LAKE VILLA DISTRICT LIBRARY

3 1981 00573 9150

Principles of Biology

Principles of Biology

Editor
Christina A. Crawford, MS Ed

SALEM PRESS

A Division of EBSCO Information Services, Inc.
Ipswich, Massachusetts

GREY HOUSE PUBLISHING

Copyright ©2017, by Salem Press, A Division of EBSCO Information Services, Inc., and Grey House Publishing, Inc.

All rights reserved. No part of this work may be used or reproduced in any manner whatsoever or transmitted in any form or by any means, electronic or mechanical, including photocopy, recording, or any information storage and retrieval system, without written permission from the copyright owner. For permissions requests, contact proprietarypublishing@ebsco.com.

∞ The paper used in these volumes conforms to the American National Standard for Permanence of Paper for Printed Library Materials, Z39.48 1992 (R2009).

Publisher's Cataloging-In-Publication Data
(Prepared by The Donohue Group, Inc.)

Names: Crawford, Christina A., 1983- editor.
Title: Principles of biology / editor, Christina A. Crawford, M.S. Ed.
Description: [First edition]. | Ipswich, Massachusetts : Salem Press, a division of EBSCO Information Services, Inc. ; [Amenia, New York] : Grey House Publishing, [2017] | Series: Principles of | Includes bibliographical references and index.
Identifiers: ISBN 978-1-68217-324-4 (hardcover)
Subjects: LCSH: Biology.
Classification: LCC QH307.2 .P75 2017 | DDC 570--dc23

FIRST PRINTING
PRINTED IN THE UNITED STATES OF AMERICA

Contents

Publisher's Note vii
Editor's Introduction ix

Activation energy 1
Active transport 4
Aging 8
Amino acids 10
Anatomy 15
Animal kingdom 18
Apes to hominids 22
Artificial organs 27
Asexual reproduction 32

Biochemical engineering 35
Biochemistry 42
Bioengineering 46
Biology 53
Bionics and biomedical engineering 55
Biophysics 60
Bioprocess engineering 65
Biosynthetics 69
Birth 74
Bone and cartilage 76
Brain 79

Cannibalism 85
Cardiology 87
Cell and tissue engineering 91
Cell communication 96
Cell organelles 100
Cell specialization 102
Cell types 104
Cellular respiration 108
Circulatory systems of vertebrates 112
Cleavage, gastrulation, and neurulation . . . 115
Cloning 119
Copulation 124
CRISPR-Cas9 127
Cryogenics 128

Death and dying 135
Defense mechanisms 138
Demographics 143
Dentistry 148
Deoxyribose nucleic acid 152
Dermatology and dermatopathology 155
Diffusion 160

Digestion 163
Digestive tract 166
Diseases 171
DNA analysis 176
DNA/RNA synthesis 182
DNA/RNA transcription 186

Egg production 191
Embryology 195
Emotions 198
Endocrine systems of vertebrates 201
Endocrinology 205
Endocytosis and exocytosis 211
Enzyme engineering 213
Eukaryotes and prokaryotes 218
Evolution: animal life 220
Evolution: historical perspective 224
Eyes 230

Fertilization 233
Forensic science 236
Fur and hair 242

Gametogenesis 247
Gastroenterology 250
Gene flow 255
Genetics 258
Geriatrics and gerontology 262
Growth 268

Hematology 273
Histology 277
Homeostasis 282
Hominids 284
Homo sapiens and human diversification 289
Human evolution analysis 296
Human genetic engineering 300
Human-computer interaction 307
Hydrophilic and hydrophobic 312
Hypnosis 316

Immune system 321
Immunology and vaccination 324
Intelligence 328

Kinesiology 333

Lactation..................................339
Lactic acid fermentation....................342
Life spans.................................344

Metabolic engineering......................349
Multicellularity...........................353
Mutations.................................360

Natural selection..........................365
Neanderthals..............................370
Nephrology................................374
Neural engineering.........................379
Neurology.................................384
Noses.....................................390
Nutrient requirements......................391

Obstetrics and gynecology...................397
Optometry.................................401
Orthopedics...............................406
Osmoregulation............................410
Osmosis...................................414

Parasitology..............................419
Pathology.................................423
Placental mammals.........................427
Polymers and monomers.....................430
Protein synthesis..........................432
Proteins, enzymes, carbohydrates, lipids,
    and nucleic acids......................436
Pulmonary medicine........................439

Reproduction..............................445
Reproductive science and engineering........449

Reproductive system of female mammals......454
Reproductive system of male mammals........457
Respiration and low oxygen.................461
Respiratory system.........................466
RNA/protein translation....................470
RNAase...................................473

Sex differences: evolutionary origin..........477
Sexual development........................481
Skin......................................484
Smell.....................................487
Stem cell research and technology...........490

The Hardy-Weinberg law of
    genetic equilibrium....................495
Thermoregulation..........................499
Tool use..................................504
Toxicology................................507
Tribology.................................510

Urology...................................515

Virology..................................521
Vision....................................525

Zoology...................................529

Glossary..................................533
The Last Twenty Years of Nobel Prize winners
    in Biological Studies...................546
Body Systems.............................548
Bibliography..............................550
Subject Index.............................587

Publisher's Note

Salem Press is pleased to add *Principles of Biology* as the sixth title in the *Principles of* series that includes Chemistry, Physics, Astronomy, Computer Science, Physical Science, and Biology. This new resource introduces students and researchers to the fundamentals of physical science using easy-to-understand language, giving readers a solid start and deeper understanding and appreciation of this complex subject.

The 112 entries range from Activation Energy to Zoology and are arranged in an A to Z order, making it easy to find the topic of interest. Entries include the following:
- Related fields of study to illustrate the connections between the various branches of biology including biochemistry, anthropology, reproduction science, physics, classical genetics, and cell biology;
- A brief, concrete summary of the topic and how the entry is organized;
- Principal terms and concepts that are fundamental to the discussion and to understanding the concepts presented;
- Basic principles that clarify the essentials of the topic
- Text that gives an explanation of how it works, important discoveries, key figures in the study of biology, or applications in fields such as medicine or industry;
- Illustrations that clarify difficult concepts via models, diagrams, and charts of such key topics as cell communication, cell and tissue engineering, DNA and RNA synthesis;
- Photographs of significant contributors to the field of physical science;
- Fascinating facts about the topic or field; sidebars devoted to important figures such as Comte de Buffon and Thomas Hunt Morgan; essays about topics such as sequestration and mutualism;
- Further reading lists that relate to the entry.

This reference work begins with a comprehensive introduction to the field, written by editor Christina Crawford, Assistant Director for Biology and Life Sciences of the Rice Office of STEM Engagement.

The book's backmatter is another valuable resource and includes:
- The Last Twenty Years of Nobel Prize winners in Biological Studies;
- Glossary;
- Diagram of Body Systems
- General bibliography; and
- Subject index.

Salem Press and Grey House Publishing extend their appreciation to all involved in the development and production of this work. The entries have been written by experts in the field. Their names and affiliations follow the Editor's Introduction.

Principles of Biology, as well as all Salem Press reference books, is available in print and as an e-book. Please visit www.salempress.com for more information.

Editor's Introduction

Principals of Biology

No matter if you are referring to a cell or a large herd of elephants, life is a symphony of natural phenome that connects all living things. In order for something to be considered living, it must possess all of the essential characteristics of life. These characteristics include being composed of cells, being able to obtain and utilize energy, growth and development, reproduction, homeostasis, evolution, and all living organisms must possess genetic material in the form of deoxyribose nucleic acid (DNA). Life gives us a reason to learn about other fields of study such as mathematics, physics and chemistry, but without physic and chemistry life would not possible. "Principals of Biology" includes multiple articles which bring to light the interdisciplinary nature of multiple scientific fields which allows life to not only be possible but more importantly understood. Aspects of life on Earth are connected in many ways. One of the most important connections is through the carbon cycle. Without the element carbon, life on Earth would not be chemically possible. The term "organic" refers to compounds being composed of carbon bonded to additional elements such as hydrogen and oxygen. Because of the carbon cycle, photosynthesis and cellular respiration are able to occur and provide the necessary oxygen for almost all forms of life and carbon dioxide needed by plants and other organism which rely on the sun as an energy source. The understanding of physics allows scientists to provide explanations of the mechanics of organ systems within the human body.

What is Biology?

Biology is the study of all forms of life. This study is not limited to organisms individually but also extends to the habitats in which those organisms live. To understand the many fields of biology, scientists overse multiple aspects of living organisms. A goal of natural biologists is to be able to provide explanations of an organism's structure, function, growth, origin, and evolution. These explanations help provide a clear picture of the organization of the natural word. The biological natural world is classified into taxonomic groups which help us understand the differences and similarities between species of plant, animal, fungi, bacteria and protist. The majority of this book focuses on the principals of biology as they relate to the classification of *Homo sapiens*, humans. By studying human fossil, cultural, and genetic evidence, researchers continue to trace development of humans. It is believed by many that the earliest form of human life appeared almost 350,000 years ago and that the modern humans appeared somewhat before 100,000 years ago, with major diversification of the human species continuing thereafter. From the basics of physiology to modern advances in medicine, *Principals of Biology* provides relevant articles which will help the reader understand the multiple aspects of biology.

Stepping stones in Biology

It was during the fifth century BCE when Greek philosophers began to study and question the existence of life. The term 'Biology' comes from the Greek word *bios* meaning life. Greek scientist Alcmaeon of Croton was interested in determining where intelligence comes from in the human body. Although his conclusions were incorrect, his observations and work provided stepping stones for philosophers such as Aristotle and Theophrastus to theorize on the structure and function of the human body.

During the 1500s Dutch eye glass maker, Zacharia Janssen, began working on a lens that might significantly improve eye sight. What he instead invested was a lens which would be incorporated into the first models of single and compound lens microscopes. Zacharia Janssen is another example of a scientist who provided stepping stones for others to succeed in the advances of biology. It was because of Janssen that Robert Hooke was able to see and coin the term 'cells' using a microscope in 1663 and Anton Van Leeuwenhoek was able to describe microscopic life ten years later in 1673.

Gregor Mendel formulated his laws of inheritance in 1866 which provided a basic explanation for how traits may be passes on from parent to offspring. These laws include independent assortment and the law of segregation which basically states that traits are linked to single genes and each allele of a gene is independently passed on from parent to offspring. In Mendelian's pattern of inheritance, each parent contributes only one of two possible alleles for a trait. According to the pattern of inheritance, the distribution of phenotypes expected for the population of offspring can be predicted when the parent genotypes are known. However, in 1908 Carl Correns discovered that not all traits follow this pattern of inheritance.

The basics discoveries within the field of biology continues to provide the ground work for multiple biologists to provide answers to questions that our ancestors did not even dream to ask. The need for advances in health, medicine, agriculture, and biotechnology allows researchers to continuously inquire on how to improve our understanding of biology and life.

FIELDS OF BIOLOGY

Biology is an extensive field of study. For convenience purposes alone, biology is subdivided into categories based on the molecule, the cell, the organism, and the population, although the majority of the subdivisions overlap in subject matter. The major fields of biology covered in the *Principal of Biology* are as follows:

- Anatomy: the study of structure and function of plants, animals, and other organisms
- Bacteriology: the study of bacteria
- Biochemistry: the study of the chemical reactions necessary for life to function
- Bioengineering: the branch of engineering which focuses biotechnology
- Biogeography: the study of the distribution of species spatially and temporally
- Bioinformatics: the use of information for the study, collection, and storage of genomic and other biological data
- Biomechanics: the study of the mechanics of living organisms
- Biomedical research: the study of health and disease
- Biophysics: the study of biological processes through physics
- Biotechnology: the study of the manipulation of living structures, including DNA modification and synthetic biology
- Botany: the study of plants
- Cell biology: the study of the cell and the molecular and chemical interactions that occur within a living cell
- Cognitive biology: the study of cognition as a biological function
- Cryobiology: the study of the effects of lower than normally preferred temperatures on living beings
- Developmental biology: the study of the processes through which an organism forms, from zygote to full structure
- Embryology: the study of the development of embryo
- Epidemiology: a major component of public health research, studying factors affecting the health of populations
- Evolutionary biology: the study of the origin and descent of species over time
- Genetics: the study of genes and heredity.
- Histology: the study of cells and tissues, a microscopic branch of anatomy
- Microbiology: the study of microscopic organisms and their interactions with other living things
- Molecular biology: the study of biology and biological functions at the molecular level, some cross over with biochemistry
- Nano-biology: the study of how nanotechnology can be used in biology, and the study of living organisms and parts on the nanoscale level of organization
- Neurobiology: the study of the nervous system, including anatomy, physiology and pathology
- Paleontology: the study of fossils and sometimes geographic evidence of prehistoric life
- Parasitology: the study of parasites and parasitism
- Pathology: the study of diseases, and the causes, processes, nature, and development of disease
- Physiology: the study of the functioning of living organisms and the organs and parts of living organisms
- Structural biology: a branch of molecular biology, biochemistry, and biophysics concerned with the molecular structure of biological macromolecules
- Synthetic biology: research integrating biology and engineering; construction of biological functions not found in nature

- Virology: the study of viruses and some other virus-like agents
- Zoology: the study of animals, including classification, physiology, development, and behavior

Molecular biology touches on biophysics and biochemistry. It is the branch of biology that deals with the structure and development of biological systems in terms of the physics and chemistry of their molecules. Botany is the science or study of plant life. Ecology, also referred to as bionomics, is the study of relationships between organisms and their environments.

Biomedical engineering further melds biomedical and engineering sciences by producing medical equipment, tissue growth, and new pharmaceuticals. An example of biomedical engineering is human insulin production through genetic engineering to treat diabetes.

Human genetic engineering is a branch of genetic engineering focusing on the understanding of human genes to produce applications that can improve human life. Genes, formulated by DNA, determine genotype. Human genetic engineering aims to alter genotypes to cause changes in physical expression of genes to engineer products, such as medications, that can cure or improve the quality of human life by addressing genetic disorders.

Current Advances in Biological research

Every year biologists are finding new ways to engineer medicine and technology with the ability to treat, cure, or prevent life-threatening illness and disorders. Scientists are currently studying ways to treat cancers through the use of nanoparticles and genetic disorders with the aid of gene editing tools.

CRISPR-Cas9 is fast, inexpensive, and the most accurate technique for editing DNA thus far. It has a wide range of potential applications, but some scientists are worried about the ethical issues of using a genetic tool of this nature. This unique technology allows scientists to remove, add, or alter sections of a person's DNA sequence and may be used to cure genetic diseases. However, it can also be used to edit reproductive cells, which could affect generation after generation of humans. CRISPR-Cas9 is still under research and currently has not been used on human genomes.

"RNAase" is a collective term that refers to a relatively new finding for an additional used of ribose nucleic acid (RNA). In the pass it was believed that all enzymes were proteins, but researchers are now discovering that some enzymes are made from the nucleic acid RNA. The functions of the RNAase enzymes are very specific. They initiate, control and terminate all stages of the processes of replication and transcription, and are therefore completely self-regulating. One of the 'RNAase' enzymes initiates opening of the DNA molecule from end to end in replication, while another opens the DNA molecule at specific locations only to retrieve the amino acid code for a specific protein. Still other RNAase enzymes control closure of the DNA strands, as well as setting the helical structure of the DNA molecule, and making all of the necessary manipulations of RNA in protein synthesis. An example of RNAase is RNAi, or RNA interference. This specific form of RNA prevents the translations of mRNA within a cell. Through the discovery of RNAi researchers are able to understand the purpose of genes by turning them off. This advance in biology is particularly exciting because it provides a form of gene therapy for diseases by turning off a gene that is causing harm to an organism.

As our level of understanding of the natural world increases, the need for students majoring in biological fields continues to increase as well. The articles in "Principals of Biology" are written to provide students and researchers with a solid foundation to study of biology. A diverse collection on topics pertaining to the history of biology, fields of biology and current advances of biology are addressed.

—*Christina A. Crawford, MS Ed*

WORK SITED

"Biology." Wikipedia. Wikimedia Foundation, 15 Feb. 2017. Web. 17 Feb. 2017.

Biomedical Engineering Society http://www.bmes.org

Center for Genetics and Society http://www.geneticsandsociety.org

Gascoigne, Bamber. HistoryWorld. From 2001, ongoing. http://www.historyworld.net

Genetics Education Center University of Kansas Medical Center http://www.kumc.edu/gec

Activation energy

FIELDS OF STUDY
Physical Chemistry; Inorganic Chemistry; Biochemistry

SUMMARY
The activation energy of a process is defined, and its importance in chemical processes is elaborated. Activation energy is a widely variable quantity in different reactions but is nevertheless characteristic of any specific reaction process.

PRINCIPAL TERMS

- **Arrhenius equation:** a mathematical function that relates the rate of a reaction to the energy required to initiate the reaction and the absolute temperature at which it is carried out.
- **catalyst:** a chemical species that initiates or speeds up a chemical reaction but is not itself consumed in the reaction.
- **chemical reaction:** a process in which the molecules of two or more chemical species interact with each other in a way that causes the electrons in the bonds between atoms to be rearranged,
- **resulting:** in changes to the chemical identities of the materials.
- **reaction rate:** how much of a particular reaction or reaction step occurs per unit time.
- **transition state:** an unstable structure formed during a chemical reaction at the peak of its potential energy that cannot be isolated and ultimately breaks down, either forming the products of the reaction or reverting back to the original reactants.

Activation Energy in Chemical Reactions

Activation energy can be thought of as a barrier that the reactants in a chemical reaction must overcome if the reaction is to proceed to the formation of products. The molecules that are involved must rearrange to form either a transition state or an intermediate that is higher in energy than the starting materials. An intermediate is a stable chemical structure formed during a reaction process that can often be captured and isolated by chemical means. Once this structure is formed, the reaction process can either progress to form products or revert back to the original reactants.

Activation energy is the energy required for a specific chemical reaction to occur. In a reaction, two reactant molecules contact each other with the energy of their ambient states. (The ambient state of a material can be thought of as its "default" state—the state it takes at one atmosphere of pressure and what is commonly considered to be room temperature.) In the case of a spontaneous reaction, the energy of the collision is sufficient to initiate the formation of the transition state or intermediate. In a nonspontaneous reaction, the energy of the molecular collision is not sufficient, and the two molecules will not interact. The input of some additional energy is required to drive the two molecules together so that the transition state or intermediate is formed and the reaction can proceed. The energy released in the transformation of reactants into products is generally sufficient to drive the reactions of other molecules in the reaction mixture.

Another way to look at activation energy is to think of it as the minimum amount of energy that two interacting molecules must gain in order to weaken bonds between atoms in both molecules so that those bonds can be rearranged. Since chemical reactions are essentially processes of breaking and making bonds, having sufficient energy to overcome the strength of the appropriate bonds is essential if there is to be any reaction between the two molecules.

Activation Energy and Reaction Rates

Reaction rates can be related directly to their activation energies. This relationship is defined by the Arrhenius equation, formulated in 1884 by the Swedish scientist Svante Arrhenius (1859–1927), who received the Nobel Prize in Chemistry in 1903. The

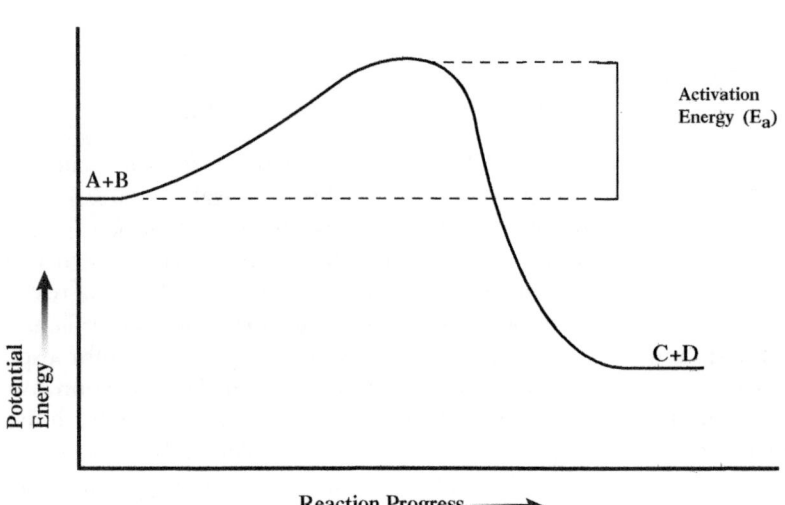

ACTIVATION ENERGY DIAGRAM

Arrhenius equation relates the rate constant of a reaction to its activation energy and the absolute temperature and has the form

$$k = Ae^{-E/RT}$$

where k is the rate constant for the reaction or process; A is the pre-exponential factor, also known in some cases as the frequency factor; E (or E_a) is the activation energy for the reaction or process; R is the gas constant; T is the absolute temperature; and the mathematical constant e is the base of the natural logarithm, so that the natural logarithm (ln) of e is equal to 1. The Arrhenius equation has been found to apply not only to chemical reactions but to physical processes as well. The relationship can be most clearly seen by plotting experimentally determined logarithmic values of k against the inverse of the absolute temperature, $1/T$. This results in a straight line plot, from which the activation energy of the reaction or process can be calculated. The pre-exponential factor A is identified as the value that the specific rate constant k would have if the activation energy E were zero (a spontaneous reaction). In that special case, the exponent $-E/RT$ would also be equal to zero, making $e^{-E/RT}$ equal 1 and thus causing the value of k to be equal to A. For different specific reactions, the value of A ranges over several orders of magnitude, but the rate constant k is determined almost solely by the value of $e^{-E/RT}$, which can range over several hundred orders of magnitude, depending on the relative values of E and T.

The course of a reaction depends on the relative difference between the energy of the reactants and that of the products. The greater this difference is, the more impetus there is for the reaction to proceed to the formation of products. This is typically illustrated in a plot of energy versus the reaction coordinate, a symbolic representation of the progress of a reaction. In the plot, the energy level of the reactants, on the left, is either higher or lower than that of the products, on the right. In between, a curved line rises from the energy level of the reactants to a maximum value before falling to the energy level of the products. The difference between the energy level of the reactants and this peak energy value represents the activation energy for the reaction, while the difference between the energy levels of the reactants and the products represents the energy released in the reaction, also called its enthalpy.

The specific rate of any individual reaction is determined by its activation energy. However, in mass quantities, the energy differences between reactants and products in the system also play a role. This can be understood by considering the Boltzmann fraction $e^{-E/RT}$, which describes the fraction of molecules in the system having energy greater than E. As energy is released from several reactions, the fraction of molecules present at any given time with sufficient energy to react increases, and more reactions can occur in any given time period. Each reaction requires the same activation energy, and the amount of energy that is available in the system to permit reactions to occur may be anything from "barely enough" to "excessive."

The activation energy of a reaction can be greatly reduced by the inclusion of a catalyst, a material that takes part in the reaction mechanism but is not consumed in the reaction. Catalysts function by forming an activated complex with the reactants, typically

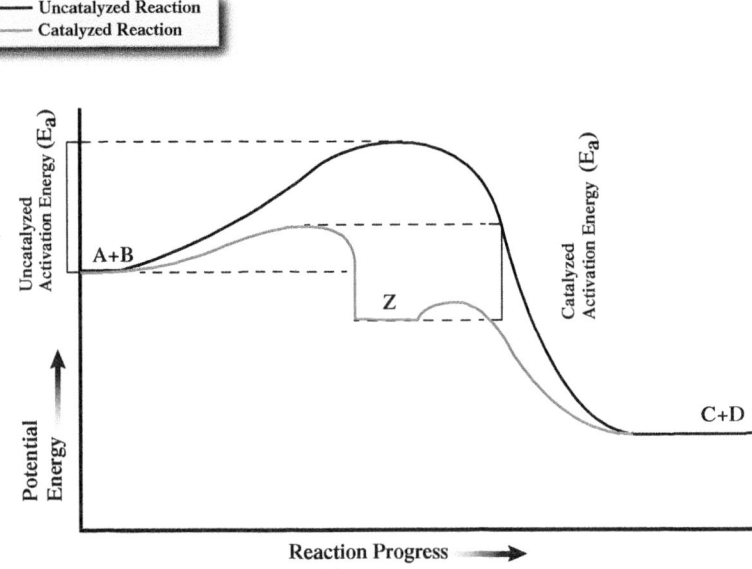

CATALYZED ACTIVATION ENERGY

constraining the reactant molecules in an orientation that they would otherwise have to achieve through collision with each other. This reduces the energy necessary to achieve that particular orientation—that is, the activation energy—so that the reaction can proceed. When the reactant molecules are constrained in the activated complex with the catalyst, bonds between certain atoms are weakened and the orientations of atomic and molecular orbitals that must interact are often brought into the proper alignment, or trajectory, for the new bonds to form between atoms.

Activation Energy in Action

The activation energy of a reaction can range from exceedingly small to very large. Two examples serve to illustrate this point. For the first, consider the addition of two parts hydrogen gas (H_2) to one part oxygen gas (O_2). This is an explosive mixture of gases, yet the two mixed gases are quite happy to coexist quietly in the same container, no matter how much is present. The introduction of an initiator such as an electrical spark, however, results in an almost instantaneous reaction to form water (H_2O), accompanied by the release of a great deal of energy. The activation energy of the reaction between hydrogen and oxygen is very low, and the amount of energy released by one reaction is more than sufficient to drive many instances of the reaction in the gas mixture, with each subsequent occurrence releasing an equal amount of energy as the enthalpy of reaction.

The second example is the so-called thermite reaction, in which iron oxide and aluminum metal react to produce aluminum oxide and iron metal. This is a spectacular reaction often demonstrated for chemistry exhibitions. Because the activation energy of the thermite reaction is very high, the reaction is very difficult to initiate and must typically be ignited by a burning piece of magnesium metal; once it has begun, however, it is essentially impossible to stop it due to the amount of energy that is released. Typically, the iron metal falls out of the reaction mixture as a white-hot liquid.

Activation Energy in Biological Systems

Activation energy applies to biochemical processes as well as to physical processes. The chirping of crickets, for example, is dependent on temperature in a manner that is in complete accord with the Arrhenius equation. In biological systems, the activation energy of processes is made a great deal lower by the catalytic action of protein molecules called enzymes. Enzymes have well-defined three-dimensional structural shapes that allow them to coordinate with other molecules in specific ways, rather like the way a key works with a lock. The coordination normally alters the three-dimensional shape of the substrate molecule or otherwise interacts with it so that specific bonds are weakened and the molecular geometry is changed such that reaction is highly favored.

Richard M. Renneboog, MSc

Further Reading

Arnaut, Luís, Sebastião Formosinho, and Hugh Burrows. *Chemical Kinetics: From Molecular Structure to Chemical Reactivity*. Oxford: Elsevier, 2007. Print.

Bell, Jerry A. *Chemistry: A Project of the American Chemical Society*. New York: Freeman, 2005. Print.

Lafferty, Peter, and Julian Rowe, eds. *The Hutchinson Dictionary of Science*. 2nd ed. Oxford: Helicon, 1998.Print.

Masterton, William L., Cecile N. Hurley, and Edward Neth. *Chemistry: Principles and Reactions*. 7th ed. Belmont: Brooks, 2012. Print.

Raymond, Kenneth W. *General, Organic, and Biological Chemistry: An Integrated Approach*. 4th ed. Hoboken: Wiley, 2014.Print.

Zumdahl, Steven S., and Susan Zumdahl. *Chemistry*. 9th ed. Belmont: Brooks Coles, 2013. Print.

ACTIVATION ENERGY SAMPLE PROBLEM

Use the Arrhenius equation to determine the activation energy at 0°C for a reaction having a specific rate constant (k) of 0.023 moles per liter per second and a pre-exponential factor (A) of 2,303 moles per liter per second. Use the gas constant.

$$R = 8.314 \frac{J}{mol\ K}$$

Answer:

Convert the temperature from degrees Celsius to kelvins, given that K = °C + 273.15:

$$K = 0 + 273.15 = 273.15$$

The Arrhenius equation is

$$k = Ae^{-E/RT}$$

Rearrange the equation using natural logarithmic (ln) relationships:

$$\ln k = \ln A + \ln(e^{-E/RT})$$

$$\ln k = \ln A - \frac{E}{RT}$$

$$\frac{E}{RT} = \ln A - \ln k$$

$$E = RT(\ln A - \ln k)$$

Substitute in the values of R ($8.314 \frac{J}{mol\ K}$), T (temperature), A (pre-exponential factor), and k (rate constant). Calculate, paying attention to the units throughout:

$$E = RT(\ln A - \ln k)$$

$$E = (8.314 \frac{J}{mol\ K} \times 273.15\ K)(\ln 2303 - \ln 0.023)$$

$$E = (8.314 \frac{J}{mol\ K} \times 273.15\ K)[7.742 - (-3.772)]$$

$$E = 26147.938 \frac{J}{mol}$$

ACTIVE TRANSPORT

FIELDS OF STUDY

Biochemistry; Molecular Biology; Genetics

SUMMARY

The process of active transport is defined, and its importance in biochemical processes is elaborated. Active transport is an essential feature of the biochemistry of living systems and helps maintain the necessary concentrations of various biochemical components and electrolytes for the proper functioning of cellular metabolism.

PRINCIPAL TERMS

- **adenosine triphosphate (ATP):** a molecule consisting of adenine, ribose, and a triphosphate chain that is used to transfer the energy needed to carry out numerous cellular processes.
- **cell membrane:** a biological membrane that forms a semipermeable barrier separating the interior of a cell from the exterior.
- **concentration gradient:** the gradual change in the concentration of solutes in a solution across a specific distance.

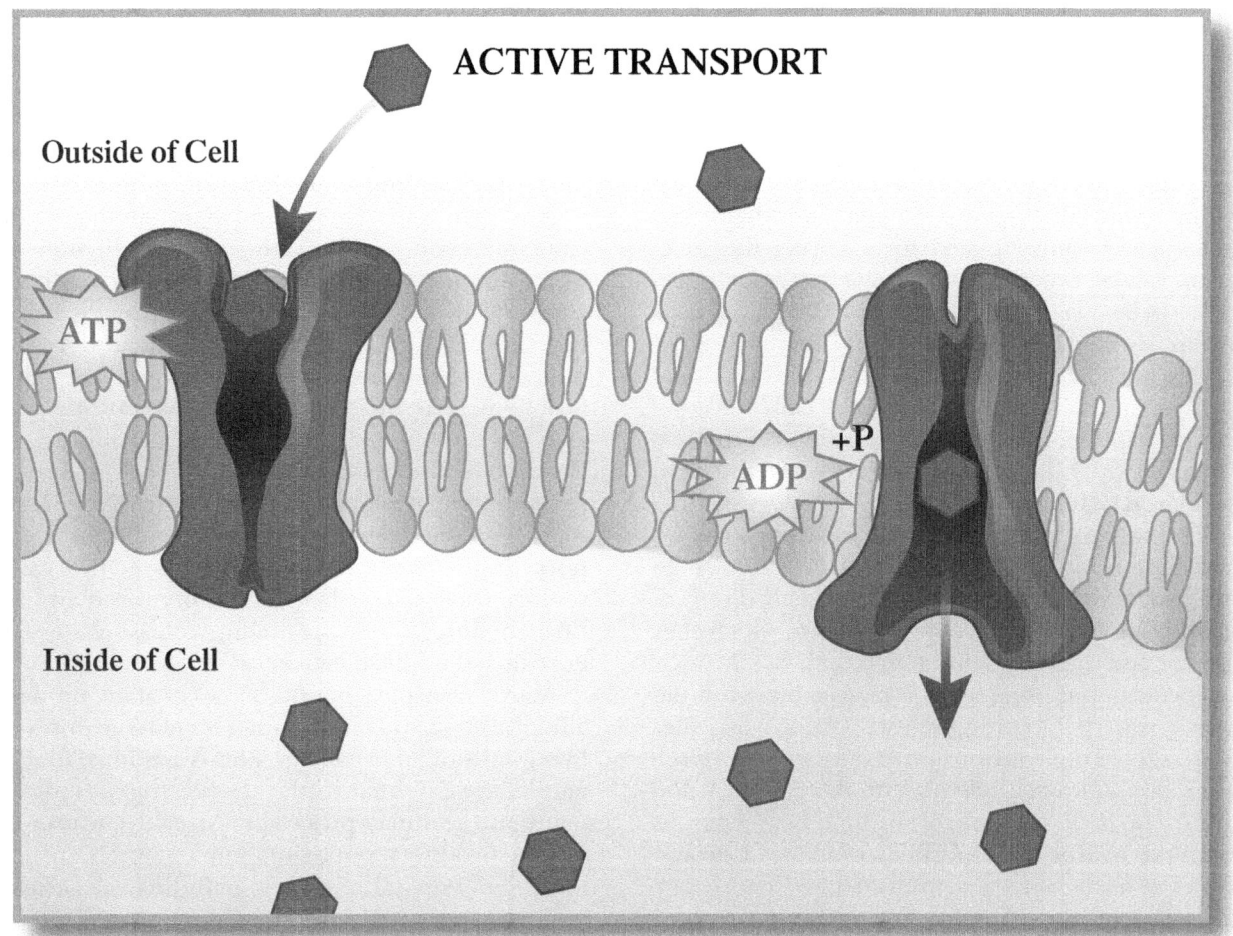

- **diffusion:** the process by which different particles, such as atoms and molecules, gradually become intermingled due to random motion caused by thermal energy.
- **passive transport:** the passage of materials through a membrane with no input of energy required.

The Mechanics of Active Transport

In living cells, biochemical processes transport materials necessary for a properly functioning metabolism through cell membranes. Passive transport does not require an input of energy to move materials across cell walls because it operates in the same direction as the concentration gradient, moving the materials from an area of high pressure to one of low pressure. Active transport can be thought of as a "shuttle service" for ions and other polar materials that cannot pass through a cell membrane by diffusion, a kind of passive transport. Instead, those entities must be physically transported across membranes by various mechanisms collectively termed pumps. A pump is a type of mediated transport system that functions to conduct ions, amino acids, glucose, and other polar compounds through the nonionic lipid bilayer, the highly nonpolar material that makes up the cell wall. Pumps always work against the concentration gradient to move materials out of regions of low concentration and into regions of higher concentration, using energy derived from biochemical reactions. The transported material is subsequently used in other biochemical reactions that return the energy used during transport.

Cell Walls and Lipid Bilayers

Long-chain fatty acids are organic molecules whose molecular structure consists of a single hydrocarbon chain terminated by a carboxylic acid functional group (–COOH). The carboxyl group is highly polar and hydrophilic, while the hydrocarbon moiety, or portion, of the molecule is very nonpolar and hydrophobic. Carboxylic acids are converted to esters by enzyme-mediated reactions with alcohols. In an ester, the carboxyl functional group retains the highly polar character that it had in its free carboxylic acid form, giving the long-chain esters, called lipids, a polar-nonpolar structure similar to that of the free carboxylic acids. When carboxylic acids are esterified with glycerol, which has three hydroxyl (–OH) functional groups, the resulting triesters are called triglycerides. Lipids and triglycerides are the principal forms in which long-chain fatty acids are found in biological systems.

The hydrocarbon chains and the carboxyl-based portions of fatty acids and their esters do not interact with each other due to their different hydrophilicities—that is, the degrees to which they attract and interact with water and other polar molecules—but they are quite capable of interacting with the corresponding portions of other molecules. The hydrocarbon chains associate preferentially with each other, as do the carboxyl portions. The basic structure of the lipid bilayer results from the hydrocarbon portions of the acids of two layers of such molecules intermingling and essentially dissolving each other. The carboxyl functions on the other ends of the one on either side of the very hydrophobic interior layer. The resulting structure is a lipid bilayer.

The walls of all animal cells are formed of lipid bilayers, allowing them to interact with water-based fluids while isolating the sensitive materials and processes that take place within each cell. The fluid inside of each cell is also water based, which necessitates some means of transporting vital polar materials from the exterior of the cell to the interior and moving extraneous materials and metabolites in the opposite direction for elimination. This movement is accomplished by active transport.

FUNCTIONS OF ACTIVE-TRANSPORT SYSTEMS

Active-transport systems serve a variety of functions in the biochemistry of living systems. Their principal function is to allow the organism to extract "fuels" and other essential materials for use in the metabolic functions that occur within cells. This is a very important function, and the nature of active transport allows cells to retain a relatively high concentration of such materials even when their concentrations outside of the cell are quite low. A second important function of active-transport systems is to regulate and maintain the organism's metabolic steady state, a balanced state in which the material and energy that the organism removes from its environment through living functions is equal to the energy and materials that it returns to the environment through those same functions. The biochemical processes of metabolism use energy and materials taken from the environment. Anabolic processes remove materials from the environment and use energy from reactions involving those materials to build and support the life of the organism. Catabolic processes remove used materials from the organism and return them to the environment, releasing the energy stored in those materials.

Active transport maintains a constant optimal amount of various inorganic elements within the living cells of an organism. Potassium ions, for example, are essential to the proper functioning of many intracellular processes. An active-transport system produces potassium-ion channels in cell walls of nerves and muscles, including the cardiac muscles. Potassium ions are delivered into the cytoplasm of the cell via these channels to replace ejected sodium ions, thus maintaining a constant ionic concentration within the cell. The system maintains a relatively high concentration of potassium ions in most aerobic cells, between 100 and 150 millimolars (mM), whether they are plant, animal, or microbial in nature and regardless of the concentration outside of the cells. (A 1 mM solution has a concentration of 0.001 moles per liter.) The potassium ions that are pumped into the cell also serve to maintain the electric potential across the cell membrane, a factor that affects the free-energy change in reactions involved in active-transport systems.

ACTIVE TRANSPORT IN ACTION

The transfer of ions across a membrane or against a concentration gradient by active transport is accompanied by a free-energy change (ΔG) that can be calculated by one of two equations. The first equation represents the free-energy change for the transfer of neutral materials against a concentration gradient. This is described by the following equations:

$$\Delta G = RT \ln \frac{c_2}{c_1}$$

$$8.314 \frac{J}{mol\ K}$$

where R is the gas constant and T is the absolute temperature in kelvins, ln is the natural logarithm function, and c_1 and c_2 are concentrations on either side of the membrane in molars, or moles per liter (M), with c_2 being greater than c_1.

The second equation, which represents the free-energy change for the transfer of electrically charged materials, needs to account for the charge on the material being transported and the difference in electric potential across the membrane. The latter is determined by the neutral nature of the lipid bilayer, which causes it to act as a capacitor, or energy-storage device, and the presence of charge as maintained by the potassium ions in the cytosol. The free-energy expression for the transport of charged species across a cell membrane is given by the following equation:

$$\Delta G = RT \ln \frac{c_2}{c_1} + ZF\Delta\Psi$$

where Z is the charge on the ion, F is the Faraday constant (96,485.3365 coulombs per mole, the electric charge on one mole of electrons), and is the difference in electric potential across the membrane in volts.

ATP and Active Transport

The energy used in active-transport systems is obtained through enzyme-mediated reactions of adenosine triphosphate (ATP). ATP molecules consist of a molecule of the nucleobase adenine that is bonded to a molecule of ribose sugar, which in turn is bonded to a triphosphate ion. A magnesium ion coordinates and stabilizes the second and third segments of the triphosphate moiety. Energy is derived from the structure by the enzymatic cleavage of the third phosphate segment from the triphosphate moiety, transforming the molecule into adensosine diphosphate (ADP), and it is restored by concatenating, or joining, a third phosphate ion to ADP to re-form ATP.

The function of muscle cells depends on the active transport of calcium ions and sodium ions, a process termed the calcium ion pump or Ca^{2+} pump. The calcium ion pump works in an organelle of muscle cells called the sarcoplasmic reticulum and is powered by ATP hydrolysis reactions mediated by the enzyme calcium adenosine triphosphatase. This process is critical to the contraction and relaxation of muscle fibers, especially heart muscles. The sarcoplasmic reticulum is a cell structure that stores and releases calcium ions to aid in this contraction and relaxation. In muscle cells, the rapid release of calcium ions from the sarcoplasmic reticulum into the cytosol, the cellular fluid outside of the organelles, triggers contraction of the muscle, while rapid removal of calcium ions from the cytosol and back into the sarcoplasmic reticulum triggers relaxation of the muscle.

The normal concentration of free calcium ions in the cytosol is between 0.1 and 0.2 micromolar (μM, or 10^{-6} moles per liter), increasing when the muscle contracts and returning to the normal value when it relaxes.

Richard M. Renneboog, MSc

Further Reading

Lafferty, Peter, and Julian Rowe, eds. *The Hutchinson Dictionary of Science*. 2nd ed. Oxford: Helicon, 1998. Print.

Lehninger, Albert L. Biochemistry: The Molecular Basis of Cell Structure and Function. 2nd ed. New York: Worth, 1975. Print.

Lodish, Harvey, et al. *Molecular Cell Biology*. 7th ed. New York: Freeman, 2013. Print.

Pelczar, Michael J., Jr., E. C. S. Chan, and Noel R. Krieg. *Microbiology: Concepts and Applications*. New York: McGraw, 1993. Print.

Reece, Jane B., et al. *Campbell Biology*. 10th ed. San Francisco: Cummings, 2013. Print.

ACTIVE TRANSPORT SAMPLE PROBLEM

Use the free-energy equation for active transport against a concentration gradient to determine the free energy associated with transporting neutral amino-acid molecules across a membrane from a concentration of 20 μM to one of 43 μM. Assume normal body temperature of 37°C. Use.

$$R = 8.314 \frac{J}{mol\,K}$$

Answer:
The materials being transported are electrically neutral. Therefore, use the equation

$$\Delta G = RT \ln \frac{c_2}{c_1}$$

Convert the temperature from °C to K:

$$K = °C + 273.15$$
$$K = 37 + 273.15 = 310.15$$

Convert the concentration values from micromolars to molars:

$$c_1 = 20\,\mu M = 20 \times 10^{-6}\,M = 0.00002\,M$$
$$c_2 = 43\,\mu M = 43 \times 10^{-6}\,M = 0.000043\,M$$

Substitute in the values of R, T, c_1, and c_2 and calculate, paying attention to the units throughout:

$$\Delta G = RT \ln \frac{c_2}{c_1}$$
$$\Delta G = (8.314 \frac{J}{mol\,K})(310.15\,K) \ln \frac{0.000043}{0.00002}$$
$$\Delta G = 1973.8 \frac{J}{mol}$$

The free energy of active transport of neutral amino acids across a concentration gradient from 20 μM to 43 μM is 1973.8 joules per mole, or 1.9738 kilojoules per mole.

AGING

FIELDS OF STUDY

Anatomy, cell biology, developmental biology, genetics, neurobiology, pathology, physiology

SUMMARY

Aging is the process of progressive and irreversible change common to all living organisms. There are striking similarities in the physical process of aging among all animal species.

PRINCIPAL TERMS

- **aging:** a process common to all living organisms, eventually resulting in death or conclusion of the life cycle
- **cognition:** ability to perceive or understand death: the cessation of all body and brain functions
- **function:** ability, capacity, performance
- **life span:** length of life from birth to death
- **longevity:** length of life

BASIC PRINCIPLES

Progressive and irreversible change has been called the single common property of all aging systems. When change is reversible or self-maintaining, such as one would see in a forest, for example, the effects of aging are often not observable. Growth of the forest is evident, but with the right conditions, trees within the forest may grow for hundreds of years in the absence of disease. Certain conditions of the forest system help to regenerate, renew, and reverse changes that happen within that system.

However, in animals some change is not reversible. The changes in the cells of the body accumulate over time and result in a steady downward trend. The end point of this trend is the death of the organism. Aging is a normal part of the life cycle. This is known to be true because aging changes within populations are rather predictable. The changes associated with aging that are seen in all animal species may occur for similar reasons. These may include chemical aging, extracellular aging, intracellular aging, and aging of cells.

Aging occurs within body systems as a result of unseen changes at the molecular and cellular levels. Although the mechanisms through which aging occurs may be understood, the causes are less clear. The fact remains that due to changes in chemical balances such as those of hormones, and to the dying of cells within the body, each of the bodily systems shows deterioration over time.

Changes that occur in domestic animals over the life span can be similar to those that occur in humans. Dogs experience the graying of their hair, a decrease in vision, and a slowing of movement with age. They also experience cataract formation, arthritis, skin problems, cancer, and diabetes. Certain breeds of animals may demonstrate a tendency toward specific illnesses or diseases. For example, German shepherds often develop hip problems, and collies commonly develop progressive arthritis that may seriously inhibit mobility by around ten years of age.

How It Works

Common Effects of Aging. There are many variations in the effects of aging among the species of animals. The life span of animals may range from a few days (among insects) to thirty years or more, with great variation depending upon many factors. Animals that live in captivity, as pets or in zoos where they are sheltered from the effects of predation, disease, and adverse climate, also tend to live significantly longer than animals in the wild.

Very little research has been done on the aging of most animal species. The reasons for this include the difficulty of observing animals over a long period of time in their natural habitat. Aging in monkeys has been studied more than that in other animals because of the notion that aging patterns may closely reflect those of humans.

Aging monkeys show changes in their circulatory systems similar to those found in humans: There is notable atherosclerosis and arteriosclerosis, or hardening of the arteries. The heart pumps less effectively, and vessels show buildup of plaque. These changes often result in cardiac problems, including heart attacks. The respiratory system also shows a decrease in elasticity. Senile emphysema has been noted. The kidneys show signs of atrophy and sclerosis in aged monkeys. The kidneys of humans may lose up to half of the functioning nephrons with advanced age and thus become less effective in filtering waste products from the body.

Physical function or capacity tends to decline with age. This is largely due to the atrophy of muscles, which is more common as the body gets older. The joints tend to become stiffer and less mobile. Range of motion may be restricted. Changes in bone density may lead to loss of teeth, osteoporosis, and subsequent fractures. Tooth loss and osteoporosis have been documented in monkeys over the age of twenty years. Pictures of such older monkeys reveal a stooped posture, with shoulders hunched forward, similar to the kyphosis observed in many older human women.

Physical function among animals has been less studied than that in humans, but certain physiological characteristics are similar. For example, survival times after severe physical injury with blood loss and trauma decreases in both humans and animals as age increases. Male monkeys do not lose reproductive capabilities until toward the end of the life span, while females have a more restricted period of time to bear offspring. Fertility among all females tends to decline with age after its peak.

The immune system functions less effectively as age increases. This leaves the body more susceptible to a range of illnesses and diseases. Neoplasms, or tumors, are most common among mammals as they age. An impaired immune system allows various types of tumors or cancers to spread more rapidly in the older body. Response to stress and ability to adapt to stressors also decline with age. For example, older mice become less able to adapt to cold temperatures.

Social roles and behaviors among animals may also change with age. Longitudinal studies on animals in the wild are scarce, so only generalities may be speculated upon. Even studies done within controlled laboratory settings yield only broad suggestions, since numbers of animals available for study are limited. Males generally tend to dominate the females in both physical strength and social ranks. Some nonhuman primates show different characteristics with advanced age. That is, some monkeys and baboons allow older males to remain part of the social group, while other species support the male leader in the group only as long as the female harem supports him, whether younger or older. Individual monkeys in stable groups have been observed to resort less frequently to aggressive behavior to maintain their status within the group.

Causes of Death. Among nonhuman primates, the leading cause of spontaneous death is digestive problems. Older animals that die do not always show advanced signs of tissue aging. Since much less research has been done on aging among animals than among humans, data about causes of death are rare. However, it appears that there is an increased probability of dying from trivial illnesses, perhaps due to decreased resistance factors, as animals age.

Predator-prey relationships among animals are particularly significant as causes of death. Thus, the effect of the environment on animal aging and death requires more investigation. Do animals age more quickly if they are objects of prey? Do animals relate to stress in ways similar to those of people, thus showing signs of wear and tear that are seen with premature aging under stress? Are there risk factors among animals that affect their life span? These are some of the questions that remain to be answered on the topic of aging among animals.

—*Kristen L. Mauk*

See also: Birth; Death and dying; Demographics; Diseases; Growth; Life spans; Natural selection.

FURTHER READING

Bowden, Douglas M. *Aging in Nonhuman Primates.* New York: Van Nostrand Reinhold, 1979. Discusses aging in monkeys, particularly the effects of disease.

Kohn, Robert R. *Principles of Mammalian Aging.* Englewood Cliffs, N.J.: Prentice-Hall, 1971. One of the few general surveys of aging in mammalian species.

Schmidt-Nielsen, Knut. *Animal Physiology: Adaptation and Environment.* New York: Cambridge University Press, 1997. A well-regarded college-level textbook on animal physiology, which covers aging in the context of the whole life of the animal.

Slater, P. J. B. *Essentials of Animal Behavior.* New York: Cambridge University Press, 1999. A basic introduction to ethology, which considers the effects of aging on behavior.

Slobodkin, Lawrence B. *Growth and Regulation of Animal Populations.* New York: Holt, Rinehart and Winston, 1961. A classic text on population analysis; covers the effects of aging on population demographics.

AMINO ACIDS

FIELDS OF STUDY

Biochemistry; Organic Chemistry

SUMMARY

The basic structure of amino acids is explained, as is their function in the creation of polypeptides and proteins. The transcription of nucleotides from DNA by RNA to create amino acid chains is also discussed.

PRINCIPAL TERMS

- **amino group:** a functional group containing a nitrogen atom bonded to two hydrogen atoms ($-NH_2$).
- **carboxyl group:** a functional group containing a carbon atom double bonded to an oxygen atom and single bonded to a hydroxyl group ($-OH$); has the formula CO_2H, typically written $-COOH$.
- **catalyst:** a chemical species that initiates or speeds up a chemical reaction but is not itself consumed in the reaction.
- **peptide bond:** a covalent bond that links the carboxyl group of one amino acid to the amine group of another, enabling the formation of proteins and other polypeptides.
- **protein:** a biological polymer consisting of one or more long chains of amino acids linked by peptide bonds in a sequence specified by an organism's DNA.

THE NATURE OF AMINO ACIDS

Strictly speaking, an amino acid is any compound whose molecular structure contains both an amino group and a carboxyl group, also called a carboxylic acid group—hence the term "amino acid." The term in general use, however, refers to the specific group of amino acids relevant to the genetic code in the DNA molecule. These twenty (or sometimes twenty-three, depending how they are classified) amino acids are called proteinogenic amino acids, which refers to the fact that they are the only amino acids used in the creation of proteins, enzymes, and other biomolecules. The three disputed amino acids

AMINO ACIDS

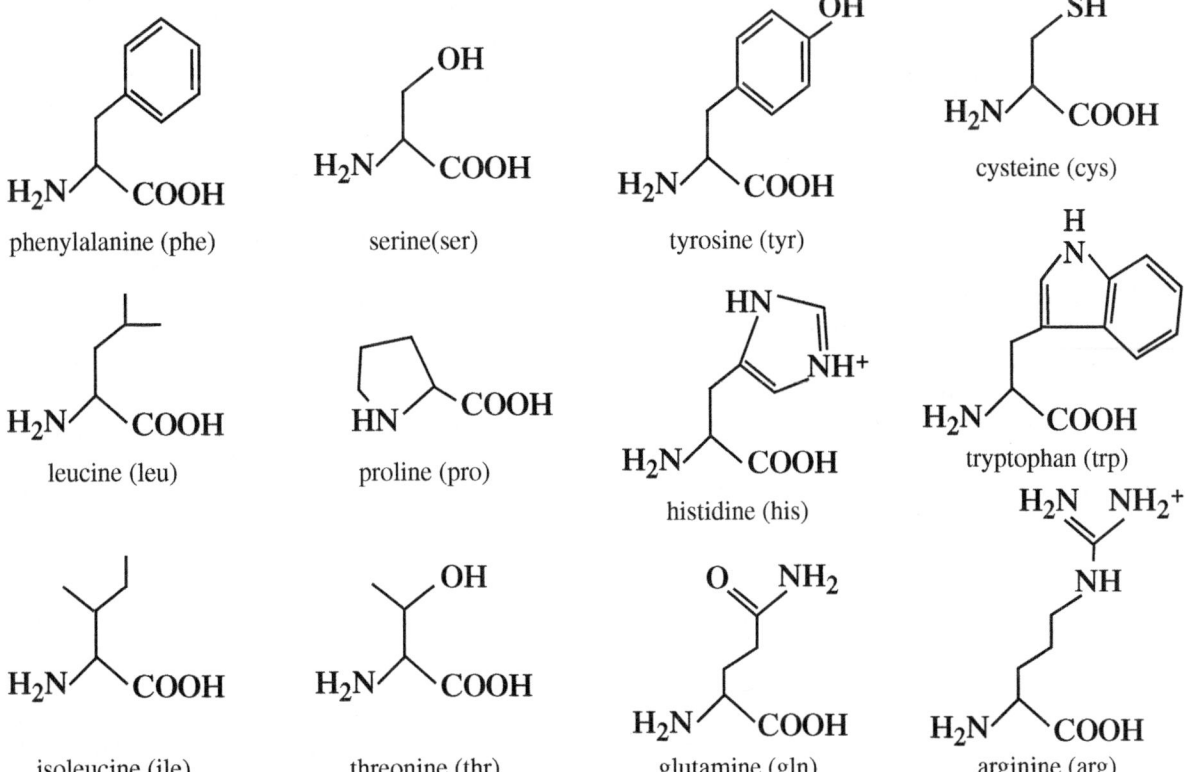

AMINO ACIDS (cont.)

[Structural diagrams of amino acids:]
- methionine (met) — START
- alanine (ala)
- asparagine (asn)
- glycine (gly)
- valine (val)
- lysine (lys)
- aspartic acid (asp)
- glutamic acid (glu)

[Reaction diagram: amide bond formation]

R–C(=O)–OH + H$_2$N–R' → R–C(=O)–N(H)–R'

amide bond formation

are selenocysteine and pyrrolysine, which are not directly coded for in the genetic code but rather synthesized by other means and incorporated later, and N-formylmethionine, which initiates protein creation in some prokaryotes but is typically removed afterward. Selenocysteine is the only one of the three found in eukaryotes.

Of the standard twenty proteinogenic amino acids, nine are deemed "essential" because they are not synthesized in human metabolism but must be acquired through diet. All proteinogenic amino acids are also called α- amino acids, meaning that their amino and carboxyl groups are both bonded to the same carbon atom, known as the α-carbon (alpha carbon). This same carbon atom is also bonded to a hydrogen atom. The fourth atom or group bonded to the α-carbon determines the identity of the amino acid.

Chemically, the amino acids have unique properties due to the presence of both a base and an acid in the same molecule. Self- neutralization, in which the acid transfers a proton to the base, readily takes place to produce a zwitterion, an electrically neutral molecule in which both a positive and a negative charge exist in separate parts of the molecule at the same time. In addition, each amino acid has a unique isoelectric point, which is the specific degree of acidity or basicity (pH) at which the amino acid has no net electrical charge. These particular characteristics are responsible for most, if not all, of the behavior of amino acids and the much larger compounds they form as proteins and enzymes.

Formation of Proteins and Enzymes

The structures of all proteins are determined by the sequence of amino acids encoded in the DNA molecule. With just one each of the twenty standard amino acids, there are thousands of trillions of possible ways to arrange them in what is called a polypeptide chain, and most proteins and enzymes contain far more than just twenty amino acids.

The synthesis of polypeptides and their formation into proteins is extremely quick, taking as little as six minutes, according to various tracer studies using radioactively labeled amino acids. The process begins with transcription, during which specific enzymes open up the double-stranded structure of a DNA molecule and assemble copies of the nucleotide sequence using RNA segments. Each segment, called messenger RNA (mRNA), carries specific sequences of three nucleotides called codons. During the next step, translation, the mRNA translates the genetic code from DNA to structures called ribosomes, composed of ribosomal RNA (rRNA), where they match up with the rRNA sequence of nucleotides. When this occurs, the codons are exposed. In the cytosol (intracellular fluid) of the cell, a third type of RNA called transfer RNA (tRNA) transfers the specific amino acid corresponding to a particular codon to the mRNA strand in the ribosome. The anticodon on the tRNA segment matches to the codon on the mRNA strand, and specific enzymes there act as a catalyst to form a peptide bond between two neighboring amino acids. A peptide bond is just the normal amide structure that forms between a carboxylic acid and an amine:

The term is used to refer specifically to an amide bond formed between amino acids in a polypeptide, as here:

TRANSLATING THE GENETIC CODE INTO PROTEINS
The DNA and RNA molecules use only four different nucleotide bases to specify the entire genetic code, yet the number of possible three- nucleotide codons that can be formed from these is more than sufficient to differentiate the twenty standard amino acids. The system is actually quite redundant, with several different codons signifying the same amino acid. There are also specific nucleotide sequences that designate the starting and ending points of a particular sequence of amino acids, and hence the protein structure that derives from that sequence. DNA itself has been known to exist since 1869, but it was not thought to have any relation to genetic information until the connection was unequivocally demonstrated in 1943. Subsequent research eventually revealed the structure and function of DNA and RNA. By preparing synthetic sequences of mRNA codons, researchers were able to determine which codons encoded for each specific amino acid in transcription and translation. The code is translated in the accompanying chart.

From this it can be seen that the same codon, AUG, indicates both methionine and the start sequence, while the three stop codons are unique. The context in which the AUG codon appears that determines whether or not it functions as a start codon or the codon for methionine.

Amino Acid	Symbol	mRNA Codon
Alanine	Ala	GCA, GCC, GCG, GCU
Arginine	Arg	AGA, AGG, CGA, CGG, CGC, CGU
Asparagine	Asn	AAC, AAU
Aspartic acid	Asp	GAC, GAU
Cysteine	Cys	UGC, UGU
Glutamic acid	Glu	GAA, GAG
Glutamine	Gln	CAG, CAA
Glycine	Gly	GGA, GGC, GGG, GGU
Histidine	His	CAC, CAU
Isoleucine	Ile	AUA, AUC, AUU
Leucine	Leu	CUA, CUC, CUG, CUU, UUA, UUG
Lysine	Lys	AAA, AAG
Methionine	Met	AUG
Phenylalanine	Phe	UUU, UUC
Proline	Pro	CCA, CCC, CCG, CCU
Serine	Ser	AGC, AGU, UCA, UCG, UCC, UCU
Threonine	Thr	ACA, ACG, ACC, ACU
Tryptophan	Trp	UGG
Tyrosine	Tyr	UAC, UAU
Valine	Val	GUA, GUG, GUC, GUU
Start codon		AUG
Stop codon		UAA, UAG, UGA

AMINO ACIDS AND PROTEINS
The sequence of amino acids in a protein molecule defines its primary structure. Since each amino acid group has a specific geometry dictated by the rules of molecular structure and bond formation, no polypeptide chain or protein can be just a linear molecule. The angles of the bonds at each atom create all sorts of twists and turns along the entire length of the polypeptide molecule, moving the various functional groups on each amino acid into positions that allow them to interact with each other. Some segments of the protein molecule form larger physical shapes, such as spirals or flattened sheets. These constitute the

Amino Acid	Symbol	mRNA Codon
Alanine	Ala	GCA, GCC, GCG, GCU
Arginine	Arg	AGA, AGG, CGA, CGG, CGC, CGU
Asparagine	Asn	AAC, AAU
Aspartic acid	Asp	GAC, GAU
Cysteine	Cys	UGC, UGU
Glutamic acid	Glu	GAA, GAG
Glutamine	Gln	CAG, CAA
Glycine	Gly	GGA, GGC, GGG, GGU
Histidine	His	CAC, CAU

Amino Acid	Symbol	mRNA Codon
Isoleucine	Ile	AUA, AUC, AUU
Leucine	Leu	CUA, CUC, CUG, CUU, UUA, UUG
Lysine	Lys	AAA, AAG
Methionine	Met	AUG
Phenylalanine	Phe	UUU, UUC
Proline	Pro	CCA, CCC, CCG, CCU
Serine	Ser	AGC, AGU, UCA, UCG, UCC, UCU
Threonine	Thr	ACA, ACG, ACC, ACU
Tryptophan	Trp	UGG
Tyrosine	Tyr	UAC, UAU
Valine	Val	GUA, GUG, GUC, GUU
Start codon		AUG
Stop codon		UAA, UAG, UGA

secondary structure of the protein. A third, or tertiary, structure results from the interaction of the various functional groups as they form bonds due to their proximity to each other. A fourth, or quaternary, structure results when two or more protein molecules combine to form a larger reactive complex.

Richard M. Renneboog, MSc

FURTHER READING

Berg, Jeremy M., John L. Tymoczko, Gregory J. Gatto, and Lubert Strye. *Biochemistry*. 8th ed. New York: W. H. Freeman, 2015. Print.

Lehninger, Albert L. *Biochemistry: The Molecular Basis of Cell Structure and Function*. 2nd ed. New York: Worth, 1975. Print.

Lodish, Harvey, et al. *Molecular Cell Biology*. 7th ed. Print.

Morrison, Robert Thornton, and Robert Neilson. Boyd. *Organic Chemistry*. 7th ed. Englewood Cliffs, N.J.: Prentice Hall, 2003. Print.

Pine, Stanley H. *Organic Chemistry*. 5th ed. New York: McGraw, 1987. Print.

Reece, Jane B., et al. *Campbell Biology*. 10th ed. San Francisco: Cummings, 2013. Print.

AMINO ACID SAMPLE PROBLEM

Given the following RNA nucleotide sequence, determine the sequence of amino acids in the protein that may be assembled from its code.

CCGAUGUGGGGGGGCGCUCUUUU
UUUGUGCGCUCUAUACACGCGGG
GCGCGCGAGAUAUAUAGAGCGCC

Answer:
The start codon, AUG, begins three units from the left end of the string. No synthesis is carried out until a start codon is encountered. Each subsequent three-unit codon then specifies the next amino acid in the sequence, until a stop codon is encountered.

The sequence begins at AUG and is followed by the codons
UGG, GGG, GGC, GCU, CUU, UUU, UUG, UGC, GCU, CUA, UAC, ACG, CGG, GGC, GCG, CGA, GAU, AUA, UAG, AGC, GCC.

The stop codon is encountered at UAG, the third codon from the end of the string. Thus, only the amino acids coded for between AUG and UAG will be synthesized.

Based on the chart, the relevant codons produce the following polypeptide string:
Trp-Gly-Gly-Ala-Leu-Phe-Leu-Cys-Ala-Leu-Tyr-Thr-Arg-Gly-Ala-Arg-Asp-Ile

---

ANATOMY

FIELDS OF STUDY

Anatomy, anthropology, developmental biology, embryology, entomology, evolutionary science, herpetology, human origins, invertebrate biology, marine biology, neurobiology, ornithology, paleontology, physiology, systematics (taxonomy), zoology

SUMMARY

Anatomy is the branch of natural science that focuses on the structural organization of living organisms. Physiology, which is closely related to anatomy, is the study of function, activities, and processes of living organisms. It is concerned with such basic activities as reproduction, growth, metabolism, excitation, and contraction as they are carried out within structures such as the cells, tissues, organs, and organ systems of the body.

PRINCIPAL TERMS

- **comparative anatomy:** the study of relationships between the anatomies of different species
- **developmental anatomy:** the study of the anatomical changes an animal undergoes in the process of growth

BASIC PRINCIPLES

One common approach to the study of anatomy is from the viewpoint of a classification system that is based on the type of organisms studied, generally plant anatomy and animal anatomy. Animal anatomy can be further subdivided into human anatomy and comparative anatomy. Other anatomy subdivisions are developmental, pathological, and surgical anatomy and anatomical art. An example of developmental is the study of embryos, and an example of pathological is the study of diseased organs. Examples of applied anatomy are surgical anatomy and anatomical art. Anatomy encompasses the following systems: musculoskeletal, nervous, circulatory, immune, respiratory, digestive and excretory, endocrine, reproductive, and integumentary. These systems differ widely among animals, but most animals need to fulfill the functions of these anatomical structures in one way or another. For simplicity's sake, anatomy of warm-blooded vertebrate creatures will be discussed here.

HOW IT WORKS

Musculoskeletal System. A muscle is a tissue composed of fibers capable of contracting and relaxing to effect bodily movement. The skeleton is the internal supporting structure of a vertebrate, composed of bone and cartilage. Skeletons are bound together by tough and relatively inelastic connective tissues called ligaments. Ligaments allow the limbs, connected by joints, to move freely. Movements of the bones of the skeleton are effected by contractions of the skeletal muscles, to which tendons attach the bones. These muscular contractions are controlled by the nervous system.

Nervous System. The nervous system in vertebrates is a network of cells, tissues and organs that regulates the body's responses to internal and external stimuli. The nervous system has two divisions: the somatic, which allows voluntary control over skeletal muscle, and the autonomic, which is involuntary and controls cardiac and smooth muscle and glands. The autonomic nervous system has two divisions: the sympathetic and the parasympathetic. These divisions tend to have opposing effects. For example, the sympathetic system increases heartbeat, and the parasympathetic system decreases heartbeat. However, the two nervous systems are not always antagonistic. For example, both nerve supplies to the salivary glands excite the cells of secretion. Voluntary movement of head, limbs, and body is caused by nerve impulses arising in the motor area of the cortex of the brain and carried by cranial nerves or by nerves that emerge from the spinal cord to connect with skeletal muscles. Movement may occur also in direct response to an outside stimulus. These involuntary responses are called reflexes. Muscular contractions do not always cause actual movement. A small percentage of the total numbers of fibers in most muscles are usually contracting. This serves to maintain the posture of a limb. This slight continuous contraction is called muscle tone.

Circulatory System. The circulatory system is composed of the heart, blood vessels, and lymphatic system of the body. Blood is pumped by the heart through the right chambers of the heart, into the lungs, where it picks up oxygen, and back into the left chambers of the heart. From these, it is pumped into the main artery, the aorta, which branches into increasingly smaller arteries until it passes through the smallest, known as arterioles. Beyond the arterioles, the blood passes through a vast number of tiny, thin-walled structures called capillaries. Here, the blood gives up its oxygen and its nutrients to the tissues and absorbs from them carbon dioxide and other waste products of metabolism. The blood completes its circuit by passing through small veins that join to form increasingly larger vessels until it reaches the largest veins, the inferior and superior venae cavae, which return it to the right side of the heart. Contractions of the heart working with the contractions of the skeletal muscle propel the blood and contribute to circulation. Valves in the heart and in the veins ensure blood flow in one direction.

Immune System. The immune system is an integrated system of organs, tissues, cells, and cell by-products (such as antibodies) that differentiates self from non-self and neutralizes potentially pathogenic organisms or substances. The body defends itself against foreign proteins and infectious microorganisms by means of a complex dual system that depends on recognizing foreign patterns. The two parts of the system are termed cellular immunity, in which lymphocytes are the effective agent, and humoral immunity, based on the action of molecules. When particular lymphocytes recognize a foreign molecular pattern, termed an antigen, they release antibodies in great numbers; other lymphocytes store the memory of the pattern for future release of antibodies should the molecule reappear. Antibodies attach themselves to the antigen and mark it for destruction by other substances in the body's defense system, such as enzymes and phagocytes. The latter are cells that engulf and digest foreign matter.

Respiratory System. The respiratory system comprises the organs involved in the intake and exchange of oxygen and carbon dioxide between an organism and the environment. Respiration is carried on by the expansion and contraction of the lungs; the process and the rate at which it proceeds are controlled by a nervous center in the brain. In the lungs, oxygen enters tiny capillaries, where it combines with hemoglobin in the red blood cells and is carried to the tissues. Simultaneously, carbon dioxide, which entered the blood in its passages through the tissues, passes through capillaries into the air contained within the lungs. Inhaling draws into the lungs air that is higher in oxygen and lower in carbon dioxide; exhaling forces from the lungs air that is high in carbon dioxide and low in oxygen. Changes in the size and gross capacity of the chest are controlled by contractions of the diaphragm and of the muscles between the ribs.

Digestive System. The digestive system is composed of the alimentary canal, along with glands such as the liver, salivary glands, and pancreas that produce substances needed in digestion. Digestion starts with the ingestion and chewing of food mixed with saliva. The food passes down the esophagus into the stomach, where the gastric and intestinal juices continue the process. Thereafter, the mixture of food and secretions, called chyme, is pushed down the alimentary canal by peristalsis, rhythmic contractions of the smooth muscle of the gastrointestinal system. The contractions are initiated by the parasympathetic nervous system and can be inhibited by the sympathetic nervous system.

In ruminants, the stomach has multiple sections, and chyme is passed back and forth several times between the stomach sections and the mouth for rechewing and redigestion. Absorption of nutrients from chyme occurs mainly in the small intestine; unabsorbed food and secretions and waste substances from the liver pass to the large intestines and are expelled as feces. Water and water-soluble substances travel via the bloodstream from the intestines to the kidneys, which absorb all the constituents of the blood plasma except its proteins. The kidneys return most of the water and salts to the body, while excreting other salts and waste products, along with excess water, as urine.

Endocrine System. The endocrine system involves internal secretions related to the function of the endocrine glands such as the thyroid, adrenal, pituitary. Hormonal secretions from these glands pass directly into the blood stream. An important part of this system, the pituitary, lies at the base of the brain. This master gland secretes a variety of hormones, including hormones that stimulate the thyroid gland and control its secretion of thyroxine, which dictates the rate at which all cells utilize oxygen; control the secretion in the adrenal gland of hormones that influence the metabolism of carbohydrates, sodium, and potassium and control the rate at which substances are exchanged between blood and tissue fluid; control the secretion in the ovaries of estrogen and progesterone and the creation in the testicles of testosterone; control the rate of development of the skeleton and large interior organs through its effect on the metabolism of proteins and carbohydrates; and inhibit insulin—a lack of insulin causes diabetes mellitus.

The posterior lobe of the pituitary secretes vasopressin, which acts on the kidney to control the volume of urine; a lack of vasopressin causes diabetes insipidus, which results in the passing of large volumes of urine. The posterior lobe also elaborates oxytocin, which causes contraction of smooth muscle in the intestines and small arteries and is used to bring about contractions of the uterus in birth. Other glands in the endocrine system are the pancreas, which secretes insulin, and the parathyroid, which secretes a hormone that regulates the quantity of calcium and phosphorus in the blood.

Reproductive System. The union of male sperm and the female ovum accomplishes reproduction. In coitus, the male organ ejaculates millions of sperm into the vagina, with some making their way to the uterus. Ovulation is the release of an egg into the uterus; the uterus is prepared for the implantation of a fertilized ovum by the action of estrogens. In some primates, if a male cell fails to unite with a female cell, other hormones cause the uterine wall to slough off during menstruation. After childbirth, prolactin, a hormone secreted by the pituitary, activates the production of milk.

Integumentary System. Skin, the natural outer covering of the body, is an important part of the integumentary system. The skin is an organ of double-layered tissue stretched over the surface of the body and protecting it from drying or losing fluid, from harmful external substances, and from extremes of temperature. The inner layer, called the dermis, contains sweat glands, blood vessels, nerve endings (sense receptors), and the bases of hair and nails. The outer layer, the epidermis, is only a few cells thick and contains pigments, pores, and ducts, and its surface is made of dead cells that it sheds from the body. The sweat glands excrete waste and cool the body through evaporation of fluid droplets; the blood vessels of the dermis supplement temperature regulation by contracting to preserve body heat and expanding to dissipate it. Separate kinds of receptors convey pressure, temperature, and pain. Fat cells in the dermis insulate the body, and oil glands lubricate the epidermis.

—Mary E. Carey

See also: Bone and cartilage; Brain; Cell types; Digestive tract; Eyes; Fur and hair; Noses; Kidneys and other excretory structures; Skin.

Further Reading

Ankel-Simons, Friderun. *Primate Anatomy*. 2d ed. San Diego, Calif.: Academic Press, 1999. Focuses on all the organ systems of primate species. Many excellent illustrations. Bone, Jesse F. Animal Anatomy and Physiology. 3d ed. Englewood Cliffs, N.J.: Prentice Hall, 1996.Aveterinary textbook that takes a systematic approach to animal anatomy.

Feher, Gyorgy. Cyclopedia *Anatomicae: More than 1,500 Illustrations of the Human and Animal Figure for the Artist*. New York: Black Dog & Leventhal, 1996. Acompendium of anatomical illustrations, focusing on musculoskeletal systems. Focuses on the comparative anatomies of humans, horses, dogs, cats, lions, sheep, cattle, hogs, camels, apes, crocodiles, and seals.

Hildebrand, Milton. *Analysis of Vertebrate Structure*. 4th ed. New York: John Wiley & Sons, 1994. A classic textbook on vertebrate morphology. Uses an organ system approach and relates morphology to evolution. Excellent illustrations, accessible to nonspecialists.

Walker, Richard. *Animal Anatomy on File Collection*. New York: Facts on File, 1990. A guide to the external and internal structure of animals, grouped thematically into eight sections (introduction, lower groups, annelids and mollusks, arthropods and echinoderms, fish, amphibians and reptiles, birds, and mammals).

ANIMAL KINGDOM

FIELDS OF STUDY

Ecology, evolutionary science, genetics, population biology, systematics (taxonomy)

SUMMARY

Among all the species that have been identified, about 75 percent are animals. Animals flourish on land, in the seas, and in the air. Today, twenty-seven distinct phyla of animals come in diverse forms and shapes, and live in almost every habitat. Together, these animals make up a crucial portion of all ecosystems.

PRINCIPAL TERMS

- **class:** the taxonomic category composed of related genera; closely related classes form a phylum or division
- **invertebrates:** animals lacking a backbone phylogeny: the evolutionary history of a group of species
- **phylum:** the taxonomic category of animals and animal-like protists that is contained within a kingdom and consists of related classes
- **species:** a group of animals capable of interbreeding under normal natural conditions; the smallest major taxonomic category
- **taxonomy:** the science by which organisms are classified into hierarchically arranged categories that reflect their evolutionary relationship
- **vertebrates:** animals with a backbone or vertebral column

BASIC PRINCIPLES

Human perception of the animal kingdom tends to focus on relatively large vertebrates. However, these large vertebrates are true minorities, accounting for just a tiny fraction of the animal world. Over 97 percent of animal species are invertebrates, the earliest animals to emerge. Insects and arthropods make up the vast majority of animal species and a huge percentage of the individual animals on earth. Most other animal phyla are also far more diverse and numerous than vertebrates. All vertebrates together constitute only part of a single phylum, Chordata. In simple terms, the small and boneless creatures called invertebrates dominate the animal kingdom. They live bountifully in diverse habitats: in pond muck, on ocean bottoms, in treetops, beneath leaf litters, and in many other environments.

Animals are easy to identify but difficult to define due to the diversity and complexity of all creatures in this kingdom. The best approach relies on a set of common characteristics that distinguish animals from individuals of other kingdoms, a field called systematics.

First, animals are multicellular (made up of many cells). Second, animals are heterotrophic, obtaining nutrients and energy by consuming other organisms. Third, animals are usually capable of sexual reproduction, although other reproductive styles may exist. Fourth, animal cells contain no cell wall. Fifth, animals are mobile during at least some stage of their lives. Finally, animals are usually capable of rapidly responding to external stimuli through their nerve cells, muscle, or contractile tissue. These six characteristics taken together distinguish animals from other living creatures.

Based upon evolutionary theories, animal phyla show trends toward increasing cellular organization and complexity. In the most ancient phylum of animals, sponges, individual cells may have specialized functions but act independently, hence are not organized into tissues or organs. Cnidarians (jellyfish and their relatives), the phylum most closely related to sponges, have well-defined tissues that coordinate movement and sensory information. Flatworms, the next phylum to emerge, have organs and organ systems, such as a reproductive system. Organ systems are also found in all the remaining, more recently emerged animal phyla. The trend toward increasing complexity goes beyond the level of cellular organization and specialization. It includes the presence and type of symmetry in body plan, the degree of development in sensory organs and brain, the presence and type of body cavity, the presence of body segmentation, and the structure of the digestive system. The members of the latest phylum, including vertebrate animals such as seals, whales, horses, and humans, also exhibit a trend toward increasing size and sophistication of the brain. Based upon these traits, animals can be grouped into twenty-seven phyla. The nine major phyla include, from simple to more complex, Porifera (sponges), Cnidaria (hydra, anemones, and jellyfish), Platyhelminthes (flatworms), Nematoda (roundworms), Annelida (segmented worms), Arthropoda (insects, arachnids, and crustaceans), Mollusca (snails, clams, and squid), Echinodermata (sea stars and sea urchins), and Chordata (primarily vertebrates).

How It Works
The Sponges, Hydra, Anemones, and Jellyfish.
Sponges (phylum Porifera) are the simplest multicellular animals that lack true tissues and organs. They resemble colonies in which single-celled organisms live together for mutual benefit. However, individual sponge cells are able to survive and function independently. All sponges, whether single-celled or colony-like, have a similar body plan. The body is perforated by numerous tiny pores, through which water enters, and by fewer large holes, through which water is expelled. Water travels within the sponge through canals where oxygen is extracted and microorganisms are filtered into cells for digestion. Some sponges can grow more than a meter in height. So far, more than five thousand species of sponges have been identified, all of which are aquatic and most of which are marine.

The phylum Cnidaria is composed of hydra, anemones, and jellyfish. Clearly more complex than sponges, cnidarians have distinct tissues, including contractile tissue that acts like muscle and nerve net that spreads through the body and controls movement and feeding behavior. However, they lack true organs and a brain. Their beautiful and diverse body shapes are variations of two basic body plans: tentacled and jellyfish-like. Tentacles attach to rocks and reach upward for grasping, stinging, and immobilizing prey. A jellyfish-like body can easily be carried by ocean currents. Cnidarians are radially symmetrical, with body parts arranged in a circle around the mouth and digestive cavity. All cnidarians are predators, but none hunt actively. They rely upon their tentacles to grasp small animals floundering by chance into contact with them. Once stimulated by contact, special cells called cnidocytes explosively inject poisonous or sticky darts into prey. The immobilized prey is forced through an elastic mouth into a digestive sac. The undigested food is expelled through the mouth. Cnidarians may reproduce sexually or asexually. Of the nine thousand or more species in this phylum, all are aquatic and most are marine. One of these, the corals, is of particular ecological importance.

Diverse Forms of Worms.
Flatworms (phylum Platyhelminthes) are more complex than cnidarians, yet are the simplest organisms with well-developed organs. Their bilaterally symmetrical bodies are an adaptation to active movement, as found in other, more complex organisms. Their sense organs, consisting of light-detecting eyespots and cells responsive to chemical and tactile stimuli, inform their bodies whether to feed, forge onward, or retreat. When the flatworm encounters smaller animals, it sucks its prey through a muscular tube called the pharynx, located in the middle of the body. Compared with more complex organisms, however, flatworms lack both respiratory and circulatory

systems. They can produce sexually or asexually. Most flatworms are hermaphroditic, possessing both male and female sex organs within one body. Examples of flatworms include parasitic tapeworms and flukes.

Roundworms (phylum Nematoda) reside in nearly every habitat on earth. Of an estimated 500,000 species, only 10,000 have been named. They are largely microscopic, although some may reach a meter in length. They have a rather simple body plan, with a tubular gut that runs from mouth to anus. A fluid-filled hydrostatic skeleton provides support and a framework against which muscles can act. They also have a tough but flexible cuticle on the outside of the body and a simple brain that processes and transmits information. They do not have circulatory or respiratory systems. Most nematodes reproduce sexually, with the male fertilizing the female by injecting sperm inside her body. Nematodes play a crucial role in breaking down organic matter in ecosystems. Some are also parasites to humans or other animals, such as hookworms that infect human feet, *Trichinella* worms that cause trichinosis, and heartworms that attack dogs' hearts.

The prominent feature of the phylum Annelida is segmentation of the body into a series of repeating units; hence they are called segmented worms. Each body compartment is controlled by separate muscles, collectively capable of far greater complexity of movement than in other worms. A well-developed closed circulatory system distributes gases and nutrients throughout the body. Primitive hearts, in essence short, expanded segments of specialized blood vessels, can contract rhythmically. A simple brain located in the head plus nerve cords along the length of the body and within each segment control movement and other activities. Among the nine thousand or so species identified, the best known examples are the earthworm and its relatives, and leeches. However, the largest annelids, the polychaetes, live primarily in the ocean.

The Arthropods, Molluscs, and Echinoderms. The phylum Arthropoda comprises insects, spiders, and crustaceans. By any standard, whether number of individuals or number of species, arthropods are the most dominant animals on earth. A mere 10 percent of animals described in this phylum constitutes one million species, including insects (class Insecta), spiders and their relatives (class Arachnida), and crabs, shrimp, and their relatives (class Crustacea). The enormous success of arthropods is due to several adaptation features. The exoskeleton allows precision movement; segmentation generates specialized and more effective organ systems; these, in turn, allow higher efficiency in gas exchange, circulation, and information processing. Most arthropods have well-developed sensory systems, including compound eyes and acute chemical and tactile senses. Of the three classes, insects are the most diverse and abundant, accounting for 850,000 species identified. Insects usually have three pairs of legs plus two pairs of wings. Their ability to fly helps them escape from predators and find widely dispersed food. Insects normally go through radical changes in body form through metamorphosis, from egg to larva to pupa and finally to winged adults that mate and lay eggs.

Spiders and scorpions are examples of the class Arachnida. They typically have eight walking legs and are mostly carnivores, living on either a liquid diet of blood (ticks and mosquitoes) or predigested prey (scorpions). Simple eyes equipped with a single lens are extremely sensitive to movement, which helps in catching prey or escaping from predators. There are about fifty thousand species of arachnids. Crab, shrimp, crayfish, and their relatives make up the class Crustacea, comprising roughly thirty thousand species. They are largely aquatic, with a wide variation in size. Except for two pairs of sensory antenna and mostly compound eyes, they are highly variable in body form.

As their name suggests, members of the phylum Mollusca—snails, clams, and squid—have a moist, muscular body supported by a hydrostatic skeleton. Some have a shell of calcium carbonate to protect their body; others escape predation by moving swiftly or by being distasteful if caught. They have an open circulatory system. Their nerve systems are more advanced than those of arthropods in that more nerves are concentrated in the brain. Reproduction is sexual; some species have separate sexes, and others are hermaphroditic. Together, there are five thousand species identified, among which clams, octopuses, oysters, scallops, snails, and squid are the most familiar. Sea stars, sea urchins, and sea cucumbers compose the phylum Echinodermata. These animals are mostly marine, and adults have radial symmetry and lack a head and distinct brain. They have very simple nervous systems, and hence move very slowly on numerous, tiny, tube feet. They feed on algae or small particles sifted from sand or water. Most species reproduce by releasing sperm and eggs into the water, where larvae develop upon fertilization.

Another distinct feature of echinoderms is their endoskeleton, a hard shell of calcium carbonate enclosed by an outer skin.

Phylum Chordata: The Tunicates, Lancelets, and Vertebrates. Animals of this phylum exhibit tremendous diversity in form and size. They include small sea squirts and lancelets (invertebrates), and birds, fish, amphibians, reptiles, and mammals (vertebrates). Members of this phylum possess four characteristics at some stage of their lives: a notochord—a stiff yet flexible rod that extends the length of the body and provides an attachment site for muscles; a dorsal, hollow nerve cord at the anterior end of the notochord that becomes a brain; specialized respiratory openings called pharyngeal gill slits; and a tail that extends past the anus. There are only two classes of invertebrates in Chordata, lancelets and tunicates, both of which are small marine animals. Lancelets reside mainly in the sandy sea bottom and live by filtering tiny food particles from the water. Sea squirts, a member of the tunicates, send out a forceful jet of water in response to touch or danger. Their filter-feeding, saclike bodies move slowly via contraction.

Vertebrates are the most conspicuous animals on earth. Their backbones and other adaptations have contributed to their success. There are seven major classes of vertebrates. Jawless fishes (Agnatha) were the earliest vertebrates to arise in the sea. Two examples are hagfishes and lampreys. The colorful hagfishes are strictly marine, living in communal burrows in mud, feeding on polychaete worms. Lampreys live in both fresh and salt water. Some lampreys are parasitic, attaching to fish with sucker-like mouths lined with rasping teeth. They live on blood and body fluids sucked from their hosts.

Cartilaginous fishes (Chondrichthyes) are skillful predators, and include sharks, skates, and rays. Their skeletons are made up exclusively of cartilage, void of bone. Many shark species have several rows of razor-sharp teeth, with back rows moving forward as front teeth are lost to action or aging. Most sharks, as most skates and rays, are shy and retiring creatures that do not attack humans. A few species, however, can be deadly when irritated. Bony fishes (Osteichthyes), spread over a wide range of aquatic habitats, are the most diverse and abundant vertebrates on earth. As suggested by their name, bones rather than cartilage make up their skeletons. Of seventeen thousand species identified, all bony fishes have bladders that help them float effortlessly. Some have lungs and modified fins that work as legs, which help them to survive periodic drying in freshwater habitats.

Carl Linnaeus, known as the father of modern taxonomy

Amphibians (Amphibia) live a double life between aquatic and terrestrial habitats. They represent the transition of life from water to land. Some adaptations, such as lungs, a three-chambered heart, and moist skin, help them live a temporary land life. However, other traits, requiring water for fertilization and juvenile development, restrict the range of amphibian habitats on land. Their double life and permeable skin have made amphibians particularly vulnerable to pollutants and environmental fouling. About 2,500 species have been identified, including frogs, toads, and salamanders. The seven thousand species of reptiles (Reptilia) identified have bodies of diverse forms. Turtles, snakes, lizards, alligators, and crocodiles are all reptiles, as well as the huge and now-extinct dinosaurs. Reptiles have more efficient lungs than amphibians, a tough, scaly skin that resists water loss and protects the body, a mechanism of internal fertilization, and a shelled egg.

The diversity of birds (Aves) is revealed through nine thousand species, including the delicate hummingbird, the endangered spotted owl, and the largest bird, the ostrich. Their ability to soar gracefully in the air depends on many anatomical and physiological traits. These features include a light body with hollow bones, light wings with feathers that also provide protection and insulation, reduced reproductive organs during non-breeding periods, a single ovary in female birds, acute eyesight, and a delicate nervous system that facilitates the extraordinary coordination and balance needed for flight. Birds, which are warm-blooded, also have four-chambered hearts that help to maintain high body temperature and a high metabolic rate, crucial for flight.

The last vertebrate class, mammals (Mammalia), is represented by some 4,500 species. In addition to being warm-blooded with high metabolic rates, mammals normally possess hair, produce milk for their offspring, assume a remarkable diversity in form, and possess more highly developed brains than any other class. The bat, cheetah, elephant, mole, monkey, seal, and whale exemplify the radiation of mammals into nearly all habitats, with their bodies finely adapted to their lifestyles.

—*Ming Y. Zheng*

APES TO HOMINIDS

FIELDS OF STUDY

Anthropology, evolutionary science, genetics, systematics (taxonomy), zoology

SUMMARY

Although the study of fossil apes has taught scientists much regarding certain aspects of primate evolutionary development, they cannot yet trace more than a general connection between the apes and hominids such as man.

PRINCIPAL TERMS

- **apes:** large, tailless, semi-erect anthropoid primates, including chimpanzees, gorillas, gibbons, orangutans, and their direct ancestors—but excluding man and his direct ancestors
- **australopithecines:** nonhuman hominids, commonly regarded as ancestral to man

FURTHER READING

Blaustein, A. "Amphibians in a Bad Light." *Natural History* 103 (October, 1994): 32-39. Thorough examination of the role of increased ultraviolet light, which is penetrating a depleted ozone layer, in recent population declines of amphibians.

Brusca, R. C., and G. J. Brusca. *Invertebrates.* Sunderland, Mass.: Sinauer Associates, 1990. A survey of invertebrates, filled with interesting reading and beautiful drawings. Very informative.

McMenamin, M. A., and D. L. McMenamin. *The Emergence of Animals: The Cambrian Breakthrough.* New York: Columbia University Press, 1990. An examination of the adaptive radiation that resulted in an explosion of animal forms at the beginning of the Cambrian period.

Morell, V. "Life on a Grain of Sand." *Discover* 16 (April, 1995): 78-86. A close look at the sand beneath shallow waters, home to incredibly diverse microscopic creatures.

Rennie, J. "Living Together." *Scientific American* 266 (January, 1992): 122. The fascinating interactions between parasites and their hosts provide insights into life on earth.

- **dryopithecines:** extinct Miocene-Pliocene apes; their evolutionary significance is unclear
- **hominid:** an anthropoid primate of the family Hominidae, including the genera *Homo* and *Australopithecus*
- **human:** a hominid of the genus *Homo*, whether *Homo sapiens sapiens* (to which all varieties of modern man belong), earlier forms of *Homo sapiens*, or such presumably related types as *Homo erectus*
- **primates:** placental mammals, primarily arboreal, whether anthropoid (humans, apes, and monkeys) or prosimian (lemurs, lorises, and tarsiers)
- **stratigraphy:** in geology, a sequence of sedimentary or volcanic layers, or the study of them—indispensable for dating specimens

BASIC PRINCIPLES

Primates are an order of the class Mammalia. The Primate order is divided into two suborders. Suborder

Prosimii (lower primates) includes lemurs, lorises, and tarsiers. Suborder Anthropoidea (higher primates), to which monkeys, apes, and humans all belong, is divided further into infraorders: Platyrrhini (flat-nosed New World monkeys, definitely not ancestral to man) and Catarrhini (down-nosed Old World monkeys, apes, and man). The infraorder Catarrhini includes two superfamilies: Cercopithecoidea (Old World monkeys) and Hominoidea (apes and man). Within the Hominoidea, finally, are three families: Hylobatidae (lesser apes), Pongidae (great apes), and Hominidae (man). As this classification suggests, it is now taken for granted that human ancestry— if it could be traced satisfactorily—would include forms that, on other genealogies, gave rise to lower and higher primates, Old World monkeys, and a series of now-extinct creatures that were ancestral to certain of the apes as well.

The lower primates (prosimians) first appeared about seventy million years ago. They still exist (as lemurs, lorises, and tarsiers) but have been declining for the last thirty million years, probably because of unsuccessful competition with their own descendants, the monkeys. Prosimians have five digits on each limb, but the digits have claws rather than nails, and the limbs are entirely quadrupedal. Prosimians also lack binocular vision, but they do have dentition anticipating the molar development of the higher primates.

THE HIGHER PRIMATE FOSSIL RECORD
The earliest evidence of any kind of higher primate— some tiny pieces of jaw found in Burma— dates from the Eocene epoch, about forty million years ago. Two creatures named *Amphipithecus* ("near ape") and *Pondaungia* ("found in the Pondaung Hills") have been proposed, each being a very primitive monkey or ape, but the evidence thus far is too sparse to ally these forms with any possible descendants. Some two million years later, in the Fayum Depression of Egypt (then a lush forest), *Apidium* and *Parapithecus* ("past ape") existed. Known only from jaws and teeth, they are the oldest known Old World monkeys presently recognized. Their dental pattern (arrangement of teeth), however, is the same as that of *Amphipithecus*. Other teeth from the Fayum, perhaps thirty-five to thirty million years old, have a different cusp pattern, more like an ape's (and a human's) than a monkey's. Possibly, then, *Propliopithecus* ("before more recent ape") is the earliest evidence of an ape line distinct from the monkey line.

The oldest apelike animal about which scientists know enough to regard it as a probable human ancestor is *Aegyptopithecus* ("Egyptian ape"), also found in the Fayum, in Oligocene deposits about thirty-two million years old. In addition to jaws and teeth, an almost complete skull and some postcranial bones (meaning those below the skull) have been recovered. Since the Fayum at that time consisted of dense tropical rain forest with little open space, *Aegyptopithecus* is assumed to have been an arboreal quadruped. (In 1871, before *Aegyptopithecus* was known, Charles Darwin had predicted that such a human ancestor existed.) It is the most primitive ape yet discovered. *Proconsul* ("before Consul," Consul being a chimpanzee in the London Zoo in 1933) and *Dryopithecus* ("forest ape") were either closely related to each other or identical. *Proconsul* appeared in Africa at the start of the Middle Miocene (about twenty million years ago) and was contemporary with *Dryopithecus* in Europe and Asia about fourteen million years ago. The relatively abundant fossils of these forms have been classified by some researchers into three species, together forming an extinct subfamily, the Dryopithecinae. One species in particular, *Dryopithecus major*, regularly left its remains on what were then the forested slopes of volcanoes; males grew significantly larger than females (a situation known as sexual dimorphism). In both of these respects, Dryopithecus resembled the modern gorilla, to which it may be ancestral. Like the gorilla, *Dryopithecus major* probably walked on its knuckles. In size, it was somewhat larger than a chimpanzee. The other two species, *Dryopithecus nyanzae* and *Dryopithecus africanus*, were smaller than *Dryopithecus major* and more like the chimpanzee. Outside Africa, *Dryopithecus* has been found from Spain to China.

THE ANCESTORS OF MODERN APES AND HUMANS
Limnopithecus ("lake ape"), found in deposits in Kenya and Uganda of about twenty-three to fourteen million years ago, is thought to be an earlier form of *Pliopithecus* ("more recent ape"). Its gibbon-like skulls, jaws, and teeth are plentiful in European sediments of Middle Miocene to Early Pliocene age—sixteen to ten million years ago. For some researchers, these two forms constitute a separate subfamily, the Pliopithecinae, which they consider to be part of the

family Hylobatidae (lesser apes). In some respects, they resembled the modern gibbon, but other aspects of their anatomy were quite different. For example, *Pliopithecus* possessed seven lumbar vertebrae, whereas gibbons (and humans) have only five. It seems to have been primarily arboreal, swinging from branch to branch. *Pliopithecus* has been known since 1837 (in France), and since then some almost complete skeletons have been recovered. *Sivapithecus* ("Siva's ape," Siva being a Hindu deity), found in India and later in Africa, is a closely related form, Miocene in age. Both the dryopithecines and the pliopithecines are often regarded as the ancestors of modern apes.

Ramapithecus ("Rama's ape," Rama being another Hindu deity), found originally as a jaw fragment in India, is remarkable for its human-looking teeth. Some researchers regard it as the earliest member of the hominid line and therefore ancestral to humans. Others, however, relate *Ramapithecus* and *Sivapithecus* to modern orangutans, seeing no direct connection to man. Though *Ramapithecus* has been recovered from Late Miocene deposits in Africa and Indian and Early Pliocene ones in India (about fourteen to ten million years ago), only teeth and jaws have been found. As a result, many opinions regarding *Ramapithecus* are highly conjectural. The most striking feature of this genus, for example, is the greatly reduced size of its canine teeth, as compared with those of earlier (as well as modern) apes. Presumably, this indicates a changed diet of some sort. However, primates also use their teeth for nondietary purposes, including weaponry and display. It has therefore been suggested that the reduced tooth size of *Ramapithecus* might indicate its having begun to use other tools or weapons; if so, none has ever been found. Another conjecture has been that climatic change brought the primates down from the trees. Once on the ground, *Ramapithecus* then developed a hunter-gatherer style of sustenance that eventually included the formation of family units (male-female bonds), tool making, and a rudimentary form of language— the beginnings of culture. Unfortunately, all that is really known about *Ramapithecus* is what can be observed from a smattering of its bones. Finally, there was *Gigantopithecus* ("giant ape"), a huge simian with protohuman teeth (clearly not ancestral to man, however) that outlasted *Dryopithecus*, *Ramapithecus*, and *Sivapithecus* to survive in Asia for almost nine million years. The largest primate that ever evolved (exactly how large is not known), it was alive in China as recently as a million years ago. Known for its immense molars, *Gigantopithecus* was apparently the only successful ground-living savanna ape. It probably competed with early hominid forms and may have been exterminated by them.

In broad outline, then, these are the fossil apes. Since much of the evidence (all of it, in several cases) consists of teeth and jawbones, it is not surprising that conjecture has played a very active part in attempts to associate this evidence with the evolution of the hominids. Before 1980, there was widespread consensus among experts with regard to an evolutionary main line extending at least from *Aegyptopithecus* through *Dryopithecus* (or *Proconsul*) and *Sivapithecus* to *Ramapithecus*, the latter being regarded as the first hominid. However, portions of two *Sivapithecus* faces, recovered from Turkey in 1980 and Pakistan in 1982, impressed researchers with their orangutan-like characteristics. Since firm ties between *Sivapithecus* and *Ramapithecus* had already been established, it began to seem that the entire lineage pointed toward the orangs rather than toward man. Another problem is that formerly accepted dating has come into question for such important branchings of the lineage as those which separated monkeys from apes and apes from man. New genetic studies having nothing to do with either fossils or stratigraphy have presented compelling (but controversial) arguments to the effect that these branchings occurred much later than hitherto believed. A third, even more serious problem is that there is virtually no pertinent fossil evidence regarding the development of simian primates into humans for a period beginning about fourteen million years ago and lasting until the appearance of the australopithecines about four million years ago. While anthropologists and biologists continue to learn more about the ancestry of modern simians, therefore, it is certainly not the case that a reliable lineage (or even a timetable) leading from other primates to humankind has been established.

The Problems of Theorizing from Fossils

The study of fossil apes is a specialization within the broader field of vertebrate paleontology, or the study of fossil bones. Like all paleontologists, therefore, paleoprimatologists are necessarily concerned with fossils and their stratigraphic occurrence. Because primates still exist, however, it is also important to study the behavior of living examples. Since behavior reflects environmental conditions, it is further necessary to reconstruct the climate, flora, and fauna of the region and time in which the fossils were found.

No complete fossil ape has ever been found. Any understanding of what they may have looked like is therefore conjectural—an extrapolation from what has been recovered to what has not. Skulls are undoubtedly the most desirable evidence, but they are not the most durable of fossils. Teeth, which constitute the hardest parts of the primate body, are preserved more often than any other part. Some kinds of fossil ape are known either exclusively or primarily from their teeth and jaws. Ape teeth differ from human teeth in two significant respects: They are generally larger (the canines especially), and the cusp patterns on their molars differ. The arrangement of teeth in an ape's jaw, moreover, is angular, like a V; in a human, the arrangement is rounder, like a U. Inevitably, whenever jawbones or molar teeth are found, an attempt is made to place them somewhere on a continuum that runs between the purely simian (ape) and the purely human. This procedure not only distinguishes primitive apes from primitive humans, and one kind of fossil ape from another, but also gives rise to inevitable conjecture as to possible anticipations of the human line.

A major difficulty with evolutionary sequences based solely upon dental evidence is that the head and body of a given species have not necessarily evolved at the same rate. One may be surprisingly apelike, the other somewhat human. Even more specifically, the fact that jaws are changing does not necessarily mean that crania (or any other specific body parts) are changing also. On the whole, scientists do not yet understand the evolution of primate anatomy well enough to interpret present evidence or reconstruct missing parts with much reliability. In the absence of factual evidence, the form taken by prevailing reconstructions at any given time may owe as much to professional politics as to objective knowledge. The controversy regarding *Ramapithecus*, which (on the basis of facial bones) moved that genus from a central position at the base of the human lineage to a similar position on a separate orangutan genealogy, has been a valuable lesson in the folly of premature commitment.

Ancient Fossils and Modern Primates
Most paleoprimatologists are to some extent modern-day primatologists also, necessarily expert in the comparative anatomy of all members of the higher primates and as knowledgeable as possible regarding their behaviors and environments. It is assumed that a changing environment (one becoming increasingly arid, for example) requires behavioral modifications and that these modifications will then create selection pressures favoring some types of anatomical variation over others; thus, one species will eventually change into (or be replaced by) another. The process is not well understood, but the primates—being so intensely studied—are often regarded as test cases for competing evolutionary theories. Some researchers stress evidence to the effect that species are always changing; others believe that species are created precipitously and then tend to endure relatively unchanged until they are abruptly superseded. All that is really known at present is that the ancient apes generally conform to a partially ascertainable progression from hypothetical earlier forms to modern-day primates—with a ten-million-year gap in between.

The Importance of Fossil Apes
The fossil apes are important for four reasons: They are an important group in their own right; they are important to the development of the mammals; they are important to the development of the primates; and, they are thought to be ancestral to humankind and therefore uniquely important among all nonhuman fossil genera. One of the unique characteristics of humans is the ability to create, preserve, and transmit knowledge. *Homo sapiens* has learned to value learning and to hoard and increase knowledge in the realization that it fortifies and enhances his existence. Humans attempt to know and understand all present-day forms of life; in order to do so, however, it must also be known how these forms came into being through time. Biology, then, is inherently evolutionary and does not sharply distinguish between plants and animals of the past and those of the present.

Insofar as an understanding of life itself is the goal of biological studies, no single form of life is inherently more important than any other. From this point of view, one would say that fossil and living apes are studied for the same reason that algae, sponges, or nematodes are. A number of biologists would maintain this view. Many others, however, believe mammals—and especially primates— to be a "higher" form of life, anatomically more complex than sponges (though not necessarily more nearly perfect) and certainly capable of more complex behaviors. No mere study of anatomy, this viewpoint suggests, can sufficiently explain a primate.

The outstanding characteristic of all primates is their intelligence. One can find surprising levels of intelligence in other animals, however: Among invertebrates, such cephalopods as the squid and the octopus have the highly developed nervous systems, senses, and brains that are normally associated with mammals. Together with some birds and social insects, all the mammals are capable of surprisingly complex behavior. Nevertheless, the higher primates constitute intellectual elite even among the mammals. Impressive as gorillas and chimpanzees can be in this respect, it is apparent that the human mind has a capacity well beyond theirs. The brains of extinct apes are seen as having been ancestral not only to those of modern apes but also to the brain of man. Scientists are fortunate in the number of fossil ape skulls that have been found, for they make the increasing mental capacity of the higher primates easy to establish. Limb bones and other less durable parts of the skeleton are much rarer. When available, they indicate the relative lengths of arms and legs; the nature of the shoulder (a key to arboreal existence); the relation of pelvis and femur (a key to posture); and the shapes, capabilities, and functions of hands and feet. Without such evidence, scientists have only conjectures based upon the presumed place of the genus in question within a supposed evolutionary sequence. When proposed sequences differ, though they are derived from the same sparse evidence, conflicting suppositions about the evolutionary sequence are at work.

—*Dennis R. Dean*

See also: Evolution: Historical perspective; Gene flow; Genetics; Hominids; *Homo sapiens* and human diversification; Human evolution analysis; Neanderthals.

FURTHER READING
Berger, Lee. *In the Footsteps of Eve: The Mystery of Human Origins.* Washington, D.C.: Adventure Press/National Geographic Society, 2000. Argues that humans originated in South Africa rather than East Africa, based on the author's own fieldwork with hominid fossils discovered in South African caves.

Campbell, Bernard. *Humankind Emerging.* 8th ed. Boston: Allyn & Bacon, 2000. See especially the chapters "Back beyond the Apes" and "The Behavior of Living Primates." Though a college-level text, this book is useful to everyone because of its comprehensive scope. Older sources should be checked against it.

Conroy, Glenn C. *Primate Evolution.* New York: W. W. Norton, 1990. A very readable book for students and general readers. Emphasizes the evolution, phylogeny, and classification of hominids, linking the fragmented fossil record with behavior and culture.

Jordan, Paul. *Neanderthal: Neanderthal Man and the Story of Human Origins.* Phoenix Mill, England: Sutton, 1999. A clearly written book for the general reader, highlighting all the ideas involved in the study of human evolution. Describes the discovery of Neanderthal Man, reconstructs the Neanderthal environment and way of life, and traces the emergence of modern humankind.

Lewin, Roger. *Bones of Contention: Controversies in the Search for Human Origins.* New York: Simon & Schuster, 1987. One rather involved chapter of this book reviews the *Ramapithecus* debate in detail.

Martin, R. D. *Primate Origins and Evolution: A Phylogenetic Reconstruction.* London: Chapman and Hall, 1990. A wide-ranging book for those interested in mammals and evolutionary biology in general. Covers classification of mammals, the fossil record of mammalian origins, continental drift and evolution, and the problems of phylogenetic reconstruction.

Szaly, Frederick S., and Eric Delson. *Evolutionary History of the Primates.* New York: Academic Press, 1979. Though not intended for beginners and somewhat dated in places, this remains a standard reference.

Tattersall, Ian, and Jeffrey H. Schwartz. *Extinct Humans.* Boulder, Colo.: Westview Press, 2000. Tattersall excels at explaining complex paleoanthropological topics for the general reader. This book presents the idea of a "bushy" human evolutionary history, with many branches, as opposed to the now-obsolete notion that humankind evolved in a single linear path. Discusses the similarities and differences among *Australopithecus*, *Paranthropus*, and *Homo*, and the importance of tools in hominid mental evolution.

Artificial organs

FIELDS OF STUDY

Biology; anatomy; biophysics; chemistry; physics; mathematics; physiology; genetics; immunology; molecular biology; organ transplantation; biomedical engineering.

SUMMARY

Artificial organs are complex systems of natural or manufactured materials used to supplement failing organs to aid in recovery, sustain failing organs until transplantation, or replace failing organs that cannot recover. Some whole organs have artificial counterparts: heart, kidneys, liver, lungs, and pancreas. Smaller body parts also have artificial counterparts: blood, bones, heart valves, joints, skin, and teeth. In addition, there are mechanical support systems for circulation, hearing, and breathing. Artificial organs are composed of biomaterials, biological or synthetic materials that are adapted for use in medical applications.

PRINCIPAL TERMS

- **biocompatibility:** absence of immune reaction or rejection against biological or synthetic materials.
- **biohybrid:** interfacing of a biological material with a synthetic material.
- **biomaterial:** biological or synthetic material that is adapted for medical use.
- **extracorporeal:** outside the body; often used to describe large mechanical support systems.
- **hemodialysis:** removal of metabolic waste products and extra water from the blood in cases of kidney failure.
- **hemoperfusion:** removal of toxins from the blood in cases of liver failure.
- **immunomodulation:** exerting an affect on the immune system, either stimulation or suppression.
- **organ failure:** state in which an organ does not perform its natural functions.

Basic Principles

Artificial organs are complex systems that assist or replace failing organs. The human body is composed of eleven major organ systems: nervous, circulatory, respiratory, immune, digestive, excretory, reproductive, endocrine, integumentary (skin), muscular, and skeletal. The nervous system transmits signals between the brain and the body via the spinal cord and nerves. The circulatory system transports blood to deliver oxygen and nutrients to the body and to remove waste products. Its organs are the heart, blood, and blood vessels. It works closely with the respiratory system, in which the lungs and trachea perform oxygen exchange between the body and the environment. The immune systems work closely with all systems to protect and promote healthy conditions. The digestive system breaks down food and absorbs its nutrients. Its organs include the esophagus, stomach, intestinal tract, and liver. The excretory system rids the body of metabolic waste in the forms of urine and feces. The reproductive system provides sex cells and in females the organs to develop and carry an embryo to term. The endocrine system consists of the pituitary, parathyroid, and thyroid glands, which secrete regulatory hormones. The integumentary system is the body's external protection system. Its organs include skin, hair, and nails. The muscular system recruits muscles, ligaments, and tendons to move the parts of the skeletal system, which consists of bones and cartilage.

Background

While he was still a medical student in 1932, renowned cardiac surgeon Michael E. DeBakey introduced a dual-roller pump for blood transfusion. It has since become the most widely used type of clinical pump for cardiopulmonary bypass and hemodialysis. Physician John H. Gibbon, Jr., of Philadelphia, developed the first clinically successful heart-lung pump. He initially demonstrated it in 1953, when he closed a hole between the atria of an eighteen-year-old girl.

In 1954, American physician Joseph Murray performed the first successful human kidney transplant from one identical twin to the other in Boston. In 1962, he performed the first kidney transplant in unrelated persons. In 1967, surgeon Christiaan Barnard performed the first successful human heart transplant in Cape Town, South Africa. The patient, a fifty-four-year-old man, lived another eighteen days. Physician Willem J. Kolff is considered to be "the father of the artificial organ." In 1967, he emigrated from the Netherlands and spent a good deal

Artificial Heart

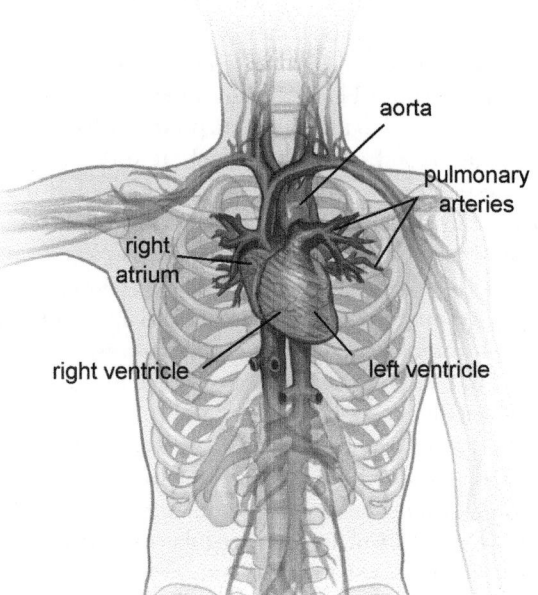

Normal Heart

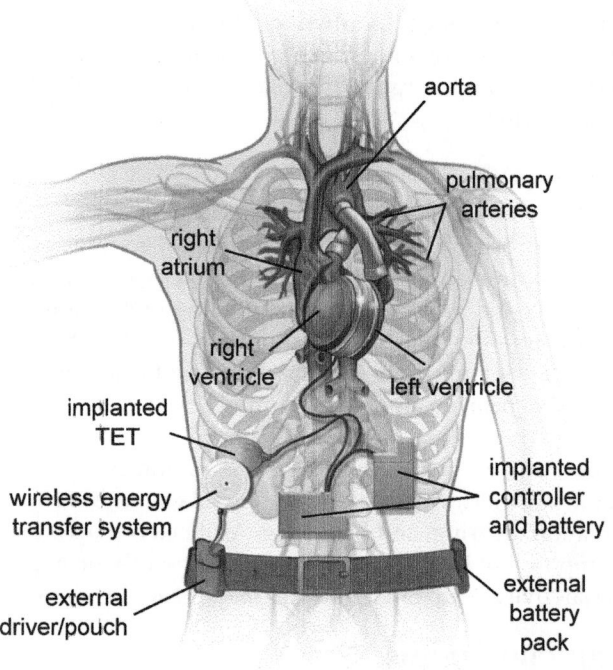

Replacement Heart

of his career at the University of Utah, where he became a distinguished professor emeritus of internal medicine, surgery, and bioengineering. He led the designing of numerous inventions, including the modern kidney dialysis machine, the intra-aortic balloon pump, an artificial eye, an artificial ear, and an implantable mechanical heart.

American physician Robert K. Jarvik refined Kolff's design into the Jarvik-7 artificial heart, intended for permanent use. In 1982, at the University of Utah, American surgeon William C. DeVries implanted it into retired dentist Barney Clark, who survived 112 days.

How It Works

The existence and performance of artificial organs depend on the collaboration of scientists, engineers, physicians, manufacturers, and regulatory agencies. Each of these groups provides a different perspective of pumps, filters, size, packaging, and regulation.

Hemodynamics. The human heart acts as a muscular pump that beats an average of 72 times a minute. Each of the two ventricles pumps 70 milliliters of blood per beat or 5 liters per minute. Blood pressure is measured and reported as two numbers: the systolic pressure exerted by the heart during contraction and the diastolic pressure, when the heart is between contractions. Hemodynamics is the study of forces related to the circulation of the blood. The hemodynamic performance of artificial organs must match that of the natural body to operate efficiently without resulting in damage. Calculations may be made using computational fluid dynamics (CFD); relevant parameters include solute concentration, density, temperature, and water concentration. In addition to artificial hearts, which are intended to perform all cardiac functions, there is a mechanical circulatory implement called a ventricular assist device (VAD) that supports the function of the natural heart while it is recovering from a heart attack or surgery. Its pumping action may be pulsatile, in rhythmic waves matching those of the beating heart, or continuous.

Mass Transfer Efficiency. The human kidney acts as a filter to remove metabolic waste products from the blood. A person's kidneys process about 200 quarts of blood daily to remove two quarts of waste and extra water, which are converted into urine and excreted. Without filtration, the waste would build to a toxic level and cause death. Patients with kidney failure may undergo dialysis, in which blood is withdrawn, cleaned, and returned to the body in a periodic, continuous, and time-consuming process that requires the patient to remain relatively stationary. Portable artificial kidneys, which the patient wears, filter the blood while the patient enjoys the freedom of mobility. Filtration systems may involve membranes with a strict pore size to separate molecules based on size or columns of particle-based adsorbents to separate molecules by chemical characteristics. Mass transfer efficiency refers to the quality and quantity of molecular transport.

Scale. The development of artificial organs requires that biological processes that can be duplicated in the laboratory be scaled up to work within the human body without also magnifying the weaknesses. Biological functions occur at the organ, tissue, cellular, and molecular levels, which are on micro- and nanoscales. In addition, machines that work in the engineering laboratory must be scaled down to work within the human body without crowding the other organs. Novel power sources and electronic components have facilitated miniaturization. Size must also be balanced with efficiency and cost. Computer-aided design software is being used to create virtual three-dimensional models before fabrication.

Biomaterials. Artificial organs are made of natural and/or manufactured materials that have been adapted for medical use. The properties of these materials must be controlled down to the nanometer scale. The biological components may serve in gene therapy, tissue engineering, and the modification of physiological responses. The synthetic materials must be biocompatible, which means that they do not trigger an adverse physiological reaction such as blood clotting, inflammatory response, scar-tissue formation, or antibody production. The biomechanics of the artificial organ, such as friction and wear, must be known and parts must be sterile before use. Biomaterials have been developed for subspecialties such as orthopedics and ophthalmics.

Regulation. The body has natural feedback systems that allow the exchange of information with the brain for optimal regulation. Artificial organs that communicate directly with the brain are still in development. The present models require sensors and data systems that may be monitored by physicians. Implanted devices must be able to be inspected without direct observation. Another aspect of regulation is the uniform manufacturing of artificial organs in compliance with performance and patient safety specifications.

APPLICATIONS

The collective knowledge of scientists, engineers, physicians, manufacturers, and regulatory agencies has produced the applications and products in the interdisciplinary realm of artificial organs.

Hemodynamics. Knowledge of hemodynamics, the study of blood-flow physics, has led to the development of artificial circulatory assistance. The ventricular assist device (VAD) supplements the contraction of the two lower chambers of the heart so the heart muscle does not have to work as hard while it is healing. The cardiopulmonary bypass pump, also known as a heart-lung machine, provides blood oxygenation and circulating pressure during open-heart surgery when the heart is stopped. A similar application called extracorporeal membrane oxygenation (ECMO) is used to assist neonates and infants in the intensive care unit and to maintain the viability of organs pending transplantation. The natural pressure generated by a healthy heart is used to send blood through versions of artificial lungs and kidneys without batteries.

Mass Transfer Efficiency. Information about molecular transport and delivery, known as mass transfer efficiency, has been applied to separation and secretion functions of artificial organs. In hemodialysis, toxins are removed from circulating blood that passes through a filter called a dialyzer. This process also removes excess salts and water to maintain a healthy blood pressure. The dialyzer is composed of a semipermeable membrane or cylinder of hollow synthetic fibers that separates out the metabolic-waste solutes in the incoming blood by diffusion into dialysate solution, leaving cleaner outgoing blood. Hemofiltration is a similar process; however, the filtration occurs without dialysate solution because instead of diffusion, the solutes are removed more quickly by hydrostatic pressure. Another separation technique in medical applications is apheresis, in which the constituents of blood are isolated. This may be achieved by gradient density centrifugation or absorption onto specifically coated beads. The therapeutic application is the absorptive removal of a specific blood component that is causing an adverse reaction in a patient, with the remaining components returned to the patient's circulatory system. The pathogenic blood component might be malignant white blood cells, excess platelets, low-density lipoprotein, autoantibodies, or plasma. The second application of apheresis is the separation of components following blood donation. Concentrated red blood cells are administered in the treatment of sickle-cell crisis or malaria. Plasmapheresis is used to collect fresh frozen plasma as well as rare antibodies and immunoglobulins.

Scale. Miniaturization of artificial organs has been facilitated by the application of smaller, more efficient batteries, transistors, and computer chips. For example, hearing aids once had to be worn with cumbersome amplifiers and batteries disguised in a purse or camera case with a carrying strap. Existing models fit completely in the ear canal and a computer chip facilitates digital rather than analogue processing for crisper sound. The artificial kidney has evolved into a wearable model that weighs 10 pounds and is seventeen times smaller than a conventional dialysis machine. Its hollow-fiber filter must be replaced once a week and its dialysate solution must be replenished daily. However, this maintenance is a trade-off that many patients are willing to make for freedom of movement. On the horizon is an artificial retina that depends on a miniature camera to transmit images. Conversely, research is under way to produce large-scale cultures of tissues on biohybrid matrices and scaffolding for transplantation.

Biomaterials. Synthetic materials are used in artificial organs. Dacron (polyethylene terephthalate) is a polyester fiber with high tensile strength and resistance to stretching whether wet or dry, chemical degradation, and abrasion. Patches of it are sewn to arteries to repair aneurysms. When tubing of it is used as an aortic valve bypass, the patient will not require subsequent blood-thinning medications. Gore-Tex (expanded polytetrafluoroethylene) is an especially strong microporous material that is waterproof. Vascular grafts made from it are supple and resist kinks and compression. It is also used for replacing torn anterior and posterior cruciate ligaments in the knee. Perfluorocarbon fluids are synthetic liquids that carry dissolved oxygen and carbon dioxide with negligible toxicity, no biological activity, and a short retention time in the body. These features make them ideal for medical applications. One of these fluids, perfluorodecalin, is typically used as a blood substitute (also called a blood extender) because it mixes easily with blood without changing the hemodynamics. It increases the oxygen-carrying capacity of the blood and penetrates ischemic (oxygen-deprived) tissues especially easily because of its small particle size. This makes it particularly useful in the healing of ulcers and burns. It is also used in conjunction with ECMO in the life support of preterm infants to increase oxygenation and to keep the lungs inflated, reducing exertion. Furthermore, it is used in the preservation of harvested organs and cultured tissue for transplantation, extending their viable storage time.

Regulation. The application of regulatory systems has allowed artificial organs to be adjusted while they are in use. Artificial cardiac pacemakers, which supplement the natural electrical pace-making capabilities of the heart to normalize a slow or irregular heartbeat, are externally programmable so that cardiologists are able to establish the optimal pacing parameters for each patient. Adjustments are made with radio frequency programming, so no further surgery is required. Contemporary hearing aids have volume controls that the wearer can adjust to suit changing surroundings. The inability to detect high- or low-pitch sounds is not a function of volume, yet pitch range can be adjusted in a hearing aid by an audiologist. Other parameters are also adjustable and the audiologist can reprogram the hearing aid as a person's hearing loss changes.

FUTURE PROSPECTS

The number of Americans older than sixty-five years of age is expected to double within the next twenty-five years. The fastest growing age group is people older than eighty-five years of age. Increasing life span of the general population is a direct result of improved health care. The shortage of donor organs is also increasing. As of 2010, more than 16,000 people were waiting for liver transplants, but each year, only about 6,500 kidneys become available. For the 93,000 people waiting for kidney transplants in 2010, only 17,000 donated kidneys became available for transplant. The need for artificial organs as a bridge to transplantation or even as a permanent substitute for failed organs is becoming increasingly urgent. Once only made of synthetic components, artificial organs are becoming biohybrid organs: a combination of biological and synthetic components. Examples include functionally competent cells enveloped within immuno-protective artificial membranes and tissues cultured on chemically constructed matrices. Experiments are underway to develop an antibacterial agent that can be incorporated into biomaterials to reduce the risk of infection from these organ surfaces. Emerging technologies also involve sensors and intelligent control systems, biological batteries and alternate power sources, and innovative delivery systems. Other areas of research include the miniaturization of artificial organs for pediatric use and the development of smaller and more efficient batteries and sensors that will be capable of more accurate communication between the artificial organ and the brain. Another goal is to incorporate wireless capabilities so the artificial organ may be programmed, monitored, and recharged remotely so the patient has increased freedom of mobility.

Bethany Thivierge, MPH

FURTHER READING

Fox, Renee C., and Judith P. Swazey. *Spare Parts: Organ Replacement in American Society*. New York: Oxford University Press, 1992. Discusses not only the progression of organ transplantation methods but also the emotional significance attached to the human body and its parts as well as the ethical concerns regarding organ replacement.

Hench, Larry L., and Julian R. Jones, eds. *Biomaterials, Artificial Organs, and Tissue Engineering*. Boca Raton, Fla.: CRC Press, 2005. Provides multiple essays and introductory topics on artificial organs and tissue engineering.

McClellan, Marilyn. *Organ and Tissue Transplants: Medical Miracles and Challenges*. Berkeley Heights, N.J.: Enslow, 2003. Explores the history of organ transplantation as well as the ensuing medical, ethical, and financial issues.

Sharp, Lesley A. *Bodies, Commodities, and Biotechnologies: Death, Mourning, and Scientific Desire in the Realm of Human Organ Transfer*. New York: Columbia University Press, 2008. Explores how organ transplantation and artificial organs have changed cultural attitudes toward the body.

WEB SITES

American Society for Artificial Internal Organs http://asaio.com

International Federation for Artificial Organs http://www.ifao.org

Society for Biomaterials http://www.biomaterials.org

See also: Bioengineering; Biomechanical Engineering; Cardiology; Cell and Tissue Engineering; Nephrology; Stem Cell Research and Technology.

FASCINATING FACTS ABOUT ARTIFICIAL ORGANS

- Aviator Charles Lindbergh worked with physician Alexis Carrel, a noted pioneer of vascular surgery, to find a means of oxygenating blood other than the lungs. They created a basic oxygen-exchange device that led to the development of the heart-lung bypass machine for artificial circulation.

- The first successful human organ transplant was performed in 1954. A kidney was transplanted from an identical twin donor.

- In the United States, the need for organ replacement therapies increases by 10 percent each year. In 2008, there were nearly 100,000 people on the national waiting list for organ transplants. Three of every four individuals on the list were waiting for a kidney, with an average wait exceeding five years. One patient waiting for an organ transplant dies every seventy-three minutes.

- About 350,000 Americans are reliant on hemodialysis because of kidney damage associated with diabetes and hypertension. Medicare spends $25 billion annually on treatments for kidney failure, which translates to spending 6 percent of its budget on 1 percent of its recipients.

- The National Organ Transplant Act of 1984 bans "valuable consideration" in exchange for providing an organ for transplantation. Thus, human organs cannot be bought or sold in the United States.

- In Spain, the law allows for the presumption of organ donation after death with consent unless the individual stated otherwise before death. As a result, the procurement rate of organ harvesting from cadavers is 35 percent higher than that in the United States. If the United States had the same presumed-consent law, nearly 14,000 more organs could be procured annually.

Asexual reproduction

FIELDS OF STUDY

Cell biology, developmental biology, reproduction science

SUMMARY

Asexual reproduction is any form of reproduction in which the fusion of haploid gametes is not the first step. There are many examples, such as the budding of a new hydra from the stalk of an existing one or the development of diploid eggs into larvae and, eventually, adults in aphids. The new organism formed is a clone of the original.

PRINCIPAL TERMS

- **clone:** an organism that is genetically identical to the original organism from which it was derived
- **diploid:** having two of each chromosome; a normal state for most animals
- **haploid:** having one of each chromosome; a normal state for animal gametes
- **parthenogenesis:** a form of asexual reproduction where the young are derived from diploid or triploid eggs produced by the mother without any genetic input from a male
- **triploid:** having three of each chromosome; an abnormal state which is unable to produce normal haploid gametes

BASIC PRINCIPLES

Although asexual reproduction is very common in organisms such as bacteria, protists, fungi, and plants, it is rarer among animals, especially among the more complex animals. In simple animals, such as sponges (phylum Porifera), the polyps of hydra, jellyfish, corals, and sea anemones (phylum Cnidaria), and many flatworms (phylum Platyhelminthes), asexual reproduction is common. Sponges can reproduce asexually when fragments break off and become established as new individuals or when mature sponges produce gemules, overwintering buds, that are produced and released by many freshwater and a few marine sponges. Cnidarian polyps frequently reproduce by budding. The buds start as small regions of less-differentiated tissue that differentiates into a new polyp. These new polyps can separate from the original to form new individuals or can remain attached and form colonies.

The largest colonies to be produced asexually by budding are those produced by the reef-building corals. In some Cnidaria, the polyps release free-floating forms of the organism called medusae (jellyfish). The medusa form of the life cycle, which was formed asexually, reproduces sexually when it matures. Many species of flatworms reproduce asexually by fragmentation. When these worms are cut into fragments, most fragments can regenerate their missing parts, and thus form several organisms from the original one. Others, like many trematode flukes, asexually reproduce by polyembryony. In this mode of reproduction, larval flukes form many juveniles of the next larval stage internally, from the mature larva's own cells. The immature larvae that are produced, which are all genetic clones of the original larva, will be released to continue the life cycle.

PARTHENOGENESIS

In higher organisms, asexual reproduction is much less frequent and far more complex. Rotifers (phylum Rotifera) are small aquatic organisms with rows of cilia around their mouths that seem to rotate as they beat. Rotifer populations are usually either mostly or entirely female. This happens because rotifers usually reproduce asexually by parthenogenesis. In this type of reproduction, the females produce diploid eggs instead of the haploid eggs that are needed for sexual reproduction. The diploid eggs have the same genes as the mother and are thus clones and mature into adults identical to their mother. In some rotifers, asexual reproduction is the only form of reproduction, and thus all of the organisms in these species are female. In other species, males are only produced during times of environmental stress, and sexual reproduction only occurs then.

Like rotifers, many populations of aphids (phylum Arthropoda), a common plant pest, are entirely or mostly female during parts of the breeding season. In spring and early summer, aphids reproduce parthenogenetically, with females producing diploid eggs that develop into adult female aphids. These eggs and the adults formed from them are clones of the original aphid. In late summer and early fall, the aphids reproduce sexually, producing haploid eggs that can be fertilized by haploid sperm from males. Even among the very complex vertebrates, a few organisms can be found that reproduce asexually. The two most studied are the whiptail lizard, native to the deserts of the American Southwest and the northwestern part of Mexico, and the gecko, found on some tropical islands of the Pacific. These lizards exist in both sexually reproducing and parthenogenetic forms. Genetic study of their chromosomes has shown that the parthenogenetic species were first formed as diploid or sometimes triploid hybrids of two sexually reproducing forms. The hybrids could not undergo normal meiosis because different chromosomes inherited from each parent could not align properly. Thus, these lizards cannot reproduce sexually. They do, however, produce diploid or triploid eggs that can be triggered to start reproduction. The progeny that are produced are exact genetic duplicates of their mothers—clones.

With asexual reproduction, organisms do not have to waste energy in sexual activity. However, the energy savings is not without a price. With the exception of the rotifers, no populations of asexually reproducing organisms have very long histories, evolutionarily speaking. Unlike most sexually reproducing populations with varying degrees of diversity, all members of an asexually reproducing population are identical, except for mutations that arose after the population's inception. This lowered diversity makes the population much less likely to be able to adapt to change More or less water, higher or lower temperatures, introduction of parasites, or disease could more easily wipe out the entire population.

—*Richard W. Cheney, Jr.*

See also: Copulation; Gametogenesis; Reproduction; Sex differences: Evolutionary origin.

FURTHER READING

Adiyodi, K. G., and Rita Adiyodi, eds. *Asexual Propagation and Reproductive Strategies*. Vol. 6, Parts A and B, in *Reproductive Biology of Invertebrates*. New York: John Wiley & Sons, 1993, 1995. Survey of the last one hundred years of research in invertebrate reproduction.

Margulis, L., and K. Schwartz. *Five Kingdoms: An Illustrated Guide to the Phyla of Life on Earth*. 3d ed. New York: W. H. Freeman, 1998. This book gives concise articles on most phyla of animals.

Richardson, S. "The Benefits of Virgin Birth." *Discover* 17 (March, 1996): 33. A brief summary of the advantages of asexuality in Pacific geckos.

Wuethrich, B. "The Asexual Life: Why Sex? Putting Theory to the Test." *Science* 281, no. 25 (September, 1998): 1980-1982. A comparison of the advantages of a sexual versus an asexual lifestyle.

B

Biochemical engineering

FIELDS OF STUDY

Biochemistry; microbiology; biotechnology; cell biology; biology; chemical engineering; chemistry; genetics; molecular biology; pharmacology; medicine; agriculture; food science; environmental science; petroleum refinement; physiology; waste management.

SUMMARY

Biochemical engineers are responsible for designing and constructing those manufacturing processes that involve biological organisms or products made by them. Biochemical engineers take commercially valuable biological or biochemical commodities and design the means to produce those commodities effectively, cheaply, safely, and in mass quantities. They do this by optimizing the growth of organisms that produce valuable molecules or perform useful biochemical processes, establishing the most effective way to purify the desired molecules, and designing the operation systems that execute these processes, while adhering to a high standard of quality, purity, worker safety, and environmental cleanliness.

PRINCIPAL TERMS

- **biofuels:** solid, liquid, or gaseous fuels derived from biomass.
- **biomass:** plant materials and animal waste used especially as a source for fuel.
- **bioreactor:** device in which industrial biochemical reactions occur with the help of either enzymes or living cells.
- **enzymes:** particular proteins or ribonucleic acids (RNAs) that accelerate the rate of chemical reactions without being consumed or changed in the process.
- **genetically engineered organisms:** biological organisms that have had their endogenous deoxyribonucleic acid (DNA) altered, usually by the introduction of foreign DNA.
- **hollow-fiber membrane bioreactor (HFMB):** cylindrical bioreactor that has a series of thin, porous, narrow, hollow tubes inside a plastic cylinder. Cultured cells grow in the spaces between the hollow tubes or fibers (extra-capillary spaces), while oxygen and nutrients continuously flow through the hollow fibers to the cells.
- **hybridoma:** cultured cell line that results from the fusion of an antibody-making B lymphocyte and a myeloma (B lymphocyte tumor cell) that secretes a monoclonal antibody.
- **monoclonal antibodies:** proteins secreted by specific cells of the immune system that precisely bind to specific sites on the surface of foreign invaders and facilitate the destruction or neutralization of the foreign invaders.
- **phage display:** test-tube selection technique that genetically fuses a protein to the outer-coat protein of a virus that infects bacteria, resulting in display of the fused protein on the outside of the virus. This allows screening of vast numbers of variants of the protein, each encoded by its corresponding DNA sequence.
- **photobioreactor:** translucent container that incorporates a light source and is used to grow small photosynthetic organisms for controlled biomass production.
- **wave bioreactor:** disposable, sterile, plastic bag bioreactor that is mounted on a rocking platform, which creates wave action inside the bag to mix the culture that grows inside it.

Basic Principles

Biochemical engineering involves designing and building those industrial processes that use catalysts, feedstocks, or absorbents of biological origin. Industrial processes used in food, waste-management, pharmaceutical, and agricultural plants are often called unit operations. Those unit operations used in combination with biological organisms or molecules include heat and mass transfer, bioreactor design and operation, filtration, cell isolation, and sterilization.

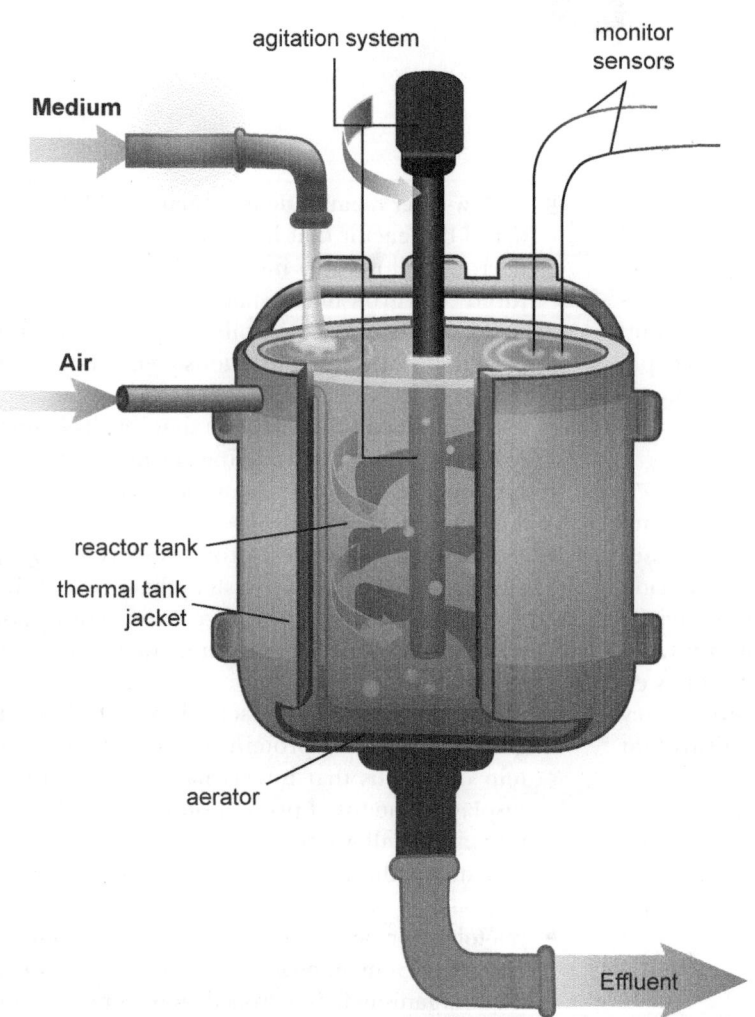

Chemical Reactor

One of the main tasks of bioengineers is to optimize the production of commercially valuable molecules by genetically engineered microorganisms. Biochemical engineers design culture containers known as bioreactors that accommodate growing cultures and maintain an environment that keeps growth at optimal levels. They also create the protocols that separate the cultured cells and their growth medium from the molecule of interest and purify this molecule from all contaminating components. Biochemical engineers do not make the genetically engineered organisms that produce or do valuable things, but instead they maximize the capacities of such organisms in the safest and most cost-effective ways.

Biochemical engineers also design systems that degrade organic or industrial waste. In these cases, bioreactors house biological organisms that receive and decompose waste. They select the right organism or mix of organisms for the job at hand, establish environments that allow these organisms to thrive, and design systems that feed waste to the organisms and remove the degradation products.

A branch of biochemical engineering called tissue engineering combines cultured cells with synthetic materials and external forces to mold those cells into organs that can serve as a replacement for diseased or damaged organs. Biochemical engineers determine the forces, materials, or biochemical cues that drive cells to form fully functional organs and then design the bioreactor and associated instrumentation to provide the proper environment and cues.

BACKGROUND

Biochemical engineering is a subspecialty of chemical engineering. Chemical engineering began in 1901 when George E. Davis, its British pioneer, mathematically described all the physical operations commonly used in chemical plants (distillation, evaporation, filtration, gas absorption, and heat transfer) in his landmark book, *A Handbook of Chemical Engineering*. Biochemical engineering emerged in the 1940's as advancements in biochemistry, the genetics of microorganisms, and engineering shepherded in the era of antibiotics. World War II created shortages in commonly used industrial agents; therefore, manufacturers turned to microorganisms or enzymes to synthesize many of the chemicals needed for the war effort. Growing large batches of microorganisms presented scaling, mixing, and oxygenation problems that had never been encountered before, and biochemical engineers solved these problems.

During the 1960's, advances in biochemistry, genetics, and engineering drove the creation of biomedical engineering, which is the application of all engineering disciplines to medicine, and separated it from biochemical engineering. During this decade, biochemical engineers developed new types of bioreactors and new instrumentation and control circuits for them. They also made breakthroughs in kinetics (the science that mathematically describes the rates of reactions) within bioreactors and whole-cell biotransformations.

The 1970's saw the development of enzyme technologies, biomass engineering, single-cell protein production, and advances in bioreactor design and operation. From 1980 to 2000 there was a virtual explosion in biochemical-engineering advances that had never been seen before. The advent of recombinant DNA and hybridoma technologies, cell culture, molecular models, large-scale protein chromatography, protein and DNA sequencing, metabolic engineering, and bioremediation technologies changed biochemical engineering in a drastic and profound way. These technologies also presented new challenges and problems, many of which are still the subject of intense research and development.

How It Works

Bioreactors. Bioreactors that utilize living cells are typically called fermenters. There are several different types of bioreactors: mechanically stirred or agitated tanks; bubble columns (cylindrical tanks that are not stirred but through which gas is bubbled); loop reactors, which have forced circulation; packed-bed reactors; membrane reactors; microreactors; and a variety of different types of reactors that are not easily classified (such as gas-liquid reactors and rotating-disk reactors). Biochemical engineers must choose the best bioreactor type for the desired purpose and outfit it with the right instrumentation and other features.

Bioreactor operation is either batch-wise or continuous. Batch-wise operation or batch cultures include all the nutrients required for the growth of cells prior to cultivation of the organisms. After inoculation, cell growth commences and ceases once the organisms have exhausted all the available nutrients in the culture medium. A modification of this type of operation is a fed-batch or semi-batch operation in which the reactants are continuously fed into the bioreactor, and the reaction is allowed to go to completion, after which the products are recovered.

Continuously operated bioreactors, use "continuous culture systems" that continuously feed culture medium into the bioreactor and simultaneously remove excess medium at the same rate. Batch-culture bioreactors work best for fast-growing biological organisms. Slow-growing organisms usually require continuous-culture bioreactors.

Several factors influence the success of bioreactor-based operations. First, choosing the right strain to make the desired product is essential. Second, the culture medium and growth conditions must optimize the growth of the chosen organism. Third, supplying the culture with adequate oxygen requires the use of agitators or stirring equipment that must operate at high enough levels to aerate the culture without severely damaging the growing cells. Fourth, the bioreactor must have sensors to measure accurately the physical properties of the culture system, such as temperature, acidity (pH), and ionic strength. Fifth, the bioreactor should also be equipped with the means to adjust these physical properties as needed. Finally, the bioreactor must be integrated into a network of peripheral equipment that allows automated monitoring and adjustment of the culture's physical factors.

Separation. Once a bioreactor makes a product, separating this molecule or group of molecules from the remaining contaminants, byproducts, and other components is an integral part of preparing that molecule for market.

There are several different separation techniques. Filtration separates undissolved solids from liquids by passing the solid-liquid mixture through solids perforated by pores of a particular size (like a membrane). If the liquid is viscous or the particle size of the solid is too small for filtration, centrifugation can separate such solids from liquids. The liquid samples are loaded into centrifuges, which spin rotors at very high speeds. This process creates pellets from the solids and separates them from liquids. Neither filtration nor centrifugation can separate dissolved components from liquids.

Adsorption and chromatography can effectively separate dissolved molecules. Adsorption involves the accumulation of dissolved molecules on the surface of a solid in contact with the liquid. The solid in most cases consists of a resin made of porous charcoal, silica, polysaccharides (complex chains of sugars), or other molecules. Chromatography runs the liquid through a stationary medium packed into a cylindrical column that has particular chemical properties. The interaction between the desired molecules and the stationary medium facilitates their isolation. Other types of separation techniques include crystallization, in which the molecule of interest is driven to form crystals. This effectively removes it from solution and facilitates "salting out," in which gradually increased salt concentrations precipitate the molecules of interest, or contaminating molecules, from a liquid solution.

Sterilization. If a culture of genetically engineered organisms is used to produce a commercially useful product, contamination of that culture can decrease the amount of product or cause the production of harmful byproducts. Therefore, all tubes, valves, the bioreactor container, and the air supplied to it during operation must be effectively sterilized before the start of any production run. Heat, radiation, chemicals, or filtration can sterilize equipment and liquids. One of the most economical means of sterilization is moist steam. Calculating the time it takes to sterilize something depends on the initial number of organisms present, the resilience of those organisms to killing with the chosen agent, the ability of the air or liquid to conduct the sterilizing agent, and length of time the organisms are exposed to the sterilizing agent.

Pharmaceuticals. Hundreds of pharmaceuticals are proteins made by genetically engineered organisms. Because these reagents are intended for clinical use, they must be produced under completely sterile conditions and are usually grown in disposable (plastic), prepackaged, sterile bioreactor systems. A variety of wave bioreactors, hollow-fiber membrane bioreactors, and variations on these devices help grow the cells that make these products. Some of the proteins made by genetically engineered cells are enzymes. Genentech, for example, makes dornase alfa, an enzyme that degrades DNA. This enzyme is made by genetically engineered Chinese hamster ovary (CHO) cells and is purified by filtration and column chromatography. Dornase alfa is administered as an inhalable aerosol to allay the symptoms of cystic fibrosis. Other therapeutic enzymes include clotting factors such as Helixate FS (native clotting factor VIII made by CSL Behring), NovoSeven (clotting factor VII made by Novo Nordisk) to treat hemophilia, and Fabrazyme or Replagal (agalsidase alfa) to treat Anderson-Fabry disease. Other pharmaceuticals are peptide hormones.

Serostim and Saizen are commercially available versions of recombinant human growth hormone. Both products are made with cultured mouse C127 cells in bioreactors. Human growth hormone is used to treat children with hypopituitary dwarfism or those who experience the chronic wasting associated with AIDS. Therapeutic proteins are normally made in the human body under certain conditions, and synthetic versions of these proteins that are made in labs can be used as medicine. For example, human cells make a protein called interferon in response to viral infections, but synthetic interferon can also be used to treat multiple sclerosis. Two synthetic forms of interferon-1β, Rebif, which is made in CHO cells by EMD Serono, and Avonex, also made in CHO cells by Biogen Idec, serve as treatments for multiple sclerosis. Alefacept (brand name Amevive), which is made by Astellas Pharma, is a fusion protein that blocks the growth of specific T cells (immune cells). No such protein exists in the human body, but alefacept is used to treat psoriasis and various cancers. These are only a few examples of the hundreds of pharmaceutical compounds made by genetically engineered organisms in bioreactors designed by biochemical engineers.

Monoclonal Antibodies. Monoclonal antibodies are Y-shaped proteins secreted by specific cells of the immune system that precisely bind to specific sites (epitopes) on the surface of foreign invaders, and act as guided missiles that facilitate the destruction or neutralization of the foreign invaders. Immune cells called B lymphocytes secrete antibodies, and the fusion of these antibody-producing cells with myelomas (B-cell tumor cells) produces a hybridoma, an immortal cell that grows indefinitely in culture and secretes large quantities of a particular antibody. Antibodies made by hybridoma cells can bind to one and only one site on a specific target and are known as monoclonal antibodies.

Monoclonal antibodies are powerful clinical and industrial tools, and by growing hybridoma cell lines in bioreactors, biotechnology companies can produce large quantities of them for a variety of applications. Mouse monoclonal antibodies end with the suffix "-omab." Tositumomab (brand name Bexxar) was approved by the Food and Drug Administration (FDA) for treatment of non-Hodgkin's lymphoma in 2003. Chimeric or humanized monoclonal antibodies, and have the suffixes "-ximab" (chimeric antibodies that are about 65 percent human) or "-zumab" (humanized antibodies that are about 95 percent human). Cetuximab (Erbitux) is a chimeric antibody that was approved by the FDA in 2004 for the treatment of colorectal, head, and neck cancers. Bevacizumab (Avastin) is a humanized antibody approved by the FDA in 2004 that shrinks tumors by preventing the growth of new blood vessels into them.

Human monoclonal antibodies are made either by hybridomas from transgenic mice that have had their mouse antibody genes replaced with human antibody genes, or by a process called phage display. Human monoclonal antibodies end with the suffix "-mumab." The first human monoclonal antibody developed through phage display technologies was adalimumab (Humira), which was approved by the FDA to treat several immune system diseases.

Tissue Engineering. Making artificial organs for transplantation represents a unique challenge. Bioreactors tend to grow cells in two-dimensional cultures, but organs are three-dimensional structures. Thus, biochemical engineers have designed synthetic scaffolds that support the growth of cultured cells and mold them into structures that bear the shape and properties of organs. They have also designed special bioreactors that subject cells to the physical conditions that induce the cells to form the tissues that compose particular organs.

People often need cartilage repair or replacement, but bone and cartilage form only when their progenitor cells are subjected to mechanical stresses and shear forces. Biochemical engineers have grown bone by seeding bone marrow stem cells on a ceramic disc imbued with zirconium oxide and loading these discs into bioreactors with a rotating bed. Cartilage biopsies are taken from the nose or knee and grown in a bioreactor in which the cells are perfused into a complex sugar called glycosaminoglycan (GAG). This engineered cartilage is then used for transplantations. Such experiments have established that nasal cartilage responds to physical forces similarly to knee cartilage and might substitute for knee cartilage. Heart muscle is grown in bioreactors that pulse the liquid growth medium through the chamber under high-oxygen tension. Blood vessels are grown in two-chambered bioreactors and contain a reservoir of smooth muscle cells and a chamber through which culture medium is repeatedly pulsed.

Food Engineering. Companies making foods that require fermentation by microorganisms or digestion of complex molecules by enzymes use bioreactors to optimize the conditions under which these reactions occur. Biochemical engineers design the industrial processes that manufacture, package, and sterilize foods in the most cost-effective manner. Starch is a polymer of sugar made by plants and is a very cheap source of sugar. To convert starch into glucose, enzymes called amylases are employed. These enzymes are often isolated from bacteria or fungi, and some are even stable at high temperatures. Degrading starch at high temperatures often clarifies it and rids it of contaminating proteins.

Lactic acid fermentation metabolizes simple sugars to lactic acid and is commonly used in the production of yogurt, cheeses, breads, and some soy products. Cheese production begins with curdling milk by adding acids such as vinegar that separate solid curds from liquid whey and an enzyme mixture called rennet that comes from mammalian stomachs and coagulates the milk. Starter bacterial cultures then ferment the milk sugars into lactic acid. Yogurt is made from heat-treated milk to which starter cultures are added. The acidity of the culture is monitored, and when it reaches a particular point, the yogurt is heated to sterilize the culture for packaging.

Ethanol fermentation converts simple sugars to ethyl alcohol and is used in the production of alcoholic beverages. The most common organism utilized for ethanol fermentation is the baker's yeast, *Saccharomyces cerevisiae*. Malted barley is the sugar source in beer production, and grapes are used to make wine. Beer production involves the extraction of wort, a sugar-rich liquid from barley, which is treated with hops to add aroma and flavor and is then fermented by yeast to form beer. For wine production, the juice from crushed grapes is fermented by yeast for from five to twelve days to generate ethanol. For most red wines and some white wines, the mixture is fermented a second time by malolactic bacteria that degrade the malic acid in the wine, which has a rather harsh, bitter taste, to lactic acid. This lowers the acidity of the wine.

Biofuel Production. Burning of fossils fuels as an energy source is not sustainable, since the supply of these fuels is finite and their combustion generates greenhouse gases such as carbon dioxide (CO_2), sulfur dioxide (SO_2), and nitrogen oxides. First-generation biofuels (biodiesel and bioethanol) utilize biomass from cultivated crops such as corn, sugar beets, and sugar cane. This results in the unfortunate consequences of tying up large swathes of farmland for fuel production and raising food prices. Second-generation biofuels come from grasses, rice straw, and bio-ethers, which are economically superior to first-generation biofuels. Third-generation biofuels show the most ecological and economic promise and come from microalgae. The oil content of some microalgae can exceed 80 percent of their dry weight, and since they use sunlight as their energy source and atmospheric CO_2 as their carbon source, microalgae can produce substantial amounts of oil with little material investment.

Microalgae can be grown in open ponds, which ties up land, or special bioreactors called photobioreactors. The fast-growing microalgae are harvested and then liquefied by microwave high-pressure reactors. Oils extracted from the algal species *Dunaliella tertiolecta* at 340 degrees Celsius for 60 minutes had physical properties comparable to fossil fuel oil.

Waste Management. The removal of pollutants from air and water provides a large global challenge to environmental engineers. While there are nonbiological ways to degrade pollution, biological strategies represent some of the most innovative and potentially effective ways to remediate pollution. To treat polluted air, it is piped through a biofilter, which consists of an inert substance called a carrier. Nutrients are trickled over the carrier, and consequently the carrier is colonized by biological organisms that can degrade the pollutant. Devices called bioscrubbers eliminate pollutants such as hydrogen sulfide (H_2S), which smells like rotten eggs, or SO_2, by dissolving the air pollutants in water and running the water into a bioreactor where the pollutants are degraded. For air pollutants that are poorly soluble in water, such as methane (CH_4) or nitric oxide (NO), hollow-fiber membrane bioreactor (HFMB) systems that house a robust population of biological organisms that can degrade gas-phase pollutants effectively treat air polluted with such molecules. Many of these same strategies can also treat polluted water.

Bioreactor landfills were designed to accelerate the degradation of municipal solid waste (MSW) in landfills. Bioreactor landfills use microorganisms to degrade solid wastes, but they also drain the water (leachate) that moves through the landfill, clean it, and recycle it back through the landfill in a process called leachate recirculation. The design of a bioreactor landfill requires extensive knowledge of the surroundings, the nature of the MSWs to be treated, and the quality of the water that becomes the leachate.

Future Prospects

Two aspects of biochemical engineering can be cause for concern to the general public. First, biochemical engineers work with genetically engineered organisms. Many people have never completely made peace with the use of such organisms, despite the fact that many of the items people consume on a regular basis, from seasonal flu vaccines and other medicines to the foods they eat, are made by genetically engineered organisms. Nevertheless, fear of genetically engineered organisms remains. For example, despite repeated tests establishing that genetically engineered foods are as safe as food from nongenetically modified crops, some people still feel the need to label genetically engineered food as Frankenfood. As long as this fear persists, the work of biochemical engineers will make some people uncomfortable. Second, biochemical engineers tend to work for large industries that are sometimes painted as inveterate polluters by environmental groups or as greedy, unconcerned capitalists by consumer-advocate groups. Since many companies abide by strict environmental standards and engage in humanitarian work, these accusations are somewhat unfair.

The development of new technologies in fields like genetic engineering, biomedicine, bioinstrumentation, biomechanics, waste management, and alternative energy development are driving new employment opportunities for biochemical engineers.

Michael A. Buratovich, MA, PhD

Further Reading

Katoh, Shigeo, and Fumitake Yoshida. *Biochemical Engineering: A Textbook for Engineers, Chemists, and*

Biologists. Weinheim, Germany: Wiley-VCH Verlag, 2009. A basic, though rather technical, textbook of biochemical engineering by two prominent Japanese biochemical engineers that contains many tables of mathematical symbols and conversions, graphs that illustrate the application of the equations presented, and problems for interested students to solve.

McNamee, Gregory. *Careers in Renewable Energy: Get a Green Energy Job.* Masonville, Colo.: PixyJack Press, 2008. This highly readable Summary of the renewable- energy job market includes more than just engineering jobs and discusses potential future employment opportunities in alternative energy sources.

Mosier, Nathan S., and Michael R. Ladisch. *Modern Biotechnology: Connecting Innovations in Microbiology and Biochemistry to Engineering Fundamentals.* Hoboken, N.J.: Wiley-AIChE, 2009. A very practical and richly illustrated and referenced guide to the advances in molecular biology for aspiring biochemical engineers.

Murphy, Kenneth M., Paul Travers, and Mark Walport. *Janeway's Immunobiology.* 7th ed. Oxford, England: Taylor & Francis, 2007. A standard immunology textbook that has an excellent section on the therapeutic use of antibodies.

Pahl, Greg. *Biodiesel: Growing a New Energy Economy.* 2d ed. White River Junction, Vt.: Chelsea Green, 2008. A popular guide to the advances in biodiesel technology and biofuel industries that also examines the food-for-fuel controversy and the issues surrounding genetically modified crops.

Vasic-Racki, Durda. "History of Biotransformations: Dreams and Realities." *Industrial Biotransformations*, edited by Andreas Liese, Karsten Seelbach, and Christian Wandrey. Weinheim, Germany: Wiley-VCH Verlag, 2000. This essay chronicles the history of using microorganisms and enzymes to synthesize commercially valuable products and the rise of biochemical engineering as an inevitable consequence of these developments.

Walker, Sharon. *Biotechnology Demystified.* New York: McGraw-Hill Professional, 2006. This is an introduction to the basics of and latest advances in molecular biology and the latest applications of these concepts to a range of discoveries, including new drugs and gene therapies.

WEB SITES

Biohealthmatics http://www.biohealthmatics.com/careers/ PID00269.aspx

National Society for Professional Engineers http://www.nspe.org

Sloan Career Cornerstone Center http://www.careercornerstone.org/pdf/bioeng/ bioeng.pdf

See also: Cell and Tissue Engineering; DNA Analysis; Stem Cell Research and Technology.

FASCINATING FACTS ABOUT BIOCHEMICAL ENGINEERING

- Capromab pendetide (ProstaScint), made by Cytogen Corporation, an engineered monoclonal antibody that specifically labels prostate cancers, was the first recombinant protein produced in hollowfiber membrane bioreactors to be approved by the Food and Drug Administration in 1996.

- Heart muscle tissue grows in bioreactors only in the presence of high oxygen concentrations, but oxygen is poorly soluble in aqueous environments. To solve this problem, tissue engineers use an artificial oxygen carrier, such as perfluorocarbon, to act as a kind of artificial hemoglobin.

- A common component of household mildew, the fungus *Aspergillus niger*, converts sugars in beets or cane molasses to citric acid, which is used to acidulate foods, beverages, and candies.

- By injecting a gel made from seaweed (calcium alginate) into rabbit leg bones, tissue engineers were able to grow engineered bone that provided viable material for bone-transplant surgeries without affecting normal bone growth.

- Tempeh, a popular fermented food eaten as a meat substitute, is made in a solid-state fermentation bioreactor by cultivating the fungus *Rhizopus oligosporus* on cooked soybeans. The fungus binds the soybeans into compact cakes that are fried and packaged to sell to the public.

- Bioartificial hollow-fiber liver devices have cultured liver cells (hepatocytes) growing in between hollow fibers. The patient's blood is passed through the fibers, where it diffuses into the extra-fiber spaces, interacts with the hepatocytes, and returns to the patient. In laboratory tests, such devices maintained liver function for a few months.

BIOCHEMISTRY

FIELDS OF STUDY

Biochemistry; Organic Chemistry

SUMMARY

The broad field of biochemistry explores the compounds and processes involved in living organisms. Biology and biochemistry are intimately related: biology does not exist without biochemical processes, and biochemistry does not exist without biological systems.

PRINCIPAL TERMS

- **biomolecule:** an organic molecule produced by a living organism.
- **carbohydrate:** an organic compound containing hydroxyl (–OH) and carbonyl (C=O) groups, often with the general formula $C_x(H_2O)_y$; includes sugars, starches, and celluloses.
- **lipid:** a type of biomolecule that is soluble in organic nonpolar solvents and generally insoluble in water; includes fats, waxes, and the major components of organic oils.
- **nucleic acid:** a biopolymer consisting of many different nucleotides bonded together; includes both DNA and RNA.
- **protein:** a biological polymer consisting of one or more long chains of amino acids linked by peptide bonds in a sequence specified by an organism's DNA.

THE BASICS OF BIOCHEMISTRY

Biochemistry is quite literally the chemistry of life. Because they are composed of matter, all living organisms are made of atoms and molecules that obey the same chemical principles as any other atoms and molecules. However, the very complexity of biological systems separates biochemistry from "basic" chemistry. In any one of the trillions of cells that compose the physical body of an animal, tens of thousands of interrelated chemical processes occur almost constantly. These processes are often cyclic in nature and extract energy from matter taken in as nutrition. For example, the process of cellular respiration uses oxygen from the atmosphere and quantities of inorganic phosphate ions to break down glucose into carbon dioxide and water, releasing the energy from the chemical bonds so it can be used to drive other biochemical processes. In green plants, the opposite process occurs: carbon dioxide from the atmosphere and water are combined in the presence of sunlight to produce glucose, which polymerizes to form starches, celluloses, and other carbohydrates. In this way, the sun is the source of all energy that maintains the vast majority of life-forms on the planet.

Deoxyribonucleic acid (DNA) and other nucleic acids are central to the processes of life. DNA molecules are composed of a series of several million nucleotides. Nucleotide sequences are the blueprint for every one of the many thousands of proteins in living beings. The living cell itself exists only as a construct resulting from the hydrophilic ("water loving") and hydrophobic ("water fearing") properties of lipids. All of these biomolecules, and an untold number of others, are the subject of the field of biochemistry.

THE MAJOR BIOCHEMICALS

Carbohydrates. Carbohydrates are compounds composed of carbon atoms, each of which is bonded to the components of a water molecule: two hydrogen atoms and one oxygen atom. Technically, the simplest carbohydrates are methanol (H–CH$_2$–OH) and ethylene glycol (HO–CH$_2$CH$_2$–OH), since the carbon atoms in each of these is bonded to both a hydrogen atom and a hydroxyl (–OH) group. However, the first truly important carbohydrate in biochemical systems is glycerol, an alcohol compound containing three hydroxyl groups. Glycerol is the first product formed during glycolysis, produced by splitting a molecule of glucose into two molecules of glycerol. This compound is also the "backbone" of the phospholipids that form the phospholipid bilayer of cell membranes.

Larger carbohydrate molecules are called saccharides, or sugars, and include fructose, glucose, lactose, and many other similar compounds. The most important of these are ribose and deoxyribose, the essential sugar components of ribonucleic acid (RNA) and DNA, respectively. In plants, photosynthesis produces glucose exclusively, mostly because of its stable six-membered ring structure. Fructose and other sugars are produced by plants and animals by

BIOCHEMISTRY

Nucleic Acids

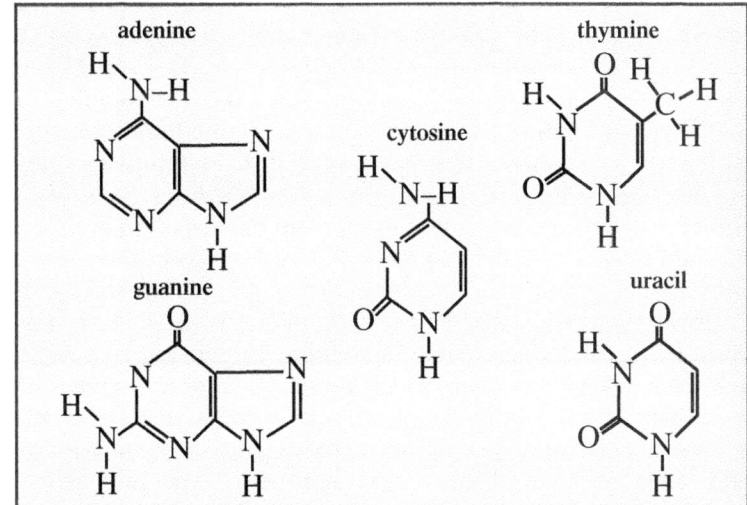

Endocrine System

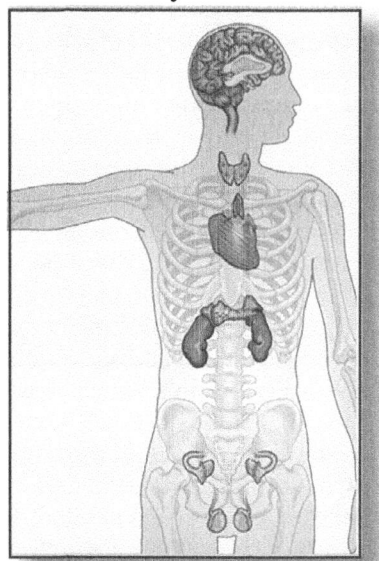

Proteins

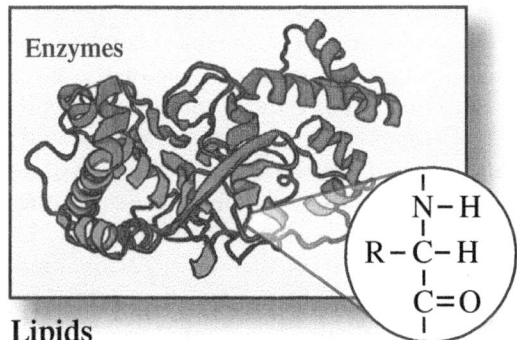

Carbohydrates

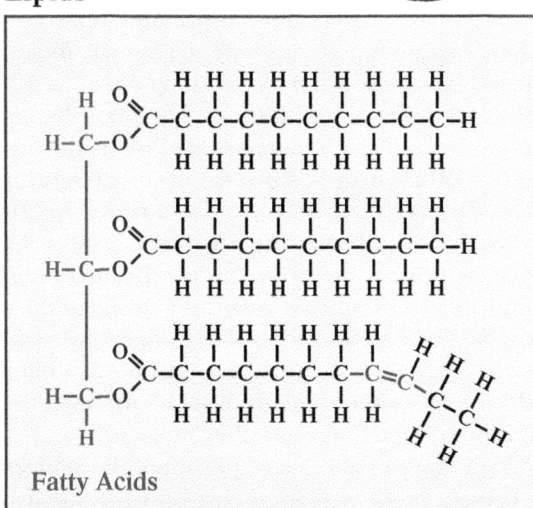

Lipids

Fatty Acids

the alteration of glucose or some other synthetic process. Glucose polymerizes into various starches and celluloses in plants and into glycogen in animals, providing the primary sources of energy in metabolism.

Proteins. All proteins in living systems are created from a limited assortment of approximately twenty amino acids. Protein molecules can be any length, and the same amino acids can repeat any number of times, so the potential variety of proteins is essentially infinite.

Amino acids bond chemically to each other via amide linkages, which are called peptide bonds when they occur in proteins. Extended chains of amino acids form a protein molecule. Proteins are the principal structural component of the many different types of animal tissues, from heart muscle to hair and fingernails. All of the proteins created and used in living organisms are synthesized by enzyme-mediated processes according to the nucleotide sequences contained in each individual DNA molecule. Therefore, in a way, all living things are simply self-replicating collections of proteins.

Enzymes, which are in fact specialized protein molecules, carry out their functions based on their molecular structure. Each protein molecule has a primary structure consisting of the sequence of amino acids from which it was assembled. The shapes of these individual molecules induce a secondary protein structure, causing some sections to twist in a spiral and others to resemble pleated sheets. A tertiary structure arises as functional groups in different parts of the protein molecule interact with each other. A final, multi-unit quaternary structure forms when two or more independent protein molecules become intertwined but not chemically connected, forming a protein complex or an enzyme complex.

Nucleic Acids. Nucleic acids are the central feature of protein synthesis. DNA carries the genetic code of the corresponding living organism. It was first isolated in 1869 but was not known to bear genetic information until 1943, when Oswald Avery (1877–1955), Colin MacLeod (1909–72), and Maclyn McCarty (1911–2005) demonstrated that introducing DNA from a virulent strain of a bacterium into a non-virulent strain could produce the virulent strain. In 1953, James D. Watson (1928–) and Francis Crick (1916–2004), with significant assistance from their laboratory manager, Maurice Wilkins (1916–2004), and based on the analytical and theoretical work of Rosalind Franklin (1920–58), first published their discovery that the DNA molecule is in the shape of a double helix.

Since then, methods and techniques for the manipulation and analysis of DNA have advanced, permitting biochemists and geneticists to obtain a much better understanding of the role of genes and chromosomes in the DNA molecule. In February 2001, the journal *Nature* published the first complete analysis of the human genome. Subsequent study of the genome has demonstrated that all modern humans are descended from a very few human populations that originated in Africa in the distant past.

As the genome was deciphered over time, scientists gained an understanding of the mechanisms by which DNA and RNA are synthesized. In DNA replication and protein transcription, certain enzymes separate the dual-strand DNA molecule into two single strands so that RNA segments can assemble against the nucleotide pattern contained in the molecule. These RNA segments are then used to synthesize either proteins or new DNA molecules. DNA and RNA are each formed from only four different nucleotide bases, but the number of possible combinations of those four is sufficient to control the functional process in its entirety.

Lipids. Lipids are superficially related to carbohydrates through the intermediary of the glycerol molecule. The category "lipid" describes several different types of molecule, including the many and various fats, oils, and greases found in the cells of living organisms. Numerous metabolic processes synthesize lipids to carry out a variety of functions. The lipids formed from fatty acids (carboxylic acids attached to long hydrocarbon chains) are the essential material of the cell walls and membranes in animal cells, forming what is called the phospholipid bilayer.

A phospholipid consists of a glycerol molecule that has been esterified by two long-chain fatty-acid molecules and a phosphate ion. The phosphate portion is a "mixed ester," composed of phosphoric acid and one of the hydroxyl groups of the glycerol molecule; the fatty-acid portions form normal carboxylate esters with the other two hydroxyl groups. Thus, the phospholipid molecule has two distinct regions, a phosphoester end that is strongly hydrophilic and a long-chain fatty-acid end that is decidedly hydrophobic. These two different regions do not interact well with each other and are instead attracted to the corresponding sections of other molecules.

When large quantities of phospholipid molecules are present in a water-based environment, the hydrophobic fatty-acid portions are forced to aggregate and effectively blend into each other. Due to the repulsive force between the negatively charged phosphate groups to which they are attached, the molecules automatically form a sandwich-like structure called a bilayer. The two outer surfaces of the bilayer consist of the hydrophilic phosphoester portions of the phospholipid molecules, while sandwiched between them is a thick, hydrophobic layer of intertwined fatty-acid chains. This bilayer fully encloses the interior contents of living animal cells, forming a membrane that allows materials to pass from one side to the other by various mechanisms. Being hydrophilic, the inner and outer surfaces can interact freely with the water-based fluids on either side of the membrane.

Metabolic Processes. The biochemical processes that living systems use to obtain energy and sustain life are collectively called metabolism. These metabolic processes fall into two distinct categories, anabolic and catabolic. Anabolic processes build up the structure of the system; examples include muscle- protein synthesis and bone construction. Catabolic processes deconstruct materials in order to extract energy for anabolic and homeostatic processes; examples include the breakdown of glycogen and fat deposits, the digestion of foods, and the elimination of waste products.

Biochemistry and Organic Chemistry. Since life, as it is known on Earth, is based on carbon compounds, the basic principles of biochemistry are the same as those of organic chemistry. Biochemistry can be thought of as the application of the principles of organic chemistry to aqueous (water-based) substances, which are normally the realm of inorganic chemistry. All of the fluids of living things, such as blood and tree sap, are water based. Accordingly, mineral salts and electrolyte ions are important factors in the maintenance of biological systems, and the field of biochemistry is as concerned with the role of inorganic substances, such as minerals and trace elements, as it is with the organic aspects of biochemical processes.

Richard M. Renneboog, MSc

FURTHER READING

Abbott, David. The Biographical Dictionary of Scientists. New York: P. Bedrick, 1984. Print.

Acheson, R.M. *An Introduction to the Chemistry of Heterocyclic Compounds.* 3rd ed. New York: Wiley, 2008. Print. Berg,

Jeremy M., John L. Tymoczko, Gregory J. Gatto, and Lubert Strye. *Biochemistry.* 8th ed. New York: W. H. Freeman, 2015. Print.

Fenichell, Stephen. *Plastic: The Making of a Synthetic Century.* New York: Harper, 2009. Print.

Hendrickson, James B., Donald J. Cram, and George S. Hammond. *Organic Chemistry.* 3rd ed. New York: McGraw, 1970. Print.

Mann, J., et al. *Natural Products: Their Chemistry and* 2nd ed. London: Chapman & Hall, 1989. Print.

Morrison, Robert Thornton, and Robert Neilson. Philadephia: Saunders College Pub., 1987. Print.

Lehninger, Albert L. *Biochemistry: The Molecular Basis* Print. *of Cell Structure and Function.* 2nd ed. New York: Robbers, James E., Marilyn K. Speedie, and Varro Worth, 1975. Print.

Lodish, Harvey, et al. *Molecular Cell Biology.* 7th ed. Rev. ed. Baltimore: Williams, 1996. Print. New York: Freeman, 2013. Print.

Vieira, Ernest R. *Elementary Food Science.* 4th ed. Gaithersburg: Aspen, 1999. Print.

BIOCHEMISTRY SAMPLE PROBLEM

Determine whether the following processes are anabolic or catabolic:
  protein synthesis
  protein digestion
  glycolysis
  photosynthesis
  respiration

Answer:

Protein synthesis is a process for building up the system; it is anabolic.

  Protein digestion breaks down proteins into component amino acids portions; it is catabolic.

  Glycolysis breaks down the glucose molecule into two molecules of glycerol; it is catabolic.

  Photosynthesis assembles glucose from carbon dioxide and water to store energy from the sun; it is anabolic.

  Respiration breaks down the glycerol from glycolysis into carbon dioxide and water to release the energy stored in its bonds; it is catabolic.

Bioengineering

FIELDS OF STUDY

Cell biology; molecular biology; biochemistry; physiology; ecology; microbiology; pharmacology; genetics; medicine; immunology; neurobiology; biotechnology; biomechanics; bioinformatics; physics; mechanical engineering; electrical engineering; materials science; buildings science; architecture; chemical engineering; genetic engineering; thermodynamics; robotics; mathematics; computer science; biomedical engineering; tissue engineering; bioinstrumentation; bionics; agricultural engineering; human factors engineering; environmental health engineering; biodefense; nanotechnology; nanoengineering.

SUMMARY

Bioengineering is the field in which techniques drawn from engineering are used to tackle biological problems. For example, bioengineers may use mechanics principles—knowledge about how to design and construct mechanical objects using the most ideal materials—to create drug delivery systems. They may work on developing efficient ways to irrigate and drain land for growing crops, or they may be involved in building artificial environments that can support life even in the harsh climate of outer space. A highly interdisciplinary, collaborative field that synthesizes expertise from multiple research areas, bioengineering has had a significant impact on many fields of study, including the health sciences, technology, and agriculture.

PRINCIPAL TERMS

- **biocompatible material:** material used to replace or repair tissues in the body, or to perform a biological function in a living organism.
- **bioinformatics:** application of data processing, retrieval, and storage techniques to biological research, especially in genomics.
- **biomechanics:** application of mechanical principles to questions of motor control in biological systems.
- **bioreactor:** tool or device that generates chemical reactions to create a product.
- **bioremediation:** use of bacteria and other microorganisms to solve environmental problems, such as neutralizing hazardous waste.
- **geoengineering:** use of engineering techniques to modify environmental or geological processes, such as the weather, on a global scale.
- **prosthetic device:** artificial part or implant designed to replace the function of a lost or damaged part of the body.
- **regenerative medicine:** therapies that aim to restore the function of tissues that have been damaged or lost through injury or disease by using tissue or cells grown in laboratories or compounds created in laboratories.
- **systems biology:** theoretical branch of bioengineering that creates models of complex biological processes or systems, using them to predict future behavior.
- **transgenic organism:** plant or animal containing genetic information taken from another species.

Basic Principles

Bioengineering is an interdisciplinary field of applied science that deals with the application of engineering methods, techniques, design approaches, and fundamental knowledge to solve practical problems in the life sciences, including biology, geology, environmental studies, and agriculture. In many contexts, the term bioengineering is used to refer solely to biomedical engineering. This is the application of engineering principles to medicine, such as in the development of artificial limbs or organs. However, the field of bioengineering has many applications beyond the field of health care. For example, genetically modified crops that are resistant to pests, suits that protect astronauts from the ultra-low pressures in space, and brain-computer interfaces that may allow soldiers to exercise remote control over military vehicles all fall under the wide umbrella of bioengineering.

Each of the sub-disciplines within bioengineering relies on different sets of basic engineering principles, but a few fundamental approaches can be said to apply broadly across the entire field. From an engineering perspective, three basic steps are involved in solving any problem: an analysis of how the system in question works, an attempt to synthesize the

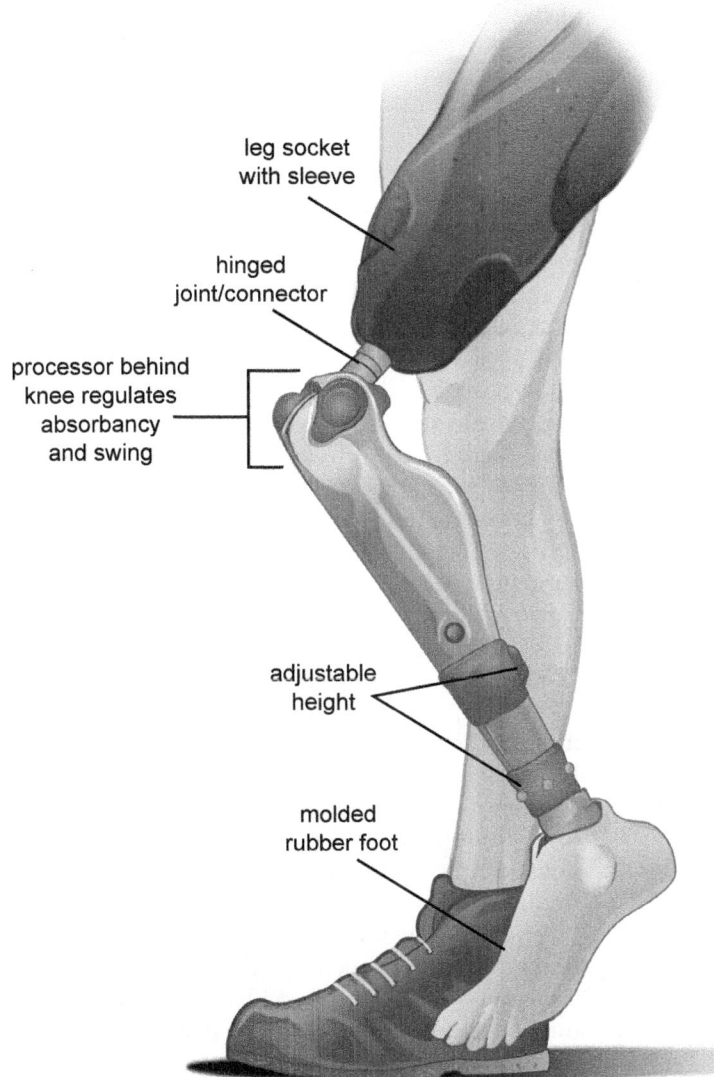

PROSTHETIC LEG (C-LEG model)

- leg socket with sleeve
- hinged joint/connector
- processor behind knee regulates absorbancy and swing
- adjustable height
- molded rubber foot

heart itself on a cellular or tissue level but also the complex dynamics of the organ's interactions with the rest of the body through the circulatory system and the immune system. They must build a device whose parts can mimic the functionality of a healthy heart and whose materials can be easily integrated into the body without triggering a harmful immune response.

Background

Principles of chemical and mechanical engineering have been applied to problems in specific biological systems for centuries. For example, bioengineering applications include the fermentation of alcoholic beverages, the use of artificial limbs (which are documented as far back as 500 BCE), and the building of heating and cooling systems that regulate human environments.

Bioengineering did not emerge as a formal scientific discipline, however, until the middle of the twentieth century. During this period, more and more scientists began to be interested in applying new technologies from electronic and mechanical engineering to the life sciences. As the United States, Japan, and Europe began to enter a period of economic recovery and growth following World War II, governments increased funding for bioengineering efforts. The cardiac pacemaker and the defibrillator, both developed during this postwar period, were two of the earliest and most significant inventions to come out of the quickly developing field. In 1966, the Engineers Joint Council Committee on Engineering Interaction with Biology and Medicine first used the term "bioengineering." At about the same time, academic institutions began to form

information gathered from this analysis and generate potential solutions, and finally an attempt to design and test a useful product. Bioengineers apply this three-stage problem-solving process to problems in the life sciences. What is somewhat novel about this approach is that it is a holistic one. In other words, it treats biological entities as systems—sets of parts that work together and form an integrated whole—rather than looking at individual parts in isolation. For example, to develop an artificial heart, bioengineers need to consider not just the structure of the specialized departments and programs of study to train professionals in the application of engineering principles to biological problems. In the twenty-first century, rapid technological advances continue to produce growth in the field of bioengineering.

How It Works

Because bioengineering is such a large and diverse field, it would be impossible to enumerate all the processes involved in creating the totality of its applications. The following are a few of the most significant examples of the types of technological tools used in bioengineering.

Materials Science. One of the most important areas of bioengineering is the intersection of materials science and biology. Scientists working in this field are charged with developing materials that, although synthetic, are able to successfully interact with living tissues or other natural biological systems without impeding them. (For example, it is vital that biocompatible materials not allow blood platelets to adhere to them and form clots, which can be fatal.) Depending on the specific application in question, other properties, such as tensile strength, resistance to wear, and permeability to water, gases, and small biological molecules, are also important. To manipulate these properties to achieve a desired end, engineers must carefully control both the chemical structure and the molecular organization of the materials. For this reason, biocompatible materials are generally made out of some kind of synthetic polymer—substances with simple and extremely regular molecular structures that repeat again and again. In addition, additives may be incorporated into the materials, such as inorganic fillers that allow for greater mechanical flexibility or stabilizers and antioxidants that keep the material from becoming degraded over time.

Biochemical Engineering. Since living cells are essentially chemical systems, the tools of chemical engineering are especially applicable to biology. Biochemical engineers study and manipulate the behavior of living cells. Their basic tool for doing this is a fermenter, a large reactor within which chemical processes can be carried out under carefully controlled conditions. For example, the modern production of virtually all antibiotics, such as penicillin and tetracycline, takes place inside a fermenter. A central vessel, sealed tight to prevent contamination and surrounded by jackets filled with coolants to control its temperature, contains propellers that stir around the nutrients, culture ingredients, and catalysts that are associated with the reaction at hand.

Genetic engineering is a subfield of biochemical engineering that is growing increasingly significant. Scientists alter the genetic information in one cell by inserting into it a gene from another organism. To do this, a vector such as a virus or a plasmid (a small strand of DNA) is placed into the cell nucleus and combines with the existing genes to form a new genetic code. The technology that enables scientists to alter the genetic information of an organism is called gene splicing. The new genetic information created by this process is known as recombinant DNA. Genetic engineering can be divided into two types: somatic and germ line. Somatic genetic engineering is a process by which gene splicing is carried out within specific organs or tissues of a fully formed organism; germ-line genetic engineering is a process by which gene splicing is carried out within sex cells or embryos, causing the recombinant DNA to exist in every cell of the organism as it grows.

Electrical Engineering. Electrical engineering technologies are an essential part of the bioengineering tool kit. In many cases, what is required is for the bioengineer to find some way to convert sensory data into electric signals, and then to produce these electric signals in such a way as to enable them to have a physiological effect on a living organism.

The cochlear implant is an example of one such development. The cochlea is the part of the brain that interprets sounds, and a cochlear implant is designed for people who are profoundly deaf. A cochlear implant uses electronic devices that capture sounds and relay them to the cochlea. The implant has four parts: a microphone, a tiny computer processor, a radio transmitter, and a receiver, which surgeons implant in the user's skull. The microphone picks up nearby sounds, such as human speech or music emerging from a pair of stereo speakers. Then the processor converts the sounds into digital information that can be sent through a wire to the radio transmitter. The software used by the processor separates sounds into different channels, each representing a range of frequencies. In turn, the radio transmitter translates the digital information into radio signals, which it relays through the skull to the receiver. The receiver then turns the radio signals into electric impulses, which directly stimulate the nerve endings in the cochlea. It is these electric signals that the brain is able to interpret as sounds, allowing even profoundly deaf people to hear.

Another example of how electric signals can be used to direct biological systems can be found in brain-computer interfaces (BCIs). BCIs are direct channels of communication between a computer and the neurons in the human brain. They work because activity in the brain, such as that produced by thoughts or sensory processing, can be detected by bioinstruments designed to record electrophysiological signals. These signals can then be transmitted to a computer and used to generate commands. For example, BCIs allow stroke victims who have lost the use of a limb to regain mobility; a patient's thoughts about movement are transmitted to an external machine, which in turn transmits electric signals that precisely control the movements of a cradle holding his or her paralyzed arm.

APPLICATIONS

Biomedical Applications. Biomedical engineering is a vast subdiscipline of bioengineering, which itself encompasses multiple fields of interest. The many clinical areas in which applications are being developed by biomedical engineers include medical imaging, cell and tissue engineering, bioinstrumentation, the development of biocompatible materials and devices, biomechanics, and the emerging field of bionanotechnology.

Medical imaging applications collect data about patients' bodies and turn that data into useful images that physicians can interpret for diagnostic purposes. For example, ultrasound scans, which map the reflection and reduction in force of sounds as they bounce off an object, are used to monitor the development of fetuses in the wombs of pregnant women.

Magnetic resonance imaging (MRI), which measures the response of body tissues to high-frequency radio waves, is often used to detect structural abnormalities in the brain or other body parts. Cell and tissue engineering is the attempt to exploit the natural characteristics of living cells to regenerate lost or damaged tissue. For example, bioengineers are working on creating viable replacement heart cells for people who have suffered cardiac arrests, as well as trying to discover ways to regenerate brain cells lost by patients with neurodegenerative disorders such as Alzheimers disease. Genetic engineering is a closely related area of biomedicine in which DNA from a foreign organism is introduced into a cell so as to create a new genetic code with desired characteristics.

Bioinstrumentation is the application of electrical engineering principles to develop machines that can sense and respond to biological or physiological signals, such as portable devices for diabetics that measure and report the level of glucose in their blood. Other common examples of bioinstrumentation include electroencephalogram (EEG) machines that continuously monitor brain waves in real time, and electrocardiograph (ECG) machines that perform the same task with heartbeats.

Many biomedical engineers work on developing materials and devices that are biocompatible, meaning that they can replace or come into direct contact with living tissues, perform a biological function, and refrain from triggering an immune system response. Pacemakers, small artificial devices that are implanted within the body and used to stimulate heart muscles to produce steady, reliable contractions, are a good example of a biocompatible device that has emerged from the collaboration of engineers and clinicians.

Biomechanics is the study of how the muscles and skeletal structure of living organisms are affected by and exert mechanical forces. Biomechanics applications include the development of orthotics (braces or supports), such as spinal, leg, and foot braces for patients with disabling disorders such as cerebral palsy, multiple sclerosis, or stroke. Prostheses (artificial limbs) also fall under the field of biomechanics; the sockets, joints, brakes, and pneumatic or hydraulic controls of an artificial leg, for example, are manufactured and then combined in a modular fashion, in much the same way as are the parts of an automobile in a factory.

Bionanotechnology. Nanotechnology is a fairly young field of applied science concerned with the manipulation of objects at the nanoscale (about 1-100 nanometers, or about one-thousandth the width of a strand of human hair) to produce machinery.

Bionanotechnological applications within medicine include microscopic biosensors installed on small chips; these can be specialized to recognize and flag specific proteins or antibodies, helping physicians conduct extremely fast and inexpensive diagnostic tests. Bioengineers are also developing microelectrodes on a nanoscale; these arrays of tiny electrodes can be implanted into the brain and used to stimulate specific nerve cells to treat movement disorders and other diseases.

Military Applications. Bioengineering applications are making themselves felt as a powerful presence on the front lines of the military. For example, bioengineering students at the University of Virginia designed lighter, more flexible, and stronger bulletproof body armor using specially created ceramic tiles that are inserted into protective vests. The armor is able to withstand multiple impacts and distributes shock more evenly across the wearer's body, preventing damaging compression to the chest. Others working in the field are creating sophisticated biosensors that soldiers can use to detect the presence of potential pathogens or biological weapons that have been released into the air.

One of the most significant contributions of bioengineering to the military is in the development of treatments for severe traumas sustained during warfare. For example, stem cell research may one day enable military physicians to regenerate functional tissues such as nerves, bone, cartilage, skin, and muscle—an invaluable tool for helping those who have lost limbs or other body parts as a result of explosives. The United States military was responsible for much of the early research done in creating safe, effective artificial blood substitutes that could be easily stored and relied on to be free of contamination on the battlefield.

Agriculture. Agricultural engineering involves the application of both engineering technologies and knowledge from animal and plant biology to problems in agriculture, such as soil and water conservation, food processing, and animal husbandry. For example, agricultural engineers can help farmers maximize crop yields from a defined area of land. This technique, known as precision farming, involves analyzing the properties of the soil (factors such as drainage, electrical conductivity, pH [acidity] level, and levels of chemicals such as nitrogen) and carefully calibrating the type and amount of seeds, insecticides, and fertilizers to be used.

Farm machinery and implements represent another area of agriculture in which engineering principles have made a big impact. Tractors, harvesters, combines, and grain-processing equipment, for example, have to be designed with mechanical and electrical principles in mind and also must take into account the characteristics of the land, the needs of the human operators, and the demands of working with particular agricultural products. For example, many crops require specialized equipment to be successfully mechanically harvested. Thus a pea harvester may have several components—one that lifts the vines and cuts them from the plant, one that strips pea pods from the stalk, and one that threshes the pods, causing them to open and release the peas inside them. Another example of an agricultural engineering application is the development of automatic milking machines that attach to the udders of a cow and enable dairy farmers to dispense with the arduous task of milking each animal by hand.

The management of soil and water is also an important priority for bioengineers working in agricultural settings. They may design structures to control the flow of water, such as dams or reservoirs. They may develop water-treatment systems to purify wastewater coming out of industrial agricultural production centers. Alternatively, they may use soil walls or cover crops to reduce the amount of pesticides and nutrients that run off from the soil, as well as the amount of erosion that takes place as a result of watering or rainfall.

Environmental and Ecological Applications. Environmental and ecological engineers study the impact of human activity on the environment, as well as the ways in which humans respond to different features of their environments. They use engineering principles to clean, control, and improve the quality of natural spaces, and find ways to make human interactions with environmental resources more sustainable. For example, the reduction and remediation of pollution is an important area of concern. Therefore, an environmental engineer may study the pathways and rates at which volatile organic compounds (such as those found in many paints, adhesives, tiles, wall coverings, and furniture) react with other gases in the air, causing smog and other forms of air pollution. They may design and build sound walls in residential areas to cut down on the amount of noise pollution caused by airplanes taking off and landing or cars racing up and down highways.

The life-support systems designed by bioengineers to enable astronauts to survive in the harsh conditions of outer space are also a form of environmental engineering. For example, temperatures around a space shuttle can vary wildly, depending on which side of the vehicle is facing the Sun at any given moment. A complex system of heating, insulation, and ventilation helps regulate the temperature inside the cabin. Because space is a vacuum, the shuttle itself must be filled with pressurized gas. In addition, levels of oxygen, carbon dioxide, and nitrogen within the cabin must be controlled so that they resemble the atmosphere on Earth. Oxygen is stored on board in tanks, and additional supplies of the essential gas are produced from electrolyzed water; in turn, carbon dioxide is channeled out of the shuttle through vents.

Geoengineering. Geoengineering is an emerging subfield of bioengineering that is still largely theoretical. It would involve the large-scale modification of environmental processes in an attempt to counteract the effects of human activity leading to climate change. One proposed geoengineering project involves depositing a fine dust of iron particles into the ocean in an attempt to increase the rate at which algae grows in the water. Since algae absorbs carbon dioxide as it photosynthesizes, essentially trapping and containing it, this would be a means of reducing the amount of this greenhouse gas in the atmosphere. Other geoengineering proposals include the suggestion that it might be possible to spray sulfur dust into the high atmosphere to reflect some of the Sun's light and heat back into space, or to spray drops of seawater high up into the air so that the salt particles they contain would be absorbed into the clouds, making them thicker and more able to reflect sunlight.

FUTURE PROSPECTS

Bioengineering is a field with the capacity to exert a powerful impact on many aspects of social life. Perhaps most profound are the transformations it has made in health care and medicine. By treating the body as a complex system—looking at it almost as if it were a machine—bioengineers and physicians working together have enabled countless patients to overcome what once might have seemed to be insurmountable damage. After all, if the body is a machine, its parts might be reengineered or replaced entirely with new ones—as when the damaged cilia of individuals with hearing impairments are replaced with electro-mechanical devices. Some aspects of bioengineering, however, have drawn concern from observers who worry that there may be no limit to the scientific ability to interfere with biological processes.

Transgenic foods are one area in which a contentious debate has sprung up. Some are convinced that the ecological and health ramifications of growing and ingesting crops that contain genetic information from more than one species have not yet been fully explored. Stem cell research is another area of controversy; some critics are uncomfortable with the fact that human embryonic stem cells are being obtained from aborted fetuses or fertilized eggs that are left over from assisted reproductive technology procedures.

One aspect of bioengineering that has been the subject of both fear and hope in the twenty-first century is the question of whether it might be possible to stop or even reverse the harmful effects of climate change by carefully and deliberately interfering with certain geological processes. Some believe that geoengineering could help the international community avoid the devastating effects of global warming predicted by scientists, such as widespread flooding, droughts, and crop failure. Others, however, warn that any attempt to interfere with complex environmental systems on a global scale could have wildly unpredictable results. Geoengineering is especially controversial because such projects could potentially be carried out unilaterally by countries acting without international agreement and yet have repercussions that could be felt all across the world.

M. Lee, MA

Further Reading

Artmann, Gerhard M., and Shu Chien, eds. *Bioengineering in Cell and Tissue Research*. New York: Springer, 2008. Examines bioengineering's role in cell research. Heavily illustrated with diagrams and figures; includes a comprehensive index and references after each section.

Enderle, John D., Susan M. Blanchard, and Joseph D. Bronzino, eds. *Introduction to Biomedical Engineering*. 2d ed. Boston: Elsevier Academic Press, 2005. A broad introductory textbook designed for undergraduates. Each chapter contains an outline, objectives, exercises, and suggested reading.

Huffman, Wallace E., and Robert E. Evenson. *Science for Agriculture: A Long-Term Perspective*. 2d ed. Ames, Iowa: Blackwell, 2006. A history of agricultural engineering research within the United States. Includes a glossary and list of relevant acronyms.

Madhavan, Guruprasad, Barbara Oakley, and Luis G. Kun, eds. *Career Development in Bioengineering and Biotechnology*. New York: Springer, 2008. An extensive guide to careers in bioengineering, biotechnology, and related fields, written by active practitioners. Covers both traditional and alternative job opportunities.

Nemerow, Nelson Leonard, et al., eds. *Environmental Engineering*. 3 vols. 6th ed. Hoboken, N.J.: John Wiley & Sons, 2009. Discusses topics such as food protection, soil management, waste management, water supply, and disease control. Each section includes references and a Bibliography.

Web Sites

Biomedical Engineering Society http://www.bmes.org
National Institutes of Health National Institute of Biomedical Imaging and Bioengineering http://www.nibib.nih.gov
Society for Biological Engineering http://www.aiche.org/sbe

See also: Artificial Organs; Biochemical Engineering; Biomechanical Engineering; Bionics and Biomedical Engineering; Bioprocess Engineering; Cell and Tissue Engineering; Human Genetic Engineering; Military Sciences and Combat Engineering; Rehabilitation Engineering.

FASCINATING FACTS ABOUT BIOENGINEERING

- Bioengineering has enabled scientists to grow replacement human skin, tracheas, bladders, cartilage, and other tissues and organs in the laboratory.

- Materials scientists and clinical researchers are working together to develop contact lenses that can deliver precise doses of drugs directly into the eye.

- By genetically engineering crops that are naturally resistant to insects, bioengineers have helped reduce the need to use harmful pesticides in industrial farming.

- Bacteria whose genetic information has been carefully reengineered may eventually provide an endless supply of crude oil, helping meet the world's energy needs without engaging in damaging drilling.

- In 2009, an MIT bioengineer invented a new way to pressurize space suits that does not use gas, making them far sleeker and less bulky than conventional astronaut gear.

- One military application of bioengineering is a robotic system that seeks out and identifies tiny pieces of shrapnel lodged within tissue, then guides a needle to those precise spots so that the shrapnel can be removed.

- Bionic men and women are not just the stuff of television and motion-picture fantasy. In fact, anyone who has an artificial body part, such as a prosthetic leg, a pacemaker, or an implanted hearing aid, can be considered bionic.

- Some bioengineers are working on developing artificial noses that can detect and diagnose disease by smell—literally sniffing out infections and cancer, for example.

- One day, it may be possible to "print out" artificial organs using a three-dimensional printer. Layer by layer, cells would be deposited onto a glass slide, building up specialized tissues that could be used to replace damaged kidneys, livers, and other organs.

- Each year, more women choose to enter the field of biomedical engineering than any other specialty within engineering.

Biology

FIELDS OF STUDY

Cell biology, conservation biology, developmental biology, invertebrate biology, marine biology, neurobiology, population biology, zoology

SUMMARY

Biology is the science that studies life and living organisms.

PRINCIPAL TERMS

- **biodiversity:** the total of all living organisms in an environment
- **deoxyribonucleic acid (DNA):** the carrier of all an organism's genetic information
- **National Institutes of Health:** the United States' governmental division that monitors and improves public health
- **xenotransplantation:** the transplantation of organs from one species to another

Biology is an extensive field subdivided into categories based on the molecule, the cell, the organism, and the population. Molecular biology touches on biophysics and biochemistry. It is the branch of biology that deals with the structure and development of biological systems in terms of the physics and chemistry of their molecules. Cellular biology is closely related to molecular biology through understanding the functions and basic structure of the cell. The cell is the smallest structural unit of an organism that is capable of independent functioning, consisting of one or more nuclei, cytoplasm, and organelles, all surrounded by a semipermeable membrane. Botany is the science or study of plant life. Ecology, also referred to as bionomics, is the study of relationships between organisms and their environments.

Developmental biology encompasses a number of issues, including gene regulation, genetics, and evolution. These are concepts of importance to vertebrates, invertebrates, and plants. A gene is a hereditary unit that occupies a particular location on a chromosome, determines a particular characteristic of an organism, and can undergo mutation. Genetics is the branch of biology that deals with heredity, especially the hereditary transmission and variation of inherited characteristics. Evolution is the theory that groups of organisms change with the passage of time, mainly as a result of natural selection, so that descendants differ morphologically and physiologically from their ancestors. Population genetics, the study of gene changes in populations, and ecology have been established subject areas since the 1930's. These two fields were combined in the 1960's to form a new discipline called population biology, which became established as a major subdivision of biological studies in the 1970's. Central to this field is evolutionary biology, in which the contributions of Charles Darwin are noted.

Microbiology is the branch of biology that deals with microorganisms. The study of bacteria, including their classification and the prevention of diseases that arise from bacterial infection, is the primary focus of microbiology. This branch is of interest not only among bacteriologists but also among chemists, biochemists, geneticists, pathologists, immunologists, and public health professionals. Parasitology is the study of parasites, organisms that feed on or in different organisms while contributing nothing to the survival of their hosts. Ethology is the scientific study of animal behavior. Animal behavioral studies have developed along two lines. The first of these, animal psychology, is primarily concerned with physiological psychology, and has traditionally concentrated on laboratory techniques such as conditioning. The second, ethology, had its origins in observations of animals under natural conditions, concentrating on courtship, flocking, and other social contacts. One of the important recent developments in the field is the focus on sociobiology, which is concerned with the behavior, ecology, and evolution of social animals such as bees, ants, schooling fish, flocking birds, and humans. Ethics is the study of the general nature of morals and of specific moral choices. Bioethics addresses such issues as animal experimentation, cloning, euthanasia, gene therapy, genetic engineering, genome projects, protection of human research subjects, organ transplants, and patients' rights. Biotechnology is the industrial use of living organisms or biological techniques developed through basic research. Biotechnology products

A diagram of a fly from Robert Hooke's innovative Micrographia, 1665

include antibiotics, insulin, interferon, and recombinant DNA, and techniques such as waste recycling.

The Office of Biotechnology Activities of the National Institutes of Health monitors scientific progress in human genetics research in order to anticipate future developments, including ethical, legal, and social concerns, in basic and clinical research involving recombinant DNA, genetic testing, and xenotransplantation (the use of organs from other species of mammals for transplants). In addition to organs donated from humans, researchers are exploring the use of partially or wholly artificial organs manufactured in the laboratory. Biodiversity focuses on such issues as conservation, extinction and depletion from overexploitation, habitat pollution, global patterns and values of biodiversity, and endangered species protection. Well known pioneer biologists include naturalist and explorer Sir Joseph Banks (1743-1820), naturalist and explorer Charles William Beebe (1877-1962), biochemist Gunter Blobel (b. 1936), environmentalist Rachel Carson (1907-1964), biochemist Stanley Cohen (b. 1922), biophysicist and codiscoverer of DNA, Francis Crick (b. 1916), naturalist and father of evolutionary theory Charles Darwin (1809-1882), zoologist Richard Dawkins (b. 1941), marine biologist Sylvia Earle (b. 1935), bacteriologist Paul Ehrlich (b. 1932), bacteriologist and discoverer of penicillin Sir Alexander Fleming (1881-1955), microscopist Antonie van Leeuwenhoek (1632-1723), botanist and taxonomist Carolus Linnaeus (1707-1723), ethologist Konrad Lorenz (1903-1989), zoologist A. S. Loukashkin (1902-1988), botanist and geneticist Barbara Mc- Clintock (1902-1992) , botanist and genetic theorist Gregor Mendel (1822-1884), endocrinologist and inventor of the birth control pill Gregory Goodwin Pincus (1903-1967), naturalist Alfred Russel Wallace (1823-1913), biophysicist and codiscoverer of DNA James Watson (b. 1928), and sociobiologist Edward O. Wilson (b. 1929).

—*Mary E. Carey*

See also: Anatomy; Animal kingdom; Ecology; Embryology; Ethology; Genetics; Marine biology; Paleoecology; Paleontology; Systematics; Veterinary medicine; Zoology.

FURTHER READING

Campbell, Neil A. *Biology*. 2d ed. San Francisco: Benjamin/Cummings, 1993. A wideranging introductory textbook, covering all the basics of biology.

Lavers, Chris. *Why Elephants Have Big Ears: Understanding Patterns of Life on Earth*. New York: St. Martin's Press, 2001. An encyclopedic exploration of form and function in animal physiology.

Meyr, Ernst. *The Growth of Biological Thought: Diversity, Evolution, and Inheritance*. Reprint. Cambridge, Mass.: Belknap Press of Harvard University Press, 1985. Aclassic exposition of the history of biology as a field of study and its philosophical evolution.

Watson, James D. *The Double Helix: A Personal Account of the Discovery of the Structure of DNA*. Reprint. New York: Simon & Schuster, 1998. An autobiographical account by the codiscoverer of the structure of DNA. Notable for its honest depiction of the process of scientific experimentation, discovery, and academic life.

Bionics and biomedical engineering

FIELDS OF STUDY

Biology; physiology; biochemistry; engineering; orthopedic bioengineering; physics; bionanotechnology; biomechanics; biomaterials; neural engineering; genetic engineering; tissue engineering; prosthetics.

SUMMARY

Bionics combines natural biologic systems with engineered devices and electrical mechanisms. An example of bionics is an artificial arm controlled by impulses from the human mind. Construction of bionic arms or similar devices requires the integrative use of medical equipment such as electroencephalograms (EEGs) and magnetic resonance imaging (MRI) machines with mechanically engineered prosthetic arms and legs. Biomedical engineering further melds biomedical and engineering sciences by producing medical equipment, tissue growth, and new pharmaceuticals. An example of biomedical engineering is human insulin production through genetic engineering to treat diabetes.

PRINCIPAL TERMS

- **biologics:** medicines produced from genes by manipulating genes and using genetic technology.
- **biomaterials:** substances, including metal alloys, plastic polymers, and living tissues, used to replace body tissues or as implants.
- **bionanotechnology:** construction of materials on a very small scale, enabling the use of microscopic machinery in living tissues.
- **bionic:** integrating biological function and mechanical devices.
- **clone:** genetically engineered organism with genetic composition identical to the original organism.
- **human genetic engineering:** genetic engineering focused on altering or changing visible human characteristics through gene manipulations.
- **prosthesis:** artificial or biomechanically engineered body part.
- **recombinant DNA:** DNA created by the combination of two or more DNA sequences that do not normally occur together.

BASIC PRINCIPLES

The fields of biomedical engineering and bionics focus on improving health, particularly after injury or illness, with better rehabilitation, medications, innovative treatments, enhanced diagnostic tools, and preventive medicine.

Bionics has moved nineteenth-century prostheses, such as the wooden leg, into the twenty-first century by using plastic polymers and levers. Bionics integrates circuit boards and wires connecting the nervous system to the modular prosthetic limb. Controlling artificial limb movements with thoughts provides more lifelike function and ability. This mind and prosthetic limb integration is the "bio" portion of bionics; the "nic" portion, taken from the word "electronic," concerns the mechanical engineering that makes it possible for the person using a bionic limb to increase the number and range of limb activity, approaching the function of a real limb.

Biomedical engineering encompasses many medical fields. The principle of adapting engineering techniques and knowledge to human structure and function is a key unifying concept of biomedical engineering. Advances in genetic engineering have produced remarkable bioengineered medications.

Recombinant DNA techniques (genetic engineering) have produced synthetic hormones, such as insulin. Bacteria are used as a host for this process; once human-insulin-producing genes are implanted in the bacteria, the bacteria's DNA produce human insulin, and the human insulin is harvested to treat diabetics. Before this genetic technique was developed in 1982 to produce human insulin, insulidependent diabetics relied on insulin from pigs or cows. Although this insulin was life saving for diabetics, diabetics often developed problems from the pig or cow insulin because they would produce antibodies against the foreign insulin. This problem disappeared with the ability to engineer human insulin using recombinant DNA technology.

BACKGROUND

In the broad sense, biomedical engineering has existed for millennia. Human beings have always envisioned the integration of humans and technology to increase and enhance human abilities. Prosthetic devices go back many thousands of years: A three-thousand-year-old Egyptian mummy, for example, was found with a wooden big toe tied to its foot. In the fifteenth century, during the Italian Renaissance, Leonardo da Vinci's elegant drawings demonstrated some early ideas on bioengineering, including his helicopter and flying machines, which melded human and machine into one functional unit capable of flight. Other early examples of biomedical engineering include wooden teeth, crutches, and medical equipment, such as stethoscopes.

Electrophysiological studies in the early 1800's produced biomedical engineering information used to better understand human physiology. Engineering principles related to electricity combined with human physiology resulted in better knowledge of the electrical properties of nerves and muscles. X rays, discovered by Wilhelm Conrad Rontgen in 1895, were an unknown type of radiation (thus the "X" name). When it was accidentally discovered that they could penetrate and destroy tissue, experiments were developed that led to a range of imaging technologies that evolved over the next century. The first formal biomedical engineering training program, established in 1921 at Germany's Oswalt Institute for Physics in Medicine, focused on three main areas: the effects of ionizing radiation, tissue electrical characteristics, and X-ray properties.

In 1948, the Institute of Radio Engineers (later the Institute of Electrical and Electronics Engineers), the American Institute for Electrical Engineering, and the Instrument Society of America held a conference on engineering in biology and medicine. The 1940's and 1950's saw the formation of professional societies related to biomedical engineering, such as the Biophysics Society, and of interest groups within engineering societies. However, research at the time focused on the study of radiation. Electronics and the budding computer era broadened interest and activities toward the end of the 1950's.

James D. Watson and Francis Crick identified the DNA double-helix structure in 1953. This important discovery fostered subsequent experimentation in molecular biology that yielded important information about how DNA and genes code

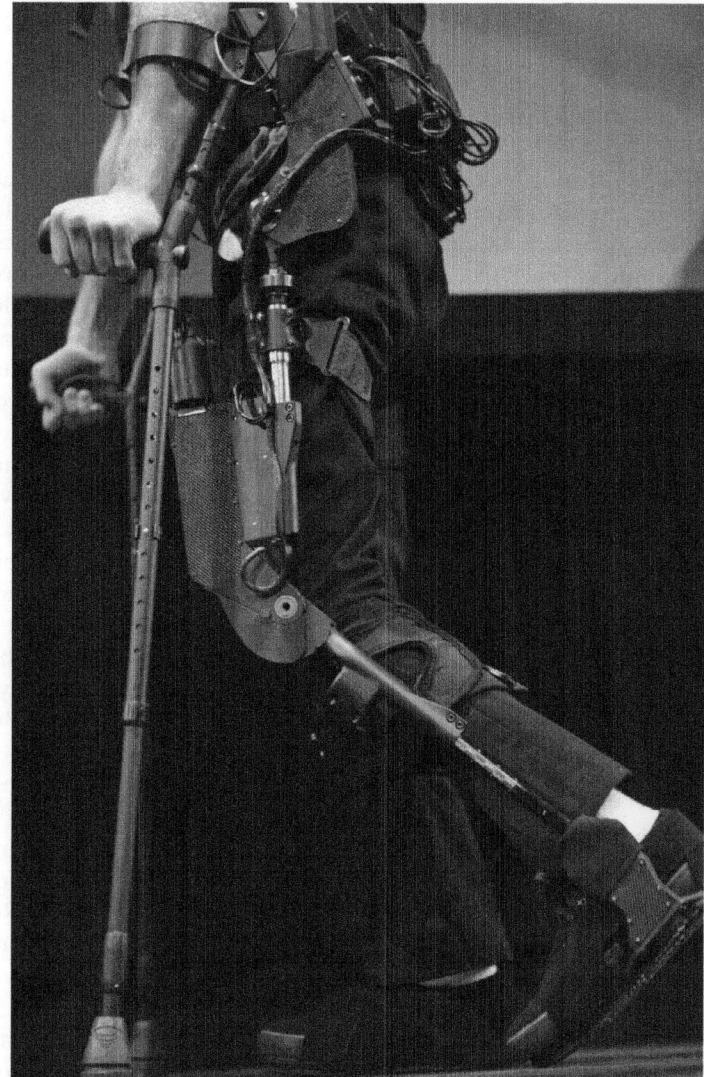

Bionics

for the expression of traits in all living organisms. The genetic code in DNA was deciphered in 1968, arming researchers with enough information to discover ways that DNA could be recombined to introduce genes from one organism into a different organism, thereby allowing the host to produce a variety of useful products. DNA recombination became one of the most important tools in the field of biomedical engineering, leading to tissue growth as well as new pharmaceuticals. In 1962, the National Institutes of Health created the National Institute of General Medical Sciences, fostering the development of biomedical engineering programs. This institute funds research in the diagnosis, treatment, and prevention of disease. Bionics and biomedical engineering span a wide variety of beneficial health-related fields. The common thread is the combination of technology with human applications. Dolly the sheep was cloned in 1996. Cloning produces a genetically identical copy of an existing life-form. Human embryonic cloning presents the potential of therapeutic reproduction of needed organs and tissues, such as kidney replacement for patients with renal failure.

How It Works

In the twenty-first century, the linking of machines with the mind and sensory perception has provided hearing for deaf people, some sight for the blind, and willful control of prostheses for amputees.

Restorative Bionics. Restorative bionics integrates prosthetic limbs with electrical connections to neurons, allowing an individual's thoughts to control the artificial limb. Tiny arrays of electrodes attached to the eye's retina connect to the optic nerve, enabling some visual perception for previously blind people. Deaf people hear with electric devices that send signals to auditory nerves, using antennas, magnets, receivers, and electrodes. Researchers are considering bionic skin development using nanotechnology to connect with nerves, enabling skin sensations for burn victims requiring extensive grafting.

Biomedical Devices. Many biomedical devices work inside the human body. Pacemakers, artificial heart valves, stents, and even artificial hearts are some of the bionic devices correcting problems with the cardiovascular system. Pacemakers generate electric signals that improve abnormal heart rates and abnormal heart rhythms. When pulse generators located in the pacemakers sense an abnormal heart rate or rhythm, they produce shocks to restore the normal rate. Stents are inserted into an artery to widen it and open clogged blood vessels. Stents and pacemakers are examples of specialized bionic devices made up of bionic materials compatible with human structure and function.

Cloning. Cloning is a significant area of genetic engineering that allows the replication of a complete living organism by manipulating genes. Dolly the sheep, an all-white Finn Dorset ewe, was cloned from a surrogate mother blackface ewe, which was used as an egg donor and carried the cloned Dolly during gestation (pregnancy). An egg cell from the surrogate was removed and its nucleus (which contains DNA) was replaced with one from a Finn Dorset ewe; the resulting new egg was placed in the blackface ewe's uterus after stimulation with an electric pulse. The electrical pulse stimulated growth and cell duplication. The blackface ewe subsequently gave birth to the all-white Dolly. The newborn all-white Finn Dorset ewe was an identical genetic twin of the Finn Dorset that contributed the new nucleus.

Recombinant DNA. Another significant genetic engineering technique involves recombinant DNA. Human genes transferred to host organisms, such as bacteria, produce products coded for by the transferred genes. Human insulin and human growth hormone can be produced using this technique. Desired genes are removed from human cells and placed in circular bacterial DNA strips called plasmids. Scientists use enzymes to prepare these DNA formulations, ultimately splicing human genes into bacterial plasmids. These plasmids are used as vectors, taken up and reproduced by bacteria. This type of genetic adaptation results in insulin production if the spliced genes were taken from the part of the human genome producing insulin; other cells and substances, coded for by different human genes, can be produced this way. Many biologic medicines are produced using recombinant DNA technology.

Medical Devices. Biomedical engineers produce life-saving medical equipment, including pacemakers, kidney dialysis machines, and artificial hearts. Synthetic limbs, artificial cochleas, and bionic sight chips are among the prosthetic devices that biomedical engineers have developed to enhance mobility, hearing, and vision. Medical monitoring devices, developed by biomedical engineers for use in intensive care units and surgery or by space and deep-sea explorers, monitor vital signs such as heart rate and rhythm, body temperature, and breathing rate.

Equipment and Machinery. Biomedical engineers produce a wide variety of other medical machinery, including laboratory equipment and therapeutic equipment. Therapeutic equipment includes laser devices for eye surgery and insulin pumps (sometimes called artificial pancreases) that both monitor blood sugar levels and deliver the appropriate amount of insulin when it is needed.

Imaging Systems. Medical imaging provides important machinery devised by biomedical engineers. This specialty incorporates sophisticated computers and imaging systems to produce computed tomography (CT), magnetic resonance imaging (MRI), and positron emission tomography (PET) scans. In naming its National Institute of Biomedical Imaging and Bioengineering (NIBIB), the U.S. Department of Health and Human Services emphasized the equal importance and close relatedness of these subspecialties by using both terms in the department's name. Computer programming provides important circuitry for many biomedical engineering applications, including systems for differential disease diagnosis. Advances in bionics, moreover, rely heavily on computer systems to enhance vision, hearing, and body movements.

Biomaterials. Biomaterials, such as artificial skin and other genetically engineered body tissues, are areas promising dramatic improvements in the treatment of burn victims and individuals needing organ transplants. Bionanotechnology, another subfield of biomedical engineering, promises to enhance the surface of artificial skin by creating microscopic messengers that can create the sensations of touch and pain. Bioengineers interface with the fields of physical therapy, orthopedic surgery, and rehabilitative medicine in the fields of splint development, biomechanics, and wound healing.

Medications. Medicines have long been synthesized artificially in laboratories, but chemically synthesized medicines do not use human genes in their production. Medicines produced by using human genes in recombinant DNA procedures are called biologics and include antibodies, hormones, and cell receptor proteins. Some of these products include human insulin, the hepatitis B vaccine, and human growth hormone.

Bacteria and viruses invading a body are attacked and sometimes neutralized by antibodies produced by the immune system. Diseases such as Crohn's disease, an inflammatory bowel condition, and psoriatic arthritis are conditions exacerbated by inflammatory antibody responses mounted by the affected person's immune system. Genetic antibody production in the form of biologic medications interferes with or attacks mediators associated with Crohn's and arthritis and improves these illnesses by decreasing the severity of attacks or decreasing the frequency of flare-ups.

Cloning and Stem Cells. Cloned human embryos could provide embryonic stem cells. Embryonic stem cells have the potential to grow into a variety of cells, tissues, and organs, such as skin, kidneys, livers, or heart cells. Organ transplantation from genetically identical clones would not encounter the recipient's natural rejection process, which transplantations must overcome. As a result, recipients of genetically identical cells, tissues, and organs would enjoy more successful replacements of key organs and a better quality of life. Human cloning is subject to future research and development, but the promise of genetically identical replacement organs for people with failed hearts, kidneys, livers, or other organs provides hope for enhanced future treatments.

FUTURE PROSPECTS

Bionics technologies include artificial hearing, sight, and limbs that respond to nerve impulses. Bionics offers partial vision to the blind and prototype prosthetic arm devices that offer several movements through nerve impulses. The goal of bionics is to better integrate the materials in these artificial devices with human physiology to improve the lives of those with limb loss, blindness, or decreased hearing.

Cloned animals exist but cloning is not a yet a routine process. Technological advances offer rapid DNA analysis along with significantly lower cost genetic analysis. Genetic databases are filled with information on many life-forms, and new DNA sequencing information is added frequently. This basic information that has been collected is like a dictionary, full of words that can be used to form sentences, paragraphs, articles, and books, in that it can be used to create new or modified life-forms.

Biomedical engineering enables human genetic engineering. The stuff of life, genes, can be modified or manipulated with existing genetic techniques. The power to change life raises significant societal concerns and ethical issues. Beneficial results such as optimal organ transplantations and effective medications are the potential of human genetic engineering.

Richard P. Capriccioso, MD, and Christina Capriccioso

FURTHER READING

Braga, Newton C. Bionics for the Evil Genius: Twenty-five Build-It-Yourself Projects. New York: McGraw-Hill, 2006. Step-by-step projects that introduce basic concepts in bionics.

Fischman, Josh. "Merging Man and Machine: The Bionic Age." National Geographic 217, no. 1 (January, 2010): 34-53. A well-illustrated consideration of the latest advances in bionics, with specific examples of people aided by the most modern prosthetic technologies.

Hung, George K. Biomedical Engineering: Principles of the Bionic Man. Hackensack, N.J.: World Scientific, 2010. Examines scientific bioengineering principles as they apply to humans.

Richards-Kortum, Rebecca. Biomedical Engineering for Global Health. New York: Cambridge University Press, 2010. Examines the potential of biomedical engineering to treat diseases and conditions throughout the world. Examines health care systems and social issues.

Smith, Marquard, and Joanne Morra, eds. The Prosthetic Impulse: From a Posthuman Present to a Biocultural Future. Cambridge, Mass.: MIT Press, 2007. Examines the developments in prosthetic devices and addresses the social aspects, including what it means to be human.

WEB SITES

American Society for Artificial Internal Organs http://www.asaio.com

Biomedical Engineering Society http://www.bmes.org

National Institute of Biomedical Imaging and Bioengineering http://www.nibib.nih.gov/HomePage

Rehabilitation Engineering and Assistive Technology Association of North America http://resna.org

Society for Biomaterials http://www.biomaterials.org

See also: Artificial Organs; Biosynthetics; Cloning; Human Genetic Engineering; Stem Cell Research and Technology.

FASCINATING FACTS ABOUT BIONICS AND BIOMEDICAL ENGINEERING

- In 2010, scientists at the University of California, San Diego, developed biosensor cells that can be implanted in the brain to help monitor receptors and chemical signals that allow cells in the brain to communicate with one another. These cells may help scientists understand drug addiction.

- Vanderbilt engineers in 2010 began testing a knowledge repository and interactive software that will help surgeons more accurately and rapidly place electrodes in the brains of people with Parkinson's disease in a procedure called deep brain stimulation. The data allow for faster surgery and the implementation of best practices.

- In 2010, scientists at Vanderbilt University developed a robotic prosthesis for the lower leg that has powered knee and ankle joints. Intent recognizer software takes information from sensors, determines what the user wants to do, and provides power to the leg.

- The Wadsworth Center at the New York State Department of Health in 2009 developed a brain-computer interface that translates brain waves into action. It allowed a patient with amyotrophic lateral sclerosis who was no longer able to communicate with others because of failing muscles to write e-mails and convey his thoughts to others.

- In 2008, a research team at the Johns Hopkins University modified chondroitin sulfate, a natural sugar, so it could glue a hydrogel (like the material used in soft contact lenses) to cartilage tissue. It is hoped that this technique may help those experiencing joint pain from oseoarthritis, in which the natural cartilage in a joint disappears.

BIOPHYSICS

FIELDS OF STUDY

Physics; physical sciences; chemistry; mathematics; biology; molecular biology; chemical biology; engineering; biochemistry; classical genetics; molecular genetics; cell biology.

SUMMARY

Biophysics is the branch of science that uses the principles of physics to study biological concepts. It examines how life systems function, especially at the cellular and molecular level. It plays an important role in understanding the structure and function of proteins and membranes and in developing new pharmaceuticals. Biophysics is the foundation for molecular biology, a field that combines physics, biology, and chemistry.

PRINCIPAL TERMS

- **circular dichroism (CD):** differential absorption of left- and right-handed circularly polarized light.
- **electromagnetic waves:** waves that can transmit their energy through a vacuum.
- **molecular genetics:** branch of genetics that analyzes the structure and function of genes at the molecular level.
- **polarized light:** light waves that vibrate in a single plane.
- **quantum mechanics:** physical analysis at the level of atoms or subatomic fundamental particles.
- **thermodynamics:** branch of physics that studies energy conversions.
- **vector:** quantity that has both magnitude and direction.

BASIC PRINCIPLES

The word "biophysics" means the physics of life. Biophysics studies the functioning of life systems, especially at the cellular and molecular level, using the principles of physics. It is known that atoms make up molecules, molecules make up cells, and cells in turn make up tissues and organs that are part of an organism, or a living machine. Biophysicists use this knowledge to understand how the living machine works. In photosynthesis, for instance, the absorption of sunlight by green plants initiates a process that culminates with synthesis of high-energy sugars such as glucose. To fully understand this process, one needs to look at how it begins—light absorption by the photosystems.

Photosystems are groups of energy-absorbing pigments such as chlorophyll and carotenoids that are located on the thylakoid membranes inside the chloroplast, the photosynthetic organelle in the plant cell. Biophysical studies have shown that once a chlorophyll molecule captures solar energy, it gets excited and transfers the energy to a neighboring unexcited chlorophyll molecule. The process repeats itself, and thus, packets of energy jump from one chlorophyll molecule to the next. The energy eventually reaches the reaction center, where it begins a chain of high-energy electron-transfer reactions that lead to the storage of the light energy in the form of adenosine triphosphate (ATP) and nicotinamide adenine dinucleotide phosphate (NADPH). In the second half of photosynthesis, ATP and NADPH provide the energy to make glucose from carbon dioxide. Biophysics is often confused with medical physics. Medical physics is the science devoted to studying the relationship between human health and radiation exposure. For example, a medical physicist often works closely with a radiation oncologist to set up radiotherapy treatment plans for cancer patients.

BACKGROUND

In comparison with other branches of biology and physics, biophysics is relatively new and, therefore, still evolving. Even though the use of physical concepts and instrumentation to explain the workings of life systems had begun as early as the 1840's, biophysics did not emerge as an independent field until the 1920's. Some of the earliest studies in biophysics were conducted in the 1840's by a group known as the Berlin school of physiologists. Among its members were pioneers such as Hermann von Helmholtz, Ernst Heinrich Weber, Carl F. W. Ludwig, and Johannes Peter Muller. This group used well-known physical methods to investigate physiological issues, such as the mechanics of muscular contraction and the electrical changes in a nerve cell during impulse transmission.

Karl Pearson

The first biophysics textbook was written in 1856 by Adolf Fick, a student of Ludwig. Although these early biophysicists made significant advances, subsequent research focused on other areas. In the 1920's, the first biophysical institutes were established in Germany and the first textbook with the word "biophysics" in its title was published. However, through the 1940's, biophysics research was primarily aimed at understanding the biophysical impact of ionizing radiation. In 1944, Austrian physicist Erwin Schrodinger published *What Is Life ? The Physical Aspect of the Living Cell*, based on a series of lectures that addressed biology from the viewpoint of a classical physicist. This cross-disciplinary work motivated several physicists to become interested in biology and thus laid the foundation for the field of molecular biology. From 1950 to 1970, the field of biophysics experienced rapid growth, tremendously accelerated by the discovery in 1953 of the double helix structure of DNA by James D. Watson and Francis Crick. Both Watson and Crick have stated that they were inspired by Schrodinger's work.

How It Works

Biophysicists study life at all levels, from atoms and molecules to cells, organisms, and environments. They attempt to describe complex living systems with the simple laws of physics. Often, biophysicists work at the molecular level to understand cells and their processes.

The work of Gregor Mendel in the late nineteenth century laid the foundation for genetics, the science of heredity. His studies, rediscovered in the twentieth century, led to the understanding that the inheritance of certain traits is governed by genes and that the alleles of the genes are separated during gamete formation. Experiments in the 1940's revealed that genes are made of DNA, but the mechanisms by which genes function remained a mystery. Watson and Crick's discovery of the double helix structure of DNA in 1953 revealed how genes could be translated into proteins.

Biophysicists use a number of physical tools and techniques to understand how cellular processes work, especially at the molecular level. Some of the important tools are electron microscopy, nuclear magnetic resonance (NMR) spectroscopy, circular dichroism (CD) spectroscopy, and X-ray crystallography. For example, the discovery of Watson and Crick's double helix model was possible in part because of the X-ray images of DNA that were taken by Rosalind Franklin and Maurice H. F. Wilkins. Franklin and Wilkins, both biophysicists, made DNA crystals and then used X-ray crystallography to analyze the structure of DNA. The array of black dots arranged in an X-shaped pattern on the X-ray photograph of wet DNA suggested to Franklin that DNA was helical.

Electron Microscopy. Electron microscopes use beams of electrons to study objects in detail. Electron microscopy can be used to analyze an object's surface texture (topography) and constituent elements and compounds (composition), as well as the shape and size (morphology) and atomic arrangements (crystallographic details) of those elements and compounds. Electron microscopes were invented to overcome the limitations posed by light microscopes, which have maximum magnifications of 500x or 1000x and a maximum resolution of 0.2 millimeter. To see and study subcellular structures and processes required magnification capabilities of greater than 10,000x. The first electron microscope was a transmission electron microscope (TEM) built by Max Knoll and Ernst Ruska in 1931. The invention of the scanning electron microscope (SEM) was somewhat delayed (the first was built in 1937 by Manfred von Ardenne) because the field had to figure out how to make the electron beam scan the sample.

NMR Spectroscopy. Nuclear magnetic resonance (NMR) spectroscopy is an extremely useful tool for the biophysicist to study the molecular structure of organic compounds. The underlying principle of NMR spectroscopy is identical to that of magnetic resonance imaging (MRI), a common tool in medical diagnostics. The nuclei of several elements, including the isotopes carbon-12 and oxygen-16, have a characteristic spin when placed in an external magnetic field. NMR focuses on studying the transitions between these spin states. In comparison with mass spectroscopy, NMR requires a larger amount of sample, but it does not destroy the sample.

CD Spectroscopy. Circular dichroism (CD) spectroscopy measures differences in how left-handed and right-handed polarized light is absorbed. These differences are caused by structural asymmetry. CD spectroscopy can determine the secondary and tertiary structure of proteins as well as their thermal stability. It is usually used to study proteins in solution.

X-Ray Crystallography. X rays are electromagnetic waves with wavelengths ranging from 0.02 to 100 angstroms (A). Even before X rays were discovered in 1895 by Wilhelm Conrad Rontgen, scientists knew that atoms in crystals were arranged in definite patterns and that a study of the angles therein could provide clues to the crystal structure. As is true of all forms of radiation, the wavelength of X rays is inversely proportional to its energy. Because the wavelength of X rays is smaller than that of visible light, X rays are powerful enough to penetrate most matter. As X rays travel through an object, they are diffracted by the atomic arrangements inside and thus provide a guideline for the electron densities inside the object. Analysis of this electron density data offers a glimpse into the internal structure of the crystal. As of 2010, about 90 percent of the structures in the Worldwide Protein Data Bank had been elucidated through X-ray crystallography.

Biophysical tools and techniques have become extremely useful in many areas and fields. They have furthered research in protein crystallography, synthetic biology, and nanobiology, and allowed scientists to discover new pharmaceuticals and to study biomolecular structures and interactions and membrane structure and transport. Biophysics and its related fields, molecular biology and genetics, are rapidly developing and are at the center of biomedical research.

Biomolecular Structures. Because structure dictates function in the world of biomolecules, understanding the structure of the biomolecule (with tools such as X-ray crystallography and NMR and CD spectroscopy), whether it is a protein or a nucleic acid, is critical to understanding its individual function in the cell. Proteins function as catalysts and bind to and regulate other downstream biomolecules. Their functional basis lies in their tertiary structure, or their three-dimensional form, and this function cannot be predicted from the gene sequence. The sequence of nucleotides in a gene can be used only to predict the primary structure, which is the amino acid sequence in the polypeptide. Once the structure-function relationship has been analyzed, the next step is to make mutants or knock out the gene via techniques such as ribonucleic acid interference (RNAi) and confirm loss of function. Subsequently, a literature search is performed to see if there are any known genetic disorders that are caused by a defect in the gene being studied. If so, the structure-function relationship can be examined for a possible cure.

Membrane Structure and Transport. In 1972, biologist S. J. Singer and his student Garth Nicolson conceived the fluid mosaic model of the plasma membrane. According to this model, the plasma membrane is a fluid lipid bilayer largely made up of phospholipids arranged in an amphipathic pattern, with the hydrophobic lipid tails buried inside and the hydrophilic phosphate groups on the exterior. The bilayer is interspersed with proteins, which help in cross-membrane transport. Because membranes control the import and export of materials into the cell, understanding membrane structure is key to coming up with ways to block transport of potentially harmful pathogens across the membrane.

Electron microscopes were used in the early days of membrane biology, but fluorescence and confocal microscopes have come to be used more frequently.

The development of organelle-specific vital stains has rejuvenated interest in evanescent field (EF) microscopy because it permits the study of even the smallest of vesicles and the tracking of the movements of individual protein molecules.

Synthetic Biology. In the 2000's, the term "systems biology" became part of the field of life science, followed by the term "synthetic biology." To many people, these terms appear to refer to the same thing, but they do not, even though they are indeed closely related. While systems biology focuses on using a quantitative approach to study existing biological systems, synthetic biology concentrates on applying engineering principles to biology and constructing novel systems heretofore unseen in nature. Clearly, synthetic biology benefits immensely from research in systems and molecular biology. In essence, synthetic biology could be described as an engineering discipline that uses known, tested functional components (parts) such as genes, proteins, and various regulatory circuits in conjunction with modeling software to design new functional biological systems, such as bacteria that make ethanol from water, carbon dioxide, and light. The biggest challenge to synthetic biologists is the complexity of life-forms, especially higher eukaryotes such as humans, and the possible existence of unknown processes that can affect the synthetic biological systems.

Drug Discovery. In the pharmaceutical world, the initial task is to identify the aberrant protein, the one responsible for generating the symptoms in any disease or disorder. Once that is done, a series of biophysical tools are used to ensure that the target is the correct one. First, the identity of the protein is confirmed using techniques such as N-terminal sequencing and tandem mass spectroscopy (MS-MS). Second, the protein sample is tested for purity (which typically should be more than 95 percent) using methods such as denaturing sodium dodecyl sulfate polyacrylamide gel electrophoresis (SDS-PAGE). Third, the concentration of the protein sample is determined by chromogenic assays such as the Bradford or Lowry assay.

The fourth and probably the most important test is that of protein functionality. This is typically carried out by either checking the ligand binding capacity of the protein (with biacore ligand binding assays) or by testing the ability of the protein to carry out its biological function. All these thermodynamic parameters need to be tested to develop a putative drug, one that could somehow correct or restrain the ramifications of the protein's malfunction.

Nanobiology. With the aid of biophysical tools and techniques, the field of biology has moved from organismic biology to molecular biology to nanobiology. To get a feel for the size of a nanometer, picture a strand of hair, then visualize a width that is 100,000 times thinner. Typically nanoparticles are about the size of either a protein molecule or a short DNA segment. Nanomedicine, or the application of technology that relies on nanoparticles to medicine, has become a popular area for research. In particular, the search for appropriate vectors to deliver drugs into the cells is an endless pursuit, especially in emerging therapeutic approaches such as RNA interference. Because lipid and polymer-based nanoparticles are extremely small, they are easily taken up by cells instead of being cleared by the body.

FUTURE PROSPECTS

The discovery of the structure of DNA set off a revolution in molecular biology that has continued into the twenty-first century. In addition, modern scientific equipment has made study at the molecular level possible and productive. Many biophysicists, especially those who have also had course work in genetics and biochemistry, are working in molecular biology, which promises to be an active and exciting area for the foreseeable future.

Organisms are believed to be complex machines made of many simpler machines, such as proteins and nucleic acids. To understand why an organism behaves or reacts a certain way, one must determine how proteins and nucleic acids function. Biophysicists examine the structure of proteins and nucleic acids, seeking a correlation between structure and function.

Once proper function is understood, scientists can prevent or treat diseases or disorders that result from malfunctions. This understanding of how proteins function enables scientists to develop pharmaceuticals and to find better means of delivering drugs to patients, and someday, this knowledge may allow scientists to design drugs specifically for a patient, thus avoiding many side effects. In addition, the scientific equipment developed by biophysicists in their research has been adapted for use in medical imaging for diagnosis and treatment. This transformation of laboratory equipment to medical equipment is likely to continue.

Biophysics applications have played and will continue to play a large role in medicine and health care, but future biophysicists may be environmental scientists. Biophysics is providing ways to improve the environment. For example, scientists are modifying microorganisms so that they produce electricity and biofuels that may lessen the need for fossil fuels. They are also using microorganisms to clean polluted water. As biophysics research continues, its applications are likely to cover an even broader range.

Sibani Sengupta, MS, PhD

Further Reading

Bischof, Marco. "Some Remarks on the History of Biophysics and Its Future." In *Current Development of Biophysics*, edited by Changlin Zhang, Fritz Albert Popp, and Marco Bischof. Hangzhou, China: Hangzhou University Press, 1996. This paper delivered at a 1995 symposium on biophysics in Neuss, Germany, examines how the field of biophysics got its start and predicts future developments.

Claycomb, James R., and Jonathan Quoc P. Tran. *Introductory Biophysics: Perspectives on the Living State*. Sudbury, Mass.: Jones and Bartlett, 2011. This textbook considers life in relation to the universe. Contains a compact disc that allows computer simulation of biophysical phenomena. Relates biophysics to many other fields and subjects, including fractal geometry, chaos systems, biomagnetism, bioenergetics, and nerve conduction.

Glaser, Roland. *Biophysics*. 5th ed. New York: Springer, 2005. Contains numerous chapters on the molecular structure, kinetics, energetics, and dynamics of biological systems. Also looks at the physical environment, with chapters on the biophysics of hearing and on the biological effects of electromagnetic fields.

Goldfarb, Daniel. *Biophysics Demystified*. Maidenhead, England: McGraw-Hill, 2010. Examines anatomical, cellular, and subcellular biophysics as well as tools and techniques used in the field. Designed as a self-teaching tool, this work contains ample examples, illustrations, and quizzes.

Herman, Irving P. *Physics of the Human Body*. New York: Springer, 2007. Analyzes how physical concepts apply to human body functions.

Kaneko, K. *Life: An Introduction to Complex Systems Biology*. New York: Springer, 2006. Provides an introduction to the field of systems biology, focusing on complex systems.

Web Sites

Biophysical Society http://www.biophysics.org

International Union of Pure and Applied Biophysics http://iupab.org

Worldwide Protein Data Bank http://www.wwpdb.org

See also: DNA Analysis; Human Genetic Engineering.

FASCINATING FACTS ABOUT BIOPHYSICS

- The concept of biophysics was first developed by ancient Greeks and Romans who were trying to analyze the basis of consciousness and perception.
- Human sight begins when the protein rhodopsin absorbs a unit of light called the quanta. This energy absorption triggers an enzymatic cascade that culminates in an amplified electric signal to the brain and enables vision.
- Crystallin, the lens protein, is made only in the lens of the human eye, and melanin, the skin pigment, is made only in skin cells, or melanocytes, even though all the cells in the human body have the genes to make crystallin and melanin.
- A technique called footprinting allows scientists to determine exactly where a protein binds on DNA and how much of the protein actually binds.
- Genes in human cells are selectively turned on and off by proteins called regulators. A defect in this regulatory mechanism can cause diseases such as cancer.

Bioprocess engineering

FIELDS OF STUDY

Biology; engineering; bioengineering; medicine; genetic engineering; molecular biology.

SUMMARY

Bioprocess engineering is an interdisciplinary science that combines the disciplines of biology and engineering. It is associated primarily with the commercial exploitation of living things on a large scale. The objective of bioprocess engineering is to optimize either growth of organisms or the generation of target products. This is achieved mainly by the construction of controllable apparatuses. Both government agencies and private companies invest heavily in research within this area of applied science. Many traditional bioprocess engineering approaches (such as antibiotic production by microorganisms) have been advanced by techniques of genetic engineering and molecular biology.

PRINCIPAL TERMS

- **biomass:** mass of organisms or organic material; traditionally refers to the biomass of plants and microorganisms.
- **bioreactor:** apparatus for growing microbial, plant, or animal cells, with a practical purpose under controlled conditions; these closed systems range from small (5- to 10-milliliter), laboratory-scale devices to larger, industrial-scale devices of more than 500,000 liters.
- **bioremediation:** use of living organisms to clean up the environment.
- **enzymes:** biological catalysts made of proteins.
- **fermentation:** metabolic reaction that is necessary to generate energy in microbial cells; used to produce many important compounds, such as alcohol and acetone.
- **fermenter:** type of traditional bioreactor (involving either stirred or nonstirred tanks) in which cell fermentation takes place; in continuous-culture fermenters, nutrients are continuously fed into the fermentation vessel so that cells can ferment indefinitely, whereas in batch fermenters, nutrients are added in batches.

BASIC PRINCIPLES

Bioprocess engineering is the use of engineering devices (such as bioreactors) in biological processes carried out by microbial, plant, and animal cells in order to improve or analyze these processes. Largescale manufacturing involving biological processes requires substantial engineering work. Throughout history, engineering has helped develop many bioprocesses, such as the production of antibiotics, biofuels, vaccines, and enzymes on an industrial scale. Bioprocess engineering plays a role in many industries, including the food, microbiological, pharmaceutical, biotechnological, and chemical industries.

People have been using bioprocessing for making bread, cheese, beer, and wine—all fermented foods— for thousands of years. Brewing was one of the first applications of bioprocess engineering. However, it was not until the nineteenth century that the scientific basis of fermentation was established, with the studies of French scientist Louis Pasteur, who discovered the microbial nature of beer brewing and wine making. During the early part of the twentieth century, large-scale methods for treating wastewater were developed. Considerable growth in this field occurred toward the middle of the century, when the bioprocess for large-scale production of the antibiotic penicillin was developed. The World War II goal of industrialscale production of penicillin led to the development of fermenters by engineers working together with biologists from the pharmaceutical company Pfizer. The fungus *Penicillium* grows and produces antibiotics much more effectively under controlled conditions inside a fermenter.

Later progress in bioprocess engineering has followed the development of genetic engineering, which raises the possibility of making new products from genetically modified microorganisms and plants grown in bioreactors. Just as past developments in bioprocess engineering have required contributions from a wide range of disciplines, including microbiology, genetics, biochemistry, chemistry, engineering, mathematics, and computer science, future developments are likely to require cooperation among scientists in multiple specialties.

How It Works

Living cells may be used to generate a number of useful products: food and food ingredients (such as cheese, bread, and wine), antibiotics, biofuels, chemicals (enzymes), and human health care products such as insulin. Organisms are also used to destroy or break down harmful wastes, such as those created by the 2010 oil spill in the Gulf of Mexico, or to reduce pollution.

A good example of how bioprocess engineering works is the development of a bioprocess using bacteria for industrial production of the human hormone insulin. Without insulin, which regulates blood sugar levels, the body cannot use or store glucose properly. The inability of the body to make sufficient insulin causes diabetes. In the 1970's, the U.S. company Genentech developed a bioprocess for insulin production using genetically modified bacterial cells. The initial stages involve genetic manipulation (in this case, transferring a human gene into bacterial DNA). Genetic manipulation is done in laboratories by scientists trained in molecular biology or biochemistry. After creating a genetically engineered bacterium, scientists grow it in a small tubes or flasks and study its growth characteristics and insulin production.

Once the bacterial growth and insulin production characteristics have been identified, scientists increase the scale of the bioprocess. They use or build small bioreactors (1-10 liters) that can monitor temperature, pH (acidity-alkalinity), oxygen concentration, and other process characteristics. The goal of this scale-up is to optimize bacterial growth and insulin production.

The next step is another scale-up, this time to a pilot-scale bioreactor. These bioreactors can be as large as 1,000 liters and are designed and built by engineers to study the response of bacterial cells to large-scale production. During a scale-up, decreased product yields are often experienced because the conditions in the large-scale bioreactors (temperature, pH, aeration, and nutrient supply) differ from those in small, laboratory-scale systems. If the pilot-scale bioreactors work efficiently, engineers will design industrial-scale bioreactors and supporting facilities (air supply, sterilization, and process-control equipment).

Bioreactor

All these stages are part of upstream processing. An important part of bioprocess engineering is the product recovery process, or so-called downstream processing. Product recovery from cells often can be very difficult. It involves laboratory procedures such as mechanical breakage, centrifugation, filtration, chromatography, crystallization, and drying. The final step in bioprocess engineering is testing of the recovered product, in which animals are often used.

Applications

A wide range of products and applications of bioprocess engineering are familiar, everyday items.

Foods, Beverages, Food Additives, and Supplements. Living organisms play a major role in the production of food. Foods, beverages, additives, and supplements traditionally made by bioprocess engineering include dairy products (cheeses, sour cream, yogurt, and kefir), alcoholic beverages (beer, wines, and distilled spirits), plant products (soy sauce, tofu, sauerkraut), and food additives and supplements (flavors, proteins, vitamins, and carotenoids). Traditional fermenters with microorganisms are used to obtain products in most of these applications. A typical industrial fermenter is constructed from stainless steel. Mixing of the microbial culture in fermenters is achieved by mechanical stirring, often with baffles. Airlift bioreactors have also been applied in the manufacturing of food products such as crude proteins synthesized by microorganisms. Mixing and liquid circulation in these bioreactors are induced by movement of an injected gas (such as air).

Biofuels. Bioprocess engineering is used in the production of biofuels, including ethanol (bioethanol), oil (biodiesel), butanol, biohydrogen, and biogas (methane). These biofuels are produced by the action of microorganisms in bioreactors, some of which use attached (immobilized) microorganisms. Cells, when immobilized in matrices such as agar, polyurethane, or glass beads, stabilize their growth and increase their physiological functions. Many microorganisms exist naturally in a state similar to immobilization, either on the surface of soil particles or in symbiosis with other organisms.

Environmental Applications. Bioprocess engineering plays an important role in removing pollution from the environment. It is used in treatment of wastewater and solid wastes, soil bioremediation, and mineral recovery. Environmental applications are based on the ability of organisms to use pollutants or other compounds as their food sources. One of the most important and widely used environmental applications is the treatment of wastewater by microorganisms. Microbes eat organic and inorganic compounds in wastewater and clean it at the same time. In this application, microorganisms are placed inside bioreactors (known as digesters) specifically designed by engineers. Engineers have also developed biofilters, bioreactors for removing pollutants from the air. Biofilters are used to remove pollutants, odors, and dust from air by the action of microorganisms. In addition, the mining industry uses bioprocess engineering for extracting minerals such as copper and uranium through the use of bacteria. Microbial leaching uses leaching dumps or tank bioreactors designed by engineers.

Enzymes. Enzymes are used in the health, food, laundry, pulp and paper, and textile industries. They are produced mainly from fungi and bacteria using bioprocess engineering. One of these enzymes is glucose isomerase, important in the production of fructose syrup. Genetic manipulation provides the means to produce many different enzymes, including those not normally synthesized by microorganisms. Fermenters for enzyme production are usually up to 100,000 liters in volume, although very expensive enzymes may be produced in smaller bioreactors, usually with immobilized cells.

Antibiotics and Other Health Care Products. Most antibiotics are produced by fungi and bacteria. Industrial production of antibiotics usually occurs in fermenters (stirred tanks) of 40,000- to 200,000- liter capacity. The bioprocess for antibiotics was developed by engineers during World War II, although it has undergone some changes since the 1980's. Various food sources, including glucose and sucrose, have been adopted for antibiotic production by microorganisms. The modern bioprocess is highly efficient (90 percent). Process variables such as pH and aeration are controlled by computer, and nutrients are fed continuously to sustain maximum antibiotic production. Product recovery is also based on continuous extraction.

The other major health care products produced with the help of bioprocess engineering are steroids, bacterial vaccines, gene therapy vectors, and therapeutic proteins such as interferon, growth hormone, and insulin. Steroids are important hormones that are manufactured by the process of biotransformation, in which microorganisms are used to chemically modify an inexpensive material to create a desired product. Health care products are produced in traditional fermenters.

Biomass Production. Biomass is used as a fuel source, as a source of protein for human food or animal feed, and as a component in agricultural pesticides or fertilizer. Baker's yeast biomass is a major product of bioprocess engineering. It is required for making bread and other baked goods, beer, wine, and ethanol. Yeast is produced in large aerated fermenters of up to 200,000 liters. Molasses is used as a nutrient source for the cells. Yeast is recovered from the fermentation liquid by centrifugation and then is dried. People also use the biomass of algae. Algae are a source of animal feed, plant fertilizer, chemicals, and biofuels. Algal biomass is produced in open ponds, in tubular glass, or in plastic bioreactors.

Animal and Plant Cell Cultures. Bioprocess engineering incorporating animal cell culture is used primarily for the production of health care products such as viral vaccines or antibodies in traditional fermenters or bioreactors with immobilized cells. Antibodies, for example, are produced in bioreactors with hollow-fiber immobilized animal cells. Plant cell culture is also an important target of bioprocess engineering. However, only a few processes have been successfully developed. One successful process is the production of the pigment shikonin in Japan. Shikonin is used as a dye for coloring food and has applications as an anti-inflammatory agent.

Chemicals. There is an on-going trend in the chemical industry to use bioprocess engineering instead of pure chemistry for production of a variety of chemicals such as amino acids, polymers, and organic acids (citric, acetic, and lactic). Some of these chemicals (citric and lactic acids) are used as food preservatives. Many chemicals are produced in traditional fermenters by the action of microbes.

FUTURE PROSPECTS

The role of bioprocess engineering in industry is likely to expand because scientists are increasingly able to manipulate organisms to expand the range and yields of products and processes. Developments in this field continue rapidly.

Bioprocess engineering can potentially be the answer to several problems faced by humankind. One such problem is global warming, which is caused by rising levels of carbon dioxide and other greenhouse gases. A suggested method of addressing this issue is carbon dioxide removal, or sequestration, based on bioprocess engineering. This bioprocess uses microalgae (microscopic algae) in photobioreactors to capture the carbon dioxide that is discharged into the atmosphere by power plants and other industrial facilities. Photobioreactors are various types of closed systems made of an array of transparent tubes in which microalgae are cultivated and monitored under illumination.

The health care industry is another area where bioprocess engineers are likely to be active. For example, if pharmaceutical applications are found for stem cells, a bioprocess must be developed to produce a reliable, plentiful source of stem cells so that these drugs can be produced on a large scale. The process for growing and harvesting cells must be standardized so that the cells have the same characteristics and behave in a predictable manner. Bioprocess engineers must take these processes from laboratory procedures to industrial protocols.

In general, the future of bioprocess engineering is bright, although questions and concerns, primarily about using genetically modified organisms, have arisen. Public education in such a complex area of science is very important to avoid public mistrust of bioprocess engineering, which is very beneficial in most applications.

Sergei A. Markov, PhD

FURTHER READING

Bailey, James E., and David F. Ollis. *Biochemical Engineering Fundamentals.* 2d ed. New York: McGraw-Hill, 2006. Covers all aspects of biochemical engineering in an understandable manner.

Bougaze, David, Thomas R. Jewell, and Rodolfo G. Buiser. *Biotechnology. Demystifying the Concepts.* San Francisco: Benjamin/Cummings, 2000. Classical book on biotechnology and bioprocessing.

Doran, Pauline M. *Bioprocess Engineering Principles.* London: Academic Press, 2009. A solid, basic textbook for students entering the field.

Glazer, Alexander N., and Hiroshi Nikaido. *Microbial Biotechnology: Fundamentals of Applied Microbiology.* New York: Cambridge University Press, 2007. In-depth analysis of the application of microorganisms in bioprocessing.

Heinzle, Elmar, Arno P. Biwer, and Charles L. Cooney. *Development of Sustainable Bioprocesses: Modeling and Assessment.* Hoboken, N.J.: John Wiley & Sons, 2007. Looks at making bioprocesses sustainable by improving them. Includes case studies on citric acid, biopolymers, antibiotics, and biopharmaceuticals.

Nebel, Bernard J., and Richard T. Wright. *Environmental Science: Towards a Sustainable Future.* 10th ed. Englewood Cliffs: Prentice Hall, 2008. Describes several bioprocesses used in waste treatment and pollution control.

Yang, Shang-Tian. *Bioprocessing for Value-Added Products from Renewable Resources: New Technologies and Applications.* Amsterdam: Elsevier, 2007. Reviews the techniques for producing products through bioprocesses and lists suitable organisms, including bacteria and algae, and describes their characteristics.

WEB SITES

Biotechnology Industry Association http://www.bio.org
International Society for BioProcess Technology http://www.isbiotech.org
Society for Industrial Microbiology http://www.simhq.org/index.aspx
U.S. Department of Agriculture http://usda.gov
U.S. Department of Energy Bioenergy http://www.energy.gov/energysources/bioenergy.htm

See also: Biochemical Engineering; Bioengineering; Food Science; Proteomics and Protein Engineering.

FASCINATING FACTS ABOUT BIOPROCESS ENGINEERING

- Citric acid, a common supplement of soft drinks, is a major product of bioprocess engineering. It is produced in fermenters by the common mold *Aspergillus niger*.
- Bioprocess engineering is used to recover gold from gold ores. The bioprocess uses bacteria in bioreactors to attack ores, releasing the trapped gold.
- Most insecticides are produced by the genetically modified bacterium *Bacillus thuringiensis* in bioreactors.
- One of the bacteria commonly used to produce antibodies is *Escherichia coli*, largely because so much is known about *E. coli* protein expression and because gene manipulation is relatively easy in this bacterium.
- Lysine, an amino acid added to animal feed, is produced in fermenters using the bacterium *Corynebacterium glutamicum*. About 700,000 tons are produced this way each year.
- Global Cell Solutions and Hamilton has developed a benchtop incubator-bioreactor for highdensity three-dimensional cell cultures. This type of bioprocess engineering may enable pharmaceutical companies to test the toxicity of their drugs without using animals.
- Metabolomics, a technique from functional genetics in which all the metabolites in a cell are analyzed and compared, allows scientists to optimize bioprocesses by improving the strains of bacteria and the medium used in fermentation.

BIOSYNTHETICS

FIELDS OF STUDY

Organic chemistry; biochemistry; bio-organic chemistry; bioinorganic chemistry; medicinal chemistry; pharmaceutical chemistry; pharmacology; analytical chemistry; nanotechnology; biomedical engineering; genetic engineering; genetics; synthetic biology; biology; molecular biology.

SUMMARY

Biosynthesis is the process of using small, simple molecules to make larger, more complex molecules, either inside the body or in the laboratory. Numerous applications for drug development and medicine include the synthesis of proteins, hormones, dietary supplements, blood products, and surgical dressings for wounds. Additional techniques to facilitate the diagnosis and treatment of disease include protein biomarkers for immune assays, the development of proteomics to analyze changes in proteins in response to a drug, the development of polyclonal and monoclonal antibodies, immunizations, and various drug delivery systems.

PRINCIPAL TERMS

- **amino acid:** building block of proteins.
- **antibody:** glycoprotein that binds to and immobilizes a substance that the cell recognizes as foreign.
- **antigen:** substance that triggers an immune response.
- **binding assay:** experimental method for selecting one molecule out of a number of possibilities by specific binding.
- **DNA (deoxyribonucleic acid):** molecule that contains the genetic code.
- **enzyme:** biological catalyst, usually a globular protein.
- **gene:** individual unit of inheritance that consists of a sequence of dna.
- **hormone:** substance produced by endocrine glands and delivered by the bloodstream to target cells, producing a desired effect.
- **hydrophilic:** property of tending to dissolve in water.
- **insulin:** hormone released from the pancreas.
- **monoclonal antibody:** antibody produced from the progeny of a single cell and specific for a single antigen.

- **peptide:** molecule formed by linking two to several dozen amino acids.
- **protein:** macromolecule formed by polymerization of amino acids.

BASIC PRINCIPLES

In general, the term "biosynthetic" refers to any type of material produced via a biosynthetic process. A biosynthetic process uses enzymes and energetic molecules to transform small molecules into larger molecules within the cells of organisms. The two types of metabolites produced from cellular biosynthetic pathways include the primary metabolites of fatty acids and DNA needed by cells and the secondary metabolites of pheromones, antibiotics, and vitamins that assist the entire organism. Additional small molecules, such as adenosine triphosphate (ATP), provide the energetic driving force for the biosynthetic pathways, and other small molecules, including enzymes, further facilitate the reactions in these pathways. Thus, there have been many possibilities for numerous types of scientists, including chemists, biochemists, biologists, and geneticists, to create innovations.

The term "biosynthetic" differs from the term "chemosynthetic," because chemosynthetic indicates the production of materials that cannot take place within a living organism. Scientists generally begin the process of developing a new medical application or dietary supplement by first isolating and characterizing the DNA of the proteins or other small molecules directly involved in the biological process. They then try to duplicate this naturally occurring biological process to produce massive quantities of the desired material, and ultimately they combine these naturally occurring processes with chemicals that can mimic the process during laboratory manufacturing processes.

BACKGROUND

The biochemical pharmacologist Hermann Karl Felix "Hugh" Blaschko was a trailblazer whose discoveries in the 1930's initiated the field of biosynthetics. His work elucidated the biosynthetic pathway for adrenaline, which is often called the fight-and-flight hormone, and encompassed the study of the enzymes important for regulation of this hormone. This work led the way toward the development of syntheses using amino acids for therapeutic applications. Another key development in biosynthetics was the discovery of the role of the amino acid L-arginine in the synthesis of creatine, an important biomolecule, by G. L. Foster, Rudolf Schoenheimer, and D. Rittenberg in 1939. Since that time, L-arginine has also been shown to be a precursor to nitrous oxide and nitric oxide, as well as a component of the urea cycle, which is important for ammonia regulation and thus influences the operation of the kidneys and other organs. Nitric oxide is important in the regulation of blood flow to muscles. These discoveries involving L-arginine have led to dietary supplements useful to bodybuilders who wish to enhance their weight-lifting performance.

Throughout the 1940's, 1950's, and 1960's, progress was made toward understanding the genetic composition of organisms, enzymes, and biosynthetic pathways. Researchers made contributions to understanding pyrimidine, galactosidase, *Escherichia coli*, and chlorophyll. Practical biosynthetic applications that were made possible by these fundamental discoveries began to manifest themselves throughout the 1970's, 1980's, and 1990's, with the development of surgical dressings, therapeutic hormones, and plant supplements for increased nutritional value.

HOW IT WORKS

General Process. Often the isolation and characterization of a specific gene responsible for producing an important enzyme or other small molecule is the first step in a lengthy process toward synthesis of a product that undergoes lengthy clinical trials before the final, approved product is ready for manufacture. Once the gene has been characterized, its DNA is further characterized to facilitate the process of peptide synthesis (the process of producing long peptides is known as protein biosynthesis). The process of peptide synthesis involves the general concepts of antigenicity, hydrophilicity, and surface probability, as well as flexibility indexes. The process involves an analysis of the peptide's characteristics, the use of software and databases to determine hydrophilicity (affinity for water), study of the antigenicity (capacity to stimulate the production of antibodies) to assist with antibody production, the study of surface probability (which determines the likelihood of inducing the formation of antibodies), the determination of the protein sequence, phosphorylation (process that activates or deactivates many protein enzymes), and then selection of two to three peptides, followed by comparison of their homology (similarity of structure).

In a general process called screening, the efficacy of an antibiotic is first tested using bacterial cultures, followed by injection of the antibiotic into laboratory animals, such as rats, rabbits, or guinea pigs; then clinical trials are conducted according to protocols established by the Food and Drug Administration (FDA). Combinatorial chemistry, a faster screening method, is often used instead. FDA-approved products are then manufactured on a larger scale.

Antibody Production. The application of a binding assay is used for isolation of the purified protein that is to be the source of an antigen. This antigen is then used as a conjugate to a carrier protein, such as kehole limpet hemocyanin (KLH), to produce a target peptide with a length of thirteen to twenty amino acids to stimulate the immune system. A carrier protein is a membrane protein that can bind to a substance to facilitate the substance's passive transport into a cell. Injection into a laboratory animal occurs next, and then the animals undergo a series of four to six immunizations separated by about twenty days. Enzyme-linked immunosorbent assay (ELISA) is used to detect antibodies. ELISA is based on the antibody-antigen binding interaction and often uses color to visually indicate the concentration of antibodies. Purification of antibodies obtained from the antiserum for specific antigen binding completes the antibody production process.

Antigen Preparation. This process is facilitated through bioinformatics analysis to choose the appropriate two to three peptides based on the protein sequence provided by a customer. KLH conjugation used for immunization, and bovine serum albumin (BSA) conjugation is carried out for screening. After immunization protocols and specific antibodies have been selected during fusion and screening, a cell can be cryo-preserved.

Combinatorial Chemistry. In combinatorial chemistry synthesis, a high-throughput screening method, the starting small molecule is attached to a type of polymeric resin, followed by different permutations of reagents, to produce large libraries containing hundreds of unique products that can be rapidly screened for enzymatic activity, specific antigen-binding, or protein-protein interactions. Often the process is controlled by a computer and completed through the application of robotics. A customer can specify antigen details, and a pharmaceutical company can design a protocol involving the general phases of preparation of antigen, immunization, fusion and screening of assays, and finally selection, purification, and production of antibodies.

APPLICATIONS

Biosensors. Biosensors are microelectronic devices that use antibodies, enzymes, or other biological molecules to interact with an optical device or electrode to record data electronically. These devices can be operated by home health care providers to transmit data obtained from blood or urine samples, for example, to a clinical laboratory some distance away.

Therapeutic Proteins. Plasmids are used to transfer human genes that provide the code for proteins important for growth hormones, blood clotting, and insulin production to bacterial cells.

Disposable Micropumps for Drug Delivery. Disposable micropumps manufactured by Acuros in Germany are capable of delivering a preset amount of liquid hormones, proteins, antibodies, or other medications. An osmotic micro-actuator, based on osmotic pressure, is used to regulate the amount of drug delivered, and there are no moving parts or power supply components.

High-throughput Screening. High-throughput screening can assay more than twenty thousand potentially useful drugs per week by using multi-well plates, standard binding assay methodologies, and robotics.

Protein Biomarker Assays. NextGen Sciences has developed a mass spectrometry method for protein biomarker assays that does not depend on antibodies but instead uses surrogate proteins to facilitate development of assays. The mass spectrometer measures the amount of surrogate peptides and applies statistical evaluation to assess each biomarker. This first stage requires that a protein be confirmed; then only these selected proteins are used for the second stage of validation of these protein biomarkers. The mass spectrometry data are used along with carbon-13 or nitrogen-15 isotopically labeled standards to calculate protein concentrations. Reporting the protein biomarkers in terms of concentration is important to allow batches containing hundreds of samples to be analyzed and validated. This technique uses proteomics (the quantitative analysis of proteins based on a physiological response) to allow for much faster development of assays than immunoassays. A wide range of at least 500 plasma proteins and 3,000 tissue proteins can be analyzed at once.

Gene Expression Databases. Gene Logic's BioExpress System is a comprehensive genome-wide gene expression database. The BioExpress System allows cells from a patient to be collected and analyzed to develop a useful biomarker profile for comparison with a database sample to indicate a therapeutic target. This process is made possible by the use of high-throughput gene expression profiling of the mononuclear cell fractions present in a blood sample. The software is capable of mining a database that has access to more than 18,000 samples containing biomarkers for the expression of the gene associated with ovarian cancer. This system is also capable of developing biomarker profiles to help diagnosis autoimmune diseases. Autoimmune diseases include rheumatoid arthritis, Crohn's disease, multiple sclerosis, systemic lupus erythematosus, and psoriasis, which affect about 20 million people in the United States.

Biosynthetic Temporary Skin Substitute. A biosynthetic skin substitute is a useful treatment for partial-thickness wounds, including skin tears, burns, and abrasions. After applying a gel to the surface of the wound, a semipermeable membrane of biosynthetic skin is used to cover the wound for protection from infection. Before the development of biosynthetic skin grafts, a physician had to choose between an allograft, which uses cadaver skin, and a xenograft, which uses tissue from another species. Biosynthetic dressings have also been developed. The dressing called Hydrofiber contains ionic silver and has been shown to prevent the spread of bacteria.

Needle-free Drug Delivery Systems. The three types of needle-free drug delivery systems are liquid, powder, and depot injections. Each of these types uses some form of mechanical compression to create enough pressure to force the medication into the skin. Although these needle-free delivery systems cost more initially and require more technical expertise because of their complexity, they also have many advantages. In addition to eliminating pain from needle injections and reducing physician visits, these needle-free delivery systems decrease the frequency of incorrect doses. They are being used to deliver anesthetics, chemotherapy injections, vaccines, and hormones.

Nanoparticles. DNA nanotechnology uses discoveries involving nanoparticles and nanomaterials to manipulate DNA's molecular recognition abilities to build tiny medical robots that mimic bond parts or function within cells.

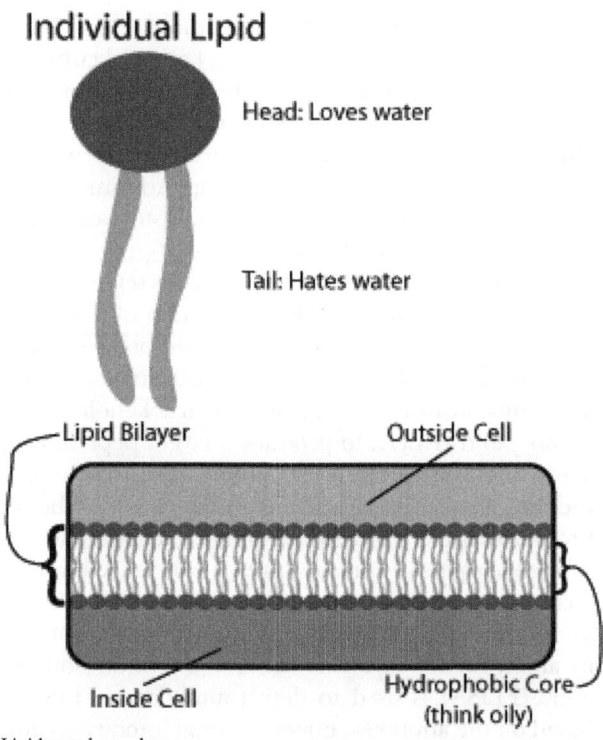

Lipid membrane layer

FUTURE PROSPECTS

The Human Genome Project has facilitated the mapping of genes, which has been instrumental to the development of vaccines to treat influenza, cervical cancer, and malaria, as well as the creation of new diagnostic tools for analysis. As a result, the pharmaceutical industry in the United States has become a multibillion-dollar industry. The generation of biosynthetic products has enhanced the lives of thousands of people through the development of treatments for many types of cancer, pneumonia, cardiovascular diseases, diabetes, tuberculosis, neurological disorders, strokes, blood disorders, and many other diseases.

Combinatorial chemistry has allowed for rapid screening of potentially successful medications that may enhance and extend the lives of many people. Normally, only one out of every 5,000 to 10,000 compounds screened makes it through the multiyear process of clinical trials to become an FDA-approved drug. However, the desire to recoup the money spent during the years of research required to bring a drug to market has caused some pharmaceutical companies to launch a product as early as possible, which has resulted in serious litigation because some drugs proved to have harmful side effects. The application of biosynthetic growth hormones for nonmedical applications, such as bodybuilding, has also caused ethical and medical controversy. However, as the global population continues to grow and the percentage of elderly persons increases, the need for the products of biosynthetic research will continue to grow.

Jeanne L. Kuhler, MS, PhD

Further Reading

Arya, Dev. *Aminoglycoside Antibiotics: From Chemical Biology to Drug Discovery.* New York: Wiley-Interscience, 2007. Describes the design and synthesis of antibiotics and the process of antibiotic resistance.

Dewick, Paul. *Medicinal Natural Products: A Biosynthetic Approach.* New York: John Wiley & Sons, 2009. Comprehensive textbook describing biosynthetic methods and processes, including new techniques in genetic engineering and isolation of genes.

Lazo, John, and Peter Wipf. "Combinatorial Chemistry and Contemporary Pharmacology." *The Journal of Pharmacology and Experimental Therapeutics* 293, no. 3 (February, 2000): 705-709. Describes the process of combinatorial chemistry. Includes experimental strategies and flow charts describing the screening of compounds.

Pettit, George. *Biosynthetic Products for Cancer Chemotherapy.* Vol. 5 London: Elsevier Science, 1985. A discussion of the fundamental processes involved with screening for antitumor agents.

Savageau, Michael. *Biochemical Systems Analysis: A Study of Function and Design in Molecular Biology.* New York: CreateSpace, 2010. Detailed textbook describing the immune system and gene regulation.

Spentzos, Dimitri. "Gene Expression Signature with Independent Prognostic Significance in Epithelial Ovarian Cancer." *Journal of Clinical Oncology* 22, no. 23 (December, 2004): 4648-4658. The research article describes the diagnosis of ovarian cancer and the use of biomarkers for detection.

Stanforth, Stephen. *Natural Product Chemistry at a Glance.* New York: Wiley-Blackwell, 2006. An introductory textbook that describes much of the organic chemistry involved in biosynthesis.

Web Sites

American Chemical Society http://acs.org

Society for Industrial Microbiology http://www.simhq.org

See also: Bioengineering; Bioprocess Engineering.

FASCINATING FACTS ABOUT BIOSYNTHETICS

- The generation of biosynthetic products has led to the development of successful treatments for many types of cancer, pneumonia, cardiovascular diseases, diabetes, tuberculosis, neurological disorders, strokes, blood disorders, and many other diseases and conditions.

- Biosynthetic corneas were used to restore vision in people with keratoconus, a condition that causes corneal scarring. These biosynthetic corneas replaced rejection-prone, scarce cadaver corneas.

- The J. Craig Venter Institute synthesized the first self-replicating synthetic bacteria cell in 2010. Synthesis of such cells may aid researchers and help develop new drugs.

- Synthetic genomics has made it possible to design and assemble chromosomes and genes and gene pathways, which may be used in creating green biofuels, pharmaceuticals, and vaccines.

- Scientists at the University of Sheffield are mapping the metabolism of the *Nostic* bacterium, which fixes nitrogen and releases hydrogen, which could be used as fuel. Once they understand the metabolic process thoroughly, they hope to be able to genetically engineer an organism that can produce hydrogen more efficiently.

- Scientists have identified biosynthetic gene clusters for many aminoglycoside antibiotics, including streptomycin, kanamycin, butirosin, neomycin and gentamicin. A full understanding of how these antibiotics work may enable scientists to get around the problem of antibiotic-resistant bacteria.

- Mass-produced biosynthetic bovine growth hormone, which when injected into dairy cows raises milk production, has been used in many developing countries. However, its use is controversial as questions have arisen regarding its effects on the health of the cows and the people who drink the milk.

Birth

FIELD OF STUDY

Reproduction science

SUMMARY

Birth is often defined as the act of being born, via parturition, from a mammal or another vivaparous organism, after pregnancy. More broadly stated, it is the beginning of awareness in an animal.

PRINCIPAL TERMS

- **gestation:** the term of pregnancy hormone: a substance produced by one organ of a multicellular organism and carried to another organ by the blood, which helps the second organ to function
- **larva:** a newly hatched form of an organism that looks very different from adults of the species and must undergo metamorphosis to the adult form
- **metamorphosis:** the form changes in a larva that turn it into the adult form motile: able to move about spontaneously oviparous: born from an externally incubated egg
- **parthenogenesis:** a process whereby a female sex cell develops without fertilization, in an organism that reproduces sexually uterus: the organ in which fertilized eggs develop during gestation
- **viviparous:** born alive after internal gestation zygote: a fertilized egg

BASIC PRINCIPLES

Animals are born in one of two ways. They may be born via parturition after an internal pregnancy (a gestation period). They may also be born from an egg that hatches externally. This includes eggs that are spawned and then fertilized externally by organisms such as fishes; those fertilized internally and laid in huge numbers to hatch on their own, as in snails and millipedes; or those laid in much smaller numbers and incubated by their parents, as in birds.

In all animals, male and female reproductive cells (gametes) unite to form a single cell, known as a zygote. The zygote then undergoes successive cellular divisions, as well as cellular differentiation, to form a new organism. In most higher animals, individuals of a species are male or female, according to the type of reproductive cells they produce. Male reproductive cells—the sperms—are motile cells, with heads containing nuclei and tails that allow them to move. Female reproductive cells— eggs or ova—are round cells many times larger than sperms. They also contain large amounts of cytoplasm located around the nucleus.

How It Works

Viviparous Birth. In viviparous organisms, a fertilized egg will develop into an incompletely finished miniature or miniatures of an adult of the same species. After fertilization of an egg, the zygote enters the uterus, undergoes both cell division and differentiation, and forms an embryo within the mother. In due time parturition (birth) occurs. Most viviparous organisms are mammals. Early in gestation the implanted dividing egg and the uterine wall become interconnected by a placenta, composed of both maternal and embryonic tissue. The placenta brings oxygen and nutrients to the embryo and carries away wastes. The transfer of nutrients uses the circulatory systems of both the mother and the embryo.

At the time of birth, hormonal changes cause the mother's birth canal to enlarge, the muscles of the uterus to rhythmically contract, and the embryo is expelled as a newborn. The overall process can be exemplified with the female gorilla. She menstruates monthly and can mate successfully at any time of year. Her gestation period is 9.5 months and yields one or two almost fully formed offspring. Gestation is much shorter in smaller primates and there are variations in the difficulty of parturition, related to the head-first entry of young into the world.

Some primates—including humans—must undergo major dilation of the uterine mouth (cervix) before parturition can begin. This allows the large head of the fetus to pass out of the body safely. In species such as monkeys, in which the head of the fetus is close to the size of the cervical opening, far less dilation is needed. In other placental mammals, the position of fetuses in the uterus and the fashion of birth differ.

Oviparous Birth. Many animals, including snails, insects, birds, lizards, and fish, lay eggs either before they are fertilized or before their young are completely developed. These organisms are termed oviparous. In the case of snails, most species are hermaphrodites. This means that each snail has both male and female sex organs. However, each individual snail usually mates with another snail of the same species, passing sperm to its partner and getting sperm from the partner. Fertilized eggs are then spawned into the water or laid on rocks or aquatic plants. The eggs hatch in two weeks to two months. Hatching is considered to be the time of birth of the young snails. In most cases, offspring hatch as miniature replicas of their parents. In insects, eggs are laid in a wide variety of places. For example, grasshoppers lay eggs in the ground or on plants. When the offspring are born, they hatch as wingless grasshopper larvae, called nymphs. Over several months the nymphs undergo metamorphosis to adult locusts. In contrast, ants, wasps, and termites lay their eggs in special chambers in their nests (or colonies). Worker termites place eggs laid by a colony's queen into hatching chambers in "nurseries." Termites are born as wormlike larvae when eggs hatch. The larvae undergo metamorphosis into workers, soldiers, or reproductives (kings or queens) as a result of being fed varied amounts of hormones obtained from queens.

Birds lay eggs in nests, that are located in a wide variety of locales depending on species. Adults then incubate the eggs by sitting on them. Offspring are born when they use a specialized egg tooth to break open their egg shells. In the case of lizards, the eggs are laid after they are fertilized. However, they are not cared for by parents. Large lizards such as alligators, crocodiles, and caimans lay eggs covered with hard, calcium-containing shells like those of bird eggs—reptiles and birds are distant relatives—in holes in the ground, where they hatch into offspring that look like adults. Most fish lay fertilized eggs on plants or on the bottom of the sea, lakes, or rivers, and leave their offspring to hatch on their own. These offspring then develop into adults.

Ovoviviparous birth. Ovoviviparous animals produce eggs in shells like those of the oviparous organisms but the eggs are hatched within the body of the mother, or by expulsion from her body. There are numerous examples of ovoviviparous organisms among animals. They include some oysters, snails, and other gastropods, as well as numerous species of sharks, and the live-bearing tropical fish such as the guppy or swordtail. The eggs of live-bearing guppies hatch internally, just before leaving the mother's body, and the young are born alive. These young fish usually leave the body of the mother head first. In all cases, the development of the egg or eggs of ovoviviparous species begins with internal fertilization of the female of the species. Then, the zygotes formed pass through many cycles of internal cell division and differentiation. Ultimately, each egg yields a miniature of the adult organism involved. However, there is no placenta formed and the zygote becomes the complete organism in processes that depend on a yolk sac for food and energy. Often, upon birth, the newborn organism has part of its yolk sac left and can survive for one or several days without eating.

—*Sanford S. Singer*

See also: Asexual reproduction; Cleavage, gastrulation, and neurulation; Fertilization; Gametogenesis; Reproduction; Reproductive system of female mammals; Reproductive system of male mammals; Sexual development.

FURTHER READING

Dekkers, Midas. *Birth Day: A Celebration of Baby Animals.* New York: W. H. Freeman, 1995. A book for children, covering conception, pregnancy, and birth in a wide range of animals. Well illustrated.

Hayes, Karen E. N. *The Complete Book of Foaling: An Illustrated Guide for the Foaling Attendant.* New York: Howell Book House, 1993. A guidebook for horse breeders. Covers the final three weeks of a horse's pregnancy through the first twelve hours of the newborn foal's life, with detailed discussion of the birthing process.

Pinney, Chris C. *Veterinary Guide for Dogs, Cats, Birds, and Exotic Pets.* Blue Ridge Summit, Pa.: Tab Books, 1992. Holds much useful data on pet-keeping, breeding, and parturition

Prine, Virginia Bender. *How Puppies Are Born: An Illustrated Guide on the Whelping and Care of Puppies.* New York: Howell Book House, 1975. A guidebook for dog breeders, with helpful illustrations of the canine birth process.

Spaulding, C. E., and Jackie Clay. *Veterinary Guide for Animal Owners: Sheep, Poultry, Rabbits, Dogs, Cats.* Emmaus Pa.: Rodale Press, 1998. Provides much interesting information on keeping animals, including gestation, parturition and related problems

> **BIRTH IN CATTLE**
>
> The birth of a calf, relatively representative of the births of artiodactyls, is simpler than that of primates. The first signal of readiness to calve is the passage of mucus and a bloody discharge. After this, the pregnant cow lays down on the ground. Within an hour or two, she gives birth to her calf or calves. In normal birth, each calf is delivered in either a front or a rear presentation. Most cows deliver their calves front end first (front presentation). Initially, a dark bulge, the amniotic sac (water bag), is seen in the birth canal. This is soon followed by the tips of the calf's hooves, toes down. These hooves are soft so that they do not injure the mother during the birthing process. However, they harden quickly. A front presentation calf is born in the "diver position," with its front legs stretched out and its head between them. The legs protrude about a foot, and then the head comes out, nose first. The head is soon followed by the shoulders, and then the rest of the calf slides out. Usually, afterbirth (the placenta) comes out as well. However, it may take several hours after calving for this to happen. Within a few hours after birth, the calf is ready to nurse and grow.

> **TYPES OF SHARK BIRTH**
>
> Reproduction in sharks is very interesting. Most often it is ovoviviparous, with the eggs fertilized internally, hatching within the female, and being born as live young. Some sharks are oviparous, laying eggs externally. Often these eggs are encased in tough shells with filaments that anchor them to rocks or sea plants, while development occurs.
>
> In still other sharks, the young develop in the uterus in a fashion very similar to that seen in mammals. In those cases, the yolk sac becomes a yolk placenta in the folds of the uterine wall. This yolk placenta brings nutrients to the embryo, as well as carrying wastes away. In such cases gestation lasts between six months and two years. In cases of live birth, the offspring of some sharks are up to three feet long, almost fully developed, and capable swimmers that eat the same prey as do the adults.

BONE AND CARTILAGE

FIELDS OF STUDY

Anatomy, biochemistry, physiology

SUMMARY

Bone is a connective tissue, much of which derives from cartilage. It forms vertebrate skeletons, which are body support networks, prevent organ damage, anchor muscles so that they can function, provide calcium to serve the body needs, and contain the sites of blood cell synthesis.

PRINCIPAL TERMS

- **articular:** pertaining to bone joints bone: the dense, semirigid, calcified connective tissue which is the main component of the skeletons of all adult vertebrates
- **calcification:** calcium deposition, mostly as calcium carbonate, into the cartilage and other bone-forming tissue, which facilitates its conversion into bone
- **cartilage:** elastic, fibrous connective tissue which is the main component of fetal vertebrate skeletons, turns mostly to bone, and remains attached to the articular bone surfaces
- **collagen:** a fibrous protein very plentiful in bone, cartilage, and other connective tissue
- **connective tissue:** any fibrous tissue that connects or supports body organs
- **osteoblast:** a bone cell which makes collagen and causes calcium deposition
- **periosteum:** the fibrous membrane which covers all bones except at points of articulation, containing blood vessels and many connections to muscles

Basic Principles

Bone is the hard substance that forms the supportive framework of the bodies of all of vertebrate organisms. This framework, the skeleton, is composed of hundreds of separate parts called bones. The bones support the bodies of vertebrates and protect their delicate internal organs, such as the brain, lungs, and liver, from injury. In addition, the muscles are attached to the bones, which act as levers to enable their function in actions as diverse as walking or swallowing. Furthermore, bone provides the calcium needs of the body, serves as the main repository for calcium storage, and contains the sites where the red blood cells are made.

Much of the bone in adult vertebrates derives from cartilage, elastic, fibrous connective tissue which is the main component of fetal vertebrate skeletons. Such bone, for example, that of the long bones, is called cartilage bone. Cartilage is an extracellular matrix made by body cells called chondrocytes. It is surrounded by a membrane, the periosteum, and much of its firmness and elasticity arises from plentiful fibrils of the protein collagen that it contains. These fibrils and their many interconnections provide mechanical stability and very high tensile strength, while allowing nutrients to diffuse into the chondrocytes to keep them alive. The blood vessels which surround the cartilage in the periosteum provide all of the needed nutrients and remove the cellular waste materials produced by life processes. The cartilage-containing skeletons of newborn vertebrates become cartilage bone by ossification, a process which includes calcification, chondrocyte destruction, and replacement by bone cells, which lay down more bone. This cartilage, called hyaline cartilage, remains at the articular sites of bones. In young vertebrates, cartilage is the site for the con tinued growth and calcification that produces the bone lengthening required for the attainment of adult size and stature. In addition to the cartilage bone, so-called membrane bone occurs exclusively in the top portion of the skull.

Bone is thought to have developed over a half billion years ago, as shown by its presence in the fossils of fishlike carnivores of that time period. In those creatures, it seems to have been formed into interconnected external plates covering their bodies as sheaths that strengthened and protected their bodies. The existence of bone only at the surfaces of these fossils has led many scientists to suppose that the first function of bone was protection, rather than body support. Be that as it may, bone has both functions in modern organisms. It is interesting to note that many of these early organisms lacked bone in their heads. It seems possible that this lack may have led to the development of the separate mechanisms for formation of membrane bone and cartilage bone, different means to the same end.

Physical Characteristics of Bone

To best serve their biofunctions, bones must be very hard, strong, and rigid, but remain supple enough to stay unbroken under normal conditions. These characteristics are provided by the collagen fibrils and insoluble calcium phosphate which make up the bones. The bones must also be light enough to allow vertebrates to move easily and remain erect. Overly heavy bones are prevented by the occurrence of two general types of bone tissue. The first of these is compact bone, the portion most familiar because it makes up the hard exterior of many bones, except at their very ends. The second bone type, cancellous bone, which appears spongy, is found at the ends of long bones and inside them. It serves to lighten the bones, acting in the same fashion as the air-filled sinuses of the skull, which diminish the overall weight of the skull without weakening it. Bones are covered on their outsides by the important fibrous membrane called the periosteum, along with cartilage. Their insides are lined by an endosteum membrane, very similar to the inner layer of the periosteum. It is also useful to think of bones in terms of woven, lamellar, and osteonic forms. These terms indicate the relative number of cells in a bone matrix region and the arrangement of collagen fibers in the region. Collagen fibers of woven bone crisscross within the bone matrix, and its bone cells are distributed randomly. In lamellar bone the collagen fibrils are more ordered and fewer bone cells are present. Osteonic bone is also well-organized. However, its cells are found in concentric rings, with narrow channels (Haversian canals) inside them. A blood vessel passes through each canal and feeds the concentric cell rings formed around it. The bone layers form from the outside in, within the internal bone cavity. This narrows its diameter more and more. A Haversian canal and its rings develop when cancellous bone is converted into compact bone.

Bones are either "long" or "short" bones. Most long bones are located in the arms and legs. They are divided into three parts: a shaft (the diapysis), the long central part of the bone; a flared portion at each bone end (the metaphysis); and a rounded bone end (the epiphysis). The short bones, designed for flexibility, include those in the skull, spine, hands, and feet. The centers of bones—medullary cavities—are most often filled with either red or yellow bone marrow. The yellow marrow is mostly fat. Red marrow is a network of blood vessels, connective tissue, and blood-cellmaking tissue. Red blood cells (erythrocytes) are made in this red marrow. Each bone has nerves that stimulate it and blood vessels that supply nutrients and take away wastes.

Bone Composition, Development, and Remodeling

Between 66 and 70 percent of bone is an inorganic mineral composite made of calcium phosphate and calcium carbonate, which is mostly hydroxyapatite. Much of the remainder of bone is the fibrous protein collagen. This mineral and protein together are called the bone matrix. Within the bone matrix are the three types of specialized cells which ensure its formation, remodeling as is needed, and continuity throughout life. The first cell type, the osteoblast, produces the bone matrix and surrounds itself with it, synthesizing collagen and stimulating mineral deposition. The second cell type, the osteocyte, is a branched cell that becomes embedded in bone matrix, is interconnected, and acts in the control of the mineral balance of the body. Finally, the osteoclast cells destroy the bone matrix whenever it is remodeled during skeleton growth or the repair of bone breaks and bone fractures.

The stepwise conversion of cartilage into bone begins when the chondrocytes of hyaline cartilage enlarge and arrange themselves in rows. This is followed by the synthesis of collagen fibers, and mineral deposition around them. Just below the inner surface of the periosteum a vascular membrane— the perichondrium—forms and supplies the osteoblasts needed for bone formation. Simultaneously, osteoclasts excavate layers through the bone layer and set the stage for the formation of additional bone.

All the bones in the bodies of the vertebrates change their sizes and shapes as these organisms pass through their lives. The processes involved are collectively called remodeling. An example of such change is the growth of the long bones in circumference as the limbs grow from puberty to adulthood. In the course of such bone growth the periosteum provides the osteoblasts required to deposit bone matrix around the bone exterior and to calcify it. At the same time the endosteum-derived osteoclasts often dissolve bone in the interior, thus enlarging the marrow cavity.

Remodeling in such cases occurs in response to biosignals including those caused by increases in the need for bone to bear additional weight or to anchor increased muscle mass. Conversely, inactivity and the lack of exercise can result in remodeling which produces diminished bone mass. The complex changes involved in bone remodeling are also controlled by vitamin D and hormones originating in the pituitary gland, the thyroid gland, and the parathyroid glands. Abnormalities in bone growth and remodeling are associated with a great many bone diseases, ranging from rickets to bone cancer.

—*Sanford S. Singer*

See also: Anatomy; Cell types; Growth; Nutrient requirements; Vertebrates.

Further Reading

Alexander, R. McNeill. *Bones: The Unity of Form and Function.* Reprint. Boulder, Colo.: Westview, 2000. Covers bones, joints, muscle attachment, and related topics in an interesting fashion.

Hukins, David W. L., ed. *Calcified Tissue.* Boca Raton, Fla.:CRCPress, 1989. Covers many topics froma molecular and structural perspective. Particularly interesting is its coverage of the bones and their calcification.

Murray, Patrick D. F. *Bones: A Study of the Development and Structure of the Vertebrate Skeleton.* Cambridge, England: Cambridge University Press, 1936. Reprint. New York: Cambridge University Press, 1985. Covers many aspects of the development and anatomy of the skeletons of the vertebrates.

Rosen, Vicki, and R. Scott Theis. *The Cellular and Molecular Basis of Bone Formation and Repair.* Austin, Tex.: R. G. Landes, 1995. This very nice book covers many topics related to bone growth, bone regeneration, other remodeling, and growth factors. Included are a great many bibliographical references and useful illustrations.

Siebel, Markus J., Simon P. Robins, and John P. Bilezikian, eds. *Dynamics of Bone and Cartilage Metabolism.* San Diego, Calif.: Academic Press, 1999. Covers many aspects of bone and cartilage metabolism and diseases. References and illustrations.

Vaughan, Janet Maria. *The Physiology of Bone.* 3d ed. Oxford, England: Clarendon Press, 1981. Covers the basis of the physiology of bone.

MEMBRANE AND CARTILAGE BONE

Membrane bone, such as that which makes up the upper portion of the skull, forms after cells of vertebrate embryo connective tissue gather together in the area where such bone is to form. This aggregation is followed by the development of small blood vessels in the area and the differentiation of the connective tissue cells into osteoblast cells, which make collagen, intracellular material, and cause the deposition of calcium.

In contrast, cartilage bone, in the long bones, for example, forms where cartilage was initially laid down in the vertebrate embryo. This kind of bone, also called the endochondrial bone, results via osteoblast formation from chondrocytes, followed by ossification. The differentiation between membrane bone and cartilage bone, which are indistinguishable after ossification, is made by the careful examination of appropriate tissue structures during the course of embryogenesis.

The conversion of cartilage into endochondrial bone is not complete until adulthood. At the ends of the immature cartilage bones, regions of actively growing cartilage (epiphysial plates) occur. In these regions, continued longitudinal cartilage growth followed by ossification leads to the lengthening of bones required for the development of newborn vertebrates into full-sized adults. At adulthood, the cells of the epiphysial plates stop reproducing. It is believed that imbalances between chondrocyte numbers and bone matrix material levels in articular cartilage probably play major roles in the genesis of arthritis.

BRAIN

FIELDS OF STUDY

Biochemistry, cell biology, histology, neurobiology

SUMMARY

The brain is the center of the body's information integration and storage, and the site that determines how an animal will react to changes in its environment. Both the brain and spinal cord receive information from sensory receptors and send messages to the muscles and glands, telling them how to respond to internal or external changes.

PRINCIPAL TERMS

- **brainstem:** lowest or most posterior portion of the vertebrate brain, including midbrain, pons, and medulla oblongata; controls "housekeeping" functions such as breathing and heartbeat
- **cell body:** the central portion of a neuron, containing the nucleus, where most processing and integration of information occur
- **cerebellum:** second largest part of the brain, manages fine muscle control and muscle memories
- **cerebrum:** largest part of most vertebrate brains, with areas that control vocalizations, vision, hearing, smell, and taste, as well as voluntary skeletal muscle movements
- **cortex:** thin layer of gray matter that covers surfaces of the cerebrum and cerebellum
- **ganglia:** clustered cell bodies of neurons that may form a brain-like center in lower animals
- **gray matter:** region of the brain or spinal cord that contains cell bodies of neurons, where information processing and storage occur
- **white matter:** region of neural tissue that contains axons of neurons that carry electrical nerve impulses from one processing center to another

BASIC PRINCIPLES

Animals are multicellular organisms that obtain their nutrients by eating or ingesting other organisms, and many have locomotor abilities. Obtaining food and avoiding being eaten are behaviors enhanced by the ability of an animal to tell what is going on in its surroundings. Using information from sensory receptors and responding to changes in the environment are generally managed by a nervous system of some sort, usually with a center where processing occurs, a brain or brainlike structure. Invertebrate nervous systems are generally very primitive and may contain only a very rudimentary brainlike structure. Some animals, however, are so structurally simple that they have no neural processing center at all.

INVERTEBRATES WITH AND WITHOUT BRAINS

Sponges (phylum Porifera) are invertebrates with no brain or nervous system of any sort, in either the sedentary adults or the free-swimming larvae. Stimuli received at the body surface produce responses (movements) directly, over the entire body, in these and related lower metazoan animals. Other primitive invertebrates such as hydra, jellyfish, corals, and sea anemones (phylum Cnidaria) have one or more nerve nets. For these radially symmetrical animals, food or danger can come from any direction in the water, and the meshlike nervous system can respond directly without a central control region. Some jellyfish also have a nerve ring that helps coordinate their movements, but no brain.

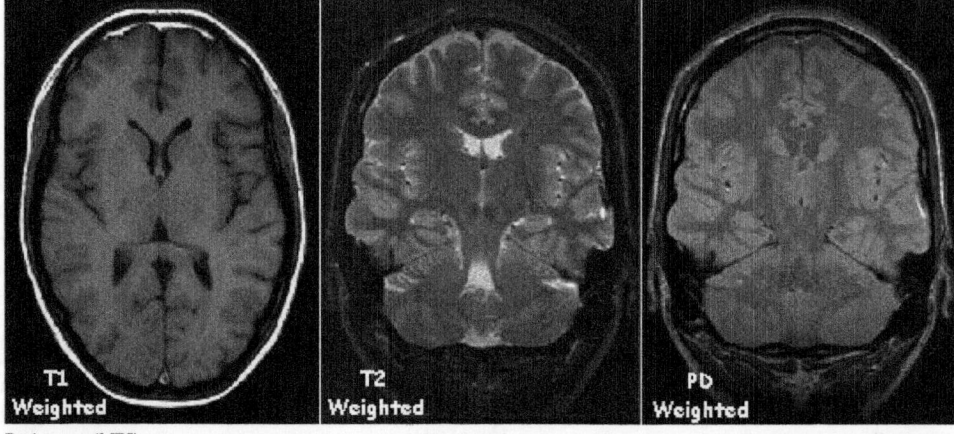

Brain scan (MRI)

Bilaterally symmetrical invertebrates include the flatworms (phylum Platyhelminthes), round worms (Nematodes), mollusks (Mollusca), segmented worms (Annelida), and insects and their relatives (Arthropoda). Most of these show cephalization, the presence of an anterior head containing the main processing center of the nervous system, specialized sensory receptors, and the mouth. Echinoderms (Echinodermata) such as starfish are bilaterally symmetrical as larvae, but develop radial symmetry as adults, when they lack a head. Some mollusks are not symmetrical as adults, despite the bilateral symmetry of the larvae. Among the flatworms, some have only nerve nets like those of cnidarians, while more complex planarians, tapeworms, and flukes generally have one or more pairs of ladderlike longitudinal nerve cords with ganglia at the head. These ganglia are clusters of cell bodies of neurons, the most primitive form of a brainlike structure. Nematodes or roundworms have a nerve ring and anterior ganglia organized around the anterior digestive tract, with nerve cords extending toward the head and tail from this center. Mollusks include clams and oysters (class Bivalva), snails and slugs (Gastropoda), and octopuses and squid (Cephalopoda). These animals have nervous systems that vary from simple and relatively uncephalized nerve rings and nerve cords, to a more centralized system with at least four pairs of ganglia.

Octopuses and squid have the most complex nervous systems of the mollusks and are the most intelligent invertebrates. The relatively large cephalopod brain contains many clustered or fused ganglia that manage sensory information from complex eyes and produce motor instructions for extremely rapid muscular responses. Giant nerve fibers in squid are the largest neurons known in any animal, up to one millimeter in diameter in a single cell, and are able to conduct rapid impulses that allow lightning-fast movements. Extensive studies of these neurons' structure and function have provided scientific insights that are also applicable to human neurons. Gastropods and cephalopods may show extremely complex behaviors, such as homing, territoriality, and learning. An octopus can have as many as thirty functional brain centers, some of which are memory banks used for experiential learning.

Annelids such as earthworms and leeches have paired cerebral ganglia near the mouth, connected by a solid ventral nerve cord to smaller paired ganglia in each body segment. Giant nerve fibers in the nerve cord allow rapid responses to escape from threats using reflex actions and patterned behavior. Earthworms can be taught to travel a maze by simple associative learning, in which repeated stimuli become linked to a specific behavior pattern, but this learning requires many repetitions and disappears within a few days if not reinforced.

Arthropods include spiders, scorpions, ticks, and mites (class Arachnida), lobsters, crabs, and shrimp (Crustacea), and insects (Insecta). The nervous system in arthropods is similar to that of annelids in its segmentation, but it is much more complex, and the anterior ganglia tend to be fused into a true brain. Many arthropods have giant neurons like those of some mollusks and annelids, capable of rapid nerve impulse transmission for efficient muscle control. Insects in particular, especially ants and bees, are capable of complex learning and very intricate social behavior. Habituation allows individuals to learn to ignore repeated stimuli that do not produce harmful effects, and cockroaches and ants can learn to run mazes.

Sea urchins, sand dollars, sea stars (starfish), and sea cucumbers are echinoderms, in which the bilaterally symmetrical larvae develop a secondary radial or biradial symmetry as they mature. The resulting radial nervous system is not greatly centralized, as there is a mouth but essentially no head. The nervous system consists of a nerve ring around the mouth that is connected to radial nerves and a nerve net. Thus, behavior generally involves only localized responses to stimuli, as along one arm of a starfish.

Evolutionary Development of the Vertebrate Brain

The location of most animals' brain or brainlike organ at the anterior or superior end of the body is important, since it places the brain at the leading end of the moving animal or at its highest point. Many sensory receptors are located in the head, and information from the eyes, ears, and nose can be rapidly received and processed if the processing center is in the same region.

In vertebrates, the nervous system is much more advanced than the primitive systems of invertebrates. The vertebrate brain is an anterior enlargement of the dorsal hollow nerve cord that develops above the notochord in all chordates. This swelling of the nerve cord allows development of a large collection of neurons that receive, process, and store information, and determine what the organism's response to that information will be. The central nervous system consists of the brain at the anterior end of the nerve cord and the spinal cord behind it, encased in a skull and vertebral column of bone or cartilage. The rest of the vertebrate nervous system is called the peripheral nervous system, with nerve fibers bundled into nerves. Clusters of the cell bodies of neurons in the central nervous system are called nuclei, while the same kind of clusters in the peripheral nervous system are called ganglia.

The components of the vertebrate embryonic brain are divided into three areas or primary vesicles, known as the forebrain (prosencephalon), midbrain (mesencephalon), and hindbrain (rhombencephalon). As development occurs, the three primary vesicles form five secondary vesicles that continue to develop into the mature brain structures. The forebrain becomes subdivided into the telencephalon, which matures into the cerebrum, and the diencephalon, which contains the thalamus and hypothalamus. The midbrain does not undergo further developmental separation. The hindbrain develops into the metencephalon, which will form the pons and cerebellum, and the myelencephalon, which becomes the medulla oblongata that is connected to the spinal cord. The lower or posterior part of the brain is called the brain stem, consisting of the medulla oblongata, pons, and midbrain, which manages the most primitive functions required for life. Higher brain functions reside in the cerebrum, particularly in the outer cortex of gray matter on its surface. The cerebellum coordinates skeletal muscle or motor activities, while the diencephalon processes and sends on sensory information to the cerebrum and cerebellum, as well as being the center of autonomic or visceral motor control.

The different classes of vertebrates are grouped into subphylum Vertebrata within phylum Chordata, with the main classes including cartilaginous fish (Chondrichthyes), bony fish (Osteichthyes), amphibians (Amphibia), reptiles (Reptilia), birds (Aves), and mammals (Mammalia). The brains of fish and amphibians are relatively primitive as compared to those of other vertebrates, with the main control over body functions handled by the medulla oblongata, the oldest part of the vertebrate brain. Olfactory lobes for processing sensations of smell and perhaps also taste, located in the cerebrum, and optic lobes for vision, in the diencephalon, are large in comparison to other parts of the brain, and responses are generally reflexive. Animals that lay eggs with shells, called amniotes and including reptiles and birds, are adapted to the rigorous requirements of life on land, and have larger, more complex brains than fish and amphibians. The amniote brain has a larger telencephalon and is able to process and store more information about the land environment, which is much more likely to vary than is a watery environment. In addition to having a larger telencephalon, the brain contains more gray matter that is closer to the brain surface in amniotes than in fish or amphibians.

Mammals and some reptiles have much or all of the surface of the cerebrum covered in gray matter, which forms a structure called the cerebral cortex.

The evolutionarily newer portion of this cortex is called the neocortex, while the older part is called the paleocortex. The paleocortex is the control center for drive-related behaviors, such as activities associated with feeding (licking, chewing, swallowing), sexual behavior, and primitive emotions (anger, fear). The limbic system occupies the paleocortex, which is sometimes called the reptilian brain, because it is the highest brain area present in reptiles and governs nearly all their behaviors. The neocortex is a "higher" control area that is well developed even in primitive mammals, but it is most completely expressed and covers the entire cerebral surface in humans.

In cetaceans (whales and dolphins) and primates, the neocortex is the center of higher learning, logical thinking, and storage of many memories. The activities of the neocortex can override the more primitive responses of the paleocortex under most conditions, but when the higher brain areas are inactive, as in alcoholic intoxication in humans or when removed surgically in experimental animals, the lower areas reassert themselves and take control, often causing inappropriate behaviors. Mammalian brains have convolutions on the surface of the cerebrum and cerebellum, with the neural cortex following and covering every "hill" and "valley" of the convolutions. This provides a much greater surface area occupied by gray matter, especially in humans, the species in which the convolutions and cerebral cortex are most extensive. Below the gray matter surface is white matter, myelinated neuron fibers that carry information from one area of gray matter to another. Deep to this white matter are basal nuclei, gray matter centers that help regulate subconscious and involuntary control of body functions.

The gray matter of the cerebrum in birds is nearly all in the deep basal nuclei, which are relatively much larger than they are in mammals, and in an overlying gray area specific to birds called the hyperstriatum. The avian brain lacks a neocortex entirely, with no equivalent of the cerebral cortex present. The area of the basal nuclei called the corpus striatum is apparently the center for complex behavior patterns, while the hyperstriatum manages learning and memory.

Humans have been found to show lateralization of the brain, where one side of the cerebrum (left) controls language production and interpretation, while the other side (right) controls spatial awareness and artistic creativity. This lateralization is not generally seen in other vertebrates, but recently it has been observed in some birds, where memories of song patterns and migratory homing directions are located in gray matter areas on specific sides of the brain.

Because the vertebrate central nervous system develops from a dorsal hollow nerve cord, the anterior end of its hollow, fluid-filled central canal enlarges into four ventricles or spaces. These are the first and second or lateral ventricles of the cerebral hemispheres, the third ventricle within the diencephalon, and the fourth ventricle associated with the pons, medulla oblongata, and cerebellum. The midbrain retains a simple canal called the cerebral or mesencephalic aqueduct that connects the third and fourth ventricle spaces.

The fluid that fills the canal and ventricle spaces is cerebrospinal fluid (CSF), produced by filtration of fluids from the blood at specialized capillary beds called choroid plexuses within the ventricles. Besides filling the hollow spaces of the central nervous system, CSF also washes over the surfaces of the brain and spinal cord in an area below the arachnoid layer, one of the central nervous system's coverings or meninges. It provides protection against traumatic injury, delivers nutrients, removes wastes, and helps regulate neurochemicals for the central nervous system.

The Primate Brain

Primates, the order of animals that includes monkeys, apes, and humans, contains species that show a higher level of brain development than most other mammals. Primate brains, especially in humans, are among the largest in the animal kingdom, compared to the body size of the animal. The primate brain retains in its structure the earlier forms and functions that have developed in lower vertebrates over evolutionary time, such as the brainstem and limbic system, but higher areas give new and more complex possibilities for learning and behavior.

Because humans are upright, bipedal walkers, the human brain is at the top of the spinal cord rather than somewhat in front of it as in other primates. The human brain weighs only about three pounds, or about 2 percent of the weight of a 150 pound individual, but that is still larger relative to body size than the brains of other primates, even chimpanzees. The cerebrum makes up about 87 percent of the volume of the brain, and the cerebellum occupies most of the remaining volume.

The diencephalon and brainstem in primates are relatively smaller than in most other mammals, compared to the entire brain size. The cerebral cortex in humans contains only six layers of cell bodies in the gray matter on the surface of the cerebrum. Many axons extending down from these cell bodies into the underlying white matter crossconnect the neurons that receive stimuli, process information, determine responses, and store memories. Specific areas of this cerebral cortex determine the body's voluntary muscle actions, or receive and analyze sensory information from the skin, muscles, and joints. Other cortical areas process incoming information about smell, taste, vision, and hearing, and compare those sensations to previous memories or store them as new memories.

The most "human" aspect of the brain is the prefrontal cortex of the cerebrum, where logical analysis, predictions of the results of specific actions, and social interactions take place, although even in monkeys and apes the front of the brain manages social awareness and behavior. Since the primate brains of apes and monkeys are so similar to those of humans, many studies of brain function have involved experimentation on these animals, humans' closest relatives. Other mammals such as mice, rats, cats, and dogs have also served as subjects of brain studies that can be related not only to their own specific behavior, but also to how the human brain works in its various component parts. Since neurons are very similar to each other, whether they come from sea slugs, squid, or mammals, experimentation using these animals has produced insight into how all brains and nervous systems work.

—*Jean S. Helgeson*

See also: Anatomy; Communication; Intelligence.

Further Reading

Calvin, William H. *The Throwing Madonna: Essays on the Brain*. Updated ed. New York: Bantam Books, 1991. One of the essays describes studies on the brain of *Aplysia*, a sea slug widely used in neurophysiological experimentation on habituation learning.

Eccles, John C. *Evolution of the Brain: Creation of the Self*. London: Routledge, 1989. The chapter on learning and memory uses modern anthropoid apes as a model for prehuman ancestral hominids of modern humans. Comparisons are made of the size of brain areas devoted to learning in apes and humans, as well as other mammals such as monkeys and even rabbits. Emphasis is on the development and characteristics of the neocortex.

Falk, Dean. *Braindance*. New York: Henry Holt, 1992. This study of the evolution of the human brain describes a comparison of the brains and behaviors of humans and other primates, including common chimpanzees, pygmy chimpanzees (bonobos), and monkeys. Areas of particular interest are the visual, motor, and premotor areas of the cerebral cortex, and the concept of brain lateralization in animals as well as humans.

Hickman, Cleveland P., Jr., Larry S. Roberts, and Allan Larson. *Integrated Principles of Zoology*. 9th ed. St. Louis: C.V. Mosby, 1993. Comparative anatomy and physiology of nervous coordination are described in this college text. The different categories of animals are also covered, with descriptions of the nervous systems of each in their individual chapter

Marieb, Elaine N. *Human Anatomy and Physiology*. 5th ed. San Francisco: Benjamin/ Cummings, 2001. This college text describes the structure and function of neurons in general, mainly dependent on animal studies that can be applied to humans as well. The human brain and its functions are similar in many ways to the general patterns of the mammalian brain and nervous system.

Mitchell, Lawrence G., John A. Mutchmor, and Warren D. Dolphin. *Zoology*. Menlo Park, Calif.: Benjamin/ Cummings, 1988. This college text covers the nervous system in general, as well as discussing the different individual categories of animals, including their nervous system specializations and behaviors.

Restak, Richard M. *The Modular Brain*. New York: Charles Scribner's Sons, 1994. Discussion of the workings of the human brain is accompanied by descriptions of animal brain and behavior experimentation as well as observations on humans. Brain activities in cats, dogs, rats, and monkeys are examined and described.

Cannibalism

FIELDS OF STUDY

Anthropology, evolutionary science, population biology, reproduction science

SUMMARY

Cannibalism is the act or practice of consuming the bodies or parts of the bodies of a species by members of the same species. Although its practice is not universal in the animal world, there are more than 1,300 kinds of cannibal animals in existence in the world today.

PRINCIPAL TERMS

- **ecosystem:** a community of organisms in relation to each other and their physical environment
- **endocannibalism:** a form of human cannibalism in which members of a related group eat their own dead
- **exocannibalism:** a form of human cannibalism in which unrelated humans are eaten

BASIC PRINCIPLES

The most persuasive reason why cannibalism takes place within a species is the need to survive. Paramount to survival is the necessity to have a diet sufficient to support development and continued existence. Competition for survival begins at birth and continues throughout the life cycle of the animal. Since most animals produce more young than can survive, it sometimes occurs that the strongest of the young feeds on the weakest. Young mantids (praying mantises), black widow spiderlings, and varieties of young salamanders, for example, often feast on their brothers and sisters as soon as they are born. In some varieties of sharks, only one or two shark pups are born from the large number of eggs that the mother shark carried during gestation; the surviving shark pups consume their brothers and sisters before birth.

Another reason for cannibalism, also related to the need to survive, is the necessity to eliminate competitors within an ecosystem. Some young tiger salamanders, when living in extremely crowded conditions, develop special structures in their mouths that enable them to eat other salamanders that are their competitors. Adult male Kodiak bears often kill and eat young cubs, especially male cubs, as a means of both supplementing their diet and eliminating future competitors. Male lions and male feral cats are also known to kill and eat the cubs of another male, thus enabling them to mate with the mother of those cubs and ensure that their own offspring will survive. Male chimpanzees are also known to engage in the practice of killing and eating infants of females that they have not impregnated.

Sometimes cannibalism is related either to the lack of enough food or to a diet deficiency. Two popular household pets, guppies and gerbils, eat their young if there is not enough food available. Female gerbils also cannibalize their own or another female's litter as a means of gaining more protein in their diet. Furthermore, livestock may be forced into cannibalism by the practice of feeding them by-products of slaughtered animals to increase their protein intake; this practice is believed to have spread bovine spongiform encephalopathy, or "mad cow disease," among cattle.

One form of cannibalism still perplexes scientists. During the mating and reproduction process, some female members of the animal kingdom kill and later consume their suitors. Black widow spiders are perhaps best known for this practice, but not all black widows eat their mates after killing them. Scientists have discovered that if the female black widow spider is not hungry, she will not consume her dead mate.

The female praying mantis, however, will always devour her mate after she has killed him.

Some animals appear especially capable of consuming members of their species if there is nothing else readily available. *Tyrannosaurus rex*, the famous prehistoric predator, was apparently in that category. Scientists have discovered that the North American *T. rex* may have devoured members of its own group in order to gain a fast and easy meal. South American horned frogs apparently feed on anything, including fellow horned frogs, that moves near them.

HUMAN CANNIBALISM

While factors that explain cannibalism among animals can also be applied to humans, there are several other possibilities to examine in the case of humans. Prehistoric humans engaged in cannibalism as a means of survival, and modernhuman cannibalism because of a natural disaster or an accident has also been recorded. There are stories of shipwrecked sailors resorting to cannibalism of their dead and even murder and cannibalism in order to survive. Perhaps two of the most famous instances of modern cannibalism forced by starvation was the 1846-1847 experience of the Donner Party in California, and the 1972 incident involving the Andes mountain crash of a plane carrying a Uruguayan soccer team. In both instances, the survivors resorted to cannibalism (but not murder) in order to withstand the peril they faced.

However, the human animal is unique in that cannibalism is also practiced as a social and religious custom not involving subsistence and survival in the normal sense. One type of human cannibalism involves a genuine reverence by relatives for their dead. Called endocannibalism, this practice is based upon the belief that eating the flesh of departed relatives shows great respect and veneration of the dead. This type was practiced among the natives of islands of the southern Pacific Ocean until it was declared illegal following World War II.

Exocannibalism has ritualistic and religious overtones as well. Native warriors of the South Pacific, popularly referred to as headhunters, ate parts of their vanquished opponents as a means of controlling them and gaining their strength. Sixteenth century South American natives mixed the eating of captured slaves with religion, making the cannibalistic ritual into a festival. Indeed, it was actions similar to these, observed by the Spanish, that gave us the word "cannibal"— Columbus incorrectly transcribed the name of the humaneating Caribs of Cuba as *Canibalis*.

—*Robert L. Patterson*

See also: Digestion; Reproduction.

FURTHER READING

Brown, Paula, and Donald Tuzin, eds. *The Ethnography of Cannibalism.* Washington, D.C.: The Society for Psychological Anthropology, 1983. A collection of essays from a 1980 symposium by the Society for Psychological Anthropology.

Elgar, Mark A., and Bernard J. Crespi, eds. *Cannibalism: Ecology and Evolution Among Diverse Taxa.* New York: Oxford University Press, 1992. Standard study of the role of cannibalism in animal evolution.

Goldman, Laurence R., ed. *The Anthropology of Cannibalism.* Westport, Conn.: Bergin and Garvey, 1999. Provides general textbook-style information

Klitzman, Robert. *The Trembling Mountain: A Personal Account of Kuru, Cannibals, and Mad Cow Disease.* New York: Plenum Trade, 1998. Tells of the author's attempt to show, after months of field study, the relation between endocannibalism and Kuru (a brain disease).

Cardiology

FIELDS OF STUDY

Biology; chemistry; anatomy; physiology; biochemistry; neurology; pharmacology; pathology.

SUMMARY

Cardiology is the study of the heart. By understanding the heart's normal functional state, applied science can address its abnormalities, designing stents and bypasses for blocked coronary vessels, replacements for faulty valves, and pacemakers to compensate for electrical abnormalities. In addition, cardiopulmonary bypass machines allow the patient to undergo open-heart surgery, and artificial hearts briefly extend the life of a patient waiting for a transplant. Automated external defibrillators for applying an electrical shock to reset a heart are becoming increasingly available in public places such as airplanes and shopping malls.

PRINCIPAL TERMS

- **angiogram:** X ray of blood vessels using radioactive dye
- **angioplasty:** surgical procedure that unblocks blood vessels
- **aorta:** largest artery that carries oxygenated blood to the body
- **arrhythmia:** irregular heartbeat
- **atrium:** one of two upper chambers of the heart
- **bradycardia:** lower than normal heart rate
- **catheter:** thin tube that fits within blood vessels and other body tunnels
- **defibrillator:** machine that delivers an electrical shock to the heart
- **echocardiogram:** image of the heart using ultrasound
- **electrocardiogram (EKG):** recording over time of the heart's electrical activity
- **pacemaker:** device implanted in the chest to regulate the heartbeat
- **stent:** implanted device that keeps arteries open for smooth blood flow
- **stethoscope:** medical device for listening to sounds inside the body such as blood flow
- **tachycardia:** higher than normal heart rate.
- **vena cava:** largest vein that delivers deoxygenated blood from the body to the heart
- **ventricle:** one of two lower chambers of the heart

Basic Principles

Cardiology is the branch of medicine that deals with the heart and the cardiovascular system. The heart is a muscular organ in the middle of the chest and consists of four chambers with valves that send blood to the lungs for oxygenation and then send the oxygenated blood throughout the body. It is essentially a pump; the muscle contracts in response to electrical signals that arise from the sinoatrial node, a patch of specialized tissue located at the top of the right atrium. Without oxygenated blood circulating to the body's tissues, a person would die.

Blood that is poor in oxygen enters the right atrium from the vena cava. The right atrium collects this oxygen-poor blood and sends it through the tricuspid valve into the right ventricle. The right ventricle sends this oxygen-poor blood through the pulmonary valve and pulmonary artery to the lungs for oxygenation. Oxygen-rich blood returns to the heart and enters the left atrium through the pulmonary vein. The left atrium collects this oxygen-rich blood and sends it though the mitral valve into the left ventricle. The left ventricle sends this oxygen-rich blood through the aortic valve to the rest of the body. A wall of muscle, called the septum, separates the left and right halves of the heart.

Background

A great deal of applied technology related to the heart began to evolve in the twentieth century. In 1903, Dutch physiologist Willem Einthoven invented the electrocardiograph; for this, he was awarded the Nobel Prize in Physiology or Medicine in 1924. Werner Forssmann performed the first human cardiac catheterization in Eberswalde, Germany, in 1929. Twelve years later, in New York, Andre F. Cournand and Dickinson W. Richards first used cardiac catheter techniques to measure cardiac output for diagnostic purposes. In 1956, these three men shared the Nobel Prize in Physiology or Medicine for their work.

In 1952, at the University of Minnesota, American surgeons F. John Lewis and C. Walton Lillehei

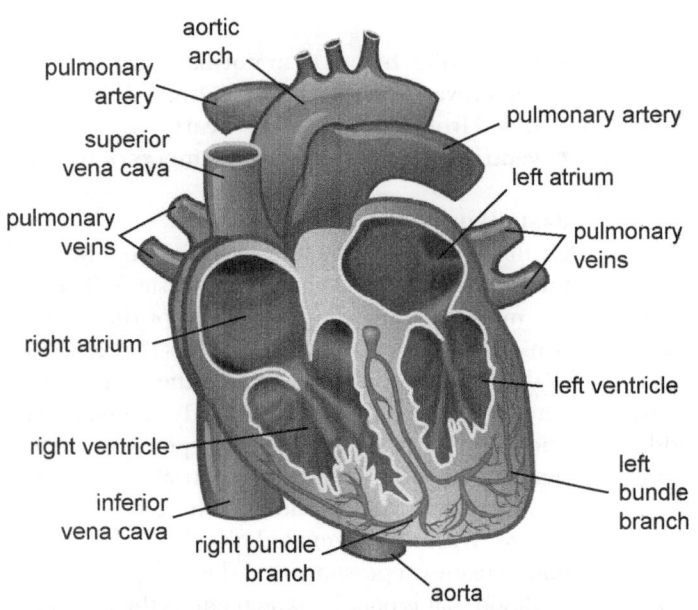

The Human Heart

successfully performed the first open-heart surgery on a human patient. After reducing the body temperature of a five-year-old girl, they repaired a hole in her heart in a ten-minute operation. In 1967, the first wholeheart transplant was performed by South African surgeon Christiaan Barnard. The patient, Louis Washkansky, died eighteen days later of pneumonia. In 1982, American surgeon William C. DeVries implanted the first permanent artificial heart, the Jarvik-7, invented by American physician Robert K. Jarvik, into Seattle dentist Barney Clark. The dentist lived for another 112 days.

How It Works

Blockage of Coronary Vessels. Blockage in the blood vessels leading to the heart may result from blood clots or fatty deposits called plaques. Blockages may be found on an electrocardiogram, echocardiogram, or angiogram. Small blockages may be managed by lifestyle changes such as increased exercise, smoking cessation, and a diet low in cholesterol. Medications are also available that lower cholesterol levels, lower blood pressure, or thin the blood, minimizing blood clot formation. Applied science has contributed to the development of surgical procedures: An artery can be kept open for adequate blood flow by the insertion of a stent, blood vessels can be expanded by balloon angioplasty, and oxygenated blood can bypass a blocked coronary artery and reach the heart through an artery from another part of the body that is used to connect the blocked artery and the aorta.

Faulty Heart Valves. A heart valve may function improperly as the result of a birth defect, an infection, or age-related changes. It may also become deformed as the result of damage and scar tissue. Properly functioning valves permit only the one-way flow of blood. One malfunction, called regurgitation, occurs when the valve does not close tightly, allowing blood to reenter the chamber from which it came. Another malfunction, called stenosis, occurs when the valve becomes thick and stiff so it does not open completely and insufficient blood passes through with each heartbeat. The third malfunction, called atresia, occurs when the valve is fused shut, not allowing blood to pass. Untreated valve malfunctions may progress to cause sudden cardiac arrest and death. Less severe valve malfunctions may be managed by lifestyle changes and medications to relieve the symptoms. More severe valve malfunctions may be surgically corrected by repair or replacement. Replacement tissue valves may come from cows, pigs, or human cadavers. They last ten to fifteen years in older, less-active patients but wear out and must be replaced sooner in younger, more active patients. Applied science has developed mechanical replacement valves that are intended to last beyond a patient's lifetime, so that only one surgery is needed. However, the risk of blood clot formation is higher, so the patient must remain on anticoagulant drugs.

Electrical Irregularities. Electrical irregularities of the heart may result in a heartbeat that is too fast (tachycardia), too slow (bradycardia), or erratic (arrhythmia). Such dysfunction may be detected using a stethoscope, feeling peripheral pulses, or generating an electrocardiogram. Treatment depends on the stability of the patient's condition. In some cases, physical maneuvers may be employed to regulate occasional palpitations. In other cases, medications that prevent arrhythmia may be prescribed; these must be taken in conjunction with anticoagulant drugs to reduce the risk of blood clot formation. In cases of chronic bradycardia, an electrical pacemaker may be implanted to deliver a shock to the heart when the heart rate falls too low.

Stethoscopes and Electrocardiograms. Electrical irregularities such as an erratic heartbeat may be detected through a stethoscope, felt in a peripheral pulse, or seen as an abnormal tracing on electrocardiography. The electrocardiography invented by Dutch physiologist Einthoven in 1903 was a large tablelike contraption. The patient put his or her hands and feet in buckets of saltwater to facilitate electrical conduction. Four decades later, the machine had become smaller and received input from wire leads on metal disks attached to the patient's wrists and ankles. Modern machines are compact and easily transported on a wheeled cart from one exam room to another. The wire leads clip to disposable, self-adhesive disks for easy placement and removal.

Pacemakers. Heartbeats that flutter instead of beating strongly and regularly may be corrected by electrical stimulation from a cardiac pacemaker. In 1950, Canadian John A. Hopps invented the first cardiac pacemaker; it was external because it was simply too heavy to be implanted into the chest. His background in electrical engineering led him to be called the father of biomedical engineering. When transistors replaced vacuum tubes, the pacemaker became less cumbersome, and in 1958, Colombian scientist Jorge Reynolds Pombo designed the first internal cardiac pacemaker.

Defibrillators. A heart that has just stopped beating or that is pumping with no apparent rhythm may be restarted with an electrical charge delivered by a defibrillator. The shock disrupts the chaotic heart action and allows the sinoatrial node to resume its regulatory function. Automated external defibrillators are becoming increasingly available for emergency situations in public places. When activated, these machines deliver audible instructions so that untrained bystanders may use them effectively.

Stents. A coronary stent is a wire-mesh tube that is inserted into an artery to improve blood flow. It is placed as part of an angioplasty to remove a blood clot or plaque deposit and remains permanently. Charles Dotter invented the first coronary stent in 1969 and implanted it in a dog. Stents were implanted in humans in Europe as early as 1986. The U.S. Food and Drug Administration (FDA) approved the Palmaz-Schatz stent, a balloon-expandable coronary stent, for human use in 1994.

Angiography, Angioplasty, and Catheters. A cardiac catheter is a long, small-diameter tube that is threaded to the heart through the femoral artery. When contrast dye is injected through the tube, the coronary arteries may be visualized on X rays, indicating the location of any blockage; this process is called coronary angiography. These catheters and specific techniques for using them were developed by American radiologist Melvin Judkins in the 1960's. One such catheter, a balloon-tipped catheter, is used to open a blocked artery. When the catheter with a balloon fitted on its tip reaches the site of a blockage, the balloon is inflated to enlarge the interior of the artery by flattening plaque deposits against the vessel wall. In 1977, German cardiologist Andreas Gruentzig performed the first balloon angioplasty in a human in Zurich, Switzerland.

When angioplasty cannot sufficiently open a closed artery, coronary artery bypass surgery may be performed. In this procedure, an artery from another part of the body is grafted to the heart to reroute blood flow around the blockage. This procedure requires open-heart surgery, for which the patient must be put on a heart-lung machine (cardiopulmonary bypass machine) to remain alive while the heart is not beating. The patient's blood is pumped from the body through the vena cava into the machine, where it is filtered, cooled, diluted with a specific solution to lower its viscosity, and oxygenated. The blood is then returned to the body through the ascending aorta. The patient is also given an anticoagulant to prevent the formation of blood clots.

Heart Valves. The cardiopulmonary bypass machine, which allows blood flow to bypass the heart and lungs, is also used in other surgical procedures, such as valve repair and replacement, repairs of septal defects and congenital heart defects, and heart transplantation. Surgeons repair or replace heart valves in 99,000 operations per year in the United States. The valves most commonly affected are those on the left side of the heart, namely the mitral and aortic valves, because they are exposed to higher blood pressure than those on the right side of the heart to pump oxygenated blood into the body.

When any one of the four heart valves requires replacement, the transplanted valve may be from a cow, pig, or deceased human, or it may be mechanical. Biological valves of animal tissue last only about ten years and are better suited for use in older, less-active patients. Mechanical valves are typically fashioned from plastic, carbon, or metal. Although they last longer than tissue valves and seldom require replacement, blood clots may form on their surface, so the recipient must take an anticoagulant for the rest of his or her life.

Artificial hearts. An artificial heart is a machine that substitutes for a heart in which both halves no longer function properly. It is generally used as a temporary measure until a healthy human heart becomes available for transplantation. However, the goal of ongoing development is to create a lightweight, durable, functional machine that does not need to be replaced and will not be rejected by the body. Various models of artificial hearts are available; some have tubes for pumping pressure and wires for electrical charging that attach to equipment outside of the body; however, later models are designed to be completely enclosed within the chest. Novel biomaterials lessen the risk of foreign-body rejection and complete enclosure reduces the risk of infection.

FUTURE PROSPECTS

Researchers are continuing their efforts to develop an artificial heart that will sustain patients for longer periods while they wait for a transplant, with the ultimate goal of implanting a mechanical heart that will not need replacement. The AbioCor artificial heart, manufactured by Abiomed of Danvers, Massachusetts, was the first self-contained implantable device; previous versions of an artificial heart required patients to remain in bed connected to machines with tubes and electrodes. The AbioCor device is powered by an external battery pack, allowing the patient to be ambulatory. In human trials, this device was implanted in patients whose life expectancy was thought to be less than 30 days. The goal was to extend their lives by an additional 30 days. This device received FDA approval on September 5, 2006. So far, fifteen patients have received an AbioCor artificial heart; the longest a recipient lived was 512 days.

A modern prototype of a fully implantable artificial heart contains cutting-edge electronic sensors, synthetic microporous skins, and other novel biomaterials.

Bethany Thivierge, MPH

FURTHER READING

Holler, Teresa. *Cardiology Essentials*. Sudbury, Mass.: Jones and Bartlett Learning, 2007. Presents cardiology with a practical clinical orientation for medical personnel working in a cardiology office.

Mueller, Richard L., and Timothy A. Sanborn. "The History of Interventional Cardiology: Cardiac Catheterization, Angioplasty, and Related Interventions." *American Heart Journal* 129, no. 1 (January, 1995): 146-172. Contains plenty of names, dates, and details of interest in cardiology history.

Murphy, Joseph G. *Mayo Clinic Cardiology: Concise Textbook*. 3d ed. London: Informa Healthcare Communications, 2006. This easy-to-read textbook features information on all aspects of cardiology.

Topol, Eric J., ed. *Textbook of Cardiovascular Medicine*. 3d ed. Philadelphia: Lippincott Williams & Wilkins, 2006. A complete, well-organized, user-friendly reference book, complete with audio and visual aids.

WEB SITES

Alliance of Cardiovascular Professionals http://www.acp-online.org

American College of Cardiology http://www.cardiosource.org

American Heart Association http://www.heart.org/HEARTORG

Heart Disease Research Institute http://heart-research.org

National Institutes of Health National Heart Lung and Blood Institute http://www.nhlbi.nih.gov

See also: Artificial Organs; Bioengineering; Bionics and Biomedical Engineering; Cell and Tissue Engineering; Geriatrics and Gerontology.

FASCINATING FACTS ABOUT CARDIOLOGY

- The average adult heart weighs 7 to 15 ounces and is slightly smaller than two clenched fists.
- Each day, the heart beats 100,000 times and pumps 2,000 gallons of blood. Most adults have a total blood volume of 10 pints, which is 1.25 gallons, in their body.
- A healthy heart beats with enough pressure to shoot blood 30 feet.
- In 1949, a crude prototype of an artificial heart was built by two Yale doctors from an erector set, small cheap toys, and mismatched household items. It kept a dog alive for more than an hour.
- A modern version of the Jarvik-7 total artificial heart has been implanted in more than eight hundred people since 1982 but each device was removed when a donor heart became available.
- In the 1980's, the external pneumatic power sources that drove artificial hearts were large and based on milking machines.
- The prevalence of heart disease increased so dramatically between 1940 and 1967 that the World Health Organization called it the world's most serious epidemic.
- Surgeons got the idea to lower the body temperatures of patients to slow their metabolism and heart rate during surgery from observing hibernating groundhogs.

CELL AND TISSUE ENGINEERING

FIELDS OF STUDY

Cell biology; cardiology; cardiac surgery; biochemistry; organic chemistry; developmental biology; physiology; transplant surgery; stem cell research; biomaterial science; drug and gene delivery; neuroscience; bioinformatics; molecular engineering; orthopedic surgery; mechanical engineering; physical therapy; biophysical tools; medical ethics; public health.

SUMMARY

Cell and tissue engineering are fields dedicated to discovering the mechanisms that underlie cellular function and organization to develop biological or hybrid biological and nonbiological substitutes to restore or improve cellular tissues. The most immediate goal of cell and tissue engineering is to allow physicians to replace damaged or failing tissues within the body. The field was first recognized as a distinct branch of bioengineering in the 1980's and has since grown to attract participation from numerous medical and biological disciplines. Engineered cellular materials may be used to grow new tissue within a patient's heart or to replace damaged bone, cartilage, or other tissues. In addition, research into the mechanisms affecting cellular organization and development may aid in the treatment of congenital and developmental disorders. Cell and tissue engineering has developed in conjunction with stem cell research and is therefore subject to debate over the ethics of stem cell research.

PRINCIPAL TERMS

- **bioartificial device:** substance, tissue, or organ that combines biological and synthetic components
- **bioengineering:** medical or biological application of engineering principles, including the process of engineering biological tissues and components from raw materials
- **biomaterial:** cellular or synthetic material that can be introduced into living tissues; often part of a medical device
- **cardiology:** branch of medicine concerned with the disorders, diseases, and function of the heart and associated systems
- **cell therapy:** introduction of cells or tissues to treat disease or other physiological disorders
- **differentiation:** processes by which cells change morphology and function to fill a specific role within an organism or tissue

Cell and Tissue Engineering

Diagram: A syringe extracts the Inner Cell Mass from a Blastocyst and deposits it into a Petri Dish, from which differentiated cell types arise: Insulin, Hormones; Neurons; White Blood Cells; Smooth Muscle; Skin, Fibers.

- **drug and gene delivery:** field of study dedicated to the methods involved in introducing medical chemicals and genes to an organism
- **extracellular matrix:** substance surrounding cellular tissues in which connecting tissues are fixed
- **growth factor:** substance that stimulates growth of a cell, tissue, or other part of an organism
- **heterologous:** derived from an organism of a different species
- **in vitro:** outside a living body or organism, in an artificial environment
- **in vivo:** within a living body or organism
- **orthopedics:** branch of medicine concerned with disorders, diseases, and injuries to the skeleton and associated tissues

- **regenerative medicine:** branch of medicine concerned with applying techniques and tools from a variety of disciplines to restore or repair damaged tissues, cells, and organs by stimulating the biological healing and regeneration processes
- **rejection:** immune response in which the host's immune system attempts to defend against cells or tissues introduced from a foreign organism
- **stem cell:** undifferentiated cell type that gives rise to specialized cells within the body; most are derived from embryonic tissues
- **transplant:** transfer of an organ, tissue, or other cellular material from one individual to another

Basic Principles

Cell and tissue engineering is a branch of bioengineering concerned with two basic goals: studying and understanding the processes that control and contribute to cell and tissue organization and developing substitutes to replace or improve existing tissues in an organism. Substitute tissues can be composed either of biological materials or of a blend of biological and nonbiological materials.

The basic goal of cell and tissue engineering is to create more effective treatments for tissue degeneration and damage resulting from congenital disorders, disease, and injury. Engineers may, for instance, introduce foreign tissues that have been modified to stimulate healing within the patient's own tissues, or they may implant synthetic structures that help control and stimulate cellular development. Another goal in cell and tissue engineering is to create tissues that are resistant to rejection from the host organism's immune system. Rejection is one of the primary difficulties in organ transplant and limb replacement surgery.

One of the basic principles of cell and tissue engineering is to use and enhance an organism's innate regenerative capacity. Engineers therefore examine the ways that tissues grow and change during development. Using cutting-edge development in genomics and gene therapy, engineers are working to develop ways to stimulate a patient's immune system and enhance healing.

Cell and tissue engineering have a wide variety of potential applications. In addition to creating new therapies, engineering principles can be used to create new methods for delivering drugs and engineered cells to target locations within a patient. The potential applications of cell and tissue engineering depend on the capability to create cultures of cells and tissues to use for experimentation and transplantation. Research on cell growth is a major facet of the bioengineering field.

Background

Cell and tissue engineering emerged from a field of study known as regenerative medicine, a branch concerned with developing and using methods to enhance the regenerative properties of tissues involved in the healing process. Ultimately, cell and tissue engineering became most closely associated with transplant medicine and surgery.

Medical historians have found documents from as early as 1825 recording the successful transplantation of skin. The first complete organ transplants occurred in the 1950's, and the first heart transplant was completed successfully in 1964.

The science of cell and tissue engineering arose from attempts to combat the problems that affect transplantation, including scarcity of organs and frequent issues involving rejection by the host's immune system. In the 1970's and 1980's, scientists began working on ways to build artificial or semi-artificial substitutes for organ transplants. Most early work in tissue engineering involved the search for a suitable artificial substitute for skin grafts.

By the mid-1980's, physicians were using semi-synthetic compounds to anchor and guide transplanted tissues. The first symposium for tissue engineering was held in 1988, by which time the field had adherents around the world. The rapid advance of research into the human genome and genetic medicine in the mid-1990's had a considerable effect on bioengineering. In the twenty-first century, cell and tissue engineers work closely with genetic engineers in an effort to create new and better tissue substitutes.

How It Works

Broadly speaking, cell and tissue engineering involves creating cell cultures and tissues that are introduced to an organism to repair damaged or degenerated tissues. There are a wide variety of techniques and specific applications for cell and tissue engineering, ranging from cellular manipulation at the chemical or genetic level to the creation of artificial organs for transplant.

Most cell and tissue engineering methods share several common procedures. First, scientists must produce cells or tissues. Next, engineers must tell the cells what to do. This can be done in a variety of ways, from physically manipulating cellular development and tissue formation to altering the genes of cells in such a way as to direct their function. Finally, engineered tissues and cells must be integrated into the body of the host organism under controlled conditions to limit the potential for rejection. Cell and tissue engineering can be divided into two main categories, in vitro engineering and in vivo engineering.

In vitro engineering is the development of cell cultures and tissues outside of the body in a controlled laboratory environment. This method has several advantages. Producing tissues in a laboratory has the potential for growing large amounts of tissue and eventually entire organs. This could help solve a major issue with transplant surgery: the scarcity of viable organs for transplantation. Scientists can more precisely control the growing environment and can therefore exert greater control over developing cells and tissues. In vitro engineering allows engineers to modify and adjust cellular properties without the need for surgery or invasive techniques.

In vitro engineering is commonly used in the creation of skin tissues, cartilage, and some bone replacement tissues. Although in vitro techniques have certain advantages, they have serious drawbacks, including a higher rejection rate for cells and tissues created in vitro. In addition, there are physiological advantages to engineering within the host organism's body, including the presence of accessible cellular nutrients.

In vivo engineering. In vivo engineering is the family of techniques that involves creating engineered cellular cultures or tissues within the host's body. It involves the use of chemicals to alter cellular function and the use of synthetic materials that interact with the host's body to stimulate or direct cellular growth.

In vivo procedures typically involve introducing only minor changes to the host's internal environment, and therefore, these tissues are more likely to be resistant to rejection. In addition, working in vivo allows engineers to take full advantage of the host's existing cellular networks and the physiological environment of the body. The body provides the essential nutrients, exchange of materials, and disposal of waste, helping create healthy tissues.

The primary disadvantages of the in vivo approach are that engineers have less direct control over the development of the cells and tissues and cannot make exact changes to the microenvironment during development. In addition, in vivo engineering does not allow for the production of mass quantities of cells and is therefore not an avenue toward addressing the shortage of available tissues and organs for transplant.

Hundreds of bioengineers are working around the world, and they have created a wide variety of applications using cell and tissue engineering research. Among the most promising applications are cell matrices and bioartificial organ assistance devices.

Cell matrices. In an effort to improve the success of tissue transplants, bioengineers have developed a method for using artificial matrices, also called "scaffolds," to control and direct the growth of new tissues. Using cutting-edge microengineering techniques and materials, engineers create three-dimensional structures that are implanted into an organism and thereafter serve as a "guide" for developing tissues.

The scaffold acts like an extracellular matrix that anchors growing cells. New cells anchor to the artificial matrix rather than to the organism's own extracellular material, allowing engineers to exert control over the eventual size, shape, and function of the new tissue. In addition, scaffolds can aid in the diffusion of resources within the growing tissue and can help engineers direct the placement of functional cells, as the scaffold can be installed directly at the site of an injury.

Matrices may be constructed from a variety of materials, including entirely synthetic combinations of polymers and other structures that are created from derivatives of the extracellular matrix. Many researchers have been designing scaffolds that dissolve as the tissues form and are then absorbed into the organism. These biodegradable scaffolds allow engineers to avoid further surgical procedures to remove implanted material.

Cellular scaffolds represent a middle ground between in vivo and in vitro engineering. Engineers can create a scaffold in a laboratory environment and can allow tissue to anchor and grow around the matrix before implantation, or they can place a scaffold in their target area within the organism and allow the organism's own cells to populate the matrix.

Scaffolds have been used successfully in cardiac repair, especially in conjunction with stem cells. A scaffold seeded with stem cells may be implanted directly into a heart valve, roughly at the site where a cardiac infarction has occurred. The scaffold then directs the growing cells toward the injured area and facilitates regeneration of damaged tissue.

Artificial matrices have also been successful in treating disorders that affect the kidney, bone, and cartilage. Researchers are hopeful that cellular scaffolds could eventually allow the creation of entire organs by coaxing cells to develop around a scaffold designed as an organ template.

Bioartificial organs. One of the major areas of research in tissue engineering is the creation of machines that assist organs damaged by disease or injury. Made from a combination of synthetic and organic materials, these machines are sometimes called bioartificial devices.

One of the most promising organ assistance devices is the bioartificial liver (BAL), which has been developed to help patients suffering from congenital liver disease, acute liver failure, and other metabolic disorders affecting the liver. The BAL consists of cells incorporated into a bioreactor, which is a small machine that provides an environment conducive to biological processes. Cells growing within the BAL receive optimal nutrients and are exposed to hormones and growth factors to stimulate development.

The bioreactor is also designed to facilitate the delivery of any chemicals produced by the developing tissues to surrounding areas. The BAL performs some of the functions usually performed by the liver: It processes blood, removes impurities, produces proteins, and aids in the synthesis of digestive enzymes. The BAL is not intended to permanently replace the liver but rather to supplement liver function or to allow a patient to survive until a liver transplant can be arranged. The bioartificial liver enables patients to forgo dialysis treatments, and some researchers hope to develop BAL devices that may function as a permanent replacement for patients in need of dialysis.

Researchers are working on bioartificial kidney devices that would aid patients with diabetes and other disorders leading to kidney failure. Again, the bioartificial kidney devices are bioreactors, using stem cells and kidney cells to perform some of the purification and detoxification functions of the kidney. Researchers are also developing bioartificial devices to treat disorders of the pancreas and the heart and to help patients suffering from nervous system or circulatory disorders. Taken as a whole, the development of organ assistance devices may be a step toward the development of bioartificial devices that can function to fully replace a patient's malfunctioning organ.

Future Prospects

Bioengineering is intended to improve daily life, both for those suffering from injury and illness and for the population at large. Cell and tissue engineers are focusing on ways to replace damaged tissues, providing, for instance, new skin where skin has been destroyed, and technology to supplement the function of essential organs. One of the ultimate goals of the industry is to create artificial organs that can fully and permanently replace damaged organs. Bioengineers are confident that in the future it will be possible to provide patients with a variety of organs including a heart, liver, or pancreas.

Although most cell and tissue engineers focus on combating physical illness and injury, bioengineering also has the potential to produce technology that will allow humans to improve their functional abilities. At some point, combinations of synthetic computer technology and biological components could be used to improve human visual capacity or to endow humans with more precise access to memory.

As a distinct discipline, bioengineering is relatively new and scientists have only begun to investigate the potential applications and discoveries possible with further research. As the field has begun to expand, so too have opportunities for scientists, engineers, and physicians interested in exploring the future of medicine and science. The bioengineering field has already created billions in revenue and is still in a state of rapid growth. Universities, hospitals, and biomedical corporations are likely to increase their investment in these emerging technologies and techniques, creating a strong and growing industry for many years to come.

Micah L. Issitt, MA

Further Reading

Chien, Shu, Peter C. Y. Chen, and Y. C. Fung, eds. *An Introductory Text to Bioengineering*. Hackensack, N.J.: World Scientific Publishing, 2008. While definitely written with advanced science students in mind, this text is one of the most basic and yet comprehensive texts available as an introduction to all types of bioengineering.

De Gray, Aubrey, and Michael Rae. *Ending Aging: The Rejuvenation Breakthroughs That Could Reverse Aging in Our Lifetime*. New York: St. Martin's Griffin, 2008. An investigation of research programs in bioengineering, nutrition, and other fields of medicine that are aimed at prolonging life. Provides interesting coverage of organ transplantation and cellular manipulation.

Mataigne, Fen. *Medicine by Design: The Practice and Promise of Biomedical Engineering*. Baltimore: The Johns Hopkins University Press, 2006. An introduction to and investigation of bioengineering and the potential future of the field. Provides discussions of issues such as bioreactors and organ replacements.

Rose, Nickolas. *The Politics of Life Itself: Biomedicine, Power, and Subjectivity in the Twenty-first Century.* Princeton, N.J.: Princeton University Press, 2006. An introduction to the moral, ethical, and political issues that surround medical engineering, genetic manipulation, and bioengineering. Addresses several prominent fields in cell and tissue engineering.

Valentinuzzi, Max. *Understanding the Human Machine: A Primer for Bioengineering.* Hackensack, N.J.: World Scientific Publishing, 2004. An accessible reference designed to give students much of the biological knowledge needed to pursue studies in bioengineering. Also provides useful information about the nature, goals, and development of the bioengineering field.

Zenios, Stefanos, Josh Makower, and Paul Yock, eds. *Biodesign: The Process of Innovating Medical Technologies.* New York: Cambridge University Press, 2010. Covers the biomedical industry, with a particular focus on the process of creating and marketing medical technology. Also provides information about the future of biotechnology and bioengineered products.

WEB SITES

Johns Hopkins University School of Medicine Institute for Cell Engineering http://www.hopkins-ice.org

Tissue Engineering International and Regenerative Medical Society http://www.termis.org

See also: Artificial Organs; Bioengineering; Biomechanical Engineering; Bionics and Biomedical Engineering; Biosynthetics; Human Genetic Engineering; Stem Cell Research and Technology.

FASCINATING FACTS ABOUT CELL AND TISSUE ENGINEERING

- In 2008, scientists from a number of European countries reported on the first procedure to install a bioengineered trachea, constructed from synthetic materials and cultures produced from stem cells, in a woman with a failing respiratory system.
- In August of 2009, a team of Italian scientists announced they were working on an innovative method to replace damaged bone with substitute bone made from wood.
- The Russ Prize, given by the National Academy of Engineering since 1999, is considered the Nobel Prize for bioengineering. Past winners have come from both the engineering and the medical fields.
- Scientists at the Fraunhofer Institute in Germany are working on creating artificial human organs that can be used to replace animal subjects in clinical experiments, allowing scientists to achieve more accurate results and to avoid costly and controversial animal trials.
- The large number of soldiers who lost limbs while serving in Iraq has prompted the U.S. Department of Defense to invest in a University of Michigan program aimed at creating prosthetic hands that can transmit touch sensations to a patient's brain.
- In 2009, scientists in Germany revealed a plan to institute an automated process to produce synthetic skin. Automation would be a first necessary step toward producing sufficient quantities of skin to meet all needs.

CELL COMMUNICATION

FIELDS OF STUDY

Biochemistry; Molecular Biology; Genetics

SUMMARY

The process of cell communication is defined, and its importance in biochemical processes is elaborated. Cell communication is an essential feature of nerve function and control of the metabolism of living systems, affecting the operation of the autonomic and sympathetic nervous systems.

PRINCIPAL TERMS

- **autocrine signaling:** a type of cell signaling in which the signaling compound is produced within a cell and delivered to receptors on the outside of the same cell.
- **endocrine signaling:** a type of cell signaling in which the signaling compound is produced in one location in the body and transported to a receptor site some distance away.
- **juxtacrine signaling:** a type of cell signaling in which the signaling compound is produced within

CELL COMMUNICATION

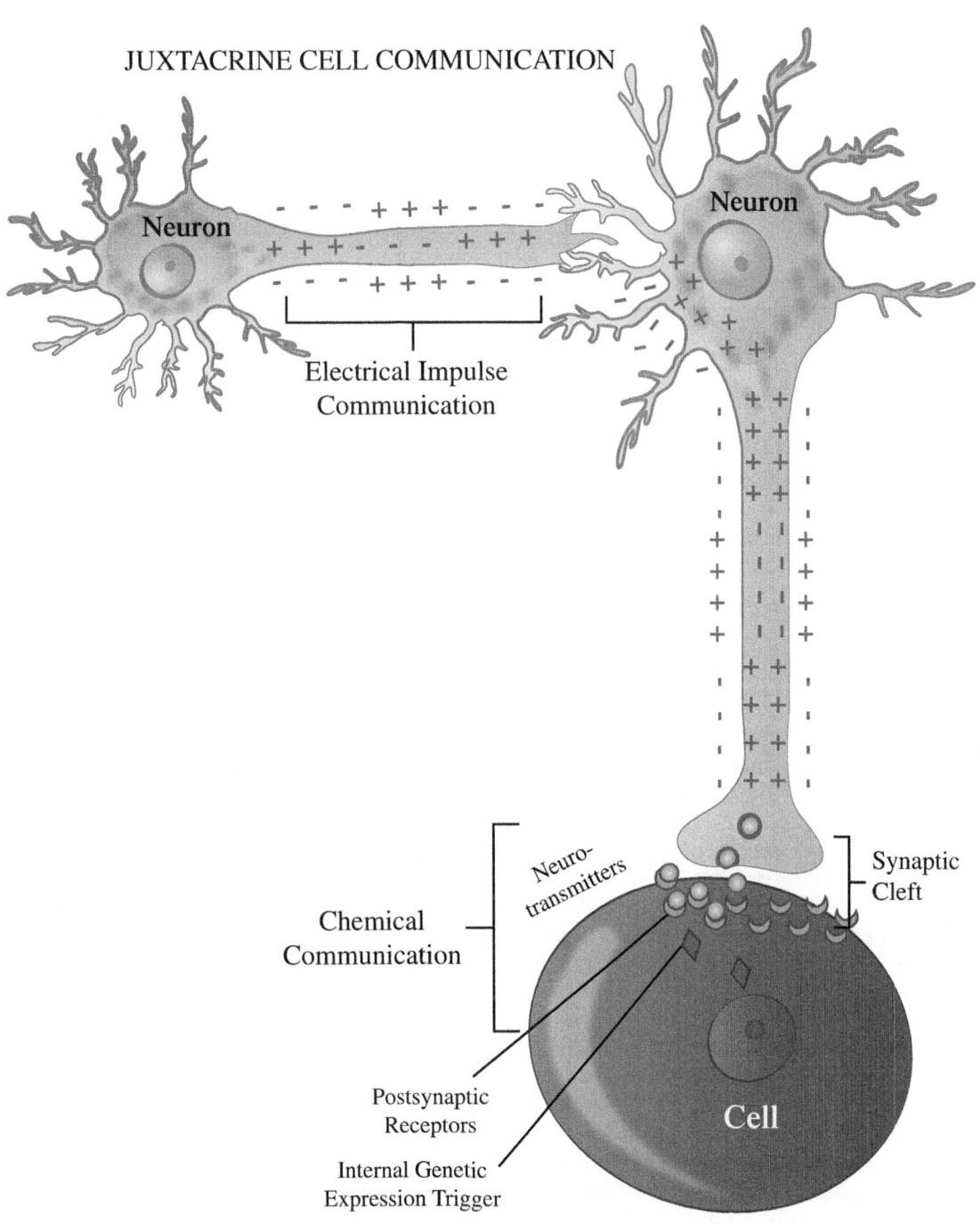

a cell and delivered to receptors in an adjacent cell via physical contact.
- **paracrine signaling:** a type of cell signaling in which the signaling compound is produced in one in the body and delivered to receptors in a nearby cell.
- **receptor:** a molecule or molecular structure, typically a protein or enzyme, that interacts only with compounds that have a matching molecular structure; the interaction normally triggers a biochemical response in the cells to which the receptor is attached.
- **transmitter:** a biochemical compound produced to trigger a specific response at a corresponding receptor site.

THE BASICS OF CELL COMMUNICATION

Cell communication is also known as cell signaling or extracellular signaling. It occurs via a complex multistep process that uses several different pathways. The process begins with the synthesis of a specific signaling compound within a cell or group of cells of the same type. After synthesis, the signaling compound is released by the cell, typically crossing the cell membrane via an active-transport mechanism. The signaling compound is then transported to its target cell. At the target cell, the signaling compound is detected by a specific protein or enzyme receptor site. The detection of the signaling compound triggers a change in cellular metabolism or gene expression in the cell to which the receptor is attached. The final step in the process is the removal of the signaling compound from the receptor and its subsequent elimination from the system.

CELL COMMUNICATION FUNCTIONS AND METHODS

Cell communication or signaling serves a number of purposes, depending on the nature of the living system in which it occurs. In single-cell organisms such as bacteria and protozoa, the production of specific cell-communication compounds coordinates the organisms for cell differentiation or mating, according to which particular reproductive pathway is appropriate. For more complex life-forms, this role is carried out by compounds called pheromones, which can range from relatively simple hydrocarbons to complex proteins.

Pheromones are typically released directly into the environment, where they are detected by other members of the same species. The sense of smell plays a significant role in the detection of pheromones; in many creatures, specific anatomical structures exist for the sole purpose of detecting pheromone molecules. Moths and butterflies, for example, have receptors in their antennae that have evolved to maximize their sensitivity to specific airborne molecules. Almost all other species have a molecule-sensitive structure called the vomeronasal organ as part of their olfactory sense.

This method of cell communication is also used by plants. The production of aroma compounds and other molecular materials for the attraction of pollinators has long been known. More recent studies have also shown that plants use chemical signals to communicate with other plants and to exert some control over their environment by inhibiting the growth of competitors or promoting the growth of companion plants.

Cell communication within organisms is a more fundamental function and is responsible for a great many essential features of living biochemical systems. Intramolecular cell communication is essential to the control of metabolic processes within the cells of organisms. An extremely complex variety of chemical reactions takes place within cells, involving tens of thousands of different proteins and other biochemicals. All of these many and varied reactions are interconnected by their necessity for maintaining the existence of the organism, and a great many are more intimately related to each other as components of various cyclic biochemical mechanisms.

One function of intracellular communication is to control tissue growth. Signaling by hormones controls anabolic processes, which are processes that extract energy from adenosine triphosphate (ATP) to power the movement of various materials across cell membranes; the construction of protein molecules from amino acids; the movement of calcium into bone structures; and many other processes that build up and maintain the physical structure of the organism. The many biochemical processes that take place within a cell's organelles (organ-like structures within the cell) and cytosol (intracellular fluid surrounding the organelles) extract energy and convert needed materials into unusable product materials that must be eliminated from the organism. Catabolic processes, also controlled by hormone signaling, are the processes that break down and eliminate byproducts of metabolism and materials that are no longer required, such as signaling molecules that have already served their function.

TYPES OF SIGNALING

Cell communication takes place between cells or within cells. Endocrine signaling is typical of hormonal processes in which the signaling compound is produced in one part of the organism, such as the endocrine glands, and carried in the bloodstream to receptor sites located in other parts of the organism at some distance from the signaling compound's point of origin. Paracrine signaling, which affects receptors located on cells that are in close proximity to the originating cell, is typical of such functions as nerve-signal transmission across neural synapses and is mediated by acetylcholine and other compounds that function as neurotransmitters and neurohormones. Like paracrine signaling, juxtacrine signaling also takes place between adjacent cells; however, juxtacrine signaling relies on physical contact. In autocrine signaling, a cell responds to its own self-synthesized signaling compounds; receptors may be within the cell or located on the outer surface of the cell membrane. This method of cell communication is typical of the action of growth hormones and is especially relevant to the growth of tumors. There is some crossover in the methods of cell communication, since many compounds can signal by more than one method.

RECEPTORS IN CELL COMMUNICATION

All methods of cell communication function by the interaction of a signaling compound with the corresponding receptor. Receptor proteins only bind to specific signaling compounds; their molecular structure also plays a role in their function. Lipophilic (literally, "fat-loving") receptor proteins found on the surfaces of cell and organelle membranes in the cytosol typically interact with fats, oils, and other hydrophobic ("water-fearing") signaling compounds. Hydrophilic ("water-loving") receptor proteins, normally found on cell surfaces, interact with polar or water-soluble signaling compounds. Activation of cell surface receptors often triggers the formation of a secondary signaling compound that delivers the signal into the cell. Secondary signaling compounds formed in response to such triggering include the cyclic forms of adenosine monophosphate (cAMP) and guanisine monophosphate (cGMP), inositol triphosphate (IP_3), and diacylglycerol (DAG).

Receptors are classified in one of four major categories. G-protein-coupled receptors (GPCRs) interact with compounds such as epinephrine, serotonin, and glucagon to activate G proteins that subsequently activate or inhibit a second messenger or ion channel, ultimately bringing about a change in the cell function. Ion-channel-linked receptors, typically activated by acetylcholine, change the conformation of the ion channel to allow the passage of specific ions across the cell membrane. A third class of receptor is the enzyme-linked receptors, which either behave as enzymes themselves or activate associated enzymes. The fourth class of receptor is nuclear receptors, which are found within cells rather than on the cell surface and mainly respond to steroid and thyroid hormones.

Signaling compounds that function in cell communication are many and varied in structure. The neurotransmitter acetylcholine is the principal signaling compound in nerve synapses. Other paracrine signaling compounds include dopamine, serotonin, and gamma-amino butyric acid (GABA). Endocrine signaling typically involves hormones as the signaling compounds. Most hormones are steroid compounds, based on the molecular structure of cholesterol, though the very important thyroid hormone thyroxine is not.

Richard M. Renneboog, MSc

FURTHER READING

Berg, Jeremy M., John L. Tymoczko, Gregory J. Gatto, and Lubert Strye. *Biochemistry*. 8th ed. New York: W. H. Freeman, 2015. Print.

Lafferty, Peter, and Julian Rowe, eds. *The Hutchinson Dictionary of Science*. 2nd ed. Oxford: Helicon, 1998. Print.

Lehninger, Albert L. Biochemistry: The Molecular Basis of Cell Structure and Function. 2nd ed. New York: Worth, 1975. Print.

Lodish, Harvey, et al. *Molecular Cell Biology*. 7th ed. New York: Freeman, 2013. Print.

Pelczar, Michael J., Jr., E. C. S. Chan, and Noel R. Krieg. *Microbiology: Concepts and Applications*. New York: McGraw, 1993. Print.

Reece, Jane B., et al. *Campbell Biology*. 10th ed. San Francisco: Cummings, 2013. Print.

> **CELL COMMUNICATION SAMPLE PROBLEM**
>
> The rate of lateral movement of a protein molecule in the lipid bilayer of a cell membrane can be described by the equation
>
> $$\text{rate} = \frac{2kT}{3\pi r \eta}$$
>
> where k is the Boltzmann constant, given in kilocalories per kelvin, with a value of $3.3 \times 10^{-27} \frac{\text{kcal}}{\text{K}}$; T is the absolute temperature in kelvins; r is the effective radius of the protein molecule; and η is the viscosity of the lipid bilayer.
>
> If the radius of the protein molecule is estimated to be 0.5 nanometers (nm), what is the rate of lateral movement at a temperature of 37°C? Use $2.4 \times 10^{-11} \frac{\text{kcal s}}{\text{cm}^3}$ for the value of η.
>
> Because the viscosity of the lipid bilayer is given in units per cubic centimeter, the radius of the protein molecule also be expressed in centimeters:
>
> $$r = 0.5 \text{ nm} = 0.5 \times 10^{-9} \text{ m} = 0.5 \times 10^{-7} \text{ cm}$$
>
> Substitute in the values of k, T, r and η and calculate, paying attention to the units throughout:
>
> $$\text{rate} = \frac{2kT}{3\pi r \eta}$$
>
> $$\text{rate} = \frac{2 \times \left(3.3 \times 10^{-27} \frac{\text{kcal}}{\text{K}}\right) \times 310.15 \text{K}}{3\pi \times \left(0.5 \times 10^{-7} \text{ cm}\right) \times \left(2.4 \times 10^{-11} \frac{\text{kcal s}}{\text{cm}^3}\right)}$$
>
> $$\text{rate} = \frac{2046.99 \times 10^{-27} \text{ kcal}}{11.31 \times 10^{-18} \frac{\text{kcal s}}{\text{cm}^2}}$$
>
> $$\text{rate} = 181 \times 10^{-9} \frac{\text{cm}^2}{\text{s}}$$

CELL ORGANELLES

FIELDS OF STUDY

Biochemistry, Molecular Biology, Genetics

SUMMARY

A cell organelle is a special subunit within a cell that performs a specific function. It can be likened to an organ in the body; even the word "organelle" means "little organ." Many types of organelles are found in eukaryotic cells. Prokaryotic cells do not have organelles in the classic sense, but some have compartments within them that are thought to function in much the same way. Some of the major organelles in a cell include the nucleus, mitochondria, Golgi apparatus (or Golgi body), and endoplasmic reticulum. Organelles are difficult to study because they are tiny and hard to examine, even under powerful microscopes.

PRINCIPAL TERMS

- **eukaryotic cell:** a complex cell on which all life more complex than a bacterium or yeast is based, characterized by a nucleus containing the cell's DNA and a number of specialized organelles, supplied with energy by mitochondria, which individually resemble the more primitive prokaryotic cell. *See also* prokaryotic cell.
- **mitochondria:** a type of organelle within eukaryotic cells, where oxygen and nutrients are converted into adenosine triphosphate, the molecule that stores chemical energy for the cell. This process, called aerobic respiration, is possible only in the presence of oxygen. Mitochondria are rod-shaped, have their own DNA, and reproduce independently within the cell, resembling some primitive prokaryotic cells. This suggests that prokaryotic cells were absorbed within the cell walls of evolving eukaryotic cells in a symbiotic relationship. Mitochondria enable cells to produce adenosine triphosphate fifteen times more efficiently than is possible by anaerobic respiration.
- **prokaryotic cell:** the earliest and most primitive type of cell, probably the first life form on Earth, lacking a cell nucleus. Most bacteria are prokaryotic. Most one-celled animals, such as

paramecium, and simple plants such as algae have the more complex eukaryotic cell.

Basic Principles

Scientists differ on what exactly defines an organelle. A "true" organelle is bound by a membrane and only exists in a eukaryotic cell. Some structures in a cell that perform certain functions but are not bound by a membrane (such as ribosomes) spark hot debate about whether they are true organelles. This article discusses the major organelles that are generally recognized as such.

Major Types of Eukaryotic Organelles and Their Functions

Nucleus: This large organelle controls all the cell's activities, including growth and metabolism. It is sometimes called the "brain" of the cell. It uses the deoxyribonucleic acid (DNA) stored within to direct these activities. The nucleolus is a structure contained within the nucleus that stores ribonucleic acid (RNA). RNA helps transmit orders from the DNA to the rest of the cell.

Mitochondria: These structures are the powerhouse of the cell. They change the molecules in food, such as glucose, into adenosine triphosphate (ATP) which fuels all cellular processes. Mitochondria vary from cell to cell based on the cell's needs. For example, cells that use a lot of energy, such as liver and muscle cells, have many mitochondria. Mitochondria are also involved in apoptosis, or death of cells that are no longer necessary.

Endoplasmic reticulum: This network of membranes is the cell's delivery system. It shares part of its membrane with the cell nucleus and winds its way throughout the cell to provide a route for materials to move through the cell to where they are needed. This structure is also involved in protein and lipid manufacture.

Golgi apparatus (or Golgi body): This structure is also a network of membranes. It takes proteins from the endoplasmic reticulum and modifies them so the cell can use them if necessary.

Marianne Moss Madsen, MS

Further Reading

Bradshaw, Ralph A., and Edward A. Dennis, eds. *Regulation of Organelle and Cell Compartment Signaling: Cell Signaling Collection*. Academic Press, 2011. Contains 55 articles about the function of nuclei and other organelles discussing how they signal each other, respond to stress, and affect cell life and death.

Clark, Jonathan. *Biology: Explaining the Cell: Cell Structure and Organelles, Cell Specialization and Function*. Amazon Digital Services, 2014. Explains the features of the cell including structure and function of organelles and their significance. Generally geared toward students with a question-and-answer format.

Dasheck, William V., and Gurbachan S. Miglani, eds. *Plant Cells and Their Organelles.* Wiley-Blackwell, 2017. Provides a comprehensive overview of plant organelles and their structure and function, describing the differences between these organelles and eukaryotic cell organelles.

FASCINATING FACTS ABOUT CELL ORGANELLES

- The cell nucleus takes up about 10 percent of the volume of a cell. It contains around 6 feet of tightly packed, but very organized, DNA.
- Mitochondria are ever-changing components of a cell. They change shape and move around the cell to where they are needed. When the cell needs more energy, they grow larger and divide, then die off as the cell needs less energy.
- Mitochondria have their own DNA. This mitochondrial DNA is inherited only from the female parent and can be used to trace back families through the maternal line. This also means that any diseases that are related to mitochondrial dysfunction are inherited through the maternal line.
- The Golgi apparatus, or Golgi body, was identified in 1897 by physicist Camillo Golgi and named for him in 1898.
- The endoplasmic reticulum comes in two different forms: Smooth endoplasmic reticulum (sER) and rough endoplasmic reticulum (rER). The sER is where fats are synthesized, sugars are metabolized, and toxins are eliminated. The rER is rough because it is lined with ribosomes, which are involved in protein synthesis.

Goodman, Steven R. *Medical Cell Biology*. Academic Press, 2007. Combines a historical aspect with current literature to explain eukaryotic cell biology using the context of organ systems and disease. Generally geared toward students.

Sadava, David E. *Cell Biology: Organelle Structure and Function.* Jones & Bartlett Publishers, 1993. Detailed classic analysis of structure and function of each cellular organelle.

WEB SITES

6 Cell Organelles, Brittanica.com, https://www.britannica.com/list/6-cell-organelles

Cellular Organelles and Structure, Eukaryotic Cells, Khan Academy, https://www.khanacademy.org/test-prep/mcat/cells/eukaryotic-cells/a/organelles-article

Organelles of Eukaryotic Cells: Windows to the Universe, https://www.windows2universe.org/earth/Life/cell_organelles.html

CELL SPECIALIZATION

FIELDS OF STUDY

I didn't see any that made sense to me.

SUMMARY

Cell specialization is a process that happens after cells divide. This process, sometimes called cell modification or cell differentiation, is where cells are modified to perform specific functions. These specialized cells are different from each other in structure, such as size or shape, and function.

PRINCIPAL TERMS

- **DNA (deoxyribonucleic acid):** the complex molecule making up genes, encoding inheritance in all living species. It is known for a unique double-helix structure, with the code in an "alphabet" of four types of molecule: cytosine pairs with guanine, while adenine pairs with thymine. DNA is increasingly used for identification, particularly in crime scenes, to determine paternity of a child, and to study inheritance of both individuals and demographic groups. Study of DNA is also leading to new treatments for genetically inherited diseases. *See also* chromosome, gene.
- **gene:** a unit of hereditary information found within a chromosome that determines the characteristics of an organism. Each gene is an ordered series of nucleotides, which are subunits of DNA, composed of a base molecule containing nitrogen, a phosphate molecule, and a pentose sugar molecule. *See also* chromosome, DNA.
- **protein:** a long chain of amino acids. There are thousands of different proteins in each cell of the human body, and since each species has slightly different proteins in its cells, there are millions of different proteins in the biosphere. A balance of all necessary proteins is essential to the continued life of any organism. Food consumption must either supply each complete protein that the human body cannot manufacture for itself or a wide variety of incomplete proteins that can be assembled into complete proteins.
- **RNA (ribonucleic acid):** a complex molecule similar to, but less complex than, DNA. RNA molecules transfer genetic information from genes in longer DNA molecules forming the chromosomes of living cells to the active metabolic proteins in a living cell. *See also* DNA.

BASIC PRINCIPLES

Specialization happens through a process called differential gene expression. All your cells have exactly the same genes, but they are turned off and on in different patterns, depending on the type of cell the body is creating. Some genes are expressed in all cells, while other genes are expressed in only a few selected cells. Unique cells called stem cells can reproduce and differentiate whenever new specialized cells are needed to replace any dead or damaged cells throughout the body. DNA inside the cell determines which type of specialized cells will be made. RNA makes the proteins that the cell needs to function and translates and transcribes the DNA code, thus influencing cell specialization.

EXAMPLES OF SPECIALIZED CELLS

Different types of cells in the body are necessary to perform specialized functions. Here are just a few kinds of the numerous types of cells that your body needs to perform specific functions.

Red blood cell: These cells contain hemoglobin, which combines with oxygen, and transport oxygen throughout the body. They have specialized into a cell with no nucleus to leave more room for hemoglobin to be contained in the cell, which allows this type of cell to transport more oxygen. They have developed a concave shape to increase surface area to enable oxygen to move more freely in and out of the cell.

Muscle cell: These cells are long and elastic to facilitate contraction and relaxation. They contain many mitochondria to provide energy to each cell so the muscle can contract.

Melanocytes: These rounded cells with long dendrites (branch-like extensions) produce and secrete melanin, the pigment that gives skin color and protects cell DNA from UV radiation. As we age, melanocytes begin to die, lightening the skin.

Sperm cells: These specialized reproductive cells have a unique shape related to their function. These cells have a head, midpiece, and a tail. The head contains all the chromosomes to be delivered from the male parent. The midpiece is packed with mitochondria to provide energy to the tail, which is the source of locomotion to help deliver the chromosomes to be united with an ovum.

Marianne Moss Madsen, MS, ND

Further Reading

Alberts, Bruce, Dennis Bray, Karen Hopkin, Alexander D. Johnson, Julian Lewis, Martin Raff, Keith Roberts, and Peter Walters. *Essential Cell Biology*. Garland Science, 2013. An easy-to-understand introduction to concepts crucial to the study of cell biology with clear writing and beautiful illustrations.

Johnson, Lori. *Cell Function and Specialization (Sci-Hi Life Science)*. Raintree, 2009. Discusses how specialized cells function with good photographs.

Milo, Ron and Rob Phillips. *Cell Biology by the Numbers*. Garland Science, 2015. Features calculations that investigate key numbers in the study of cell biology such as sizes, concentrations, rates, energies, etc.

Pollard, Thomas, D. MD, William C. Earnshaw, Jennifer Lippincott-Schwartz PhD, and Graham Johnson MA PhD, CMI. *Cell Biology*. Elsevier, 2016. Cover key principles of cellular function with clear, concise text and visually interesting illustrations, diagrams, and charts.

Romano, Amy. *Cell Specialization and Reproduction: Understanding How Cells Divide and Differentiate (Library of Cells)*. Rosen Publishing Group, 2005. Gives both a high-level overview and a deep look into how specialized cells, tissues, and organs develop; covers both mitosis and meiosis in depth.

Web Sites

Bozeman Science, Cellular Specialization, http://www.bozemanscience.com/044-cellular-specialization/.

Carmel Clay Schools, Cell Specialization, http://www1.ccs.k12.in.us/teachers/downloads/cms_block_file/50931/file/51165

Texas Gateway, Cell Specialization and Differentiation, https://www.texasgateway.org/resource/cell-specialization-and-differentiation

---

FASCINATING FACTS ABOUT CELL SPECIALIZATION

- Red blood cells have no nucleus, which makes them small enough to squeeze into tiny capillaries. Your body replaces 2 million red blood cells every second.
- An adult human has over 200 different types of specialized cells throughout its body.
- Many factors influence cell specialization, such as available nutrients, salinity, temperature, and hormones.
- Sperm cells can move about 3 mm per hour and wave their tail more than 1000 times to move half an inch.

Cell types

Fields of Study

Anatomy, developmental biology, genetics, histology, invertebrate biology, zoology

Summary

Living cells are the basic units that make up the animal body. Cells are variable in size, shape, structural components, and function. Understanding the functions of cells provides insights into various diseases.

Principal Terms

- **adenosine triphosphate (ATP):** a molecule produced in the cell that provides energy for cell processes
- **amino acid:** the subunit that makes up larger molecules called proteins
- **cytoplasm:** the living portion of the cell that is contained within the cell membrane deoxyribonucleic acid (DNA): the molecular structure within the chromosomes that carries genetic information
- **differentiation:** the process during development in which specialized cells acquire their characteristic structures and functions
- **gamete:** the sex cells of an animal; each gamete contains only one chromosome from each available pair of chromosomes found in normal body cells
- **gene:** the part of the chromosome that includes the DNA and is the carrier of heredity nucleus (pl. nuclei): a central cell structure that controls the activity of the cell because of the genetic material it contains
- **organelle:** a subcellular structure found within the cytoplasm that has a specialized function
- **protein:** a substance made up of amino acids; proteins are the chief building blocks of cellular structures

Basic Principles

The cells that make up the body of an animal are the most basic units of life. Living cells use nutritional sources for energy to maintain structure as well as to maintain life processes such as growth and reproduction. Life requires structure, and cells have the minimal architectural design that enables them to retain life and to pass life on to future generations of cells. Typically, cells are joined in the animal body to form larger structures called tissues. By definition, a tissue, such as connective or muscle tissue, is an aggregate of similar cells and intercellular materials that are combined to perform a common function. An organ, such as the skin or the biceps, is frequently composed of several types of tissues.

Cell Structure

There are essential structural characteristics of animal cells that enable them to maintain life. Most fundamentally, there must be a border that limits the physical portion of the cell from its environment or surroundings. In animal cells, that border is called the cell membrane. Within the membrane lies the cytoplasm (literally, the plasm of the cell); outside the membrane is the environment from which the cell must extract its nutritional needs and into which the cell must pass the waste products that result from the numerous chemical reactions that are continually occurring inside the cell.

Structurally, the cell membrane consists of a double layer of lipid molecules. These molecules are bipolar: The head end of each molecule has an attraction, or affinity, for water molecules (it is hydrophilic), while the tail end of the molecule tends to repel water molecules (it is hydrophobic). In the formation of the membrane, these molecules are found parallel to one another and arranged in a double row. Their heads face the inside of the cell in the inner row, and their heads face their environment in the outer row. The hydrophobic tails form the interior of the membrane and provide an effective barrier that prevents the free passage of water and water-soluble substances. Interspersed among the lipid molecules, like boats in the waves of a lake, are numerous protein molecules. Some of these proteins, called transmembrane proteins, extend entirely through the lipid bilayer and have ends exposed to both the interior and exterior of the cell.

Others, called integral proteins, are found only on one side of the membrane and extend only part way into the lipid bilayer. Finally, some proteins, called peripheral proteins, are attached to the outside or inside surfaces of the membrane. Transmembrane proteins may function as channels or pores allowing specific ions or molecules to pass through them. Integral proteins may also function as transmembrane carriers in that they can bind to specific products outside the cell—such as a certain amino acid—and then flip-flop across the membrane to the inner side and release the product into the cell's interior. Peripheral proteins on the outer surface of the membrane may serve as identification markers for other cells. The membrane itself functions to maintain the integrity of the cell's cytoplasm by holding essential things inside and preventing the cell from drying out.

The cytoplasm is a general term for the viscous fluid within the cell's membrane. In the cytoplasm are numerous small, specialized structures called cell organelles. An essential structure for living cells is the nucleus, which contains genetic information in the form of DNA. DNA determines a particular cell's structure and function. Other organelles in the cytoplasm are designed for specific processes such as manufacturing energy, building structural proteins, or storing cell products prior to exporting them. Some organelles are enclosed within their own membranes, which are structurally very similar to the membrane forming the cell's outer boundary.

THE NUCLEUS AND ITS CONTENTS

The nucleus is an essential control center of the cell. It is enclosed within a unique membrane— the nuclear envelope—that has relatively large pores, through which large information-carrying molecules called ribonucleic acid (RNA) pass from the nucleus to the cytoplasm. Resident within the nucleus is a dark material called the chromatin. During cell division, the chromatin material condenses into clearly observable structures called chromosomes. When the cell divides into two, each daughter cell contains equal numbers of chromosomes from the original maternal cell. The chromatin material within the cell's nucleus is made up of two major types of material— DNA and associated proteins.

The DNA is organized into discrete packets of information called genes, which form the backbone of the chromosomes. The genes determine the characteristics of the specific cell or organism. The DNA-associated proteins regulate the gene's expression. At times, the genes may express themselves by replicating their information into chemical messengers called RNA. The RNA may diffuse out through the nuclear pores into the cytoplasm. Within the cytoplasm, RNA may bind to an organelle called the ribosome, which produces new protein molecules. Normally, the nucleus also contains one or more very dark round structures called nucleoli. The nucleoli assist in the production of the cell's ribosomes. A single nucleolus is made up of protein and RNA.

A particular cell may also contain hundreds of mitochondria, also typical organelles. The mitochondria are complex, double-membraned structures, and they are found throughout the cytoplasm of the cell. Mitochondria are the energy producers of the cell. These oval-shaped structures contain numerous enzymes that stimulate energy-producing reactions. The net sum of these reactions results in the formation of high-energy ATP molecules. These molecules diffuse throughout the cell to various other organelles and release their energy, thereby fueling most cell processes.

Lysosomes are cell organelles that are round in shape and are enclosed within a membrane; they contain many different enzymes. The enzymes function to break down, or digest, many substances into simpler substances that the cell can use. The endoplasmic reticulum (ER) consists of membrane-enclosed spaces in the cytoplasm. These complex membrane arrays are extensions of the outer cell membrane. Some of this membrane system is covered with ribosomes that function to produce protein. These parts of the ER are called rER, for rough ER or ribosomal ER. Other ER portions, that lack ribosomes, have a smooth appearance and are called smooth ER (sER). Another cellular organelle, the Golgi apparatus, appears as a bunch of flattened bags. This organelle is usually not far distant from the ER. After the rER produces a protein product, the Golgi apparatus further processes the product and packages it for cellular export.

CELL SIZE AND FUNCTION

Animal cells have great variations in size. Among the smallest are some bacteria, which may be about 0.1 micron in diameter. (One micron equals 0.001 millimeter.) The largest known cell is the single-cell ostrich egg, which is about seventy-five millimeters in diameter—more than 750,000 times larger than the smallest bacterium. Somatic cells, the cells that make up the body structures of animals, are typically more intermediate in size. For example, the human red blood cell is about seven microns in diameter, while an intestinal epithelial cell is about thirty microns in diameter. Most animal cells are approximately the same size and typically have diameters between ten and twentyfive microns.

An essential characteristic for cell survival is the ratio of the cell's surface area to the volume of the cell, or its surface-to-volume ratio. Typically, cells with small diameters have large surface-tovolume ratios, while large cells have small surfaceto- volume ratios. Many of the substances that are needed for the cell's survival, such as oxygen or nutrients, enter the cell through the surface membrane by simple diffusion. The cells that have high rates of metabolism tend to be very small and have larger surface-to-volume ratios. Alternatively, larger cells either have lower metabolic rates or have specialized shapes such as numerous membrane enfoldings to optimize diffusion of essential materials into their interiors. In multicellular animals, cells are differentiated. Differentiated cells are those that have a specialized modification in their structure to enable them to perform a specific task. Thus, a striated muscle cell contains numerous myofibrils that shorten as the cell contracts, while a glandular cell may contain numerous secretory granules filled with products for export.

Animal cells can be classified on the basis of the number of chromosomes that they contain. Somatic cells, or body cells of an organism, are called diploid, because they contain the total number of chromosome pairs that is characteristic for that organism. For example, somatic cells in mosquitoes contain four pairs of chromosomes (or eight individual chromosomes). Each chromosome in the pair contains genetic information for the same genes that the other paired chromosome contains.

Alternatively, sex cells or gamete cells—sperm and egg cells—contain a haploid number of chromosomes, which consists of a single chromosome from each of the possible pairs. Thus, a human sperm cell contains a total of twenty-three individual chromosomes, whereas a human skin cell contains a total of forty-six individual chromosomes. Another way to classify animal cells is to consider the primary way that they use proteins. Some animal cells are primarily protein-secreting and manufacture much of their proteins for body use outside the cell that produced the protein. An example of this is the pancreatic acinar cell, which produces numerous digestive enzymes (proteins) and secretes them into the pancreatic duct for transport into the digestive tract, where the enzymes break down complex foods into simpler forms. Other animal cells are primarily proteinretaining cells in whichmuch of the manufactured protein is retained for use by the cell itself. An example of this is the keratinocyte, the prominent cell type found in the skin. This cell produces a large amount of the protein keratin, which re- mains stored in the cell's cytoplasm. Because of the presence of the keratin, the skin is able to maintain its waterproofing, protective function. Without keratin, the skin would lose body water.

CELL SHAPES AND TYPES

Animal cells have a wide variety of shapes. Individual cells that are mobile within an aqueous environment tend to be spherical in shape. The neutrophil, a type of white blood cell, is an example of this. Cells that are mobile within tissues and that migrate from one area to another often have long cytoplasmic extensions. An example of this is the macrophage, which has long, changeable, armlike processes (projections) that enable it to migrate through tissue spaces and ingest bacteria thatmay be found there. Other migratory cells have a tail called a flagellum. Sperm cells, for example, use their flagella to propel themselves through the female reproductive tract to reach the egg.

Some cells have branching, stationary cytoplasmic processes through which information molecules move. Often, these processes form a complex network with similar cells such as the network of neuron cells in the brain. Body tissues are made of cells that have tight cell-to-cell connections between their membranes. These cells may provide a covering or form the wall of a particular structure. Protective cells, such as skin cells, are often many layers thick. A typical animal's body is composed of about two hundred recognizable different cell types. Most of these are variations of four main categories: epithelial cells, connective cells, movement cells, and message cells.

Epithelial cells form a continuous layer over surfaces that are external or internal to the body. Skin is an external protective tissue that is made of many such layers of cells. The inner lining of the digestive tract, for the most part, consists of a single layer of epithelial cells that absorb and secrete materials.

Connective cells provide the structural support of the animal body. Examples of these cells are fibrocytes, found in the dermal layer of the skin, and osteocytes, found in the matrix of bone.

Cells responsible for body movement are typified by the muscle cells. Muscle cells that are attached to bone and cause limb movement are called striated muscle cells. Muscle cells that are found in the walls of body organs such as the stomach are called smooth muscle cells; contraction of these cells causes the contents of the stomach to be mixed and stirred. Cardiac muscle cells are found in the heart, and they contract to force the movement of blood throughout the circulatory system.

Message or conveyance cells are very diverse. Branching nerve cells have long processes that carry information molecules (for example, from the spinal cord to the finger). Red blood cells carry oxygen from the lungs to the body cells. Gamete (sperm or egg) cells transfer genetic information from one organism to the next generation of organisms.

STUDYING CELLS

Ever since the 1600's, when Antoni van Leeuwenhoek first used a simple magnifying lens system to study the structure of single-celled life forms, microscopes have been an important tool in cytology (the study of cells). Several types of microscopes are commonly used to study cells, whether alive or preserved.

Light (bright-field) microscopes with two sets of magnifying lenses are most commonly used to study animal cells today. These microscopes have magnification powers ranging from about forty to two thousand times. In their natural state, most cells are essentially colorless. In order to make microscopic viewing more effective, cells are often stained so that their structures are more readily visible. A phase-contrast microscope is a modified light microscope that can produce visible images from quite transparent objects. This type of microscope is frequently used to study unstained or living cells. Its magnification range is similar to the light microscope.

A transmission electron microscope (TEM) uses electrons instead of light rays to visualize objects. A TEM passes its electron beam through very thinly sectioned cells. The density of the stained cellular structures absorbs electrons in a differential fashion so that an image corresponding to the cell's architecture can be visualized. The TEM can magnify cellular structures from 1,000 to 250,000 times. Consequently, this type of microscopy is often used to study the small subcellular organelles.

Another popular technique used to study animal cells involves growing them in cultures outside an animal's body. This technique, called in vitro cell culture, involves obtaining a group of living cells from an animal, usually in the form of pieces of tissue. An enzyme solution is commonly used to digest the cell-to-cell connections and to produce a suspension of free individual cells. These cells are then placed in petri dishes along with a liquid medium that contains essential nutrients. Some types of animal cells, especially fibrocytes found in the connective tissues, are easily cultured with this technique. These cell cultures, if properly maintained, will grow and reproduce for generations. Experimenters can use such cell cultures to investigate how living cells function and respond to varied environmental influences. Since cells are the basic units of life, an understanding of their function is essential in comprehending the way living organisms function. Many diseases are caused by a malfunctioning group of cells. For example, a group of cells normally undergoes an orderly sequence of growth and reproduction. At times, however, some cells become disordered and began to multiply rapidly without stopping. This may be caused by an abnormality that appears in the genetic code or by the presence of a virus that takes over the genetic controls of the cell. This situation is typical of some types of cancer.

As scientists learn more about how cells live and why cells die, they will gain valuable insights into the aging process and may thereby increase the span of life. As differentiation is better understood, scientists may be able to change mature cells or even replace them if they become damaged, destroyed, or simply worn out with age.

—*Roman J. Miller*

See also: Asexual reproduction; Biology; Cleavage, gastrulation, and neurolation; Embryology; Fertilization; Gametogenesis; Genetics; Multicellularity; Mutations; Osmoregulation; Reproduction.

FURTHER READING

Loewy, Ariel G., et al. *Cell Structure and Function: An Integrated Approach.* 3d ed. Philadelphia: Saunders, 1991. A college textbook that covers the major facts and theories of cell biology, as well as genetic manipulation and genetic analysis in their relationship with structural and biochemical approaches to the cell. Suitable for beginning students with a background in elementary organic chemistry.

Prescott, David M. *Cells: Principles of Molecular Structure and Function.* Boston: Jones and Bartlett, 1988. This college-level textbook on classic cell biology is one of the best written on the topic. Cytology is covered well, as is other molecular biology information. Chapters dealing with cellular evolution and specialized cells are especially well done and will be of interest to the reader who wants further information.

Sadava, David E. *Cell Biology: Organelle Structure and Function.* Boston: Jones and Bartlett, 1993. A textbook on the biochemistry, molecular biology, and structure of eukaryotic cells and organelles. For students with college-level introductory biology and chemistry. Illustrated.

Starr, Cecie, and Ralph Taggart. *Biology: The Unity and Diversity of Life.* 9th ed. Pacific Grove, Calif.: Brooks/Cole, 2001. Chapter 5, "Cell Structure and Function: An Over- view," gives a well-written, easily understood synthesis of common characteristics of all cells. Special emphasis is placed on cell organelles and cell surface specializations.

Telford, Ira R., and Charles F. Bridgman. *Introduction to Functional Histology.* 2d ed. Grand Rapids, Mich.: HarperCollins, 1995. This introductory histology text with excellent diagrams and photographs shows the relationships between individual animal cells and the tissues they form. Beginning with basic animal cell biology, the text moves to cover the major organ systems. The frequent use of line drawings to complement electron microscope photographs makes their interpretation much more apparent.

CELLULAR RESPIRATION

FIELDS OF STUDY

Biochemistry; Molecular Biology

SUMMARY

The process of cellular respiration is defined, and its importance in biochemical processes is explained. Cellular respiration controls both the uptake of gases required for maintenance of the living processes and the elimination of by-product gases that would poison the living biochemical system if not removed.

PRINCIPAL TERMS

- **adenosine triphosphate (ATP):** a molecule consisting of adenine, ribose, and a triphosphate chain that is used to transfer the energy needed to carry out numerous cellular processes.
- **aerobic respiration:** a form of cellular respiration that requires oxygen in order to generate energy from glucose.
- **anaerobic respiration:** a form of cellular respiration that does not require oxygen in order to generate energy from glucose.
- **electron transport chain:** a series of oxidation and reduction (redox) reactions in which the electrons released by oxidation are transferred from one molecule to the next, ultimately enabling the production of ATP.
- **Krebs (citric acid) cycle:** a cyclic series of biochemical reactions that completes the conversion of glucose into carbon dioxide, water, and adenosine triphosphate (ATP), consuming and then regenerating citric acid in the process.

UNDERSTANDING CELLULAR RESPIRATION

Cellular respiration is the process that underlies the act of breathing. On the macroscopic scale, living aerobic organisms take in oxygen from the atmosphere through aerobic respiration, normally by inhaling air into the lungs and then exhaling air that has become enriched with carbon dioxide, which is a waste product of metabolic processes. This is not

CELLULAR RESPIRATION

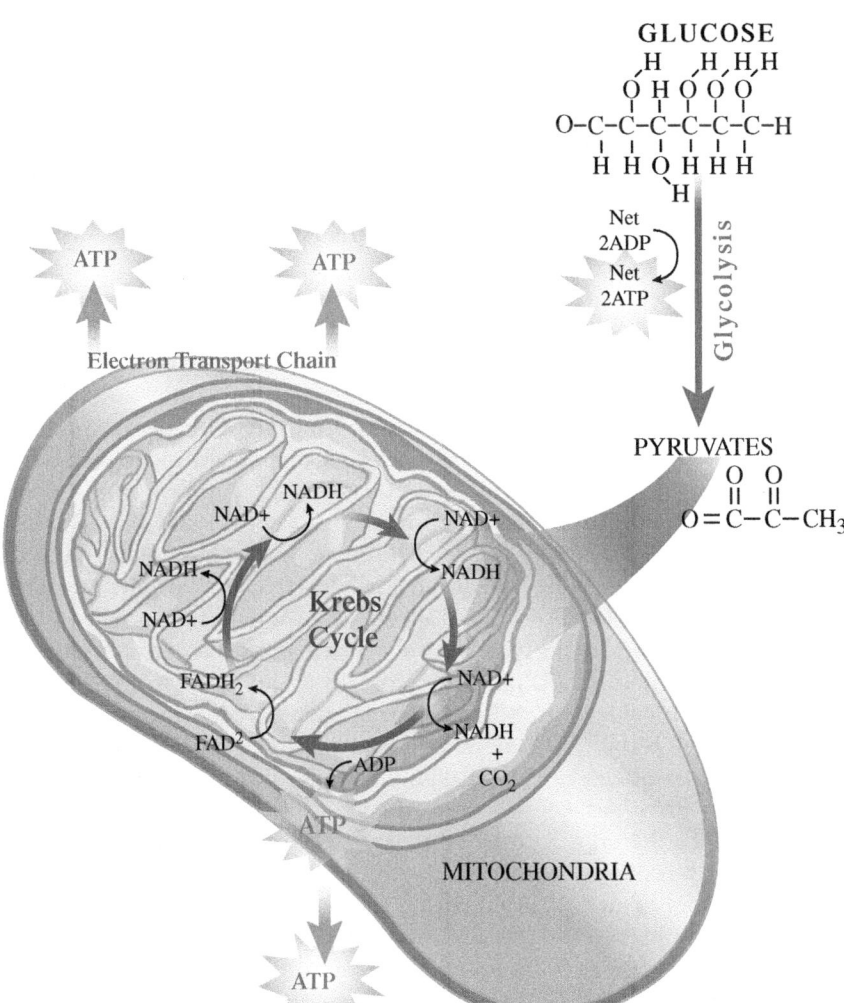

Respiration in plants provides the carbon dioxide used in photosynthesis to produce glucose and releases oxygen that is formed as a by-product of photosynthesis.

At the cellular level, during respiration, oxygen being transported in the blood is taken into eukaryotic cells, where it is used to generate adenosine triphosphate (ATP), a substance that carries the energy necessary for many enzyme-mediated biochemical processes.

CELLULAR RESPIRATION FUNCTIONS

Cellular respiration is the fundamental process by which ATP is generated in biological systems. In chlorophyll-bearing plant cells, sunlight is required to begin the process of photosynthesis, which both consumes and produces ATP. Through photosynthesis, carbon dioxide and water are combined to form the simple carbohydrate sugar glucose. The net production of ATP is used to support various aspects of plant metabolism. The photosynthetic reaction also produces oxygen, which is then released into the atmosphere, becoming available to take part in respiration-based biochemical reactions by other organisms. The glucose that is formed through photosynthesis is the principal foodstuff that supports biological systems, whether as a simple sugar or in more complex molecular forms. Once it is ingested, digestive processes break down the more complex forms into the simple sugar, which is subsequently absorbed and used as a high-energy compound. The energy produced by the the only means by which organisms take in oxygen, however. Fish extract oxygen from water by passing it over their gills, while some amphibians and insects are able to absorb oxygen from the air through their skin.

By contrast, anaerobic organisms do not require oxygen to maintain their life functions and in fact may cease to live if oxygen is present; rather, they survive by practicing anaerobic respiration. Such organisms include bacteria and deep-ocean dwellers that have evolved to depend on chemosynthetic rather than photosynthetic processes for their energy needs. Plants, especially green plants, also rely on respiration as part of their continuing living processes.

biochemical breakdown of glucose is extracted via the linked processes of glycolysis and the Krebs cycle, also known as the citric acid cycle.

GLYCOLYSIS

Glycolysis begins when a molecule of glucose is chemically bonded to an inorganic phosphate group by the enzyme hexokinase, producing glucose 6-phosphate (the number indicates the position of the carbon atom to which the phosphate group has been attached). This step consumes one molecule of ATP. The third segment of the triphosphate group on the ATP molecule is cleaved off and transferred to the glucose molecule, leaving adenosine diphosphate (ADP). The conversion of ATP to ADP is the means by which the high energy contained in the phosphate-anhydride bond is released to do work in biochemical systems. Conversely, converting ADP to ATP stores that energy for later use.

The glucose 6-phosphate is converted by another enzyme to fructose 6-phosphate, which is then bonded to a second phosphate group from another ATP molecule. The symmetrical structure of fructose 6-phosphate is then cleaved to produce two molecules of glyceraldehyde 3-phosphate, each of which is then bonded to a second phosphate group to produce 1,3-bisphosphoglycerate. At this point, the glycolysis process begins the net generation of ATP, as the enzyme phosphoglycerate kinase transfers a phosphate group from each of the two 1,3-bisphosphoglycerate molecules to two ADP molecules, thus forming two molecules of ATP. The 3-phosphoglycerate produced by this reaction is isomerized to 2-phosphoglycerate the two molecules of phosphoenolpyruvate undergo reaction with the enzyme pyruvate kinase, which transfers the two remaining phosphate groups to two more molecules of ADP to generate two more molecules of ATP, leaving two units of pyruvate, the anion (negatively charged ion) of pyruvic acid. The process of glycolysis uses energy obtained from the conversion of two units of ATP to ADP before converting four units of ADP into ATP. In other words, glycolysis doubles the amount of ATP avail-able for use in other energetic processes.

THE KREBS CYCLE

The Krebs cycle, also known as the citric acid cycle and the tricarboxylic acid cycle, is directly connected to glycolysis through the pyruvate produced in the final step of the process. Each unit of pyruvate loses a molecule of carbon dioxide, leaving an acetyl group that is transferred to a substance called coenzyme A and then converted to phosphoenolpyruvate. Finally, (CoA) to produce acetyl-CoA.

The cycle begins with the interaction of acetylCoA with a molecule of oxaloacetate to produce a molecule of citrate. Dehydration converts the citrate into *cis*-aconitate, which is an isomer of aconitic acid from which a hydrogen ion has been removed. The *cis*-aconitate is then rehydrated to produce the iso-meric molecule isocitrate. A chemical reaction called decarboxylation takes place, in which a carboxyl group (–COOH) is removed from the isocitrate, thus releasing carbon dioxide. The decarboxylation, mediated by a coenzyme called nicotinamide adenine dinucleotide (NAD+), converts the isocitrate to α-ketoglutarate. A second reaction with NAD+ and CoA decarboxylates the α-ketoglutarate and transfers the succinyl residue to CoA to produce succinyl-CoA.

The succinate anion is released from suc-cinyl-CoA through a reaction that converts guanosine diphosphate (GDP) to guanosine triphosphate (GTP). When the succinate is dehydrogenated by flavin adenine dinucle-otide (FAD), the product is the fumarate anion, which is then hydrated to the malate structure. Oxidation of malate by NAD+ then produces the oxaloacetate that begins the cycle again.

ELECTRON TRANSPORT CHAIN

Although it is possible for biological systems to extract energy from glucose without the involvement of oxygen, a great deal more energy and ATP is generated by the aerobic decomposition of glucose into carbon dioxide and water. This occurs by a series of processes that includes not only glycolysis and the Krebs cycle but also the electron transport chain.

Through repeated iterations of the Krebs cycle, the six carbon atoms of each glucose molecule are converted to six corresponding molecules of carbon dioxide. Two carbon atoms are eliminated when the two pyruvate units are decarboxylated to form acetyl-CoA, two more are eliminated when isocitrate is converted to α-ketoglutarate, and the remaining two when α-ketoglutarate is converted to succinyl-CoA.

The electron transport chain is the major generator of ATP during the metabolism of glucose. The process is a complex series of oxidation-reduction (redox) reactions, which involve molecular oxygen and such substances as flavin mononucleotide, ubiquinone, or various cytochromes. An oxidation reaction is one in which a reactant loses one or more electrons, while a reduction reaction is one in which one or more electrons are gained.

In the cytosol, the liquid inside the cells, glycolysis produces a net quantity of two units of ATP directly, in addition to two units of NADH (the reduced form of NAD+). In the mitochondria, enclosed units within eukaryotic cells, the rest of the glycolytic process occurs, beginning with the conversion of pyruvate to acetyl-CoA, which is accompanied by the formation of two more units of NADH. As the Krebs cycle continues, two more units of ATP are produced, as well as six more units of NADH and two of $FADH_2$ (the reduced form of flavin adenine dinucleotide, or FAD). NADH and $FADH_2$ are formed during the electron transport chain through a mitochondrial process known as oxidative phosphorylation. About three units of ATP are produced for each unit of NADH and two for each unit of $FADH_2$. Overall, a total of roughly forty units of ATP are produced, while four are consumed, for a net production of about thirty-six units of ATP per molecule of glucose that is metabolized.

ANAEROBIC RESPIRATION IN ACTION

Anaerobic respiration, a term that has come to mean any type of metabolic process that breaks down molecules into smaller units and uses an electron transport chain to produce ATP, typically extracts energy from glucose via alcoholic fermentation. In the absence of oxygen, fermentation by prokaryotic cells begins with glycolysis to produce two units of pyruvic acid. Two units of ATP and two units of NADH are produced in this process. The pyruvic acid is then decarboxylated to produce two units of carbon dioxide and two units of acetaldehyde. Subsequent reduction of the acetaldehyde by NADH produces two units of ethanol (ethyl alcohol) for a net gain of two units of ATP. Anaerobic respiration by this means is thus an energy-poor method that is generally restricted to a few groups of bacteria and yeasts.

CELLULAR RESPIRATION SAMPLE PROBLEM

What is the yield of ATP when pyruvate, fructose 1,6-biphosphate, and phosphoenolpyruvate are each completely oxidized to carbon dioxide? Assume that glycolysis, the citric acid cycle, and oxidative phosphorylation in the electron transport chain are fully involved.

Answer:

Recall that oxidative phosphorylation generates three ATP for each NADH and two ATP for each $FADH_2$ that is produced in glycolysis and the Krebs cycle. Use a chart of the glycolysis process and the Krebs cycle to identify the necessary steps from the material in question and tabulate the ATP output from that starting point.

Pyruvate:
  Pyruvate to acetyl-CoA produces 2 NADH and 6 ATP.
  Citric acid to α-ketoglutaric acid produces 2 NADH and 6 ATP.
  Succinic acid to fumaric acid produces 2 FADH2 and 4 ATP.
  Malic acid to oxaloacetic acid produces 2 NADH and 6 ATP.

$$6 + 6 + 4 + 6 = 22$$

Therefore, a total of twenty-two ATP are produced from oxidation of pyruvate.

Fructose 1,6-Biphosphate:
  Fructose 1,6-biphosphate to 1,3-diphosphoglycerate produces 2 NADH and 6 ATP.
  1,3-diphosphoglycerate to 3-phosphoglycerate produces 2 ATP.
  Phosphoenolpyruvate to pyruvate produces 2 ATP.
  Add the ATP produced by the oxidation of pyruvate.

$$6 + 2 + 2 + 22 = 32$$

Therefore, a total of thirty-two ATP are produced from oxidation of
fructose 1,6-biphosphate.

Phosphoenolpyruvate:
  Phosphoenolpyruvate to pyruvate produces 2 ATP.
  Add the ATP produced by the oxidation of pyruvate.
$$2 + 22 = 24$$

Therefore, a total of twenty-four ATP are produced from oxidation of phospho enolpyruvate.

A second form of anaerobic respiration, lactic acid fermentation, is well known to anyone who has "felt the burn" during exercise. As in alcoholic fermentation, the process begins with glycolysis, producing two units each of ATP, NADH, and pyruvic acid. In this process, however, the pyruvic acid is not decarboxylated but rather reduced by the NADH to two units of lactic acid. Again, only two units of ATP are obtained by this process. When insufficient oxygen is present in muscle tissue, lactic acid is produced faster than it can be removed from the muscle tissue, and the familiar burning sensation in the muscle results.

Richard M. Renneboog, MSc

FURTHER READING

Berg, Jeremy M., John L. Tymoczko, Gregory J. Gatto, and Lubert Strye. *Biochemistry*. 8th ed. New York: W. H. Freeman, 2015. Print.

Lehninger, Albert L. Biochemistry: The Molecular Basis of Cell Structure and Function. 2nd ed. New York: Worth, 1975. Print.

Lodish, Harvey, et al. *Molecular Cell Biology*. 7th ed. New York: Freeman, 2013. Print.

Pelczar, Michael J., Jr., E. C. S. Chan, and Noel R. Krieg. *Microbiology: Concepts and Applications*. New York: McGraw, 1993. Print.

Reece, Jane B., et al. *Campbell Biology*. 10th ed. San Francisco: Cummings, 2013. Print.

CIRCULATORY SYSTEMS OF VERTEBRATES

FIELDS OF STUDY

Anatomy, histology, zoology

SUMMARY

Vertebrates' circulatory systems consist of a closed system of blood vessels with a centrally placed heart, which receives oxygen-poor blood from the body and pumps it to the organs of respiration, where it is oxygenated and returned to the body's tissues. It is here that the oxygen and nutrients are unloaded and carbon dioxide and excretory products are picked up.

PRINCIPAL TERMS

- **aorta:** the major arterial trunk, into which the left ventricle of the heart pumps its blood for transport to the body
- **artery:** a blood channel with thick muscular walls which transports blood from the heart to various parts of the body
- **atria:** the two chambers of the heart, which receive venous blood from the body (via the right atrium) or oxygenated blood from the lungs (left atrium)
- **capillaries:** the very fine vessels in various tissues, which connect arterioles with venules; it is here that the exchange between blood and the extracellular fluid takes place
- **cardiac output:** the amount of blood ejected by the left ventricle into the aorta per minute
- **diastole:** relaxation (filling with blood) of the heart chambers
- **pacemaker:** a specialized group of cardiac muscle cells in the right atrium which initiates the heartbeat; also called the sinoatrial node
- **systole:** contraction (emptying of blood) of the heart chambers
- **valves:** specialized, thickened groups of muscle cells in the heart chambers, major arterial trunks, arterioles, and veins which prevent backflow of blood

BASIC PRINCIPLES

Cells, the units of the animal body, need a constant supply of blood. Blood effects the transport of important materials needed for metabolic, synthetic, and degradative activities, supplying energy and materials necessary for growth, repair of worn-out components of cells, reproductive activity, and other functions of the body. Among the many products that blood transports through a system of closed channels are oxygen, nutrients, metabolic wastes, heat, and hormones. The circulatory system links all tissues with one another and with the external environment to and from which many of these materials are transported. Basically, the circulatory system of vertebrates consists of two parallel systems of blood vessels: One, the arterial system, actively transports blood and its constituents from a central pumping station, the heart; the other, the venous system, more or less passively brings the blood back to the heart. The two systems branch again and again until they ramify all tissues. In the extracellular space of tissues, the finest branches of each system, called arterioles and venules, are connected by means of a network of fine capillaries that allow the movement of blood in one direction, from arterial into the venous system, in which the valves prevent any backflow of blood. A head of pressure, generated in the heart, pumps the blood in this direction, facilitating the transport of substances as well as their movement and filtration out of the capillary membranes and into the extracellular fluid.

Circulatory Systems of Fish, Amphibians, and Reptiles

The simplest level of organization of the circulatory system of vertebrates is seen in fishes. The heart in fishes consists of two chambers, an atrium (auricle) and a ventricle. The oxygen-poor, carbondioxide-rich blood returning from the body via a system of veins is first received by an enlarged vein, the sinus venosus, prior to entering the atrium. The atrium empties its blood into the thick-walled, muscular ventricle, which then pumps it into an enlarged artery, the conus arteriosus. The blood then passes through a major arterial trunk, the ventral aorta, going directly to the gills. The arteries in gills branch profusely and are connected via capillaries with other arteries. In the capillary bed, the blood becomes oxygenated and provides nutrients to the tissue. The oxygenated blood then flows to the head and the rest of the body, and from there returns to the heart through the venous system. In preparation for their journey to land, ancient aquatic vertebrates had to evolve lungs for aerial breathing and had to evolve a complementary circulatory system. As demands for oxygen for a terrestrial existence increased, greater blood pressure and a new way of oxygenating blood were in order. The atrium became divided into two, the right one receiving the deoxygenated blood returning from the body and the left one receiving the oxygenated blood from the lungs (which replaced gills). The deoxygenated blood, entering the right part of the single ventricle, is pumped into the pulmonary artery, all the way to the lungs. The left part of the ventricle, receiving oxygenated blood from the left atrium, pumps it into the body. This three-chambered heart is present in amphibians and most reptiles. The oxygenated and deoxygenated bloods mix partially in the ventricle. In some amphibians, flaps and partial valves tend to prevent such mixing. Reptiles have a partition between the right and left parts of the ventricle, which is complete in alligators, crocodiles, and turtles.

Circulatory Systems of Birds and Mammals

Later reptiles, birds, and mammals developed four-chambered hearts. This complete division of the heart into two separate right and left pumps enables birds and mammals to achieve high speeds. One pumping circuit, the pulmonary, receives blood from the body and pumps it to the lungs; the other pumping circuit, the systemic, receives oxygen-rich blood from the lungs and pumps it into the systemic circulation. Valves within the heart prevent the blood from flowing through it in the opposite direction. The contractile tissue of the heart consists of muscle cells that receive sympathetic and parasympathetic nerve impulses. The vertebrate heart is myogenic; that is, all of its muscle cells and fibers possess an inherent capacity to contract (electrically depolarize) rhythmically; however, all these fibers are under the control of a group of specialized heart muscle cells which have a lower threshold for depolarization than other heart muscle cells: the pacemaker. In fish, amphibians, and reptiles, the pacemaker is located in the wall of the sinus venosus (the first heart chamber before the atrium). In higher vertebrates, which lack a sinus venosus, the pacemaker is found in the wall of the atrium and is called the sinoatrial node. The wave of electrical depolarization initiated here is conducted through the atrioventricular node via a special group of fibers called the Bundle of His, which branch out into the ventricular muscle. The depolarization enters and traverses the atrioventricular node only relatively slowly but spreads down the atrioventricular bundle and its branches much more rapidly than it could travel through ordinary ventricular muscle. This regulates the sequence of contraction of the heart chambers: The atria contract first and the ventricles later, each group of muscles contracting approximately in unison. Since the pulmonary (right) circuit is much shorter than the systemic circuit, it contains less blood volume and offers less frictional resistance to blood flow; also, the right ventricle has muscular walls that are less thick than those of the left ventricle, which has to pump large volumes of blood to the entire body via the systemic circuit.

After the two ventricles are completely filled (a condition referred to as diastole), they contract simultaneously (called systole). During systole, the maximum arterial pressure is generated; during diastole (just before systole), arterial pressure decreases to a minimum. The pulmonary side of the heart contains the funnel-shaped valve between the atrium and the ventricle known as the atrioventricular valve, the right one having three flaps, or cusps (and hence named the tricuspid valve), and the left one (the bicuspid or mitral valve) having two. The free edges of these cusps hang down into the ventricular cavities and are anchored by tendonlike cords of connective tissue called chordae tendinae, each of which is attached to the ventricular wall by a lump called a papillary muscle. The pulmonary artery and the aorta originate at the base of the right and left ventricles, respectively, each having a semilunar valve at its origin. Each of these valves opens in the direction of the blood flow and prevents the backflow of blood. The ventricular contraction and the resulting turbulence in the blood produce the long, low-pitched "lub" sound that can be detected with a stethoscope. The sudden closure of the semilunar valves is similarly perceived to emit a relatively short, high-pitched "dup" sound.

BLOOD VOLUME AND BLOOD VESSELS
The volume of blood that is pumped by the heart each minute is called the minute-volume, whereby the heart beats (contractions) per minute (cardiac stroke rate) eject a typical quantity of blood per beat. This rate is altered by the body's activity and by the volume of blood returning to the heart from the veins each minute. If the venous blood volume is adequate, then increase in stroke rate can increase minute-volume. The increased stroke rate, however, involves a decrease in the ventricular filling time, and as a result, the ventricles do not fill completely. Thus, the stroke volume is decreased; at rapid heart rates, even the minutevolume may be decreased, so that it offsets the stroke rate. During systole, the ventricles do not empty completely. A small residual volume of blood remains in them. An increased venous return may cause more complete filling and emptying of the ventricles, thus increasing the cardiac output without changing the stroke rate.

The vessels at various points in the circulatory path differ anatomically and functionally. The great arteries have thick walls heavily lined with smooth muscle and contractile tissue to enable them to transport blood under pressure from the heart to peripheral tissues. The arteries become smaller and thinner-walled as they branch out toward the periphery. The systemic arteries deliver blood to the microcirculatory beds of the tissues and organs. These "capillary beds" consist of microscopic arterioles, capillaries, and venules. The contraction (vasoconstriction) and relaxation (vasodilation) of the smooth muscles in the terminal branches of the arteries play an important role in regulating blood flow in the capillary bed. Control of the arteriole muscles is mediated by sympathetic neurotransmitters, hormones, and local effects. From the arterioles, the blood enters the capillaries, minute vessels whose walls consist of a single layer of cells, facilitating transfer of oxygen and nutrients to the tissues and the loading of metabolic waste and carbon dioxide, all via the extracellular fluid. Their density depends on the need of the particular tissue for nutrients and oxygen. The capillaries drain into small, thin-walled but muscular vessels called venules, whence the blood begins its return to the heart through the veins. The veins have elastic walls but are without muscles. The venous vasculature serves as a reservoir, storing about 60 percent of the blood.

STUDYING VERTEBRATE CIRCULATION
Circulatory systems of vertebrates have been studied since ancient times through dissection and observation of animal and human cadavers: The heart can be cut open to examine its chambers and their structures, and the body wall can be cut open from the ventral side to expose the circulatory organs. Preserved, dissected animals, including fish, amphibians, reptiles, and mammals, are available from suppliers for students of anatomy who wish to conduct their own dissections. The venous systems of these animals are dyed blue and the arterial systems are dyed red. Plastic models of the circulatory system can be purchased for classroom use.

Scientists are also interested in microcirculation, or circulation at the capillary level. One can fasten a live frog on a frog board and observe the capillaries in the frog's foot web under a microscope. The movement of the red blood cells into the capillary is observed; it is slow and intermittent. The blood flow is regulated by the central nervous system (the vasomotor center in the medulla), as well as by local conditions (such as levels of carbon dioxide, acidity, histamine, temperature, and inflammation). One can then immerse the foot in hot or cold water and observe the resulting change in blood flow. Histamine can be applied to cause vasodilation, which can be controlled by epinephrine. Drops of dilute hydrochloric acid can be applied to the foot to cause vasodilation and inflammation. Thus, it is clear from the foregoing discussion that the heart and circulatory system are of vital importance to the health of an animal.

—*M. A. Q. Khan*

See also: Digestion; Endocrine systems of vertebrates; Reproductive system of female mammals; Reproductive system of male mammals; Respiration and low oxygen; Respiratory system.

FURTHER READING

Berne, R., and M. Levy. *Physiology*. 4th ed. St. Louis: C. V. Mosby, 1998. An advanced undergraduate textbook. The chapter on the cardiovascular system provides an introduction to controls over circulatory function. A good source of information on anatomical details and a good introduction to technical terms.

Hill, Richard W., and Gordon Wyse. *Animal Physiology*. 2d ed. New York: Harper&Row, 1987. A textbook for advanced undergraduates. The chapter on circulation deals lucidly with the needs, evolution, and functioning of circulatory systems in animals. Covers all groups of invertebrate and vertebrate animals.

Raven, P. H., and G. B. Johnson. *Biology*. 4th ed. Boston: McGraw-Hill, 1996. An introductory biology text for college students. The chapter on circulation provides a good description of circulatory systems in animals, with helpful diagrams and illustrations. Good for the beginner.

Robinson, T. F., et al. "The Heart as a Suction Pump." *Scientific American* 254 (June, 1986): 84-91. An excellent introduction to the heart and its functioning, for high school students, college freshmen, and general readers. Provides basic information at a nontechnical level.

CLEAVAGE, GASTRULATION, AND NEURULATION

FIELDS OF STUDY

Anatomy, cell biology, invertebrate biology, physiology, reproduction, zoology

SUMMARY

A single-celled zygote goes through the processes of cleavage, gastrulation, and neurulation to become a many-celled embryo. By learning about normal development, scientists are finding ways to prevent abnormal development.

PRINCIPAL TERMS

- **archenteron:** the primitive gut cavity formed by the invagination of the blastula; the cavity of the gastrula
- **blastula:** an early stage of an embryo which is shaped like a hollow ball in some animals and a small, flattened disc in others; contains a cavity called the blastocoel
- **cleavage:** the process by which the fertilized egg undergoes a series of rapid cell divisions which result in the formation of a blastula
- **gastrulation:** the transformation of a blastula into a three-layered embryo, the gastrula; initiated by invagination germ layers: the embryonic layers of cells which develop in the gastrula: ectoderm, mesoderm, and endoderm
- **invagination:** the turning of an external layer into the interior of the same structure; formation of archenteron
- **morula:** a solid ball or mass of cells resulting from early cleavage divisions of the zygote

- **neurulation:** the process by which the embryo develops a central nervous system notochord: a fibrous rod in an embryo which gives support; a structure that will later be surrounded by vertebrae zygote: the fertilized egg; the first cell of a new organism

BASIC PRINCIPLES

A fertilized egg divides into many smaller cells, which then undergo rearrangement and differentiation to form the embryo of a new individual. The division of the one-celled zygote into smaller and smaller cells is called cleavage. The cellular rearrangement is known as gastrulation, and the proliferation and movement of cells into position to form the beginnings of the central nervous system is termed neurulation. The significance of these events lies in the fact that a single cell with genetic information from two parents is transformed into a multicellular structure with three germ layers that will give rise to all the organs and systems of the body.

CLEAVAGE

After fertilization, the resultant zygote undergoes many rapid cell divisions. The cleavage process results in smaller and smaller cells, called blastomeres. The cell divisions are by mitosis, which produces identical chromosomes in each new cell. When between sixteen and thirty-two cells have been formed, the structure is called amorula, from the Latin for "mulberry," which it resembles. The morula stage is short-lived because, as soon as it is formed, processes are initiated that bring it to the next stage, known as the blastula. A cavity begins to form in the center of the morula as water flows in and pushes out the cells. The new cavity is called the blastocoel and the embryonic stage the blastula. Cleavage continues until the blastula consists of hundreds of cells but is still no larger than the original zygote. The blastula is the terminal cleavage structure. The egg, much larger than an average cell, has been fertilized and sub- divided into hundreds of normal-sized cells. The blastomeres all appear to be similar to one another, but studies have shown that the individual cells are already destined for the tissues they will become.

The principles of cleavage are the same in all vertebrate groups, but the mechanics differ according to the amount of yolk in the egg. Eggs with large amounts of yolk undergo only partial cleavage, because the yolk retards the cytoplasmic division. In birds, reptiles, and many fishes, the yolk is so dense that the cytoplasm and nucleus are crowded into a small cap or disk on one side of the cell. The cleavage divisions all occur in this small area, resulting in a flattened blastula atop the large inert yolk.

Eggs with but a moderate amount of yolk, such as amphibian eggs, are able to cleave completely. Because division proceeds more slowly through the part of the cell where yolk has accumulated, the cleavage is uneven. The cells are formed more slowly on the yolky side and are larger and fewer in number. The blastocoel is smaller and displaced to the side, with less yolk. The side with smaller blastomeres will develop into the embryo and is called the animal hemisphere. The side containing larger amounts of yolk is called the vegetal hemisphere and will provide nutrients for the embryo. Eggs with very little yolk undergo total and equal cleavage divisions. The blastula has a large, centrally located blastocoel, and blastomeres are uniform in size. Starfish and the primitive chordate amphioxus undergo this kind of cleavage.

They are often used to demonstrate the successive cleavage stages which are more easily seen in the absence of yolk. Though mammalian eggs do not have large amounts of yolk, their development is similar to that of birds. The outer layer of cells of the morula develop into a membrane, called the trophoblast, that surrounds the embryo. The embryo forms fromcells in the inner region known as the inner cell mass. Alarge, fluid-filled blastocoel forms within the trophoblast, giving rise to the term "blastocyst" for the mammalian blastula. The inner cell mass develops atop the blastocoel as the bird embryo on the yolk.

GASTRULATION

Gastrulation is the next process in embryonic development and consists of a series of cell migrations that result in cellular rearrangement. The final gastrula will have three embryonic germ layers destined to give rise to all body structures and systems.

The first step in gastrulation is an indenting or invagination in the blastula at a spot known as the dorsal lip. Cells begin to move over the lip and drop into the interior, forming the lining of a new cavity, the archenteron, or primitive gut. Continued inward movement of cells forms a middle layer between outer cells and inner ones which have dropped in through the opening, or blastopore. The three embryonic germ layers have now been formed, and they are called ectoderm, mesoderm, and endoderm.

In animals with little egg yolk, such as the starfish, gastrulation begins when a few cells lose their adhesiveness and drop into the blastocoel. That causes a dent or depression in that area. Cells move in and deepen the depression, forming the archenteron. As the archenteron expands, the inner blastocoel shrinks and is finally obliterated. This process may be visualized as punching in the side of a hollow rubber ball with one's finger. The hole the finger makes is the blastopore; the new cavity formed by the hand represents the archenteron; and the original space inside the ball represents the blastocoel. The indentation forms two cell layers, and a third one is formed as cells continue to move in and take position between the inner and outer layers.

The outer ectoderm is destined to become epidermis and nerve tissue. The inner endoderm will form digestive glands and the lining of the digestive and respiratory systems. The middle germ layer, the mesoderm, will give rise to bone, muscle, connective tissue, and the cardiovascular and urinary systems. Additional mesoderm forms a rodlike structure known as the notochord, which lies in the roof of the archenteron. The notochord is a distinctive characteristic of chordates and gives embryonic support. The mesoderm lateral to the notochord will segregate into paired masses known as somites, each with prospective skin, bone, and nerve segments.

Gastrulation in blastulas with moderate quantities of yolk, such as amphibians have, proceeds similarly, but the archenteron is displaced toward the animal hemisphere and is filled with yolk cells. The early stages are similar to those in starfish. Gastrulation in birds and mammals is initiated in a manner different from that in starfish and amphibians, because of the discoidal configuration of the blastula. Both groups have incomplete cleavage with embryonic development on a disklike area on one side of the egg. The upper cells of the disk separate from the lower ones, forming two layers, the epiblast and the hypoblast. After the two layers are formed, a thickening occurs in one quadrant of the blastula and soon becomes noticeable as a distinct streak, the primitive streak. The streak becomes grooved, and cells fromeither side begin to migrate to the groove and sink down through it. The cells then move into position between the epiblast and hypoblast. The three embryonic germ layers have been formed.

The primitive groove in the gastrula is considered homologous to the blastopore in the starfish and amphibians. After the germ layers have been established, cells continue to move in to the new cavity, the archenteron, and form a mesodermal notochord in the roof of the archenteron.

Neurulation

Neurulation is the final stage of early embryonic development. Studies have shown that the notochord induces the neurulation process to begin. Cells just above the notochord are induced to proliferate and thicken, forming a neural plate. After the neural plate is formed, a buckling occurs in it, forming a depression known as the neural groove. Modern microscope techniques have revealed microfilaments and microtubules lying beneath the surface of the plate. Contraction of the microfilaments and elongation of microtubules appear to cause cell buckling and folding of the plate. The neural groove deepens at its cephalic end, and folds on either side continue to grow higher until they actually touch each other, forming an enclosed tube, the neural tube. At the same time that the neural tube is forming, the head is growing forward and tissue is folding beneath it so it projects forward free from the surface. Brain differentiation begins with the enlargement of the anterior end of the neural tube. The undilated caudal portion will give rise to the spinal cord. The brain forms several constrictions, so that three bulges appear. These will become the three embryonic brain divisions: the forebrain, the midbrain, and the hindbrain.

Upon completion of the three brain divisions, the embryo undergoes forward flexion of the forebrain and a lateral torsion so that the embryo comes to life with its left side on the yolk. A final caudal flexion causes the embryo to take its typical C-shaped configuration. Extraembryonic membranes form from tissue outside the embryo to provide oxygen, nutrients, and waste storage. In birds, an outer chorion and amnion fuse to form a membrane with a large blood supply which provides for the exchange of oxygen and carbon dioxide between the embryo and the atmosphere. The allantois is a membranous sac to contain waste secretions.

In mammals, the outer chorion becomes extensively vascularized on one side and interconnects with the uterus to form the placenta. Nutrient and waste exchange between mother and baby take place in the placenta. The amnion forms a fluidfilled sac that lies closely around the embryo and cushions it. The allantois is not needed for waste storage and is not well developed.

EMBRYOLOGY

Humans have always been intrigued by the processes of gestation and birth. Aristotle questioned whether the embryo unfolds from a preformed condition and then enlarges to adult proportions or progressively differentiates from simple to complex form. Not until the eighteenth century were actual observations made of a developing embryo. The chick egg was the first to be studied, because of its large size. Early studies were descriptive, as each stage of the embryo was observed and carefully described. It was found that development does proceed from simple form to forms increasingly complex.

In the late nineteenth century, great interest developed in evolutionary theory, and comparative embryology became the focal point of studies. Clues were sought for possible evolutionary relationships between organisms. The theory emerged that embryonic stages reflect the evolutionary past of an organism.

The twentieth century has seen the explosion of experimental embryology and multiplication of knowledge. Cleavage of the large fertilized egg was first observed in the eighteenth century, but not until the late twentieth century did the mechanics begin to be understood. With improved microscope techniques, a ring of microfilaments can be seen just below the egg cell surface. These protein filaments have contractile qualities, and it was thought perhaps they lined up around the equator to contract and squeeze the cell in two. To test this hypothesis, a drug which causes microfilament subunits to break down was added to the cell culture. It was found that cell division was inhibited, suggesting that microfilaments are involved in the division process. Removal of astral rays also hindered cleavage. Each new discovery answers some questions and raises more. Embryologists have questioned how blastomeres all formed from the same cell could differentiate into many kinds of cells and tissues. Some of the earliest experiments in embryology involved separating the first two daughter cells to demonstrate that each could form two complete individuals. How and when cells differentiate continues to be a challenge to researchers. The substance in cells which predisposes them to differentiate between one another is still not understood.

It has been discovered that each part of the embryo surface is already divided into prospective organ areas by the blastula stage. Fate maps have been constructed by marking certain areas on the blastula with vital stains and observing the structures into which they develop. Since the early days of experimental embryology, researchers have performed all kinds of operations on embryos, marking areas and observing their movement, transplanting cells from one area to another, exchanging cell nuclei and removing portions. These experiments have led to many discoveries and better understanding of the complicated developmental process. When one considers the multitude of complex events that must take place in the development of a new individual from a single cell, it might seem impossible that the entire developmental process could occur without a slip.

Malformation usually begins during early development. Deformities may arise from inherited mistakes in the genetic code or from the harmful influence of external factors such as radiation, poor nutrition, or infection. Studies of cell migration in the embryo have led to ideas for procedures to inhibit tumor cell migration. Knowledge of normal cell development is helping to find ways to prevent abnormal cell development.

—*Katherine H. Houp*

See also: Asexual reproduction; Birth; Brain; Cell types; Fertilization; Gametogenesis; Growth; Hormones and behavior; Hormones in mammals; Multicellularity; Mutations; Placental mammals; Reproduction; Reproductive system of female mammals.

FURTHER READING

Horder, T. J., J. A. Witkowski, and C. C. Wylie, eds. *A History of Embryology.* Cambridge, England: Cambridge University Press, 1986. A sourcebook that covers the history of embryology from 1818, when the first human abnormalities were described, until the 1943 production of radioisotopes at Oak Ridge, which are used in marking and tracing development. Describes the contributions of scientists from around the world, with interesting sidelights. Includes an extensive Bibliography on all aspects of embryology.

Johnson, Leland G., and Rebecca L. Johnson. *Essentials of Biology*. Dubuque, Iowa: Wm. C. Brown, 1986. An introductory-level college text, well illustrated and including outlines, major concepts, key terms, essays, summaries, questions, suggested readings, and a glossary. The chapter on reproduction and development gives a concise Summary of embryonic development in each vertebrate group and has an informative essay on animal cloning.

Mathews, Willis W. *Atlas of Descriptive Embryology*. 5th ed. Upper Saddle River, N.J.: Prentice Hall, 1998. A paperback manual intended for laboratory work in an intermediate college course. The manual includes a complete series of large photomicrographs of developmental stages. A different form of development is noted in eggs with differing amounts of yolk. Acomplete series of cross sections is shown for seaurchin, amphioxus, frog, chick, and pig embryos. Cross sections include an illustration of a whole mount showing the exact location of each section, helping to integrate the three-dimensional aspect of the embryo. The complete series is helpful for laboratory identification of microscopic embryology sections. A thorough glossary is included.

Oppenheimer, Steven B ,, and Edward J. Carroll. *Introduction to Embryonic Development*. Upper Saddle River, N.J.: Pearson Education, 2004. Print. An intermediate-level college text which gives extensive coverage to the embryological stages in primitive chordate and vertebrate classes. Molecular and cellular aspects of development are emphasized, and the discussion of molecular genetics is informative. Extensive coverage of the topics of cleavage, gastrulation, and neurulation. Illustrated, glossary, references.

Starr, Cecie, Ralph Taggart, Christine A , Evers, and Lisa Starr. *Biology : The Unity and Diversity of Life*. 14th ed. N.p.: n.p., 2016. Print. An introductory-level college text that uses the principles of evolution and energy flow as a conceptual framework for each chapter. Clear writing style and color illustrations on every page make this an attractive and informative text. Gives a concise overview of the early embryological stages and describes experiments that have led to the understanding of mechanisms of development.

Cloning

FIELDS OF STUDY

Biology; genetics; biotechnology; animal husbandry; aquaculture; veterinary medicine; developmental biology; embryology; biochemistry; theriogenology; agriculture; horticulture; botany; cell biology; conservation biology; viticulture; enology; medicine; pharmacology; microbiology; pomology; toxicology; zoology.

SUMMARY

Cloning is any type of biological reproduction that produces offspring that are genetically identical to their parents. Cloning occurs naturally, since many organisms routinely reproduce through natural cloning processes. Artificial cloning technologies include molecular cloning, which reproduces large quantities of discrete segments of DNA; reproductive cloning, which uses assisted reproductive technologies to produce animals that share the same desirable genetic characteristics as another living or previously existing organism; and therapeutic cloning, which uses the same techniques as reproductive cloning but instead derives useful cell lines from cloned embryos.

PRINCIPAL TERMS

- **clone:** organism whose genetic information is identical to the donor organism from which it was created, or a macromolecule that is an exact replicate of another macromolecule
- **embryonic stem cell:** stem cell made from the inner cell mass of very young mammalian embryos
- **enucleation:** microsurgical technique that removes nuclei from cells
- **genome:** sum total of the dna stored in the cells of an organism

- **parthenogenesis:** biological process whereby an egg initiates embryonic development without having first undergone fertilization
- **pharming:** use of genetic engineering to express cloned genes that encode useful pharmaceutical products in host animals or plants
- **pluripotent:** ability of a cell to differentiate into any fetal or adult cell type
- **restriction endonuclease:** special enzyme that cuts dna at a specific sequence motif
- **somatic cell nuclear transfer:** implantation of nuclei from somatic (body) cells into an egg to make a cloned embryo
- **transgenic organism:** biological entity that has had a foreign gene inserted into its genome. commercially available transgenic organisms are often called genetically modified organisms (GMOs)

BASIC PRINCIPLES

Cloning is a means of producing biological organisms, cells, or DNA molecules that are genetically identical to their progenitors. There are natural forms of cloning and three main types of artificial cloning: molecular, reproductive, and therapeutic cloning.

Natural mechanisms of cloning occur in organisms such as bacteria that simply split or fragment into identical copies of themselves. In other organisms, reproductive cells, or gametes, undergo a process called parthenogenesis, in which they initiate development without the benefit of fertilization. Cloning is uncommon in mammals, but rarely, early mammalian embryos undergo a form of cloning called twinning, in which the embryo splits into two embryos, which develop into genetically identical twins.

Molecular cloning, also known as recombinant DNA technology or DNA cloning, involves the transfer of an isolated fragment of DNA from an organism of interest to a host cell that replicates it. Such isolated DNA fragments are known as cloned DNA or genes.

Reproductive cloning uses assisted reproductive technologies to generate animals with the same nuclear genome as another animal. The particular procedure used during reproductive cloning is called somatic cell nuclear transfer (SCNT). Cloned embryos are gestated in the womb of a surrogate mother until they come to term. Cloned organisms are not genetically modified organisms but are simply produced through a type of assisted reproduction.

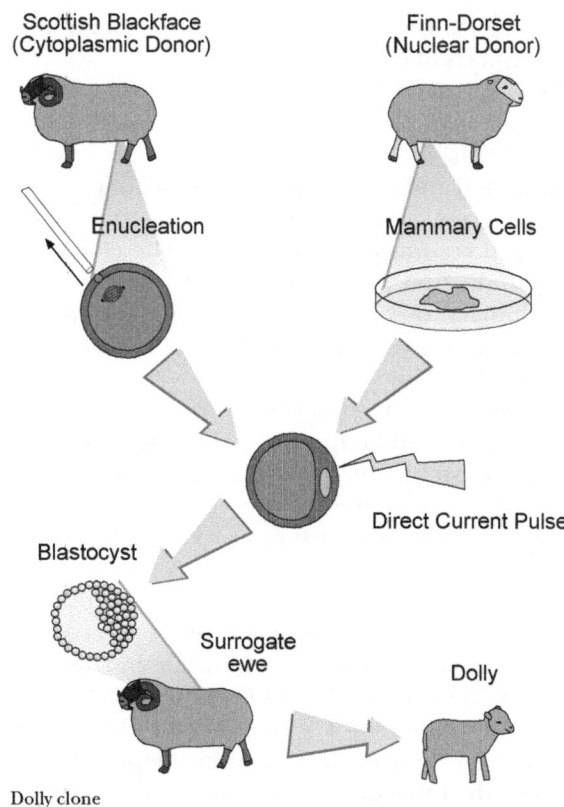

Dolly clone

Therapeutic cloning uses the same procedures as reproductive cloning; however, instead of transferring the cloned embryo into the womb of a surrogate mother, the embryo is further manipulated in the laboratory to make cell cultures of embryonic cells for basic or clinical research

BACKGROUND

Sea urchins were the first animal cloned in the laboratory. In 1894, Hans Dreisch isolated sea urchin embryo cells and watched them develop into small, separate larvae. In 1902, Hans Spemann used the same procedure, embryo splitting, to isolate cells from salamander embryos, which also developed into identical adult salamanders. In 1903, U.S. Department of Agriculture employee Herbert Webber coined the word "clon" for asexually produced cells or organisms, which later evolved into "clone." This term comes from the Greek *klon*, which means "trunk" or "branch." Horticulturists have used this term for more than a century, since an entire new plant can grow from a cutting, resulting in a plant that is genetically identical to the plant from which the cutting was taken.

In 1928, Spemann cloned salamanders by transferring the nucleus, the subcellular compartment that houses the chromosomes, from one salamander embryo into the egg of another. Since Spemann's seminal experiments, scientists have adapted nuclear transfer technology to clone other organisms. In 1952, frogs were cloned, and in 1963, the Chinese embryologist Tong Dizhou cloned a carp to produce the first cloned fish. During the 1980's and 1990's, sheep, cows, and mice were cloned. However, all these animals were cloned by using nuclei from embryos. In 1996, Ian Wilmut and his team at the Roslin Institute in Edinburgh, Scotland, cloned a sheep from an adult cell, demonstrating that adult cells could serve as the source of genetic material for animal clones. This technological feat was followed by the cloning of goats, mules, gaurs (an endangered species), horses, pigs, mouflons (a wild sheep), mice, rats, dogs, cats, water buffalos, camels, rabbits, deer, wolves, and African wildcats, and even embryos from nonhuman primates and humans.

How It Works

Molecular Cloning. To clone a gene, the DNA of the model organism is selectively fragmented by enzymes called restriction endonucleases (REs) and inserted into another piece of DNA called a cloning vector. Cloning vectors are either small circles of DNA called plasmids, bacterial viruses, or bacterial or yeast artificial chromosomes. They ferry the DNA fragments from the genome of the model organism into a host cell (either a bacterium or yeast). This population of host cells collectively carries the entire genome of the model organism in small fragments, and is called a gene library.

To isolate a gene from a gene library requires a probe, which is a fragment of DNA or RNA of any length that has a sequence that is complementary to the sequence of the gene that is to be isolated. Probes can be made synthetically or can come from the genes of closely related organisms. By screening the gene library with the probe, the gene of interest is cloned, which simply means to isolate it from all the other sequences found in the genome of the model organism.

Alternatively, scientists can synthesize small strands of DNA called primers, whose sequences are complementary to different locations in the gene. These primers can be used to specifically amplify the gene from the library by means of a polymerase chain reaction (PCR). A polymerase chain reaction makes large quantities of the gene of interest from a very small amount of starting material, and the amplified DNA can also be cloned into a cloning vector or analyzed directly.

Organisms that express cloned genes make many useful pharmaceuticals such as human insulin, growth hormone, clotting factors, fertility drugs, and vaccines. Cloned genes are also used to genetically screen individuals for genetic diseases. Pharmacologists even use cloned genes for pharmacogenetics, which screens patients for the presence of gene variants that can profoundly affect the efficacy and toxicity of particular drugs. This allows clinicians to tailor treatment to the exact genetic makeup of the patient to maximize treatment efficacy and minimize side effects. Such a strategy is called personalized medicine. Cloned genes are also used in gene therapy, which delivers cloned genes into the bodies of patients who suffer from genetic diseases in an attempt to cure them. Patients with cancer and inherited deficiencies of the immune system, blindness, and blood-based defects have been treated with gene therapy protocols. In agriculture, the introduction of cloned genes into plants that are used as food crops has generated transgenic crops. These crops display several advantageous traits: reduced dependence on agrochemical applications (for example, Bt-corn and herbicide-resistant crops), increased nutritional value (for example, Golden Rice), increased resistance to environmental stresses, and reduced spoilage (for example, the Flavr Savr tomato).

Reproductive Cloning. To clone an animal, mature eggs are isolated from females of the animal species that is to be cloned. The egg is enucleated by piercing it with a microscopically narrow (0.0002-inch-wide) glass tube that is used to vacuum out the egg nucleus. The enucleated egg is fused with a cell from the body of the animal to be cloned and activated with either chemicals or an electric current. This procedure is called somatic cell nuclear transplantation (SCNT). After activation, the egg divides and grows like a newly formed embryo. However, if the animal is a mammal, the embryo can survive only for a limited period of time before it must implant into the inner layer of the mother's womb. Therefore, a surrogate female from the same species of the animal to be cloned, or a closely related species, is made pseudo-pregnant by feeding her hormones, and the embryo is released into her receptive womb, where it implants. Barring any technical or biological mishap, the cloned embryo will develop, and the process will result in a live birth.

When farmers identify food animals with desirable traits, they typically breed those animals as much as possible to improve the genetic quality of their herds and flocks. However, such prize animals inevitably die. Propagating these animals by reproductive cloning and mating them to as many animals as possible preserves the exceptional genetic content of a prize animal and allows it to produce far more offspring. This significantly raises the genetic quality of the flock or herd, and commercial dissemination of such cloned animals to other farmers raises the overall genetic quality of food animals. Reproductive cloning also eliminates the need for artificial insemination, which is often expensive and inconvenient.

Cloning effectively maintains high-quality animal stocks. Reproductive cloning of only the healthiest and most productive animals increases their numbers and improves the gene pool (sum total of genetic diversity) and overall health of food animals.

This results in safer and healthier food and reduces the use of growth hormones, antibiotics, and other chemicals in the raising of animals. In the field of conservation biology, the numbers of endangered species are often increased by captive breeding programs. However, not all endangered species can effectively breed in captivity. Reproductive cloning can aid in the preservation of those organisms that do not reproduce in captivity. Cloning can also resurrect genetic material from dead animals and potentially expand the gene pool of endangered species. In 2001, scientists at the University of Teramo, Italy, cloned the European mouflon, an endangered sheep, from cells sampled from a dead animal.

When combined with other reproductive technologies, cloning can help save endangered species. Cloned animals also serve as excellent research models. Because each cloned animal is genetically identical, experiments on cloned animals are devoid of differences caused by heterogeneous genetic backgrounds.

Genetic manipulation of cloned animals allows researchers to modify genes of interest and more completely analyze their contribution to development and disease. Modifying particular genes of cloned animals also generates model systems for particular genetic diseases. Cloned, transgenic mice and cloned knockout mice, which have had a specific gene inactivated, are examples of the vast usefulness of such model systems.

Of enormous interest is modifying the genomes of cloned animals so that they can produce clinically and pharmaceutically significant products. By genetically modifying pigs, it is possible to make cloned pigs that contain organs that are fit for transplantation into humans (xenotransplantation). Also, producing antibodies, clotting factors, or even vaccines in the blood or milk of farm animals provides a means to mass-produce potentially expensive pharmaceutical agents at a fraction of the normal cost. This process is called pharming.

Therapeutic Cloning. To make embryonic cell cultures, cloned embryos are made by means of somatic cell nuclear transplantation. They are then either disassembled in the laboratory and used to establish embryonic cell cultures or gestated in a surrogate mother to the fetal stage, at which time the fetus is aborted, and cells from the fetus are used to establish fetal cell cultures.

By culturing specific cells from cloned embryos, scientists can make embryonic stem cell (ESC) cultures. During mammalian development, two distinct cell populations form after the first few days of embryonic development. The trophoblast, or the flattened, outer layer of cells, will eventually form the placenta and its associated structures. The inner cell mass (ICM) is the round, inner clump of cells that develop to form the embryo proper and a few structures associated with the placenta. If ICM cells are isolated and cultured on feeder cells, a layer of nondividing skin cells that secrete a cocktail of growthpromoting chemicals, the ICM cells will grow and spread over the surface of the culture dish. Such a culture is an embryonic stem cell culture, and these cells are pluripotent, which means that they can differentiate into any cell type in the adult body.

Therapeutic cloning has tremendous potential for numerous clinical applications. Embryonic stem cells (ESCs) made from therapeutic cloning procedures are pluripotent. Therefore, injured, diseased, or failing tissues or organs could potentially be replaced by tissues or organs manufactured from embryonic stem cells in the laboratory or fetal cells from cloned fetuses. Furthermore, embryonic stem cells made from cloned embryos, or any tissues or organs fashioned from these cells, would not be regarded by the patient's body as foreign. Experiments in laboratory animals have shown that such scenarios are possible. Therapeutic cloning, coupled with embryonic stem cells technology, could christen a new era of regenerative medicine.

Embryonic stem cells from cloned embryos have toxicological applications. Toxicologists typically use laboratory animals or cultured cells to gauge the biological effects of natural or industrially produced molecules on human beings. Unfortunately, laboratory animals show limited utility as a model for human toxicology, and cultured cells do not represent the response of an organ or tissue to foreign molecules. Furthermore, neither of these model systems can assess the individual responses people will have to such molecules, because the genetic variation between individual humans causes differential responses to drugs, toxins, or environmental pollutants. However, cultured embryonic stem cells from cloned embryos can test the biological effects of drugs or environmental pollutants on cells made from a specific person. In addition, because these cells can be differentiated into various tissues and even organs, they can be used to evaluate the individual and tissue-specific responses people might have to particular drugs or pollutants.

Future Prospects

Despite the reservations of some people, cloning is a part of everyday life. Many of the foods Americans consume contain some genetically engineered products. Physicians prescribe medicines, give vaccines, and apply other biological products made by genetically engineered microorganisms on a quotidian basis. Given the inroads molecular cloning has already made into people's lives, it is unlikely that people would suffer any revulsion from eating meat from cloned cattle or sheep or having their lives saved by the transplantation of an organ that came from a cloned pig.

People would also probably not protest seeing cloned versions of endangered species at their local zoos. Nevertheless, many people have raised concerns over cloning technologies. First, conservation biologists have suggested that cloning endangered species does not address the habitat destruction and environmental degradation that pushed these species to near extinction in the first place. Second, cloning only makes one species and does not re-create an ecosystem. For example, cloning cannot recapitulate a coral reef or an old growth forest. Thus, it is the wrong solution for the problem.

Genetically modified organisms have become the focal point of concern for several environmental activism groups. Such groups oppose GMOs because they believe that the cloned genes inserted into them can spread to other species and cause severe environmental disruption and that genetically engineered foods have not been sufficiently tested and are potentially dangerous to human health.

The most contentious aspect of cloning technologies is human genetic engineering and reproductive cloning. Transhumanists are some of the most energetic proponents of human cloning and genetic enhancement. As a movement, Transhumanism regards infirmity, disease, aging, and death as undesirable and unnecessary and views science and technology as the means to defeat human limitations. Transhumanists' main argument for human cloning is that reproductive freedoms extend to everyone, and therefore, every human being has an inherent right to clone himself or herself.

Opponents of human cloning object to the manufacturing of human beings. Cloned children are made to be identical to someone else and therefore will always live in the shadow of the original person and never be completely the person they choose to be. These unreasonable expectations can psychologically damage them and violate their human dignity and individuality. Cloning would also alter the concept of human nature and therefore undermine the very foundation of liberal democracy. In the future, the argument over cloning will not dissipate, but cloning research will certainly advance and provide more and more examples of the utility of this remarkable technology.

Michael A. Buratovich, MA, PhD

Further Reading

Alexander, Brian. *Rapture: A Raucous Tour of Cloning, Transhumanism, and the New Era of Immortality.* New York: Basic Books, 2004. A reporter examines the fringe groups that support human cloning and genetic enhancement and finds people who want to defeat the effect of entropy and live forever.

Fukuyama, Francis. *Our Posthuman Future: Consequences of the Biotechnology Revolution.* New York: Picador, 2003. A historian's admonition of the consequences of the biotechnology revolution and its potential to abolish human rights and erode the foundations of liberal democracy.

Mitchell, C. Ben, et al. *Biotechnology and the Human Good.* Washington D.C.: Georgetown University Press, 2007. A distinctly Christian assessment of the application of biotechnology to humans that remains optimistic but cautious and concerned.

Shanks, Pete. *Human Genetic Engineering: A Guide for Activists, Skeptics, and the Very Perplexed.* New York: Nation Books, 2005. A helpful explication of the science behind cloning, coupled with stern warnings against it, by a noted social activist.

Silver, Lee. *Challenging Nature: The Clash Between Biotechnology and Spirituality.* New York: Harper Perennial, 2006. A Princeton stem cell scientist explains the science behind biotechnology and stem cells. He offers some rather harsh critiques of more conservative thinkers who do not agree with his optimistic views of genetic enhancement and embryonic stem cells.

_____. *Remaking Eden: How Genetic Engineering and Cloning Will Transform the American Family.* New York: Harper Perennial, 2007. A very readable introduction to the science of cloning and genetic engineering by a noted mammalian embryologist, who believes that humans should be cloned and that people should welcome the profound changes that it will invoke within human societies.

Wilmut, Ian, Keith Campbell, and Colin Tudge. *The Second Creation: Dolly and the Age of Biological Control.* New York: Farrar, Straus and Giroux, 2000. The two researchers who made Dolly team up with a noted British science writer to give a personal but rigorous explanation and thoughtful examination of cloning. Contains a helpful glossary of terms.

WEB SITES

Human Cloning Foundation http://www.human-cloning.org

MedlinePlus Cloning http://www.nlm.nih.gov/medlineplus/cloning.html

National Human Genome Research Institute Cloning http://www.genome.gov/25020028

See also: Bioengineering; Reproductive Science and Engineering; Stem Cell Research and Technology.

FASCINATING FACTS ABOUT CRYOGENICS

- Scientists at Advanced Cell Technology used fetal heart muscle cells from cloned cow fetuses to reverse the effects of heart attacks in adult cows.
- The first cloned cat, CC (CopyCat), made at Texas A & M University in 2001, has a completely different personality than the donor cat. Even though CC is genetically identical to her donor, she is shy and timid whereas the donor cat is outgoing and playful.
- In 2008, BioArts International held an essay contest that invited people to argue why their dog should be cloned. The winner was Trakr, a German Shepherd police dog, who discovered the last survivor of the September 11, 2001, terrorist attacks on the World Trade Center in New York City.
- By cloning vaccines into plants, scientists have made edible vaccines against digestive diseases such as cholera, the Norwalk virus, some food poisonings, and enterotoxigenic *Escherichia coli*. These vaccines are not injected but rather eaten.
- Ingo Potrykus and Peter Beyer invented Golden Rice in the 1990's. This genetically engineered strain of rice produces beta-carotene, a precursor for vitamin A biosynthesis, which is not found in normal rice in appreciable quantities. Children who live in countries where rice is the main food staple are at higher risk for vitamin A deficiency, and Golden Rice was developed to help prevent this deficiency. Subsequent development has increased the nutritional value of Golden Rice even further. Even though the makers of Golden Rice want to give it to farmers completely free of charge, opposition to genetically modified organisms has prevented it from ever being cultivated for food.

COPULATION

FIELDS OF STUDY

Anatomy, developmental biology, embryology, reproduction science

SUMMARY

Copulation, also called coitus or sexual intercourse, is the process by which sperm are placed directly into the female reproductive system by the male reproductive system, thus assuring that sperm and ova are in close contact for fertilization. It is also a critical adaptation for land-based reproduction, since on land, unlike in the water, sperm released into the environment cannot swim to ova that have also been released. The behaviors leading up to copulation are also important in assuring that gametes of the same species meet for fertilization.

PRINCIPAL TERMS

- **amplexus:** a form of pseudocopulation seen in amphibians, where the male mounts and grasps the female so that their cloacae are aligned, and eggs and sperm are released into the water in close proximity and at the same time
- **cloaca:** a common opening for the reproductive, urinary, and digestive systems
- **heat:** that part of the estral cycle when the female is receptive to male copulatory behavior
- **semen:** fluid produced by the male reproductive system that contains the sperm

BASIC PRINCIPLES

Animals have many diverse strategies to ensure that eggs and sperm are in close enough proximity for fertilization to take place. One widespread process that is seen in many different phyla is copulation. Copulation is seen in many aquatic phyla and is the rule in terrestrial phyla. In most forms of copulation, the male reproductive system has an intromittent organ, often called a penis, which deposits sperm into the female reproductive system. Once there, sperm can travel the short distance to the eggs. In some organisms, pseudocopulation is seen. In hermaphroditic oligochaetes, such as earthworms (*Lumbricus terrestris*), two worms align in opposite directions so that their genital pores are applied to the openings of the seminal receptacles of their partners. Semen released by the genital pores flows into the seminal receptacles where it is stored. Fertilization, however, is actually external. The worms build cocoons where they lay their eggs and then deposit the stored sperm. Amphibian amplexus is also a form of pseudocopulation. Here, the male frog clasps the female in such a way that their cloacae are in close proximity. Sperm are not deposited in the female's cloaca, however. Instead, both sperm and eggs are released into the aquatic environment for external fertilization.

COPULATORY ORGANS

True copulation takes many forms. In several invertebrates, such as a few flatworms (some Acoela, Rhabdocoela, and Polycladida) and the bedbugs (Cimicidae), hypodermic injection is sometimes seen. In this form of copulation, the female has no external gonopore and the intromittent organ punctures the epidermis and deposits sperm in the underlying body tissue. This sperm must then migrate through the intercellular spaces to the female reproductive organs for fertilization of the eggs to occur. In most organisms, however, the male does not have to pierce the female's epidermis, but instead deposits sperm in an already-present opening of the female's reproductive system.

In birds and some reptiles (such as the tuatara, *Sphenodon punctatus*), the male does not have a true intromittent organ. Instead, the male must manipulate the female during mounting so that their cloacae are pressed against each other. During this "cloacal kiss," the male ejaculates sperm into the female's cloaca. In some bird species, a false penis is present in the male. These organs are not connected to the ducts of the male reproductive system and thus do not serve as intromittent organs. There is speculation that they may provide a necessary stimulation to the female during copulation.

In some fish, fins are modified for semen delivery. In guppies and their allies (Poeciliidae), gonopodia, modified anal fins, are used for insemination. Each gonopodium is a hollow, tubelike structure formed from the paired anal fins of the male. When mating, the male inserts his gonopodium directly into the female's gonopore. Usually, not all the sperm are used to fertilize this batch of eggs, and the rest is stored in the oviduct walls for future fertilizations. Other fish (such as the Coodeidae) have the anal fins modified into andropodia, which are cup-shaped structures that direct the flow of semen into the female without the andropodia actually entering the female's gonopore. Sharks (Elasmobranchii) have modified pelvic fins called claspers, which the male directs into the female's cloaca for insemination. Each shark has two claspers and, depending on species, either the one closer to the female or both are inserted for copulation.

Males of mammals, some reptiles, and many arthropods also have intromittent organs that deposit sperm directly into the female reproductive tract. In these copulations, by either female behavior or male manipulation, the opening of the female reproductive tract must be exposed. In many organisms, the male mounts a squatting or otherwise stationary female. Male snakes and lizards (Squamata) have two intromittent organs called hemipenes. Males and females line up side by side and the male uses the hemipenis closer to the female to inseminate her. Many arthropods often go through intricate body contortions to bring the male's penis in proper position for mating. This may be the common rear-mounting pattern, but can also be face-to-face or tail-to-tail. In many animal species, insertion of the penis is followed by one or more thrusting movements that lead to ejaculation.

COPULATORY BEHAVIORS

Among animals, both the lengths of time per copulation and the frequencies of copulation vary widely. When a female lion comes into heat, the male will remain near her, copulating up to one hundred times a day for periods up to ten days. Each copulation, however, lasts for just a few seconds. Other animals may copulate only once, but the copulation may be prolonged. Canid females do not usually remain stationary for mating. To remedy this, once the male mounts, his penis becomes further engorged and this effectively locks him to the female long enough to ejaculate even if she tries to get away. Other animals have hooks and barbs on their penises that may also help to lock them to a female for prolonged copulation. In some animals, prolonged copulation can last several hours. This may be a mechanism to prevent other males from fertilizing the same female. Females can also play a role in prolonging copulation. In some water mites (*Arrenerus* sp.), the female gonopore can be opened or closed by means of chitinous plates. The smaller male inserts his intromittent organ into the female's gonopore, which then closes, trapping the male. Although sperm transfer is thought to occur in the first few minutes, the female may swim off dragging the male with her for several hours.

Copulation can be dangerous to males. In the domestic honeybee (*Apis mellifera*), a swarm of drones pursue the unmated queen. In-air copulation occurs as a drone inserts his endophallus into the queen's sting chamber. After ejaculation, a small part of the drone's phallus remains inside the queen and the drone falls to the ground and soon dies. Several more drones mate and die until the queen's spermatheca is filled. Male spiders have to be very careful when copulating. If a male does not leave the female's web immediately after depositing his sperm, the female may envenomate and then eat him. The female praying mantis (*Stagmomantis carolina*) have also been known to begin feeding on the heads of males with which they are copulating. Luckily, the headless male can continue to deposit sperm.

—*Richard W. Cheney, Jr.*

See also: Asexual reproduction; Fertilization; Reproduction; Reproductive system of female mammals; Reproductive system of male mammals.

FURTHER READING

Davey, K. G. *Reproduction in the Insects*. San Francisco: W. H. Freeman, 1965. A compendium of insect reproduction.

Hayssen, Virginia Douglass, Ari van Tienhoven, and Ans van Tienhoven. *Asdell's Patterns of Mammalian Reproduction: A Compendium of Species-Specific Data*. Ithaca, N.Y.: Comstock, 1993. A brief look at mating in mammals, divided taxonomically.

Sieglaff, D. "Most Spectacular Mating." In *University of Florida Book of Insect Records*, edited by T. J. Walker. Gainesville: University of Florida Department of Entomolgy and Nematology, 1999. Also at the Web site http://gnv.ifas.ufl.edu/~tjw/recbk.htm. An electronically published paper on bee mating.

Smith, R. L. ed. *Sperm Competition and the Evolution of Animal Mating Systems*. Orlando, Fla.: Academic Press, 1984. A thorough look at mating patterns in animals.

ECHIDNA MATING

The female echidna, or spiny anteater (Tachyglossus aculeata), an Australian monotreme mammal, makes it rather difficult for the male to access her cloaca. As a female comes into heat, she is followed closely by one or more males that will try to mate with her. When she is ready to mate, she lies flat on the ground on her abdomen, often grabbing a tree or other support with her forelegs. This directs her cloaca toward the ground, certainly not exposed toward the male. If there are no other males present, the single enterprising male must dig a ditch on one side of the female. When it is deep enough, he can descend into the ditch, turn on his back and insert his tail beneath the female's tail. This positions his penis to enter her cloaca and deposit semen. When more than one male is present, they dig a semicircular rut on both sides of the female's tail, in which rut the males jostle for position until one finally maneuvers his tail under the female's. If all goes well the female lays a single small egg a few weeks later that she incubates in an abdominal pouch until hatching.

CRISPR-Cas9

FIELD OF STUDY

Molecular biology

SUMMARY

CRISPR-Cas9 is a fast, inexpensive, accurate way of editing DNA. It has a wide range of potential applications, but some scientists worry about the way this tool might be used. This unique technology allows scientists to remove, add, or alter sections of a person's DNA and could be used to cure genetic diseases. However, it can also be used to edit reproductive cells, which could affect generation after generation of humans.

PRINCIPAL TERMS

- **chromosome:** a basic unit of heredity in living cells. Each chromosome is composed of proteins and DNA, which carry thousands of genes. In a healthy, normal, human cell, there are twenty-three pairs of chromosomes. In sexual reproduction, one chromosome in each pair comes from the father, the other from the mother. See also DNA, gene.
- **DNA (deoxyribonucleic acid):** the complex molecule making up genes, encoding inheritance in all living species. It is known for a unique double-helix structure, with the code in an "alphabet" of four types of molecule: cytosine pairs with guanine, while adenine pairs with thymine. DNA is increasingly used for identification, particularly in crime scenes, to determine paternity of a child, and to study inheritance of both individuals and demographic groups. Study of DNA is also leading to new treatments for genetically inherited diseases. See also chromosome, gene.
- **enzyme:** a biological catalyst, any protein molecule within a living organism that speeds up biochemical reactions to a rate that will sustain life. The effect may speed up metabolic reactions by a factor of one million, compared with what would occur chemically outside the body. Names and classification of enzymes are regulated by the International Commission on Enzymes. Most enzymes are named by adding -ase to the root of a corresponding substrate, the molecule an enzyme acts upon. Sucrase catalyzes the hydrolysis of sucrose into glucose and fructose. A living cell has a unique set of 3,000 enzymes, each defined by the cell's DNA.
- **gene:** a unit of hereditary information found within a chromosome that determines the characteristics of an organism. Each gene is an ordered series of nucleotides, which are subunits of DNA, composed of a base molecule containing nitrogen, a phosphate molecule, and a pentose sugar molecule. See also chromosome, DNA.
- **RNA (ribonucleic acid):** a complex molecule similar to, but less complex than, DNA. RNA molecules transfer genetic information from genes in longer DNA molecules forming the chromosomes of living cells to the active metabolic proteins in a living cell. See also DNA.

BASIC PRINCIPLES

CRISPR. CRISPR stands for "Clustered Regularly Interspaced Short Palindromic Repeats." These are segments of prokaryotic DNA with short base sequences that are repeated over and over. Each repetition of the DNA is followed by spacer DNA, short segments of DNA that are left over from exposure to foreign DNA.

Cas-9. Cas-9 is an enzyme that is produced by CRISPR that can bind with DNA and cut it, thereby editing a gene.

How CRISPR and Cas9 Works Together. CRISPR-Cas9 works by causing a mutation in the DNA of a cell. Cas9 acts like a pair of scissors that cuts DNA at a specific chosen location in the cell's DNA so that it can be changed or removed. Guide RNA (gRNA) is used to bind to the DNA and guide Cas9 to the previously selected part of the gene to make sure that Cas9 is cutting the DNA at the correct place. After Cas9 cuts across both strands of DNA, the cell recognizes that its DNA is damaged and begins to repair it.

APPLICATIONS

Gene editing has great potential as a tool for helping people who have conditions that are genetically related, from conditions such as cancer to conditions like high cholesterol. However, gene editing of reproductive cells has caused many people to question it. Any change made in reproductive cells will be passed on from generation to generation, causing people to wonder about the long-term implications and consequences.

Marianne Moss Madsen, MS

FURTHER READING

Doudna, Jennifer A. and Samuel H. Sternberg. *A Crack in Creation: Gene Editing and the Unthinkable Power to Control Evolution.* Houghton Mifflin Harcourt, 2017. Written by one of the scientists who discovered this earthshaking technology; focuses on whether or not to actually use this method to change our DNA and the promises and perils of this gene-editing tool.

Enriquez, Juan and Steve Gullans. *Evolving Ourselves: Redesigning the Future of Humanity—One Gene at a Time.* Current, 2016. Discusses the rapidly changing field of altering human evolution; shows the inner workings of innovative molecular biology and how this will affect who humans become in the future.

Kozubek, James. *Modern Prometheus: Editing the Human Genome with Crispr-Cas9.* Cambridge University Press, 2016. Discusses the potential for gene editing, including ethical and legal implications; tells the story across a 50-year timeline, including stories of the scientists involved in the process.

Lipkin, Steven Monroe and John Luoma. *The Age of Genomes: Tales from the Front Lines of Genetic Medicine.* Beacon Press, 2016. Focuses on the real-life stories of patients who may be helped by this type of gene editing in an easy-to-read and accessible way.

Saboowala, Hakim. *CRISPR Cas 9: An Enzymatic Scissor for Specific Site Modification of Genome.* Amazon Digital Services, 2016. Discusses how the components of CRISPR-Cas9 can be combined in multiple ways to edit the genome.

WEB SITES

McGovern Institute for Brain Research at MIT, Genome Editing with CRISPR-Cas9, https://www.youtube.com/watch?v=2pp17E4E-O8.

New England Bio Labs, CRISPR/Cas9 and Targeted Genome Editing: A New Era in Molecular Biology, https://www.neb.com/tools-and-resources/feature-articles/crispr-cas9-and-targeted-genome-editing-a-new-era-in-molecular-biology.

Your Genome, Facts, What is CRISPR-Cas9? http://www.yourgenome.org/facts/what-is-crispr-cas9.

FASCINATING FACTS ABOUT THE CRISPR-CAS9

- Some types of bacteria have a built-in gene editing system that is similar to the CRISPR-Cas9 system. They use this system to respond to pathogens like viruses by snipping out a piece of the bacteria and keeping it so that they recognize the virus the next time it attacks.
- Geneticists use gene mutation to study its effects and discover what the function of that particular gene is.
- Scientists have used chemicals or radiation to alter genes, but CRISPR-Cas9 is faster, less expensive, and more reliable than these other methods.
- Currently, gene editing of reproductive cells is illegal in many countries.

CRYOGENICS

FIELDS OF STUDY

Astrophysics; cryogenic engineering; cryogenic electronics; nuclear physics; cryosurgery; cryobiology; high-energy physics; mechanical engineering; chemical engineering; electrical engineering; cryotronics; materials science; biotechnology; medical engineering; astronomy.

SUMMARY

Cryogenics is the branch of physics concerned with creation of extremely low temperatures and involves the observation and interpretation of natural phenomena resulting from subjecting various substances to those temperatures. At temperatures near absolute zero, the electric, magnetic, and thermal properties of most substances are greatly altered, allowing useful industrial, automotive, engineering, and medical applications.

PRINCIPAL TERMS

- **absolute zero:** temperature measured 0 Kelvin (−273 degrees Celsius), where molecules and atoms oscillate at the lowest rate possible.
- **cryocooler:** device that uses cycling gases to produce temperatures necessary for cryogenic work.
- **cryogenic processing:** deep cooling of matter using cryogenic temperatures so the molecules and atoms of the matter slow or almost stop movement.
- **cryogenic tempering:** onetime process using sensitive computerization to cool metal to cryogenic temperatures then tempering the metal to enhance performance, strength, and durability.
- **cryopreservation:** cooling cells or tissues to subzero temperatures to preserve for future use.
- **evaporative cooling:** process that allows heat in a liquid to change surface particles from a liquid to a gas.
- **heat conduction:** technique where a substance is cooled by passing heat from matter of higher temperature to matter of lower temperature.
- **Joule-Thomson effect:** technique where a substance is cooled by rapid expansion, which drops the temperature. So named for its discoverers, British physicists James Prescott Joule and William Thomson (Lord Kelvin).
- **Kelvin temperature scale (K):** used to study extremely cold temperatures. On the Kelvin scale, water freezes at 273 K and boils at 373 K.
- **superconducting device:** device known for its electrical properties and magnetic fields, such as magnetic resonance imaging (MRI) in medicine.
- **superconducting magnet:** electromagnet with a superconducting coil where a magnetic field is maintained without any continuing power source.
- **superconductivity:** absence of electrical resistance in metals, ceramics, and compounds when cooled to extremely low temperatures.
- **superfluidity:** phase of matter, such as liquid helium, that is absent of viscosity and flows freely without friction at very low temperatures.

Basic Principles

Cryogenics comes from two Greek words: *kryo*, meaning "frost," and *genic*, "to produce." This science studies the implications of producing extremely cold temperatures and how these temperatures affect substances such as gases and metal. Cryogenic temperature levels are not found naturally on Earth. The usefulness of cryogenics is based on scientific principles. The three basic states of matter are gas, liquid, and solid. Matter moves from one state to another by the addition or subtraction of heat (energy). The molecules or atoms in matter move or vibrate at different rates depending on the level of heat. Extremely low temperatures, as achieved through cryogenics, slow the vibration of atoms and can change the state of matter. For example, cryogenic temperatures are used in the liquefaction of atmospheric gases such as oxygen, nitrogen, hydrogen, and methane for diverse industrial, engineering, automotive, and medical applications.

Sometimes cryogenics and cryonics are mistakenly linked, but use of subzero temperatures is the only thing these practices share. Cryonics is the practice of freezing a body right after death to preserve it for a future time when a cure for fatal illness or remedy for fatal injury may be available. The practice of cryonics is based on the belief that technology from cryobiology can be applied to cryonics. If cells, tissues, and organs can be preserved by cryogenic temperatures, then perhaps whole body can be preserved for future thawing and life restoration. Facilities exist for interested persons or families, although the cryonic process is not considered reversible as of this writing.

Background

The history of cryogenics follows the evolution of low-temperature techniques and technology. Principles of cryogenics can be traced to 2500 BCE, when Egyptians and Indians evaporated water through porous earthen containers to produce cooling. The Chinese, Romans, and Greeks collected ice and snow from the mountains and stored it in cellars to preserve food. In the early 1800's, American inventor Jacob Perkins created a sulfuric-ether ice machine, a precursor to the refrigerator. By the mid-1800's, William Thomson, a British physicist known as Lord Kelvin, theorized that extremely cold temperatures could stop the motion of atoms and molecules. This became known as absolute zero, and the Kelvin scale of temperature measurement emerged.

Scientists of the time focused on liquefaction of permanent gases. By 1845, the work of British

Cryogenics

physicist Michael Faraday accomplished liquefaction of permanent gases by cooling immersion baths of ether and dry ice followed by pressurization. Six permanent gases— oxygen, hydrogen, nitrogen, methane, nitric oxide, and carbon monoxide—still resisted liquefaction. In 1877, French physicist Louis-Paul Cailletet and Swiss physicist Raoul Pictet produced drops of liquid oxygen, working separately and using completely different methods. In 1883, S. F. von Wroblewski at the University of Krakow in Poland, discovered oxygen would liquefy at 90 Kelvin (K) and nitrogen at 77 K. In 1898, Scottish chemist James Dewar discovered the boiling point of hydrogen at 20 K and its freezing point at 14 K. Helium, with the lowest boiling point of all known substances, was liquefied in 1908 by Dutch physicist Heike Kamerlingh Onnes at the University of Leiden. Onnes was the first person to use the word "cryogenics." In 1892, Scottish physicist James Dewar invented the Dewar flask, a vacuum flask designed to maintain temperatures necessary for liquefying gases, which was the precursor to the Thermos. The liquefaction of gases had many important commercial applications, and many industries use Dewar's concept in applying cryogenics to their processes and products.

The usefulness of cryogenics continued to evolve, and by 1934 the concept was well established. During World War II, scientists discovered that metals became resistant to wear when frozen. In the 1950's, the Dewar flask was improved with the multilayer insulation (MLI) technique for insulating cryogenic propellants used in rockets. Over the next thirty years, Dewar's concept led to the development of small cryocoolers, useful to the military in national defense. The National Aeronautics and Space Administration (NASA) space program applies cryogenics to its programs. Cryogenics can be used to preserve food for long periods—this is especially helpful during natural disasters. Cryogenics continues to grow globally and serve a wide variety of industries.

Cryogenics is an ever-expanding science. The basic principle of cryogenics that the creation of extremely low temperatures will affect the properties of matter so the changed matter can be used for a number of applications. Four techniques can create the conditions necessary for cryogenics: heat conduction, evaporative cooling, rapid-expansion cooling (Joule-Thomson effect), and adiabatic demagnetization.

CREATING LOW TEMPERATURES

With heat conduction, heat flows from matter of higher temperature to matter of lower temperature in what amounts, basically, to a transfer of thermal energy. As the process is repeated, the matter cools. This principle is used in cryogenics by allowing substances to be immersed in liquids with cryogenic temperatures or in an environment such as a cryogenic refrigerator for cooling. Evaporative cooling is another technique employed in cryogenics. Evaporative cooling is demonstrated in the human body when heat is lost through liquid (perspiration) to cool the body via the skin. Perspiration absorbs heat from the body, which evaporates after it is expelled. In the early 1920's in Arizona during the summers, people hung wet sheets inside screened sleeping porches. Electric fans pulled air through the sheets to cool the sleeping space. In the same way, a container of liquid can evaporate, so the heat is removed as gas; the repetitive process drops the temperature of the liquid. An example is reducing the temperature of liquid nitrogen to its freezing point.

The Joule-Thomson effect occurs without the transfer of heat. Temperature is affected by the relationship between volume, mass, pressure, and temperature. Rapid expansion of a gas from high to low pressure results in a temperature drop. This principle was employed by Dutch physicist Heike Kamerlingh Onnes to liquefy helium in 1908 and is useful in home refrigerators and air conditioners.

Adiabatic demagnetization uses paramagnetic salts to absorb energy from liquid, resulting in a temperature drop. The principle in adiabatic demagnetization is the removal of the isothermal magnetized field from matter to lower the temperature. This principle is useful in application to refrigeration systems, which may include a superconducting magnet.

Cryogenic Refrigeration

Cryogenic refrigeration, used by the military, laboratories, and commercial businesses, employs gases such as helium (valued for its low boiling point), nitrogen, and hydrogen to cool equipment and related components at temperatures lower than 150 K. The selected gas is cooled through pressurization to liquid or solid forms (dry ice used in the food industry is solidified carbon dioxide). The cold liquid may be stored in insulated containers until used in a cold station to cool equipment in an immersion bath or with sprayer.

Cryogenic Processing and Tempering.

Cryogenic processing or treatment increases the length of wear of many metals and some plastics using a deepfreezing process. Metal objects are introduced to cooled liquid gases such as liquid nitrogen. The computer-controlled process takes about seventy-two hours to affect the molecular structure of the metal. The next step is cryogenic or heat tempering to improve the strength and durability of the metal object. There are about forty companies in the United States that provide cryogenic processing.

Applications

Early applications of cryogenics targeted the need to liquefy gases. The success of this process in the late 1800's paved the way for more study and research to apply cryogenics to developing life needs and products. Examples include applications in the auto and health care industries and in development of rocket fuels and methods of food preservation. Cryogenic engineering has applications related to commercial, industrial, aerospace, medical, domestic, and defense ventures.

Superconductivity Applications. One property of cryogenics is superconductivity. This occurs when the temperature is dropped so low that the electrical current experiences no resistance. An example is electrical appliances, such as toasters, televisions, radios, or ovens, where energy is wasted trying to overcome electrical resistance. Another is with magnetic resonance imaging (MRI), which uses a powerful magnetic field generated by electromagnets to diagnosis certain medical conditions. High magnetic field strength occurs with superconducting magnets. Liquid helium, which becomes a free-flowing superfluid, cools the superconducting coils; liquid nitrogen cools the superconducting compounds, making cryogenics an integral part of this process. Another application is the use of liquefied gases that are sprayed on buried electrical cables to minimize wasted power and energy and to maintain cool cables with decreased electrical resistance.

Health Care Applications. The health care industry recognizes the value of cryogenics. Medical applications using cryogenics include preservation of cells or tissues, blood products, semen, corneas, embryos, vaccines, and skin for grafting. Cryotubes with liquid nitrogen are useful in storing strains of bacteria at low temperatures. Chemical reactions needed to release active ingredients in statin drugs, used for cholesterol control, must be completed at very low temperatures (–100 degrees Celsius). Highresolution imaging, like MRI, depends on cryogenic principles for the diagnosis of disease and medical conditions. Dermatologists uses cryotherapy to treat warts or skin lesions.

Food and Beverage Applications. The food industry uses cryogenic gases to preserve and transport mass amounts of food without spoilage. This is also useful in supplying food to war zones or natural disaster areas. Deep-frozen food retains color, taste, and nutrient content while increasing shelf life. Certain fruits and vegetables can be deep frozen for consumption out of season. Freeze-dried foods and beverages, such as coffee, soups, and military rations, can be safely stored for long periods without spoilage. Restaurants and bars use liquid gases to store beverages while maintaining the taste and look of the drink.

Automotive Applications. The automotive industry employs cryogenics in diverse ways. One is through the use of thermal contraction. Because materials will contract when cooled, the valve seals of automobiles are treated with liquid nitrogen, which shrinks to allow insertion and then expands as it warms up, resulting in a tight fit. The automotive industry also uses cryogenics to increase strength and minimize wear of metal engine parts, pistons, cranks, rods, spark plugs, gears, axles, brake rotors and pads, valves, rings, rockers, and clutches. Cryogenic-treated spark plugs can increase an automobile's horsepower as well as its gasoline mileage. The use of cryogenics allows a race car to race as many as thirty times without a major rebuild on the motor compared with racing twice on an untreated car.

Aerospace Industry Applications. NASA's space program utilizes cryogenic liquids to propel rockets. Rockets carry liquid hydrogen for fuel and liquid oxygen for combustion. Cryogenic hydrogen fuel is what enables NASA's workhorse space shuttle to get into orbit. Another application is using liquid helium to cool the infrared telescopes on rockets.

Tools, Equipment, and Instrument Applications. Metal tools can be treated with cryogenic applications that provide wear resistance. In surgery or dentistry, tools can be expensive, and cryogenic treatment can prolong usage. Sports equipment, such as golf clubs, benefits from cryogenics as it provides increased wear resistance and better performance. Another is the ability of a scuba diver to stay submerged for hours with an insulated Dewar flask of cryogenically cooled nitrogen and oxygen. Some claim musical instruments receive benefits from cryogenic treatment; in brass instruments, a crisper and cleaner sound is allegedly produced with cryogenic enhancement.

Other Applications. Other applications are evolving as industries recognize the benefits of cryogenics to their products and programs. The military have used cryogenics in various ways, including infrared tracking systems, unmanned vehicles, and missile warning receivers. Companies can immerse discarded recyclables in liquid nitrogen to make them brittle, then these recyclables can be pulverized or grinded down to a more eco-friendly form. No doubt with continued research, many more applications will emerge.

Future Prospects

The economic and ecological impact of cryogenic research and applications holds global promise for the future. In 2009, Netherlands firm Stirling Cryogenics built a cooling system with liquid argon for the ICARUS project, which is being carried out by Italy's National Institute of Nuclear Physics. In China, the Cryogenic and Refrigeration Engineering Research Centre (CRERC) focuses on new innovations and technology in cryogenic engineering. Both private industry and government agencies in the United States are pursuing innovative ways to utilize existing applications and define future implications of cryogenics. Although cryogenics has proved useful to many industries, its full potential as a science has not yet been realized.

Marylane Wade Koch, MSN., R.N.

Further Reading

Hayes, Allyson E., ed. *Cryogenics: Theory, Processes and Applications.* Hauppauge, N.Y.: Nova Science Publishers, 2010. Details global research on cryogenics and applications such as genetic engineering and cryopreservation.

Jha, A. R. *Cryogenic Technology and Applications.* Burlington, Mass.: Elsevier, 2006. Deals with most aspects of cryogenics and cryogenic engineering, including historical development and various laws, such as heat transfer, that make cryogenics possible.

Schwadron, Terry. "Hot Sounds From a Cold Trumpet? Cryogenic Theory Falls Flat." *New York Times*, November 18, 2003. Explains how two Tufts University researchers studied cryogenic freezing of trumpets and determined the cold did not improve the sound.

Ventura, Gugliemo, and Lara Risegari. *The Art of Cryogenics: Low-Temperature Experimental Techniques.* Burlington, Mass.: Elsevier, 2008. Comprehensive discussion of various aspects of cryogenics from heat transfer and thermal isolation to cryoliquids and instrumentation for cryogenics, such as the use of magnets.

Web Sites

Cryogenic Society of America http://www.cryogenicsociety.org

Help Mary Save Coral http://www.helpmarysavecoral.org/obe

National Aeronautics and Space Administration Cryogenic Fluid Management http://www.nasa.gov/centers/ames/research/technology-onepagers/cryogenic-fluid-management.html

National Institute of Standards and Technology Cryogenic Technologies Project http://www.nist.gov/mml/properties/cryogenics/index.cfm

---

FASCINATING FACTS ABOUT CRYOGENICS

- American businessman Clarence Birdseye revolutionized the food industry when he discovered that deep-frozen food tasted better than regular frozen food. In 1923, he developed the flash-freezing method of preserving food at below-freezing temperatures under pressure. The "Father of Frozen Food" first sold small-packaged foods to the public in 1930 under the name Birds Eye Frosted Foods.

- In cryosurgery, super-freezing temperatures as low as −200 Celsius are introduced through a probe of circulating liquid nitrogen to treat malignant tumors, destroy damaged brain tissue in Parkinson's patients, control pain, halt bleeding, and repair detached retinas.

- Cryogenics can be used to save endangered species from extinction. Smithsonian researcher Mary Hagedorn is using cryogenics to establish the first coral seed banks: She's collecting thousands of sample species and freezing them for the future. Hagedorn refers to this as an insurance policy for natural resources.

- The Joule-Thomson effect, discovered in 1852 by James Prescott Joule and William Thomson (Lord Kelvin), is responsible for the cooling used in home refrigerators and air conditioners.

- Helium's boiling point, 173 Kelvin, is the lowest of all known substances.

- Surgical tools and implants used by surgeons and dentists have increased strength and resistance to wear because of cryogenic processing.

- Cryogenic processing is 100 percent environmentally friendly with no use of harmful chemicals and no waste products.

- In 1988, microbiologist Curt Jones, who studied freezing techniques to preserve bacteria and enzymes for commercial use, created Dippin' Dots, a popular ice cream treat, using a quickfreeze process with liquid nitrogen.

D

DEATH AND DYING

FIELDS OF STUDY

Cell biology, ethology, genetics, marine biology, population biology, physiology, wildlife ecology, zoology

SUMMARY

Animals die from old age, disease, their encounters with predators, starvation, human use, human-made disasters, and pollution created by industrial and agricultural chemicals. Some nonhuman animals appear to have a concept of dying.

PRINCIPAL TERMS

- **cetaceans:** plant-eating marine mammals, such as whales, dolphins, and porpoises marine mammals: part of the class of mammals that adapted to life in the sea
- **myocarditus:** inflammation of the heart muscle
- **persistent organic pollutants (POPs):** chemicals that remain in the environment for a very long time and can be found at long distances from where they are used or released; they are nearly all of human origin
- **pinnipeds:** flipper-footed marine mammals, such as sea lions, fur seals, true seals, walruses
- **sirenians:** plant-eating dugongs and manatees

BASIC PRINCIPLES

The life span of all species in the animal kingdom depends upon genetic composition, environmental conditions, and the amount of energy expended throughout their lifetime. The natural life span varies from one species to another. Insects generally have the shortest lives; the adult mayfly lives only a few hours, the fruit fly lives from thirty to forty days. At the other extreme, a giant tortoise may live up to 177 years and the quahog clam can live up to 220 years. Human life expectancy has increased substantially since the beginning of the twentieth century. In 1998, it ranged from seventy-five to eighty years in the United States, Canada, Western Europe and Australia, to fifty-five years in most African countries.

Environmental conditions affecting the genetically determined life span and hastening death include the number and ferocity of predators, viral, bacterial and fungal disease, poisons and pollutants, changing climate, and the rise of carbon dioxide in the air. Humans have contributed to the annihilation of many species and placed others close to extinction by either deliberately or accidentally destroying animal habitats and by overhunting wildlife. In 1973, the United States Congress passed the Endangered Species Act, to protect endangered animals and their habitats. The Environmental Protection Agency (EPA) monitors the fate of endangered species. If pesticide use adversely affects the habitat of an endangered species, the EPA can prohibit it.

How It Works

Pollutants. Maritime oil spills, which kill huge numbers of marine life and birds, affect wildlife species differently. Birds especially are sensitive internally and externally to the effects of crude oil and its refined products. If they become coated with oil and their feathers collapse and mat, the insulating properties of their feathers and down change, making them vulnerable to hypothermia. They become vulnerable to predators and can suffer from dehydration, drowning, and starvation.

Cetaceans, sirenians, and pinnipeds, who depend on air and have amphibious habits, are all susceptible to the effects of oil spills. Like birds, they can suffer from hypothermia. Due to ingesting the oil during grooming and feeding, they suffer from organ dysfunction, congested airways, damaged lungs, gastrointestinal ulceration and hemorrhaging, and eye and skin lesions. Sea turtles are particularly vulnerable during their breeding season, as their nesting sites are on beaches and their eggs may become contaminated by the oil. Newly hatched turtles would have to move over the oiled beach to the water. Among the most deadly oil spills was the wreck in Alaska of the supertanker *Exxon Valdez*, in March 1989, which resulted in more than thirty thousand sea birds dying.

Persistent organic pollutants (POPs) have been related to many behavioral problems in birds, marine mammals, and fish, as well as in humans. Studies of humans exposed through food to POPs show a possible relationship to disruptions of the immune system. This finding has been used to explain why more seals and whales are dying and getting stranded. High levels of cancers in fish have been attributed to another class of potential POPs. Environment Canada reported that when POP levels were reduced, population declines in some birds reversed.

Viral and Bacterial Disease. Nonhuman animals, like humans, are vulnerable to viruses. Livestock contract highly contagious and serious diseases. Among the more commonly known are foot and mouth disease, which affects hoofed animals; scrapie, which affects sheep; and bovine spongiform encephalopathy (BSE), known also as mad cow disease, which occurs in cattle. BSE appears to jump species; humans who contract BSE can develop Creutzfeldt-Jacob disease, a fatal brain disorder. Pigs contract swine fever. Capripox occurs in sheep. In Africa, Rift Valley fever kills livestock and humans. Poultry can contract avian influenza and Newcastle disease, both of which spread rapidly, killing more than 90 percent of infected birds. Because rabies is fatal in animals and humans, many countries require a quarantine period for animals entering the country. In wildlife, trapping has been used to prevent the spread of rabies. However, usually the healthier animals are caught in traps and not the sick, who are less active, are symptomatic, and debilitated, and who are more likely to deviate from their normal behavior.

Organized Animal Fighting, Animal Farming, and Sport Hunting. Humans are voracious predators of other species, killing nonhuman animals for food, clothing, sport, scientific experimentation, and financial gain. Many animals meet their death through organized fighting: bull fighting in Spain and Mexico, and cock fighting in the United States and several Asian cultures, which usually results in the death of one or both roosters. Dog fighting, although banned in most of the United States, is still held clandestinely.

Animal agriculture is the largest food industry in the United States. Animals reared for slaughter are frequently housed in crowded conditions in large buildings, which are ideal for disease. Under natural conditions, chickens can live for as long as fifteen to twenty years. In a modern egg factory hens live about a year and a half. Each year in the United States, about two-thirds of the eighty million pigs raised for slaughter live their lives in a confinement system, as do about half of the ten million milking cows and heifers raised. When birds are debeaked and calves and pigs weaned prematurely, they can die from the shock. Slaughtering is sometimes undertaken without safeguards in place to prevent unnecessary pain.

About 7 percent of the United States' population legally hunt animals. Sport hunting of polar bears in areas of Canada eventually led to such a substantial loss of bears that the local government banned hunting in 2002. In 2001, the government of British Columbia placed a moratorium on the hunting of grizzly bears. The whale population initially decreased because of hunting. The blue whale, which once numbered 200,000, was estimated to be 10,000 in 2001. Marine mammals also become accidentally entangled in fishing nets and collide with boats. Dolphins died at a considerable rate due to tuna fishing methods until U.S. legislation prohibited the method and the number caught in nets was reduced dramatically. Manatees, who move slowly and sometimes sleep near the surface of the water, are particularly vulnerable to being fatally hit by motor boats.

Scientific Experimentation. Using animals in scientific experiments has been widely sanctioned throughout the world for testing consumer products, disease prevention and/or progression techniques, the effects of noxious agents, and psychological theories of behavior. An animal rights movement developed in the late 1970's and early 1980's to protest this use of animals, who not only died during and following the experimental procedures but were also subjected to extreme pain and injury. Industrial manufacturers and scientists were urged to find alternate methods of safety testing and conducting experiments. The Johns Hopkins Center for Alternatives to Animal Testing was founded in 1981, while In Defense of Animals (http://www.idausa.org) grew out of challenges to the University of California's research. It grew into one of the foremost animal advocacy organizations in the United States. The ethical question raised by animal rights groups is whether nonhuman animals should be treated as independent sentient beings and not as a means to human ends.

Beyond the ethical issues raised by philosophers, such as Tom Regan and Peter Singer, are the questions concerning the emotional life of animals and whether animals experience grief and have a concept of death. Marc Hauser, an animalcognition researcher, maintains that animals, lacking a capacity for empathy, sympathy, shame, guilt, and loyalty, are without self-awareness or an awareness of what another of their species experiences, and therefore are incapable of having a deep understanding of death.

Researcher Cynthia Moss, at the Amboseli Elephant Research Project in southern Kenya, takes a different view. From her field observations, she maintains that elephants have a concept of death. They recognize one of their own carcasses or skeletons, always react to the body of a dead elephant, and have been seen putting dirt on a dead elephant's body and covering it with branches and palm fronds. Healthy elephant mothers whose young calves have died look lethargic for many days afterward, trailing behind their family. Wild animals and birds, as well as animals in captivity and animal pets, have been seen reacting to the loss of a mate or companion that can be interpreted as mourning behavior and grief.

—*Susan E. Hamilton*

See also: Aging; Birth; Cannibalism; Demographics; Diseases; Emotions; Immune system; Life spans.

FURTHER READING

DeWaal, Frans B. M. *Good Natured: The Origins of Right and Wrong in Humans and Other Animals.* Cambridge, Mass.: Harvard University Press, 1996. De Waal is a zoologist and ethnologist. His provocative book examines morality in animals.

Gould, James L., and Carol Grant Gould. *The Animal Mind.* New York: Scientific American Library, 1994. A fascinating, well-illustrated inquiry into animal intelligence.

Hauser, Marc D. *Wild Minds: What Animals Really Think.* New York: Henry Holt, 2000. An exploration

---

FOOT AND MOUTH DISEASE

Foot and mouth disease (FMD—also known as hoof and mouth disease), an acute viral disease, is one of the most contagious animal diseases, affecting ungulates: cattle, sheep, pigs, and goats, wild and domestic cloven hoofed animals, elephants, hedgehogs and rats. It causes fever, followed by the development of blisters and sores on the feet and in the mouth. Pigs and sheep may suddenly become lame. There is a high mortality rate in young animals due to myocarditis.

The virus thrives in moist conditions, is airborne and endemic in parts of Asia, Africa, the Middle East, and North America. Transmission occurs directly or indirectly from contact with an infected animal, contaminated foodstuffs, or from a human who has attended an infected animal. It may be picked up on contaminated roads by the wheels of vehicles. Until they are disinfected, vehicles and implements from places where infected animals may have been present are sources of infection.

Prevention and control involves protecting disease-free zones with border animal movement control and surveillance, slaughtering infected, recovered, and FMD-susceptible contact animals, disinfecting premises, cars, clothes, and implements, destroying cadavers, litter, and susceptible animal products in the infected area, and introducing quarantine measures. After two initial vaccinations, one month apart, of an inactivated virus vaccine, immunity is provided for six months, depending on the antigenic relationship between the vaccine and outbreak strains.

An outbreak in 2001 in Great Britain was vastly different from a previous outbreak in 1967 due to the speed and geographical scale of the spread of the infection and the species involved. The outbreak necessitated slaughtering thousands of cattle. Despite the strenuous measures employed, the disease spread from Britain to parts of Europe and Ireland. Experts agreed the outbreak was unprecedented internationally.

of the intellectual and emotional lives of animals and how researchers examine animal skills and cognition.

Levine, Herbert M. *Animal Rights*. Austin, Tex.: Steck-Vaughn, 1998. Offers arguments pro and con on many issues concerning animal rights. Mason, Jim, and Peter Singer. *Animal Factories*. New York: Harmony Books, 1990. A stringent and discerning examination of the manufacturing of animals for food and profit.

Moussaieff Masson, Jeffrey, and Susan McCarthy. *When Elephants Weep: The Emotional Lives of Animals*. New York: Delacorte Press, 1995. Full of engaging anecdotes, this scholarly, insightful book is a delight to read.

Regan, Tom. *The Case for Animal Rights*. Berkeley: University of California Press, 1983. Regan's book describes his philosophy, as an animal rights leader, that nonhuman animals have moral rights and that recognition of these rights requires changes in how we treat them.

Singer, Peter. *Writings on an Ethical Life*. New York: HarperCollins, 2000. Includes extracts from this Australian philosopher's scholarly writings on animal liberation.

Suhowatsky, Gary. "The Role of Trapping in Wildlife Disease." http://articles.animal concerns.org/arvoices/archive/trapping_disease.html. Testimony delivered before the New York State Assembly Subcommittee on Wildlife in March, 1977.

DEFENSE MECHANISMS

FIELDS OF STUDY

Biochemistry, ecology, entomology, ethology, invertebrate biology, marine biology

SUMMARY

All organisms represent a potential resource for their predators. Several have evolved ingenious ways to prevent themselves from becoming their predator's next meal.

PRINCIPAL TERMS

- **aposematic coloration:** brightly colored warning coloration that toxic species use to advertise their distastefulness to would-be predators
- **autotomy:** the self-induced release of a body part mimicry: a type of defense in which an organism gains protection from predators by looking like a dangerous or distasteful species
- **predation:** broadly defined, any interaction in which one organism consumes another living organism, including herbivory (predation on plants), parasitism (predation by small organisms), and familiar predation (where one animal kills and eats another animal)
- **secondary metabolite:** a biochemical that is not involved in basic metabolism, often of unique chemical structure and capable of serving a defensive role for the organism
- **sequester:** to store a material derived from elsewhere. In defenses, some predators sequester defensive properties from their prey to defend themselves from their own predators
- **symbiosis:** "living together"; a term that describes the association between two species in which one species typically lives in or on the other species. Parasitism is a common type of symbosis

BASIC PRINCIPLES

All organisms are composed of fixed carbon, biomolecules, and mineral nutrients, and therefore represent energy and nutrient resources for consumers. To be successful in life, animals must avoid, tolerate, or defend themselves against natural enemies such as predators, parasites, and competitors. The term "defense" can be attributed to any trait that reduces the likelihood that an organism, or part of an organism, will be consumed by a predator. There are several categories of defenses that have evolved in animals, including structural defenses, chemical defenses, associational defenses, behavioral defenses, autotomy, and nutritional defenses. Animals often possess more than one type of defense, thereby having backup plans in case the first line of defense fails. The number of defenses devised by organisms is a reflection of the strong selective pressure exerted by predators.

How It Works

Structural Defenses. Structures that defend animals can act as external shields: sharp spines located externally or internally, skeletal materials that make tissues too hard to bite easily, or weaponry such as horns, teeth, and claws. External structures that protect vulnerable soft tissues include the chitonous exoskeleton of crustaceans, the calcareous shells of corals, mollusks, and barnacles, the tests (skeletal plates) of echinoderms, the tough tunic of ascidians, and the hard plates of armadillos. The pretty shells that tourists collect along beaches were once used to protect a soft, delicate animal that lived inside the shell. Hard, protective shells remain after the animal dies and can be used by other animals for protection. For example, small fishes will retreat into empty conch shells when they feel threatened by predators, and hermit crabs live inside empty snail shells to protect their soft, vulnerable abdomens. Some animals cover their bodies with sharp structures that puncture predators that try to bite them. The porcupine is a good example of a mammal that uses this defensive strategy. Porcupines are covered with tens of thousands of long, pointed spines, or quills, growing from their back and sides. The quills have needle-sharp ends containing hundreds of barbs that make the quills difficult to remove. Sea urchins are also covered with long, sharp spines that deter would-be predators. Urchins can move their spines, and will direct them toward anything that comes in contact with them, such as a predator. While porcupines and urchins are covered with multiple spines, stingrays defend themselves from enemies by inflicting a wound with a single barbed spine. The wound is extremely painful, giving these rays their common name.

Predators have sharp claws and teeth that help them grasp, subdue, and consume their prey. These same structures, used offensively in hunting, can also be used to protect themselves from their own predators. Small predators such as badgers, raccoons, and foxes can fend off larger predators such as wolves and mountain lions with their weaponry. Rather than risk injury, the larger predators will avoid a fight with the smaller predator and seek a less risky meal, such as a rabbit or mouse.

Chemical Defenses. Both plants and animals defend themselves by using compounds that are distasteful, toxic, or otherwise repulsive to consumers. Most defensive compounds are secondary metabolites of unique structures, but can also include more generic compounds such as sulfuric acid or calcium carbonate. Secondary metabolites get their name because they are not involved in basic metabolic pathways such as respiration or photosynthesis (that is, primary metabolic reactions), not because they are of secondary importance. Indeed, many organisms probably could not survive in their natural environment without the protection of their secondary metabolites.

Stink bugs get their names because of the smelly secondary metabolites they release from pores located on the sides of their thorax. These smelly compounds repel predators, and may even indicate toxicity to the predator. These insects are common garden pests that are usually controlled with chemical pesticides. However, it appears that the eggs of stink bugs are not defended against roly-poly pill bugs, which can control stink bug numbers (and hence, garden damage) by preying on eggs.

Bombardier beetles take chemical defenses a step further, erupting a boiling hot spray of chemicals in the direction of a predator. To accomplish this, the bombardier beetle has a pair of glands that open at the tip of its abdomen. Each gland has two compartments, one that contains a solution of hydroquinone and hydrogen peroxide, and the other that contains a mixture of enzymes. When threatened by a predator, the bombardier beetle squeezes the hydroquinone and hydrogen peroxide mixture into the enzyme compartment, where an exothermic reaction that produces quinone takes place. The large amount of heat generated brings the quinone mixture to its boiling point, and it is forcefully emitted as a vapor toward the threat. An average bombardier beetle can produce about twenty loud discharges of repulsive, hot chemicals in quick succession. Chemical defenses are common among small, slow animals such as insects, sponges, cnidarians, and sea slugs, which might be limited in their ability to flee from predators. However, chemical defenses are rather rare among large, fast animals.

One of the few mammals that uses chemical defenses is the black-and-white-striped skunk. Most people are familiar with the smelly chemical brew emitted from these animals, as it is distinctly detectable along roads when skunks get hit by cars, and can be detected up to a mile from the location where a skunk sprays. These mammals hold their smelly musk in glands located below their tail, and squirt the liquid through ducts that protrude from the anus. When threatened by a predator, the skunk raises its tail and directs its rear end toward the predator. A predator that has had prior experience with a skunk might retreat from this display, but if the predator is persistent at harassing the skunk, the striped mammal will deliver a spray of smelly chemicals that usually sends the predator running. The musk also causes intense pain and temporary blindness if it gets in the eyes of the predator.

Associational Defenses. Associational defenses occur when a species gains protection from a natural enemy by associating with a protective species, such as when humans gain protection from enemies by keeping a guard dog on their property. Types of protection provided to the defended species through this co- evolution can be structural, chemical, or aggressive. Small animals can avoid predators by using a defended species as habitat. For example, small fishes defend themselves by associating with sea urchins, gaining protection by hiding among the sharp spines. Some species of shrimp inhabit the cavities and canals of sponges. Sponges are known to be chemically and structurally defended against most predators, with the exception of angel fishes and parrot fishes. Finally, much of the diverse coral reef fauna seeks protection among the cracks and the crevices in the reef. Reefs, slowly built by coral animals, are the largest structures ever made by living organisms, and serve a protective role for thousands of species that inhabit reefs.

Associational defenses can also be chemically mediated. For example, bacteria that grow symbiotically on shrimp eggs produce secondary metabolites that protect the egg from a parasitic fungus. The numerous examples of sequestration of chemical defenses can be categorized as associational defenses, as they involve associating with chemically defended prey.

An organism might even be defended by protective species that aggressively attack would-be predators, especially if the protected species is a resource for the aggressive defender. For example, humans are protected by guard dogs because dogs view people as a resource that provides them with food, water, and shelter. Stop feeding the dog, and it is likely to look elsewhere for somebody to protect. There are several nonhuman examples of aggressive defensive associations, especially among ants. Aphids are insects that feed on the sugary phloem stream of plants. In the process of feeding and processing phloem, the aphids secrete large amounts of honeydew, which the ants harvest and consume; that is, aphids provide ants with a resource. Ants tend to aphids in the same way that dairy farmers tend to their cows. The ants carry aphids to prime feeding locations, defend aphids from predators, and periodically "milk" the aphids of their honeydew by stroking them with their antennae.

Defensive Behaviors. Being chemically defended does not protect an animal from being accidentally eaten. Therefore, chemically defended animals often advertise the fact that they are nasty to avoid such accidents. This advertisement is often in the form of outlandish colors and patterns that flaunt the animal's distastefulness to predators. Using bright warning patterns is called aposematic coloration. One problem with aposematic coloration is the training of predators: Bright coloration is only useful if the predator understands the warning. Otherwise, the coloration simply makes the animal a conspicuous prey item. An interesting way that different species with aposematic coloration share the cost of training naïve predators is through mimicry. A predator that eats an individual of species A (assume species A is bright red with blue stripes) and vomits shortly thereafter may learn to avoid things that are red with blue stripes, though at the cost of that first individual's life. This educated predator will now avoid other members of species A, and any other organism that looks like species A (the mimic), whether the mimic is toxic or not. If the mimic is toxic, the system is termed Müllerian mimicry. If the mimic is a palatable species that looks like a toxic model, the system is termed Batesian mimicry.

Mimicry is common within groups of closely related organisms (for example, snakes, butterflies, and bees) which are already similar in appearance. However, mimicry can also occur even when the model and mimic are distantly related. For example, there are caterpillars that mimic the head of a snake, moths that mimic the eyes of a cat, and beetles, moths, and flies that mimic stinging bees and wasps.

Autotomy. Sometimes, despite the best defenses, a predator will get hold of a prey. If this happens, some animals are able to sacrifice a portion of their body to the predator, with the hope that the remaining parts will survive, and perhaps even regrow the lost parts. This ability to lose a body part intentionally is called autotomy.

Many lower animals, such as sponges, cnidarians, and worms, have great regeneration abilities, and can regrow body parts well. In fact, these animals can even use regeneration as a form of asexual reproduction: Break the animal into four parts, and the parts will generate four complete individuals. Sea cucumbers, in addition to being chemically defended, are able to eviscerate (autotomy of intestines) when harassed by a predator. These are not fast animals, so this action does not allow them to escape, but it might satisfy (or disgust) the predator enough to make it lose interest in the sea cucumber. Losing a large portion of its digestive tract interferes with feeding, but the sea cucumber can regenerate those parts of the gut that were eviscerated, restoring itself to original function. Sea cucumbers also play an important role in a defensive association with the pearlfish. When the pearlfish feels threatened, it locates the anus of a sea cucumber, then backs into its intestine, where it hides until the danger has passed.

The regenerative ability of higher animals is generally less than that of lower animals. However, autotomy does occur even in some vertebrates. Lizards are well known for their ability to release the tips of their tails when grabbed by a predator. The predator is distracted, and perhaps satisfied, by the wiggling piece of flesh, and in the meantime, the remainder of the lizard scampers off to safety. Geckos release skin instead of tails. The part of the skin that is grabbed by the predator is released, enabling the gecko to break free and escape.

Nutritional Defenses. Some animals, such as corals, jellyfish, anemones, and gorgonians (phylum Cnidaria), possess a type of combined structural and chemical defense in the form of specialized stinging cells called nematocysts. When nematocysts are stimulated, they rapidly discharge a barb that punctures the skin of a predator, often releasing toxic chemicals at the same time. The stinging sensation that people get when they swim into a jellyfish is caused by nematocysts. Some of these jellyfish stings are so potent that they can result in death. Not only do many predators avoid jellyfish because they posses nematocysts, but predators may avoid jellyfish because they are jellylike, being composed of more than 95 percent water.

It takes time and effort for predators to locate, handle, ingest, and digest prey. If the prey item is basically a bag of seawater (as jellyfish are), then predators might not bother eating these nutrientdeficient animals. Thus, these animals are "nutritionally" defended. Nutritional defenses are also used by plants, but they are generally not an avail- able strategy for animals other than jellyfish, as most animal tissue is relatively nutritious.

—*Greg Cronin*

See also: Communication; Death and dying; Nutrient requirements.

Further Reading

Cloudsley-Thompson, John L. *Tooth and Claw: Defensive Strategies in the Animal World*. London: J. M. Dent&Sons, 1980.Areadable volume that covers defense mechanisms in great detail and with many examples.

Edmunds, Malcolm. *Defence in Animals*. Burnt Mill, England: Longman, 1974. Technical, comprehensive guide to antipredator defenses, and the evolutionary arms race between predator and prey. Contains photographs, illustrations, and quantitative results from experiments.

Evans, David L., and Justin O. Schmidt, eds. *Insect Defenses: Adaptive Mechanisms and Strategies of Prey and Predators*. Albany: State University of New York Press, 1990. An edited volume that examines themanyways that the most successful group of organisms on earth deals with predators.

Kaner, Etta. *Animal Defenses: How Animals Protect Themselves*. Toronto: Kids Can Press, 1999.Acolorfully illustrated book aimed at adolescents. It places defenses of animals in context of human behavior.

McClintock, James B., and Bill J. Baker, eds. *Marine Chemical Ecology*. Boca Raton, Fla.: CRC Press, 2001. This technical volume is the most current, comprehensive book on marine chemical ecology. The book provides cellular, physiological, organismal, evolutionary, and applied perspectives creating a high-resolution snapshot of the field at the start of the twenty-first century.

Owen, Denis. *Survival in the Wild: Camouflage and Mimicry*. Chicago: University of Chicago Press, 1980. A look at animals that appear to be something other than what they are. Some try to look like their background to avoid detection, others try to appear like a dangerous animal, while others are brightly colored to advertise nastiness. Easy reading with numerous illustrations and photographs.

SEQUESTRATION

The production of chemical defenses is often assumed to be expensive because it requires resources that might otherwise have been used for growth and reproduction. One way for a species to avoid such cost is to sequester compounds produced by its prey. Sequestration was first discovered in the monarch butterfly. Monarch caterpillars feed on milkweeds and sequester the plant's cardenolides. Milkweeds produce toxic cardenolides that deter most vertebrate herbivores, but monarchs have evolved the ability to tolerate and sequester these compounds. When blue jays were fed monarchs, the birds soon regurgitated and learned to associate this unpleasant response with eating monarchs. To help advertise their chemical nastiness and perhaps to increase the learning response of birds, monarchs evolved aposematic coloration. Other species of butterflies, such as the viceroy and the queen butterflies, mimic monarchs, affording them protection from predators that have already learned to avoid orange and black butterflies.

Nudibranchs are a class of sea slugs that lack the protective shell that most of their gastropod relatives possess. Because they lack the physical defense of a shell, they must protect themselves from other mechanisms, and have ingenious ways of doing so. Most nudibranchs do not produce their own defenses. Rather, they sequester the defenses of their prey. For example, aeolid nudibranchs are famous for sequestering functional nematocysts from their cnidarian prey. They transfer nondischarged nematocysts from their gut to their skin, thereby protecting themselves from predators such as fish. Nudibranchs also sequester secondary metabolites from their invertebrate prey, which effectively protect them from predatory fishes.

ACACIA-ANT MUTUALISM

Acacia ants are found throughout Central America and only inhabit the hollow thorns of acacia trees. The ants harvest and feed on nectar from extrafloral nectaries and special leaf tips, called Beltian bodies, produced by the acacia trees. In a sense, these ants are herbivores of acacia trees. However, in return for the shelter and food provided by the tree, the ants defend the tree from other herbivores and competitors.

When an herbivore begins feeding on the acacia tree, the ants release an alarm order that signals the colony to attack the herbivore with painful bites and stings. The busy ants also remove competing vegetation that comes in contact with the tree or that grows near the trunk of the tree. The weeding activity of the ants results in a circle of cleared ground surrounding the acacia tree that also protects the tree from the damaging effects of fire. This mutualistic relation is obligate, meaning that each species is absolutely dependent on the other: The tree cannot survive without the ants and vice versa. Other species take advantage of the acacia-ant mutualism: Some birds preferentially nest in acacia trees, presumedly because the ants deter egg predators as well as herbivores. How the birds avoid attack by the ants is not well understood.

Demographics

FIELDS OF STUDY

Environmental science, population biology, zoology

SUMMARY

Demography is the study of the numbers of organisms born in a population within a certain time period, the rate at which they survive to various ages, and the number of offspring that they produce. Many different patterns of birth, survival, and reproduction are found among organisms in nature.

PRINCIPAL TERMS

- **cohort:** a group of organisms of the same species, and usually of the same population, that are born at about the same time fecundity: the number of offspring produced by an individual
- **life table:** a chart that summarizes the survivorship and reproduction of a cohort throughout its life span
- **mortality rate:** the number of organisms in a population that die during a given time interval
- **natality rate:** the number of individuals that are born into a population during a given time interval
- **population:** a group of individuals of the same species that live in the same location at the same time
- **survivorship:** the pattern of survival exhibited by a cohort throughout its life span

BASIC PRINCIPLES

No animal lives forever. Instead, each individual has a generalized life history that begins with fertilization and then goes through embryonic development, a juvenile stage, a period in which it produces offspring, and finally death. There are many variations on this general theme. Still, the life of each organism has two constants: a beginning and an end. Many biologists are fascinated by the births and deaths of individuals in a population and seek to understand the processes that govern the production of new individuals and the deaths of those already present. The branch of biology that deals with such phenomena is called demography.

The word "demography" is derived from Greek; *demos* means "population." For many centuries, demography was applied almost exclusively to humans as a way of keeping written records of new births, marriages, deaths, and other socially relevant information. During the first half of the twentieth century, biologists gradually began to census populations of naturally occurring organisms to understand their ecology more fully. Biologists initially focused on vertebrate animals, particularly game animals and fish. Beginning in the 1960's and 1970's, invertebrate animals, plants, and microbes also became subjects of demographic studies. Studies clearly show that different species of organisms vary greatly in their demographic properties. Often, there is a clear relationship between those demographic properties and the habitat in which these organisms live.

How It Works

Demographic Parameters. When conducting demographic studies, a demographer must gather certain types of basic information about the population. The first is the number of new organisms that appear in a given amount of time. There are two ways that an organism can enter a population: by being born into it or by immigrating from elsewhere. Demographers generally ignore immigration and concentrate instead on newborns. The number of new individuals born into a population during a specific time interval is termed the natality rate. The natality rate is often based on the number of individuals already in the population. For example, if ten newborns enter a population of a thousand individuals during a given time period, the natality rate is 0.010. A specific time interval must be expressed (days, months, years) for the natality rate to have any meaning.

A second demographic parameter is the mortality rate, which is simply the rate at which individuals are lost from the population by death. Losses that result from emigration to a different population are ignored by most demographers. Like the natality rate, the mortality rate is based on the number of individuals in the population, and it reflects losses during a certain time period. If calculated properly, the natality and mortality rates are directly comparable, and one can subtract the latter from the former to provide an index of the change in population size over time. The population increases whenever natality exceeds mortality and decreases when the reverse is true.

The absolute value of the difference denotes the rate of population growth or decline. When studying mortality, demographers determine the age at which organisms die. Theoretically, each species has a natural life span that no individuals can surpass, even under the most ideal conditions. Normally, however, few organisms reach their natural life span, because conditions are far from ideal in nature. Juveniles, young adults, and old adults can all die. When trying to understand the dynamics of a population, it makes a large difference whether the individuals are dying mainly as adults or mainly as juveniles.

Patterns of Survival. Looking at it another way, demographers want to know the pattern of survival for a given population. This can best be determined by identifying a cohort, which is defined as a group of individuals that are born at about the same time. That cohort is then followed over time, and the number of survivors is counted at set time intervals. The census stops after the last member of the cohort dies. The pattern of survival exhibited by the whole cohort is called its survivorship. Ecologists have examined the survivorship patterns of a wide array of species, including vertebrate animals, invertebrates, plants, fungi, algae, and even microscopic organisms. They have also investigated organisms from a variety of habitats, including oceans, deserts, rain forests, mountain peaks, meadows, and ponds. Survivorship patterns vary tremendously. Some species have a survivorship pattern in which the young and middle-aged individuals have a high rate of survival, but old individuals die in large numbers. Several species of organisms that live in nature, such as mountain sheep and rotifers (tiny aquatic invertebrates), exhibit this survivorship pattern. At the other extreme, many species exhibit a survivorship pattern in which mortality is heaviest among the young. Those few individuals that are fortunate enough to survive the period of heavy mortality then enjoy a high probability of surviving until the end of their natural life span. Examples of species that have this pattern include marine invertebrates such as sponges and clams, most species of fish, and parasitic worms. An intermediate pattern is also observed, in which the probability of dying stays relatively constant as the cohort gets older. American robins, gray squirrels, and hydras all display this pattern.

These survivorship patterns are usually depicted on a graph that has the age of individuals in the cohort on the x axis and the number of survivors on the y axis. Each of the three survivorship patterns gives a different curve when the number of survivors is plotted as a function of age. In the first pattern (high survival among juveniles), the curve is horizontal at first but then swings downward at the right of the graph. In the second pattern (low survival among juveniles), the curve drops at the left of the graph but then levels out to form a horizontal line. That curve resembles a backward letter J. The third survivorship pattern (constant mortality throughout the life of the cohort) gives a straight line that runs from the upper-left corner of the graph to the lower right (this is best seen when the y axis is expressed as the logarithm of the number of survivors). In the first half of the twentieth century, demographers Raymond Pearl and Edward S. Deevey labeled each survivorship pattern: Type I is high survival among juveniles, type II is constant mortality through the life of the cohort, and type III is low survival among juveniles. That terminology became well entrenched in the biological literature by the 1950's. Few species exhibit a pure type I, II, or III pattern, however; instead, survivorship varies so that the pattern may be one type at one part of the cohort's existence and another type later on. Perhaps the most common survivorship pattern, especially among vertebrates, is composed of a type III pattern for juveniles and young adults followed by a type I pattern for older adults. This pattern can be explained biologically. Most species tend to suffer heavy juvenile mortality because of predation, starvation, cannibalism, or the inability to cope with a stressful environment. Juveniles that survive this hazardous period then become strong adults that enjoy relatively low mortality. As time passes, the adults reach old age and ultimately fall victim to disease, predation, and organ-system failure, thus causing a second downward plunge in the survivorship curve.

Patterns of Reproduction. Demographers are not interested only in measuring the survivorship of cohorts. They also want to understand the patterns of reproduction, especially among females. Different species show widely varying patterns of reproduction. For example, some species, such as octopuses and certain salmon, reproduce only once in their life and then die soon afterward. Others, such as humans and most birds, reproduce several or many times in their life. Species that reproduce only once accumulate energy throughout their life and essentially put all of it into producing young. Reproduction essentially exhausts them to death. Conversely, those that reproduce several times devote only a small amount of their energy into each reproductive event.

Species also vary in their fecundity, which is the number of offspring that an individual makes when it reproduces. Large mammals have low fecundity, because they produce only one or two progeny at a time. Birds, reptiles, and small mammals have higher fecundity because they typically produce a clutch or litter of several offspring. Fish, frogs, and parasitic worms have very high fecundity, producing hundreds or thousands of offspring. A species' pattern of reproduction is often related to its survivorship. For example, a species with low fecundity or one that reproduces only once tends to have type I or type II survivorship. Conversely, a species that produces huge numbers of offspring generally shows type III survivorship. Many biologists are fascinated by this interrelationship between survivorship and reproduction. Beginning in the 1950's, some demographers proposed mathematically based explanations as to how the interrelationship might have evolved as well as the ecological conditions in which various life histories would be expected. For example, some demographers predicted that species with low fecundity and type I survival should be found in undisturbed, densely populated areas (such as a tropical rain forest). In contrast, species with high fecundity and type III survival should prevail in places that are either uncrowded or highly disturbed (such as an abandoned farm field). Ecologists have conducted field studies of both plants and animals to determine whether the patterns that actually occur in nature fit the theoretical predictions. In some cases the predictions were upheld, but in others they were found to be wrong and had to be modified.

Age Structures and Sex Ratios. Another feature of a population is its age structure, which is simply the number of individuals of each age. Some populations have an age structure characterized by many juveniles and only a few adults. Two situations could account for such a pattern. First, the population could be rapidly expanding, with the adults successfully reproducing many progeny that are enjoying high survival. Second, the population could be producing many offspring that have type III survival. In this second case, the size of the population can remain constant or even decline. Other populations have a different age structure, in which the number of juveniles only slightly exceeds the number of adults. Those populations tend to remain relatively constant over time. Still other populations have an age structure in which there are relatively few juveniles and many adults. Those populations are probably declining or are about to decline because the adults are not successfully reproducing. Since most animals are unisexual, an important demographic characteristic of a population is its sex ratio, defined as the ratio of males to females. While the ratio for birds and mammals tends to be 1:1 at conception (the fertilization of an egg), it tends to be weighted toward males at birth, because female embryos are slightly less viable. After birth, the sex ratio for mammals tends to favor females, because young males suffer higher mortality. The posthatching ratio in birds tends to remain skewed toward males, because females devote considerable energy to producing young and suffer higher mortality. As a result, male birds must compete with one another for the opportunity to mate with the scarcer females.

The Age-Specific Approach. To understand the demography of a particular species, one must collect information about its survivorship and reproduction. The best survivorship data are obtained when a demographer follows a group of newly born organisms (this being a cohort) over time, periodically counting the survivors until the last one dies. Although that sounds relatively straightforward, many factors complicate the collection of survivorship data; demographers must be willing to adjust their methods to fit the particular species and environmental conditions. First, a demographer must decide how many newborns should be included in the cohort. Survivorship is usually based on one thousand newborns, but few studies follow that exact number. Instead, demographers follow a certain number of newborns and multiply or divide their data so that the cohort is expressed as one thousand newborns. For example, one may choose to follow five hundred newborns; the number of survivors is then multiplied by two. Demographers generally consider cohorts composed of fewer than one hundred newborns to be too small. Second, methods of determining survivorship are much more different for highly motile organisms, such as mammals and birds, than for more sedentary ones, such as bivalves (oysters and clams). To determine survivorship of a sedentary species, demographers often find some newborns during an initial visit to a site and then periodically revisit that site to count the number of survivors. Highly motile animals are much more difficult to census because they do not stay in one place waiting to be counted. Vertebrates and large invertebrates can be tagged, and individuals can be followed by subsequently recapturing them. Some biologists use small radio transmitters to follow highly active species. The demography of small invertebrates such as insects is best determined when there is only one generation per year and members of the population are all of the same age-class. For such species, demographers merely count the number present at periodic intervals. Third, the frequency of the census periods varies from species to species. Short-lived species, such as insects, must be censused every week or two. Longer-lived species need be counted only once a year. Fourth, the definition of a "newborn" may be troublesome, especially for species with complex life cycles. Demographic studies usually begin with the birth of an infant. Some would argue, however, that the fetus should be included in the analysis because the starting point is really conception. Manysedentary marine invertebrates (sponges, starfish, and barnacles) have highly motile larval stages, and these should be included in the analysis for survivorship to be completely understood. Parasitic roundworms and flatworms that have numerous juvenile stages, each found inside a different host, are particularly challenging to the demographer.

The Time-Specific Approach. The survivorship of long-lived species, such as large mammals, is really impossible to determine by the methods given above. Because of their sheer longevity, one could not expect a scientist to be willing to wait decades or centuries until the last member of a cohort dies. Demographers attempt to overcome this problem by using the age distribution of organisms that are alive at one time to infer cohort survivorship. This is often termed a "horizontal" or "time-specific" approach, as opposed to the "vertical" or "age-specific" approach that requires repeated observations of a single cohort. For example, one might construct a timespecific survivorship curve for a population of fish by live-trapping a sufficiently large sample, counting the rings on the scales on each individual (which for many species is correlated with the age in years), and then determining the number of one-year-olds, two-year-olds, and so on. Typically, demographers who use age distributions to infer age-specific survivorship automatically assume that natality and mortality remain constant from year to year. That is often not the case, however, because environmental conditions often change over time. Thus, demographers must be cautious when using age distribution data to infer survivorship. Methods for determining fecundity are relatively straightforward. Typically, fertile individuals are collected, their ages are determined, and the number of progeny (eggs or live young) are counted. Species that reproduce continually (parasitic worms) or those that reproduce several times a year (small mammals and many insects) must be observed over a period of time.

Demographers usually want to determine whether the production of new offspring (natality) balances the losses attributable to mortality. To accomplish this, they construct a life table, which is a chart with several columns and rows. Each row represents a different age of the cohort, from birth to death. The columns show the survival and fecundity of the cohort. By recalculating the survivorship and fecundity information, demographers can compute several interesting aspects of the cohort, including the life expectancy of individuals at different ages, the cohort's reproductive value (which is the number of progeny that an individual can expect to produce in the future), the length of a generation for that species, and the growth rate for the population.

Uses of Demography. Demographic techniques have been applied to nonhuman species, particularly by wildlife managers, foresters, and ecologists.Wildlife managers seek to understand how a population is surviving and reproducing within a certain area, and therefore to determine whether it is increasing or decreasing over time.With that information, a wildlife biologist can then estimate the effect of hunting or other management practice on the population. By extension, fisheries biologists can also make use of demographic techniques to determine the growth rate of the species of interest. If the population is determined to be increasing, it can be harvested without fear of depleting the population. Alternatively, one can conduct demographic analyses to see whether certain species are being overfished.

An often unappreciated benefit of survivorship analyses is that they can help ecologists pinpoint factors that limit population growth in an area. This maybe especially important in efforts to prevent rare animals and plants from becoming extinct. Once the factor is identified, the population can be appropriately managed. Increasing amounts of public and private money are allocated each year to biologists who conduct demographic studies on rare species.

—*Kenneth M. Klemow*

See also: Birth; Death and dying; Life spans.

Further Reading

Begon, Michael, John L. Harper, and Colin R. Townsend. *Ecology: Individuals, Populations, and Communities.* 3d ed. Boston: Blackwell Scientific, 1996. A college-level introductory ecology textbook. Covers demography by asking the question "what is an individual?" The discussion of life tables includes many perspectives not found in other texts.

Begon, Michael, Martin Mortimer, and David J. Thompson. *Population Ecology: A Unified Study of Animals and Plants.* 3d ed. Cambridge, Mass.: Blackwell Science, 1996. An up-to-date and readable textbook for those with no background in population ecology. The text is divided into three sections: Single-Species Interactions, Interspecific Interactions, and Synthesis. The first section contains information most relevant to the study of demographics. References.

Brewer, Richard. *The Science of Ecology.* 2d ed. Fort Worth, Tex.: Saunders College Publishing, 1994. This clearly written textbook is aimed at upper-level undergraduates and was written by an author who has experience researching both plants and animals. Contains a succinct, well-presented discussion of life tables and survivorship curves. Some of the evolutionary applications of demography are also treated.

Elseth, Gerald D., and Kandy D. Baumgardner. *Population Biology.* New York: Van Nostrand, 1981. Intended for graduate students and advanced undergraduates, provides rigorous mathematical treatment of population biology. Discussion of demography includes detailed discussions about age structure, the calculation of population growth rates from demographic data, the evolution of demographic traits, and sex ratios.

Gotelli, Nicholas J. *A Primer of Ecology.* 2d ed. Sunderland, Mass.: Sinauer Associates, 1998. Covers the essential mathematical models of ecology. Makes the assumptions and predications of the models explicit, and provides empirical examples of each. Each chapter includes problems and solutions. Written for undergraduate and graduate students.

Hutchinson, G. Evelyn. *An Introduction to Population Ecology.* New Haven, Conn.: Yale University Press, 1978. Presents an engaging account of the study of populations by a very well-respected ecologist. Chapter 2, "Interesting Ways of Thinking About Death," provides the reader with a perspective on survivorship analysis that includes examples from humans, vertebrates, invertebrates, and plants. Also discusses the interrelationships between fecundity and other demographic properties.

Smith, Robert Leo. *Elements of Ecology.* 4th ed. San Francisco, Calif.: Benjamin/Cummings, 2000. This textbook does an excellent job of covering the breadth of topics found in modern ecology. The treatment of the topics is unusually complete and provides a close linkage between theoretical principles and numerous examples from specific ecological situations. Covers population age structure, as well as natality, mortality, and survivorship.

Wilson, Edward O., and William Bossert. *A Primer of Population Biology.* Sunderland, Mass.: Sinauer Associates, 1977. This classic handbook has been used extensively by students wishing to master the basics of population biology. Demography is discussed in a way that emphasizes the estimation of population growth rates from survivorship and reproduction data, as well as the determination of age distributions.

Dentistry

FIELDS OF STUDY

Biology; health science; physiology; anatomy; pharmacology; biochemistry; chemistry; mathematics; microbiology; physics.

SUMMARY

Dentistry involves the diagnosis, treatment, and prevention of disorders and diseases of the teeth, mouth, jaw, and face. Dentistry includes instruction on proper dental care, removal of tooth decay, teeth straightening, cavity filling, and corrective and reconstructive work on teeth and gums. Dentistry is recognized as an important component of overall health. Practitioners of dentistry are called dentists. Dental hygienists, technicians, and assistants aid dentists in the provision of dental care.

PRINCIPAL TERMS

- **appliance:** removable restorative or corrective dental or orthodontic device.
- **bite:** contact of the upper and lower teeth; also known as occlusion.
- **caries:** tooth decay; also known as cavities.
- **cleaning:** removal of plaque and tartar from the teeth, generally above the gum line.
- **enamel:** hard ceramic that covers and protects the exposed part of the tooth.
- **gingiva:** soft, pink tissue surrounding the base of the teeth; also known as gums.
- **permanent teeth:** the thirty-two teeth that appear after the loss of the primary teeth, beginning around the age of six years; also known as adult teeth.
- **plaque:** sticky film of food particles, bacteria, and saliva that forms on the teeth and can eventually turn into tartar.
- **primary teeth:** the first set of teeth that appear between the ages of six months and one year that help children learn to speak and chew; also known as baby teeth or deciduous teeth.
- **pulp:** soft, inner part of the tooth that contains nerves and blood vessels.
- **root:** part of the tooth that is embedded in the gums.
- **tartar:** hard deposit that adheres to teeth and attracts plaque; also known as dental calculus.

BASIC PRINCIPLES

Dentistry is a branch of medicine that focuses on diseases and disorders of the teeth, mouth, face, and oral cavity. Dentistry includes examining the teeth, gums, mouth, head, and neck to evaluate dental health. The examination may include a variety of dental instruments, imaging techniques, and other diagnostic equipment. Dentistry involves diagnosing oral or dental diseases or disorders and formulating treatment plans. Dentistry is instrumental in teaching patients about the importance of maintaining oral health and instructing patients on proper oral hygiene techniques.

Although dentistry is an independent health care field, it is not entirely detached from other health care services and collaboration between dentistry and other health care providers ensures positive outcomes for patients. Dentists often see patients more often than physicians and may be the first to diagnose systemic diseases, including inflammatory conditions, autoimmune diseases, and cardiovascular risk factors. Dentists also work with pharmacists to prescribe the best antibiotics or anesthetics for dental patients, as well as to understand how certain medications affect dental care and oral health.

Dentistry not only prevents and treats serious oral health disorders but also provides cosmetic services to enhance facial features and correct signs of aging. Dentistry strives to promote oral health as a part of overall health and applies principles of basic medicine, pharmacology, and psychology to dental care.

BACKGROUND

Before the seventeenth century, dental care was crude, unrefined, and most often provided by physicians. Through the eighteenth and nineteenth centuries, dentistry emerged as its own medical discipline, and most dentists trained through apprenticeships. Pierre Fauchard is credited as the founder of modern dentistry. In 1728, the French surgeon published *Le Chirurgien Dentiste: Ou, Traité des dents* (*The Surgeon Dentist: Or, Treatise on the Teeth*, 1946), which summarized all available knowledge of dental of dentures.

In 1840, Horace Hayden and Chapin Harris established the world's first dental school, the Baltimore College of Dental Surgery, in Baltimore, Maryland. In 1867, Harvard University became the first university to

Dentistry

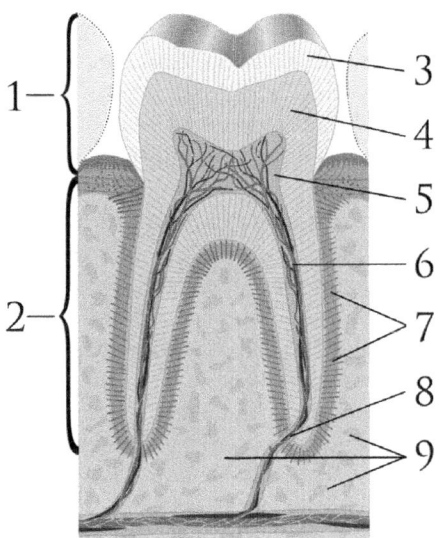

A sagittal cross-section of a molar tooth; 1: crown, 2: root, 3: enamel, 4: dentin and dentin tubules, 5: pulp chamber, 6: blood vessels and nerve, 7: periodontal ligament, 8: apex and periapical region, 9: alveolar bone

establish a university-affiliated dental school. Several scientific milestones transformed dentistry in the nineteenth century. In 1844, Horace Wells administered nitrous oxide to a patient before a tooth extraction, becoming the first dentist to use anesthesia. In 1890, dentist Willoughby Dayton Miller connected microbes to the decay process, extending the germ theory to dental disease. In 1898, William Hunter introduced the term "oral sepsis" to the profession of dentistry and called attention to the contaminated practices and instruments used by dentists. In 1918, radiology was added to dental school curricula, and by the 1930's, most dentists in the United States were using X rays as part of routine dental diagnostics.

Advances in science and technology, including the sequencing of the human genome and the arrival of the digital age, have revolutionized dentistry, rendering it nearly unrecognizable when compared with nineteenth-century dentistry and improving the diagnostic and treatment capabilities within the field.

Dental Tools

Many common dental tools are available for home use as part of a daily oral care routine. The most basic of dental tools is the toothbrush. Toothbrushes come in a variety of sizes, shapes, and stiffness. Patient age and oral condition determine the best toothbrush for each individual. Toothbrushes usually consist of a plastic handle with nylon bristles that remove food, bacteria, and plaque that can lead to tartar and dental caries. Toothpaste is usually added to a toothbrush to aid in cleaning the teeth and freshening the mouth. Toothpaste is available in a variety of flavors and compositions and may contain polishing or bleaching agents. Dental floss is another basic tool used to remove food and debris from between the teeth. Floss is available in waxed and unwaxed formulations and in a variety of widths and thicknesses. Mouthwash is a rinse that prevents gum disease. Mouthwash is available in many flavors, but all types reduce the number of germs in the mouth that cause gingivitis.

More sophisticated dental tools are used by dentists during dental examinations and procedures. A routine dental cleaning removes stains on the teeth, as well as tartar that brushing and flossing cannot remove. Polishing the teeth aids the dentist in visualizing the teeth and makes it more difficult for plaque to accumulate on the surface of the teeth. Mirrors, scrapers, scalers, and probes are essential in-office dental tools.

Dental Therapy and Devices

Countless therapies and devices are available to diagnose, prevent, and treat disorders and diseases of the teeth and mouth. Extraction, previously the mainstay of dentistry, involves simply removing the affected tooth. Fillings are used to replace a portion of a tooth that is missing or decayed. Fillings are often made of gold or silver but may also be made of composite resins or amalgam depending on the size and location of the filling. A dental implant is the extension or replacement of a tooth or its root by inserting a post made of metal or other material into the bone to support a new artificial tooth. Crownwork involves covering a damaged tooth with porcelain or other alloy to restore the tooth's original size and shape. A denture is a removable prosthetic appliance that replaces missing teeth. Dentures may replace all or just some teeth. In contrast, a bridge is a tooth-replacement device that cannot be removed. A bridge is made of one or more artificial crowns that are cemented to adjacent teeth.

Orthodontic appliances are necessary to correct and prevent irregularities in the alignment of the teeth, face, and jaw. Braces are among the most common orthodontic appliances, along with headgear and retainers. Conventional braces have metal brackets that are attached to the outer surfaces of the teeth. Wires are attached to the brackets, and manipulation of the wire allows movement and rotation of the teeth into the desired position. The braces may be attached to headgear to help move teeth or secure them into position. Retainers are often worn after braces are removed to maintain the new position of the teeth. Retainers may be permanent or removable. Removable retainers consist of a wire attached to a resin base that is worn at all times (except during meals) to hold the teeth in place for up to several years after braces are removed. A permanent retainer is a metal wire attached to the tongue side of the lower teeth that can maintain the desired position of severely crowded or rotated teeth.

General Dentistry

Most dentists practice dentistry as general practitioners. In addition, the American Dental Association recognizes nine specialties within the field of dentistry, each of which requires additional education or training beyond dental school. A general practitioner of dentistry deals with the overall maintenance of patients' teeth, gum, and mouth health. Ideally, general dentistry is preventive in nature, focusing on the maintenance of oral health and hygiene to avoid the occurrence of disorders and diseases of the mouth. Dentists who practice general dentistry encourage regular checkups to ensure proper functioning of the mouth and teeth. A general dentist will provide individualized treatment plans that include dental examinations, tooth cleanings, and X rays or other diagnostic tests to prevent or treat disorders of the mouth as early as possible. General dentists also repair and restore injuries of the teeth and mouth that result from decay, disease, or trauma. All dentists are able to prescribe medicines and treatments to diagnose, prevent, or treat diseases of the mouth and teeth.

Orthodontics

The largest specialty within dentistry is orthodontics. Orthodontics focuses on straightening teeth and correcting misalignment of the bite, usually using braces and retainers. Misalignment of the teeth or bite can cause eating or speaking disorders, making orthodontics an important part of overall health. Also, orthodontics may be aesthetic in nature, focusing on improving the structure and appearance of the teeth, mouth, and face to improve a patient's self-esteem. Most orthodontic patients are children because corrective procedures of the teeth are most effective when started early. However, an increasing number of adults are seeking orthodontic care, owing to the development of new methods and techniques in orthodontics that allow minimal discomfort and improved healing.

Oral and Maxillofacial Surgery

Commonly referred to as oral surgery, oral and maxillofacial the diagnosis and treatment of disorders of the teeth, mouth, face, and jaw. An oral surgeon may remove damaged or decayed teeth under intravenous sedation or general anesthesia; place dental implants to replace missing or damaged teeth; repair facial trauma, including injuries to soft tissues, nerves, and bones; evaluate and treat head and neck cancers; alleviate facial pain; perform cosmetic surgery of the face; perform corrective and reconstructive surgery of the face and jaw; and correct sleep apnea.

Pedodontics

Also known as pediatric dentistry, pedodontics focuses on dental care and oral hygiene of children and adolescents. Pediatric dentists apply the principles of dentistry to the growth and development of young patients, oral disease prevention, and child psychology. Some pediatric dentists also specialize in the treatment of patients with developmental or physical disabilities. Pediatric dentists emphasize proper oral hygiene, beginning with baby teeth, because healthy teeth allow for proper chewing and correct speech. Pediatric dentists also stress the importance of proper nutrition for its role in oral health, as well as overall growth and development. Early dental care facilitates lifelong oral health.

Periodontics

The field of dentistry called periodontics studies the bone and connective tissues that surround the teeth. Periodontics also involves the placement of dental implants. Periodontists prevent, diagnose, and treat periodontal disorders and infections, including gingivitis and periodontitis. Most periodontal diseases are inflammatory in nature, as are some cardiovascular diseases, and a connection between these two disease states has prompted physicians and periodontists to work together to treat patients at risk for either condition.

Prosthodontics

Also known as prosthetic dentistry, prosthodontics is the specialized field of dentistry that focuses on restoring and replacing teeth with dental implants, bridges, dentures, and crowns. Although general dentists can perform simple restoration or replacement of teeth, prostodontists handle severe or extreme cases of tooth loss because of trauma, disease, congenital defects, and age.

Endodontics
The field of dentistry that studies abnormal tooth pulp and focuses on the prevention, diagnosis, and treatment of diseases of the tooth pulp is called endodontics. Endodontic treatment is also known as root canal therapy. Endodontic therapy may also include surgery necessary to save a diseased tooth. Endodontists are often able to treat the diseased or damaged inside of a tooth instead of extracting it completely.

Oral and Maxillofacial Pathology
In oral and maxillofacial pathology, the principles of dentistry are applied to investigating the causes and effects of diseases of the mouth, head, and neck. Oral pathologists are trained to diagnose and treat such diseases, as well as to expose the connection between oral disease and systemic disease.

Oral and Maxillofacial Radiology
The use of advanced imaging techniques to diagnose and treat disorders of the mouth, teeth, head, and neck is known as oral and maxillofacial radiology. An oral and maxillofacial radiologist is a dentist who uses radiographic images to diagnose disease and guide treatment plans. Radiologists may use X rays, computed tomography (CT) scans, magnetic resonance imaging (MRI), ultrasound, and positron emission tomography (PET) to visualize the oral cavity or maxillofacial regions. Specialized sialography images the salivary glands. Intraoral radiographs are used routinely by general dentists as part of regular dental checkups.

Dental Public Health
The field of dental public health is involved in the epidemiology of dental diseases and applies the principles of dentistry to populations rather than individuals. Dental public health specialists have been involved in promoting fluoridation of drinking water and examining the links between commercial mouthwash and cancer. Dental health specialists assess the oral health needs of communities, develop programs to teach and promote oral health, and implement policies and regulations to address oral health issues.

Future Prospects
The connection between oral health and overall health has led to an increase in oral home care as well as in professional dentistry services. Patients seek dental care for routine maintenance of oral health and cosmetic procedures to improve the appearance of the face and mouth. In the future, dentistry will increasingly play a fundamental role in people's overall health and wellness. From preventing childhood tooth decay and age-related tooth loss to improving self-esteem through a brighter, straighter smile, dentistry has evolved from a fearful, painful process of tooth extraction to a respected field of medicine that is associated with comfortable care and daily hygiene.

Dentistry of the future will emphasize less painful therapy and disease prevention. It will seek to identify at-risk groups and to provide services to underserved populations to improve dental public health, which will have lasting benefits in education and overall disease morbidity and mortality. Emerging research is focused on mouthwashes that prevent the buildup of plaque on teeth, vaccines that prevent decay and dental caries, and long-lasting pellets that deliver a continuous dose of fluoride to the teeth. Braces may soon be replaced or aided by small, battery-operated paddles that deliver an undetectable electric current to the gums to rearrange bone and tissue structures of the mouth. Lasers will replace existing surgical techniques, allowing for pain free treatment of dental disease. Further, new enzymes and plastics are emerging as options for tooth restoration and dental diagnostics.

Dentistry will continue to be a collaborative and interdisciplinary practice that meets the growing and changing needs of dental health.

Jennifer L. Gibson, D.P.

Further Reading
Kendall, Bonnie. *Opportunities in Dental Care Careers.* Rev. ed. New York: McGraw-Hill, 2006. A review of the educational requirements and professional expectations for all specialties of dentistry and dentalrelated careers.

Picard, Alyssa. *Making the American Mouth: Dentists and Public Health in the Twentieth Century.* New Brunswick, N.J.: Rutgers University Press, 2009. Presents a history of dentistry as well as essays on issues such as

dental hygiene, dental economics, and the American diet.
Pyle, Marsha, et al. "The Case for Change in Dental Education." *Journal of Dental Education* 70, no. 9 (September, 2006): 921-924. The American Dental Education Association's Commission on Change and Innovation in Dental Education examines the need for change in dental education. It takes into account the financial expense of a dental education and the professional responsibilities of meeting all individual and public health needs.
Rossomando, Edward F., and Mathew Moura. "The Role of Science and Technology in Shaping the Dental Curriculum." *Journal of Dental Education* 72, no. 1 (January, 2008): 19-25. Offers a history of the changing dental school curricula in the United States and offers perspectives for the future of dentistry education.
Wynbrandt, James. *The Excruciating History of Dentistry: Toothsome Tales and Oral Oddities from Babylon to Braces.* New York: St. Martin's Press, 1998. An entertaining history of the development of the dental profession, offering humorous anecdotes and macabre tales of the profession.

WEB SITES
American Dental Association http://www.ada.org
American Dental Education Association http://www.adea.org

See also: Anesthesiology; Cardiology.

DEOXYRIBOSE NUCLEIC ACID

FIELDS OF STUDY

Animal Breeding and Husbandry; Biosynthetics; Cell and Tissue Engineering; Cloning; DNA Analysis; DNA Sequencing; Forensic Science; GM Foods; GMOs; Genomics; Human Genetic Engineering; Pharmacology; Plant Breeding and Propagation; Proteomics; Reproductive Science and Engineering; Stem Cell Research and Technology; Toxicology; Virology; Wildlife Conservation; Xenotransplantation

SUMMARY

DNA is the most complex biopolymer known, yet it is constructed from just four different nucleotides attached to a long backbone structure of deoxyribose sugar molecules bonded to phosphate groups. The order of nucleotides in DNA is the blueprint for all enzymes and proteins that are involved in the biochemistry of a living organism. It has a double helix structure produced by two complimentary single strands of DNA. The DNA molecule is reproduced by the process of replication, while proteins are produced from the nucleotide sequences in the process of transcription.

PRINCIPAL TERMS

- **amino acids:** biological molecules that serve as the building blocks of proteins and enzymes. Amino

---

FASCINATING FACTS ABOUT DENTISTRY

- In the mid-1850's, a textbook on dental surgery taught that asbestos could be placed under the filling of a sensitive tooth because asbestos is unable to conduct heat or electricity.
- The earliest evidence of dental caries dates back 100 million years to the Cretaceous period; dinosaur and fish fossils from this period show signs of dental decay.
- Toothpicks were in use at least 3,000 years ago and were made of wood, metal, thorns, or porcupine quills; ornate metal toothpicks were a sign of wealth in ancient Egypt.
- Saint Appollonia, a Christian martyr whose teeth were removed by her Roman captors, is the patron saint of dental pain sufferers, and her intercession is thought to bring healing to all oral pain and afflictions.
- The structure and anatomy of the teeth and mouth are unique to each individual, and dental records and examinations are used to identify victims of accidents, terrorism, or natural disasters. Paul Revere was the first dentist to suggest using bridgework to identify remains—namely, a Revolutionary War general—and he became a pioneer of forensic dentistry in the United States.
- More than 90 percent of systemic diseases exhibit oral manifestations; oral signs and symptoms may come before, after, or at the same time as signs and symptoms elsewhere in the body but are often the first signs of systemic illness.

acids are incorporated into proteins by transfer RNA, according to the genetic code contained in DNA. The majority of amino acids have names ending with -ine, and are complex arrangements of atoms of carbon, nitrogen, hydrogen, and oxygen. *See also* enzyme, protein.

- **chromosome:** a basic unit of heredity in living cells. Each chromosome is composed of proteins and DNA, which carry thousands of genes. In a healthy, normal, human cell, there are twenty-three pairs of chromosomes. In sexual reproduction, one chromosome in each pair comes from the father, the other from the mother. *See also* DNA, gene.
- **enzyme:** a biological catalyst, any protein molecule within a living organism that speeds up biochemical reactions to a rate that will sustain life. The effect may speed up metabolic reactions by a factor of one million, compared with what would occur chemically outside the body. Names and classification of enzymes are regulated by the International Commission on Enzymes. Most enzymes are named by adding -ase to the root of a corresponding substrate, the molecule an enzyme acts upon. Sucrase catalyzes the hydrolysis of sucrose into glucose and fructose. A living cell has a unique set of 3,000 enzymes, each defined by the cell's DNA.
- **forensic:** the application of science to legal concerns. The analysis of crime scenes, firearms, DNA, and the pathology of dead bodies are common subjects of forensic investigation, but dentists, toxicologists, psychiatrists, engineers, and practitioners in many other fields can also be called upon.
- **gene:** a unit of hereditary information found within a chromosome that determines the characteristics of an organism. Each gene is an ordered series of nucleotides, which are subunits of DNA, composed of a base molecule containing nitrogen, a phosphate molecule, and a pentose sugar molecule. *See also* chromosome, DNA.
- **protoplasm:** the living substance of a cell, including the content of the cell membrane and the substance within the cell—a transparent gelatinous material composed of inorganic substances (90 percent water with mineral salts and gases such as oxygen and carbon dioxide), and organic substances (proteins, carbohydrates, lipids, nucleic acids, and enzymes). Protoplasm outside the cell nucleus is called cytoplasm.

BASIC PRINCIPLES

A biopolymer is a large molecule formed within a biological system, a living organism, by the repetitive connection of a large number of smaller molecules having identical or similar chemical structures or properties. Typically, polymers are formed by the repeated head-to-tail bonding of smaller molecules to produce very long bio molecules, in which the number of repeating subunits, or monomers, is some whole number greater than one and ranging into the hundreds of thousands. This type of polymerization produces linear polymers. If a monomer has an additional functional group that will allow a polymerization reaction to occur from the same molecule, then cross-linking can occur in which different polymer chains become bonded together to form a three-dimensional network. Biopolymers obey the same basic principles as synthetic polymers, but are based on biomolecules such as sugars and amino acids. Their formation within biological systems is mediated by enzymes.

Different monomeric units can copolymerize to produce more complex polymers, and Deoxyribose Nucleic Acid (DNA) is the most complex biopolymer known. The essential structure of the DNA molecule was recognized in 1953 by Watson and Crick, based on X-ray diffraction photographs obtained by Maurice Wilkins and Rosalind Franklin. This demonstrated the now familiar 'double helix' of complimentary strands of DNA, or 'duplex DNA'. The molecular structure of a single DNA strand is a unique combination of deoxyribose sugar molecules bonded head-to-tail through intermediate phosphate groups. To each deoxyribose sugar moiety is attached one of the four purine and pyrimidine bases adenosine, thymine, cytosine and guanine, forming the corresponding nucleotide. In the complimentary strand of duplex DNA, the adenosine nucleotide couples only to the thymine nucleotide of the other strand, and the cytosine nucleotide couples only to the guanine nucleotide of the other strand. The functional groups of the different bases match up to each other in such a way that their interactions cause the formation of a very stable coordination of the two strands to each other. The structure is often described using a zipper as an analogy.

DNA AND PROTEINS

DNA contains the structural blueprint for each and every protein in the human body. The order of the nucleotides in a segment of DNA determines the identity of the corresponding protein that will be produced through interaction with proteins that moderate the opening of the DNA molecule at that location. The pattern is transcribed by the formation of a complimentary strand of RNA, which then is used by enzyme complexes in the protoplasm to construct the desired protein from free amino acids in the cytoplasm. Enzyme proteins control the formation of DNA, while DNA determines the identity of the proteins that carry out that function. It is a fascinating cyclic relationship. Specific segments of the DNA molecule define gene and chromosome regions that determine human heredity.

THE UNIQUE CHARACTER OF DNA

Every one of the myriad proteins that are produced in a living organism, believed to be between 10^7 and 10^9 in humans, is blueprinted in the order of the nucleotides in the DNA molecule. Despite this multiplicity, all are described by a sequence that uses just four nucleotide bases, and that sequence is unique to each individual organism. The exact duplication of DNA to produce a new organism results in the formation of a 'clone' having the identical characteristics of the original. Normally, however, sexual reproduction, whether in plants or animals, blends the DNA of two different individuals, to form a new individual with its own unique DNA. The individual pattern of the nucleotide structure in DNA is unique to every individual species, and to every individual member within a species. By determining the sequence of nucleotides in a sample of DNA, an individual can be identified with absolute certainty. This methodology has been used in forensic analysis both to identify the perpetrator of a crime, and to prove that a specific person is in fact innocent of a crime.

REPLICATION AND TRANSCRIPTION

Replication is the process by which new copies of DNA are produced during cell division. Various specific enzymes act on a strand of DNA to 'unzip' the two strands and assemble a complimentary strand of ribose nucleic acid (RNA) for each strand. RNA is based on a ribose sugar backbone rather than deoxyribose, and uses a uracil nucleotide in place of a thymine nucleotide. Each of the RNA strands is then replaced nucleotide by nucleotide to produce the new strand of DNA. Overall, one duplex DNA molecule separates into two single DNA strands, and when each single strand has formed its complimentary strand, the end result is that two molecules of duplex DNA have been formed from one. Transcription is the process in which specific portions of the DNA molecule are accessed, rather than the entire molecule. The nucleotide sequence in the specific portion of the DNA molecule is copied by a form of RNA and carried out into the cell protoplasm and cytoplasm. Other forms of RNA identify the three-nucleotide codons that correspond to specific amino acids and assemble those amino acids into a specific protein. The entire processes of replication and transcription are controlled from start to finish by a variety of specific enzyme proteins.

Richard M. Renneboog MSc

See also: RNAase.

FURTHER READING

Watson, James D. and Berry, Andrew *DNA: The Secret of Life* New York, NY: Knopf Doubleday, 2009.

Butler, John M. *Fundamentals of Forensic DNA Typing* Burlington, MA: Academic Press, 2010.

Rosenfeld, Israel, Ziff, Edward and Van Loon, Borin *DNA. A Graphic Guide to the Molecule That Shook the World* New York, NY: Columbia University Press, 2011.

FASCINATING FACTS ABOUT DNA

- DNA was first isolated from cells in 1859, but the structure of the molecule was not determined until 1953.
- DNA can be easily recovered from strawberries by mashing the fruit in a mild detergent solution. Stirring with a thin stick or glass rod collects a white substance that is strawberry DNA.
- Although small enough to fit inside the nucleus of a cell with plenty of room to spare, a DNA molecule would be more than six feet long if stretched out in a straight line.
- The nucleotide sequence of DNA, using A for adenosine, C for cytosine, T for thymine, and G for guanine, is written in the form ATGCTTCAGCG... To write out the complete structure of a strand of human DNA in this form would require about one million closely printed pages.

Kornberg, Arthur and Baker, Tania A. *DNA Replication* 2nd ed., Sausalito, CA: University Science Books, 2005.

Matisoo-Smith, Elizabeth and Horsburgh, K. Ann *DNA for Archaeologists* Walnut Creek, A: West Coast Press, 2012.

Dermatology and dermatopathology

FIELDS OF STUDY

Medicine; pathology; surgery; surgical pathology; biology; histology; chemistry; physics; immunodermatology; pediatric dermatology; cosmetic dermatology; surgical dermatology; veterinary dermatology; Mohs micrographic surgery.

SUMMARY

Dermatologists diagnose and treat medical conditions of the skin, including acne, rosacea, psoriasis, warts, hair loss, and various forms of skin cancer. Dermatopathologists analyze the mechanisms of skin diseases and perform microscopic diagnoses based on the tissue samples submitted by dermatologists. Skin disorders have a high prevalence and can affect patients of all ages, from neonates to elderly people. Because of the great variety and dynamic nature of the lesions, specialties focusing on skin are among the most complex in medicine.

PRINCIPAL TERMS

- **botulinum toxin:** neurotoxin produced by the bacterium *Clostridium botulinum;* commonly known as Botox, its trade name.
- **epidermis:** upper (outer) skin layer.
- **flow cytometry:** technique for separating and counting cells or chromosomes by suspending them in fluid and passing them by a focused light.
- **immunohistochemistry:** antibody-based method of detecting a specific protein in a tissue sample.
- **keratinocyte:** common epidermal cell that synthesizes keratin and changes while moving upward from basal to superficial layers.
- **macule:** flat, colored skin area that measures less than 10 millimeters in diameter.
- **melanocyte:** epidermal cell that produces the skin pigment melanin.
- **papule:** solid, raised spot on the skin that measures less than 10 millimeters in diameter.
- **plaque:** broad, raised area of skin.
- **pustule:** small skin swelling filled with pus.
- **retinoids:** class of compounds chemically related to vitamin A.

BASIC PRINCIPLES

Dermatology is the branch of medicine dedicated to the diagnosis, treatment, and prevention of diseases and conditions of the skin, the hair, the nails, and mucous membranes. A subspecialty of pathology and dermatology, dermatopathology focuses on studying the mechanisms of skin diseases and on the microscopic examination of cutaneous tissue. Dermatologists assess the appearance and distribution of any abnormalities in the skin, identifying primary and secondary lesions. These lesions can manifest in numerous forms, including macules, papules, plaques, nodules, pustules, vesicles, wheals (hives), scales, fissures, and scars. The patient may complain of itchiness (pruritus), pain, or hair loss, or may be uncomfortable with the appearance of a skin area. If a diagnosis is not readily apparent, the dermatologist performs a skin biopsy. A dermatopathologist examines the tissue under a microscope and renders a pathological diagnosis.

The skin is the largest and most visible organ of the human body, with essential functions in storage, absorption, thermoregulation, vitamin D synthesis, and protection against pathogens. It is readily accessible to the examiner; however, the potential abnormalities are numerous and the differential diagnoses extensive, rendering dermatology one of the most complex medical disciplines. Although the field has been morphologically oriented for centuries, advances in molecular medicine and genetics have opened new opportunities for understanding the pathogenesis of skin diseases and for improved diagnosis strategies. An evolving interrelationship with other disciplines such as plastic surgery and endocrinology has been expanding the frontiers of this medical specialty.

Background

People have been concerned with the health and appearance of their skin throughout history. Egyptian physicians used arsenic applications to treat skin cancer and sandpaper to smooth scars. Queen Cleopatra was known for her cosmetic knowledge. Geoffrey Chaucer's *The Canterbury Tales* (1387-1400) and William Shakespeare's plays contain numerous references to unsightly skin afflictions, such as boils, carbuncles, and scabs. Not surprisingly, their appearance is frequently a metaphor for character flaws.

Some of the first skin treatments were undoubtedly borrowed from the plant world, making use of leaves, flowers, and roots. The juice of the aloe vera, for example, is an ancient and effective remedy that continues to be used for some skin conditions. For centuries, physicians treated a wide range of afflictions, from rashes to wounds, using oils, powders, and salves they mixed themselves. Sunlight was used by European physicians in the eighteenth and nineteenth centuries to treat psoriasis and eczema.

Starting in the nineteenth century, a true revolution in biology galvanized the progress of skin sciences. The terms "dermatology" and "dermatosis" were introduced. In the late 1800's, dermatologists began using a variety of chemicals to smooth facial wrinkles and scars. Cryosurgery and electrosurgery came into use. Soon after the development of the laser in the 1950's, dermatologists used it to treat skin conditions. The surge of innovations has continued, making dermatology an exciting and rapidly evolving specialty.

How It Works

Skin diseases and conditions affect patients of all ages and ethnicities. Physicians may specialize in a specific age group, such as children, or a category of conditions. Some dermatologists focus on cosmetic disorders of the skin and may be certified to perform procedures such as injections of botulinum toxin, chemical peels, and laser therapy. Others concentrate on skin cancers or immunological conditions. Regardless of the focus of a dermatologist's practice, the day-to-day work can be divided into three main areas: diagnosis, treatment, and management.

Diagnosis. Dermatologists obtain the patient's medical history and assess his or her status. They examine the affected skin and adjacent areas to determine the nature and extent of the lesions. A frequently used method is dermoscopy (or epiluminescent microscopy), which employs a quality magnifying lens and a powerful lighting system to allow a close examination of the skin's structure. It is useful in evaluating pigmented skin lesions and can facilitate the diagnosis of melanoma.

Some skin conditions are more readily diagnosable than others. Acne and psoriasis, for example, often do not necessitate further tests. The lesions, however, may be of an ambiguous nature or potentially malignant. In these cases, the physician takes a tissue sample (for example, a biopsy or nail clippings) and submits it, usually with a differential diagnosis, to a laboratory. There, the sample undergoes a dermatopathological evaluation.

Dermatopathologists interpret tissue samples on specially prepared slides using light, fluorescent, and sometimes electron microscopy. They first determine how the specimen was obtained (for example, a punch or shave biopsy), then establish if the condition appears to be infectious, inflammatory, degenerative, or neoplastic (benign or malignant). Often, consultation with other dermatopathologists and the attending dermatologist or primary care physician is necessary. Additional sections of the specimen may be required before a diagnosis can be rendered and the report sent to the clinician. The work needs to be extremely thorough; no part of the microscopy slide can be left unexamined. Ancillary methods used by dermatopathologists include immunohistochemistry and flow cytometry.

Additional tests that may be undertaken in the dermatologist's office include a potassium hydroxide examination for fungi, bacterial stains, fungal and bacterial cultures, skin scrapings for scabies, patch tests (for contact allergies), and blood tests.

Treatment. Once the diagnosis has been made, treatment options are considered and discussed at length with the patient or caregiver. Dermatopathol- may involve medications to be administered externally or internally, injections, or surgical procedures. Punch biopsy, shave biopsy, electrodesiccation and curettage, blunt dissection, and simple excision and suture closure are the basic techniques that dermatologists master. They are also familiar with more sophisticated techniques, such as Mohs micrographic surgery and, if appropriate, may refer patients to physicians who perform these techniques.

Management. Skin conditions can be lifelong problems. Eczema, acne, and psoriasis are only a few of many conditions that require regular visits to the dermatologist. Managing the patient's condition often takes the form of control rather than cure.

APPLICATIONS

Dermatologist diagnose and treat many disorders and diseases. The most common examples of disorders treated are infections, inflammatory diseases, papulosquamous diseases, and tumors.

Infections. Several categories of pathogens cause infections with cutaneous manifestations. *Staphylococcus aureus* and group A beta-hemolytic streptococci account for most skin and soft tissue infections, such as impetigo, folliculitis, cellulitis, and furuncles. Syphilis is an infectious disease caused by the bacterium *Treponema pallidum*. Primary syphilis, acquired by direct contact with a skin or mucosal lesion, manifests with a cutaneous ulcer (chancre). Warts are benign epidermal tumors caused by numerous types of human papillomaviruses (HPVs). These viruses infect epithelial cells of the skin, mouth, and other areas, causing both benign and malignant lesions. Herpesvirus infections are caused by herpes simplex virus 1 (HSV1) and herpes simplex virus 2 (HSV2), distinguishable by laboratory tests. HSV1 is generally associated with oral infections, and HSV2 causes genital infections. The lesions appear as grouped vesicles on a red base.

The agents that induce superficial fungal infections include dermatophytes (responsible for tinea, or ringworm) and *Candida* species yeasts.

Inflammatory Diseases. Eczema is the most common inflammatory disorder. It manifests with itchiness and exhibits three clinical stages: acute (redness and vesicles), subacute (redness, scaling, fissuring, and scalded appearance), and chronic (thickened skin). There are numerous types of eczemas, including atopic dermatitis (in patients with personal or family history of allergies) and contact dermatitis (allergy to a common material such as nickel or poison oak).

Acne is a common disorder with important psychosocial effects. It occurs in predisposed individuals when sebum production increases. Proliferation of the microorganism *Propionibacterium acnes* in the sebum alters it and causes pore clogging. Lesions are noninflammatory (comedones, also known as blackheads and whiteheads) or inflammatory (papules, pustules, or nodules). The extent and severity of the lesions varies, from a few comedones to the strongly inflammatory acne conglobata.

Papulosquamous Diseases. The group of disorders known as papulosquamous diseases are characterized by scaly papules and plaques. Psoriasis, an immune-mediated skin and joint inflammatory disease, develops when inflammation primes basal stem keratinocytes to proliferate excessively. Initial red, scaling papules coalesce to form round-oval plaques. The scales are adherent, silvery white, and show bleeding points when removed (Auspitz sign). Inflammatory arthritis is present in some patients.

Tumors. The two most common skin cancers are basal cell carcinoma (BCC) and squamous cell carcinoma (SCC). Approximately 80 percent of nonmelanoma skin cancers are the basal cell type, and 20 percent are the squamous cell type.

Basal cell cancer, the most common invasive malignant skin tumor in humans, represents more than 90 percent of skin cancers in the United States. The patient typically has a bleeding or scabbing sore that heals and subsequently recurs. The tumor advances by direct extension and destroys normal tissue but rarely metastasizes. The cells of basal cell carcinoma resemble those of the basal epidermal layer. They have a large nucleus and develop an orderly line around the periphery of tumor nests (palisading).

Squamous cell carcinoma is the second most common cancer among light-skinned individuals. The relationship to ultraviolet radiation is stronger and the chances of metastasis much higher than for basal cell carcinoma. Actinic keratosis, the most common precursor of squamous cell carcinoma, begins on sun-exposed skin as isolated or multiple flat, pink-brown, rough lesions. Abnormal squamous cells originate in the epidermis from keratinocytes and proliferate indefinitely.

Melanocytes. Skin melanoma either begins on its own or develops from a preexisting lesion, such as a mole (nevus). One of the most aggressive tumors, melanoma can metastasize to any organ, including the brain and heart. Individuals who sunburn easily or who experienced multiple or severe sunburns have a twofold to threefold increased risk for developing skin melanoma. The goal of specialists and patients alike is to recognize melanomas as early as possible in their development. Compared with common acquired melanocytic nevi, malignant melanoma tend to have four characteristics:

asymmetry, border irregularity, color variation, and diameter enlargement (ABCD). These four characteristics are the primary criteria for clinical melanoma recognition. Changes in the shape and color of a mole are important early signs and should always arouse suspicion. Ulceration and bleeding are late signs; at this stage, the chance of cure diminishes greatly.

Important Treatment Modalities. Common ways of dealing with dermatological problems are topical treatments (such as ointments and creams) and oral treatments (drugs taken by mouth). Any bodily injury, irritation, or trauma that eliminates water, lipids, or protein from the epidermis compromises its function. Restoration of the normal epidermal barrier can often be accomplished using mild soaps and emollient creams or lotions. The often-cited dermatologic adage is "If it is dry, wet it; if it is wet, dry it." Consequently, wet compresses are a frequently used remedy. A multitude of other topical treatments are available, from antibiotic, antiviral, or steroid ointments applied to treat infectious diseases or eczema to vitamin D derivative creams for psoriasis and retinoid creams for acne. Drugs can also be taken orally to treat a variety of conditions such as acne and autoimmune disorders.

Surgical and Cosmetic Procedures. Dermatologists use several techniques to obtain skin biopsies. Most procedures are done in the doctor's office, and each technique has specific indications. Punch biopsies are employed for most superficial inflammatory diseases and skin tumors (except melanoma). Shave biopsies are used for superficial benign and malignant tumors. Deep inflammatory diseases and malignant melanoma benefit from excisions.

Electrodesiccation and curettage (ED&C; also known as scrape and burn) is an important technique for removing a variety of superficial skin lesions, such as cancerous growths and genital warts. The physician uses a sharp dermal curette to cut away the growth and a needle-shaped electrode that delivers an electric current to remove any remaining material and to stop the bleeding.

Blunt dissection is a fast, elementary, usually nonscarring surgical procedure used to remove warts and other epidermal tumors. Unlike ED&C and excision, it does not disturb normal tissue. Small, superficial, nonmalignant lesions may be quickly and efficiently frozen with liquid nitrogen, administered with a spray or sterile contact probe. Cryosurgery for malignant lesions, however, requires experience and sophisticated equipment with thermocouples that measure the depth of freeze. This minimally invasive technique is also successfully employed for common lesions, including genital warts, actinic keratoses, and certain infectious conditions.

An important surgical breakthrough occurred in the 1930's, when physician Frederic Mohs developed a microscope-guided method of tracing and removing basal cell carcinomas. These—and other tumors—may not grow in a well-circumscribed fashion but instead extend in fingerlike projections. Thin layers of tissue are removed, and all margins of the specimen are mapped to determine whether any tumor remains. This tissue-sparing technique has high cure rates. Chemical peeling of facial skin uses a caustic agent to achieve a controlled, chemical burn of the epidermis and the outer dermis. Skin regeneration results in a fresh and orderly epidermis with ablation of fine wrinkles and pigmentation reduction.

In liposuction surgery, fat is removed through half-inch incisions using small- diameter cannulae. Multiple to-and-fro movements mechanically disrupt the fat and create tunnels. The loosened fat is removed by strong suction.

Photothermolysis is based on the property of a chromophore (melanin, hemoglobin, tattoo ink) in a target tissue to strongly absorb a selected laser wavelength and generate heat. It removes the target tissue while producing only a local thermal injury, resulting in less injury to the surrounding tissue and lowered risk of scarring. Vascular lesions, for example, can be treated in this manner, including port-wine stains, benign tumors, and spider veins in legs. In vascular lesions, the targeted chromophore is hemoglobin. Specific types of lasers can be used to treat benign pigmented lesions with a predominant epidermal component such as freckles and tattoos. In addition, numerous laser-based devices can remove unwanted hair.

Other common techniques and devices include the use of intense pulsed light for resurfacing (to treat vascular lesions and acne) and light-activated drugs in photodynamic therapy (for precancerous and cancerous cells, acne, rosacea, or skin enhancement). One of the most popular nonsurgical cosmetic procedures is injections of botulinum toxin (Botox). This neurotoxin blocks the release of the chemical messenger acetylcholine, effectively causing chemical denervation. The injections reduce facial lines caused by hyperfunctional muscles.

Future Prospects

The burden of skin diseases on society is significant. According to a 2004 study by the American Academy of Dermatology Association and the Society for Investigative Dermatology, the annual cost of skin diseases in the United States is about $39 billion; direct medical costs account for $29 billion and indirect costs related to lost productivity make up the remaining $10 billion. At any given time, one in three people in the United States suffers from an active cutaneous condition. The most prevalent disorders are herpes simplex, shingles, sun damage, eczema, warts, and hair and nail conditions. The incidence of melanoma is on the rise. The main reasons for this high level of skin disease are increased exposure to the sun during recreational activities and the atmospheric changes brought on by pollutants that result in increased radiation. Understanding the biology of skin tumors, especially melanoma, has become a priority of research efforts worldwide. New therapeutic agents such as antibodies and immunomodulators offer hope for stubborn medical conditions such as psoriasis, still an incurable disease in need of good long-term therapeutic approaches. Biological treatments are on their way to bringing relief. Stem cells hold promise for tissue regeneration. Advances in understanding the pathogenesis of various disorders have led to improved management and to a reduced risk of incorporating nonevidencebased components into dermatological practice. The close cooperation between dermatologists, pathologists, rheumatologists, and surgeons enhances the quality and efficiency of care. The ever-increasing preoccupation with young, healthy skin has fueled an unprecedented explosion in the popularity of cosmetic procedures. More important, skin diseases with significant aesthetic, psychological, and social consequences have prompted dermatologists to implement and refine numerous cosmetic techniques involving peeling, botulinum toxin, hyaluronic acid, and lasers. These techniques have enabled many categories of patients with skin disorders to lead a normal social life.

Mihaela Avramut, MD, PhD

Further Reading

Bickers, D. R., et al. "The Burden of Skin Diseases, 2004: A Joint Project of the American Academy of Dermatology Association and the Society for Investigative Dermatology." *Journal of the American Academy of Dermatology* 55, no. 3 (September, 2006): 490-500. Summary of the well-documented study assessing the prevalence and economic burden of skin diseases and how they effect quality of life.

Bolognia, Jean, et al., eds. *Dermatology*. 2d ed. 2 vols. St. Louis, Mo.: Mosby Elsevier, 2008. A basic textbook that covers nearly all aspects of dermatology, from cancers to cosmetic procedures.

Ferri, Fred. *Ferri's Fast Facts in Dermatology: A Practical Guide to Skin Diseases and Disorders.* Philadelphia: Saunders/Elsevier, 2011. A handbook for the diagnosis of dermatological disorders.

FASCINATING FACTS ABOUT DERMATOLOGY AND DERMATOPATHOLOGY

- Under normal conditions, the top layer of skin on an adult human sheds every twenty-four hours, and the skin completely renews itself in three to four weeks.
- Throughout the centuries, people with leprosy have been ostracized by their communities. Although modern medicine has made diagnosis and treatment of leprosy easy, the stigma associated with the disease remains and presents an obstacle to self-reporting. About one hundred patients are diagnosed each year in the United States.
- The use of botulinum toxin is not limited to the treatment of wrinkles. It has also been used as a remedy for muscle spasms, migraines, strabismus (lazy eye), and other conditions.
- Vitiligo, a skin disorder that affects one in every two hundred people, causes patches of skin that lack pigment and are prone to sun damage but not to skin cancer. A gene mutation responsible for increasing the risk of developing vitiligo also decreases the risk of skin malignancy.
- Researchers are studying noninvasive techniques for removing adipose tissue that could help eliminate localized fat deposits in individuals of average weight. These include exposing fat cells beneath the skin to ultrasound waves or low temperatures.
- Scientists have created artificial skin with biomechanical properties similar to real skin using biomaterials such as fibrin (from blood), agarose (from seaweed), chitosan (from crustacean shells), and collagen.

Habif, Thomas P. *Clinical Dermatology*. 5th ed. St. Louis, Mo.: Mosby Elsevier, 2010. Leading manual with excellent photographs, online access, multiple appendixes, and an online differential diagnoses (DDX) mannequin for lesion localization.

Hall, Brian J., and John C. Hall. *Sauer's Manual of Skin Diseases*. 10th ed. Philadelphia: Lippincott, Williams & Wilkins, 2010. Accessible textbook includes numerous color photographs, diagnostic algorithms, and a dictionary-index. Has an accompanying Web site.

Pilla, Louis. "Cosmetic Versus Medical Dermatology: A Widening Gap?" *Skin and Aging* 11, no. 6 (June, 2003). Analysis of the interplay between medical and cosmetic dermatology in modern practices.

WEB SITES

American Academy of Dermatology http://www.aad.org

European Academy of Dermatology and Venereology http://www.eadv.org

European Society for Dermatological Research http://www.esdr.org

Society for Investigative Dermatology http://www.sidnet.org

See also: Geriatrics and Gerontology; Pathology.

DIFFUSION

FIELDS OF STUDY

Molecular Biology; Physical Chemistry; Chemical Engineering

SUMMARY

The process of diffusion is defined, and its underlying principles are discussed. Diffusion is presented in various contexts, including fundamental characteristics of living cells and the enrichment of uranium. In all applications and occurrences, the basic principles of the diffusion process are the same.

PRINCIPAL TERMS

- **Brownian motion:** the continuous, random motion of particles in a fluid medium, caused by impacts with the molecules that make up the medium.
- **concentration gradient:** the gradual change in the concentration of solutes in a solution across a specific distance.
- **equilibrium:** the state that exists when the forward activity of a process is exactly equal to the reverse activity of that process.
- **osmosis:** the passage of solvent molecules through a semipermeable membrane from a region of low solute concentration to one of higher concentration; also the primary mechanism by which water moves through cell walls.
- **semipermeable membrane:** a membrane that allows the passage of a material, such as water or another solvent, from one side to the other while preventing the passage of other materials, such as dissolved salts or another solute.

VISUALIZING DIFFUSION

The process of diffusion can be easily modeled using an array of marbles of two different colors and a box. Place the marbles in the box in a single layer, with all the marbles of one color on one side and the marbles of the second color on the other. When the box of marbles is vibrated, marbles of one color will become dispersed throughout those of the other color. When the rate and amplitude of vibration are low, the interspersion of the marbles will be slow, but as either the rate or the amplitude is increased, the marbles will intersperse much more rapidly. At the atomic and molecular levels, diffusion functions in essentially the same manner.

DIFFUSION OF ATOMS AND MOLECULES

Atoms and molecules are constantly in motion, even at the absolute zero of temperature. For the most part atoms in the solid state oscillate around their equilibrium positions but the presence of lattice vacancies and other defects allows a certain amount of hopping. In the liquid state atomic movements are larger and somewhat more difficult to describe. In the gaseous state one can treat atomic motion as

DIFFUSION

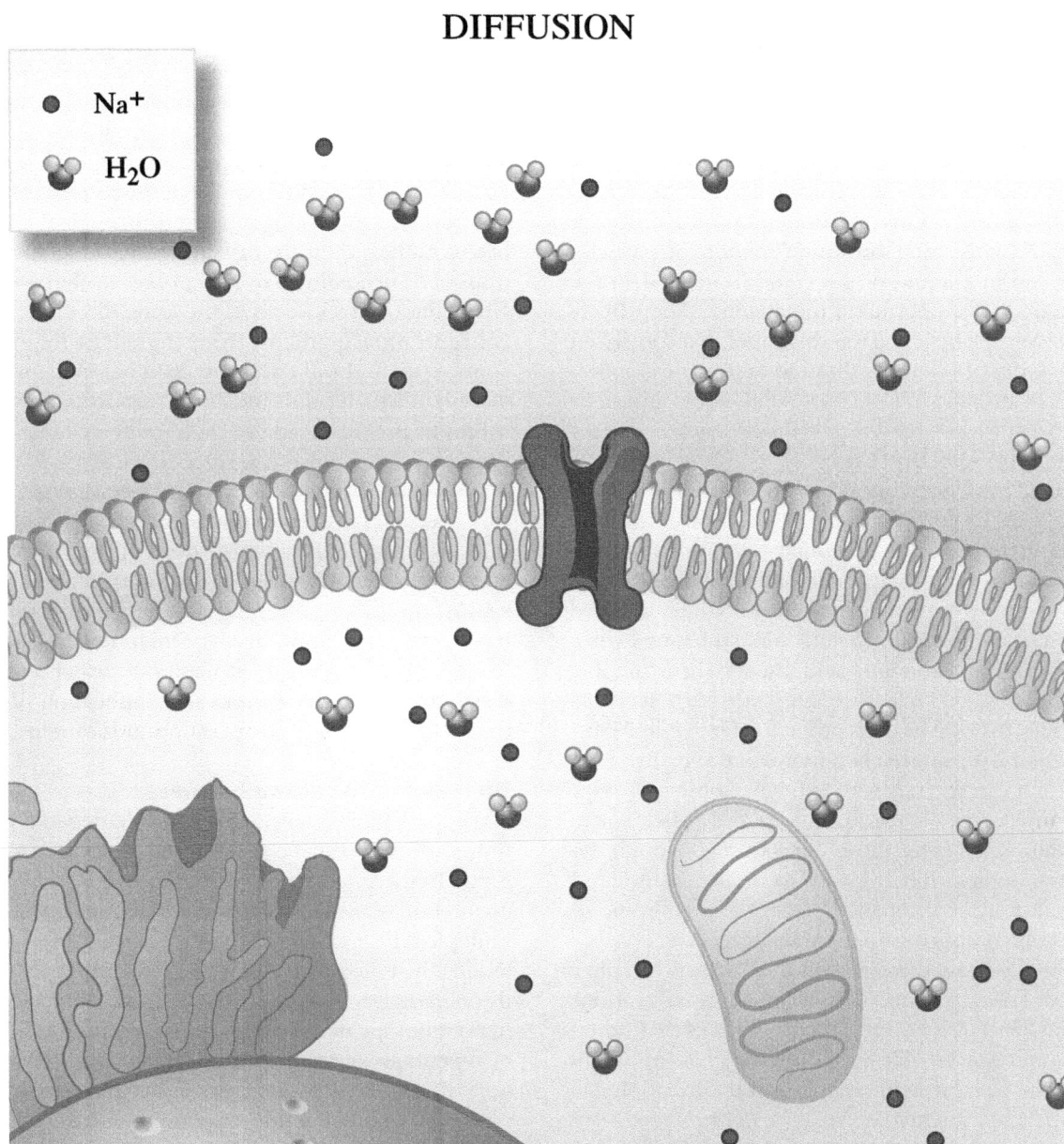

Location	Concentration Calculations	Relative Na⁺ Concentration	Concentration Gradient	Na⁺ Movement
Inside Cell	$\dfrac{22\ Na^+}{35\ Molecules} = 63\%\ Na^+$	High Concentration		
Outside Cell	$\dfrac{16\ Na^+}{37\ Molecules} = 43\%\ Na^+$	Low Concentration		

largely in straight lines with occasional collisions. In al thee phases of matter the average translational kinetic energy at absolute temperature T is equal to $3RT/N_A$ where R is the gas constant, ant M the molecular mass in kilograms. As a result, the mean distance a molecule or particle will travel in a time t obeys $\langle r^2 \rangle = Dt$, where D is the diffusion coefficient.

In 1905, Albert Einstein showed how the diffusion coefficient governing Brownian motion of particles suspended in a liquid or gas could be related to the bulk transport properties of the material. Many historians of science believe it was this paper that convinced the majority of scientists that the atomic description of matter was valid in all phases: solid, liquid and gas.

APPLICATIONS OF DIFFUSION

The most important application of diffusion occurs naturally in the cell membranes of living systems. The plasma membranes of all animal cells are composed of a lipid bilayer, primarily containing phospholipids. They are hydrophilic (attracted to water) on the phosphate end of the molecule, while the long hydrocarbon chains of the fatty acid are very hydrophobic (repelled by water) in nature. The hydrocarbon chains essentially dissolve in each other, forming a dense hydrocarbon layer sandwiched between two hydrophilic layers that can interact with proteins, ions, and other polar molecules. Sugars, proteins, and other essential compounds enter the cell by passing through the cell membrane, while by-products of metabolism and respiration must also pass through the membrane in order to be transported away and eliminated.

Molecules pass through cell membranes by either active or passive transport. In active transport, various proteins bind to ions and other molecules that otherwise would not be able to diffuse through the membrane due to their high polarity and the hydrophobic nature; these proteins, known as carrier proteins, form a channel in the membrane through which the polar materials can pass. The difference in composition between the fluid within the cell and the interstitial fluid outside of it forms a concentration gradient. Active transport processes move materials against a concentration gradient, from a region of lower concentration to one of higher concentration. This enables the transfer of signaling compounds between neurons in the nervous system, the passage of metabolic by-products and water through the nephrons of the kidneys, and similar actions in many other such systems.

In strictly chemical systems, diffusion is somewhat more straightforward. The process involves a semipermeable membrane, which is a porous membrane in which the pores are large enough to allow the passage of solvent molecules, typically water, but not large enough to allow the passage of dissolved materials, such as ions. Water molecules that pass through the pores are not actively carried through the membrane; rather, they are driven by the molecular motions of diffusion in an overall process called osmosis. Normally, osmosis works in the same direction as the concentration gradient, from higher to lower concentration, and will continue until equilibrium is attained, at which point the concentrations of the two solutions on either side of the semipermeable membrane are equal and the concentration gradient has disappeared. In the process of reverse osmosis, pressure is applied to the solution of higher concentration to overcome the osmotic pressure of the membrane and drive solvent molecules through the membrane against the concentration gradient. Reverse osmosis has numerous applications, including obtaining fresh potable water from salt-laden seawater and concentrating solutions without the application of heat that would alter or destroy compounds of value.

DIFFUSION OF NUCLEAR ISOTOPES

Since the kinetic energy of a molecule depends on its mass, diffusion can be used in the separation of isotopes. In the uranium enrichment process, the volatile compound uranium hexafluoride (UF_6) passes through multiple diffusion stages. At each stage, molecules containing the isotopes uranium-234 and uranium-235 pass through slightly more quickly than those containing the heavier isotope uranium-238. After a sufficient number of repetitions of the diffusion process, the concentration of uranium-235 increases relative to that of uranium-238, rendering the uranium more suitable for use as a fuel source for nuclear reactors. It is also the method of choice for generating the more highly enriched weapons-grade uranium.

Richard M. Renneboog, MSc

FURTHER READING

Askeland, Donald R., Wendelin J. Wright, D. K. Bhattacharya, and Raj P. Chhabra. *The Science and Engineering of Materials*. Boston: Cengage Learning, 2016. Print.

Bailey, James E., and David F. Ollis. *Biochemical Engineering Fundamentals*. 2nd ed. New York: McGraw, 1988. Print.

Berg, Jeremy M., John L. Tymoczko, Gregory J. Gatto, and Lubert Strye. *Biochemistry*. 8th ed. New York: W. H. Freeman, 2015. Print.

Lodish, Harvey, et al. *Molecular Cell Biology*. 7th ed. New York: Freeman, 2013. Print.

Pelczar, Michael J., Jr., E. C. S. Chan, and Noel R. Krieg. *Microbiology: Concepts and Applications*. New York: McGraw, 1993. Print.

Silbey, Robert J., Robert A. Alberty, and Moungi G. Bawendi. *Physical Chemistry*. 5th ed. Hoboken: Wiley, 2012. Print.

DIFFUSION SAMPLE PROBLEM

A reverse osmosis apparatus has been set up to extract freshwater from a saline solution. The concentration of sodium chloride (NaCl) on one side of the semipermeable membrane is 4.7 grams per liter, while the water on the other side has no sodium chloride at all. The membrane is 1 millimeter thick. Calculate the concentration gradient of sodium chloride across the membrane.

Answer:

The concentration gradient is described by the relationship $\frac{\Delta c}{\Delta x}$ where Δc is the change in concentration and Δx is the distance over which the concentration has changed.

First, calculate the concentration of NaCl in the saline solution in terms of moles per liter, using the atomic masses of sodium and chlorine:

$$23 \frac{g}{mol} + 35.5 \frac{g}{mol} = 58.5 \frac{g}{mol}$$

Convert 4.7 grams of NaCl into moles:

$$\frac{4.7 \text{ g}}{58.5 \text{ g/mol}} = 4.7 \text{ g} \times \frac{1 \text{ mol}}{58.5 \text{ g}} = 0.08 \text{ mol}$$

The concentration of NaCl in the saline solution is 0.08 moles per liter, or 0.08 M. Subtract the concentration of NaCl in the recovered freshwater (0 M) from the concentration in the saline solution to determine Δc:

$$0.08 \text{ M} - 0 \text{ M} = 0.08 \text{ M}$$

The distance across the membrane is 1 millimeter, or 0.001 meters. Calculate the concentration gradient across the membrane:

$$\frac{\Delta c}{\Delta x} = \frac{0.08 \text{ M}}{0.001 \text{ m}} = 80 \frac{M}{m}$$

The concentration gradient of NaCl is 80 moles per liter per meter. This means that the sodium chloride concentration would change by 80 moles per liter if the membrane were 1 meter thick.

Digestion

FIELDS OF STUDY

Biochemistry; Physical Chemistry; Forensic Chemistry

SUMMARY

The processes of digestion are defined and described in their various contexts. Digestion generally indicates that a system is undergoing chemical changes over time with the input of both thermal energy and physical agitation to maintain a constant elevated temperature.

PRINCIPAL TERMS

- **catabolic reaction:** a metabolic reaction in cells that breaks down large molecules into smaller ones, resulting in a release of energy.
- **decomposition reaction:** a chemical reaction in which a single reactant breaks apart to create several products with smaller molecular structures.
- **enzyme:** a protein molecule that acts as a catalyst in biochemical reactions.

DIGESTION

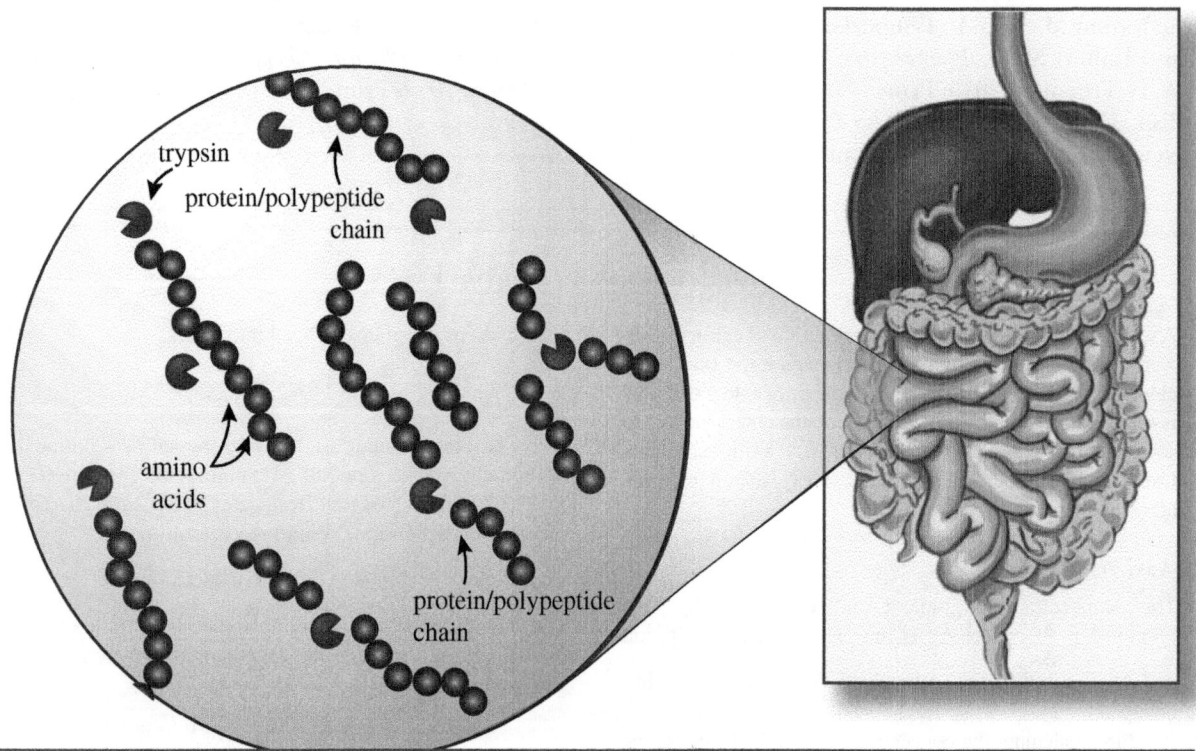

Digestion of protein molecules in the small intestine occurs when trypsin binds to the carbon of a carboxyl group on one amino acid and nitrogen of the next amino acid in a polypeptide chain. Trypsin breaks the polypeptide bond, producing smaller polypeptide chains.

- **hydrochloric acid:** a corrosive aqueous solution of hydrogen chloride, present in the common digestive fluid of the stomach.

Visualizing Digestion

Digestion refers to several similar processes that occur in different practical contexts. In all cases, starting materials are broken down chemically into smaller molecules derived from the component pieces of a larger molecular structure. Digestion can be demonstrated by grinding up various food items and placing the materials in a jar with a solution of hydrochloric acid, a key component of the gastric acid found in the stomach. After some time, the mixture will have become a more or less homogenous liquid. This closely approximates the process of digestion as it occurs in the human stomach.

Forms of Digestion

Digestion can take a number of forms. Chemists carrying out preparations in laboratories often set a reaction mixture to digest over a heat source such as a steam bath, essentially slow cooking it to promote a desired reaction. This method might be used when a reaction produces a product in high yield that isomerizes slowly into the desired target molecule at an elevated temperature. (Isomerization is when a molecule transforms into another molecule containing the same atoms in a different configuration.) The process by which this isomerization is carried out is a form of digestion.

Physiological digestion is the most familiar form of the process, as it is the process by which living creatures break down foods into their basic chemical components for use in metabolic processes. Digestion can actually begin with the preparation of food, as techniques such as heating cause various physical and chemical changes. For example, heating vegetables produces steam as the liquids within plant cells evaporate, and the resulting pressure ruptures the cell walls, making the vegetables tender. Freezing has the same effect, but the action is caused by ice crystals forming and puncturing the cell walls.

Once food is taken into the mouth, the physiological process of digestion begins. Saliva contains the enzyme amylase, a protein that promotes the breakdown of starches into their component sugar molecules. The action is fairly rapid, and the release of simple sugars in the process explains why certain foods become sweeter as they are chewed. Amylase, like all other enzymes, is a protein molecule whose three-dimensional shape forms an active site into which only certain molecular structures can fit. When that happens, the amount of energy necessary to cause changes to the chemical bonds within the substrate (the molecule acted upon by the enzyme) is significantly reduced, and specific reactions are promoted. In digestion, these are always decomposition reactions that break large food molecules into smaller components such as sugars, amino acids, and fatty acids. Accordingly, digestion is a catabolic reaction process, by which energy is obtained from the chemical breakdown of various compounds.

The process continues when the masticated (chewed) food is swallowed and enters the stomach. There, gastric acid and various enzymes are secreted through the stomach lining while, at the same time, muscles surrounding the stomach rhythmically squeeze the stomach and its contents. This churns the mixture in the stomach so that the interaction of the food materials with the digestive materials is maximized. The gastric acid in the stomach acts to break down the peptide bonds in proteins and the ester bonds in fats to release amino acids and long-chain fatty acids, respectively.

Enzyme Activity
Enzyme activity in digestion, as in all other aspects of metabolism, is strictly regulated by the biochemical environment in which it functions. Enzymes specialize in carrying out specific actions. In the enzymatic hydrolysis (chemical breakdown via the addition of water) of proteins and other polypeptides, for example, the enzymes known as trypsin and thrombin function to cleave amide bonds between individual amino acid residues in the molecule. Amino acids are essentially carbon chains with a carboxylic acid group ($-COOH$) on one end and an amine group ($-NH_2$) on the other. Trypsin will cleave the amide bond on the carboxyl side of arginine (Arg) and lysine (Lys) residues, but thrombin will cleave only the amide bond between arginine and glycine (Gly) residues when they occur in particular amino acid sequences. A reaction first with trypsin and then with thrombin will produce the same result, since there is no longer an Arg-Gly linkage in any of the fragments and no further cleavage can take place.

Learning from Digestion
In addition to helping human beings take in the nutrients necessary for survival, the process of physiological digestion can reveal a wealth of information to those who study it. Digestion is a particularly useful source of information in forensic investigations. When a person dies, the physiological process of digestion effectively stops. By examining the stomach contents to determine the extent of digestion that has occurred, a forensic scientist can estimate how much time has elapsed since the person last had something to eat or drink. This information, as well as further analysis of the partially digested materials, can provide valuable clues to the activities of the deceased person prior to his or her demise and allow investigators to isolate the cause of death.

Richard M. Renneboog, MSc

Further Reading
Bailey, James E., and David F. Ollis. *Biochemical Engineering Fundamentals.* 2nd ed. New York: McGraw, 1988. Print.

Berg, Jeremy M., John L. Tymoczko, Gregory J. Gatto, and Lubert Strye. *Biochemistry.* 8th ed. New York: W. H. Freeman, 2015. Print.

Lehninger, Albert L. *Biochemistry: The Molecular Basis of Cell Structure and Function.* 2nd ed. New York: Worth, 1975. Print.

Lodish, Harvey, et al. *Molecular Cell Biology.* 7th ed. New York: Freeman, 2013. Print.

Reece, Jane B., et al. *Campbell Biology.* 10th ed. San Francisco: Cummings, 2013. Print.

Wang, Zhaocai, Jian Tan, Dongmei Huang, Yingchao "A Biological Algorithm to Computation." *Applied Mathematics and Computation.* Volume 244 (2014), pp. 183-190.

DIGESTION SAMPLE PROBLEM

Given a polypeptide segment with the amino acid sequence Arg-Thr-Lys-Arg-Gly-Lys-Gly-Gly-Ser-Lys-Ser, what products will be formed when the material is treated with the enzyme thrombin? With trypsin? With trypsin and then thrombin? Recall that trypsin cleaves the peptide bond on the carboxyl side of arginine and lysine residues, while thrombin cleaves only the peptide bond between arginine and glycine residues. (By convention, polypeptide sequences are written so that the amide group of each amino acid is on its left side and the carboxylic acid group is on the right.)

Answer:

The reaction of the amino acid sequence with thrombin will cleave the peptide bond between arginine and glycine to yield the following fragments:

Arg-Thr-Lys-Arg

Gly-Lys-Gly-Gly-Ser-Lys-Ser

The reaction of the original sequence with trypsin will cleave the peptide bonds between arginine and threonine (Thr), lysine and arginine, arginine and glycine, lysine and glycine, and lysine and serine (Ser) to yield the following fragments:

Arg

Thr-Lys

Arg

Gly-Lys

Gly-Gly-Ser-Lys

Ser

DIGESTIVE TRACT

FIELDS OF STUDY

Cell biology, histology, physiology

SUMMARY

The digestive tract is the group of organs where nutrients brought into the animal from the environment are broken down to molecules that can be absorbed into the animal's body. Knowledge of the digestive tract and its functions allows an appreciation of the importance of correct diet and the causes of gastrointestinal diseases.

PRINCIPAL TERMS

- **absorption:** the movement of nutrients out of the lumen of the gut into the body bile salts: organic compounds derived from cholesterol that are secreted by the liver into the gut lumen and that emulsify fats
- **digestion:** the process by which larger organic nutrients are broken down to smaller molecules in the lumen of the gut
- **duodenum:** the first part of the small intestine, where it joins the stomach
- **enterocytes:** the cells that line the lumen of the small intestine
- **lumen:** the central opening through the digestive tract, which is continuous from the mouth to the anus
- **lymphatic vessels:** very thin tubes that carry water, proteins, and fats from the gut to the bloodstream
- **mucosa:** the lining of the inner wall of the gut facing the lumen
- **pancreas:** an organ derived from the gut that secretes digestive enzymes; it is connected to the gut by a duct through which its secretions enter the gut
- **plexus:** a group of nerve cells and their connections to one another
- **sphincter:** a ring of muscle that can close off a portion of the gut

BASIC PRINCIPLES

Digestion is the process by which food is broken down into molecules that are small enough to be absorbed into the body. Digestion takes place in the digestive tract of animals. The digestive tract is a continuous tube that acts on ingested food in a sequential manner. Each part of the digestive tract is adapted to reduce the size of food particles, either mechanically or enzymatically, until they are small enough to be absorbed into the body. Consideration of the mechanisms of food intake in lower animals will illustrate the evolution of complexity as an adaptation to the changing environments of these animals.

DIGESTIVE TRACTS OF SIMPLE ANIMALS

Sponges are primitive water-dwelling animals that are attached to a fixed point in the water. They bring food into their bodies from currents of water containing particulate food passing through openings in their outer wall. These currents are created by movements of flagella on cells called choanocytes. Food enters the choanocytes by phagocytosis. Phagocytosis is the process in which a cell surrounds a particle with extensions of its cell membrane until the particle is completely surrounded and thus becomes enclosed within a small sac, or vesicle, within the cell. Intracellular enzymes then digest the food particles dissolved in the fluid into their component molecules, which then become available to the metabolic systems in the cytoplasm of the cell. Some cells, called amoebocytes, carry the food particles to other cells in the sponge by crawling through the spaces between the cells. Their travel is by amoeboid mo- tion, in which the cell sends out an extension, called a pseudopod, and then follows it.

This method of feeding and digestion is adequate for a sponge because most of the sponge cells are in close contact with the water currents in which it lives. Thus, these cells can have direct access to food carried in the water currents. The cells that are not in close contact with water currents can be adequately supplied by the amoebocytes. The digestion, or breakdown, must be carried on inside the cells by cytoplasmic enzymes because if these enzymes were released to the extracellular surface, they would be washed away. The coelenterates, such as jellyfishes or hydras, are more advanced than sponges and require a more elaborate digestive mechanism. These waterdwelling animals are either attached to a surface or float in the water currents. Thus, like sponges, they are dependent on food carried in the water. These animals, however, can eat live prey as well as particulate food. They are equipped with tentacles that can reach out and trap smaller animals and paralyze them with poisoned darts called nematocysts. The tentacles then bring the food into a distinct body cavity, the gastrovascular cavity, through its one opening. The digestive cavity is, at least partially, not in direct contact with the water currents around the animal. Digestion can take place through extracellular enzymes secreted by the cells lining this cavity. The resulting molecules are then absorbed through the cell membranes. Amoebocytes also function in these animals. These animals have limited motion through musclelike cells and this motion moves fluid within the gastrovascular cavity, thus carrying fluid to all parts of the animal.

Flatworms are more advanced than sponges and coelenterates and live in a moist, but not watery, environment. They have a distinct nervous and muscular system and can move to search for food. Their digestiv e tract, as that of sponges, has only a single opening. Food is pushed into this opening by the muscle action of the first part of the digestive tract, which can be protruded to the outside of the animal. Digestion is extracellular and carried to the rest of the animal through muscle contractions of the digestive tract. The digestive tract is highly branched and extends to all parts of the animal.

DIGESTIVE TRACTS OF MORE COMPLEX ANIMALS

Animals more highly evolved than flatworms, including roundworms, insects, fish, mammals, and birds, have a functionally similar digestive tract. These animals all have similar requirements, which have necessitated further, more efficient digestion. These animals are more active and thus must ingest more food. Their digestive tracts have two openings, allowing a continuous digestion: Food enters at one end and is excreted at the other. In contrast, an animal with only one digestive opening cannot excrete and ingest at the same time. The greater size of these more-evolved animals also requires that digested food be absorbed into the circulatory system so that distribution to the rest of the body cells is quick. The absorbing portion of the gut is therefore surrounded by blood vessels.

Further adaptations have required sophisticated specializations of the digestive tract. These include initial chewing devices that can mechanically reduce the size of food so that it can be swallowed. Parts of the digestive system have evolved to store food until it can be efficiently digested. This adaptation allows animals to eat sporadically, when food is available, and allows time for other activities, such as hunting or hiding. Other portions of the digestive tract have become specialized to secrete powerful enzymes that sequentially break down the molecules in food to smaller and smaller molecules. Last, the terminal portions of the digestive tract retain food and extract any remaining nutritional value and eliminate the rest at a convenient time. Many of these adaptations required the formation of a space, called a coelom, between the digestive tract and the rest of the body. This space allows the gut to coil and thus become much longer than the animal, with a resulting increase in the surface area available for digestion and absorption.

The mouth, or buccal cavity, is designed for the entry of food into the digestive tract. The lips and tongue are highly sensitive to the texture and taste of food. They are capable of very precise movements because their musculature is supplied with an extensive nerve supply. The tongue can move laterally, up and down, and in and out, because it has both longitudinal and circular muscles. Movements of the jaws during chewing (mastication) cause the teeth to crush and tear food in the mouth. The teeth have an outer covering of very tough enamel, which protects them against abrasion. Some animals have teeth that grow throughout their life and replace the worn-out ends. Salivary glands in the sides or base of the jaw secrete saliva through ducts that empty into the mouth at the sides of or under the tongue. Saliva has the primary function of lubricating and wetting chewed food. Saliva contains an enzyme, called salivary amylase, which begins the digestion of starch, although the digestion is greatly slowed after the food enters the acidic stomach lumen.

After the food has been reduced to small particles and mixed with saliva, it is swallowed (deglutition). Swallowing is partly a reflex action, controlled by a center in the base of the brain. The tongue rises to the roof of the mouth, pushing the rounded mass of chewed food, called a bolus, into the opening of the esophagus. Further propulsion is created by contraction of the area between the mouth and the esophagus, called the pharynx. The esophagus is a muscular tube leading to the stomach. Contractions of muscles which encircle the esophagus cause a moving ring of contraction, called peristalsis, which propels the bolus into the stomach.

MUCOUS LAYERS

The wall of the gastrointestinal tract is similar throughout its length. The layers, from lumen outward, are the mucosa, submucosa, submucosal nerve plexus, circular muscle, myenteric nerve plexus, longitudinal muscle, and the thin connective tissue covering called the serosa. The stomach and intestines are suspended from the back wall of the abdominal cavity by a sheet of connective tissue called the mesentery. Nerves, blood vessels, and lymphatic vessels reach the gut in the mesentery.

The cavity of the gut, or lumen, is lined by a single sheet of cells called the mucosa. The mucosa contains a wide variety of cell types. Most of the mucosa is composed of a cell type which is called columnar epithelium because the cells are longer than their diameter. Mucous, or goblet, cells secrete mucus, which is the viscous slippery material that protects the cells of the gut against mechanical abrasion and chemical attack. Other cells secrete enzymes into the lumen. Hydrochloric acid is secreted by parietal cells in the stomach. Other cells in the small intestine secrete basic bicarbonate ion. These cells provide the degree of acidity or basicity appropriate to the different regions of the gut. Other cells are adapted to absorb nutrients from or to secrete fluid into the lumen of the gut.

There are also many endocrine cells in the mucosa. These cells secrete hormones into the blood when they are stimulated by nerves or by the contents of the gut. These hormones control the degree of motility or secretion of the gut and the metabolic and physiological responses of the body following feeding. Indeed, the digestive system is the largest endocrine gland in the body. These same hormones are found in the brain, where they act as neurotransmitters, and in other endocrine glands. They have numerous functions revolving around the digestive tract. Some of these hormones can increase or decrease hunger. Others prepare the body for the nutrients that will be absorbed from the digestive tract so that the nutrients can be efficiently utilized.

Certain hormones can be released by different types of nutrients in the lumen of the digestive tract. Other hormones can be released through the action of nerves when food is eaten.

The layer next to the mucosa, called the submucosa, is composed of fibrous connective tissue. It provides a mechanical support for the mucosa and also contains the nerve and blood supply leading to and from the mucosa. The lymphatic vessels draining the mucosa also travel through the submucosa.

Nerve and Muscle Layers

The next, more external layer, is a sheet of nerves, called the submucous (Meissner) plexus. These nerves send fibers inward to the mucosa and also outward to the other layers. They respond to the luminal contents and to other nerves and hormones. There are as many nerves in the gut as there are in the spinal cord. They are an intrinsic nervous system of the gut—that is, they begin and end in the gut. They are considered a separate category along with the autonomic (involuntary) and somatic (voluntary) nervous systems. The next layer of the gut wall is a layer of visceral smooth muscle oriented circularly around the circumference of the gut. Contraction of these muscles causes a ring of contraction that may or may not move down the intestine. Next, there is another layer of nerves called the myenteric (Auerbach) plexus. Both nerve plexuses are responsible for controlling and integrating the functions of the intestine. Motility of the muscles of the gut, absorption of salt, water, and nutrients, and blood flow are all regulated by these nerves. The outermost layer of the gut is composed of visceral smooth muscle oriented longitudinally along the gut. Contractions of these muscles shorten the length of the gut.

There are also rings of smooth muscle, called sphincters, which control the movement from one part of the gut to the adjacent part. These sphincters are found between the esophagus and the stomach, the stomach and small intestine, the small and large intestine, and the large intestine and the outside.

Food that enters the stomach is partially digested by the enzyme pepsin, which is secreted by the chief cells of the gastric mucosa. Pepsin begins the digestion of protein. The hydrochloric acid secreted by the parietal cells has the functions of activating the pepsin and killing bacteria. The most necessary function of the stomach is storage of food (now reduced to a semiliquid state called acid chyme, or chyme) and slowly propelling it into the small intestine. Additionally, the stomach secretes a substance called intrinsic factor, required for absorption of vitamin B12, which promotes red blood cell formation.

Ruminants, such as cattle and sheep, have the end of the esophagus and the beginning of the stomach modified into large chambers, called the rumen and reticulum, in which food is stored. These portions of the stomach are alkaline because of the enormous volume of basic saliva secreted by the animal. Bacterial digestion of the chyme occurs in these chambers. In addition, the contents can be regurgitated into the mouth and this cud then chewed further. After the cud is chewed and reswallowed, it bypasses the previous chambers and enters a third chamber, called the omasum, where it is churned by muscular contractions. Finally, it enters the abomasum, which is similar to the stomach of other animals. Birds have specialized adaptations of the stomach, called the crop and gizzard. The crop is a large structure at the beginning of the stomach that stores food until it enters the stomach. The gizzard is a muscular portion of the stomach that grinds the food. This grinding by the gizzard is necessary because birds have no teeth. Frequently, birds will ingest small stones, which are stored in the gizzard and help grind the food.

The Small Intestine

The stomach empties into the small intestine. The first portion of the small intestine is called the duodenum, the middle portion the jejunum, and the terminal portion the ileum. There are two large organs that are connected to the duodenum through ducts that empty into its lumen. These organs are the liver and the pancreas. The liver secretes bile salts, which are necessary to emulsify fats into small particles for absorption. Bile salts are stored in the gallbladder between meals. The gallbladder is connected, by a branch, to the duct leading from the liver to the duodenum. The pancreas secretes basic bicarbonate, which helps neutralize stomach acids that enter the duodenum. The pancreas also secretes many different digestive enzymes, which break down proteins, fats, carbohydrates, nucleic acids, and other large molecules. Thus, as soon as chyme enters the duodenum, it is immediately mixed with digestive enzymes and bile salts that entered the lumen from the pancreatic and bile ducts.

The chyme is mixed and propelled along the small intestine by longitudinal and circular muscle contractions. These contractions continually mix the chyme with the pancreatic enzymes and bile salts and present the digested molecules to the mucosal surface, where further digestion takes place. Most of the mucosal cells, called enterocytes, produce enzymes and absorb nutrients. Enterocytes are continuously formed in mucosal pits, called crypts. They migrate up tiny fingerlike projections, called villi, which protrude into the lumen of the gut. It takes about three days for the enterocyte to travel from the base of the crypt to the tip of the villi, and then it is sloughed into the lumen. The villi are thought to increase the surface area of the gut on which digestion and absorption take place. The enterocytes produce enzymes that are attached to the mucosal surface of the cells. These enzymes are responsible for the final stages of digestion, producing the smallest molecules, which are now in a form that can be absorbed by the intestine. Because the digestion takes place on the cells' surface, it is called contact digestion.

After molecules are in their completely digested form, they are absorbed by enterocytes, which transport them from the lumen of the gut to the circulatory or the lymphatic system. Most organic nutrients, such as amino acids, fats, and glucose, are absorbed in the first half of the small intestine, the duodenum and jejunum. Salt, water, and bile salts are absorbed primarily in the ileum. Absorption is virtually complete as long as the digestive system is functioning normally. Usually, the main problems that arise during gastrointestinal disorders are associated with malabsorption of fats. Fats require bile salts to be emulsified. Emulsification is necessary for enzymes to break down fats and also to reduce the final size of the fat microdroplet that results. If any step in this process is not functioning well, then the fats come out of suspension in the intestine and are excreted. The final contents of the small intestine consist mostly of salts, water, indigestible fiber, and the debris from sloughed enterocytes. The small intestine empties into the large intestine, where some bacterial digestion occurs, which produces mostly small fatty acid molecules. The debris from these bacteria add to the bulk of the undigested material. Muscle contraction propels these feces through the large intestine until it is eliminated by defecation. Sphincters control the final evacuation.

STUDYING THE DIGESTIVE TRACT

The structural features of the digestive tract can be determined by classical techniques of anatomical dissection and histological examination of the cellular characteristics of the different sections of the digestive tract. The secretions and the digestive steps can be determined by sampling the luminal contents. The sampling can be done by passing a tube through the digestive tract until the end reaches the desired portion and then withdrawing a sample for biochemical analyses.

Motility can be measured by attaching a balloon to a tube passed into the digestive tract and measuring the changes in pressure from muscle contractions. Absorption can be measured by perfusing a solution of known composition from one opening in a double tube and collecting the solution remaining after it has passed through the gut lumen from a second opening.

Motility of the intestine or the presence of obstructions that prevent the passage of food along the gastrointestinal tract can be observed by X-ray techniques. A liquid substance, such as a barium suspension, which is opaque to X rays, is swallowed. A series of X rays is taken, or continuous monitoring by an X-ray camera is used. Obstructions can be visualized from the buildup of barium above the blockade. The speed of movement can be estimated to determine if the overall motility of the gastrointestinal tract is abnormal. X rays can also be used to determine directly the presence of abnormal structures such as gallstones, which form in the bile ducts, or tumors. The bile duct and gallbladder system can be visualized with X rays by administering a radioopaque dye that is secreted by the liver into the duct system. The overall integrity of the gastrointestinal tract can be determined by ingesting inert substances of different molecular sizes and determining if they appear in the blood. Normally, only relatively small molecules can penetrate the very tight mucosal lining of the gut, unless they are nutrients of the body. The penetration of larger molecules across the mucosa indicates leaks resulting from damage to the gastrointestinal lining.

—*David Mailman*

See also: Anatomy; Cannibalism; Digestion; Nutrient requirements.

FURTHER READING

Arms, Karen, and Pamela S. Camp. *Biology*. 4th ed. FortWorth, Tex.: Saunders College Publishing, 1995.Aclear and well-illustrated general biology text. Presents the functions of the digestive tract in different phyla of animals and specifically of higher animals.

Ganong,William F. *Review of Medical Physiology*. 19th ed. Stamford, Conn.: Appleton and Lange, 1999. An advanced text, but has well-illustrated and concise explanations of digestive functions and structures. The structures of the digestive tract, their functions, and the control mechanisms are thoroughly covered. The text is very useful because the topics are presented in small sections and subsections.Well indexed for locating information.

Guyton, Arthur C. *Textbook of Medical Physiology*. 10th ed. Philadelphia:W. B. Saunders, 2000. An advanced-level medically oriented physiology text. Strong emphasis on disease mechanisms as well as normal structure and function.

Johnson, Leonard R., and David H. Alpers, et al., eds. *Physiology of the Gastrointestinal Tract*. 3d ed. 2 vols. New York: Raven Press, 1994. Acomprehensive, up-to-date survey of the state of knowledge in gastrointestinal function. Sections cover regulation and growth; motility; salivary, gastric, and pancreatic secretion; and digestion and absorption. Aimed at professionals in the field, but understandable by graduatelevel students.

Johnson, Leonard R., and Thomas Gerwin, eds. *Gastrointestinal Physiology*. 6th ed. St. Louis: C. V. Mosby, 2000.Apopular, readable textbook for graduate and medical students. Highlighted key terms, summaries, review questions. Illustrated.

Tortora, Gerard J., and Nicholas P. Anagnostakos. *Principles of Anatomy and Physiology*. 9th ed. New York: JohnWiley & Sons, 2000. An introductory-level text. Includes detailed illustrations of anatomy and cellular structures of the digestive tract and simple explanations of function. Points out the relationship to normal life and pathologic events.

DISEASES

FIELDS OF STUDY

Anatomy, histology, immunology, physiology

SUMMARY

Diseases may be caused by infectious agents, such as bacteria or viruses, by parasites, by inheritance, by developmental mistakes, or as a result of aging. The major way in which animals are able to overcome disease is through the activity of the immune system.

PRINCIPAL TERMS

- **bacteria:** single-celled microorganisms that are often the cause of infectious diseases in animals
- **developmental disorders:** diseases caused by embryonic or fetal mistakes in normal development
- **diseases of aging:** loss of functions required for health due to age-related degeneration of tissues
- **genetic diseases:** disorders caused by lack of enzymes or structural proteins caused by mutations
- **immune system:** system that produces antibodies and cells that attack foreign substances and pathogens that invade the body
- **parasites:** protozoans, fungi, or animals that survive by obtaining nourishment froma living host, frominside the host or on its surface
- **prions:** infectious proteins that cause neurological diseases such as "mad cow disease" viruses: noncellular infectious agents that must enter a host cell to infect it

BASIC PRINCIPLES

A disease is essentially any disturbance in the structure or function of the body, and may be accompanied by characteristic and well-defined areas of damage, or lesions, of specific tissues. The study of diseases is called pathology, and the causes of a disease are referred to as its etiology. An agent that can cause a disease is a pathogen. A particular disease has characteristic objective physical signs that may be seen by the examining physician or other outside observer, such as redness or swelling. Subjective feelings associated with disease, which can be described by an affected human but not by an animal or outside observer, are symptoms that might include pain or weakness. A set of signs and symptoms associated with a particular disease is called a syndrome. Diseases are often diagnosed by recognizing the signs and symptoms that characterize a particular disease and are not seen together in other diseases.

INFLAMMATION AND IMMUNITY

Diseases related to tissue damage may be caused by pathogens such as bacteria and viruses, or by trauma, toxins, heat or cold, or exposure to chemical pollutants. When the body first responds to damage, certain cells release chemicals that produce a localized, nonspecific reaction called inflammation. Histamine causes blood vessels to dilate and become leaky, so inflamed areas show redness, swelling, pain, and heat, and may have a loss of function. Inflamed tissues also send out chemical messages that call in wandering cells (neutrophils and macrophages) from the blood and connective tissues to engulf pathogens and cellular debris by phagocytosis. This response is usually restricted to a localized area where tissue damage has occurred, and the area may be walled off from the rest of the body to restrict movement of pathogens or toxins

The immune system involves other cells from the blood, lymph, and connective tissues called lymphocytes, which respond to specific antigens, usually small fragments of proteins that are not part of the "self" antigens the lymphocytes recognize. A bacterium may have thousands of different antigens associated with it, fragments produced by a macrophage during processing of the bacterium after phagocytosis. Each lymphocyte can respond against only one kind of antigen, but millions of different kinds of lymphocytes in each individual produce a widespread immune response specific to each foreign antigen associated with the invading pathogens, their toxins, or other materials.

The first time the immune system encounters a foreign antigen, its primary response is slow, and a disease may result from a pathogen's metabolic effects. Eventually, the immune response generates activated lymphocytes and antibodies that kill the bacteria or the virally infected cells to end the disease process. Memory lymphocytes are also produced that will respond against the same antigen if needed later. When the animal recovers, it will usually be immune to a second infection by the same disease-producing agent. The ability to resist a second infection is called immunological memory, and it may last for the life of the individual, as long as the memory lymphocytes live. Modern disease prevention techniques use immunizations to prevent the first experience of disease caused by a pathogen. In immunization, a derivative of the pathogen is injected into the individual to produce the slow primary response, so that memory is generated and the individual will be immune when the same agent is encountered naturally in the environment. In some cases, a booster shot must be given regularly to maintain memory and immunity to that agent, as in repeated immunizations of pets against rabies.

In some cases, however, diseases are caused by an overreaction by the immune system and the inflammation that it helps to generate. Allergies, for example, are not directly caused by pollen or dust particles, but by the body's responses to these allergy-producing antigens, or allergens. An allergic reaction is an immediate hypersensitivity response that may just cause an irritating, itchy swelling of the mucous membranes or skin, or maybe extreme and even life threatening. In highly allergic individuals responding to allergens, the respiratory passages close, blood vessels leak fluid into the tissues, and death can result in a hyperallergenic process called anaphylactic shock. Much more often, though, both the inflammatory and immune responses are protective, causing the destruction of invading pathogens or other foreign materials that get into the body past the barriers of the skin and mucous membranes. The extent to which the immune system protects against disease can be seen when it is not functioning, as in humans who have acquired immunodeficiency syndrome (AIDS) and die of infections or cancer that would be prevented by a fully functioning immune system. Viruses similar to the human immunodeficiency virus (HIV) cause related immunodeficiency diseases and leukemias or anemias in animals as well, including simian immunodeficiency virus (SIV) in monkeys and feline leukemia virus (FeLV) in cats or murine leukemia viruses (MuLVs) in mice. Studying these related viruses has been very important in scientists' understanding of HIV.

BACTERIAL DISEASES

Most inflammatory diseases are caused by pathogenic or infectious microorganisms. Infections occur when pathogens enter the body, replicate, and cause metabolic changes or toxic damage. Pathogens are often easily spread directly from an infected host to uninfected individuals, or may be spread indirectly through air, food, water, soil, or other materials. Infectious and other diseasecausing agents include bacteria, fungi, viruses, prions, protozoans, and parasitic animals.

Bacteria are unicellular, prokaryotic organisms that have much simpler cells than those of multicellular organisms such as animals, which are eukaryotic (having a nucleus). Some bacteria are pathogenic in humans and other animals. Bacteria can reside normally on the skin or in the digestive tracts of animals without harmful effects, comprising the "normal flora" of those animals. A commonbacterium, *Escherichia coli*, is found in the colons of humans and many other animals, but different strains may cause diseases in some hosts while being harmless in others. *E. coli* strain O157:H7 causes no disease in cattle, but has caused deadly infections in children who ate undercooked hamburger contaminated with fecal material fromthe infected cattle. *Salmonella* are bacilli common in and on animals such as chickens (and their eggs) and turtles, but can cause food poisoning in humans. In some cases, bacteria that are normal on the skin or in internal organs of an animal may cause diseases when their number increases abnormally or when they are transferred to other body regions. Certain bacteria can cause disease in both animals and humans, such as the anthrax bacterium *Bacillus anthracis*, which may be used in weapons for germ warfare. Infectious anthrax spores can persist in soil or on contaminated wool, leather, or other animal products for many years, then can enter through skin or lungs to cause a lethal disease of animals or humans. Wild animal populations may serve as reservoirs of bacteria that can cause epidemics in both humans and animals, such as plague bacteria in modern-day prairie dogs of the American Southwest. The same bacteria, *Yersinia pestis*, were also a major problem in medieval Europe, when the Black Plague was carried to humans by dying rats and their fleeing fleas.

Viral Diseases

Viruses are not cells, and so are not actually microorganisms, but are still infectious agents. All viruses can infect only certain cells of particular host organisms, and they must be inside host cells to replicate. Viral infections can be latent (the cell is infected but uninjured); in some latent infections the virus is totally silent and inactive, while in others new virus particles may be released into the tissue fluid surrounding the cell. In cytopathogenic infections, host cells may increase in size and number, causing tissue enlargement (hyperplasia), or the cells may die. Some latent viral infections can become activated to produce disease, cell death, and viruses that infect other cells or organisms. Activation of latent viruses occurs when the host's immune defenses are diminished by stress or illness. Antibiotics are ineffective against viruses, whether they are free in tissue fluids or inside host cells, because the actions of antibiotics are directed against bacteria. Drugs that counteract viral infections often kill the infected cells, and may cause toxic reactions in the host as a result. Most viral infections can be handled at some level by the immune system, but in some cases the viruses may be present in the host organism for the rest of the host's life span.

Prion and Fungal Diseases

A very unusual kind of infectious pathogen is an abnormal protein called a prion, not associated with eitherDNAor RNA. Neurodegenerative diseases such as scrapie in sheep and bovine spongiform encephalopathy (BSE, or mad cow disease), as well as kuru and Creutzfeld-Jacob disease in humans, are prion diseases. The abnormal protein is a refolded form of a normal brain protein, and prions cause the normal proteins to fold into the prion form in brains of infected individuals. Prions are very stable and are not destroyed by heat, light, or acid, so prions eaten in animal proteins can cause rapid or very delayed disease in the individual that ingested the prions. Animal tissues have been banned in animal feed in Europe and elsewhere, and many Europeans have stopped eating beef because of outbreaks of mad cow disease in England.

Fungi are plantlike parasites without chlorophyll, subdivided into yeasts and molds. They may cause skin infections or localized or systemic internal infections. Yeasts are single cells that are much more complex than bacteria, while molds have branching filaments (hyphae) extending into host cells to obtain nutrients. Fungi that colonize the skin (dermatophytes) may cause diseases such as mange or ringworm, while yeasts infect mucous membranes or other moist surfaces. The immune response can often control or eliminate fungal infections, but such infections can be lethal in immunocompromised hosts.

Parasites may enter the body or remain on the body surface as they obtain nutrients froma living animal host. The condition of having parasites is called infestation rather than infection, but disease generally results from the parasites' effects on the host. Diseases caused by protozoan parasites include amebic dysentery, giardiasis, toxoplasmosis, malaria, and pneumocystis pneumonia. Important parasitic worms are roundworms, tapeworms, and flukes, which infest nearly all animals, and often have multiple host species during their complex life cycles. Worms may reside in the digestive tract, heart, liver, lungs, eyes, lymphatic system, skeletal muscle, or other organs and systems, where they can cause malnutrition, tissue damage, and death of the host. Blood-sucking insects such as lice and mosquitoes, or other arthro- pods such as ticks and mites, may act as vectors that transmit protozoan parasites, bacteria, or viruses to a host animal during blood removal.

Arthropod-vectored diseases with animal reservoir hosts have caused many major diseases in humans as well, such as malaria, yellow fever, hantavirus diseases, sleeping sickness, and several forms of encephalitis. All animals are subject to infestation by many parasites against which the immune system responds; reduction of parasite load in humans and domestic animals in industrialized societies is thought to be related to the increased incidence of allergy in both humans and pets.

Genetic and Congenital Diseases

Diseases that result from genetic abnormalities may be present at birth, or may not become apparent until later in life when metabolic processes fail to function because of inherited errors. Both single gene mutations and chromosomal abnormalities may occur, or diseases can be caused by interaction of genetic predisposition and environmental factors. Congenital diseases are present at birth, and can be genetic or developmental in cause. Developmental problems may be associated with nutrient deficiency, intrauterine injury, inadequate placental support for the fetus, or environmental agents such as radiation, toxins, or pathogens.

Genetic problems tend to occur particularly in inbred lines of animals, such as purebred dogs, where breeding selection for desirable characteristics also inadvertently produces recessive inherited diseases such as hip dysplasia and deafness. Some congenital defects are considered desirably exotic in companion animals, such as curled ears or stubby tails in cats, droopy ears in rabbits, or short legs, flattened faces, or lack of hair in dogs.

Metabolic, Neoplastic, and Degenerative Diseases

Metabolic disorders include those that are strictly genetic, and those that have a combined etiology involving inheritance and environment. Disturbances of metabolism can include changes in endocrine functions or metabolic imbalances when an enzyme is missing due to a genetic mutation. The enzyme's substrate would then build up and cause damage, while the enzyme's product would not be formed, also causing problems. Neoplastic diseases are characterized by abnormal cell division and tumors, enlarged growths that may be benign or malignant. Benign tumors do not spread throughout the body, but are usually enclosed in a dense connective tissue capsule. While their cells remain relatively normal aside from their unrestrained growth, benign tumors may grow to enormous size and cause death by compressing organs or blocking passageways.

Malignant tumors contain many more abnormal cells, which can leave the primary tumor site and form secondary tumors, especially in the liver, bone marrow, lungs, or brain, that usually cause death in cancer conditions. Malignant cells lose their original characteristics and become much less specialized and less efficient in using nutrients and energy, causing the body to waste away. Degenerative diseases are associated with the aging process, when body tissues lose the ability to repair themselves effectively. The immune system also becomes less functional in combating foreign antigens or even recognizing the difference between self and foreign antigens, thus attacking self antigens by mistake. In some cases, degenerative diseases are not directly linked to aging, but to damage caused by pathogens, toxins, nutrient deficiency, or even nutrient excess. As normal tissues are damaged by the standard wear and tear of life, repairs become less effective and scar tissue replaces normal tissues such as muscle or liver.

Nervous system damage is particularly problematic, since neurons are unable to replicate in mature animals, and lost cells are not replaced, producing sensory reception, muscle control, and memory loss. Problems caused by diseases and trauma lead to continued loss of function over time in aging animals, eventually reaching a point that repairs can no longer bemadeor infections resisted, and the animal dies.

—*Jean S. Helgeson*

See also: Aging; Biology; Cell types; Death and dying; Genetics; Immune system; Life spans; Mutations; Nutrient requirements.

Further Reading

Campbell, Neil A., Jane B. Reece, and Lawrence G. Mitchell. *Biology*. 5th ed. Menlo Park, Calif.: Benjamin Cummings/AddisonWesley Longman, 1999. This college text discusses bacteria, viruses, fungi, and invertebrate animals in several chapters, discussing their structures and functions in general. It also covers the defensive mechanisms of the body, including inflammation and the immune system.

Crowley, Leonard V. *An Introduction to Human Disease: Pathology and Pathophysiology Correlations*. 5th ed. Sudbury, Mass.: Jones and Bartlett, 2001. While written for nursing students concerned with the human body, this book discusses many general aspects of diseases and their causes. The latter part of the text covers specific diseases of different organ systems and how they can be treated.

Fox, Michael W. *The Animal Doctor's Answer Book*. New York: Newmarket Press, 1984. This general audience paperback written by a scientific director of the U.S. Humane Society lists questions and answers from his newspaper column. It mainly concerns health and diseases in all sorts of pets and even wild animals, but aspects of animal behavior and careers in animal care are also discussed.

Koneman, Elmer W., et al. *Color Atlas and Textbook of Diagnostic Microbiology*. 5th ed. Philadelphia: Lippincott, 1997. The first two chapters in this medical text are introductory and cover basic bacteriology and the diagnosis of infectious diseases. Later chapters consider everything about bacteria, viruses, fungi, and parasites that cause diseases in humans and animals. Color plates show what these pathogens look like in culture and in wounds.

Norris, June, ed. *Diseases*. 2d ed. Springhouse, Pa.: Springhouse, 1997. Written for nursing and allied health professions students, this text examines infections, trauma, neoplasms, and other disorders of every body system. Many tables cover causes, assessment, diagnostic tests, and treatments for these disorders.

BACTERIAL INFECTIONS AND ANTIBIOTIC RESISTANCE

Bacteria are prokaryotes (cells without nuclei) that have three basic shapes, called cocci (spheres), bacilli (rods), and spirilla (corkscrews). Most have a cell wall surrounding the plasma membrane and may also have a protective capsule. The cell wall may stain gram-positive (purple) or gram-negative (pink), distinctions that help identify the bacteria, determine what disease they cause, and indicate antibiotics that may be effective in treating infections they produce. Antibiotics are antimicrobial drugs that usually work by interfering with the structure or function of bacterial cell membranes, cell walls, or proteins. Since bacterial cells are so different from those of the infected host animal, antibiotics usually do not affect the host cells. Antibiotics are often given to domestic agricultural animals to increase their rates of growth and productivity. Unfortunately, mutations in normal bacteria can make them resistant to antibiotics, and this resistance can be passed from one bacterium to others, so that many drug-resistant strains of bacteria have resulted, some of which cause diseases that may become untreatable.

FINDING NEW VIRUSES

Viruses are smaller and less complex than cells, and generally contain only a protein coat surrounding either deoxyribonucleic acid (DNA) or ribonucleic acid (RNA) as the genetic material. Some viruses may also have a covering membrane derived from the plasma membrane of the host cell from which they budded out during replication. DNA viruses include adenoviruses (respiratory infections); herpesviruses (cold sores, genital herpes, "pox" diseases, mononucleosis); papillomaviruses (certain warts); and hepatitis B virus. Viruses with genetic information in RNA include hepatitis A and C viruses; myxoviruses (measles, mumps, influenzas); picornaviruses (polio, respiratory infections); rabies virus; and retroviruses (AIDS, some leukemias). Viruses that infect animals without causing disease may jump species to infect humans or mutate to produce major disease epidemics, such as influenzas from recombinant bird or pig viruses, or the AIDS and Ebola viruses. In some cases, the animal source or reservoir is known, while often it is unknown. Many emerging diseases in human populations have originated in animals native to rain forest jungles or other habitats that humans have invaded and taken over, encountering previously unknown viruses.

DNA ANALYSIS

FIELDS OF STUDY

Biochemistry; biotechnology; molecular biology; molecular genetics; population genetics; forensic science; statistics.

SUMMARY

DNA analysis involves the use of scientific tools to access the information found in DNA to identify its source, whether some infectious agent, another organism of interest, or a particular individual, such as in forensic applications. Medical applications of this technology include the search for mutations associated with genetic disorders and the design of probes that are able to diagnose these disorders in a timely fashion.

PRINCIPAL TERMS

- **base pair:** single unit of double-stranded DNA.
- **Combined DNA Index System (CODIS):** Federal Bureau of Investigation (FBI) database that contains the DNA profiles of more than 8 million convicted violent offenders.
- **DNA (deoxyribonucleic acid):** nucleic acid that is the genetic component of all living cells. Nucleic acids are polymers of nucleotides.
- **nucleotide:** chemical composed of a nucleoside (a nitrogen-containing purine or pyrimidine base linked to a five-carbon sugar such as ribose or deoxyribose) and a phosphate or polyphosphate group.
- **polymerase chain reaction (PCR):** technique used to produce a large amount of DNA from very small quantities.
- **restriction fragment length polymorphism (RFLP):** early method of DNA analysis involving the cleavage and separation of DNA.
- **short tandem repeat (STR):** repeating element in DNA that is from one to six base pairs long; also known as microsatellite.
- **single nucleotide polymorphism (SNP):** point mutation in DNA that differs among individuals.
- **Southern blotting:** process used to detect the presence of a particular DNA sequence in a sample. DNA fragments are separated using gel electrophoresis, then transferred, or blotted, onto a membrane, where they are exposed to a hybridization probe (a fragment of DNA that will hybridize to a complementary sequence).
- **variable number of tandem repeats (VNTR):** repeating element in DNA that is tens to hundreds of base pairs long; also known as minisatellite or long tandem repeat.

BASIC PRINCIPLES

DNA analysis is, in the strictest sense of the term, an actual observation of the length or sequence of a portion of DNA. The length of a fragment of DNA can be determined using gel electrophoresis. This technique involves placing DNA onto a semisolid support, or gel, and applying an electric current to the gel so that DNA migrates toward the positive pole. The migration of DNA in gel electrophoresis is proportional to its mass, which is, in turn, proportional to its length. Determining the actual sequence of bases in a strand of DNA is much more complicated.

Although DNA analysis has at times been equated with genetic testing, the two are not always the same. Certain types of genetic testing developed in the 1960's did not technically involve DNA analysis. Amniocentesis, which allowed for Down syndrome testing, actually involved chromosomal analysis following the creation of a karyotype (organized profile of a person's chromosomes). Similarly, genetic testing for phenylketonuria (PKU) and Tay-Sachs disease originally involved enzyme assays, not an analysis of the defective genes themselves. Thus, actual DNA analysis did not begin in earnest until the mid-1970's.

BACKGROUND

Although the double helical structure of DNA was first described in 1953 by American molecular biologist James D. Watson and British biophysicist Francis Crick, more than twenty years passed before scientists developed methods of comparing DNA for the purpose of identification. In 1974, British molecular biologist Joseph Sambrook described the differentiation of human tumor viruses following cleavage by

DNA analysis

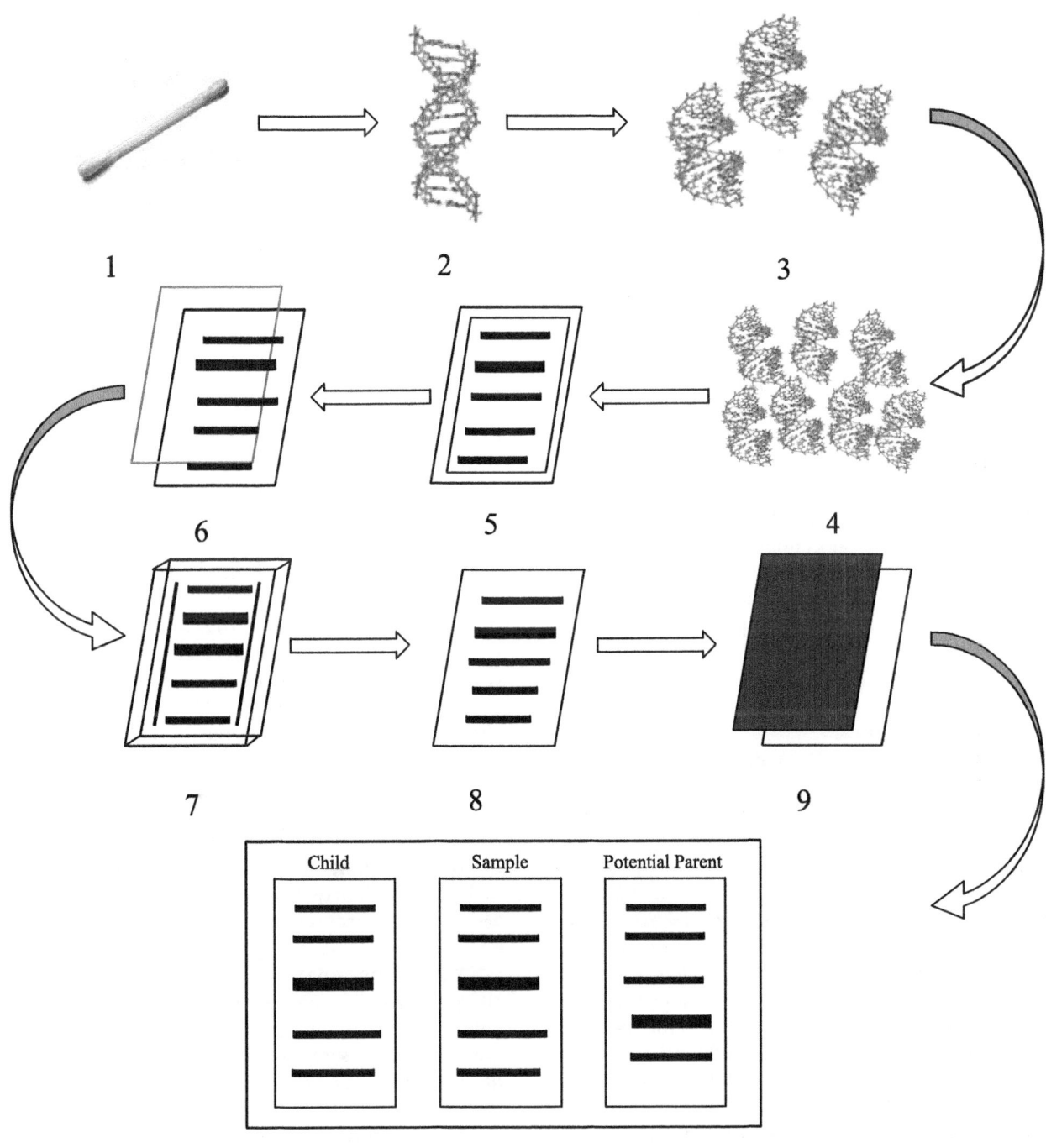

1: A cell sample is taken- 2: DNA is extracted from sample 3: Cleavage of DNA by restriction enzyme- 4: Small fragments are amplified by the Polymerase Chain Reaction 5: DNA fragments are separated by electrophoresis 6: The fragments are transferred to an agar plate 7: On the agar plate specific DNA fragments are bound to a radioactive DNA probe 8: The agar plate is washed free of excess probe 9: An X-ray film is used to detect a radioactive pattern 10: The DNA is compared to other DNA samples a restriction endonuclease (an enzyme that cleaves DNA at a specific nucleotide sequence). He noticed that different-sized bands of DNA were visible following their separation on a gel. This discovery formed the basis for what has become known as restriction fragment length polymorphism (RFLP). Although the DNA of viruses and even bacteria could be analyzed directly by RFLP, the restriction enzyme cleavage patterns of higher organisms were of sufficient complexity that only a subset of the bands that were produced could be analyzed. DNA analysis of higher organisms was made possible in 1975 by the development of the Southern blotting technique by British biochemist Edwin Southern. The following decade saw two ideas that would revolutionize the field of DNA analysis. In 1983, American biochemist Kary Mullis developed the polymerase chain reaction (PCR), a process that enabled the amplification across several orders of magnitude of small amounts of starting sample DNA in the laboratory. Then, in 1985, British geneticist Alec Jeffreys realized that human DNA was peppered with regions of repeating sequences, or variable number tandem repeats (VNTRs), and that comparisons of these regions could create a unique DNA fingerprint for any given individual.

PROBES AND PRIMERS
Most of the methods of DNA analysis take advantage of DNA's natural tendency to form a double helix. RFLP analysis has long been paired with Southern blotting. This procedure involves binding a small synthetic fragment of DNA to a region of interest that is contained within one or more of the bands of DNA that have been separated by gel electrophoresis and then blotted onto some type of membrane. This binding is made possible by the complementary nature of the DNA bases, the fact that adenine forms hydrogen bonds to pair with thymine and that cytosine pairs with guanine, a concept called Watson-Crick base pairing after the discoverers of DNA structure. The synthetic DNA, called a probe, is designed to contain about twenty complementary nucleotides to the sequence of interest. Binding of this probe to the blotted DNA is called hybridization. Originally, DNA probes were labeled with a radioactive marker to enable the detection of their position on a membrane, but later probe labeling has included nonradioactive alternatives such as fluorescent dyes.

The polymerase chain reaction (PCR) also involves the binding of small synthetic fragments of DNA to a region of interest, but in this process, two such fragments bind to opposite strands of the target DNA. These fragments, although identical in structure to the probes described earlier, are called primers because they are used to prime a DNA synthesis reaction. Also, the binding reaction, which involves the same process of complementary bases coming together to form hydrogen bonds, is referred to as annealing. It had been previously discovered that DNA could be made in the laboratory by taking a given single strand of DNA and adding a specific primer, the four types of nucleotides, and purified DNA polymerase (the enzyme normally involved in the polymerization process), but Mullis's insight in the early 1980's was that this process could be converted into a chain reaction that produced large amounts of DNA. By adding two primers instead of one and by using double-stranded DNA as a target, twice as many molecules of DNA could be created, but this necessitated a step in which the DNA had to be heated to near-boiling temperatures to separate, or melt, the two strands of the double helix. Mullis reasoned that a DNA polymerase that had been purified from a thermophilic microbe would be able to survive this heating step. This would allow the stringing together of a number of cycles with three different temperatures—one each for annealing, polymerization, and DNA melting—without having to add more DNA polymerase enzyme. Thus, after each cycle, the amount of DNA double helix would be doubled, resulting in more than a billion molecules of DNA after thirty cycles, even if the cycle started with only one strand of DNA.

DNA POLYMORPHISM
DNA analysis takes advantage of the intrinsic variability that exists among organisms as well as among members of the same species. Polymorphism, a word derived from the Greek for "many forms," is used to describe this variability, the simplest form of which is a single nucleotide polymorphism (SNP). Single nucleotide polymorphisms, which are also referred to as point mutations in genetics, are detectable by RFLP analysis only when they occur within the recognition sequence for a particular restriction enzyme because the enzyme fails to cleave the altered sequence. RFLP analysis also readily detects deletions or insertions of

DNA sequences that have occurred between restriction enzyme cleavage sites. What Jeffreys realized in the 1980's was that most restriction fragment length variation in humans was not caused by large insertions or deletions of unique DNA sequences but by a variation in the number of repetitive DNA elements that were found in tandem with one another. He did not, however, use the PCR method that was being developed at the time because a practitioner of PCR must know the precise sequences that flank a site of interest to design the primers used in this procedure. Instead, Jeffreys performed Southern blotting using a probe designed to hybridize with the about fifteen-nucleotide-long sequence that he was studying. This probe specifically labeled the regions of the membrane that contained these variable number tandem repeats (VNTRs). For this contribution, Jeffreys has been called the father of DNA fingerprinting.

Subsequent analysis of various regions in human DNA has taken advantage of PCR to produce results, focusing on even smaller tandem repeats with repeating units that are only one to six nucleotides in length. Discovered in 1989, these short tandem repeats (STRs) were eventually found to outnumber variable number tandem repeats by nearly one hundredfold, being found at more than 100,000 sites in human DNA. As more and more of these STR sites were characterized over time, primers that annealed to their flanking sequences were designed to amplify the repeat area in question.

APPLICATIONS

Although the tools involved in DNA analysis are often used in basic research such as determining the evolutionary relationships between organisms, much of the application of this technology involves analysis for the purposes of identification. Although identification could potentially include any organism of interest, the primary focus of DNA analysis has been disease-causing viruses and microorganisms along with humans. Ever since Sambrook and colleagues first applied RFLP analysis to differentiate between two strains of viruses, viral epidemiology has remained an important application for tools such as PCR. For example, around the beginning of the twenty-first century, nucleic acid amplification testing (NAAT) was developed to detect the viral load of the human immunodeficiency virus (HIV). The procedure is a faster and more effective way to test for the presence of HIV in a person. NAAT has also been applied as a diagnostic test for certain bacterial infections. Other PCR-based methods have been adapted to test for bacterial contamination of foods as well as of hospital areas and supplies. In most cases, the identification of the precise strain of virus or microbe present is unnecessary because the physical presence of an infectious agent, not its detailed classification, is of interest. Tandem-repeat-based methods of identification are largely useless when analyzing such infectious agents because these agents tend to lack such repetitive DNA sequences. Because the DNA sequence of the entire genome (the complete set of DNA found in a particular organism) of most known infectious agents has been determined, it is possible to design primers that will specifically amplify DNA from a given target species.

In some cases, as in life-threatening illnesses, potential epidemics, and acts of bioterrorism, the speed at which an infectious agent is identified is critical to saving lives. For such applications, a type of PCR called real-time PCR has been developed. Rather than waiting to run gel electrophoresis after the full thirty or so cycles of a traditional PCR reaction have been completed, real-time PCR measures the production of a fluorescent-tagged product in real time, during the early phases of the reaction. This allows for an agent to be detected in minutes rather than hours.

Crime Scenes and Beyond. The best-known use of DNA analysis is probably in the area of forensics. The first case in which Jeffreys applied DNA fingerprinting was an immigration dispute. In 1983, British authorities had denied a thirteen-year-old boy entry into the country, claiming that his passport was forged and that his stated mother, a British subject, was not his biological mother. The dispute continued until 1985, when Jeffreys was able to apply his new technique to prove that the maternal relationship stated on the passport was indeed correct. Since that time, maternity tests have been vastly outnumbered by paternity tests, but the principle used in both types of parental testing remains the same.

The first use of DNA fingerprinting in a criminal case occurred in 1986, when it was used to exonerate a suspect accused of the rape and murder of a teenage girl near Leicester, England. Later, the same technique was used to identify the real killer. Since this early case, evidence from DNA fingerprinting has helped convict thousands of criminals. The source of DNA is blood in about half of all cases; other common sources are semen and hair. DNA analysis also plays an important role in the identification of human remains following disasters, acts of terrorism, and war.

Limitations of PCR. Following the advent of PCR, the amount of forensic sample required for analysis was reduced significantly. The original DNA fingerprinting procedure developed by Jeffreys required a blood sample about the size of a quarter, but later methods needed only a few cells swabbed from a person's cheeks to perform an analysis. Although PCR requires much less starting material than RFLP analysis and is also a more rapid procedure to perform, it does have a number of limitations. The first limitation, that flanking DNA sequences must be known ahead of time, was largely overcome as more and more human short tandem repeats were characterized along with the DNA that surrounded them. A second limitation is that the method is so sensitive that it is prone to contamination by outside sources. Because even a single fragment of DNA can be amplified into large amounts on a gel, care must be taken not to introduce foreign DNA from an investigator's hair or fingertips. A third limitation is that only a single area, or locus, of DNA can be analyzed at one time. To overcome this limitation, a procedure called multiplex PCR has been developed. This method simultaneously employs a number of primers that have been labeled with fluorescent tags. These can be identified during the subsequent gel electrophoresis step based on their specific labels.

Medical Applications. Besides using DNA analysis to identify infectious agents, the medical community has begun to use this technique to study genetic disorders. However, common methods of DNA analysis cannot identify most genetic disorders, with the exception of a class of disorders called trinucleotide repeat expansion disorders. This rare class of disorders, which includes Huntington's disease as well as fragile X syndrome, is readily detectable using PCR amplification of the short tandem repeats that contribute to the disorders in question. A more common class of genetic disorders results from point mutations in genes and can therefore be linked to particular single nucleotide polymorphisms in the human genome. Unfortunately, single nucleotide polymorphisms are not detectable by PCR and will show up in RFLP analysis only if they occur in the restriction enzyme recognition site itself, which is a rare occurrence. The identification of genetic disorders is therefore largely dependent on determining the actual sequence of the DNA, still a technically challenging and expensive undertaking despite progress that has been made since the inception of the Human Genome Project DNA sequencing program in the 1990's.

Methods involved in DNA sequencing include many of the same principles as other forms of DNA analysis. A single primer is labeled with a fluorescent dye and mixed with a target sequence in the presence of a thermostable DNA polymerase. This procedure does not amplify the DNA as in PCR but results in primer extension for a certain length along the target sequence. Another difference from PCR is that modified nucleotides are added to this mixture so that the primer extension is halted whenever these particular nucleotides are incorporated into a growing DNA strand. Four separate tubes are used in this method, one for each of the four DNA bases. Once these four reactions are separated by electrophoresis, the order of bases can be determined using computer software that monitors the relative migration of the bands that occurs from each of the four reaction tubes.

FUTURE PROSPECTS

Single nucleotide polymorphisms (SNPs), although not used extensively in forensic applications, potentially contain valuable information that can be of use to crime scene investigators. For example, the presence of particular SNPs may indicate a perpetrator's race, while others could indicate hair color. One disadvantage of SNPs, besides the relative difficulty of identifying them, is that many more of them are needed to provide a unique identification (compared to the number of short tandem repeats needed for PCR). Because most SNPs are biallelic, they contain one base or another but generally not all four possible bases, and it is estimated that as many as fifty would have to be analyzed to obtain the same level of confidence as provided by the thirteen STR loci contained in CODIS. This may not prove as difficult as it sounds because it is estimated that there are probably about 10 million SNP sites scattered throughout the human genome. If accurate, that would mean that SNPs outnumber short tandem repeats to the same degree that short tandem repeats outnumber variable number tandem repeats.

DNA sequencing in some form or another is likely to continue to play an increasing role in DNA analysis. The cost of DNA sequencing probably will drop as it becomes more prevalent and increasingly automated. Although the first human genome sequence was produced at a cost of billions of dollars, scientists have set a goal of reducing the cost of DNA sequencing to about one thousand dollars. At the same time, scientists are developing a number of methods that allow SNPs to be determined without first finding the sequence of the 99.7 percent of DNA bases that do not exist as SNPs. These methods include directed hybridizations, ligations, primer extensions, or nuclease cleavages that specifically involve SNPs while leaving the rest of the DNA alone.

With any increase in the involvement of DNA sequencing in forensics comes the likelihood that debate will intensify concerning privacy issues regarding the use of sequence information. Unlike commonly used methods of PCR analysis, SNP determination will reveal certain details about suspects that could be open to abuse. Ethical issues involving the use and dissemination of DNA data will have to be resolved as the methods of DNA analysis continue to evolve.

James S. Godde, PhD

Further Reading

McClintock, J. Thomas. *Forensic DNA Analysis: A Laboratory Manual.* Boca Raton, Fla.: CRC Press, 2008. Examines the various methods of DNA analysis and DNA fingerprinting.

Nakamura, Yusuke. "DNA Variations in Human and Medical Genetics: Twenty-five Years of My Experience." *Journal of Human Genetics* 54 (2009): 1-8. A historic perspective on the progression of DNA analysis techniques with a particular emphasis on human disease characterization.

Pereira, Filipe, Joao Carneiro, and Antonio Amorim. "Identification of Species with DNA-Based Technology: Current Progress and Challenges." *Recent Patents on DNA and Gene Sequence* 2 (2008): 187-200. Contains an excellent table comparing methods of DNA analysis, along with a helpful flowchart. Also contains clear descriptions of each method, including diagrams.

Roper, Stephan M., and Owatha L. Tatum. "Forensic Aspects of DNA-Based Human Identity Testing." *Journal of Forensic Nursing* 4 (2008): 150-156. A straightforward description of all pertinent methods and applications of DNA analysis, including simple diagrams as well as a glossary of terms.

FASCINATING FACTS ABOUT DNA ANALYSIS

- In addition to Southern blotting, two other forms of blotting are performed in molecular biology. These involve blotting of either RNA or protein for analysis and are named Northern blotting and Western blotting, respectively, as a humorous homage to Edwin Southern.

- Ironically, Kary Mullis, who developed the method used to screen for the viral load of HIV in humans, is among the handful of scientists who reject the scientific evidence that HIV is the cause of acquired immunodeficiency syndrome (AIDS).

- Colin Pitchfork, the first criminal convicted of murder based on DNA fingerprinting evidence, initially evaded arrest by telling a friend that he was terrified of needles and paying that friend to submit a blood sample for him.

- DNA extracted from blood actually comes from the white blood cells, not the red, even though the latter outnumber the former by a ratio of 700:1. Human red blood cells lose their nuclei, and therefore their DNA, during development.

- The first human genome sequence produced was actually a mosaic of DNA sequences from various anonymous donors. In 2007, J. Craig Venter, the head of a private company involved in the Human Genome Project, became the first individual to have his entire genome sequenced. Venter said his company had largely used his own DNA in the sequencing efforts that they had contributed to the project.

- According to the latest estimates, the Bureau of Justice Statistics at the U.S. Department of Justice reports that tens of thousands of requests for DNA analysis are backlogged at any given time because of the high demand for this service. This represents the highest percentage of backlogged requests for any type of analysis performed by crime laboratories under their jurisdiction.

Rudin, Norah, and Keith Inman. *An Introduction to Forensic DNA Analysis*. 2d ed. Boca Raton, Fla: CRC Press, 2002. Discusses forensic DNA analysis from both the medical and legal standpoints. Examines the advantages and limitations of the various techniques.

Watson, James D., and Andrew Berry. *DNA: The Secret of Life*. New York: Alfred A. Knopf, 2006. This comprehensive introduction to DNA has the famous biologist Watson as one of its authors.

WEB SITES

Association of Forensic DNA Analysts and Administrators http://www.afdaa.org

The DNA Initiative Advancing Criminal Justice Through DNA Technology http://www.dna.gov

Federal Bureau of Investigation CODIS http://www.fbi.gov/hq/lab/html/codis1.htm

International Society for Forensic Genetics http://www.isfg.org

See also: Human Genetic Engineering.

DNA/RNA SYNTHESIS

FIELDS OF STUDY

Biochemistry; Genetics; Molecular Biology

SUMMARY

The basic process of DNA and RNA synthesis is described, and its importance in living biochemical systems is discussed. Also described are modern advances such as artificial methods for the synthesis of DNA and RNA and their applications.

PRINCIPAL TERMS

- **complementary strand:** one of the two strands of nucleotides that make up a DNA molecule, with each nucleotide in one strand corresponding to the position of its complementary nucleotide (cytosine for guanine, adenine for thymine, and vice versa) in the other.
- **deoxyribonucleic acid (DNA):** a large molecule formed by two complementary strands of nucleotides that encodes the genetic information of all living organisms.
- **gene expression:** the process by which RNA copies genes, which are specific segments of the DNA molecule, and uses the information to synthesize either proteins or other types of RNA.
- **nucleotide:** the basic structural component of DNA and RNA, consisting of a ribose (in RNA) or deoxyribose (in DNA) sugar molecule bonded to a phosphate group and one of five nucleobases: cytosine, adenine, guanine, thymine (DNA only), or uracil (RNA only).
- **polymerase chain reaction:** a laboratory method in which a very small amount of DNA can be replicated thousands or even millions of times, using free nucleotides and an enzyme called DNA polymerase.
- **ribonucleic acid (RNA):** a category of large molecules, typically consisting of a single strand of nucleotides, that perform various functions in cells, including the transcription of DNA molecules and the transfer of specific genetic information for protein synthesis.

UNDERSTANDING DNA AND RNA SYNTHESIS

It is tempting to oversimplify the formation of deoxyribonucleic acid (DNA) by likening it to zipping up a zipper, but that is perhaps the easiest way to visualize the process. Indeed, DNA synthesis is diagrammed in virtually all biochemistry texts as such. The analogy is even further simplified by associating the "teeth" of the zipper with the purine and pyrimidine nucleotides from which the molecular structures of DNA and ribonucleic acid (RNA) are formed.

A nucleotide is formed when a purine or pyrimidine base and a phosphate group are chemically bonded to a sugar molecule. Only five different purine and pyrimidine bases are utilized in constructing DNA or RNA nucleotides. In DNA nucleotides, these are the bases adenine, cytosine, guanine, and thymine, while RNA uses the base uracil instead of thymine in its nucleotides. The different bases are indicated by the first letter of their names: A, C, T, G, and U. There are different mnemonic devices for recalling the complementarity of the different bases. One is that the curved letters C and G go together, as

DNA AND RNA SYNTHESIS

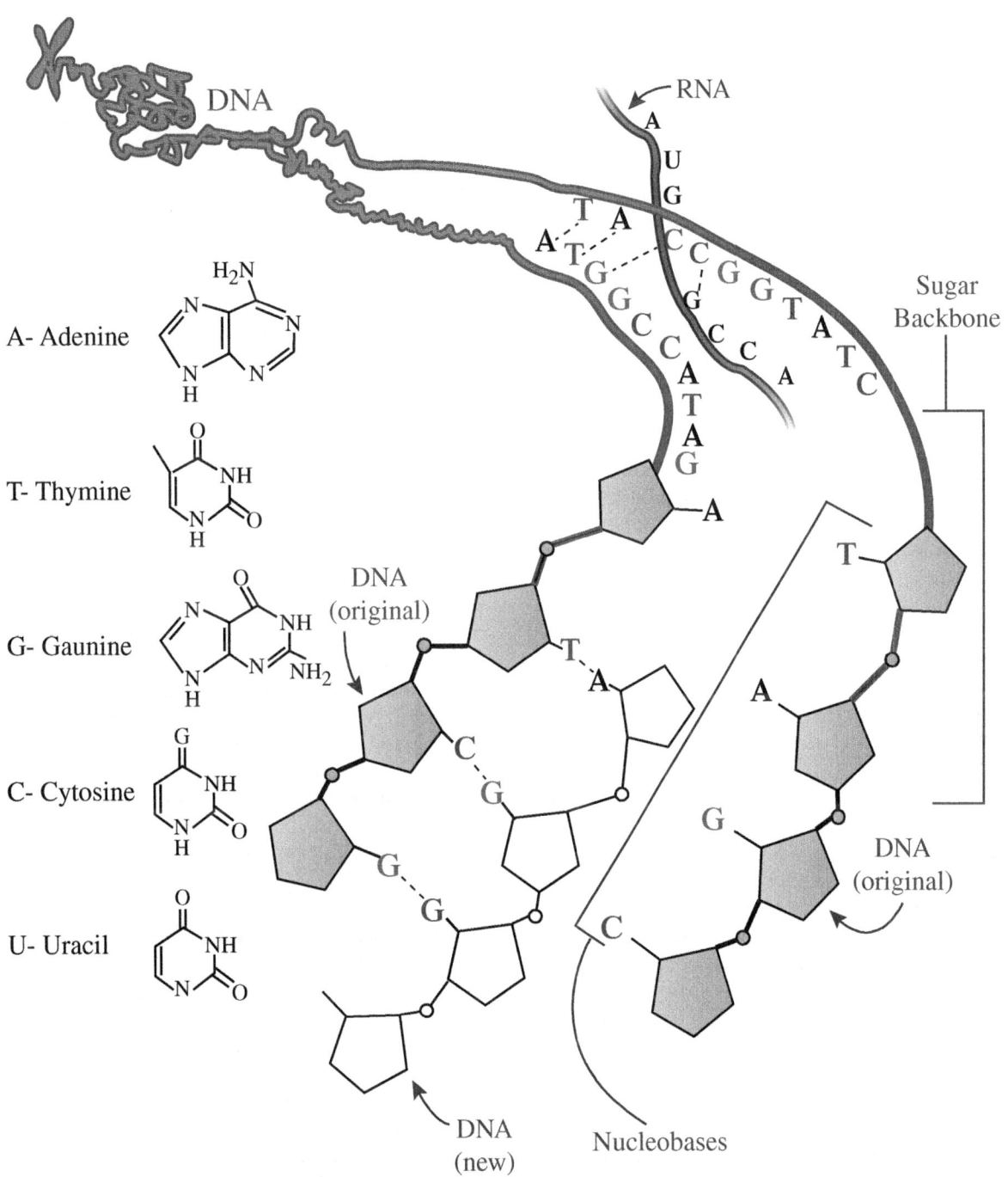

do the pointed letters A and T. Another is an easily remembered phrase such as "Cary Grant Ate Tacos." In fact, any number of such devices can be used to suit an individual's personal preference.

The second major difference between DNA and RNA is the nature of the sugars with which the nucleotides are constructed. In RNA nucleotides, the sugar molecule is ribose, a five-carbon simple sugar related to fructose. Sugar molecules are carbohydrates, indicating that each carbon atom in the molecule is chemically bonded to both an H atom and an –OH group. These are the components of the water molecule, so the term indicates that each carbon (*carbo-*) atom in the molecule is hydrated (*-hydrate*). In DNA nucleotides, however, the sugar is deoxyribose. The name indicates that the molecule lacks an oxygen atom that is part of the ribose sugar molecular structure. This difference in the structure of the sugar portion of the nucleotide is subtle, but of essential importance because it alone determines whether the molecule, and its role in the biochemical process of life, is DNA or RNA. In both DNA and RNA, the sugar molecules are in the form of a five-member ring structure made up of one oxygen atom and four of the five carbon atoms.

The third component of DNA/RNA nucleotides is the phosphate group, PO_4^{3-}, or Pi, generally referred to as "inorganic phosphate" when in that form and "phosphate" when bonded to another biomolecule, such as adenosine in adenosine triphosphate (ATP). The bonds between an oxygen atom in the phosphate group and other molecules are stable, even though they are deemed "high energy" bonds. In respiration and glycolysis (the decomposition of glucose), the bond between the third and second phosphate group in the triphosphate component is utilized both to store energy by its formation and to release energy when that bond is cleaved. Both DNA and RNA are large molecules. Their respective molecular structures consist of a long, biopolymer chain of alternating sugar molecules and phosphate groups. In both the DNA and RNA structures, each sugar component has a purine or pyrimidine base molecule bonded to the carbon atom on one side of the ring oxygen atom, and the phosphate group bonded to a carbon atom on the other side of the ring oxygen atom. It is at this point that the difference between ribose and deoxyribose sugar becomes vitally important. The RNA molecule consists of a single strand made up of adenine, cytosine, guanine, and uracil nucleotides. The DNA molecule, however, consists of two complementary strands of adenine, cytosine, guanine, and thymine nucleotides. These match up with each other as the base portions of the nucleotides in one strand and connect to the corresponding nucleotides in the other strand. The end result is that a DNA molecule is considerably bigger than an RNA molecule, and it has the form of a double helix as the two component strands coil around each other. The RNA molecule consists of a single nucleotide strand that assumes different shapes according to its role in transcription and gene expression.

DISCOVERY AND ANALYSIS OF DNA AND RNA MOLECULES

DNA was isolated from cell nuclei as early as 1869, but the fact that it bears genetic information was not known until 1943, when Oswald Avery, Colin MacLeod, and Maclyn McCarty demonstrated that introducing DNA from a virulent strain of the *pneumococcus* bacterium into a nonvirulent strain could produce the virulent strain. In 1953, James D. Watson and Francis Crick, with significant assistance from their laboratory manager Maurice Wilkins and using the analytical and theoretical work of Rosalind Franklin, first published the discovery that the DNA molecule has a double helix form, an image now so well recognized. Since then, methods and techniques for the manipulation and analysis of DNA have advanced, permitting biochemists and geneticists to obtain a better understanding of the role of genes and chromosomes in the DNA molecule. In 2001, the journal *Nature* published the first complete analysis of the human genome, which demonstrated, among other things, that all humans alive today are descended from a very few human populations that originated in Africa in the distant past. As the genome was deciphered over time, an understanding of the mechanisms by which DNA and RNA are synthesized was also obtained.

FORMATION OF DNA AND RNA IN LIVING CELLS: FOUR BASIC RULES

Both DNA and RNA are produced by copying a pre-existing DNA strand according to the base pairings of adenine to thymine and cytosine to guanine. In order to carry out this process, the DNA molecule must first "unzip," allowing nucleotide fragments to form a complementary RNA strand in which uracil nucleotides replace thymine nucleotides. From this complementary RNA strand, a duplicate of the original DNA strand is assembled from other fragments. This process can be thought of as making a mold from one half of the DNA molecule and then using it to cast a copy of the original.

Second, both RNA and DNA strands grow in one direction only. The phosphate group of each nucleotide is situated at the 5' position of the sugar molecule, the number indicating a specific location in the molecular structure according to the conventions for naming organic molecules. At the 3' position of each sugar molecule, there is a free hydroxyl (–OH) substituent that can form the phosphate ester bond with another nucleotide. Formation of a DNA or an RNA strand always proceeds from the 5'-position in one nucleotide to the 3'-position in the next nucleotide as nucleotides are added in sequence.

Third, both DNA and RNA are synthesized by very specific enzymes in polymerase chain reactions. New strands of RNA are produced only by RNA polymerases, and new DNA strands are produced only by DNA polymerases in DNA/ RNA polymerase reactions. Through the process of transcription, a new RNA strand is produced as RNA polymerases transcribe the nucleotide pattern of the parent DNA strand. RNA polymerase enzymes are able to initiate the formation of a new strand by coordinating to an appropriate site on a duplex strand of DNA (the double helix form of the molecule), where they temporarily separate the two strands of the DNA molecule and begin the process of assembling a new RNA strand from the corresponding nucleotides. DNA polymerases are not able to initiate the formation of a new strand directly. Instead, the process requires the formation of a primer, a DNA or an RNA segment that is bound to the parent DNA strand and is acting as the template for the new DNA strand. Both RNA and DNA polymerases comprise several different proteins, each of which carries out a specific function or transformation.

Fourth, synthesis of a new duplex DNA strand proceeds only from a particular formation known as a "growing fork." Specific enzymes function to open the duplex strand, allowing other enzymes to assemble the matching complementary strands from the appropriate nucleotides. As the new strands are formed, still other enzymes function to rejoin the strands as the growing fork progresses along the length of the template duplex DNA molecule. An important aspect of duplex DNA replication is that, because the strands only grow in one direction, the directions of growth on the two branches of the growing fork are opposite to each other. The new strand on the "leading" branch of the fork grows continuously, nucleotide by nucleotide. The new strand on the "trailing" branch of the fork is assembled instead in bits and pieces from various nucleotide segments.

Amplifying DNA for Analysis

Among the many techniques and methods that have developed for the manipulation of DNA samples—and that are especially important for the science of DNA "fingerprinting"—probably the most important is DNA amplification. By this method, an extremely minute sample of DNA, as might be obtained from just a few hair follicles found at a crime scene, for example, undergoes repeated replications so that enough of the DNA is present to produce a clear fragmentation pattern that is the "fingerprint" of that particular DNA. This methodology has been used to convict criminals who might otherwise have gone free, as well as to free individuals from prison who had been wrongfully convicted of crimes they did not commit.

Richard M. Renneboog, MSc

Further Reading

Berg, Jeremy M., John L. Tymoczko, Gregory J. Gatto, and Lubert Strye. *Biochemistry*. 8th ed. New York: W. H. Freeman, 2015. Print. "The Human Genome." *Nature* 409.6822 (2001): 813–958. Print.

Lafferty, Peter, and Julian Rowe, eds. *The Hutchinson Dictionary of Science*. 2nd ed. Oxford: Helicon, 1998. Print.

Lehninger, Albert L. *Biochemistry: The Molecular Basis of Cell Structure and Function*. 2nd ed. New York: Worth, 1975. Print.

Lodish, Harvey, et al. *Molecular Cell Biology*. 7th ed. New York: Freeman, 2013. Print.

Pelczar, Michael J., Jr., E. C. S. Chan, and Noel R. Krieg. *Microbiology: Concepts and Applications*. New York: McGraw, 1993. Print.

Thro, Ellen. *Genetic Engineering: Shaping the Material of Life*. New York: Facts On File, 1995. Print.

> **DNA/RNA SYNTHESIS SAMPLE PROBLEM**
>
> In 2005, a fossilized leg bone of Tyrannosaurus rex was found to contain viable tissue. Suppose a fragment of a DNA strand were to be recovered from such material and found to have the nucleotide order AGTTCGCGGAAC-TATTCG. What is the nucleotide order of the complementary strand in duplex DNA?
>
> What is its RNA complement?
>
> **Answer:**
> Recall the mnemonic device "Cary Grant Ate Tacos" (or whatever mnemonic you wish to use to signify that C/G and A/T are complementary nucleotides):
> The order of the "found" DNA fragment is
> AGTTCGCGGAACTATTCG
>
> Place the complementary nucleotide below each one in the series, as
>
> AGTTCGCGGAACTATTCG
>
> TCAAGCGCCTTGATAAGC
>
> This is the complementary sequence that would exist in a duplex DNA molecule. To generate the RNA complement, recall that RNA uses uracil (U) instead of thymine (T). The RNA complement is then easily defined by substituting U for T in the DNA complement, yielding
>
> UCAAGCGCCUUGAUAAGC
>
> (Note that listing the full sequence of a human DNA molecule in this form would require a book of approximately one million closely printed pages. It is rather unlikely that T. rex could be cloned from the above fragment.)

DNA/RNA TRANSCRIPTION

FIELDS OF STUDY

Biochemistry; Genetics; Molecular Biology

SUMMARY

The process of DNA/RNA transcription is defined, and its importance in the biochemistry of living systems is discussed. The process of transcription is essential in living systems for the synthesis of proteins in eukaryotic cells, as well as a variety of other functions that require the extraction of genetic information from DNA.

PRINCIPAL TERMS

- **complementary strand:** one of the two strands of nucleotides that make up a DNA molecule, with each nucleotide in one strand corresponding to the position of its complementary nucleotide (cytosine for guanine, adenine for thymine, and vice versa) in the other.
- **enzyme:** a protein molecule that acts as a catalyst in biochemical reactions.
- **gene expression:** the process by which RNA copies genes, which are specific segments of the DNA molecule, and uses the information to synthesize either proteins or other types of RNA.
- **RNA polymerase:** the enzyme responsible for initiating gene transcription in order to assemble and replicate strands of RNA.
- **transcription factor:** a protein that binds to DNA in order to initiate, regulate, or block gene transcription.

THE STRUCTURE OF DNA VERSUS RNA

The molecular structure of DNA is often likened to a zipper, with the two halves of the molecule matching up in a way that resembles the two halves of a zipper fitting together. However, the actual structure is much more complicated. A DNA molecule contains two complementary strands made up of specific combinations of nucleotides. Each nucleotide in a strand of DNA is composed of a molecule of the sugar deoxyribose bonded to a phosphate group and a nitrogenous base. Only four bases are used in DNA molecules: adenine, thymine, guanine, and cytosine.

Structurally, a DNA strand consists of a very long chain of alternating deoxyribose-and-phosphate units, forming what can be considered the backbone of the molecule, with the various bases appended to the deoxyribose. The complementary strand of a

DNA AND RNA

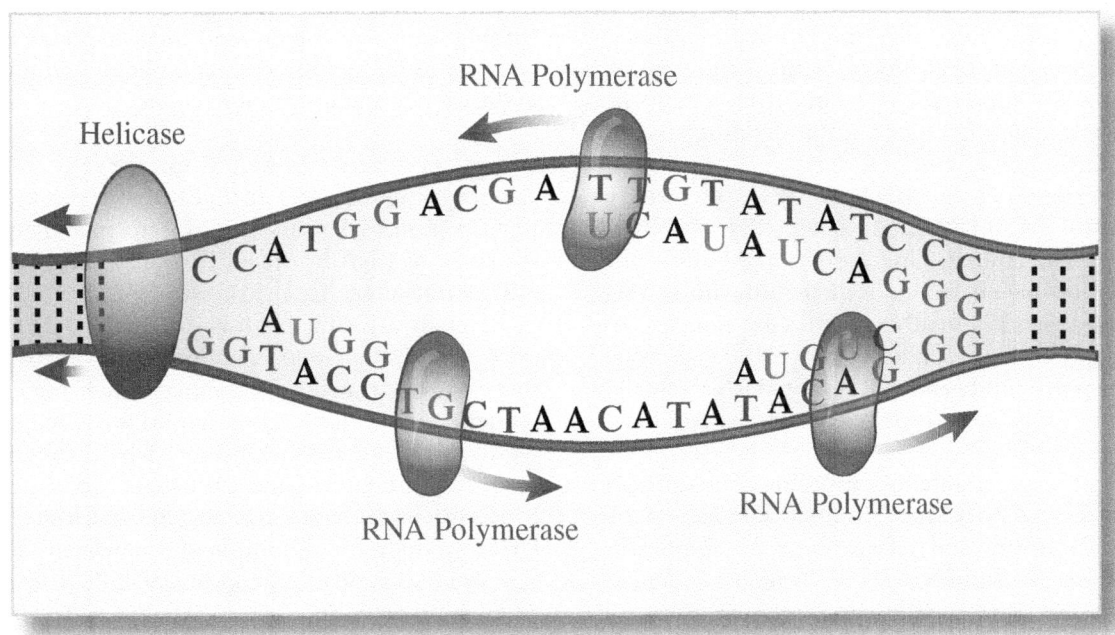

molecule of duplex DNA has the same basic structure but with a different sequence of bases. In duplex DNA, the bases form specific pairs, with adenine complementary to thymine and guanine complementary to cytosine.

The structure of the RNA molecule is very similar to that of a single DNA strand. There are two essential differences between them, however: the sugar component of the RNA molecule is the sugar ribose, not deoxyribose, and the base uracil is used in place of thymine. The difference in the sugar component is what allows DNA to form the familiar double-helix structure, which RNA cannot do.

The DNA molecule carries the genetic information that defines the identity of biological organisms. The sequence of nucleotides in each DNA strand specifies the order in which amino acids are to be assembled into proteins, the basic components of all known life. Each cell in an organism must have a DNA molecule in its nucleus—or, in the case of a prokaryote, its intracellular fluid—in order to produce the proteins and other compounds that are essential to its existence. The mechanism by which the instructions for protein assembly are translated and put into action is the DNA transcription process, which is the fundamental first step in the process of gene expression. Transcription can be described in simplified terms as RNA making a mold of the nucleotide sequence found in the parent DNA molecule and using it to synthesize new proteins.

THE TRANSCRIPTION PROCESS

Transcription is the copying of the nucleotide pattern in a strand of DNA by an enzyme called RNA polymerase. An enzyme is a protein that carries out a specific chemical function, which is determined by the relative locations of various atoms and functional groups within its three-dimensional structure. The chemical names of virtually all enzymes have the suffix *-ase*, as in lipase, transcriptase, and polymerase. Others that were first named as proteins rather than enzymes have names that end with *-in*, such as trypsin and pepsin. The various enzymes that participate in the transcription process, other than RNA polymerase, are called transcription factors.

With very few exceptions, the genetic information of the duplex DNA strand is transcribed from only one of the two strands. Due to the complementary nature of the two strands, the nucleotide sequence in the new RNA strand is identical to the

DNA strand that was not transcribed, save for the substitution of uracil for thymine, and both are the reverse of the DNA strand that served as the template. Because of this, the nontemplate strand is alternately called the "coding strand" or the "sense strand," while the template strand from which the RNA is assembled is called the "noncoding strand" or the "antisense strand."

Transcription from DNA to RNA involves a number of types of RNA. Messenger RNA (mRNA) copies genetic information from a DNA molecule to replicate the nucleotide sequence in the coding strand. Transfer RNA (tRNA) carries the amino acids specified by the nucleotide sequence to the growing end of a polypeptide chain. Ribosomal RNA (rRNA) forms the ribosome, which is where polypeptide assembly takes place.

In a eukaryote, transcription begins when an RNA polymerase attaches to an appropriate location on the duplex DNA strand and helicase enzymes temporarily separate the two strands of the DNA molecule at that location. The RNA polymerase begins to assemble nucleotides and attach them to the noncoding strand in the appropriate sequence, building up a hybrid RNA-DNA duplex strand. Due to the structural differences between the ribose and deoxyribose sugars and the complementary pairing of adenine with uracil instead of thymine, this hybrid duplex strand is not stable, so when assembly is complete, the RNA strand separates as mRNA. The mRNA strand then moves to the ribosomes formed by rRNA, where tRNA units carrying different amino acids are matched to the mRNA strand in the order specified. The amino acids are joined to one another with peptide bonds, forming the primary structure of the particular protein that has been encoded. The overall process of synthesizing proteins from genetic information contained in DNA is called "translation," with transcription being the initial step in the overall process.

Because DNA and RNA both use only four nucleotides each to specify structure—adenine, cytosine, guanine, and either thymine or uracil—and proteins are synthesized from twenty different amino acids, individual amino acids are specified by unique three-nucleotide sequences called "codons." Since the four nucleotides can produce sixty-four distinct combinations, most amino acids correspond to more than one codon, and some codons serve other purposes, such as initiating protein formation or signaling a stopping point. The codons between a start codon and a stop codon constitute what is called a "reading frame" for a specific nucleotide sequence. It is possible for multiple reading frames to overlap and the same sequence to code for different amino acids, depending on where in the sequence transcription begins.

Unraveling the Genetic Code

When biochemists recognized the role of mRNA in transcription, it became possible to investigate the structure of DNA in detail. By assembling synthetic mRNA molecules from just one type of nucleotide base, such as uracil—thus forming polyuridylic acid, or poly(U)—and examining the polypeptides that result, it can be determined which nucleotide sequences code for certain amino acids in protein synthesis. Using this technique, which they developed in 1961, biochemists Marshall Nirenberg and J. Heinrich Matthaei discovered that the codon UUU produces the amino acid phenylalanine—the first time an individual codon was linked to a specific amino acid. Similar experiments with synthetic poly(A) (polyadenylic acid) and poly(C) (polycytidylic acid) determined that the codons AAA and CCC code for lysine and proline, respectively. Poly(G) (polyguanylic acid) was found to form an unusable stacked structure that did not translate into protein synthesis. Synthetic codons of mixed nucleotide units also revealed which codons function as "start" and "stop" signals in protein synthesis and which ones code for the same amino acids.

The sequence of nucleotides in the structure of a DNA molecule is known as a "genome." If one were to transcribe the entire human genome as a sequence of nucleotides, using A, C, T, and G, the result would fill approximately one million densely typed pages. In February 2001, the science journal *Nature* published its report of the first complete analysis of the human genome. Further study of the genome has revealed, among other things, that all humans in the world today are descended from a mere handful of populations that originated in Africa; that the DNA of humans and chimpanzees only differs by approximately 2 percent; and that many modern humans, particularly those of Asian and European descent, carry Neanderthal genes—in some cases as much as 4 percent of their DNA.

Richard M. Renneboog, MSc

Further Reading

Berg, Jeremy M., John L. Tymoczko, Gregory J. Gatto, and Lubert Strye. *Biochemistry*. 8th ed. New York: W. H. Freeman, 2015. Print.

"The Human Genome." *Nature* 409.6822 (2001): 813–985. Print.

Lodish, Harvey, et al. *Molecular Cell Biology*. 7th ed. New York: Freeman, 2013. Print.

Pelczar, Michael J., Jr., E. C. S. Chan, and Noel R. Reece, Jane B., et al. *Campbell Biology*. 10th ed.

San Krieg. *Microbiology: Concepts and Applications*. New Francisco: Cummings, 2013. Print. York: McGraw, 1993. Print.

DNA/RNA TRANSCRIPTION SAMPLE PROBLEM

Some researchers think that if they can recover viable DNA from mammoths that were preserved in permafrost, they may be able to clone such a creature. Suppose a sample of mammoth DNA yields an intact DNA strand in which the template strand has the following nucleotide sequence:

AAGTGCACCTGGTATATCCAGTGTCAT

What sequence of nucleotides would be found in the mRNA produced from this DNA fragment?

Answer:

In the transcription of DNA to RNA, thymine nucleotides (T) in DNA coordinate with adenine nucleotides (A) in RNA, and adenine nucleotides in DNA coordinate with uracil (U) nucleotides in RNA. Cytosine (C) coordinates with guanine (G), and guanine with cytosine, in both cases. Therefore, the complementary mRNA sequence would be

UUCACGUGGACCAUAUAGGUCACAGUA

If you are adventurous, use a table of amino acid codons to determine what amino acid sequence a protein made from this mRNA would have.

human genome. Further study of the genome has revealed, among other things, that all humans in the world today are descended from a mere handful of populations that originated in Africa; that the DNA of humans and chimpanzees only differs by approximately 2 percent; and that many modern humans, particularly those of Asian and European descent, carry Neanderthal genes—in some cases as much as 4 percent of their DNA.

EGG PRODUCTION

FIELDS OF STUDY

Poultry/animal science; reproduction; food technology; biology; physiology; business management.

SUMMARY

The egg production field includes farm production of shell eggs for direct consumption and further processing of eggs for use in products of the food industry. Egg production includes the development of highly productive strains of laying hens, advances in technology in the production and processing of eggs, and business models that permit the efficient production and marketing of eggs.

PRINCIPAL TERMS

- **candling:** inspecting the internal quality and embryonic development of eggs by shining a bright light through them.
- **chalaza:** stringlike attachment that anchors the yolk to the center of an egg.
- **in-line production:** using a single location for production and packaging of eggs.
- **line:** group of related chickens that have similar production characteristics.
- **off-line production:** using different locations for the production and processing of eggs.
- **pullet:** immature female chicken destined for egg production.
- **salmonella:** genus of bacteria that can contaminate eggs, causing serious illness to humans who consume the eggs.
- **vertical integration:** ownership by a single firm of multiple companies in order to cover all stages of egg production, from the raw materials through distribution, including feed mills, hens, buildings, egg-processing facilities, and transportation vehicles.

BASIC PRINCIPLES

Egg production in the United States has undergone a remarkable transformation. Before the twentieth century, hens ran loose around the farmyard, largely fending for themselves. Around the late 1800's, farm flocks came into being, and egg production became a serious part of the farm enterprise. Hens were given their own housing and provided with feeders, waterers, roosts, and nests, as well as a fenced-in yard. The farm flock system allowed for applying important management principles, such as proper feeding, breeding, and egg collection. The next advance took place around the 1960's with the emergence of farms that specialized in egg production. The farmer-manager could then focus entirely on egg production and use the latest in management and feeding techniques and production stock. Later in the twentieth century, egg producers became vertically integrated, with all aspects of production and marketing under the control of the same firm. The farmer-producer became just one part of the entire system.

Egg production involves genetic research to develop strains of highly productive hens; proper management of growing pullets to maximize their potential as laying hens; the use of advanced technology in buildings, equipment, feeding, and lighting for maximal egg production at minimal cost; and the development of new egg products for the consumer. It can also involve support services such as feed mills and transportation. Modern intensive production practices involving millions of birds have come under criticism as factory farming and have raised questions of animal welfare that must be addressed by the producer.

BACKGROUND

Chickens were probably domesticated from red junglefowl in Southeast Asia. Genetic studies suggest multiple sites of domestication, including China and India. Archaeological studies indicate that chickens were present in the Americas before the time of the Spanish conquistadores.

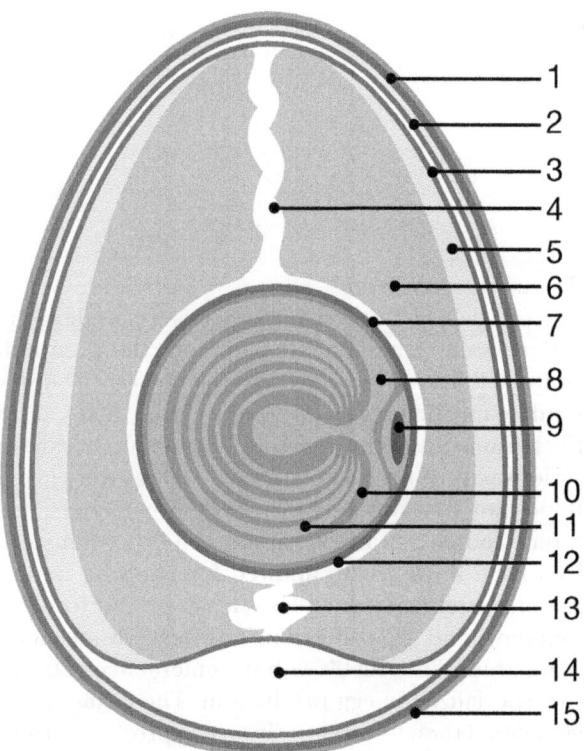

Schematic of a chicken egg: 1. Eggshell 2. Outer membrane 3. Inner membrane 4. Chalaza 5. Exterior albumen 6. Middle albumen 7. Vitelline membrane 8. Nucleus of pander 9. Germinal disc (nucleus) 10. Yellow yolk 11. White yolk 12. Internal albumen 13. Chalaza 14. Air cell 15. Cuticula

The modern egg industry is a result of a series of technological advances. In the 1870's, incubators began to be used commercially to hatch chickens, rapidly increasing the number of commercial hatcheries. Poultry breeders applied scientific principles to develop improved breeds and strains of chickens for egg production. Land-grant colleges engaged in research in poultry nutrition and feeding. This led to improved management practices and more efficient production of eggs. Better understanding and treatment of diseases, together with improved sanitation and ventilation, allowed for the creation of confinement systems.

Improved distribution systems and the development of new egg products led to greatly increased consumption of eggs, reaching a maximum of 402 eggs per capita in 1945. Health concerns about the cholesterol content of eggs and changes in lifestyle led to declines in consumption to a low of 230 eggs per capita in 1991. However, after the publication of scientific studies that stated that consuming eggs does not raise blood cholesterol, consumption of eggs began to increase, reaching 248 eggs per capita in 2008.

How It Works

Egg production begins with the selection and development of breeding stock. Many breeds of chickens have developed over time, but for commercial purposes, the laying hen (layer) must be highly productive and efficient in converting feed into eggs. These criteria are met by the white leghorn breed, which is light in body weight, is highly active, and produces a white egg. A very few breeding companies dominate the supply of egg production chicks, and they have their own specialized lines or strains of breeders. The white leghorn has been overwhelmingly adopted by the egg industry, but other breeds are used in markets that prefer a brown egg. Traditionally, this has involved using heavy breeds, such as the New Hampshire or Rhode Island red. The development of specialized lines and crossbreeds has resulted in brown-egg layers that are almost as efficient in feed conversion as the white leghorn. In many countries, including European nations, brown eggs are preferred over white eggs. The breeders must be kept in floor management systems to facilitate the breeders' mating.

From Egg to Layer. Fertilized eggs are transported to commercial incubators for incubation and hatching. After a few days of incubation, the eggs are candled to test for fertility and for viable embryo development. An infertile egg is clear, and a developing embryo shows blood-vessel development. Typically, the eggs are moved to a separate hatching incubator for the final three days of incubation. After hatching, the chicks are vaccinated and sexed, as only the female chicks are useful for egg production. Debeaking (removal of part of the beak) is performed at this time, or after the chicks are seven to ten days old.

The pullets are raised in confinement either on the floor or in cages; outside range rearing is seldom used by commercial breeders. A lighting program is essential for proper development of the pullets. One-day-old chicks receive twenty-three hours of light per day, and for the rest of the growing period, they receive a minimum of ten hours of light per day. They are transferred to laying houses at around sixteen weeks of age. Hens usually begin to lay eggs when they are five months old and continue to lay for about twelve more months.

Egg Production

Several types of management systems are commonly used by egg producers: cages, floor systems, or free-range systems. Cages are used for more than 98 percent of production operations for a variety of reasons. They allow increased population density in the poultry houses, and they are more labor efficient, as feeding, watering, egg collection, and manure removal can all be mechanized. Floor or noncage systems keep hens on litter floors inside buildings that hold feeders, waterers, roosts, and nests. This was the most common management system before the adoption of cage systems. Free-range systems allow hens access to an outdoor yard when weather permits.

The term "organic eggs" refers not so much to a management system but to the feed the hens receive. The feed must be totally vegetarian, the grains used must be pesticide-free, and the hens must not receive hormones or antibiotics.

Because most laying hens produce eggs in windowless houses, artificial lighting is provided. In fact, in all systems, lighting is essential to stimulate the pituitary gland to secrete hormones that help initiate and sustain egg production. Various lighting programs have been developed, but a typical program increases lighting from ten hours at twenty-four weeks of age to seventeen hours at thirty-two weeks and maintains this lighting period until the end of the laying cycle. The length of the lighting period should never be decreased during the laying cycle. The number of eggs produced per hen during a laying cycle can range from 180 to 200 eggs in tropical climates to 250 to 300 eggs in more temperate climates.

In cage systems, after the eggs are laid, they are transported via a conveyor belt to an egg-processing facility, where they are washed, graded for size, and either packed in flats to be shipped to a retail store or broken for further processing.

Breeding Stock. The Institut de Sélection Animale (ISA) holds a dominant position in the egg production industry as it supplies breeding stock for 50 percent of the world's egg production industry. The company began as Hendrix in the Netherlands, where it still has its headquarters. ISA expanded by purchasing many well-known and respected laying-hen breeding companies, including Babcock, J. J. Warren, Kimber, Shaver, Dekalb, Hisex, and Bovans. Many of these companies began as family-owned businesses in the early part of the twentieth century. Many strains of white and brown egg layers under the names of the original companies are sold as day-old chicks. The chicks destined as breeders must have a good egg-production capability, but good fertility is essential.

Laying Stock. Laying stock is also sold by ISA and other breeders as day-old chicks. ISA has strains of white and brown egg layers that are companions to its breeding stock. High egg production and excellent feed efficiency are essential characteristics for these strains.

Ducks for Egg Production. Ducks have never been popular for egg production in the United States and, like quail eggs, are only a niche market. However, ducks are commonly used in Asia for egg production. The Khaki Campbell breed is best known for egg production, and Metzer Farms sells a hybrid duck that produces eggs at a rate similar to the best chicken egg strains. Duck eggs are larger, have a more deeply pigmented yolk, and have firmer albumen than chicken eggs. Compared with chicken eggs, duck eggs have a higher cholesterol content, tend to pick up off-flavors more readily, and are more susceptible to contamination.

Shell Eggs. Eggs are most commonly marketed in the form in which they are laid, still in their shell. There is no difference in nutritional value between white and brown eggs, and although white eggs have a slightly thicker shell than brown eggs, brown eggshells have a stronger structure, so there is no difference in tendency to break. As the laying cycle nears its end, eggs tend to get bigger with thinner shells, leading to a greater tendency for breakage. When eggs are laid, they are coated with a protective layer called a cuticle. This cuticle is often removed during washing. The shell contains many pores, which nature intended for gaseous exchange for the developing embryo, but which also provide an entry point for bacteria.

The yolk consists of 32 to 36 percent lipids and around 16 percent protein. The lipids include triglycerides (fats), phospholipids, and cholesterol. Triglycerides contain various types of fatty acids. The fatty acid content of yolk can vary according to the diet fed to the hens. A popular modern egg product contains a high content of omega-3 fatty acids, typically 350 milligrams compared with a normal content of 60 milligrams. The eggs also have a lower content of saturated fat, as well as a somewhat lower content of cholesterol. The hens are fed flaxseed to produce these eggs. These eggs have purported health benefits and command a higher price.

Eggs are graded by weight and quality. Egg-processing machinery separates eggs by weight, which can range from jumbo to peewee. Eggs can be grade AA, A, or B in quality. Quality in eggs is determined by candling or breaking them out and measuring albumin height. Grade AA eggs are freshly laid, have a thick, cloudy albumin, and a small air cell. Most eggs in supermarkets are grade A because some time has passed since their laying. Grade A eggs have a larger air cell, and the albumin is clear but thinner. The yolk is more defined in candling but free of defects. Both AA and A eggs can be sold as shell eggs, while grade B eggs are used for further processing. Grade B eggs have poorer quality albumin and minor discoloration or minor blood or meat spots.

Liquid Egg Products. Grade B eggs or other eggs not needed for the shell egg market go to an egg-breaking plant. After breaking, the liquid products obtained include whole egg, egg white, and egg yolk. These products are destined for the food industry and are unlikely to be found in retail stores.

Dried Egg Products. The incentive for developing the technology for drying eggs in the United States began in the 1930's with the availability of large quantities of eggs from China at a very low cost. The industry got a boost during World War II when the military needed dried eggs. Dried eggs have several advantages over shell eggs or liquid eggs: They can be stored at low cost, take less space to store, are not susceptible to spoilage caused by bacteria, are easier to handle in a sanitary manner, and have lower transportation costs. Dried eggs are used extensively in many products, including bakery foods and mixes, mayonnaise and salad dressings, ice cream, pastas, and convenience foods. Most dried egg products are obtained by spray drying, but before drying, the sugars are removed from the eggs by fermentation or enzymatic treatments. These processes are necessary to avoid reactions of glucose with proteins or phospholipids in the eggs that can result in poor baking qualities or off flavors. The dried egg products are derived from egg white, egg yolk, whole egg, or blends of whole egg or yolk with carbohydrates such as sucrose or syrups.

Specialty dried egg products include a scrambled egg mix that has good storage capability and low-cholesterol egg products. Most low-cholesterol egg products contain egg white, with nonfat milk, vegetable oil, and pigments substituting for yolk. The final composition is similar to that of a whole egg.

FUTURE PROSPECTS

The modern cage system of egg production is a marvel of efficiency and low cost. However, the nature of the system has been brought to the attention of animal welfare activists. The hens are kept in very crowded conditions (typically 67 square inches per hen) and are not able to perform their natural or instinctive behaviors, such as sleeping on roosts, laying eggs in nests, and taking a dust bath. Animal activists say that this is not humane. However, egg producers reply that hens kept in cage systems are healthier than those raised in other systems, noting that their productivity is higher. Animal science departments have been aware of these criticisms and have developed a new field of farm animal welfare. Animal welfare can be studied scientifically in a manner that is objective, reliable, and reproducible. However, the demand for answers to animal welfare issues may be outpacing the results of scientific studies. This has resulted in legislation banning the use of cages for egg production in Europe and the passing of Proposition 2 in California. The California legislation will probably phase out cage use in the state, which producers say will increase production costs 40 to 70 percent and drive egg producers out of the state because they will no longer be competitive. Egg consumption fell because eggs have a high level of cholesterol, but consistent research has shown that egg consumption will not increase blood cholesterol in healthy people. Persons with heart disease may want to consult their physician as their bodies may handle cholesterol differently. The image of eggs suffered, and egg producers must convince the public of the egg's nutritive value if egg consumption is to reach or approach its 1945 peak. A problem with eggs is possible salmonella contamination. If the shells are contaminated with salmonella, proper washing can eliminate this hazard, but if hens become infected with salmonella during the growing period, the eggs are internally contaminated. In August, 2010, more than 500 million eggs produced by Wright County Egg and Hillandale Farms of Iowa were recalled because of possible salmonella contamination. Programs are being developed to certify hens in large flocks as being salmonella-free.

David Olle, MS

FURTHER READING

Bell, Donald D., William Daniel Weaver, and Mack O. North. *Commercial Chicken Meat and Egg Production.*

5th ed. Norwell, Mass.: Kluwer Academic, 2002. An essential guide for those interested in the poultry industry. This edition emphasizes managerial aspects.

Clancy, Kate. *Greener Eggs and Ham: The Benefits of Pasture-Raised Swine, Poultry, and Egg Production.* Cambridge, Mass.: Union of Concerned Scientists, 2006. The Union of Concerned Scientists looks at egg production, poultry, and pigs and presents an alternative to the intensive production methods in predominant use.

National Agricultural Statistics Service. *U.S. Broiler and Egg Production Cycles.* Washington, D.C.: USDA National Agricultural Statistics Service, 2005. A governmental document providing information on egg production cycles and chickens for those in the poultry industry.

Stedelman, William, and Owen Cotterill. *Egg Science and Technology.* 4th ed. New York: Haworth Press, 1995. Long recognized as the most comprehensive handbook on the egg-processing industry.

WEB SITES

American Egg Board http://www.aeb.org
American Poultry Association http://www.amerpoultry-assn.com
Institut de Sélection Animale http://www.isapoultry.com
United Egg Producers http://www.unitedegg.org
United Egg Producers Certified http://www.uepcertified.com
U.S. Poultry and Egg Association http://www.poultryegg.org

FASCINATING FACTS ABOUT EGG PRODUCTION

- Eggs provide a unique source of balanced nutrients, including protein, essential fatty acids, vitamins, and minerals. The protein is of such high value that it is used as a standard to measure the quality of other food proteins.
- World consumption of eggs is increasing at about 8 percent per year because of higher living standards and the introduction of efficient production methods.
- Consolidation of egg farms has resulted in around three hundred producers supplying most of the nation's eggs. These producers are primarily located in the five top egg production states: Iowa, Ohio, Indiana, Pennsylvania, and California.
- Blood spots in egg yolk do not mean that the egg is fertilized. They are caused by a broken blood vessel on the surface of the yolk as the egg is forming.
- If a carton of eggs bears a U.S. Department of Agriculture grade, it must also have a Julian date, which is the date of packing. A sell-by date, if it appears, can be no more than thirty days after the date of packing, and a use-by date can be no more than forty-five days after packing.
- Pasteurized eggs have been exposed to heat to destroy bacteria. These are the best choice for recipes that call for partially cooked or raw eggs.

EMBRYOLOGY

FIELDS OF STUDY

Cell biology, developmental biology, embryology, reproduction science

SUMMARY

Embryology is the study of the development of animals, from the formation of gametes until birth, hatching, or metamorphosis.

PRINCIPAL TERMS

- **cleavage:** cell division in the early embryo that, unlike division in adults, involves little or no growth between divisions
- **fertilization:** the process by which the egg and sperm unite to form the zygote gametes: the haploid cells, ova and spermatozoa, that fuse to form the diploid zygote
- **gastrula:** the stage of development during which the endoderm (gut precursor) and the mesoderm

(muscle and connective tissue precursor) are internalized
- **haploid:** having only one of each kind of chromosome
- **zygote:** the single cell formed when gametes from the parents (ova and sperm) unite, a one-celled embryo

BASIC PRINCIPLES

For thousands of years, humans have wondered how they and other organisms came to be. By 340 BCE, Aristotle had described the development of the chicken in the egg, but since most early embryos are too small to be seen by an unaided eye, his and later descriptions of development started with larger, more formed embryos. That did not change very much until the late 1600's, when development of the microscope gave a glimpse of life too small to be seen unmagnified.

By the early eighteenth century, the developmental patterns of many organisms had been observed and described. There was, however, still much disagreement about how the early stages progressed. The majority of scientists believed in the theory of preformation, which said that a preformed embryo was present in the gametes. There were two main factions among the preformationists.

The ovists believed that inside the egg was a tiny, fully formed organism that was stimulated to grow by the seminal fluid. Their opponents, the spermists, believed that the fully formed miniature organism was in the sperm and was nourished in its growth by the ovum. Thus seventeenth and eighteenth century drawings of sperm and of eggs often show fully formed bodies within. By the end of the eighteenth century, more and more scientists were deserting preformation in favor of the theory of epigenesis, first proposed by Caspar Wolff in 1789, which stated that development occurs through growth and remodeling of embryonic cells. Karl Ernst von Baer, who had published a collection of his observations and the observations of others, proposed that general features that are common to large groups of taxonomically related organisms appear earlier in development than more specialized features of individual species. After Darwin published his evolutionary theories, Müller, Haekel, and other proponents of Baer's law and of evolution proposed that the embryonic development of an organism (ontology) mirrored its evolution (phylogeny). Although this has been shown not to apply to all organisms or to all developmental sequences, it can be seen in the development of many embryos.

During the late nineteenth and early twentieth centuries, scientists' understanding of embryonic development increased dramatically as they began applying recently discovered knowledge in evolution, genetics, and cell biology to embryology. Edwin Ray Lankester and Hans Speeman were two prominent scientists who studied comparative embryonic development at that time. Also at that time, the new science of experimental embryology began as Wilhelm Roux and G. Schmidt manipulated the cells of amphibian embryos and began to discover how and why development occurred. Today, new discoveries in biology and chemistry are applied to the study of embryonic development.

HOW IT WORKS

Gametogenesis. The formation of gametes, eggs and sperm, is usually considered the beginning of embryology. In sperm formation, two things need to occur, reduction of chromosomes to the haploid state and maturation of the cytoplasm. During the first part of spermatogenesis, immature cells, called spermatogonia, form four haploid cells, called spermatids, by meiosis. Spermatids then go through a maturation process in which they become streamlined and motile. They also develop an acrosome that has enzymes needed to penetrate the egg. Like sperm, eggs must become haploid and mature, but both the timing and maturation are quite different. Maturation of the cytoplasm often begins before meiosis. All the cytoplasm of the early embryo comes from the egg, so immature ova are aided by various helper cells that increase each ovum's cytoplasm and add food stores, called yolk. The amount of yolk varies considerably, from mammals that have no yolk, to birds that have huge amounts. Depending on the species, meiosis can begin at any time during cytoplasmic maturation and can be a continuous process or have one or more pauses. In sea stars and many other organisms, meiosis is complete before fertilization, while in others, such as nematodes, the egg matures fully and is released by the ovary before any meiosis begins. Sperm penetration then triggers the onset of meiosis.

Fertilization and Development. Once sperm have reached the egg, the acrosomal enzymes must digest the various protective layers that surround the egg, and recognition structures on the surface of the sperm must be complementary to recognition structures on the egg cell membrane. The sperm's nucleus then enters the egg and fuses with the haploid egg nucleus. This forms a diploid cell called the zygote. Interestingly, when a sperm first penetrates the egg, the polarity of the cell changes and chemicals are released by the membrane, which make it impossible for other sperm of that species to enter the same egg.

Following fertilization, a period known as cleavage begins. During this time, cells divide rapidly with little or no growth between cell divisions. Cells become smaller and more numerous. At the end of cleavage, a structure called the blastula is formed. In some animals, such as echinoderms, amphibians, and nonvertebrate chordates, the blastula is a hollow ball of cells. In higher vertebrates, the blastula is a flat, dishshaped structure, often called the blastodisc. In mammals, the blastula is called a blastocyst, and consists of a hollow ball of cells, called the trophoblast, and a group of internal cells, called the inner cell mass. During gastrulation, surface cells become internalized to form the three germ layers—ectoderm, mesoderm, and endoderm— that are seen in most animal embryos. A second internalization, this time of some ectodermal cells, forms the beginning of the central nervous system. After this neurulation, the various body organs begin to form from the three germ layers. As these changes progress, cells become less general and more specialized, a process called differentiation.

Once the major organs have differentiated, the embryo matures and grows, a process usually called gestation. The time it takes for embryonic

---

SIR EDWIN RAY LANKESTER

Born: May 15, 1847; London, England

Died: August 15, 1929; London, England

Fields of study: Anatomy, embryology, evolutionary science, marine biology

Contribution: Lankester elucidated the structures and developmental anatomy of many marine organisms and was also a champion of Darwin's theories of evolution.

From childhood, Edwin Ray Lankester was exposed to some of the greatest scientific minds of his day. His father, Edwin Lankester, was an eminent physician, numbering among his friends Charles Darwin and Thomas Henry Huxley, who were frequent visitors to the Lankester home. Thesemenfirst exposed the younger Lankester to scientific thought, and he became their ardent disciple. After completing his university studies, Edwin Ray Lankester trained in Vienna, Leipzig, and Naples, before becoming a fellow at Oxford. He was a professor at University College, London (1874-1890), and at Oxford (1890-1898). During this time he continued his studies on animal development. He was the first to show that movement of cells through the primitive streak was by delamination (a movement of individual cells) and not by invagination (a term he coined for the inward migration of a layer of cells). He also coined the terms blastopore, for the opening formed during gastrulation, and stomadeum, for the embryonic region that gives rise to the mouth. While he was at University College, he was the prime mover in the founding of the Marine Biological Laboratory at Plymouth.

He left his professorship at Oxford in 1898 to become director of the natural history division of the British Museum, from which post he retired in 1907 to devote more time to his writing. He was a prolific writer, who wrote both for his scientific colleagues and for the popular press. He made biology, especially evolution and natural history, accessible to the nonscientist in weekly newspaper columns that were later collated and published in book form. Another of his books, *Extinct Animals* (1905), was the first book to introduce dinosaurs to the general public. He despised what he thought was quackery and was one of the first scientists to dismiss spiritualists and mediums as quacks. He was also vehemently opposed to the Lamarkian idea of the inheritance of acquired characteristics. Because of his tenacity in debate, he was often likened to a bulldog.

In 1911, a partial human skeleton was discovered near Piltdown in Sussex. Although the cranium was quite human in appearance, the lower jaw was more apelike. Many crude stone tools and bone fragments of extinct animals were also found in the same area. Lankester and many other scientists thought that this might be the skeleton of a species ancestral to modern humans. Sadly, the man who exposed spiritualists and Lamarkism had himself fallen for a hoax. Many years after its discovery, the Piltdown man skull was shown to be a composite of parts of a human cranium and a juvenile orangutan's jawbone that had been modified to look a bit more human. Lankester never knew of his mistake. He died in 1929, long before the hoax was unmasked.

—*Richard W. Cheney, Jr.*

development varies considerably. In chickens and small rodents, the process takes about three weeks; in humans, it takes approximately nine months, while in elephants, the process can take almost two years. Some organisms emerge in very immature states that require more development. Amphibians and arthropods hatch as feeding larvae that must grow before they begin a metamor- phosis that leads to the adult. Marsupials are also born at a very immature stage and must complete their embryonic development inside the mother's pouch.

—*Richard W. Cheney, Jr.*

See also: Asexual reproduction; Cleavage, gastrulation, and neurulation; Copulation; Fertilization; Gamatogenesis; Reproduction; Reproductive system of female mammals; Reproductive system of male mammals; Sexual development.

FURTHER READING

Bronson, F. H. *Mammalian Reproductive Biology*. Chicago: University of Chicago Press, 1989. A very complete look at mammalian development. One chapter gives a brief overview of development for each mammalian order.

Carlson, B. *Patten's Foundations of Embryology*. 6th ed. New York: McGraw-Hill, 1996. A comprehensive look at animal development with emphasis on vertebrate development.

Hartl, Daniel L., and ElizabethW. Jones. *Essential Genetics*. 2d ed. Sudbury, Mass.: Jones and Bartlett, 1999. The chapter on "The Genetic Control of Development" is relevant to embryology.

Kumé, Matazo, and Katsuma Dan. *Invertebrate Embryology*. Translated by Jean C. Dan. Belgrade, Yugoslavia: NOLIT Publishing House for the U.S. Department of Health and Human Services, 1968. Extensive compendium of invertebrate development by phylum.

EMOTIONS

FIELDS OF STUDY

Developmental biology, ethology, evolutionary science

SUMMARY

The idea that animals experience a wide range of emotions continues to be a controversial concept among scientists. Although it is a popular belief supported by numerous anecdotal reports, scientific researchers often stringently dismiss such evidence as signs of anthropomorphism.

PRINCIPAL TERMS

- **amygdala:** subcortical brain structure related to emotional expression anthropomorphism: attributing human characteristics to animal behavior
- **dopamine:** neurotransmitter involved in movement and reward systems
- **field observations:** observing behavior in naturalistic settings
- **limbic system:** brain structures related to the regulation of emotions
- **oxytocin:** hormone involved with pleasure during bonding
- **primary emotions:** emotions related to innate motivations
- **secondary emotions:** emotions with a strong social component

BASIC PRINCIPLES

In attempting to prove the existence and extent of emotions in animals, researchers have struggled with the question ofhowto identify and measure feelings in various species. For many scientists, it is nonsense to speak of animal emotions without the capacity to objectively define and measure them. Such scientists have an aversion to the nonscientific tendency to ascribe humanlike characteristics to animals. Anthropomorphism is the term used to describe this tendency.

DEFINING AND COMMUNICATING EMOTIONS

Defining emotions can be difficult even in humans. Psychologists view emotions as organized psychological and physiological reactions to change in one's relationship to the world. An emotion is a positive or negative transitory experience that is felt with some intensity. Emotional reactions are partly subjective experiences and partly objectively measurable patterns of behavior and physiological arousal. The subjective experiences can include how a person appraises a situation and what actions result from that appraisal. For example, when a student receives a passing grade on an extremely difficult exam, she may experience joy after appraising the situation as a success. Even with this appraisal, however, humans cannot decide to experience joy or some particular emotion. The subjective aspects of emotions are triggered by the thinking self and felt as happening to the self. Objective aspects of emotions include learned and innate physiological responses and expressive displays. The expressive displays include smiles, frowns, and squinting of the eyes. The innate physiological responses are biological adjustments needed to perform the actions generated by the emotional experience. For example, if anger develops in a person, heart rate increases in order to supply additional oxygen to the muscles.

Since animals do not have the capacity of speech, any inner states cannot be expressed directly to a scientific observer. Consequently, field observations of behavior are often used to infer emotions in animals. There are problems, however, in assessing emotions through behavioral manifestations. It becomes difficult or impossible to attribute an emotion to an act with many possible motivations. If a dog chews on the shoes of an owner who is out on a date, does this indicate jealously, anger, boredom, or merely a poorly trained pet?

Historically, animals have been seen from a mechanistic perspective as being without the capacity for humanlike emotions. Behaviorism dictated that instincts and patterns of reinforcement in the environment provided the motivation for the behavior of animals. For centuries, Christian religions also promoted the idea that animals lacked humanlike emotions. The role of animals was to serve the needs of humans. The concept of "speciesism" suggested that only humans were capable of emotions because of their special place in creation. Charles Darwin was one of the first scientists to study animal emotions, and to utilize field observations to ascribe emotions to animals. In his book, *The Expression of the Emotions in Man and Animals* (1872), Darwin stressed the communicative aspects of emotion. Positive inner states were expressed through a signal for sociability, while aggressiveness indicated a desire for isolation. He believed that species developed special social signals to indicate how they would react to a social encounter. Yet the behavioristic view of animals continued to dominate the debate about animal emotions. Over a hundred years later, Jane Goodall, in her book *The Chimpanzees of Gombe: Patterns of Behavior* (1986), was criticized by the scientific community for suggesting that chimpanzees had personalities and experienced excitement and joy.

PRIMARY AND SECONDARY EMOTIONS

Today even the most critical scientists accept the fact that many animals experience a core group of emotions that are similar to those found in humans. Making the distinction between primary and secondary emotions, there exists some agreement about the basic emotions of fear and aggression. The primary emotions, such as fear, involve instinctual tendencies that are essential to survival. Fear permits escape from dangerous situations or predators. The fight-or-flight response is an instinctual pattern of behavior found in response to danger. The primary emotions, which are instinctual or hardwired into many species, can be demonstrated quite easily. When a specific stimulus is presented to an animal, a predictable response takes place. For example, if the shadow of a hawk is projected on the ground among a group of chickens, the birds will respond with "fear" and attempt to get under cover.

It is the realm of secondary emotions that creates the most controversy between those with opposing views about the extent of animal emotions. Expressions of love, grief, or jealousy may be commonplace among humans, but it is debatable whether they can be inferred in animals. Grief is commonly reported during field observations of various animals. The behaviors of elephants, chimpanzees, sea lions, and geese suggesting grief in response to the loss of a mate or offspring have been well documented. The dolphin who carries a dead baby around for several days is inferred to be experiencing both grief and love. Love has been attributed to animals such as swans or geese because of lifelong bonds that are established with a mate. Critics of these interpretations point out that animals may behave as if they are grieving or in love, yet there is no way of knowing whether this is an accurate reflection of their inner states. A central issue about the capacity of animals to experience a wide range of secondary emotions involves the ability to show self-consciousness. If an animal is able to be aware of its own inner states, it would then have the capacity to infer the mental states of others. With self-awareness comes the capacity for sympathy and empathy.

THE BIOLOGY OF EMOTIONS
The scientists examining the biology of emotions have discovered some similarities between the brains of humans and animals that help to explain the basic primary emotions. Emotions seem to arise from the parts of the brain that are located below the cortex and are part of the limbic system. These regions of the brain have remained intact across many species throughout evolution. So far, the amygdala has been identified as the central site of emotion. This almond-shaped structure is at the center of the brain. Neuroscientists have found that rats will show a pattern of fear when a particular section of the amygdala is stimulated. If the amygdala is damaged, a rat will not show the normal behavioral responses to danger, such as freezing or running away. The rat with a damaged amygdala also will not demonstrate the accompanying physiological reactions to danger, such as increased heart rate or blood pressure. Research with humans has highlighted the amygdala's critical role in the learning of emotional associations and the recognition of emotional expressions in other individuals. Magnetic resonance imaging studies have shown that the amygdala shows activation to fearful stimuli. In humans, the brain is also involved in the control of emotional facial expressions. Smiles that occur spontaneously as a result of genuine happiness are involuntary. The extrapyramidal motor system, which depends on subcortical areas, governs involuntary smiles and fear reactions.

The chemistry of the brain also plays an important part in animal and human emotions. The neurotransmitter dopamine is released in copious amounts during periods of pleasure and excitement. Researchers have found that rats experience an increase of dopamine when engaging in activities that appear to suggest play. Research has also shown that if dopamine production is blocked in rats through the administration of a dopamineblocking agent, the rat's play activity disappears. The effects of the hormone oxytocin have been studied in smallmammalsand appear to be related to sexual activity and bonding behaviors. In humans, oxytocin is released in mothers who are nursing their infants and is considered to aid in the mother-child bond. Researchers have investigated the role of oxytocin in bonding among voles. If a female vole is injected with oxytocin, the animal will quickly select a mate. When a female vole is given a drug to block oxytocin, however, mate selection never takes place.

Many scientists contend that it is illogical to believe that emotions appear suddenly in humans. If evolution takes place through the process of natural selection, the emotions found in humans would be present in early evolutionary ancestors. The similarities in brain anatomy and chemistry between animals and humans would then support the idea that some basic emotions exist in various species. Darwin believed that some facial expressions in humans are universal. These expressions are genetically determined and evolved as the most effective at telling others something about how a person is feeling. Research with infants shows the innate capacity to grimace in pain or to smile in pleasure. For the most basic emotions, people in all cultures show similar facial responses to similar emotional situations. For example, anger is linked with a facial expression recognized by almost all cultures. Perhaps it is this line of reasoning from the evolutionary context that provides the strongest support for the existence of a wide range of emotional reactions in animals.

—*Frank J. Prerost*

See also: Apes to hominids; Brain; Communication; Evolution: Historical perspective.

FURTHER READING
Bekoff, Marc. *The Smile of a Dolphin*. New York: Discovery Books, 2000. Aseries of true stories about animal emotions are presented under the headings of love, grief, joy, aggression, anger, and fellow feelings. The research completed by over fifty scientists in the realm of animal emotions is discussed. The book also includes a number of attractive color pictures.

Griffin, Donald R. *Animal Minds*. Chicago: University of Chicago Press, 1992. This book was written to counteract the behavioristic tradition of JohnWatson and B. F. Skinner. The book is a key to the understanding of cognitive ethology, and emphasizes the richness of the animal mind.

Marshall, Elizabeth. *The Hidden Life of Dogs*. Boston: Houghton Mifflin, 1993. As an anthropologist and ethologist, the author provides unique insights into the behavior of dogs. The behavior of a pack of dogs over the course of thirty years is documented.

Masson, Jeffrey M. *When Elephants Weep.* New York: Delacorte Press, 1995. The author is a strong advocate for the recognition of emotions in animals. After discussing the impediments in the scientific community to the serious study of animal emotions, the author presents groupings of emotions expressed by animals. He presents numerous examples to support the wide range of emotions he attributes to animal behavior.

Panksepp, Jaak. *Affective Neuroscience: The Foundation of Human and Animal Emotions.* New York: Oxford University Press, 1998. This book presents a very thorough and scientific exploration of the neurochemicals associated with emotions. The author shows how the basic emotions and motivational processes are controlled by brain chemistry.

Sheldrake, Rupert. *Dogs That Know When Their Owners Are Coming Home.* New York: Crown, 1999. The author is a scientist and philosopher who presents numerous incidents of unusual abilities found in dogs. The information on animal empathy is particularly informative.

THE MIRROR TEST

In order to determine whether animals have the capacity for self-awareness, researchers commonly utilize the mirror test. The mirror test involves a basic procedure to test for self-consciousness. In one basic procedure, after an animal is anesthetized, patches of a bright dye are placed on different parts of the face. After awakening, the animal's reflection is shown in a mirror. If the subject touches the dyed parts of its face, self-consciousness is inferred. Many animals, including primates, elephants, birds, and dolphins, have been tested with this procedure. Only chimpanzees, orangutans, and humans have consistently shown the results indicative of self-recognition.

Researchers concluded that most species cannot conceive of a self. Correctly identifying one's own reflection indicates senses of identity and of personal awareness. Without this capacity an animal would not be able to identify or model the mental states of other creatures or sympathize when witnessing the suffering of another animal. Humans routinely make inferences about what other people are feeling because of this capacity for self-awareness. Apparently, only a few other primates have the rudimentary capacity for self-consciousness, yet even those who pass the mirror test do not show a consistent pattern of being able to empathize with others. In general, some researchers have been critical of the mirror test, stating that chimpanzees may have clever minds, but they are blank minds. Chimpanzees may be able to learn, memorize, and problem solve sufficiently to pass the mirror test, but be unable to utilize their situation to take into account the experiences of others.

ENDOCRINE SYSTEMS OF VERTEBRATES

FIELDS OF STUDY

Developmental biology, reproduction science

SUMMARY

Endocrinology is a relatively young branch of physiology. It deals with chemical messengers, carried in the blood, that stimulate specific responses. The responses are agents in the control of growth, development, metabolism, osmotic and ionic regulation (hydromineral metabolism), reproduction, and control of hormone secretion itself.

PRINCIPAL TERMS

- **feedback:** in endocrinology, this usually refers to one hormone controlling the secretion of another that stimulates the first, usually in the form of negative feedback, in which the second hormone inhibits the first
- **gland:** a tissue composed of similar cells that produce a hormone
- **hormone:** a blood-borne chemical messenger receptor: a protein molecule on or in a cell that responds to the hormone by binding to it and initiating a series of events that compose the response
- **target:** cells that contain hormone receptors

Basic Principles

Endocrine systems have been known only since the early twentieth century. The first known hormone was discovered around 1902 when William Bayliss and Ernest Starling discovered, in dogs, secretin, a hormone that stimulates pancreatic exocrine secretion in response to acid in the small intestine. Since that date, dozens of other hormones have been discovered, which control all aspects of growth, metabolism, and reproduction. The endocrine system consists of glands that secrete chemical substances called hormones in response to specific signals. The hormones are secreted into the blood stream, where they travel to specific target cells or tissues, which contain specific receptors that allow the hormones to bind, initiating the response. Classically, hormones were thought to belong to two very different groups, the polypeptide (small protein) hormones and the steroid (cholesterol-like) hormones. It is now known that hormones can be composed of several different kinds of molecules, including fats (prostaglandins) and even gases (nitric oxide).

Proteinlike hormones bind receptors found on external cell membranes to stimulate second messengers, such as cyclic adenosine monophosphate (cAMP), which activate enzymes and other cellular substances to produce a response. Steroid hormones enter target cells and bind intracellular receptors. The hormone-receptor complexes migrate to the nucleus and activate gene expression, which results in the response. This, like descriptions of many concepts in biology, is an oversimplification, and many hormones appear to work by a combination of the two mechanisms.

How It Works

Endocrine Control Systems. The endocrine secretions are controlled by the nervous system through a complex chain of command. Receptors around the body monitor sensory signals and alert the brain, which then relays the information to specific cells in the median eminence of the hypothalamus. For example, temperature receptors in the skin detect cold and inform the brain of potential body cooling. The brain then relays the information to cells in the hypothalamus, which secrete a molecule called thyrotropin releasing hormone into a blood vessel called the hypothalamo-hypophysial portal vessel. This blood vessel delivers the releasing hormone to the anterior pituitary gland, which in turn secretes a hormone called thyroid-stimulating hormone (TSH), or thyrotropin, into the blood. The TSH travels to the thyroid gland to stimulate the secretion of thyroid hormones, which stimulate metabolism in liver, muscle, and other cells. Heat produced as a by-product of metabolism warms the body. Some hormones are under dual control. Growth hormone (somatotropin) is stimulated by a releasing hormone called somatocrinin and inhibited by somatostatin. There are about seven anterior pituitary hormones that are controlled by similar mechanisms. Adrenocorticotropic hormone (ACTH) is controlled by corticotropin-releasing hormone. Melanocyte-stimulating hormone (MSH) and prolactin are under dual control by both releasing hormones and inhibiting hormones.

The gonadotropins—follicle-stimulating hormone (FSH) and luteinizing hormone (LH)— are under the control of a single releasing hormone called gonadotropin releasing hormone. All of these control systems are subject to feedback loops which usually involve negative feedback (for example, TSH secretion being inhibited by thyroid hormone), but positive feedback loops exist (estrogen feeding back positively to stimulate LH secretion).

Hormones Controlling Growth, Development, and Metabolism

The major control of growth is carried out by somatotropin (STH) from the anterior pituitary. STH does not act directly, however. Cells in the liver respond to STH to produce somatomedin, which stimulates bone growth and muscle production. Prolactin, a protein similar to STH, stimulates breast development in female mammals. In an interesting case of hormone evolution, thyroid hormone stimulates amphibian metamorphosis (tadpole to frog transition); however, in warmblooded vertebrates, this same hormone has evolved to stimulate metabolism for the purpose of heat production in birds and mammals. Several hormones stimulate metabolism for different reasons. Epinephrine (adrenaline), in addition to elevating blood pressure, mobilizes glucose from glycogen, in response to stress. Steroid hormones, also produced in the adrenal glands, stimulate the production of glucose from noncarbohydrate molecules (gluconeogenesis). The stimulus for this is prolonged stress, for example, starvation. These glucocorticoids, such as cortisol and corticosterone, evolved early and are very important in combating stresses resulting from migration among birds and even fish. The pancreatic hormones insulin and glucagon also effect energy metabolism. These two proteins regulate blood sugar, fat, and protein levels. After eating, insulin stimulates transport of these molecules into liver, fat, and muscle cells and then stimulates the incorporation of the simple molecules, such as glucose, amino acids, and fatty acids, into larger storage molecules, such as glycogen, protein, and fats.

Glucagon has opposite actions. After a prolonged period without food intake, glucagon stimulates breakdown of complex molecules, such as glycogen and fats, into simple molecules, which are released into the blood and made available to metabolizing cells. These two hormones act independently of the pituitary and respond directly to blood-borne signals such as glucose concentration. This regulation ensures a steady delivery of nutrients to metabolizing cells in animals who only eat intermittently.

Control of Water and Salt Balance

The state of hydration and salt levels in the body are of critical importance to vertebrate animals. Dehydration has obvious severe detrimental consequences. The salt composition of body fluids is equally important. For enzymes and other proteins such as antibodies and even hormones to function properly, salt concentrations (ionic concentrations) must be maintained. For example, blood levels of sodium and potassium must be maintained at approximately 145 and 4 millimolers, respectively, in most vertebrates. These levels are lower in amphibians (100 and 2 millimolers). Water content of the body is controlled primarily by a posterior pituitary hormone called antidiuretic hormone (ADH). When the body becomes dehydrated, both concentration receptors and volume receptors in the brain trigger the secretion of ADH. This small peptide then stimulates thirst and water retention in the kidneys. In amphibians, it also stimulates water absorption by the skin and urinary bladder. Asteroid hormone produced in the adrenal glands called aldosterone stimulates the kidney and large intestine to conserve sodium. The kidneys also excrete increased amounts of potassium in response to aldosterone. Aldosterone secretion is stimulated by angiotensin II. When blood sodium levels decrease, there is a consequent loss of water and thus body fluid volume. Pressure receptors in the kidneys trigger the release of renin, which initiates a complex series of enzymatic reactions in the blood leading to the appearance of angiotensin II, which stimulates the secretion of aldosterone. When blood pressure increases, aldosterone secretion decreases, sodium excretion increases, and other hormones appear which also help eliminate sodium. Pressure receptors in the heart cause that organ to secrete atrial natriuretic peptide (ANP). ANP inhibits sodium conservation in the kidneys. Increased pressure in blood vessels activates nitric oxide synthetase, which produces nitric oxide locally. Nitrous oxide increases the excretion of sodium by the kidneys.

Calcium and phosphate are also controlled by hormones. The parathyroid hormones respond directly to blood calcium concentrations. When calcium levels are low, parathyroid hormone (PTH) is secreted into the blood to stimulate three centers. In bone, PTH mobilizes calcium to elevate blood levels of this ion. Because mobilization of bone also elevates phosphate, which can be toxic at high concentrations, the kidneys become important. PTH stimulates the kidneys to increase calcium conservation and potassium excretion. PTH also stimulates uptake of calcium in the small intestine. Vitamin D enhances the action of PTH. Working antagonistically to PTH, calcitonin, produced in the thyroid gland, responds directly to high blood calcium to move this ion into bone.

Digestive Hormones

The digestion and assimilation of food are also controlled by hormones. In meat-eating animals, beginning in the stomach, stretch and the presence of protein stimulate the secretion of gastrin into blood vessels in the wall of the stomach. This gastrin stimulates the secretion of hydrochloric acid into the lumen of the stomach to digest protein. When the partially digested food enters the small intestine for the completion of digestion and assimilation, a slightly alkaline pH is required.

The walls of the small intestine detect the acidity and secrete another pair of hormones into blood vessels. Secretin travels to the pancreas and stimulates sodium bicarbonate secretion. The sodium bicarbonate travels through the common bile duct to the small intestine, where it neutralizes the acid. Gastric inhibitory polypeptide travels to the stomach to inhibit acid secretion and stomach contractions. Another peptide, cholecystokininpancreozymin (CCKPZ), responds to fats and proteins in the small intestine and is thus secreted into the blood. This hormone travels to the gallbladder, causing it to contract and release its bile through the common bile duct to aid digestion of fats in the small intestine. CCKPZ also stimulates secretion of a whole host of enzymes by the pancreas.

These enzymes also move through the commonbile duct to the small intestine to aid in the digestion of carbohydrates, fats, and protein. At least two other digestive hormones have been discovered that are not well understood at present. Motilin is secreted by the small intestine and stimulates stomach muscle contractions. Vasoactive intestinal polypeptide also is secreted by the small intestine and it, in turn, stimulates sodium bicarbonate secretion by the walls of the small intestine. Both hormones are of obvious benefit, but key details of their function, such as what triggers their secretion, are not clearly understood. It is important to realize that all of the hormones of the stomach and small intestine are secreted into the blood vessels in the walls of the organs, not into their lumens.

Reproductive Hormones
The two pituitary hormones that are involved in reproduction are called the gonadotropins, FSH and LH. These hormones are identical in males and females. The gonadal hormones differ between the two sexes. Females produce estrogens and progesterone in their ovaries. Males produce androgens (primarily testosterone) in the testes. The mammalian menstrual cycle has two components. Both the ovarian cycle and the uterine cycle proceed simultaneously and last approximately four days in rats, sixteen days in sheep, and twenty-eight days in humans. The length and pattern of the cycle vary with species. For the sake of comparison, the human cycle is described here.

The first five days of each cycle is called the menstrual period, and during this period the built-up walls of the uterus (resulting from the previous cycle) are shed and discharged through the vagina. At this time the concentrations of FSH and LHin the blood are about the same. From the close of the menstrual period until ovulation is the follicular cycle. FSH stimulates the ovaries to begin the growth and maturation of an egg-containing follicle. This follicle produces estrogen. Estrogen feeds back negatively on FSH, causing its levels in the blood to drop. At the same time, estrogen is feeding back positively on LH, causing its levels to rise.

At the midpoint of the ovarian cycle, LH peaks and causes the now mature follicle to burst and eject an egg (ovum) into the oviduct. The ruptured follicle now becomes a corpus luteum and continues to secrete estrogen, but also begins to secrete progesterone. The estrogen, and now the progesterone, stimulate the walls of the uterus to thicken and produce glandular tubes and blood vessels. This goes on for the final half of the cycle, which is called the follicular phase in the ovaries and the proliferative phase in the uterus. If fertilization of the ovum in the oviduct fails to occur during this period, a hormone, probably a prostaglandin, builds up in the corpus luteum, causing it to stop producing estrogen and progesterone. With the loss of these two steroids, the thickened wall of the uterus is shed and the menses flows during the first five days of the next cycle.

Male reproductive endocrinology is much different. The first striking difference is that, although the pituitary hormones FSH and LH are the same, the patterns of secretion are different. Instead of the cyclic peaks found in females, males secrete constant levels of gonadotrophins. FSH stimulates sperm production and maturation in the seminiferous tubules of the testes. LH stimulates testosterone secretion by the interstitial cells of the testes. Testosterone helps FSH to stimulate sperm maturation. This androgen also stimulates such primary sex characteristics as penis and epididymal growth during puberty. The epididymis is a tubular structure that stores sperm in preparation for ejaculation. Testosterone also stimulates secondary sex characters, such as the deepening of the voice and development of muscle mass that manifest during puberty in humans.

—*Daniel F. Stiffler*

See also: Circulatory systems of vertebrates; Digestion; Fertilization; Gametogenesis; Growth; Reproduction; Reproductive system of female mammals; Reproductive system of male mammals; Sexual development; Thermoregulation.

FURTHER READING

Bentley, P. J. *Comparative Vertebrate Endocrinology*. 3d ed. New York: Cambridge University Press, 1998. A very complete treatment of vertebrate hormones from all of the vertebrate classes.

Norman, A. W., and G. Litwack. *Hormones*. 2d ed. San Diego, Calif.: Academic Press, 1997. An excellent source of information on the biochemistry and molecular biology of the endocrine system.

Norris, D. O. *Vertebrate Endocrinology*. 3d ed. San Diego, Calif.: Academic Press, 1997. An up-to-date, comprehensive text that presents the hormones, their glands, and targets in great detail.

Yen, S. C., R. B. Jaffe, and R. L. Barbieri. *Endocrinology: Physiology, Pathophysiology, and Clinical Management.* 4th ed. Philadelphia: W. B. Saunders, 1999. Integrates normal endocrinology, the diseases of the hormones, and their treatments.

ENDOCRINE CONTROL OF PREGNANCY, PARTURITION, AND LACTATION

During an ovarian cycle, a pregnancy can occur if the ovum becomes fertilized. The events that occur during pregnancy are also controlled by hormones. In mammals, fertilization normally occurs in the oviduct. When the fertilized embryo is implanted in the uterus, it is nourished as a result of the glandular and vascular buildup of the uterine wall. This buildup is maintained by estrogen and progesterone secreted by the corpus luteum. As the embryo grows, it begins to produce a portion of the placenta; the other portion is produced by the uterus. Cells in the embryonic chorion of the placenta produce chorionic gonadotropin, which "rescues" the corpus luteum and prevents its becoming deactivated by a prostaglandin produced by the ovary as a signal to deactivate the corpus luteum if fertilization has not occurred. During the latter stages of a pregnancy, the placenta takes over the production of estrogen and progesterone in some species. During pregnancy, chorionic gonadotropin will be present in the urine. During the last days of pregnancy, the ovary begins to produce a new hormone called relaxin. Relaxin is a very ancient hormone; it has been found in sharks and birds as well as mammals. Relaxin softens the ligaments connecting the bones in the birth canal. It also stimulates uterine contractions. The uterine contractions begin the process of labor, resulting in birth. The signal for the onset of labor is a fetal hormone, cortisol. Cortisol crosses the placenta to the maternal blood, where it causes the synthesis of another prostaglandin. The strongest support for this conclusion comes from experiments in which the adrenal glands were removed from fetal goats; the result of such in utero surgery is significantly delayed parturition (birth). This uterine prostaglandin intensifies the uterine contractions to initiate labor and thus parturition.

Following birth, the suckling of the infant at the mother's mammary glands stimulates the release of the posterior pituitary hormone oxytocin. Oxytocin stimulates contraction of muscles in the mammary glands to eject milk. Oxytocin also stimulates uterine contraction to pass the placenta (afterbirth) through the vagina. Prolactin, which stimulates mammary development during pregnancy, has nothing to do with milk ejection during lactation. Prolactin is also present in lower vertebrates that lack mammary glands. In fish, prolaction stimulates sodium conservation. In amphibians, prolactin decreases skin water permeability and increases sodium uptake across the skin. In some salamanders, prolactin initiates "water drive," which is the migration of adults back to water for reproduction. In pigeons and similar birds, prolactin stimulates the production of crop-sac milk, which the mother regurgitates to feed her young; a process remarkably similar, in effect, to lactation.

ENDOCRINOLOGY

FIELDS OF STUDY

Bariatrics; diabetes medicine; internal medicine; laboratory medicine; neuroendocrinology; obstetrics and gynecology; pediatric endocrinology; radiology reproductive endocrinology; thyroid medicine.

SUMMARY

Endocrinology is a medical field focused on the diagnosis and treatment of abnormalities of the endocrine system. The endocrine system consists of glands that produce hormones: the adrenal gland, hypothalamus, ovaries, pancreas, parathyroid glands, pituitary gland, testes, and thyroid gland. These hormones control metabolism (utilization of food by the body), reproduction, and growth. Endocrinology is practiced by medical doctors with specialized training in that field. Physicians in other fields (such as obstetricians, gynecologists, and internists) may devote some of their practice to endocrinology. Some endocrinologists specialize in one area of endocrinology (such as neuroendocrinology or pediatric endocrinology). Conditions treated by endocrinologists include diabetes, hypertension (high blood pressure), inadequate growth, infertility, obesity, osteoporosis (weak bones), menopause, metabolic disorders, and thyroid disorders.

Endocrinology

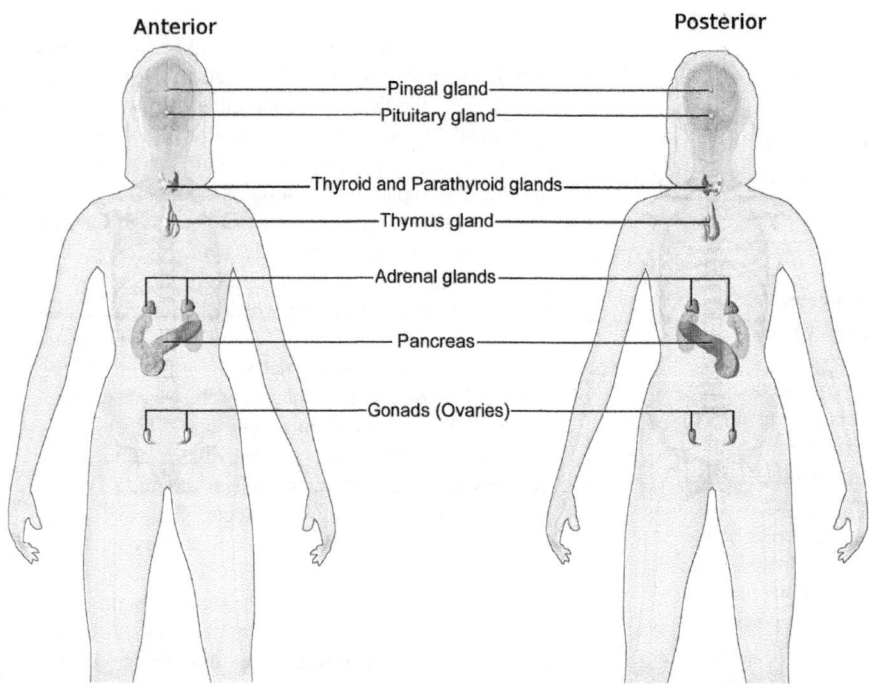

Endocrine System (Female).

PRINCIPAL TERMS

- **adrenal glands:** glands that are situated above each kidney and produce hormones, which respond to stress, such as cortisol and adrenaline.
- **hypothalamus:** portion of the brain that contains a variety of specialized cells; an important function of the hypothalamus is the linkage of the brain to the endocrine system through the pituitary gland.
- **ovaries:** paired organs adjacent to the uterus, which release eggs (ova) for reproduction and produce a variety of hormones, primarily estrogen and progesterone.
- **pancreas:** organ located in the upper abdomen, which produces hormones that regulate blood sugar levels (insulin, glucagon, and somatostatin) and secretes digestive enzymes, which pass into the small intestine.
- **parathyroid glands:** small glands, usually located within the thyroid gland, which produce parathyroid hormone; this hormone regulates calcium levels in the bloodstream and bones.
- **pituitary gland:** small gland, located at the base of the brain, which is sometimes referred to as the master gland because it produces hormones that stimulate or suppress the secretions of other endocrine glands.
- **testes:** male reproductive organs, which produce sperm and male sex hormones, such as testosterone.
- **thyroid gland:** butterfly-shaped organ located in the neck; it controls the metabolic rate, protein production, and the sensitivity of the body to other hormones.

Basic Principles

Endocrinology is a medical field dealing with the endocrine system, which is a complex system of organs that secrete hormones into the bloodstream. Hormones are chemicals that are released from cells in one location and affect cells located elsewhere in the body. To respond to these chemical messengers, a cell must possess a receptor to the hormone. Hormones control many bodily functions, including metabolism, reproduction, and growth. Diseases of the endocrine system often involve the abnormal production of a hormone or a cell's resistance to the effects of a hormone. For example, excess thyroid hormones produce a condition known as hyperthyroidism in which metabolism is increased. Patients with hyperthyroidism have increased nervousness, irritability, tremors, and a rapid heart rate. Individuals with hypothyroidism (inadequate level of thyroid hormones) have fatigue, poor muscle tone, constipation, and dry skin. Either thyroid condition may be caused by an abnormal functioning of the thyroid gland. These conditions may also be secondary to an abnormal functioning of the pituitary gland or hypothalamus, both of which regulate the thyroid gland. The hypothalamus produces a hormone, thyrotropin-releasing hormone (TRH), which causes the pituitary gland to release thyroid-stimulating hormone (TSH), which signals the thyroid gland to release thyroid hormones. Patients with diabetes usually have inadequate levels of insulin, which controls glucose (sugar) metabolism. Some cases of diabetes are caused by insulin resistance, a condition in which the cells do not respond well to insulin circulating in the bloodstream.

Background

Endocrinology is derived from the Greek words *endo* (within), *krīnō* (to separate), and *logia*, which supplies the suffix "-ology" (referring to a field of knowledge). Endocrinology originated in 200 BCE in China, when pituitary and sex hormones were isolated from the urine for medicinal purposes. In the Western world, an organ basis for pathology did not develop until the nineteenth century. In 1841, the German physician Friedrich Henle described "ductless glands" that secrete products directly into the bloodstream. In 1902, William Bayliss and Ernest Starling discovered secretin, which they described as a hormone. They defined a hormone as a chemical that is produced in an organ, then travels via the bloodstream to a distant organ and exerts a specific function on it. The field of endocrinology is based on the replacement of inadequate levels of a hormone with purified extracts. In cases in which hormone levels are unusually high, treatment involves lowering the hormonal level through surgical removal of a portion of the gland or destruction of some of the gland's cells by radiation. For example, hyperthyroidism is commonly treated with radioactive iodine, which concentrates in the thyroid gland and destroys some of the cells that produce hormones.

How It Works

Endocrinology is a medical specialty, and usually patients are referred from other physicians (such as an internist or family physician) for evaluation. The endocrinologist examines the patient and makes a diagnosis. If the referring physician has made a preliminary diagnosis, the endocrinologist often orders further tests to confirm it. The endocrinologist must be well versed in biochemistry and clinical chemistry to properly interpret these tests.

The endocrinologist frequently relies on the radiologist for diagnosis and treatment of an endocrine disorder. This involves the use of imaging equipment such as ultrasound and scintigraphy. Scintigraphy is a two-dimensional visualization of a radionuclide in the body. A radionuclide is a radioactive substance that is taken up by an endocrine grand. Computed tomography (CT) and magnetic resonance imaging (MRI) are also used to visualize organs such as the thyroid and adrenal glands. After diagnosing an endocrine condition, the radiologist might be called on to treat the condition with the injection of a radionuclide. The treatment levels of radiation are much higher than the diagnostic level and are designed to destroy cells producing excessive amounts of a hormone. Once a diagnosis is made, a course of treatment must be developed. Initial treatment might include referral to a surgeon for excision of a tumor or an abnormally functioning organ. Radiotherapy is employed to treat certain endocrine conditions such as hyperthyroidism (overactive thyroid) and Cushing's syndrome (overactive adrenal glands). Drug therapy is an option in some cases. Many diseases of the endocrine system are chronic and require lifelong treatment. A classic example is diabetes, which can develop in children and young adults (type 1 diabetes) and in adults (type 2 diabetes). Milder forms of type 2 diabetes can often be treated with medication; however, type 1 diabetes and severer forms of type 2 diabetes require insulin injections. Patients requiring insulin must be educated as to the importance of controlling their blood sugar level with self-administered insulin injections and of frequent monitoring of their blood glucose (sugar) through blood sampling. Diabetics are more prone to many conditions such as cardiovascular disease and loss of vision. Diabetics whose condition is under good control are less likely to develop serious health problems. Obesity greatly increases the risk of developing type 2 diabetes; therefore, a weight-loss program can sometimes return blood glucose levels to normal levels. Some diseases are congenital (present at birth) or occur at a young age; the pediatrician plays a crucial role in diagnosing an endocrine problem in young patients. Inadequate levels of many hormones can severely affect a child's development, both physically and mentally. Children with endocrine abnormalities are often referred to a pediatric endocrinologist.

Couples with an infertility problem may seek the help of a reproductive endocrinologist. The problem may be caused by either male or female factors. Some of these problems are nonendocrine in origin (for example, Fallopian tubes blocked from an infection); however, many infertile women can greatly increase their chances of becoming pregnant with assisted reproductive technology (ART), which involves the administration of hormones and other medications to stimulate and regulate the ovulation process. Gynecologists practice endocrinology related to the female reproductive system. A common problem dealt with by gynecologists is a menstrual irregularity. When women approach the menopause, which is caused by a drop in the level of female hormones (estrogen and progesterone), they often develop distressing symptoms, such as hot flashes, dry skin, and depression. Many women do not consult an endocrinologist at that time. Instead, they seek the advice of a gynecologist, family physician, or an internist. That physician will often diagnose and treat the condition; however, the patient may occasionally be referred to an endocrinologist.

Applications

Most hormones can be administered orally or by injection, skin patch, vaginal cream, or nasal inhalation. Some hormones cannot be administered orally. Hormones can be derived from animal or human sources. Some can be retrieved from urine, and others are obtained from animal organs harvested at slaughterhouses. In many cases, hormones have been synthesized in the laboratory.

Hormones derived from animal or human sources must be subjected to bioassay. Bioassay involves administering a hormone sample to an animal and measuring its effect. By this process, the pharmaceutical manufacturer can adjust the hormonal level of that batch to a standardized level. Thus, an individual ingesting a hormone can be assured that the dose is uniform from day to day. Synthesized hormones are easier to standardize; however, they still may require a bioassay. They are less likely to contain impurities, which could produce an adverse effect, including an allergic reaction.

Many products on the market are used for diabetes, thyroid disease, osteoporosis, infertility, and the menopause. A significant market exists for performance-enhancing products, although athletes are banned from using hormones in this manner. Specialized surgical procedures exist for medical conditions with an endocrine component, such as morbid obesity. Surgery for morbid obesity is known as bariatric surgery. An increasing proportion of surgical procedures are being done with laparoscopy, which has the advantage of a small incision and quicker recovery.

Medical Laboratory. Endocrinology is a field that uses laboratory services to a greater degree than many other specialties. Some medical laboratories contain specialized equipment to accurately measure specific hormonal levels. Inasmuch as this equipment can be quite expensive and may measure only one specific hormone, these services are usually found in specialty laboratories that analyze samples from a large geographic area.

Diabetes. Many products are marketed for diabetic patients. Meters that can reliably measure blood glucose levels with minimal discomfort are a necessity for all diabetics. Syringes for injection are also essential for management of the condition. Also available are insulin pumps that can be programmed to administer the appropriate insulin dosage throughout a twenty-four-hour period. The patient can adjust the rate at any time depending on any variance from the normal routine or an abnormal blood glucose reading.

Thyroid Disease. In addition to laboratory tests to measure thyroid levels, diagnosis of thyroid disease often involves radioactive iodine (iodine 131), which is manufactured in specialized laboratories. A small amount of the substance is given by injection or in tablet form for diagnosis. Diagnosis is facilitated by radiologists using a specialized device, a gamma camera. Treatment of hyperthyroidism involves the use of a much higher dosage of iodine 131, which destroys thyroid cells. Hyperthyroidism can also be treated with antithyroid drugs, such as methimazole and propylthiouracil. These drugs become concentrated in the thyroid gland and block production of thyroid hormones. For the immediate treatment of the symptoms of hyperthyroidism, beta-blockers such as propranolol, atenolol, and metoprolol can be used. These medications lower metabolism; however, they do not alter thyroid hormone levels in the circulation. Surgical removal of a portion of the thyroid gland is sometimes done. It is usually reserved for pregnant patients, children with an adverse reaction to antithyroid medications, and patients with a very large thyroid gland.

Osteoporosis. Osteoporosis is a weakening of the bones that can be triggered by an imbalance of a number of hormones. It commonly occurs in women at the time of the menopause. Osteoporotic bone is susceptible to fracture and disfiguring deformities. Specialized equipment, using specialized X-ray equipment or other imaging modalities, can measure the degree of osteoporosis. These devices are marketed not only for radiology facilities but also for physician offices and other health care facilities. Dual energy X-ray absortiometry (DEXA) scanning is the most common procedure used to measure the amount of calcium and other minerals present in bone. The amount of minerals present is known as the bone mineral density (BMD). The most commonly scanned areas are the hip and spine. If a patient is diagnosed with osteoporosis, hormonal and nonhormonal products are available to lessen the severity of or reverse osteoporosis. For women, estrogen preparations can slow or halt the progression of osteoporosis; however, they cannot reverse it. Bisphosphonates, pharmaceuticals that can reverse osteoporosis, are available.

Infertility. Reproductive endocrinologists are major users of specialized products. Assisted reproduction technology (ART) involves stimulating the ovaries to produce ova (eggs), extracting the ova, fertilizing the ova in the laboratory, culturing the ova (multiple cell stage or embryonic stage), then implanting the ova in the uterus. Medications such as clomiphene are used to stimulate the ovaries to release eggs; hormones, such as human menopausal hormone (HMG) and gonadotropins are also used to stimulate the ovaries. The menstrual cycle is often regulated with hormones such as estrogen and progesterone. Specialized equipment is used to extract the ova from the patient's ovaries. The process involves inserting a needle through the vaginal wall and into an ovarian follicle; it is conducted under the guidance of ultrasound. The extracted ova are placed in a culture medium and fertilized with sperm supplied by the husband or donor. The fertilized ova are then placed in specialized incubators for growth to the desired embryonic stage. The ova are then inserted into the uterine cavity for development. Ova and embryos are placed in vials and frozen with liquid nitrogen. At a future date, the embryos are thawed and inserted into the uterus.

Menopause. A number of products are on the market for replacement of hormones lost after the menopause. They include oral medication, injections, and transdermal skin patches (the hormones are absorbed through the skin and pass into the bloodstream). Many nonhormonal products are also available to treat menopausal symptoms.

Performance-Enhancing Hormones. Performance-enhancing hormones have been used by many athletes, both professional and amateur. Although most of these products are banned because they confer an unfair advantage on their user, their use continues, supported by a thriving black market enterprise. Hormones that can enhance athletic performance are anabolic steroids (male hormones), which promote muscle growth and strength, and erythropoietin (EPO), which increases red blood cell production. Stronger muscles allow a baseball player to hit more home runs and increases a cyclist's speed and endurance. Increased red cell production from EPO raises the blood's oxygen-carrying capacity.

The athlete who uses these hormones may not only be disqualified but also experience adverse effects. Anabolic steroids change cholesterol levels. They increase the level of low-density lipoprotein (LDL, or bad cholesterol) and decrease the level of high-density lipoproteins (HDL, or good cholesterol), which can raise a person's risk of developing cardiovascular disease and having a heart attack. Furthermore, these drugs can cause direct damage to the heart and liver.

A prominent example of an athlete who was disqualified is cyclist Floyd Landis (although he continues to deny using anything). Landis was the overall leader of the 2006 Tour de France, a grueling multistage bicycle race, but in stage 16, he lost eight minutes to the second-place rider, and most experts believed he could not make up the time. The following day, he made a dramatic comeback and went on to win the event. A mandatory urine test taken after stage 17 revealed high levels of testosterone, and he was stripped of his title.

FUTURE PROSPECTS

Research in endocrinology examines how hormones work in the body and how they relate to diseases and conditions. This research aims at improving the medical treatment of endocrine conditions. Some researchers are focusing on the development of synthetic hormones, which do not have the problems associated with animal-derived products. Others are looking at possible cures, in the form of transplanted or regrown organs or other ways for the body to manufacture hormones. Organ transplantation has made significant progress since the 1960's. Research continues on transplanting endocrine organs, such as the pancreas for the treatment of diabetes. The major problem faced by transplanted organs is rejection of the organ by the recipient's body. Therefore, research is focusing on developing better medications to combat rejection and also on the creation of artificial endocrine organs. Although an artificial pancreas is still a distant possibility, insulin pumps have been developed to continuously administer insulin to diabetics. These devices are likely to become more sophisticated, possibly with sensors to monitor glucose levels and administer the proper insulin dosage. A controversial topic, the subject of much debate on medical, political, and religious grounds, is embryonic stem cell research. These cells, derived from early embryos, have the potential to develop into any organ within the human body. For example, in theory, a diabetic could grow a new pancreas, which would produce insulin. Proponents of stem cell research allude to a future in which a patient could grow a new adrenal gland, a paraplegic could walk again, and a child with cystic fibrosis could be cured. Opponents of the research claim that it involves the destruction of an embryo and therefore a human life. Proponents counter that researchers use only surplus embryos—those that are destined for destruction. Treatment with stem cells is still in its infancy, but the ability to grow a replacement endocrine organ would cure many endocrine diseases.

Robin L. Wulffson, MD, FACOG

FURTHER READING

American Diabetes Association. *American Diabetes Association Complete Guide to Diabetes.* 5th ed. Alexandria, Va.: Author, 2011. Provides information to help diabetics manage their disease. Begins with a discussion of the causes and effects of diabetes. Contains a glossary, an appendix on self-monitoring and injection techniques, and a list of resources and organizations.

Borer, Katarina T. *Exercise Endocrinology.* Champaign, Ill.: Human Kinetics, 2003. Looks at the role of hormones in exercise and athletic performance. Topics include regulation of hydration and fuel use during exercise, gender and performance, biological rhythms, and exercise as a stressor.

Gardner, David, and Dolores Shoback. *Greenspan's Basic and Clinical Endocrinology.* 9th ed. New York: McGraw-Hill Medical, 2011. Examines the molecular biology of endocrine glands and discusses metabolic bone disease, pancreatic hormones and diabetes mellitus, hypoglycemia, obesity, geriatric endocrinology, and many other diseases and disorders.

Hadley, Mac E., and Jon E. Levine. *Endocrinology.* 6th ed. Upper Saddle River, N.J.: Prentice Hall, 2007. Presents explanations of basic concepts and applications. Focuses on how glands and hormones control physiological processes.

Lebovic, Dan I., John D. Gordon, and Robert N. Taylor. *Reproductive Endocrinology and Infertility: Handbook for Clinicians.* Arlington, Va.: Scrub Hill Press, 2005. A ready reference for endocrinologists treating conditions and disorders related to reproduction. Information from textbooks, articles, and endocrinologists was gathered and analyzed to provide evidence-based approaches and strategies.

Potter, Daniel A., and Jennifer S. Hanin. *What to Do When You Can't Get Pregnant: The Complete Guide to All the Technologies for Couples Facing Fertility Problems.* New York: Marlowe, 2005. A thorough guide for couples with fertility problems.

Skugor, Mario, and Jesse Bryant Wilder. *The Cleveland Clinic Guide to Thyroid Disorders.* New York: Kaplan, 2009. Skugor, an endocrinologist, teamed with

---

FASCINATING FACTS ABOUT ENDOCRINOLOGY

- Gigantism is a condition marked by excessive growth because of the secretion of a growth hormone by a pituitary tumor. Trijntje Keever, born in 1616 in the Netherlands, suffered from the condition. She was the tallest woman in recorded history. When she died at age seventeen from cancer, her height was 8 feet, 4 inches.

- A deficiency of growth hormone results in dwarfism. General Tom Thumb (the stage name of Charles Sherwood Stratton), born in 1883, probably suffered from the condition. His height at the age of eighteen was 2 feet, 8.5 inches. Stratton became wealthy as a performer in P. T. Barnum's circus.

- Within a year of giving birth, 5 to 10 percent of women develop hypothyroidism secondary to postpartum thyroiditis. Initially, thyroid hormone levels may rise, then either return to normal or drop to hypothyroid levels. Of those women who become hypothyroid, about 20 percent will require lifelong treatment.

- Premarin is an estrogen preparation that is sometimes prescribed to women at the time of the menopause as hormone replacement therapy. The name is derived from "pregnant mare's urine," the source of the hormonal preparation.

- Girls stop growing in height at puberty because estrogen closes the epiphyses (growth plates), located in the arm and leg bones.

- Newborn girls sometimes have vaginal bleeding; this is caused by a drop in the level of estrogen, which the fetus was exposed to while in the uterus.

writer Wilder to present detailed information on thyroid diseases and treatment options.

Thacker, Holly. *The Cleveland Clinic Guide to Menopause.* New York: Kaplan, 2009. Thacker, a physician at the Center for Specialized Women's Health at the Cleveland Clinic, offers safe treatments for the menopause and explains myths and facts regarding hormonal replacement therapy.

WEB SITES

American Association of Clinical Endocrinologists http://www.aace.com

American Diabetes Association http://www.diabetes.org
American Thyroid Association http://www.thyroid.org
Pediatric Endocrine Society http://www.lwpes.org/index.cfm
Society for Reproductive Endocrinology and Infertility http://www.socrei.org

See also: Bionics and Biomedical Engineering; Cell and Tissue Engineering; Geriatrics and Gerontology; Obstetrics and Gynecology; Reproductive Science and Engineering.

ENDOCYTOSIS AND EXOCYTOSIS

FIELDS OF STUDY

Biochemistry; Genetics; Molecular Biology

SUMMARY

Endocytosis and exocytosis are described as complementary processes that move materials into and out of cell cytoplasm through the cell membrane. Different types of endocytosis are described, and some of their applications are discussed.

PRINCIPAL TERMS

- **cell membrane:** a biological membrane that forms a semipermeable barrier separating the interior of a cell from the exterior.
- **endosome:** an intracellular compartment that sorts and transports material taken into a cell via endocytosis.
- **phagocytosis:** a type of endocytosis in which solid particles are taken into the cytoplasm of a cell through the cell membrane.
- **pinocytosis:** a type of endocytosis in which extracellular fluid and any substances it may contain are taken into the cytoplasm of a cell through the cell membrane.
- **vesicle:** a small bubble within a cell, surrounded by a double layer of lipids.

COMPLEMENTARY CELLULAR PROCESSES

The two processes of endocytosis and exocytosis are effectively mirror images of each other. In endocytosis, materials are brought into a cell from outside, while in exocytosis, materials are transported out of a cell. These processes are unlike active transport, which functions by opening existing channels in the membrane to allow specific materials to pass through. A model of endocytosis can be seen when two drops of a liquid such as water or mercury come together and combine into a single drop. The outer surfaces of the drops combine to form a slightly larger surface, while the interiors combine and are enclosed within the surface of the new drop. Similarly, a model of exocytosis can be seen when a large drop breaks apart into two or more smaller drops.

ENDOCYTOSIS AND EXOCYTOSIS IN ACTION

Endocytosis and exocytosis are the basic processes by which substances such as enzymes, insulin, carbohydrates, and a variety of other materials are moved about within and without various cells. They are complementary actions in many cases, as endocytosis is used to bring materials into a cell for specific operations to be performed, while exocytosis is used to eliminate the products and by-products of those operations. They are also the processes by which entities such as viruses invade and infect cells.

Endocytosis begins when a material is brought to the cell membrane. A trigger such as a hormone initiates invagination, which is the folding of the membrane to form a hollow pocket around the material. The membrane then closes around the invagination, forming a vesicle that contains the material. The vesicle can thus be thought of as a bubble of the external material within the cytoplasm of the cell. The wall

ENDOCYTOSIS AND EXOCYTOSIS

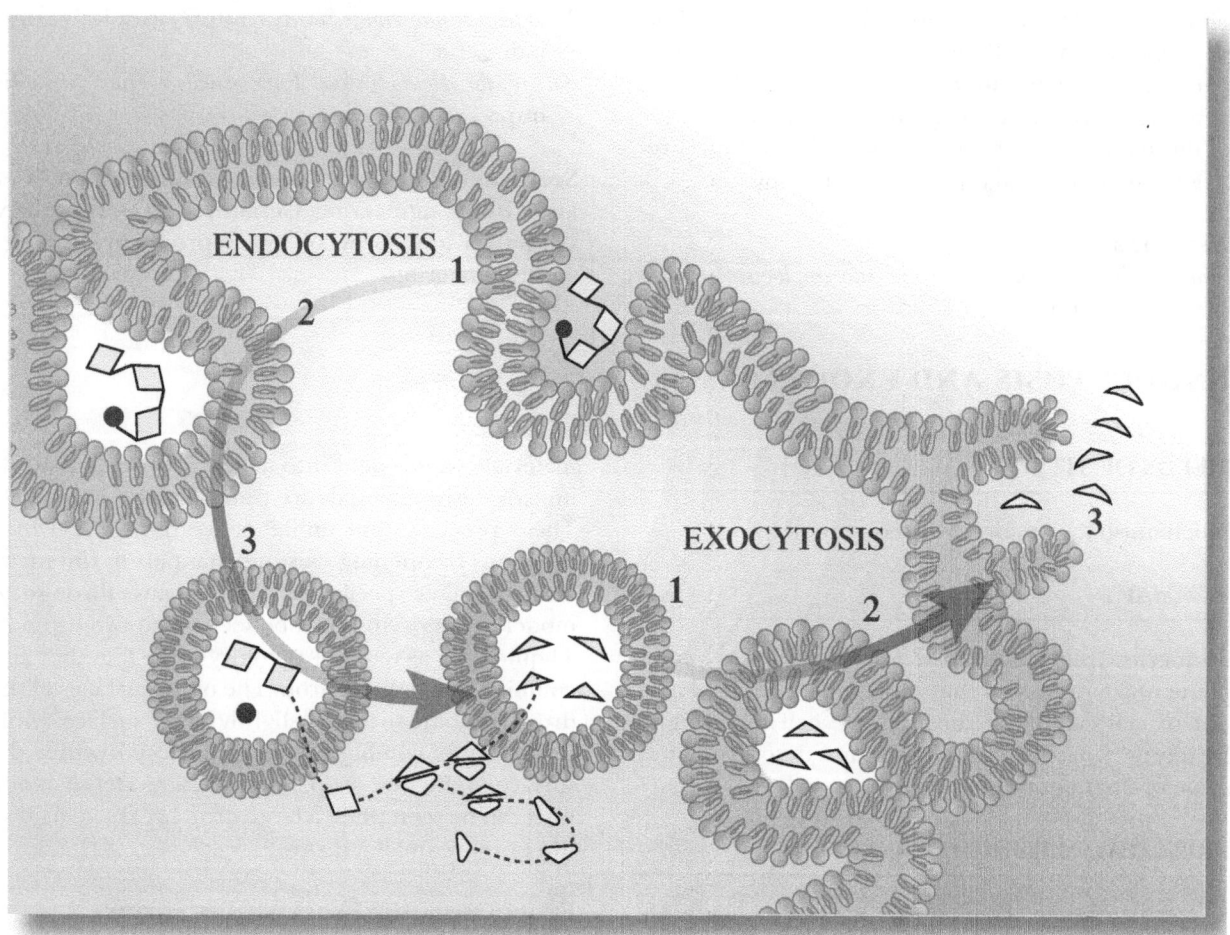

of the vesicle consists of the same material as the cell membrane. A protective coating of the protein clathrin typically forms around the vesicle to allow it to exist safely within the cytoplasm.

When the vesicle arrives at its destination within the cell, such as the Golgi complex or an endosome, it joins to the membrane of that site and essentially grows an opening through the membrane in order to eject its contents. Endosomes are compartments within the cell that serve various functions, including sorting materials introduced via endocytosis and forming pathways for the transport of those materials within the cell. Endosomes are first formed when vesicles lose their protein coatings and start to merge; the resulting structure is called an "early endosome." Over time, early endosomes mature into late endosomes, then eventually either mature further to form lysosomes or fuse with existing lysosomes.

There are three basic types of endocytosis. In phagocytosis, cells such as white blood cells extrude themselves about solids such as cellular debris and foreign objects to bring them within the cell cytoplasm, where they can be digested and converted into waste products that can be eliminated. In pinocytosis, cells encapsulate and bring in a bulk quantity of extracellular fluid. In receptor-mediated endocytosis, only specific materials, such as proteins or virus particles (virions), are brought into a cell.

The process of exocytosis within the cell, in which materials are encapsulated and transported through the cell membrane, functions in a similar manner, but in reverse. Mature virions, for example, are eliminated from within infected cells by a receptor-mediated process, while other materials are carried out of the cell and deposited into the extracellular fluid by encapsulation and extrusion at the cell membrane.

Endocytosis and Exocytosis in Biological Systems

Single-celled organisms especially depend on these twin processes for the ingestion of food and the elimination of waste. The amoeba and the paramecium are perhaps the most familiar of such organisms. On encountering a suitable particle of a foodstuff, the amoeba seems to flow around the particle to enclose and engulf it within itself. Once the particle is inside the organism, digestive processes go to work, decomposing it into usable molecules, such as simple sugars, fats, and amino acids. As the various chemical transformations take place, unusable materials are accumulated. These become encapsulated for elimination, and the amoeba then seems to flow away from the waste material. In actuality, the particle was taken into the amoeba's cell interior by endocytosis, and the waste materials were eliminated by exocytosis.

Richard M. Renneboog, MSc

Further Reading

Berg, Jeremy M., John L. Tymoczko, Gregory J. Gatto, and Lubert Strye. *Biochemistry*. 8th ed. New York: W. H. Freeman, 2015. Print.

Lehninger, Albert L. Biochemistry: The Molecular Basis of Cell Structure and Function. 2nd ed. New York: Worth, 1975. Print.

Lodish, Harvey, et al. *Molecular Cell Biology*. 7th ed. New York: Freeman, 2013. Print.

Pelczar, Michael J., Jr., E. C. S. Chan, and Noel R. Krieg. *Microbiology: Concepts and Applications*. New York: McGraw, 1993. Print.

Reece, Jane B., et al. *Campbell Biology*. 10th ed. San Francisco: Cummings, 2013. Print.

Roberts, Michael, Michael Jonathan Reiss, and Grace Monger. *Advanced Biology*. Cheltenham: Nelson,

---

ENDOCYTOSIS SAMPLE PROBLEM

Briefly explain why endocytosis always requires fluidity in the structure of the cell membrane.

Answer:

In the process of endocytosis, the cell membrane must have fluidity so that it can invaginate to form a pocket. When the pocket has been formed, the cell membrane must be able to close up around the pocket and heal over to re-form the cell membrane and create a stable vesicle. This requires the cell membrane to have a fluid character that allows it to "flow" back together and form an unbroken layer.

Enzyme engineering

FIELDS OF STUDY

Biology; biochemistry; biotechnology; chemistry; food science; genetics; medicine; microbiology; pharmacology.

SUMMARY

Catalysts accelerate the rate of chemical reactions without being essentially changed, and enzymes are biological catalysts that accelerate the rate of reactions that occur in living systems. Enzyme engineering identifies enzymes that have potentially useful catalytic activities and chemically or structurally modifies them to increase their activity, change their substrate specificity, change the types of reactions they catalyze, or change the properties of enzymes and the manner in which they are regulated. Engineered enzymes can generate completely novel molecules or new, improved ways to synthesize useful molecules.

PRINCIPAL TERMS

- **active site:** pocket or cleft in an enzyme surrounded by amino acids that specifically bind the substrate and catalyze its conversion to the product.
- **amino acids:** chemical building blocks of proteins that contain a nitrogen-containing amino group

(-NH2), a carboxylic acid (-COOH), and a functional group that varies between amino acids and determines their chemical properties.
- **catalyst:** substance that increases the rate of a reaction without being consumed or permanently changed in the process.
- **cofactors:** non-amino acid molecules that are associated with enzymes and necessary for enzymatic function.
- **enzyme:** proteins, groups of proteins, or ribonucleic acids (rnas) that are produced by living organisms and catalyze biochemical reactions.
- **protease:** enzyme that degrades proteins.
- **substrate:** molecule or molecules physically engaged by an enzyme to accelerate the chemical reaction that consists of the conversion of the substrate or substrates to the product or products.
- **transition state:** intermediate chemical structure formed during the process of a chemical reaction that represents the highest energy state of the reaction and usually degenerates to form the product.

BASIC PRINCIPLES

Enzymes are widely used as catalysts in several industrial ventures, ranging from food to synthetic chemistry to many other industrial processes. However, enzymes often show insufficient substrate selectivity, poor stability, and catalytic activities that are not robust enough for industrial use. To remedy this shortcoming, enzyme engineering builds new enzymes or modifies existing enzymes to give them novel, useful properties, or the ability to catalyze valuable chemical reactions. Engineered enzymes come in several forms. Semisynthetic enzymes (synzymes) have specific amino acids that have been chemically modified. These modifications can significantly alter the activity, specificity, or properties of the enzyme.

Directed evolution subjects the gene that encodes the enzyme to multiple rounds of mutation. The variant enzymes generated by these mutagenic genes are then sieved by some kind of selection scheme that identifies those mutant forms that display the desired characteristics or activities.

A cheaper strategy is rational design. Rational design uses detailed knowledge of the structure of proteins to identify those regions that are essential for its function and properties. By changing only those amino acids thought to be necessary for the modification of that function, the enzyme is potentially tailored for a new function with little investment.

Enzyme engineers use catalytic antibodies or abzymes. Antibodies are Y-shaped proteins made by vertebrate immune systems that bind to specific chemicals. Abzymes bind to chemicals and force them into the transition state of a chemical reaction, which accelerates the formation of the product from the reactants.

BACKGROUND

Enzyme engineering arose only after advances in several other fields made it possible to determine the primary amino acid sequences and three-dimensional structure of enzymes and directly manipulate them at the molecular level. Swedish biochemist Pehr Victor Edman gave birth to protein sequencing in 1950, when he designed the Edman degradation reactions that can determine the primary amino acid sequence of proteins. In 1958, English biochemist John Cowdery Kendrew used X-ray crystallography to solve the three-dimensional structure of the muscle oxygen-storing protein myoglobin. In the 1970's, American biochemist Herbert Wayne Boyer and American geneticist Stanley Norman Cohen pioneered molecular cloning techniques that gave scientists the means to clone genes and insert them into bacteria for propagation.

The first studies in enzyme engineering examined the effects of mutations on enzyme active sites. Betalactamase, the enzyme used by bacteria to degrade beta-lactam antibiotic (penicillin, ampicillin, and amoxicillin) was one of the first enzymes examined by enzyme engineering. In 1978, Canadian chemist Michael Smith and his colleagues invented site-directed mutagenesis, which gave biochemists a much better way to place targeted mutations into the genes that encode enzymes and thereby change their primary amino acid sequence. In 1986, the laboratories of Peter Schultz (University of California, Berkeley) and Richard Lerner (Research Institute of Scripps Clinic) made the first catalytic antibodies that could split ester bonds.

HOW IT WORKS

Semisynthetic Enzymes. Enzymes that are modified by chemical means are known as semisynthetic enzymes. There are two main ways to produce semisynthetic enzymes: atom replacement or group attachment. Atom replacement exchanges one atom within an enzyme for a different atom. Such replacements can modify enzyme activity or change the substrate specificity of the enzyme. Group attachment involves the use of particular chemical reagents to attach particular molecules to enzymes. Attaching additional molecules to enzymes can also markedly change enzyme activity and substrate specificity.

Directed Evolution. Directed evolution randomly changes amino acids in a protein without prior knowledge of the exact function of each amino acid. The first step, diversification, takes the gene that encodes the enzyme of interest and replicates it many times while using a copying machinery that is inherently error-prone. This introduces random mutations into the gene and creates a large collect of gene variants that are usually grown in bacteria.

The second step, selection, tests or screens these enzyme variants for a desired property. Once the desired variants are identified, they undergo the third step, amplification, which replicates the identified variants and sequences them in order to determine which mutations produced the desired properties. Collectively, these three steps constitute one round of directed evolution, and the vast majority of such experiments require multiple rounds. The goal is to find those variant enzymes that show the most desired characteristics to the greatest extent. Directed evolution studies suffer from the need to make huge numbers of mutants that produce no discernable effect, since up to 90 percent of all mutants made are uninformative.

Semirational Design. This enzyme engineering strategy employs sophisticated computer programs that assemble all the available structural information of the enzyme under study and predict how the mutations introduced into different locations within the enzyme might affect its activity. The enzyme engineer then notes the predicted changes that will potentially generate the desired property changes and uses this information to conduct targeted mutagenesis experiments.

Targeted mutagenesis experiments introduce mutations into specific locations of a protein. Once these mutations are made, the variant enzyme with the engineered changes is tested to determine if it has the specific properties the enzyme engineer was hoping to produce in the enzyme. These approaches combine structural information with rational design. Two computer programs that make such predictions include Protein Sequence-Activity Relationship or ProSAR and Combinatorial Active-Site Saturation Test, otherwise known as CASTing.

Rational Design. If a great deal of structural information about the enzyme in question is available, then that structural information informs which amino acids should be changed. Many rational design attempts have not succeeded because of uncertainties regarding protein structure.

De Novo Design. A computer builds an enzyme around the transition state of a reaction from scratch. The computer begins by designing the active site by placing specific amino acids in strategic positions so that they efficiently bind the transition state of the chemical reaction and stabilize it. The program then constructs a protein backbone that supports and properly positions the active-site amino acids but still provides a coherent protein structure that is predictably stable under the desired conditions.

This particular strategy suffers from gaps in the ability to predict protein structure accurately and correlate this ideal structure with enzymatic activity. For example, two enzymes (retro-aldol enzyme and a Kemp elimination catalyst) were built completely from scratch by using computer programs. However, both enzymes required further optimization by directed evolution to achieve maximum activity.

Catalytic Antibodies. The immune system of some vertebrates makes Y-shaped proteins that specifically bind to and neutralize foreign substances that invade the body. Immunizing laboratory animals with stable analogues of the transition states of various reactions directs the immune systems of those animals to synthesize antibodies that cannot only bind particular chemical reactants but force them into the transition state of the reaction, which subsequently forms the product.

APPLICATIONS

Pharmaceutical Production. Beta-lactam and cephalosporin antibiotics are commonly prescribed to combat various illnesses. Both of these drugs kill bacteria by inhibiting the synthesis of the bacterial cell wall. Beta-lactam antibiotics include such widely recognized drugs as penicillin, ampicillin, and amoxicillin, whereas cephalosporin antibiotics include such popularly used antibiotics as Ceftin (cefuroxime), Kephlex (cephalexin), and Ceclor (cefaclor).

Unfortunately, with repeated use, bacteria can become resistant to commonly used antibiotics, and making new, improved antibiotics is essential to treat some of the more recent and aggressive infectious diseases. To make new cephalosporin antibiotics, enzyme engineers have used enzymes called acylases to convert simple starting chemicals into various versions of these drugs. By engineering these acylase enzymes, pharmaceutical companies have been able to make new cephalosporin and beta-lactam antibiotics that have novel properties and can kill bacteria that are resistant to older drugs.

Enzymes as Medicines. When a person is cut, blood oozes from the damaged tissue. Fortunately, blood clotting (also known as coagulation) eventually stanches this blood flow. Blood clotting is an essential part of wound healing, but it is also a very highly regulated event. The formation of blood clots inside undamaged blood vessels clogs those vessels and leads to heart attacks if clots form inside the vessels that surround the heart, or a stroke, if they occur within vessels that surround the brain. The human body has ways to destroy unnecessary clots.

An enzyme called tissue plasminogen activating factor (TPA) activates other enzymes in the body that degrade harmful clots. Commercially available, native TPA is called Alteplase, which has a half-life in the bloodstream of four to six minutes. Engineered forms of TPA are also clinically available. Reteplase, a shortened version of TPA (consists of 357 of the 527 amino acids of Alteplase), has a longer half-life (thirteen to sixteen minutes). Tenecteplase, which has two amino acid changes (substitutes asparagine$_{103}$ with a threonine and asparagine$_{114}$ with glutamine), has an even longer half-life of twenty to twenty-four minutes.

Engineered enzymes are also used in enzyme-replacement therapies. Several genetic diseases, known as lysosomal storage diseases, result from the inability to make functional versions of enzymes that degrade various biological molecules. The accumulation of these molecules kills brain cells and causes the death of the patient. Engineered enzymes used in enzymereplacement therapies include Cerezyme (imiglucerase, used to treat Gaucher's disease), Naglazyme (galsulfase, used to treat mucopolysaccharidosis VI), Myozyme (alglucosidase alfa, used to treat Pompe disease), and Aldurazyme (laronidase, used to treat mucopolysaccharidosis I).

Enzyme Immobilization. By attaching enzymes to surfaces, embedding them in gel matrices, hollow fibers, or cross-linking them to each other, enzymes are immobilized on insoluble surfaces. This increases their stability, simplifies their recycling, and increases the tolerance of enzymes to high levels of substrate and products. Detergent enzyme preparations, such as Alcalase, immobilize the protease subtilisin by attaching it to insoluble particles. Attaching the enzyme to inert material increases its reuse as it degrades proteinaceous matter.

Making Enzymes Soluble in Organic Solvents. Enzymes usually work in water, but many reactions between organic chemicals occur in organic solvents. Although Russian chemist Alexander Klibanov showed that several enzymes are active in organic solvents, many enzymes are neither soluble in organic solvents nor work properly in such environments. Attaching a molecule called polyethylene glycol (PEG) to some enzymes makes them soluble and active in organic solvents and allows them to make things such as polyester, peptides (small proteins), esters (sweet-smelling things found in foods), and amides (nitrogen-containing compounds). Such modified enzymes also have clinical uses. For example, the enzyme asparaginase can kill cancer cells but is toxic, unstable, and some patients have severe allergies to it. PEG-treated asparaginase is not as toxic as the native enzyme, is much more stable, and does not cause allergy. PEG-asparaginase is used to treat tumors in humans.

Abzymes. A notable variety of reactions are catalyzed by catalytic antibodies that range from forming or breaking carbon-carbon bonds, rearrangements, hydrolysis of various bonds, transfer of chemical groups, and even an industrial reaction called the Diels-Alder reaction. However, abzymes are very expensive and tedious to make, and their catalytic activity is well below that of enzymes. Yet they do provide tailor-made catalysts when no other such reagent exists.

FUTURE PROSPECTS
Because modified enzymes can make certain products more cheaply, the public response to modified enzymes is generally positive. However, the genetically modified organisms (GMOs) that are used to produce these enzymes give many people pause, since the introduction of GMOs into the environment may have long-term consequences that are presently unrecognized. Strict government regulation that forbids the release of GMOs into the environment without approval allays most of these concerns, but some people are still troubled by the use of GMOs to make products that they eventually end up eating or using in some other manner. Enzyme engineering is one of the

up-and-coming fields in chemistry and biochemistry. Since the 1990's, the use of enzymes in industrial and academic chemistry has greatly increased. There are many advantages to using enzymes in that they can act outside cells and under mild conditions that minimize troublesome side effects, are environmentally innocuous, compatible with other enzymes, and are very efficient, though highly selective catalysts. The largest drawback of using enzymes is that the right enzyme is sometimes not available to catalyze the desired reaction. Enzyme engineering can eliminate this significant drawback. Furthermore, as biochemists achieve a more profound understanding of protein structure, cheaper and faster ways of doing enzyme engineering, such as rational design, become more successful and practical. This will shorten the time required for enzyme engineering experiments and reduce its cost. Companies are already looking intently at enzyme engineering as a significant investment for their research and development departments.

Michael A. Buratovich, MA, PhD

FURTHER READING

Arnold, Frances H., and George Georgiou, eds. *Directed Enzyme Evolution: Screening and Selection Methods.* Totowa, N.J.: Humana Press, 2010. Laboratory protocol book that describes, in great detail with figures and graphs, some rather ingenious techniques for screening mutant clones of enzyme genes.

_____. *Directed Evolution Library Creation: Methods and Protocols.* Totowa, N.J.: Humana Press, 2010. Encyclopedic collection of protocols for generating libraries of randomly mutagenic enzyme genes in bacteria, with tables, graphs, and some figures.

Faber, Kurt. *Biotransformations in Organic Chemistry: A Textbook.* 5th ed. New York: Springer-Verlag, 2004. A very clear, useful textbook on the uses of enzymes in chemistry that includes a chapter on engineered enzymes.

Park, Sheldon J., and Jennifer R. Cochran, eds. *Protein Engineering and Design.* Boca Raton, Fla.: CRC Press, 2010. Covers the broader field of protein engineering—methods of developing altered proteins for novel applications—in two sections: one

---

FASCINATING FACTS ABOUT ENZYME ENGINEERING

- Hemophilia is a genetic disease characterized by an inability to form blood clots. Because blood clotting requires the coordinated and sequential activity of a host of blood-based enzymes called clotting factors, hemophilia patients are treated with infusions of purified clotting enzymes. Unfortunately, some patients form antibodies against clotting factors, which abrogates the efficacy of such treatments. However, enzyme engineers have made a truncated version of clotting factor VIII (B-domain deleted recombinant factor VIII or BDDrFVIII) that is smaller than the native enzyme but just as active. This protein is not recognized by the immune system nearly as often as the native clotting factor VIII.

- Another clotting factor called factor VII (FVII) can bypass the need for other clotting factors if it is present in high enough concentrations. FVII is made in an inactive form that is only activated when the front tip of the protein is removed. Enzyme engineers have made a form of FVII that does not require activation called recombinant FVIIa (rFVIIa; the "a" stands for active). Infusions of rFVIIa can help clot the blood of hemophilia patients who have antibodies against other clotting factors and can also successfully treat patients who suffer from uncontrolled bleeding from trauma, surgery, anticoagulant drugs, or pregnancy.

- Metabolic engineering manipulates enzymes that direct metabolic pathways. Such manipulation can generate organisms that synthesize industrially useful compounds or degrade environmental pollutants. Animal feed production uses a microorganism called

- *Methylophilus methylotrophus* to convert methanol to animal protein, but by inserting genes from the common intestinal bacterial *Escherichia coli* into *M. methylotrophus*, this metabolically engineered organism can make much more protein from the same initial mass of methanol, thus increasing the overall efficiency of animal protein production and decreasing production costs.

- Antibody-directed enzyme prodrug therapy (ADEPT) uses an enzyme to a human antibody that tightly binds to a tumor-specific surface protein. When these antibodies bind to cancer cells, the enzyme converts an anticancer drug that is present in an inactive form (a prodrug) into a cancer-killing chemical. Because the prodrugs are activated only at the surface of the tumors, they cause few side effects and effectively kill tumors. As of 2011, ADEPT is in clinical trials.

on experimental protein engineering and the other on computational design. Includes discussion of enzyme engineering using both rational and combinatorial approaches.

Scheindlin, Stanley. "Clinical Enzymology: Enzymes As Medicine." *Molecular Interventions* 7, no. 1 (February, 2007): 4-8. An absorbing and readable Summary of the use of engineered enzymes in clinical diagnoses and treatments.

WEB SITES

International Enzyme Engineering Symposium http://www.enzymeengineering.ege.edu.tr

National Institutes of Health Recombinant DNA Advisory Committee http://oba.od.nih.gov/rdna_rac/rac_about.html

See also: Metabolic Engineering.

EUKARYOTES AND PROKARYOTES

FIELDS OF STUDY

Biochemical Engineering; Cell and Tissue Engineering; Drug Testing; Environmental Microbiology; Histology; Industrial Fermentation; Pathology; Pharmacology; Sewage Engineering; Toxicology; Virology; Zymurgy and Zymology

SUMMARY

Energy captured during the formation of glucose during photosynthesis is released during the processes involved in respiration in the human body. The glycolytic pathway breaks down the glucose molecule into carbon dioxide and water, and uses the energy released by the chemical changes to drive the processes of metabolism. When oxygen is not available, pyruvic acid from glucose is broken down to lactic acid rather than degraded further. Buildup of lactic acid in overworked muscles causes a burning sensation. Prokaryotic cells derive metabolic energy from glucose anaerobically via the lactic acid fermentation process, and are unable to grow and produce adenosine triphosphate (ATP) in the presence of oxygen. Eukaryotic cells require oxygen for growth and the production of ATP.

PRINCIPAL TERMS

- **eukaryotic cell:** a complex cell on which all life more complex than a bacterium or yeast is based, characterized by a nucleus containing the cell's DNA and a number of specialized organelles, supplied with energy by mitochondria, which individually resemble the more primitive prokaryotic cell. *See also* prokaryotic cell.
- **fermentation:** a biological process for breaking down complex organic compounds into simpler compounds. One of the most familiar in human history is the conversion by yeast of sugar to carbon dioxide, alcohol, and water. Fermentation also occurs in cells, including animal muscle cells, breaking down glucose to produce lactic acid, lactate, carbon dioxide, and water, as well as adenosine triphosphate, a source of energy. It is less efficient than cellular respiration but occurs when muscles are short of oxygen. Many anaerobic bacteria ferment sugars: Lactobacillus ferment milk to produce yogurt. Fermentation also produces lactic acid in a variety of foods, such as sauerkraut and sourdough bread.
- **microbe:** any microscopic form of life, also called a microorganism, particularly bacteria, protozoa, fungi, or virus. Most commonly, this term refers to pathogenic microscopic life—those that cause infection, disease, decay, sepsis, or gangrene. However, biologists are identifying an increasing number of microbes that are beneficial, even essential to life, including a variety of those found in the human intestine.
- **mitochondria:** a type of organelle within eukaryotic cells, where oxygen and nutrients are converted into adenosine triphosphate, the molecule that stores chemical energy for the cell. This process, called aerobic respiration, is possible only in the presence of oxygen. Mitochondria are rod-shaped, have their own DNA, and reproduce independently within the cell, resembling some primitive prokaryotic cells. This suggests that prokaryotic cells were absorbed within the cell walls of evolving eukaryotic cells in a symbiotic relationship. Mitochondria enable cells to produce

adenosine triphosphate fifteen times more efficiently than is possible by anaerobic respiration.
- **nucleus (cell):** an organelle within each living eukaryotic cell that acts as a control center, storing genes on chromosomes, producing messenger RNA molecules (which transfer code for essential proteins from genes in the chromosomes), producing ribosomes, and organizing replication of DNA, including complete copies for cell division.
- **prokaryotic cell:** the earliest and most primitive type of cell, probably the first life form on Earth, lacking a cell nucleus. Most bacteria are prokaryotic. Most one-celled animals, such as paramecium, and simple plants such as algae have the more complex eukaryotic cell.

Basic Principles of Cell Theory

Living organisms of all kinds are characterized as an assembly of individual cells that exist in a cooperative or symbiotic relationship with each other. The individual assemblages of cells of different types. The various unique assemblies of cells to form individual organisms produces the immense variety of members of the plant and animal kingdoms. The essential feature of the structures is the cell, all of which are defined as either a prokaryotic cell or a eukaryotic cell. The fundamental difference between the two types of cell is very distinct. A eukaryotic cell has a distinct internal structure, including a nucleus (cell), while a prokaryotic cell does not. The nucleus of a eukaryotic cell contains the DNA that defines the identity of the organism, and exists as a separate structure within the actual cell. Accordingly, all cells in the human body are eukaryotic cells. A prokaryotic cell lacks this internal definition, and in a way can be thought of as a nucleus without a surrounding cell. Eukaryotic cells contain a number of other internal structures that add to their complexity. This includes various membrane structures, organelles and fibers. A list of the major components of a eukaryotic cell includes the cell nucleus, the mitochondria, rough and smooth forms of a structure called the endoplasmic reticulum, and vacuoles. All eukaryotic cells also contain peroxisomes, while plant cells contain glyoxisomes and chloroplasts that are absent from animal cells. Mitochondria carry out the energy-producing functions of the cells. Various biochemical transformations take place at the endoplasmic reticulum. Vacuoles serve as storage sites for nutrients and wastes. All protist, fungus, animal and plant cells are eukaryotic cells. Prokaryotic cells, on the other hand, are exclusively one-celled microbes or bacteria.

Prokaryotes

The prokaryotes include all bacteria, and are generally divided into the classes of eubacteria and archaebacteria. The eubacteria are the photosynthetically active blue-green algae, or cyanobacteria. The archaebacteria are typically found in "unusual" environments. For example, the archaebacteria known as methanogens, for their production of methane as a byproduct, are found in swamps where the water and soil are free of oxygen, and in the digestive tracts of bovines such as cows. The methane is produced as a byproduct of anaerobic fermentation by methanogenic bacteria. Another branch of archaebacteria is the halophiles, bacteria that require high salt concentrations in their environment. Still another is known as the thermoacidophiles, requiring hot acidic sulfur springs in order to survive. Almost all prokaryotic cells, all of which are microbes, are "obligate anaerobes," and do not grow or function in the presence of oxygen. Prokaryotes are simple in structure, being essentially just a fluid-filled sack formed by a semipermeable phospholipid bilayer membrane. This basic structure allows the passage of gases such as methane and carbon dioxide across the membrane, but utilizes a number of different protein structures embedded in the membrane to transport sugars, amino acids and ions across the membrane.

Eubacteria

The prokaryote class of eubacteria have a slightly different structure, and are classed as either Gram-negative or Gram-positive, according to their response to the Gram staining method. Gram-positive eubacteria have a cell wall layer and an adjacent plasma membrane, and give a positive response to the Gram staining method. Gram-negative eubacteria have an additional structural layer between the cell wall itself and the internal plasma membrane. This additional layer provides a higher degree of structural rigidity, and apparently prevents the Gram stain from penetrating. They therefore give a negative response to the Gram staining method.

Richard M. Renneboog MSc

See also: Cell Organelles.

FURTHER READING

Favor, Lesli J. Eukaryotic and Prokaryotic Cell Structures. Understanding Cells With and Without a Nucleus New York, NY: Rosen Publishing Group, 2005.

Campbell, Mary K. and Farrell, Shawn O. Biochemistry 8th ed., Stamford, CT: Cengage Learning, 2015.

Hall, Brian K. and Hallgrimsson, Benedikt Strickberger's Evolution 4th ed., Sudbury, MA: Jones and Bartlett, 2008.

Weeks, Benjamin S. Alcamo's Microbes and Society 3rd ed., Sudbury, MA: Jones and Bartlett Learning, 2012.

Clark, David P. Molecular Biology Burlington, MA: Elsevier, 2010.

FASCINATING FACTS ABOUT PROKARYOTES AND EUKARYOTES

- The methane produced by methanogenic bacteria in cattle has been responsible for untoward pyrotechnic events in poorly ventilated cattle and dairy barns, while the methane produced by methanogenic bacteria in swamps is believed to be responsible for the phenomenon of glowing 'swamp gas'.
- Many of the colored deposits seen in geysers and acidic hot springs are due to the presence of thermophilic bacteria.
- It is now believed that most digestion in the human gut is carried out by the trillions of bacteria that inhabit the human gut. The bacteria break down food materials to their molecular components, often altering their molecular structure in the process, and these nutrients are then transported across the membranes of the digestive tract and into the blood stream for use by the many different cells in the body.
- Fermentation processes that utilize prokaryotic bacteria have become industrially important for the synthesis of many different pharmaceutical candidate compounds, as a source of plastics and biofuels. Bacterial pools are also used to clean up heavy metal toxic wastes from mining operations and for the treatment of sewage.

EVOLUTION: ANIMAL LIFE

FIELDS OF STUDY

Developmental biology, evolutionary science, genetics, human origins, zoology

SUMMARY

Evolution is the change in the gene pool of a population over time by such processes as natural selection, genetic drift, mutation, and migration.

PRINCIPAL TERMS

- **gene:** the basic unit of heredity
- **gene flow:** the movement of genes from one population to another
- **genetic drift:** change in gene frequencies in a population owing to chance
- **interbreeding:** the mating of closely related individuals, which tends to increase the appearance of recessive genes
- **migration:** the movement of individuals, resulting in gene flow, changing the proportions of genotypes in a population
- **mutation:** alteration in the physical structure of the DNA, resulting in a genetic change that can be inherited
- **natural selection:** the process of differential reproduction in which some phenotypes are better suited to life in the existing
- **environment and thus are more likely to survive**
- **speciation:** the formation of new species as a result of geographic, physiological, anatomical, or behavioral factors
- **species:** the basic category of biological classification representing a group of potentially or actually interbreeding natural populations which are reproductively isolated from other such groups
- **taxon (pl. taxa):** group of related organisms at one of several levels such as the family Canidae, the genus *Canis*, or the species *Canis lupus*

Basic Principles

Current consensus holds that the so-called big bang—the high-temperature, high-density event that marked the beginning of the universe—occurred some fifteen billion years ago, with the sun and Earth formed about four and a half billion years ago. Four billion years ago, the relatively newly created sun shone with only 70 percent of its current strength. The atmosphere had no free oxygen. No bacteria, no viruses, no plants, and no animals were in existence. Subsequently, as chemical processes are assumed to have created oxygen and an organic "soup," microbial life in the form of the simplest cells without a nucleus, prokaryotes, developed out of this primordial ooze. These bacteria were the only living organisms for about two billion years. After that time, about 1.5 billion years ago, more complex cells with nuclei, eukaryotes, appeared. Thus, in all, for some 5.5 billion years, bacteria were the only existing animal organisms.

The Beginnings of Animal Life

Eventually, the two kingdoms, the botanical and the zoological, started to diverge and millions of animals came into being, multicelled and with specialized body parts, distinguished from plants primarily but not exclusively by their methods of feeding, locomotion, and reproduction. Most of the phyla seem to have appeared during the Precambrian period, an immense span of geological time that ended about 590 million years ago. Few fossils have remained from these prehistoric times, but the most explosive period for the development of life was the Cambrian, some 590 to 505 million years ago. In a relatively short span of ten million years, all the animal phyla currently known came into being, perhaps encouraged by an increase in oxygen in the seas, where animal life began. Eventually, animals developed a nervous system enabling them to control their movements more appropriately, as well as sense organs to help them find suitable food. At the margins, however, the dichotomy between the botanical and the animal world has remained ambiguous, since there are many microorganisms which defy clear-cut classification At times, these difficult cases are known as Protista, or Protists. In all forms of life, including the animal kingdom, no phylum has been produced by a single evolutionary event. Nor have different animal orders appeared as a result of sudden evolutionary changes. Rather, all have come about, whether in gradual or punctuated manner, by the cumulative effect of small steps in different directions. This, at least, is the theory posited by Charles Darwin's explanation of evolution.

The time line of the animal kingdom is closely connected with this evolutionary chronology. Darwin's theory of the origin of species through natural selection, nearly universally accepted in the scientific community but at times opposed on Biblical creationist grounds and in some circles of the lay community, has it that animals, like plants, have changed since the beginning of life on earth and are still evolving today. In this view, there was no sudden creation of all species. Instead, over long periods of time, new species have evolved from isolated populations of existing species. These came to occupy new niches separate from the niches of the original species. Thus, all current species are changed descendants of others that existed previously. If there are fewer apparent links between phyla than between families further down the classification ladder, the reason is that phyla have had a longer history and so have experienced more opportunities for the elimination of intermediate forms. From an evolutionary viewpoint, the difference between species down an evolutionary line are even more recent than those between families, and so on. On average, it takes about 500,000 generations for one species to evolve into another. For species to have survived in their environments—with simultaneous changes in ecology, climate, and flora—many animal forms are now more complex and efficient than their ancestors used to be.

Genes and Evolution

Life began in the seas, so for animals to live in freshwater, let alone on dry land or in the air, many obstacles had to be overcome. The impetus to conquer these inhospitable realms came from competition, according to Darwin's theory of the survival of the fittest. Those fauna that managed to surmount the problems were the ones that underwent waves of adaptive measures and evolved into new kinds of animals.

It was only in the twentieth century that it was discovered that the characteristics of a species are passed on from parents to offspring by genes. Genes provide cells for particular features, such as webbed feet. With such characteristics inherited by offspring from parents, there is a resemblance among generations. However, at times a parent may produce quite a different offspring because of genetic change. The young, in turn, may replicate this difference in their own descendants in a process known as mutation, which may occur spontaneously for unknown reasons, but also may occur due to known causes, such as exposure to radiation. At times, mutations may be useful in allowing the species to adapt better to its environment; for instance, darker moths have a better chance of survival in a forest than lighter ones because the latter are more visible to their predators than the former. Other mutations may be harmful, such as larger size that slows down a species, making its flight from danger more difficult. Whatever the case, within a time frame of 1 to 10 million years, animals may remain the same, may evolve, or may become completely extinct, as did the dinosaurs about 60 million years ago, after being dominant for some 350 million years.

ADAPTATION TO ENVIRONMENT

Natural selection leading to a new species may be accelerated when members of the original species move to a new environment, whether voluntarily or driven by the elements. Separated populations may develop different traits as they adapt to their new condition. Eventually, they will become sufficiently different to be unable to produce offspring with members of their original population. This process has repeated itself many times, over millions of years, and accounts for the large diversity in the animal world, not to mention the additional diversity consequent on artificial breeding by humans, widely observed among domesticated animals such as horses, cattle, and dogs.

Animals occupied new environments as species living in water moved to the land or later to the air. Thus, the step from fish to amphibian was essentially one from living in water for the whole life to living on land for the adult stage of the life cycle. The step from amphibian to reptile was one of increasingly proficient adaptation to land life at all stages of the life cycle.

Birds and mammals evolved in different directions from the reptiles, the first in adaptation to an arboreal and finally a flying life and the second as a further advance in the maintenance of an even and high body temperature—homeothermy—by combining an insulating external layer such as hair with a variety of physiological thermostats. Events in geology, climate, and flora also determined the geographic distribution of species. Thus, marsupials are currently found almost entirely in Australasia and South America. The tiger exists only in India and Southeast Asia. The lion is restricted mainly to Africa. This pattern reflects the way in which these groups have evolved in relation to the physical world.

New animal groups evolve into many different forms, especially when they become dominant. For instance, when mammals came to occupy the dominant position, some became meat-eaters while others became vegetarians; some became smaller while others became larger; some became runners while others ended up as burrowers or flyers; still others returned to the water. This trend allowed the descendants of the original type to exploit a much greater range of environments and resources. Essentially, those species whose sense organs or brain morphology and functions improved the most ended up being dominant—primates in general and humans in particular.

TIME FRAME

The exact time of the origins of animals during earth's evolutionary history is not known because the early species were soft-bodied, at first single-celled and later multicellular life forms, that did not fossilize well. Fossils are the best material evidence of archaic times. Fossils do not appear earlier than 650 to 500 million years ago, not only because the animal life of the time was inappropriate for fossilization, but also because continued crustal shifts in the ensuing eons disturbed the very early rock formations. Accordingly, fossil evidence is unavailable for the entire early history of animals, which must consequently remain speculative. Current taxonomic interrelationships suggest the early history, and taxonomic diagrams may be regarded as presumptive evolutionary diagrams as well. However, a ball of carbon discovered in a cavity etched in a rock some 3.86 billion years old suggests that some life on earth was already possible at that time.

Knowledge is also limited by the fact that, even though over a million different species of animals have been identified, it is suspected that a similar number remain to be discovered or became extinct before such identification could be made. In the United States alone, some forty species of birds, about thirty-five species of mammals, and twenty-five other species have become extinct in the last two hundred years alone—less than a blip on earth's time scale—as a result of human activities such as the destruction of animal habitats through urbanization, the clearing of land for agricultural purposes, pollution, the introduction of new species from other parts of the world which turned out to be predatory to domestic specimens, hunting, and especially human population growth. It is widely predicted that climatic change triggered by greenhouse gases will continue, even enhance, this process, thereby endangering more animal species. Whatever the future, however, evolutionary biologists estimate that some 99 percent of all species that have ever lived on earth are now extinct. Despite these and other caveats, here is a very approximate timetable of the evolution of animal life:

Life Form	Date of Emergence
Simplest single-celled Protozoa	3.5 billion years ago
Invertebrates evolving from Protozoa	670 to 640 million years ago
First vertebrates evolving from invertebrates	500 million years ago
First mammals	200 million years ago
Hominids (modern man) from apelike hominoids	200,000 to 25,000 years ago

Translated in Terms of a Single Year	
January 1	Big Bang
March 22	Bacteria, the first living animals
November 9	Invertebrates
November 22	Vertebrates
December 16	Mammals
December 28	Primates, the highest order of mammals
December 31, a few minutes before midnight	Modern man, the dominant primate

Evolution of Existing and Extinct Human Species and Australopithecines

The root of the hominid evolutionary tree is still imperfectly known. The earliest australopithecine species, *Australopithecus anamensis*, is believed to be over four million years old, by which time that branch had diverged from African apelike ancestors. This species was followed by the *Australopithecus afarensis* nearly 3.5 million years ago. Much later came *Homo habilis*, called "skillful man" since they could presumably produce primitive tools, some two million years ago. They were followed by *Homo erectus*, "upright man," about one million years ago. Finally, *Homo sapiens*, "knowing man," emerged about 200,000 years ago. In the meantime, the australopithecine branch, after evolving through a number of intermediate species such as *A. africanus*, *A. aethiopicus*, and *A. robustus*, died out about one million years ago. To date, the earliest unearthed fossil, that of Lucy, a three-foot-tall female discovered in Ethiopia, is about four million years old. Modern humans are believed to have radiated out of Africa into Asia and Europe. Subsequently, cultural evolution became more prominent than biological evolution, but as modern humans evolved over the last four million years to their current condition, they developed manipulative skills, bipedalism, a change from specialized to omnivorous feeding habits, and especially, a threefold increase in cranial capacity from *H. afarensis* to *H. neanderthalensis*, together with behavior appropriate to the control of the environment.

Although humans are not the only animals capable of conceptual thought, they have refined and extended that ability until it has become their hallmark. Thus, thanks to the symbolic language of *Homo sapiens*, modern humans make possible the accumulation of experience from one generation to the next. Such cultural evolution is possessed by few, if any other animal species. It is for this reason that humans, more than other animals, have found ways to mold and change their environment according to need rather than in response to environmental demands. Because of this ability and humans' control of technology, the species has more say about their biological future than any other.

—*Peter B. Heller*

See also: Apes to hominids; Evolution: Historical perspective; Gene flow; Genetics; Hardy-Weinberg law of genetic equilibrium; *Homo sapiens* and human diversification; Natural selection; Sex differences: Evolutionary origins.

Further Reading

Carroll, Sean B., Jennifer K. Grenier, and Scott D. Weatherbee. *From DNA to Diversity: Molecular Genetics and the Evolution of Animal Design.* Malden, Mass.:

Blackwell Science, 2001. Covers general principles of the genetic basis of morphological change, including the history of animal evolution, model system developmental genetics, genetic regulatory mechanisms, and case studies of evolutionary change. Color diagrams and images, glossary.

Conway-Morris, Simon. *The Crucible of Creation: The Burgess Shale and the Rise of Animals.* Los Angeles: The Getty Center for Education in the Arts, 1999. A fascinating study of one of the richest fossil deposit sites on earth, focusing on what can be learned of evolution from fossil remains.

Lavers, Chris. *Why Elephants Have Big Ears: Understanding Patterns of Life on Earth.* New York: St. Martin's Press, 2001. Explores the evolving interaction of form and function in animals.

Long, John A. *The Rise of Fishes: Five Hundred Million Years of Evolution.* Baltimore: The Johns Hopkins University Press, 1996. A detailed history of fish evolution. Color photographs, glossary.

EVOLUTION: HISTORICAL PERSPECTIVE

FIELDS OF STUDY

Ecology, embryology, evolutionary science, genetics, paleontology, population biology, zoology

SUMMARY

Evolution is the process of change in biological populations. Historically, it is also the theory that biological species undergo sufficient change with time to give rise to new species.

PRINCIPAL TERMS

- **adaptation:** the possession by organisms of characteristics that suit them to their environment or their way of life
- **catastrophism:** a geological theory explaining the earth's history as resulting from great cataclysms (floods, earthquakes, and the like) on a scale not now observed
- **Darwinism:** branching evolution brought about by natural selection
- **essentialism (typology):** the Platonic-Aristotelian belief that each species is characterized by an unchanging "essence" incapable of evolutionary change
- **genotype:** the hereditary characteristics of an organism
- **Geoffroyism:** an early theory of evolution in which heritable change was thought to be directly induced by the environment
- **Lamarckism:** an early evolutionary theory in which voluntary use or disuse of organs was thought to be capable of producing heritable changes
- **scale of being (chain of being):** an arrangement of life forms in a single linear sequence from "lower" to "higher"
- **uniformitarianism:** a geological theory explaining the earth's history using processes that can be seen at work today

BASIC PRINCIPLES

Evolution is the theory that biological species undergo sufficient change with time to give rise to new species. The concept of evolution has ancient roots. Anaximander suggested in the sixth century BCE that life had originated in the seas and that humans had evolved from fish. Empedocles (fifth century BCE) and Lucretius (first century BCE), in a sense, grasped the concepts of adaptation and natural selection. They taught that bodies had originally formed from the random combination of parts, but that only harmoniously functioning combinations could survive and reproduce. Lucretius even said that the mythical centaur, half horse and half human, could never have existed because the human teeth and stomach would be incapable of chewing and digesting the kind of grassy food needed to nourish the horse's body.

For two thousand years, however, evolution was considered an impossibility. Plato's theory of forms (also called his "theory of ideas") gave rise to the notion that each species had an unchanging "essence" incapable of evolutionary change. As a result, most scientists from Aristotle to Carolus Linnaeus in the eighteenth century insisted upon the immutability of species. Many of these scientists tried to arrange all species in a single linear sequence known as the scale of being (also called the chain of being and the *scala naturae*), a concept supported well into the nineteenth century by many philosophers and theologians as well. The sequence in this scale of being was usually interpreted as a static "ladder of perfection" in God's creation, arranged from higher to lower forms.

The scale had to be continuous, for any gap would detract from the perfection of God's creation. Much exploration was devoted to searching for "missing links" in the chain, but it was generally agreed that the entire system was static and incapable of evolutionary change. Pierre-Louis Moreau de Maupertuis, in the eighteenth century, and Jean-Baptiste Lamarck were among the scientists who tried to reinterpret the scale of being as an evolutionary sequence, but this single-sequence idea was later replaced by Charles Darwin's concept of branching evolution. Georges Cuvier finally showed that the major groups of animals had such strikingly different anatomical structures that no possible scale of being could connect them all; the idea of a scale of being lost most of its scientific support as a result.

THE STRUGGLE TO CONCEPTUALIZE EVOLUTION

The theory that new biological species could arise from changes in existing species was not readily accepted at first. Linnaeus and other classical biologists emphasized the immutability of species under the Platonic-Aristotelian concept of essentialism. Those who believed in the concept of evolution realized that no such idea could gain acceptance until a suitable mechanism of evolution could be found. Many possible mechanisms were therefore proposed. Étienne Geoffroy Saint-Hilaire proposed that the environment directly induced physiological changes, which he thought would be inherited, a theory now known as Geoffroyism. Lamarck proposed that there was an overall linear ascent of the scale of being but that organisms could also adapt to local environments by voluntary exercise, which would strengthen the organs used; unused organs would deteriorate. He thought that the characteristics acquired by use and disuse would be passed on to later generations, but the inheritance of acquired characteristics was later disproved. Central to both these explanations was the concept of adaptation, or the possession by organisms of characteristics that suit them to their environments or to their ways of life. In eighteenth century England, the Reverend William Paley and his numerous scientific supporters believed that such adaptations could be explained only by the action of an omnipotent, benevolent God. In criticizing Lamarck, the supporters of Paley pointed out that birds migrated toward warmer climates before winter set in and that the heart of the human fetus had features that anticipated the changes of function that take place at birth. No amount of use and disuse could explain these cases of anticipation, they claimed; only an omniscient God who could foretell future events could have designed things with their future utility in mind.

The nineteenth century witnessed a number of books asserting that living species had evolved from earlier ones. Before 1859, these works were often more geological than biological in content. Most successful among them was the anonymously published *Vestiges of the Natural History of Creation* (1844), written by Robert Chambers. Books of this genre sold well but contained many flaws. They proposed no mechanism to account for evolutionary change. They supported the outmoded concept of a scale of being, often as a single sequence of evolutionary "progress." In geology, they supported the outmoded theory of catastrophism, an idea that the history of the earth had been characterized by great cataclysmic upheavals. From 1830 on, however, that theory was being replaced by the modern theory of uniformitarianism, championed by Charles Lyell. Charles Darwin read these books and knew their faults, especially their lack of a mechanism that was compatible with Lyell's geology. In his own work, Darwin carefully tried to avoid the shortcomings of these books.

DARWIN'S REVOLUTION IN BIOLOGICAL THOUGHT

Darwin brought about the greatest revolution in biological thought by proposing not only a theory of branching evolution but also a mechanism of natural selection to explain how it occurred. Much of Darwin's evidence was gathered during his voyage around the world aboard HMS *Beagle*. Darwin's stop in the Galápagos Islands and his study of tortoises and finchlike birds on these islands are usually

225

credited with convincing him that evolution was a branching process and that adaptation to local environments was an essential part of the evolutionary process. Adaptation, he later concluded, came about through natural selection, a process that killed the maladapted variations and allowed only the well-adapted ones to survive and pass on their hereditary traits. After returning to England from his voyage, Darwin raised pigeons, consulted with various animal breeders about changes in domestic breeds, and investigated other phenomena that later enabled him to demonstrate natural selection and its power to produce evolutionary change. Darwin's greatest contribution was that he proposed a suitable mechanism by which permanent organic change could take place. All living species, he said, were quite variable, and much of this variation was heritable. Also, most organisms produce far more eggs, sperm, seeds, or offspring than can possibly survive, and the vast majority of them die. In this process, some variations face certain death while others survive in greater or lesser proportion. Darwin called the result of this process "natural selection," the capacity of some hereditary variations (now called genotypes) to leave more viable offspring than others, with many leaving none at all. Darwin used this theory of natural selection to explain the form of branching evolution that has become generally accepted among scientists.

Darwin delayed the publication of his book for seventeen years after he wrote his first manuscript version. He might have waited even longer, except that his hand was forced. From the East Indies, another British scientist, Alfred Russel Wallace, had written a description of the very same theory and submitted it to Darwin for his comments. Darwin showed Wallace's letter to Lyell, who urged that both Darwin's and Wallace's contributions be published, along with documented evidence showing that both had arrived at the same ideas independently. Darwin's great book, *On the Origin of Species by Means of Natural Selection*, was published in 1859, and it quickly won most of the scientific community to a support of the concept of branching evolution. In his later years, Darwin also published *The Descent of Man and Selection in Relation to Sex* (1871), in which he outlined his theory of sexual selection. According to this theory, the agent that determines the composition of the next generation may often be the opposite sex. An organism may be well adapted to live, but unless it can mate and leave offspring, it will not contribute to the next or to future generations.

ACCEPTANCE OF DARWINISM IN THE TWENTIETH CENTURY

In the early 1900's, the rise of Mendelian genetics (named for botanist Gregor Mendel) initially resulted in challenges to Darwinism. Hugo de Vries proposed that evolution occurred by random mutations, which were not necessarily adaptive. This idea was subsequently rejected, and Mendelian genetics was reconciled with Darwinism during the period from 1930 to 1942. According to this modern synthetic theory of evolution, mutations initially occur at random, but natural selection eliminates most of them and alters the proportions among those that survive. Over many generations, the accumulation of heritable traits produces the kind of adaptive change that Darwin and others had described. The process of branching evolution through speciation is also an important part of the modern synthesis.

The branching of the evolutionary tree has resulted in the proliferation of species from the common ancestor of each group, a process called adaptive radiation. Ultimately, all species are believed to have descended from a single common ancestor. Because of the branching nature of the evolutionary process, no one evolutionary sequence can be singled out as representing any overall trend; rather, there have been different trends in different groups. Evolution is also an opportunistic process, in the sense that it follows the path of least resistance in each case. Instead of moving in straight lines toward a predetermined goal, evolving lineages often trace meandering or circuitous paths in which each change represents a momentary increase in adaptation. Species that cannot adapt to changing conditions die out and become extinct.

STUDYING EVOLUTION

Evolution is studied by a variety of methods. The ongoing process of evolution is studied in the field by ecologists, who examine various adaptations, including behavior and physiology as well as anatomy. These adaptations are also studied by botanists, who examine plants; zoologists, who examine animals; and various specialists, who work on particular kinds of animals or plants (for example, entomologists, who study insects). Some investigators capture specimens in the field, then bring back samples to the laboratory in order to examine chromosomes or analyze proteins using electrophoresis. Through these methods, scientists learn how the ongoing process of evolutionary change is working today within species or at the species level on time scales of only one or a few generations.

The long-term results of evolutionary processes are studied among living species by comparative anatomists and embryologists. Extinct organisms are studied by paleontologists, scientists who examine fossils. Biogeographers study past and present geographic distributions. All these types of scientists make comparisons among species in order to determine the sequence of events that took place in the evolutionary past. One method of reconstructing the branching sequences of evolution is to find homologies, deep-seated resemblances that reflect common ancestry. Once the sequences are established, functional analysis can be used to suggest possible adaptive reasons for any changes that took place. The sequences of evolutionary events reconstructed by these scientists represent the history of life on the earth. This history spans many species, families, and whole orders and classes, and it covers great intervals of past geologic time, measured in many millions of years.

THE HISTORICAL CONTEXT OF EVOLUTIONARY THEORY

The historical development of evolutionary theory should be viewed in two contexts: that of biological science and that of cultural history. The concept of evolution had been talked about for many years before 1859 and was usually rejected because no suitable mechanism had gained widespread acceptance. The fact that the phenomenon of natural selection was independently discovered by two Englishmen shows both that the time was ripe for the discovery and that the circumstances were right in late nineteenth century England.

Evolutionary biology is itself the context into which all the other biological sciences fit. Other biologists, including physiologists and molecular biologists, study how certain processes work, but it is evolutionists who study the reasons why these processes came to work in one way and not another. Organisms and their cells are built one way and not another because their structures have evolved in a particular direction and can only be explained as the result of an evolutionary process. Not only does each biological system need to function properly, but it also must have been able to achieve its present method of functioning as the result of a long, historical, evolutionary process in which a previous method of functioning changed into the present one. If there were two or more ways of accomplishing the same result, a particular species used one of them because found it easier to evolve one method rather than another. Everything in biology is thus a detail in the ongoing history of life on the earth, because every living system evolves. Living organisms and the processes that make them function are all products of the evolutionary process and can be understood only in that context. As biologist Theodosius Dobzhansky once said, "Nothing in biology makes sense, except in the light of evolution."

—*Eli C. Minkoff*

See also: Apes to hominids; Gene flow; Genetics; Hardy-Weinberg law of genetic equilibrium; *Homo sapiens* and human diversification; Human evolution analysis; Natural selection; Sex differences: Evolutionary origins.

FURTHER READING

Bowler, Peter J. *Evolution: The History of an Idea*. Rev. ed. Berkeley, Calif.: University of California Press, 1989. A comprehensive history of the evolutionary theory for both specialist and nonspecialist.

_____. *Life's Splendid Drama: Evolutionary Biology and the Reconstruction of Life's Ancestry, 1860-1940*. Chicago: University of Chicago Press, 1996. A history of evolutionary morphology and its relationship with paleontology and biogeography. Covers scientific debates over the emergence of vertebrates, the origins and extinctions of animal species, and the role and influence of Darwin. Biographical appendix, Bibliography.

Brandon, Robert N. *Concepts and Methods in Evolutionary Biology*. New York: Cambridge University Press, 1996. A collection of essays spanning two decades, addressing problems in the philosophy of biology, particularly the conception of relative adaptedness and the principle of natural selection.

Darwin, Charles R. *On the Origin of Species by Means of Natural Selection: Or, the Preservation of the Favoured Races in the Struggle for Life*. London: John Murray, 1859. This is the original edition, still worth reading. It is better than the more widely reprinted sixth edition, in which Darwin's more forceful statements were toned down as a response to criticism that is no longer greatly valued by biologists. Some knowledge of zoology, geology, and geography would definitely increase any reader's understanding and appreciation of this book. Darwin provided no Bibliography, but some modern editors have supplied one.

Dobzhansky, Theodosius G. *Genetics of the Evolutionary Process*. New York: Columbia University Press, 1970. Although somewhat technical in places, this book is extremely well written, and a careful reader should be able to understand it all without any formal background. It is an excellent (and very detailed) outline of the evolutionary process in terms of genetic changes. It contains much information about the genus *Drosophila* (fruit flies), on which Dobzhansky was an expert. Very comprehensive Bibliography.

Gould, Stephen J. *Ever Since Darwin*. Reprint. New York: W. W. Norton, 1992.

_____. *The Flamingo's Smile*. New York: W. W. Norton, 1985.

_____. *The Panda's Thumb*. Reprint. New York: W.W. Norton, 1992. These three books all consist of essays reprinted (and occasionally updated) from *Natural History* magazine. All are well written and directed to a general audience; no previous background is assumed. Although Gould has

---

COMTE DE BUFFON (GEORGES-LOUIS LECLERC)

Born: September 7, 1707; Montbard, France

Died: April 16, 1788; Paris, France

Fields of study: Anthropology, biology, botany, chemistry, geology, mathematics, paleontology, zoology

Contribution: The greatest naturalist of the eighteenth century, Buffon popularized zoology and botany through his publications. He tried to separate science from religious and metaphysical ideas and rejected teleological reasoning and the idea of God's direct intervention in nature.

Born to a noble family in Dijon, Buffon was admitted to the French Academy of Sciences in 1734 and the Académie Française in 1753. In 1739 he was appointed director (Intendant) of the Jardin du Roi, the royal botanical garden. Between 1749 and his death, he published (with associates) thirty-six volumes of the Histoire Naturelle (44 vols., 1739-1804; A Natural History, General and Particular, 10 vols., 1807) which took in the formation of the earth, geology, paleontology, zoology, and botany. The last eight volumes were posthumous. A master of style, his books were among the most popular in the eighteenth and early nineteenth centuries.

Buffon believed that the study of Earth was a necessary prerequisite to botany and zoology, and wrote two important texts on geology and paleontology. The first, Théorie de la terre, was published in 1749, the other, Époques de la nature, in 1778. From experiments on the cooling of globes, he estimated the age of the earth to be 85,000 years, significantly at variance with his contemporaries' estimation of an origin around 4000-6000 BCE. Buffon's cosmogony replaced the intervention of God by a cause whose effects are in accord with the laws of mathematics.

Buffon postulated that new varieties of plants and animals (including humans) were produced in nature by external geographical influences. Such influences could also cause degeneration. He was against classifying nature, as "... everything that can be, is." He gave "species" a purely biological definition: animals that by means of copulation perpetuate themselves and preserve their similarity. He thought families were artificial creations made by man. He therefore thought Linnaeus' classification of plants based on sexual characters was too rigid. He arranged animals in order of their utility to humans (later he rearranged them according to distinctive characteristics) and believed that some forms might have degenerated from others over time—thus, the ass might be a degenerate form of the horse. Buffon thus alluded to a form of "evolution" where physical characteristics produced by external influences could be passed down the generations.

Buffon tried to separate science from religious and metaphysical ideas, and rejected teleological reasoning and the idea of God's direct intervention in nature. His theories went against the accepted theological belief of immutability of species and he was reprimanded by the Faculty of Theology at the University of Paris for this. Buffon apologized but did not change his views.

Buffon was made a count in 1773 by Louis XV. He was greatly respected by his contemporaries and was made a member of almost every learned society in Europe.

—*Ranès C. Chakravorty*

occasionally supported unorthodox viewpoints, most of the views represented here have become accepted into the mainstream with the passage of time. Gould's easy, familiar style makes for lively reading and he uses esoteric cases and seemingly inconsequential details to make important points about evolution in general. The bibliographies are wide-ranging but are confined to the topics of the individual essays.

Grant, Verne. *The Evolutionary Process: A Critical Study of Evolutionary Theory*. 2d ed. New York: Columbia University Press, 1991. A comprehesive and critical review of modern evolutionary theory. Focuses on whole organisms and general principles rather than molecular changes and mathematical models.

Rose, Michael A. *Darwin's Spectre: Evolutionary Biology in the Modern World*. Princeton, N.J.: Princeton University Press, 1998. Written for the general reader. Outlines the fundamental ideas of evolutionary biology and its influence in other fields such as agriculture, medicine, and eugenics.

Wills, Christopher, and Jeffrey Bada. *The Spark of Life: Darwin and the Primeval Soup*. Cambridge, Mass.: Perseus, 2000. Describes theories of the origins of terrestrial life and its evolution. Written for the general reader.

JEAN-BAPTISTE DE LAMARCK

Born: August 1, 1744; Bazentin-le-Petit, Picardy, France

Died: December 18, 1829; Paris, France

Fields of study: Evolutionary science, invertebrate biology, paleontology, zoology

Contribution: Lamarck established the division of animal life into the vertebrate and invertebrate categories, and he formulated an evolutionary theory based on the premise that acquired traits are inheritable.

After studying briefly for the priesthood at the Jesuit seminary in Amiens, Jean-Baptiste de Monet, chevalier de Lamarck, served as an army officer in the Seven Years' War. Following an accident in 1768, he began to study botany and medicine. In 1778, he published his three-volume Flore français (French plants), which was widely used as a manual of identification. Lamarck was then employed as assistant botanist at the royal botanical gardens of Paris, and he was also appointed to the prestigious Academy of Sciences. Count Georges-Louis de Buffon engaged him as tutor to his son, which allowed him to tour European botanical gardens for two years. When the Jardin des Plantes (the National Museum of Natural History) was founded in 1793, he was placed in charge of the collection of invertebrates (a term that he coined).

During the early nineteenth century, Lamarck published numerous books about invertebrates, paleontology, and biological evolution. His Système des animaux sans vertèbres (1801; system of invertebrate animals) presented a systematic basis for the classification of the lower animals. His Hydrogéologie (1802; Hydrogeology, 1964) interpreted the history of the earth as a series of inundations, each resulting in organic deposits that built up the continents. The book was especially noteworthy for its recognition of the vastness of geologic time.

Lamarck was not the first to propose a theory of biological evolution, but his theory was more systematic and coherent than previous versions. He gave the clearest explanation for his theory in Philosophie zoologique (1809; Zoological Philosophy, 1873), presenting a two-part process. First, a change in the environment forced organisms to change their behavior. If particular organs were used, they would increase in size and strength; in contrast, disuse or disease would weaken and shrink organs. Second, Larmark argued that such changes would be inherited, so that the characteristics of a species would change gradually over many generations. Lamarck's scientific work culminated in an exhaustive study, Histoire naturelle des animaux sans vertèbres (1815-1822; natural history of invertebrate animals).

Charles Darwin acknowledged the great contribution of Lamarck's work, and Darwin's own theory of natural selection never entirely rejected the possibility that some acquired traits might be inherited. With advancements in the science of heredity during the twentieth century, the concept of Lamarckian inheritance has been largely abandoned. Late in the twentieth century, nevertheless, Edward J. Steele and other biologists found evidence that the acquired immunities of organisms might be passed on to their offspring.

—*Thomas Tandy Lewis*

Eyes

FIELDS OF STUDY

Anatomy, biochemistry, cell biology, neurobiology, physiology

SUMMARY

The eyes are the sensory organs that allow animals to visualize the world around them. Many different types of eyes have evolved, from very simple light-gathering eye spots to the complex compound eyes of insects. Each type of eye is used to convert a light signal into a useful piece of information for the animal

PRINCIPAL TERMS

- **binocular vision:** the ability to utilize image information from both eyes to form a single image with depth information
- **chromophore:** the molecule which interacts with opsin; absorption of light changes the interaction and starts the phototransduction cascade
- **ommatidium:** individual unit of the multifaceted compound eye
- **opsin:** a membrane-bound protein, or pigment, which absorbs light
- **optic nerve:** the main nerve taking information from the eyes to higher processing areas
- **photoreceptor:** cell containing membranes which house light-sensitive pigments
- **retina:** the light-sensitive membrane at the back of the eye

BASIC PRINCIPLES

There are many different kinds of eyes in the animal kingdom, each type with its own set of benefits for a particular species. The most basic function of the eye is to act as a light-sensitive organ, to detect the presence or absence of light. In addition to this, elaborations are often made, such as the ability to form an actual image on a retina, or the ability to control the amount of light entering the eye; each elaboration is made to provide the visual needs of individual species. The animal kingdom comprises eyes from the very rudimentary to exceptionally complex. The two basic types of eye are the simple eye and the compound eye. The simple eye is made up of a single light-sensitive region, whereas the compound eye comprises several such elements.

Simple Eyes. The most elementary simple eye consists of a photosensitive membrane, or eye spot. Light should enter the membrane from only one direction, and so eyes have a pigmented backing to stop entry of light from thewrong side. In order to give the light entering a sense of directionality, the photosensitive membrane is often shaped like a cup; this basic eye type is called a pigmented cup eye. An example of an animal which possesses a simple eye of this nature is the cephalopod, *Nautilus pompilius*, which is known as a living fossil as it is thought to exemplify the behavior and physiology of ancient organisms. *Nautilus* has a pinhole eye with a pigmented backing; the pinhole aperture is a primitive method to restrict the amount of light entering the eye. This eye has no other formal optics. An elaboration of this type of eye would be to add a spherical lens; this would allow the light entering to be focused and to form an image on the photosensitive membrane. In many animals, the lens can change shape (become thinner or fatter) using surrounding musculature, in order to properly focus the image on the retina; this ability is called accommodation. The lens changes the angle at which light bends, and hence the focal length of the eye. The focal length of the eye is the distance from the eye to the part of the retina on which the image is focused. Other animals actually move the entire lens forward or backward in order to accommodate. One drawback that comes along with spherical lenses is that the entire image is rarely totally focused on the retina because, as a result of the spherical shape of the lens, the light rays become focused over several focal lengths (this is called spherical aberration). To correct this, many lenses have a refractive index gradient across their length; basically, the density of the lens is different across its length, which causes the light rays to bend differently and corrects for spherical aberration, allowing proper focusing of the entire image on the retina. In spiders, a different kind of optics is found; instead of a lens, these creatures focus light rays with their corneas. The structure, unlike the lens, is fixed and does not change shape to accommodate.

Compound Eyes. Compound eyes are the most abundant type of eye in the animal kingdom, and can be thought of as being derived from the simple pigmented-cup type of eyes. The simplest compound eye consists of several pigmented-cup-type units, each of which samples a different angle of visual space. Found only in invertebrates, such as certain types of worm, this kind of eye gives very poor quality images, but does give the animal a sense of the direction from which light is coming. A more complicated version of this eye is the apposition compound eye, found in many insects, where each cup has its own optics. Each individual unit is called an ommatidium, comprising the rhabdom (containing the light-sensitive cells), directly contacting a light-focusing apparatus (either a lens or a cornea). In order to fit as many ommatidia into the eye as possible, the facets are hexagonal.

Another elaboration is the superposition compound eye; this eye has a space between the cornea (or lens) and the rhabdom of each ommatidia. As such, this allows light from many corneal facets to converge onto each rhabdom. This increases the sensitivity as compared to the apposition eye. Several mechanisms are used to bend the light from each ommatidium; the simplest is the reflecting superposition eye, found in shrimp, which uses a series of mirrors along the edge of each facet to reflect the light onto the rhabdoms.

PHOTORECEPTORS

The photoreceptive element common to all eyes differs from species to species, from the cupshaped retinas of simple eyes to the complex rhabdom structures of compound eyes. In simple eyes, the light-sensitive element is platelike, with projections containing flat layers or discs of membranes. In vertebrates, the two photoreceptor types are the cones (cone-shaped projections), which have layers of photosensitive membrane, and the rods (rod-shaped projections), which contain free-floating discs. Compound eyes have rhabdomeric microvillar photoreceptors; the microvilli are fingerlike structures that project from the rhabdomeres and are light-sensitive. Within each of these structures lies the actual light-sensitive pigment, a membrane-bound protein known as opsin. In the rhabdoms, the orientation of the opsin molecule is parallel to the axis of the microvilli; this fact aids in the perception of polarized light. There are many types of opsin protein, and they can be categorized according to which wavelength of the light spectrum they preferentially absorb. The opsin protein interacts with a Vitamin A-derived molecule called the chromophore. Its chemical name is 11- cis retinal. The absorption of a photon of light changes the chemical interaction between the opsin and the chromophore, and it is this chemical change which initiates the cascade of events that leads to a nerve signal, known as phototransduction.

The actual biochemistry of phototransduction is very complex, but involves the amplification of the signal of reception of light, and its transformation into a nerve impulse. The passage of the information through the nervous system is also very complicated and details are different in different species, but in general, the receptors converge onto axons and the nerve impulse travels through the optic nerve for higher processing. Of the two types of photoreceptor existing in vertebrates, rods are very light-sensitive and are used in situations where light is limited (scotopic conditions). Cones are less sensitive and are used in situations where light is abundant (photopic conditions). The vertebrate retina almost always possesses both rods and cones, although some nocturnal animals or animals living in an environment where light is scarce, for example deep-dwelling fishes, have allrodretinas. Mammals have rod-dominated retinas, which are probably remnants of the time millions of years ago when mammals were nocturnal. There are several kinds of cones, which are characterized according to which opsins they carry within the membrane layers; this dictates which wavelengths of light the cone preferentially absorbs. Absorption of light by combinations of at least two different cone types will allow a species color vision, given the cognitive ability to process such information.

Another important feature of eyes is their position on the head relative to each other. Most vertebrates and some invertebrates have two eyes, allowing them a certain amount of binocular vision depending on the angle between them. In fact, both eyes project a slightly different image onto each retina, but the brain is able to compute this as a single image. This results in a larger field of view and also a sense of depth; these are definite advantages, for example for a hunting animal, which needs to know how far it is from the prey target.

—Lucy A. Newman

OPTIMIZATION OF EYES

In order to adapt to the particular needs of individual species over evolutionary time, simple eyes have had to change and become optimized to deal with unique circumstances. Two major aspects of superior vision are resolution and sensitivity. Resolution, or spatial acuity, refers the ability to tell that two points are separate, and not just one point. Thus, better resolution is beneficial to an animal that needs to resolve images at a distance. To increase resolution, one can build more receptors per retina. However, there is a limit to this; to fit in more photoreceptors, they would have to become smaller (thinner) and at some point they would be too small to trap photons efficiently. This can be counteracted by making the receptors longer to catch more photons of light. Also, with too many receptors the nervous system becomes overloaded; for this reason, throughout most of the retina, many photoreceptors converge onto a single ganglion. Nevertheless, there are regions of high resolution at which each photoreceptor has its own ganglion; this region of the retina is called the fovea and is best described in vertebrate eyes. Increasing the eye size also increases the resolution, since it increases the amount of light getting to each receptor. Some species need to maximize sensitivity of their eyes, for example, animals who are active at night. To do this, the aperture needs to be large. This allows for maximum light entry into the eye. To the same end, the receptors could be made bigger, as could the entire eye. Thus, nocturnal animals, such as the opossum, have much larger eyes and apertures than diurnal animals, such as primates, which use a smaller eye and a smaller lens to give better optics.

See also: Anatomy; Bone and cartilage; Brain; Digestive tract; Eyes; Immune system; Noses; Reproductive system of female mammals; Reproductive system of male mammals; Respiratory system; Skin.

Further Reading

Baylor, D. "How Photons Start Vision." *Proceedings of the National Academy of Science* 93 (January, 1996): 560-565. This review explores the biochemistry of phototransduction, and is fairly detailed.

Dowling, J. E. *The Retina: An Approachable Part of the Brain.* Cambridge, Mass.: Belknap Press of the Harvard University Press, 1987. This work approaches the retina as a part of the brain, and discusses much of the neurobiological research available on this topic.

Goldsmith, T. "Optimization, Constraint, and History in the Evolution of Eyes." *Quarterly Review of Biology* 65, no. 3 (September, 1990): 281-320. Discusses the evolution and adaptation of eyes and vision.

Hubel D. H. *Eye, Brain, and Vision.* 2d ed. New York: Scientific American Library, 1995. Approaches vision from a neurobiological standpoint; discusses in detail processing of visual information by the brain.

Solomon, E., L. Berg, D. Martin, and C. Villee. *Biology.* 3d ed. Fort Worth, Tex.: Saunders College Publishing, 1993. Chapter 41 covers image formation and, briefly, phototransduction.

OTHER FEATURES OF PHOTORECEPTORS: PIGMENTS AND OIL DROPLETS

Some vertebrate and invertebrate photoreceptors have been found to possess filtering mechanisms. The classic example of this is oil droplets in the cones of bird eyes. The cone possesses a droplet of oil positioned just behind the outer nuclear layer of the retina, where the photoreceptor layer containing the opsin molecules is located. The droplet is carotenoid in nature and as such can be colored blue, yellow, orange, or red. As light travels through this droplet, it is filtered in intensity and in wavelength. Thus, the droplet acts as a selective spectral filter, only allowing certain wavelengths of light to get through the pigment to be absorbed at the photoreceptor level. It therefore increases the spectral sensitivity of individual receptors, and could enhance the color vision of such animals. Since most of the carotenoid oil droplets block out the shorter wavelengths of light, it is thought that they could also be a mechanism for protecting the receptors from the damage of ultraviolet light. In invertebrates, a similar filtering system has been found, although it is not in the form of an oil droplet but in the form of pigmented vesicle bundles. In the same way, these vesicles filter the incoming light, narrowing the absorbance spectrum of the photoreceptor layer above them.

Fertilization

FIELDS OF STUDY
Cell biology, developmental biology, embryology, reproduction science

SUMMARY
Fertilization occurs when the genetic information in a haploid sperm combines with the genetic information of the haploid ovum to form the diploid zygote.

PRINCIPAL TERMS
- **corona radiata:** the layers of follicle cells that still surround the mammalian egg after ovulation
- **vitelline envelope:** the protective layers that form around the egg while it is still in the ovary
- **zona pellucida:** mammalian protective layer analogous to the vitelline envelope

Assuring Eggs and Sperm are in Proximity
Animals have many mechanisms to ensure that sperm and eggs are in close proximity. This can be a major concern for aquatic organisms with external fertilization, and many release gametes in the millions or even billions to assure that at least some sperm reach the appropriate eggs. To increase the chances of a meeting between samespecies gametes, animals often have specialized mating behaviors. Corals are among those animals that release their gametes into the water and depend on currents to bring egg and sperm together. This is not, however, as random as it may seem. As the first coral releases its gametes, it also releases hormones that induce nearby corals of the same species to release their gametes. These also release the same chemicals with their gametes, and soon there are clouds of eggs and sperm, and the chances of a proper meeting are increased dramatically. One species of polychaete annelid, *Eunice viridis*, or the palolo worm, has another method of assuring male and female gametes are in the same place. In this species, sexually mature worms called epitokes swarm together at the ocean's surface in response to the lunar cycle. Females then secrete a hormone that induces males to release sperm, and the sperm induce the females to shed eggs. Many fish go through elaborate courtship rituals, during which males and females release gametes at a specific point, thus assuring that egg and sperm are together. Other fish build nests where females lay eggs and males deposit sperm. Frogs and toads usually breed in the water, but the female will only release her eggs when the male is clasped to her back in amplexus. Thus, sperm are deposited on the eggs as they are being laid.

Males of other species place sperm directly in the female's reproductive tract. The male octopus has a special tentacle that is used to place one of his sperm packets in the mantle cavity of the female. Some salamander males deposit their sperm packets on the substrate during a squat dance courtship ritual. Females also do the squat dance and pick up the packet with the lips of their cloacae. In some species of water mites, females mount a special saddle-shaped extension of the males' abdomen. The male squats to deposit a sperm packet, moves ahead slightly and then squats again when the opening of the female's reproductive system is over the packet, forcing the packet into her reproductive system. Another interesting way to assure fertilization is seen in the sea horse. In these animals, a female deposits her eggs into a pouch on the male's abdomen and the male releases sperm into the pouch at the same time. The most common way to introduce sperm into a female's reproductive tract is through copulation, where the male ejaculates sperm directly into the female's reproductive tract. The motile sperm then travel to the egg. For a sperm to gain full motility, it usually must undergo a little-understood process called capacitation.

Penetration
Once eggs and sperm are in close proximity, the sperm must begin to penetrate the egg's protective layers. All eggs have at least one protective layer

233

outside the cell membrane. Called the vitelline envelope in most organisms, it is synthesized in the ovary and composed primarily of polysaccharides and glycoproteins. The oviducts and uterus often secrete other protective layers around the egg. In some instances, the sperm must also penetrate these layers, for example, the jelly layers that surround sea urchin and frog eggs. In other instances, the egg is fertilized before these layers are added, as is the case with the many protective layers that surround bird and reptile eggs. A protective layer made up of cells is seen in most mammals, since the egg is released by the ovary with cells of the cumulus oophorus still attached. For the sperm to penetrate these layers, its acrosome must contain the appropriate enzymes to lyse (disintegrate) the chemicals that block its way. The acrosomal reaction must also take place in order to expose the digestive enzymes of the acrosome. This reaction depends on changes in membrane permeability to ions and subsequent changes in pH.

Once through the protective layers, the sperm makes contact with the egg's plasma membrane. If the sperm and egg are of the same species, sperm receptor molecules on the egg membrane attach to complementary molecules, called bindins, on the sperm membrane and the two membranes fuse. If the bindins on the sperm do not complement the receptors on the egg, there is no fusion and fertilization does not continue, thus preventing most interspecies crosses. However, closely related species often have bindins and receptors sufficiently alike to allow some fertilization to proceed. The products of these interspecific matings are hybrids, such as the mule.

Once the first sperm fuses with the egg, mechanisms to prevent polyspermy, the fertilization of an egg by more than one sperm, are put into place. The first block to polyspermy is common to most animals studied: a very quick and only temporary depolarization of the plasma membrane. In sea urchins, the resting membrane potential of the egg plasma membrane is approximately 70 millivolts, the inside being more negative than the outside. Fusion of the sperm plasma membrane with the egg cell membrane causes a rapid influx of sodium ions. The positive charges neutralize negative charges in the egg until the membrane potential is raised to +10 millivolts. All this happens in less than five seconds, and lasts for about one minute before the egg cell has actively transported enough sodium out of the cell to repolarize it. While the cell is depolarized, no further sperm membranes can fuse with the egg membrane. This is often referred to as the fast or temporary block to polyspermy and seems to occur in all animals thus far studied. The fast block also sets into motion the slow or permanent block to polyspermy. The changed membrane potential of the fast block and the release of nitrous oxide by the sperm allows cells to release calcium ions from storage. The initial calcium ion release causes the egg to release nitrous oxide, which then increases the egg's release of calcium ions. The release of calcium ions induces the cortical reaction by which cortical granules move to the surface of the cell, fuse with the cell membrane, and empty their contents into the space between the cell membrane and the vitelline envelope. In sea urchins, the first acrosomal enzymes released break the bonds between the cell membrane and the vitelline envelope. In the presence of water, other chemicals released by the cortical granules swell, lifting the vitelline envelope away from the cell membrane. Finally, other enzymes released by the cortical granules alter the vitelline envelope, knocking off any attached sperm and causing the release of peroxide ions, which harden the envelope, making it impermeable to sperm. This impermeable barrier is renamed the fertilization membrane. The released peroxide may also provide another benefit. Any sperm that had penetrated the vitelline envelope before it hardened would be killed by the peroxide and would thus not lead to polyspermy. In other animals studied, although cortical granules do empty their contents into the perivitelline space, the permanent block to polyspermy does not seem to involve the same extensive changes to the vitelline envelope (or *zona pellucida* in mammals) that are seen in the sea urchin. In large, yolky eggs, some polyspermy does occur, but the extra sperm remain in the yolk and never reach the egg nucleus for fusion.

Cell Metabolism and Meiosis

Concomitant with the cortical reactions is an increase of metabolism in the egg, which will be necessary for nuclear fusion and cleavage. In species where the egg has not completed meiosis, it does so at this time. Which parts of the sperm enter the egg is dependent on the species. In many mammals, the entire sperm enters, while all but the tail enters in echinoderms. In other organisms, the head with the nucleus and centrioles seem to be the only things that enter. There is no evidence that any parts of the sperm other than the nucleus and centrioles are used by the zygote, and other parts that enter most probably degenerate and their components are recycled.

Studies on the mitochondria of sperm indicate that soon after entering the egg, the sperm's mitochondria are tagged by ubiquitin, the first step in breakdown and recycling. After entry, the sperm nucleus imbibes water and is converted into the male pronucleus. At the same time, the egg nucleus becomes the female pronucleus. In most animals, the male pronucleus and the female pronucleus fuse to form the diploid zygote nucleus. In some nematodes, mollusks, and annelids, however, the pronuclei remain separate until after the first cleavage division. In a few others, like the copepod *Cyclops,* the pronuclei divide separately for several cleavage divisions.

The fusion of the sperm with the egg nucleus affects many other cellular processes. One of the most interesting is the displacement of some cytoplasmic constituents. These constituents of the egg determine the fate of cells derived from the parts of the egg in which they were located and probably determine the plane of bilateral symmetry. Sperm attachment and entry often causes shifts in the position of the viscous cortical and subcortical cytoplasm, where many of the fate-determining chemicals are located.

—*Richard W. Cheney, Jr.*

See also: Asexual reproduction; Cleavage, gastrulation, and neurulation; Copulation; Gametogenesis; Reproduction; Reproductive system of female mammals; Reproductive system of male mammals; Sexual development.

FURTHER READING

Balinsky, B. *An Introduction to Embryology.* 5th ed. Philadelphia: Saunders College Publishing, 1981. A thorough chapter on "Fertilization and the Beginning of Embryogenesis."

Bronson, F. H. *Mammalian Reproductive Biology.* Chicago: University of Chicago Press, 1989. A very complete look at mammalian development. One chapter gives a brief overview of development for each mammalian order.

Carlson, B. *Patten's Foundations of Embryology.* 6th ed. New York: McGraw-Hill, 1996. A comprehensive chapter on fertilization with emphasis on vertebrates.

Kumé, Matazo, and Katsuma Dan. *Invertebrate Embryology.* Translated by Jean C. Dan. Belgrade, Yugoslavia: NOLIT Publishing House for the U.S. Department of Health and Human Services, 1968. Extensive compendium of invertebrate development by phylum.

IN VITRO FERTILIZATION OF ENDANGERED SPECIES

Techniques of in vitro fertilization were first developed to aid couples who had not been able to conceive through normal sexual relations. In this technique, eggs were surgically removed from the mother and mixed with the father's sperm in the laboratory. If fertilization took place, one or more embryos were introduced into the mother's uterus in the hope that an embryo would implant and develop into a full-term infant. Almost immediately, these techniques were used in other animals, especially endangered species. It offered many advantages over natural reproduction. In pairs that showed little sexual interest in each other, eggs and sperm could be extracted, mixed in the laboratory, and viable embryos could be introduced into the female's uterus. Also, if there was little genetic diversity in a zoo population, sperm from a donor at another location could be sent and used. By the end of the twentieth century, in vitro fertilization was being coupled with surrogate motherhood. Here, after the embryos are formed, they are introduced into the uteri of females of similar, but not endangered, species. This increases the number of uteri available for the endangered species' reproduction.

Forensic science

FIELDS OF STUDY

Chemistry; biology; biochemistry; mathematics; microbiology; physics.

SUMMARY

Forensic science is commonly defined as the application of science to legal matters. Although forensic science incorporates numerous disciplines, ranging from accounting to psychology, in the traditional sense, forensic science refers to the scientific analysis of evidence collected at crime scenes, which is also known as "criminalistics." Pattern evidence, such as fingerprints, bullets, and tool marks, is often compared visually, and chemical evidence (such as illicit drugs) and biological evidence (such as DNA, blood, and bodily fluids) are analyzed and compared using scientific instruments.

PRINCIPAL TERMS

- **class evidence:** evidence that can be identified as belonging to a group containing many members, all with similar characteristics or features.
- **criminalistics:** refers to the analysis of pattern, chemical, and biological evidence; often used interchangeably with forensic science.
- **DNA (deoxyribonucleic acid):** nucleic acid that contains the genetic code and is present in nearly every cell in the body.
- **illicit drug:** substance or drug that is prohibited by federal or state laws because of its undesirable effects or high risk of abuse.
- **impression evidence:** evidence, such as fingerprints, tire tracks, and footprints, formed when an object leaves behind a characteristic marking on a surface.
- **individualizing evidence:** evidence that can be identified as belonging to a group containing only itself as a member.
- **latent print:** fingerprint residue that is not easily visible to the naked eye and must be treated, either physically or chemically, to be observed.
- **toxicology:** study of drugs and poisons and their effect on the body.
- **trace evidence:** general term for microscopic pieces of evidence that are transferred by contact between people and objects. examples include hairs and fibers, as well as fragments of paint and glass.

Basic Principles

Forensic science is the application of scientific principles to the analysis of numerous types of evidence, most commonly evidence collected at a crime scene. Crime scene investigators, usually police officers, collect evidence at the crime scene and submit it to a crime laboratory for analysis by forensic scientists.

Crime laboratories contain different sections, each of which specializes in a particular type of analysis, such as controlled substances, DNA, firearms and tool marks, latent prints, questioned documents, toxicology, and trace evidence. The type of analysis conducted depends on the type of evidence as well as the circumstances of the crime. A single piece of evidence may be analyzed in more than one section. For example, a firearm may be analyzed in the latent prints and DNA sections, as well as the firearms and tool marks section.

Following analysis, forensic scientists may be summoned to present their findings in a court of law. Forensic scientists present their analysis and interpretation of the evidence before a judge and jury, who are charged with determining the guilt or innocence of the defendant. The unbiased, accurate analysis presented by the forensic scientist is an integral part of the criminal proceedings.

Background

Forensic science aims to determine identifying or individualizing characteristics to link people, places, and objects. In the late 1880's, French criminologist Alphonse Bertillon developed a method of identifying humans based on eleven physical measurements including height, head width, and foot length. However, limitations in this method soon became apparent. In 1880, Scottish scientist Henry Faulds published an article in *Nature* that discussed the use of fingerprints as a means of identification. In 1892, Sir Francis Galton published *Fingerprints*, proposing a system of classifying fingerprint patterns. That same year, an Argentine police officer, Juan Vucetich, used fingerprint evidence that resulted in the arrest and conviction of a murder

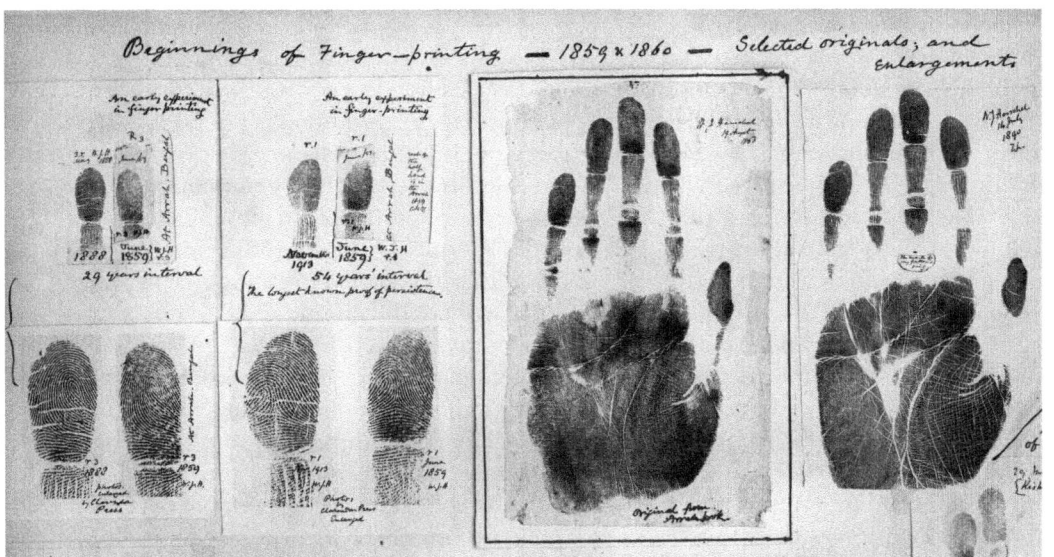
Fingerprints taken by William Herschel 1859/60.

suspect. From 1896 to 1925, Sir Edward Henry, a police official in British India, developed the Henry Classification System for fingerprints, which was based on the pattern on each finger and the two thumbs.

In the late nineteenth and early twentieth centuries, advances were being made in other areas that have become integral to forensic science. Spanish-born French scientist Mathieu Joseph Bonaventure Orfila, often considered as the pioneer of forensic toxicology, is credited with developing and improving methods for the detection of arsenic (a common poison in the nineteenth century) in the body. French scientist Edmond Locard developed the hypothesis that "every contact leaves a trace," which implies that whenever two objects make contact, there is an exchange between them. This hypothesis became known as Locard's exchange principle and is the foundation of modern trace evidence analysis. In the 1920's, the comparison microscope, which analyzes side-by-side specimens, was developed by American chemist Philip Gravelle and popularized by forensic scientist Calvin Goddard. This microscope enabled significant advances in many areas of forensic science, particularly firearms, tool marks, and trace evidence.

A major scientific breakthrough in the 1980's revolutionized the field of forensic DNA analysis. British geneticist Sir Alec Jeffreys developed DNA profiling, enabling individuals to be identified from samples of blood and other body fluids left at a crime scene. The development of the polymerase chain reaction by American biochemist Kary Mullis in 1983 allowed DNA profiling to be conducted on degraded and very small samples of DNA, making it possible for forensic scientists to test a wider range of evidence. As the field of forensic science evolves, newly developed technologies and instrumentation allow evidence to be analyzed and compared in an increasingly rapid, objective, and reliable manner. Forensic science is a truly dynamic field, constantly seeking further improvements and advancements in its analytical methodologies.

How It Works

Forensic science incorporates numerous subdisciplines, but the most common types of analysis conducted by crime laboratories are the analysis of illicit drugs, biological evidence, latent prints, firearms, footprints, tire marks, tool marks, and trace evidence. Latent prints, footprints, tire marks, and tool marks are considered pattern evidence. The patterns of an unknown sample (usually from the crime scene) and a known sample are visually compared to find similarities between the two. Samples can also be analyzed, either chemically or biologically, with scientific instruments. Some of the more common methods of testing are infrared spectroscopy, ultraviolet/visible microspectrophotometry, gas chromatography-mass spectrometry, and electrophoresis.

The major role of the forensic scientist is to analyze submitted evidence for the purposes of characterization and identification. For example, a blue fiber collected from the scene may be submitted to the trace evidence section, where forensic scientists characterize the fiber (for example, by its dimensions, color, cross-sectional shape) and then identify the type of fiber (for example, nylon, polyester, acrylic). Furthermore, when a known sample is available (such as fibers from the suspect's clothing), forensic scientists compare it with the unknown sample (collected from the crime scene) to determine if the two most likely originated from a common source. This process of characterization, identification, and comparison requires multiple stages of analysis, ranging from visual examination to instrumental analysis.

Infrared Spectroscopy. In infrared spectroscopy, the chemical structure of a sample is determined based on how the sample interacts with infrared radiation. Chemical bonds can absorb infrared radiation of a specific energy, which causes the bond to vibrate. Additionally, each bond can vibrate in different ways. Therefore, when infrared radiation is introduced, chemical bonds within the sample absorb different energies, and the results are shown in the form of an infrared spectrum. The spectrum is essentially a graph of radiation transmitted versus wave number, which is related to the energy of the radiation. Additionally, transmission can be mathematically converted to absorbance such that the spectrum can be displayed as absorbance versus wave number. The infrared spectrum of a sample displays numerous absorptions, each corresponding to a particular type of chemical bond and a particular type of vibration. The infrared spectrum of a sample is unique to that sample, and therefore, this technique can be used to definitively identify compounds.

The technique of infrared spectroscopy is commonly used in the controlled substance and the trace evidence sections of the crime laboratory. This technique can identify illicit drugs present in unknown samples, the type of fiber found at a crime scene or on a person, the polymer present in a paint chip, or the organic compounds present in explosive residues. The evidence is prepared for analysis in several ways, depending on the type of sample. Solid samples of illicit drugs can be mixed with potassium bromide and pressed into a pellet, which is then placed in the spectrometer. Infrared radiation is passed through the sample, which will absorb at characteristic energies depending on its chemical structure. The transmitted radiation is collected and the infrared spectrum is generated. Because potassium bromide does not absorb infrared radiation, the subsequent infrared spectrum shows only contributions from any drug present in the sample. For opaque samples, such as fibers or paint chips, attenuated total reflectance-infrared (ATR-IR) spectroscopy is more commonly used. The sample is positioned over a crystal, and pressure is applied to ensure good contact between the sample and crystal. Infrared radiation is passed through the crystal, and because of the close contact, the radiation penetrates a small depth into the sample. Certain energies are absorbed depending on the chemical bonds within the sample, resulting in the characteristic spectrum of the sample.

The infrared spectrum of the questioned sample can be compared to a database containing infrared spectra for known standards (drugs, fibers, paints, and so on) to identify the unknown sample. However, care must be taken when comparing a spectrum to spectra in a database. Although the spectrum of a given compound is unique, it can vary slightly depending on the instrument used to analyze the sample and standard. Rather than relying on a database search, it is often preferable to analyze the unknown sample and known standards on the same day, using the same instrument, to allow for a direct comparison of spectra.

Although samples can be rapidly analyzed using infrared spectroscopy, the technique works best for relatively pure samples. If impurities are present in the sample and they also absorb infrared radiation, the resulting spectrum contains contributions from both the sample and the impurities. This can complicate interpretation of the spectrum and subsequent identification of the sample.

Ultraviolet/Visible Microspectrophotometry. Infrared spectroscopy and ultraviolet/visible microspectrophotometry are both based on the principle of the interaction of radiation with a sample. However, ultraviolet/visible microspectrophotometry is typically used to compare the dye or pigment composition of samples. The technique is used to determine the color of a sample and identify subtle differences in color that cannot be seen with the naked eye.

A microspectrophotometer consists of a microscope with a spectrometer attached, which allows the analysis of microscopic pieces of evidence. The sample is viewed under the microscope, and ultraviolet and/or visible radiation is introduced. Depending on the chemical structure of the sample, wavelengths of light will be absorbed, reflected, or transmitted. The transmitted light is collected in the spectrophotometer, and the intensity of each wavelength is measured. Results are displayed in the form of a spectrum that is a graph of transmittance (or absorbance) versus wavelength. Subtle differences in color between two samples are observed as differences in wavelengths of light transmitted or absorbed in the corresponding spectra. Such differences are caused by differences in chemical composition between the two samples, and therefore, comparison of the resulting spectra can be used to determine if the two samples are similar in color.

The comparison and analysis of colored samples is often undertaken using microspectrophotometry. This technique is used in the trace evidence and questioned documents sections to compare the dye or pigment composition of fibers, paints, and inks.

Methods for sample preparation vary depending on the type of sample to be analyzed. Fibers are flattened and mounted on a microscope slide with a drop of immersion oil. Paint samples require more involved preparation, particularly for transmission spectra. The paint chip must be cut into a section so thin that light can be transmitted through it. Spectra of inks can be obtained directly if the paper is sufficiently thin to allow transmission. Otherwise, the ink must be removed from the document. This can be done by removing a small sample of the paper containing the ink and immersing the paper in a solvent to extract the ink. The resulting ink solution is placed on a microscope slide, and the solvent is allowed to evaporate, leaving a residue of ink for analysis. However, this is a destructive procedure because the document is damaged in removing the sample. Alternatively, a piece of clear tape can be placed on an area of the document that contains the ink. When the tape is lifted off, particles of ink adhere to the tape. These particles can be removed from the tape and transferred to a microscope slide for analysis. The document is minimally damaged using this procedure. Although microspectrophotometry offers a rapid means to investigate the dye or pigment composition of certain samples, no extensive spectral databases are readily available. Therefore, the technique is more useful when known samples are available and the color of the unknown and known samples can be compared directly, based on spectral interpretation.

Gas Chromatography-Mass Spectrometry. In any chromatography technique, sample mixtures are separated based on differences in interaction between a mobile phase and a stationary phase. In gas chromatography (GC), the mobile phase is a gas, and the stationary phase is a liquid coated on the inner walls of a very thin column. Liquid samples are typically introduced into the system and carried, in the mobile phase, through the stationary phase. Sample components that have a stronger attraction for the stationary phase spend longer in that phase, and components with less attraction spend less time in that phase and move more quickly through the system. The time it takes for sample components to travel through the system and reach the detector is known as the retention time.

In gas chromatography-mass spectrometry (GCMS), the detector is the mass spectrometer, which contains three major components: the ion source, the mass analyzer, and the detector. On emerging from the GC column, sample components enter the ion source, where each component is first ionized. The resulting ion is known as the molecular ion. This ion is unstable because of its high energy, so it breaks down, or fragments, into smaller ions. Molecular ions and fragment ions then enter the mass analyzer, where the ions are separated according to their mass-to-charge ratio. The separated ions enter the detector, where the number of ions of each mass-to-charge ratio is counted. Results are displayed in the form of a mass spectrum, which is a graph of intensity versus the mass-to-charge ratio. Because molecules break down, or fragment, in a predictable manner, the mass spectrum can be used to determine the structure of the original sample component. Furthermore, because the fragmentation pattern is unique to a molecule, the mass spectrum can be used to definitively identify the component.

On analyzing a sample by GC-MS, two pieces of information are obtained. First, from gas chromatography, a chromatogram is obtained, which is a graph of detector response versus retention time. Each separated component in the sample mixture is shown as a peak on the chromatogram. Components that take longer to reach the detector have greater attraction for the stationary phase and have longer retention times. Additionally, for each separated component, the mass spectrum is also obtained, which can be used to definitively identify the component.

As with infrared spectroscopy, gas chromatography-mass spectrometry is commonly used in the controlled substances and trace evidence sections, as well as in the toxicology section, for the determination of drugs and poisons in body fluids. GC-MS is advantageous over infrared spectroscopy in that samples containing impurities can still be identified because of the separation abilities of gas chromatography. For example, gas chromatography analysis of a drug mixture containing methamphetamine and caffeine separates the two components. In the resulting chromatogram, two peaks are observed: one for methamphetamine and one for caffeine. The mass spectrum of each peak is also obtained, which can be used to definitively identify each component. In most cases, samples must be in liquid form for GC-MS analysis. This is achieved by adding a suitable solvent to the sample and analyzing the resulting solution. For body fluid or tissue samples, a solid phase extraction or liquid-liquid extraction is necessary to isolate any drugs and poisons from additional components present in the fluids or tissues.

Solid samples can be analyzed using pyrolysis GCMS. In this case, a pyrolysis unit is attached to the gas chromatography inlet. Solid samples (for example, paint chips or fiber fragments) are placed in a small quartz tube and introduced into the pyrolysis unit, which rapidly heats the sample to a very high temperature. The sample is broken down and vaporized in the pyrolysis unit, and then carried in the flow of carrier gas onto the gas chromatography column, where the sample components are separated. Before analyzing the sample, it is important to demonstrate that the GC-MS system is free from contamination. This is usually done by injecting a volume of the solvent used to prepare the sample. If the solvent and instrument are not contaminated, the resulting chromatogram should show no peaks. For pyrolysis GC, the empty quartz tube is analyzed to demonstrate that there is no contamination in the tube or instrument.

Because the mass spectrum rather than the retention time is unique to a sample component, the spectrum of an unknown sample is compared to a suitable database of spectra. However, there may be slight differences between the database spectrum and the spectrum obtained for the unknown sample, depending on the instrument used to collect the spectra. It is often preferable to prepare and analyze a known standard in the same way as the unknown sample and then compare the corresponding mass spectra.

Electrophoresis. Although electrophoresis is also used to separate sample mixtures, the technique is not considered a chromatographic technique because no mobile phase is involved. Instead, sample mixtures are separated based on differences in migration under the influence of an applied electrical potential. Therefore, electrophoresis is used for the analysis of samples that have an electric charge. Although there are different types of electrophoresis, capillary electrophoresis is most commonly used for DNA profiling purposes. In this technique, a capillary column is filled with a polymer, and the ends of the column are immersed in reservoirs containing a buffer solution. The reservoirs also contain electrodes to allow the application of the electric potential. The sample is introduced to one end of the column, and the sample components move through the column under the influence of the applied potential. Separation occurs based on differences in the migration rate of the components through the column, which depends on size and charge. Separated components pass through a detector at the other end of the column, producing an electropherogram. The electropherogram shows the migration time of the separated components. Smaller components move more quickly, reaching the detector before larger components and have shorter migration times.

DNA profiling makes the most use of electrophoresis. Typically, blood, semen, saliva, or another body fluid from the crime scene is used to generate a DNA profile, which is compared with profiles generated from known samples. If known samples are not available, the generated DNA profile can be compared to a database of profiles. The Federal Bureau of Investigation (FBI) maintains a database of DNA profiles submitted by crime laboratories across the United States. This database, the combined DNA index system (CODIS), contains profiles from crime scenes, convicted criminals, and missing persons. Modern DNA profiling is based on the characterization of short tandem repeats (STRs) that are regions (loci) on the chromosome that repeat at least twice within the DNA. For profiling, the number of repeats at each location on the chromosome is determined. To do this, the DNA is first amplified via the polymerase chain reaction (PCR), in which the double-stranded DNA is split into two single strands and a mixture of enzymes and primers are used to replicate specific STR regions of the DNA. In the United States, STRs at thirteen loci are typically considered. The reaction is repeated many times, generating exact copies of the STRs. Because of this amplification procedure, profiles can be obtained from very small samples of DNA.

The STRs are analyzed using electrophoresis, most commonly capillary electrophoresis, which allows rapid and automated analysis. The STR mixture is separated based on differences in migration rate through the capillary column, which is related to the size of the STR. The resulting electropherogram displays a series of peaks that correspond to the STRs at each loci. Additionally, for each STR, there are two variants, one inherited from the mother and one from the father; therefore, the electropherogram actually shows a pair of peaks at each loci. A match in the number of STRs for both variants at all loci is considered strong evidence that the unknown and known samples originate from the same person. Because DNA is unique to an individual, this is one type of evidence that is considered individualizing rather than class evidence.

FUTURE PROSPECTS

Although great advances have been made in forensic science, many more have yet to be achieved. In 2009, the National Research Council published *Strengthening Forensic Science in the United States: A Path Forward*, a report on forensic science in the United States. The report highlighted several deficiencies in the field and recommended improving education, training, and certification for forensic scientists as well as developing standardized procedures and protocols for evidence analysis and reporting. Additionally, the report recommended research into the reliability and validity of many of the procedures used for evidence analysis. The report concluded that more research is necessary, not only to improve existing practices but also to develop new technologies that can be implemented in forensic science laboratories. It called for the development of a national institute of forensic science that would have many objectives, including the development of standards for certification for forensic scientists and accreditation of forensic laboratories, along with improving education and research in the field.

Ruth Waddell Smith, PhD

FURTHER READING

Bertino, Anthony J., and Patricia N. Bertino. *Forensic Science: Fundamentals and Investigations*. Mason, Ohio: South-Western Cengage Learning, 2009. Examines the tests and techniques used for the scientific analysis of various evidence types, including hairs and fibers, DNA, handwriting, and soil.

Brettell, Thomas A., John M. Butler, and José R. Almirall. "Forensic Science." *Analytical Chemistry* 79, no. 12 (2007): 4365-4384. A review of forensic science applications used in common disciplines.

Embar-Seddon, Ayn, and Allan D. Pass, eds. *Forensic Science*. 3 vols. Pasadena, Calif.: Salem Press, 2008. Extensive coverage of forensics, including historical events, famous cases, and types of investigations, evidence, and equipment.

Houck, Max M., and Jay A. Siegel. *Fundamentals of Forensic Science*. 2d ed. Burlington, Mass.: Academic Press, 2010. An introduction to forensic science and common techniques used for the analysis of physical, biological, and chemical evidence.

James, Stuart H., and Jon J. Nordby, eds. *Forensic Science: An Introduction to the Scientific and Investigative Techniques*. 3d ed. Boca Raton, Fla.: CRC Press, 2009. Discusses mass spectrometry techniques in relation to forensic applications, including forensic toxicology, controlled substance identification, and DNA analysis.

Kobilinsky, Lawrence, Thomas F. Liotti, and Jamel Oeser-Sweat. *DNA: Forensic and Legal Applications*. Hoboken, N.J.: Wiley-Interscience, 2005. Presents an overview of DNA analysis, including the historical perspective, scientific principles, and laboratory procedures.

Rudin, Norah, and Keith Inman. *An Introduction to Forensic DNA Analysis*. 2d ed. Boca Raton, Fla.: CRC Press, 2002. Contains an overview of DNA analysis, beginning with its history and examining the principles on which it is based.

Saferstein, Richard. *Criminalistics: An Introduction to Forensic Science*. 10th ed. Upper Saddle River, N.J.: Prentice Hall, 2011. Provides an introduction to forensic science, detailing the techniques to analyze physical, biological, and chemical evidence.

WEB SITES

American Academy of Forensic Sciences http://www.aafs.org

American Forensic Association http://www.americanforensics.org

Association of Forensic DNA Analysts and Administrators http://www.afdaa.org

National Institute of Justice Forensic Sciences http://www.ojp.usdoj.gov/nij/topics/forensics/welcome.htm

See also: DNA Analysis; Toxicology.

FASCINATING FACTS ABOUT FORENSIC SCIENCE

- Forensic entomology involves the study of insects that invade a body after death to determine the time that has elapsed since the person's death.
- The saliva on a discarded cigarette contains enough DNA to identify the person who smoked it.
- The first use of fingerprint evidence to solve a criminal case was recorded in Argentina in 1892. Police official Juan Vucetich used a bloody fingerprint found at the crime scene to prove that two boys were murdered by their own mother.
- "Forensic" comes from the Latin *forensic*, which means "of the forum." The "forum" relates to the law courts in ancient Rome. In modern times, forensic science is defined as science relating to the law.
- In 1981, a German publishing company purchased what were thought to be Adolf Hitler's diaries. However, forensic document examiners proved that the diaries were fake, based on the presence of a paper-whitening agent that was not used in paper manufacturing until at least 1954.
- Marie Lafarge was the first person to be found guilty of murder based on toxicology evidence. Although she poisoned her husband with arsenic, initial testing did not find any arsenic in his body. However, when French scientist Mathieu Joseph Bonaventure Orfila repeated the tests, he found arsenic in the man's body and proved that the initial testing was inaccurate.
- In 1995, O. J. Simpson was cleared of murdering his wife, Nicole Brown, and her friend Ronald Goldman, despite DNA evidence identifying blood at the crime scene as belonging to the former football player. Furthermore, DNA from Simpson, Brown, and Goldman was found in a leather glove found at the scene.

FUR AND HAIR

FIELDS OF STUDY

Physiology, zoology

SUMMARY

Fur is the hairy covering of the skin of a mammal. This covering is called fur when its individual hairs are fine and spaced closely together; when the hair is soft, kinky, and matted together, it is called wool, which grows in a fleece, and when it consists of coarse, stiff hairs, it is called bristles, or when pointed, spines or quills. Its primary role is as protection, insulation, and ornamentation.

PRINCIPAL TERMS

- **cortex:** the main part of a hair, made of pigment-containing cells, surrounding a central medulla
- **cuticle:** the outermost layer of a hair, made of scales
- **epidermis:** the dead, outermost portion of the skin follicle: the saclike organ from which a hair grows; its blood vessels nourish the hair
- **keratin:** a tough fibrous protein, seen in large quantities in epidermal structures such as hair
- **medulla:** the innermost layer of a hair
- **shaft:** the main hair part, made of dead cells arranged in a complex fashion

BASIC PRINCIPLES

The term fur is used in several ways. All are related to its being the hairy covering of the skin of a mammal. In its most common usage, fur is the dense, hairy body covering of a mammal. However, even the sparse hair covering on the arms and legs of humans may be viewed as fur, since it is the hairy covering of the skin of a mammal. Each individual hair in a mammal is a threadlike epidermis outgrowth. Collectively, hairs form the body coverings (or pelage) of all mammals. Each hair is made mostly of a fibrous, sulfur-containing protein called keratin. The pelage is most correctly called fur when its individual hairs are fine and spaced closely together. In cases where the hair is soft, kinky, and matted together, the pelage is wool, which grows in a fleece. Coarse, stiff hairs are bristles, and when pointed (as in porcupines) they are spines or quills. Hair grows on most parts of the bodies of mammals. As their body covering, its primary role is as protection, insulation, and ornamentation, like bird feathers and reptile scales.

The Structure of a Hair

Any hair consists of two main parts, the root and the shaft. The shaft is the hair part outside of the skin. It contains the hair's unattached end. Hair shaft cross-sections range from round to flattened. Due to their composition, round and flat hairs are straight and curly, respectively. Hair shafts consist of dead epithelial cells, arranged in columns which surround a central medulla (or core), covered with flat scales. The scales—a hair's cuticle— overlap like roof shingles, but their free ends face upward, away from the hair's root. Beneath the cuticle is a second layer of dead cells, the cortex, which surrounds a central core made of yet other dead cells (the medulla).

The cortex, which makes up most of each hair, consists of many dead, longitudinally arranged, keratin-rich, spindle-shaped cells which are tightly attached to each other. Hair color results from pigments in cortex cells and light reflected from the medulla. The medulla is less dense than the cortex and its cells are only loosely attached to one another.

At every hair's base is a saclike hair follicle. The hair grows from the bottom of the follicle, nourished by blood vessels in a structure called a papilla. The papilla extends into the follicle and into the hair's root. A tiny muscle is attached to each hair follicle. Action of the nervous system can cause the muscle to contract to make hair "stand on end."

Hair Growth and Replacement

A hair forms from cells that grow from the surface of the papilla, which means that it grows from the root, not the free end. As new cells develop, they push forward old ones, which become part of the shaft. Hair growth continues as long as follicle and papilla are functional. The lifetime of a hair from start of growth until it is shed depends upon the organism which produces it. When an old hair falls out a new one takes its place.

Hair follicles produce hairs in cycles of hair growth, in which the hair follicle and the shaft pass through a complex series of morphological changes. During hair growth, the follicle penetrates into the dermis, and cells of the shaft are joined together. In addition, the follicle's melanocyte cells deposit pigment into shaft cells. Once a hair shaft attains its characteristic length, the follicle contracts and a "dead" hair protrudes from it. The growth period of a single hair ranges from three years in humans to around two weeks in rodents.

Hairs are continually replaced, or shed, throughout the life of a mammal. However, their development and loss occur asynchronously, so mammals are never completely naked of pelage. In rodents, replacement is in waves, across the body. In primates, each follicle passes through the growth cycle, independent of those around it. Hormones control hair growth; however, there are other, as yet unknown components that must also affect the process, because hormones are carried in the blood, and if they acted alone they would simultaneously affect all follicles in the body. This would be disastrous, because if all follicles grew in together, at the same rate, all hairs would be shed at the same time. Then the mammal involved would have naked periods where it was deprived of hair's protection and insulation. Continuous growth of hairs can also be hazardous. For example, in merino sheep, long growth phases produce long-stranded wool. However, if these sheep strayed, were not minded well, or were sheared irregularly, they might starve to death from becoming entangled in underbrush.

Hair Origin and Function

Hair in the pelage of contemporary mammals acts mainly to insulate against temperature variation. It has been proposed that mammal hair evolved into pelage from "prehair" which had the same functions. One theory of hair origin is that it evolved from epidermal mechanoreceptors, a concept supported to some extent by the existence of sinus hairs in mammals. These hairs, whiskers (vibrissae) in mammals such as felines and rodents, have blood-filled sinuses in the skin around the follicle. This tissue, together with associated nerve fibers, engenders mechano-reception, which facilitates nocturnal movement. However, most mammal body hairs lack nerves and only insulate and protect. The basis for hair evolution from vibrissae is therefore unclear.

Certainly, mammalian hair has other functions today. For instance, it can serve the unusual protective function of the quills of porcupines, and perhaps it may more generally serve to attract mates. It seems possible that the change of prehairs into attractive pelage drove development into these main contemporary forms.

Protecting Endangered Fur-Bearers

The desire on the part of humans for fur garments has led to atrocities committed on many mammal species which have gorgeous pelage. One of many examples is clubbing young fur seals to death. Beyond that, many species have been hounded to near extinction by hunters. Classic examples of such endangered species are the big cats, such as tigers and leopards, and rodents, such as the beaver. Many other mammals are threatened species, likely to become endangered in the foreseeable future.

Fortunately, several organizations have sought to protect these fur-bearing mammals. Efforts of organizations such as the World Wildlife Fund have focused public opinion and led to animal conservation legislation. Preeminent are the Endangered Species Act of 1973, and a 1977 Convention, signed by eighty nations, including the United States. These actions have led to agreement that furs will not move interstate or between signatory countries without proof that the species from which they were harvested is not threatened or endangered. This bodes well for the future of hair.

—*Sanford S. Singer*

Further Reading

Deems, Eugene F., Jr., and Duane Pursley. *North American Furbearers: A Contemporary Reference*. Baltimore: International Association of Fish and Wildlife

---

WOOL AND WOOLENS

Wool denotes soft, curly fleece fibers from domesticated sheep. It differs from hair in that its scales are more numerous, smaller, and pointy. Curliness makes wool very resilient. This high tensile strength and elasticity help wool fabrics to retain their shape. These properties, wool's lightness, and its fine insulating ability make wool fabrics desirable. Wild sheep have sparse, woolly undercoats and coarse hair, useless in fabrics. The hair has been bred out in domestic breeds. Sheep fleeces are usually shorn annually, from late spring to early summer. The wool is cut off very close to the skin and removed in one piece, weighing nine to eleven pounds. Wool from different fleece parts varies in its fiber length, fineness, and structure. The shoulder and sides yield the best fibers. Merino sheep yield the best overall wool. It makes up 40 percent of all wool produced commercially. Crossbreeds of merino sheep and strains that produce longer, coarser wool yield most of the rest. Some apparel wool comes from alpacas, goats, and llamas. Woolen cloth manufacture begins by pulling fleeces apart and choosing the best fibers for given uses. Next, fibers are cleaned to remove lanolin and dried-on sweat. The clean fibers are disentangled and drawn straight by carding, which entails passing them between rotating cylinders to yield a thin film or web. Web processing varies, depending on whether it will be used in tweed or worsted yarn. Tweeds, woven from bulky yarn made of short, randomly arranged fibers, are thick and fuzzy. Worsted fabrics such as gabardines are woven from web made of longer, thinner fibers, tightly twisted for smoothness.

THE UNUSUAL PORCUPINE

Porcupines, whose name derives from a French word meaning "spiny pig," are animals that possess very unusual hair coverings. Included in these coverings are quills, which are used for defense against predators. A coat of regular hairs serves the porcupine as thermal insulation.

An adult porcupine possesses between twenty-five thousand and thirty-five thousand quills. Each of the quills is actually a group of long, stiff hairs that have grown together very tightly. The quills range in length from three to eighteen inches, depending upon the porcupine species and the body positions of the quills. Quills have sharp barbs on their ends. When a porcupine is attacked or otherwise disturbed, it arches its back and makes the quills stand up straight. Attacking predators are very likely to impale themselves painfully on the quills, which pull out of the porcupine's body quite easily. A porcupine can also swing its quill-filled tail at a predator. When the tail connects with the aggressor, quills enter the flesh of that animal deeply and are bound to do damage. When a quill hits an attacker's eye, it may cause blindness. If quills enter its jaw, they may prevent eating and lead to the predator's death by starvation. Porcupines grow new quills, as needed.

Agencies, 1983. Presents a great deal of information on fur and fur-bearing animals.

Robbins, Clarence R. *Chemical and Physical Behavior of Human Hair.* 3d ed. New York: Springer-Verlag, 1994. The first chapter of this text covers the morphological and macromolecular structure of hair.

Robertson, James, ed. *Forensic Examination of Hair.* London: Taylor & Francis, 2000. This book on the forensics of human hair has two solid, basic chapters on hair physiology and growth and on its microscopic examination. Spearman, R. I. C., and P. A. Riley, eds. *The Skin of Vertebrates.* New York: Academic Press, 1980. The proceedings of an international congress, including sound coverage of skin and hair.

Stanford Environmental Law Society. *The Endangered Species Act.* Stanford, Calif.: Stanford University Press, 2001. Describes endangered species law and legislation, with a solid Bibliography.

G

Gametogenesis

FIELDS OF STUDY

Cell biology, developmental biology, embryology, genetics, reproduction science

SUMMARY

Gametogenesis is the process of sex cell formation. It includes the events that lead to a reduction in chromosome number so that the sex cells will have one-half the chromosomes that are found in normal body cells.

PRINCIPAL TERMS

- **diploid:** the number of chromosomes or the amount of genetic material normally found in the nucleus of body cells; this number is constant for a particular species of animal
- **gamete:** a sex cell; the egg or ovumin the female and the sperm in the male
- **haploid:** one-half of the diploid number; the number of chromosomes or the amount of genetic material found in a gamete
- **meiosis:** reduction division of the genetic material in the nucleus to the haploid condition; it is the process used by animal cells to form the gametes
- **oogenesis:** gamete formation in the female; it occurs in the female gonads, or ovaries
- **spermatogenesis:** gamete formation in the male; it occurs in the male gonads, or testes
- **spermiogenesis:** the structural and functional changes of a spermatid that lead to the formation of a mature sperm cell

Basic Principles

Sexual reproduction is the predominant mode of reproduction in animals. Sexual reproduction involves the production of gametes: the eggs and sperm. In most animals, these gametes are produced in specialized organs called gonads (ovaries and testes). The sex cells in most animals are separate—that is, each individual animal contains either testes or ovaries, but not both. Such animals are said to be dioecious. In dioecious animals, the sex cells from two different individuals (one male and one female) will fuse together in a process known as fertilization to form the offspring. The advantage of sexual reproduction seems to be in its potential to produce variability in the gametes and therefore in the new organism.

Gametes are highly specialized cells that are adapted for reproduction. These egg and sperm cells develop by a process of gametogenesis, or gamete formation. Sperm cells are relatively small cells that are specialized for motility (movement); egg cells are larger, nonmotile cells that, in many species, contain considerable amounts of stored materials that are used in the early development of the zygote (fertilized egg).

How It Works

In animals, gametogenesis consists of two major events. One involves the structural and functional changes in the formation of the gamete. The other involves the process of meiosis. Animal body cells normally contain the diploid amount of genetic material. Each species of animal has a characteristic diploid number that remains the same from generation to generation. Because fertilization involves the fusion of the egg and the sperm, bringing together each cell's set of genetic material, some mechanism must reduce the amount of genetic information in the gamete, or it would double every generation. Meiosis is a special nuclear division whereby the genetic material is reassorted and reduced to form haploid cells. Therefore, gametes are haploid, and gamete fusion during fertilization reestablishes the diploid content in the zygote.

Sperm. Sperm are highly motile cells that have reduced much of their cellular contents and are little more than a nucleus. Sperm are produced in the testes from a population of stem cells called spermatogonia. Spermatogonia are large diploid cells that reproduce by an equal division process called meiosis. Spermatogenesis is the process by which these relatively unspecialized diploid cells will become haploid cells; it is a continuous process that occurs throughout the sexually mature male's life. When a spermatogonium is ready to become sperm, it will stop dividing mitotically, enlarge, and begin the reduction division process of meiosis. These large diploid cells that begin to divide meiotically are known as primary spermatocytes.

The first step in the division process involves each primary spermatocyte dividing to form two secondary spermatocytes. Each secondary spermatocyte continues to divide, and each forms two spermatids. These spermatids are haploid cells. For each primary spermatocyte that undergoes spermatogenesis, four spermatids are formed. The spermatids are fairly ordinary cells; they must go through a process that will form them into functional sperm. The transformation process of a spermatid into a sperm is called spermiogenesis and involves several changes within the cell. The genetic material present in the nucleus begins to condense, while much of the cytoplasm and its subcellular structures are lost. The major exception to this latter event is the retention of mitochondria, cytoplasmic structures involved in energy production. The mature sperm has three main structural subdivisions: the head, the neck (or midpiece), and the tail. All are contained within the cell's membrane. The oval head has two main parts, the haploid nucleus and the acrosome. The acrosome comes in various shapes but generally forms a cap over the sperm nucleus. The acrosome functions differently in various animals, but generally its functions are associated with the fertilization process (union and subsequent fusion of egg nucleus and sperm nucleus).

Acrosomes contain powerful digestive enzymes (organic substances that speed the breakdown of specific structures and substances) that allow the sperm to reach the egg's membrane. The midpiece of the sperm contains numerous mitochondria, which provide the energy for the sperm's movement. The tail, which has the same general organization as flagella or cilia (subcellular structures used for locomotion or movement of materials), uses a whiplike action to propel the sperm forward during locomotion. The structural changes that occur during spermiogenesis are meant to streamline and pare down the sperm cell for action of a special sort and of a limited duration.

The sperm's function is to "swim" to the egg, to fuse with the egg's surface, and to introduce its haploid nucleus into the egg's interior.

Eggs. The female gamete, the egg or ovum, is produced by a process known as oogenesis. This process occurs in the female gonads, the ovaries. At first glance, oogenesis and spermatogenesis appear to be very similar, but there are some striking differences. The major similarity is that both processes form gametes, which contain genetic material that has been reduced to the haploid condition. To understand oogenesis, one must consider that its goal is to produce a cell that is capable of development. The mature egg in all animal cells is large in comparison with other cells, particularly with the sperm. There are two important features of the egg that must be considered: the presence of a blueprint for development and the means to construct an embryo from that blueprint. In other words, the egg must be programmed and packaged during oogenesis. The programming refers to the information that is coded within the structure of the egg. This information includes the genetic material as well as the cytoplasmic information. Together, the nucleus and cytoplasm provide the egg with the potential to transform a simple cell into a complex preadult form. Since it is within the egg that this transformation occurs, the programming must be within the organization of the egg, and the directions for development must be within that organization. The packaging refers to the presence of all the material necessary to build embryonic structures, to nourish this developing embryo, and to provide its energy until it can obtain nourishment on its own.

As happens in spermatogenesis, the potential eggs are formed from unspecialized stem cells, in this case called oogonia. Oogonia contain the diploid amount of genetic material and divide by the process of mitosis. At some point in their life, oogonia stop dividing mitotically, enlarge, and prepare to become eggs—that is, they begin meiosis. The cell that begins this reduction division process is called the primary oocyte. Each primary oocyte divides into two cells— one large cell, the secondary oocyte, and a very small cell, the first polar body. The secondary oocyte continues the final reduction phase of meiosis and forms two cells, one large one (the ovum) and one very small one (the second polar body). The first and second polar bodies are nonfunctional by-products of meiosis. The one functional cell, the mature egg, contains most of the cytoplasm of the primary oocyte and one-half of its genetic material. In many animals (primarily the vertebrates), all oogonia present in the ovaries enter meiosis at the same time; the initial events of oogenesis are synchronous within the animal. Oogenesis in many animals is not a continuous process, as is spermatogenesis in the male. Rather, the primary oocytes in the first stages of reduction division may remain inactivated for a long time—in some cases, for several decades. Therefore, in female animals with this format of oogenesis, a primary oocyte population is maintained, and eggs will mature as they are needed.

Thus far it appears that the egg's formation differs from the sperm's in three ways. First, in many female animals, there is a limited number of primary oocytes capable of going on to form eggs; second, this egg formation is not necessarily a continuous process; third, one primary oocyte yields one mature egg at the end of meiosis. Although these are three very important differences, there are other distinctly egg events that deal with the developmental programming and the packaging of materials in this potential gamete.

Eggs and RNA. Little is known about the egg's storage of developmental directions or the actual programming of information, but developmental and molecular biologists are beginning to elucidate events that occur during oogenesis that are concerned with function of the egg. One such event, fairly widespread among the animal kingdom, is the formation of so-called lampbrush chromosomes during oogenesis. The chromosome's backbone unravels at many sites so that regions, composed of specific genes, loop outward from the backbone. These loops give the chromosome its distinctive lampbrush-like appearance. Large amounts of a nucleic acid known as messenger ribonucleic acid (mRNA) are being made on each loop. This mRNA is then processed and sent into the developing eggs's cytoplasm, where most of it will be stored for use during early development. After fertilization, these maternal (egg-derived) mRNAs can be used to make specific proteins necessary for the embryo.

Another event present in some developing eggs is the mass production of another type of RNA known as ribosomal RNA (rRNA). Most of this rRNA will also be stored until fertilization. After fertilization, these rRNA particles will help form cytoplasmic structures called ribosomes (the sites of protein synthesis). In addition to these egg products, many animal eggs must become filled with yolk. Yolk is the general term that covers the major storage of material in the egg. Because the maternal proteins (yolk and other protein components) and nucleic acids (various RNAs) form the bulk of the egg cytoplasm, they have profound influences on the development of the embryo. In particular, the positions of maternal mRNAs, ribosomes, and proteins affect the organization of the embryo. It is evident, then, that the maternal genetic information and the arrangement of the products of this information provide crucial developmental information that will control much of the course of embryonic development. Therefore, the egg contributes considerably more than a haploid nucleus to the zygote.

STUDYING GAMETOGENESIS

There are several approaches to the study of gametogenesis. Early biologists employed cytological techniques (methods of preparing cells for the study of their structure and function) and microscopy to study gamete formation. These early studies were, in fact, observations of the actual events themselves. Although these early descriptive approaches gave much information about the cells involved at each stage of gamete formation, they did not provide any information about the control mechanisms for this process. Biochemical studies have contributed to the understanding of certain regulatory substances and how they function in gametogenesis. By enhancing or inhibiting the presence of these regulatory substances in the organism, investigators have been able to elucidate many of the normal events of gametogenesis. Beginning at puberty, the hormones (substances released from endocrine glands, generally functioning to regulate specific body activity) of the hypothalamus, the pituitary gland, and the gonads interact to establish and regulate gametogenesis in the organism. Gonadotropin-releasing hormone (GnRH) from the hypothalamus stimulates the release of follicle-stimulating hormone (FSH) and luteinizing hormone (LH) from the anterior portion of the pituitary gland. All three of these hormones are necessary for spermatogenesis and oogenesis.

Surgical removal of the mammalian pituitary gland (hypophysectomy) in the male leads to degeneration of the testes. Testicular function can be restored in these hypophysectomized animals by administering the hormones FSH and LH. These studies suggest that FSH andLHare necessary for normal functioning of the testes. LH appears to stimulate the release of testosterone (male hormone) by certain cells (Leydig cells) of the testes. Both testosterone and FSH are necessary for spermatogenesis, but the exact role that each of these hormones plays in male sexual physiology has yet to be determined.

Oogenesis in the female has been the subject of intense investigation. At the beginning of each ovarian cycle, from puberty to menopause, one primary oocyte present in the female's ovaries is activated to continue the process of gamete formation. Release of GnRH from the hypothalamus at the beginning of each cycle stimulates the anterior portion of the pituitary gland to release FSH. FSH, in turn, affects the ovaries: It stimulates a primary oocyte to mature to the point that it can be released from the ovary, as a secondary oocyte, and it causes certain cells (follicle cells) in the ovary to produce estrogens, female hormones. High estrogen levels will cause the pituitary to inhibit FSH release, a negative feedback mechanism, and stimulate LH release. These estrogen-mediated events occur at approximately the middle of the ovarian cycle. LH also affects the ovaries. LH, however, is responsible for ovulation (the release of the oocyte from the ovaries) and for the formation of a cellular structure called the corpus luteum. LH also stimulates the corpus luteum to produce progesterone, another female hormone. Eventually, high levels of progesterone will inhibit LH release from the pituitary gland, and the cycle begins anew.

—*Geri Seitchik*

See also: Aging; Asexual reproduction; Cell types; Cleavage, gastrulation, and neurulation; Embryology; Fertilization; Growth; Reproduction; Reproductive system of female mammals; Reproductive system of male mammals; Sex differences: Evolutionary origins.

FURTHER READING

Epel, D. "The Program of Fertilization." *Scientific American* 237 (November, 1977): 128-140. A Summary of the events of fertilization and an excellent account of the world of the egg and the sperm. It explains the events that occur after the formation of the gametes. Written for the college-level reader with some background in general biology

Kinne, Rolf K. H., ed. *Oogenesis, Spermatogenesis, and Reproduction*. New York: Karger, 1991. Covers gametogenesis in fish.

Sadler, R.M. *The Reproduction of Vertebrates*. New York: Academic Press, 1973. A survey of reproductive patterns and mechanisms of reproductive control in the major vertebrate groups. Useful to the student who wants a comparative presentation of all aspects of reproductive events.

Van Blerkom, Jonathan, and PietroM. Motta, eds. *Ultrastructure of Reproduction: Gametogenesis, Fertilization, and Embryogenesis*. Boston: Kluwer, 1984. A collection of essays focusing on the use of electron microscopy to understand reproductive processes such as gametogenesis.

GASTROENTEROLOGY

FIELDS OF STUDY

Biology; chemistry; mathematics; physiology; health; nutrition; anatomy; nephrology; neurology; hematology; internal medicine; endocrinology and metabolism; oncology; nuclear medicine; infectious disease; geriatric medicine; pediatric medicine.

SUMMARY

The medical specialty known as gastroenterology focuses on conditions and diseases involving the digestive tract and related organs. The esophagus, stomach, duodenum, and large and small intestines are part of the gastrointestinal system. Related organs include the liver, gallbladder, and pancreas. Conditions addressed by the gastroenterologist include gastric ulcers, heartburn, abdominal discomfort, nausea and vomiting, constipation, diarrhea, inflammatory bowel diseases, hepatitis, nutritional deficiencies, and gastric cancers. Endoscopy, a diagnostic tool that allows visualization of the digestive tract, is an integral component of patient care. Gastroenterologists must be skilled in all aspects of the digestive process, and they must be caring, compassionate, and excellent team leaders.

PRINCIPAL TERMS

- **colon:** large intestine.
- **colonoscopy:** procedure using a thin tube to project images of the interior of the large intestine and rectum to a monitor.
- **endoscopy:** procedures using a thin, flexible fiber-optic tube with a camera to project images of the gastrointestinal tract back to a monitor.
- **gastrointestinal tract:** organs associated with the digestion and elimination of food, including the esophagus, stomach, duodenum, small intestine, large intestine (colon), rectum, and anus.
- **hepatitis:** inflammation of the liver caused by several different types of viruses.
- **motility:** movement of food through the intestines.
- **physiology:** study of the function of organs.
- **sigmoidoscopy:** procedure to inspect the lining of the last third of the colon, or descending colon.

Basic Principles

Gastroenterology is a medical specialty focused on the diagnosis and treatment of conditions involving the digestive tract and associated organs. Gastroenterologists have a thorough understanding of the normal function of the gastrointestinal tract, which encompasses the esophagus, stomach, small and large intestines, and the rectum. This specialty also includes other organs associated with the gastrointestinal tract, such as the pancreas, gallbladder, bile ducts, and liver. Training includes knowledge of the normal movement of food through the digestive tract (motility), the absorption of nutrients as foods move through the intestines, and the removal of waste products. A detailed knowledge of normal digestive processes allows gastroenterologists to evaluate abnormal conditions and plan a course of

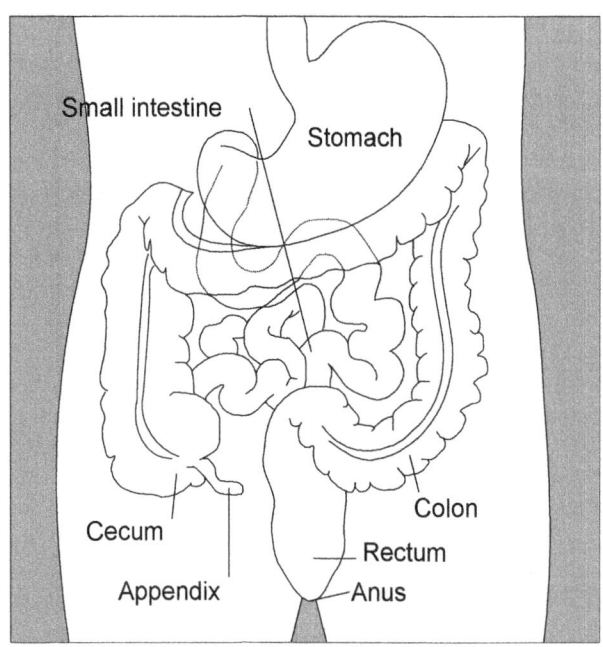

Diagram of the stomach, colon and rectum.

treatment. Gastroenterological training stresses diagnostic procedures, such as endoscopies, combined with thorough patient interviews regarding signs and symptoms. Colorectal cancer screenings are an important part of the practice. Gastroenterologists may specialize further and narrow their practice to children or the elderly, or they may become board certified in surgery. Not all gastroenterologists are surgeons, but most gastroenterologists perform minimally invasive procedures, such as removal of colon polyps. Gastroenterologists specializing in bariatric surgery generally perform procedures such as gastric bypass or lapband surgery for weight loss.

Background

Gastroenterology is a relatively new medical specialty. Until the mid-1980's, gastroenterology was considered a subspecialty of internal medicine and required only an additional one or two years of training. The introduction of endoscopy, which uses a flexible fiberoptic tube to explore the digestive tract, transformed gastroenterology and allowed the field to emerge as a separate subspecialty. The addition of hepatology, or the study of liver conditions, to gastroenterology and the rise in hepatitis further defined the field. Although gastroenterologists are first trained as internists, several more years of training in gastroenterology are required. Gastroenterologists may specialize further in gastric conditions particular to children. The field of gastroenterology has developed unevenly across developed and developing countries. Standardized training is lacking, with nations requiring different curricula. Whereas gastroenterology has emerged as a separate subspecialty in North America, the field is still considered part of internal medicine in Europe. Fellowships are required, but they last only one or two years and consist of onthe-job training. Some developing countries have no formal training guidelines or firm requirements regarding entrance and exit examinations.

PATIENT CARE

First and foremost, gastroenterologists are concerned with patient care. Most gastrointestinal conditions are chronic, allowing gastroenterologists to develop long-term relationships with their patients. Detailed medical histories are often the key to diagnosis, so gastroenterologists must be compassionate and able to listen closely to the patient's complaints.

The physical examination is an important aspect of diagnosis and treatment. Procedures such as colonoscopies and sigmoidoscopies are uncomfortable and may be embarrassing for patients, so the gastroenterologist must be skilled in relaxing patients and providing appropriate sedation.

Conditions Treated. Although they focus on the gastrointestinal system, gastroenterologists diagnose and treat a wide range of symptoms and conditions. Gastric and peptic ulcers and gastroesophageal reflux disease (GERD) are diagnosed and treated by gastroenterologists, as are gallstones. Gastroenterologists are called on to diagnose and manage abdominal pain and discomfort, hemorrhoids and bloody stools, constipation, diarrhea, nausea, and vomiting. They treat intestinal conditions, such as inflammatory bowel disease, ulcerative colitis, Crohn's disease, and diverticulitis. Gastroenterologists screen for colon and rectal cancer and remove rectal polyps. Gastroenterologists detect reasons for unexplained weight loss and poor absorption of nutrients. They also treat liver diseases such as hepatitis and jaundice.

Management Duties. Regardless of the setting, gastroenterologists are expected to handle a number of management duties in addition to their clinical caseload. The gastroenterologist in private practice manages staff and ensures that the office is run efficiently. All gastroenterologists consult with patients and other medical professionals, who offer their medical expertise. Experienced gastroenterologists train students.

Imaging Techniques. Imaging techniques are the cornerstone of the field. Endoscopy is the use of a thin, flexible, lighted tube that sends real-time video to a monitor. For upper endoscopy, also called an upper GI, the patient is sedated and the endoscopy tube is fed down through the throat into the stomach and duodenum. The gastroenterologist examines the lining of the esophagus and stomach for ulcers or other abnormalities. Tissue samples may be taken for biopsy, or the gastroenterologist can remove a foreign object that has been swallowed.

Colonoscopy, or a lower GI, uses endoscopy to examine the colon and rectum for abnormalities. Patients must prepare for the procedure by cleansing their bowels of waste materials, usually by drinking water mixed with a substance that results in rapid elimination of the contents of the bowels. Preparation is usually done at home. For the colonoscopy itself, the patient is sedated. The imaging instrument, called a colonoscope, is inserted into the rectum through the anus. As the gastroenterologist guides the tube up through the intestine, carbon dioxide gas is used to inflate the colon to allow for better imaging. As in an upper endoscopy, the gastroenterologist examines the intestinal wall for ulcers, polyps, or inflammation. If polyps are present, the gastroenterologist removes the growths by inserting miniature cutting tools through the colonoscope.

Endoscopy. Endoscopy has become the foundation of gastroenterology. Endoscopy is an imaging technique that uses a thin, flexible fiberoptic tube. As the tube is fed through the patient's gastrointestinal tract, a small camera transmits the images to a screen for viewing. The gastroenterologist manipulates the tube and camera to visually inspect the esophagus, colon, or intestines for abnormalities, lesions, or ulcers. Gastroenterologists receive extensive training on endoscopy procedures, including proper sedation of patients, and on interpretation of the images. Basic endoscopy training includes upper endoscopy, sigmoidoscopy, and colonoscopy.

Upper endoscopy examines the esophagus, stomach, and duodenum for ulcers, precancerous growths, foreign bodies, and other conditions causing pain, nausea or vomiting, bleeding, unexplained weight loss, or anemia. Gastroenterologists also receive training on using endoscopy to diagnose and dilate a narrow esophagus and to stop bleeding along the gastrointestinal tract.

Colonoscopy and sigmoidoscopy allow the gastroenterologist to visualize different portions of the colon. Colonoscopy encompasses the entire colon whereas sigmoidoscopy is limited to the last section of the colon and the rectum. Although sigmoidoscopies are quicker to perform, colonoscopies are preferred for cancer screening because they examine the entire colon. Both procedures require the patient to use a colon-cleansing product beforehand. During the procedure, the gastroenterologist examines the lining of the colon and removes polyps, or precancerous lesions.

Colorectal Cancer Screening. Colorectal cancer is the second leading cause of cancer death in the United States, according to the National Cancer Institute. Early detection through screening is an important tool in successfully treating the disease. Colonoscopy and sigmoidoscopy, along with tests measuring the presence of blood in the stool, are the tools used to screen for colon and rectal cancers. By examining the intestinal walls, the gastroenterologist detects and removes polyps, or precancerous lesions, and identifies abnormalities.

Pediatric Gastroenterology. Pediatric gastroenterology is an established subspecialty of pediatrics, whereas gastroenterology branches off from internal medicine. As such, pediatric gastroenterologists have a residency in pediatrics, and they take a separate subboard certification exam. Gastroenterologists in this subspecialty focus on gastrointestinal and nutrition-based conditions in infants, children, and adolescents. With additional training, pediatric gastroenterologists may narrow their specialization to liver transplantation, motility disorders, pancreatic diseases, endoscopic techniques for children, or nutrition.

All gastroenterologists form long-term relationships with their patients, but this is especially important for those specializing in pediatrics. Many pediatric gastric conditions are chronic and will follow the child throughout life. After the patient reaches adolescence, the pediatric gastroenterologist transfers care of the patient to a colleague; therefore, the pediatric specialist must have excellent communication skills. Pediatric gastroenterologists must consider the effect of a patient's condition on normal growth and development. Pediatric conditions often involve specialists from several disciplines, which means that the pediatric gastroenterologist must function well in teams.

Gastroesophageal Reflux Disease. The regurgitation of stomach acids into the esophagus, which causes a burning sensation in the chest known as heartburn, can lead to gastroesophageal reflux disease (GERD). Normally, a round muscle closes off the opening at the base of the esophagus leading to the stomach. Occasionally, the muscle fails to close completely, and the stomach contents wash up into the esophagus, burning the lining and causing the characteristic burning sensation. GERD may occur in adults or children.

GERD may be treated with over-the-counter antacids or other agents, but severe cases with persistent symptoms are best treated by a gastroenterologist. Upper endoscopy is useful for diagnosing GERD and evaluating treatment options. Prolonged GERD may damage the lining of the esophagus and cause bleeding ulcers.

Gastric and Peptic Ulcers. An erosion of the lining of the stomach is called a gastric ulcer; an erosion on the duodenum is called a peptic ulcer. They are caused by infection with the bacterium *Helicobacter pylori* or the use of certain medications, such as nonsteroidal anti-inflammatory drugs (NSAIDs) but are not believed to be caused by stress. Smoking may worsen ulcers. Gastroenterologists played an important role in discovering *H. pylori* as a cause of ulcers. Gastroenterologists diagnose ulcers by their symptoms, which include a dull or burning pain, weight loss, or vomiting, and a blood test for *H. pylori*. Upper endoscopy may be used if the patient has bleeding or if the ulcer blocks food from leaving the stomach. The gastroenterologist prescribes antibiotics and antacids to treat ulcers.

Parenteral Nutrition. Patients with serious medical conditions may be unable to eat or tolerate feedings introduced into the digestive tract. Parenteral nutrition—the injection of nutrients intramuscularly, intravenously, or subcutaneously—helps avoid complications of malnutrition. However, parenteral nutrition may be harmful in some instances, and it is up to the gastroenterologist to determine when the technique should be used. For example, parenteral nutrition has little effect in most patients following surgery, although it reduces postoperative complications in patients following surgery for esophageal or stomach cancer. Infants who are unable to eat and adults with prolonged malabsorption conditions benefit from parenteral nutrition.

Long-Term and End-of-Life Care. Patients in long-term care facilities and those with terminal illnesses receiving end-of life care often suffer from digestive difficulties. Patients may be unable to eat properly and become malnourished. Patients receiving opioid medications for pain relief may develop constipation as a side effect of the drugs. Opioid-induced constipation causes additional pain and discomfort to patients already suffering from end-stage illnesses. Gastroenterologists are skilled at finding alternatives to taking patients off the analgesic medications, which would relieve abdominal discomfort but would leave the patients in severe pain.

Food Allergies. Gastroenterologists may be helpful in determining specific food allergies or intolerances (for example, lactose intolerance) that lead to gastric disorders. Patients with food allergies or intolerances are often seen by primary care physicians, who consult with gastroenterologists and allergists to determine the cause of the symptoms. Gastrointestinal signs of food allergy include constipation, colic, or gastroesophageal reflux in infants, or a severe reaction immediately after eating the food.

Obesity. Gastroenterologists are uniquely positioned to assist overweight patients because many conditions that require gastroenterological care result from excess weight. The gastroenterologist faces the challenge of recommending a realistic diet and exercise plan that the patient will follow. After obtaining a detailed patient history that includes a psychiatric evaluation, the gastroenterologist must determine the amount of weight the patient must lose to achieve a healthy body mass index, balanced with the patient's level of commitment to changing lifestyle habits and losing weight.

Several treatment options are available to the gastroenterologist depending on the patient's present weight, the desired amount of weight loss, and the patient's outlook. Gastroenterologists can assist patients with dietary plans using portion-controlled servings of low-fat foods and in developing realistic exercise plans to increase physical activity and energy use. The gastroenterologist helps patients develop weight-loss goals and provides support and guidance. The doctor may prescribe medications such as sibutramine (Meridia) or orlistat (Xenical). If the patient has been unable to lose weight through diet and exercise, has no unusual surgical risks, and is at risk for obesity-related complications, such as heart disease or diabetes, the doctor may perform bariatric surgery.

Research. Gastroenterology is a fertile field for research. Gastroenterologists with an interest in research perform their own studies on the cause and diagnosis of gastrointestinal disorders. They also work with pharmaceutical companies developing medications for GERD and ulcers, Crohn's disease, ulcerative colitis, inflammatory bowel disease, and other disorders, including gastrointestinal cancers.

Gastroenterology is an active field that continues to expand. As the population ages, more people will require cancer screenings and polyp removal. Conditions such as ulcers, GERD, and hepatitis are diagnosed with increasing frequency. The general population is not only growing older but also heavier. Obesity is a growing health concern. Gastroenterologists are integral in managing complications of obesity such as gallstones and GERD, and they serve as important partners in teaching overweight patients proper nutrition and weight-loss strategies, including gastric banding procedures. In developing countries and regions with unsanitary conditions, gastroenterologists are needed to combat health concerns such as diarrhea and malnutrition. Gastroenterologists can provide basic nutritional counseling and assist patients in fighting parasitic or infectious diseases that compromise the absorption of nutrients.

Cheryl Pokalo Jones, BA

FURTHER READING

Butcher, Graham. *Gastroenterology: An Illustrated Colour Text*. Philadelphia: Elsevier Health Sciences, 2003. Full-color clinical photographs and detailed line drawings illustrate gastroenterological and liver diseases.

Collins, Paul. *Gastroenterology: Crash Course*. Philadelphia: Elsevier Health Sciences, 2008. Offers basic definitions and explanations of all aspects of gastroenterology in an easily understandable manner.

Grendell, James H., Scott L. Friedman, and Kenneth R. McQuaid. *Current Diagnosis and Treatment in Gastroenterology*. 2d ed. New York: Lange Medical Books, 2003. This comprehensive reference discusses all gastroenterological conditions, including hepatic, pancreatic, and biliary conditions.

Travis, Simon P. L., et al. *Gastroenterology*. 3d ed. Malden, Mass.: Blackwell, 2005. A concise, informative manual on gastroenterology with a global perspective.

WEB SITES

American College of Gastroenterology http://www.acg.gi.org

American Gastroenterological Association http://www.gastro.org

American Society of Gastrointestinal Endoscopy http://www.asge.org

National Institute of Diabetes and Digestive and Kidney Diseases National Digestive Diseases Information Clearinghouse http://digestive.niddk.nih.gov

See also: Geriatrics and Gerontology.

FASCINATING FACTS ABOUT GASTROENTEROLOGY

- As many as 70 million people in the United States are affected by a digestive disease, and about 9 percent of these people require hospitalization. An estimated 234,000 Americans die annually from digestive diseases, including gastric cancers.
- The United States has 11,704 certified gastroenterologists.
- Peptic ulcers affect 14.5 million Americans, resulting in more than 875,000 physician visits and 2 million prescriptions. Up to 20 percent of the population suffers from heartburn, or gastroesophageal reflux, at least once weekly.
- Colon polyps are small growths on the interior wall of the colon, some of which may become cancerous. Adenomas, or precancerous polyps, take about ten years to change from a polyp to a cancerous tumor.
- Women tend to have a higher incidence of constipation than men, perhaps because the colon empties slower in women than in men. Women also have less pressure in the anal sphincter compared with men, which enables men to withstand the urge to defecate longer than women.
- Colon cancer has few symptoms in the early stages, making screening important. Warning signs that may indicate colon cancer are blood in the stool, change in bowel habits (including formation of the stool), or abdominal pain.

GENE FLOW

FIELDS OF STUDY

Anthropology, ecology, genetics

SUMMARY

In biology, "migration" refers to the movement of a member or many members of a species of animal or plant from one geographic location to another. If other members of the same species interbreed with the migrating species, either along the way or in the new location, an exchange of genes occurs. Biologists call this exchange "gene flow" and consider it to be fundamental to the evolutionary process.

PRINCIPAL TERMS

- **allele:** one of a group of genes that occurs alternately at a given locus
- **deme:** a local population of closely related living organisms
- **fossil:** a remnant, impression, or trace of an animal or plant of a past geologic age that has been preserved in the earth's crust
- **gene pool:** the whole body of genes in an interbreeding population that includes each gene at a certain frequency in relation to other genes
- **mutation:** a relatively permanent change in hereditary material involving either a physical change in chromosome relations or a biochemical change in the codons that make up genes population: a grouping of interacting individuals of the same species
- **speciation:** the process whereby some members of a species become incapable of breeding with the majority and thus form a new species
- **species:** a category of biological classification ranking immediately below the genus or subgenus,

comprising related organisms or populations capable of interbreeding

Basic Principles

Prior to the nineteenth century, religious dogmatism retarded the activities of most scientists investigating the origins and nature of life by insisting on the immutability of species created by God. Despite mounting fossil evidence that many species of flora and fauna that once inhabited the earth had disappeared and that many extant species could not be found in the fossil record, pre-nineteenth century naturalists could find no viable explanation (other than divine intervention) for the disappearance of life-forms and their replacement by other forms. Then, in 1859, Charles Darwin published his epochal *On the Origin of Species*, which proposed the theory that all contemporary life-forms have evolved from simpler forms through a process he called "natural selection."

Many individuals before Darwin had proposed theories of evolution, but Darwin's became the first to be widely accepted by the scientific community. His success resulted from the careful and objective presentation of an overwhelming amount of evidence showing that species can and do change, and his concurrent promulgation of a convincing explanation of the mechanism that produces that change—natural selection. Since Darwin, scientists have modified and added new concepts to his theory, especially concerning the ways in which species change (evolve) over time. One of those new concepts, which was only dimly understood in Darwin's lifetime, is the importance of genetics in evolution, especially the concepts of migration and gene flow.

Genes and Gene Exchange

Genes are elements within the germ plasm of a living organism that control the transmission of a hereditary characteristic by specifying the structure of a particular protein or by controlling the function of other genetic material. Within any breeding population of a species, the exchange of genes is constant among its members, ensuring genetic homogeneity. If a new gene or combination of genes appears in the population, it is rapidly dispersed among all members of the population through inbreeding. New alleles may be introduced into the gene pool of a breeding population (thus contributing to the evolution of that species) in two ways: mutation and migration. Gene flow is integral to both processes. A mutation is the appearance of a new gene or the almost total alteration of an old one. The exact causes of mutations are not completely understood, but scientists have demonstrated that they can be caused by radiation. Mutations occur constantly in every generation of every species. Most of them, however, are either minor or detrimental to the survival of the individual and thus are of little consequence. A very few mutations may prove valuable to the survival of a species and are spread to all of its members by migration and gene flow.

When immigrants from one population interbreed with members of another, an exchange of genes between the populations ensues. If the exchange is recurrent, biologists call it "gene flow." In nature, gene flow occurs on a more or less regular basis between demes, geographically isolated populations, and even closely related species. Gene flow is more common among the adjacent demes of one species. The amount of migration between such demes is high, thus ensuring that their gene pools will be similar. This sort of gene flow contributes little to the evolutionary process, since it does little to alter gene frequencies or to contribute to variation within the species. Much more significant for the evolutionary process is gene flow between two populations of a species that have not interbred for a prolonged period of time.

Populations of a species separated by geographical barriers often develop very dissimilar gene combinations through the process of natural selection. In isolated populations, dissimilar alleles become fixed or are present in much different frequencies. When circumstances do permit gene flow to occur between two such populations, it results in the breakdown of gene complexes and the alteration of allele frequencies, thereby reducing genetic differences in both. The degree of this homogenization process depends on the continuation of interbreeding between members of the two populations over extended periods of time.

Hybridization

The migration of a few individuals from one breeding population to another may, in some instances, also be a significant source of genetic variation in the host population. Such migration becomes more important in the evolutionary process in direct proportion to the differences in gene frequencies—for example, the differences between distinct species. Biologists call interbreeding between members of separate species "hybridization." Hybridization usually does not lead to gene exchange or gene flow, because hybrids are not often well adapted for survival and because most are sterile. Nevertheless, hybrids are occasionally able to breed (and produce fertile offspring) with members of one or sometimes both the parent species, resulting in the exchange of a few genes or blocks of genes between two distinct species. Biologists refer to this process as "introgressive hybridization." Usually, few genes are exchanged between species in this process, and it might be more properly referred to as "gene trickle" rather than gene flow.

Introgressive hybridization may, however, add new genes and new gene combinations, or even whole chromosomes, to the genetic architecture of some species. It may thus play a role in the evolutionary process. Introgression requires the production of hybrids, a rare occurrence among highly differentiated animal species. Areas where hybridization takes place are known as contact zones or hybrid zones. These zones exist where populations overlap; in some cases of hybridization, the line between what constitutes different species and what constitutes different populations of the same species becomes difficult to draw. The significance of introgression and hybrid zones in the evolutionary process remains an area of some contention among life scientists. Biologists often explain, at least in part, the poorly understood phenomenon of speciation through migration and gene flow—or rather, by a lack thereof. If some members of a species become geographically isolated from the rest of the species, migration and gene flow cease. The isolated population will not share in any mutations, favorable or unfavorable, nor will any mutations that occur among its own members be transmitted to the general population of the species. Over long periods of time, this genetic isolation will result in the isolated population becoming so genetically different from the parent species that its members can no longer produce fertile progeny should one of them breed with a member of the parent population. The isolated members will have become a new species, and the differences between them and the parent species will continue to grow as more ages pass. Scientists, beginning with Darwin himself, have demonstrated that this sort of speciation has occurred on the various islands of the world's oceans and seas.

Studying Gene Flow

Scientists from many disciplines are currently studying migration and gene flow in a variety of ways. For decades, ornithologists and marine biologists have been placing identifying tags or markers on members of different species of birds, fishes, and marine mammals to determine the range of their migratory habits in order to understand the role of migration and subsequent gene flow in the biology of their subjects. These studies have led, and will continue to lead, to important discoveries. Most studies of migration and gene flow, however, relate to human beings.

Many of the important discoveries concerning the role of gene flow in the evolution of life come from the continuing study of the nature of genes. A gene, in cooperation with such molecules as transfer ribonucleic acid (tRNA) and related enzymes, controls the nature of an organism by specifying amino acid sequences in specific functional proteins. In recent decades, scientists have discovered that what they previously believed to be single pure enzymes are actually groups of closely related enzymes, which they have named "isoenzymes" or "isozymes." Current theory holds that isozymes can serve the needs of a cell or of an entire organism more efficiently and over a wider range of environmental extremes than can a single enzyme. Biologists theorize that isozymes developed through gene flow between populations from climatic extremes and enhance the possibility of adaptation among members of the species when the occasion arises. The combination and recombination of isozymes passed from parent to offspring are apparently determined by deoxyribonucleic acid (DNA). Investigation into the role of DNA in evolution is one of the most promising avenues to an understanding of the nature of life.

A classic example of the importance of understanding migration and gene flow in the animal kingdom is the spread of the so-called killer bees. In the 1950's, a species of ill-tempered African bee was accidentally released in South America. The African bees mated with the more docile wild bees in the area; through migration and gene flow, they transmitted their violent propensity to attack anything approaching their nests. As the African genes slowly migrated northward, they proved to be dominant.

Further research into migration and gene flow promises to provide information indispensable to the attempt to unravel the mysteries of life. Coupled with the concept of mutation, gene flow is a crucial component of evolution.

—*Paul Madden*

See also: Demographics; Evolution: Animal life; Evolution: Historical perspective; Genetics; Hardy-Weinberg law of genetic equilibrium; Natural selection.

FURTHER READING

Ammerman, A. J., and L. L. Cavalli-Sforza. *The Neolithic Transition and the Genetics of Populations in Europe.* Princeton, N.J.: Princeton University Press, 1984. This book is rather technical in its language, but it presents a thoughtful discussion of the influence of migration and gene flow on the Neolithic revolution and the complex workings of gene flow on human evolution.

Bailey, Jill. *Evolution and Genetics: The Molecules of Inheritance.* New York: Oxford University Press, 1995. An encyclopedia of the current understanding of genetics and evolution.

Cavalli-Sforza, Luigi Luca. *Genes, Peoples, and Languages.* Berkeley: University of California Press, 2000. Although this book focuses on humans and their languages, Cavalli-Sforza's introductory sections on genetics and gene flow are exceptionally clear and well written, accessible to nonspecialists, and applicable to all animal species.

Crow, J. F., and Motoo Kimura. *An Introduction to Population Genetics.* New York: Harper & Row, 1970. An excellent starting place for those whose knowledge of migration and gene flow and their influence on evolution is limited and who wish to learn more about the subject. The book is relatively free of the technical jargon that can make some biological texts difficult for the nonbiologist.

Endler, John A. *Geographic Variation, Speciation, and Clines.* Princeton, N.J.: Princeton University Press, 1977. Endler's book is valuable primarily because of an excellent chapter on gene flow and its influence on the evolutionary process. Endler sees evolution as a very slow and gradual process in which gene flow and small mutations cause massive change over long periods of time.

Hoffmann, Ary A., and Peter A. Parsons. *Evolutionary Genetics and Environmental Stress.* New York: Oxford University Press, 1991. Another excellent starting place for those whose knowledge of migration and gene flow and their influence on evolution is limited and who wish to learn more about the subject. The book is relatively free of the technical jargon that can make some biological texts difficult for the nonbiologist.

Raup, D. M., and D. Jablonski, eds. *Patterns and Processes in the History of Life.* New York: Springer-Verlag, 1986. Raup and Jablonski's book is a compilation of articles from the Dahlem workshop, "Patterns and Processes in the History of Life," held in Berlin in 1985. The articles in the book discuss the evidence concerning the role of migration and gene flow in the evolutionary process. Thought-provoking.

GENETICS

FIELDS OF STUDY

Biochemisty, cell biology, developmental biology, embryology, evolutionary science, reproduction science

SUMMARY

Genetics is the study of inheritance of characteristics fromone generation to the next. Humans have been studying genetics since prehistoric times with the first selective breeding of wolves for companion animals. In the 1800's, an Austrian monk, Gregor Mendel, described the basic laws that govern the inheritance of genetic traits. In the twentieth century, the field of molecular genetics was created as biologists determined the actual chemical makeup of genes.

PRINCIPAL TERMS

- **allele:** alternative forms of a single gene chromosome: a long strand of DNA with supporting proteins, that contains many genes
- **deoxyribonucleic acid (DNA):** the chemical polymer that is the genetic material of multicellular organisms
- **gene:** factors in cells that are responsible for an observable characteristic of an organism
- **genome:** all of the genetic material of an organism
- **genotype:** the actual genetic makeup of an organism
- **mutation:** any heritable change in the genetic material
- **phenotype:** the observable characteristics of an organism (for example, black fur color in a cat)

Basic Principles

Before any recorded history, ancient man chose alert pups from a litter of wolves for breeding. This practice of selectively breeding the wolves that were good companions eventually gave rise to the domesticated dog. The oldest undisputed dog bones known, excavated from a twenty-thousand-year-old Alaskan settlement, demonstrate that prehistoric humans knew that traits could be passed from one generation to the next, and that selectively breeding animals (or plants) could produce an organism that possessed desired characteristics. This practice of deliberate breeding is known as artificial selection.

Humans have practiced artificial selection on numerous animals, including pigs, cattle, goats, and sheep. Homer and other Greek poets wrote about selective breeding, and part of the wealth of the ancient city of Troy was attributed to its expertise in horse breeding. Although humans had some control over the traits of domesticated animals through selective breeding, the results of matings were not always predictable, and nothing was known about the mechanism through which traits were passed from one generation to another until the mid-1800's.

Mendelian Genetics

Gregor Mendel, an Austrian monk, is the undisputed father of the science of genetics. Working with garden peas, Mendel analyzed thousands of breeding experiments to describe laws that governed the inheritance of traits. Though Mendel studied a plant, his laws for the inheritance of traits apply to all sexually reproducing organisms, including humans.

Mendel chose seven distinct traits to study in his garden peas: flower color, plant height, seed shape, seed color, pod shape, pod color, and flower position. He concluded that each of these traits was determined by a single, discrete factor called a gene. For instance, there was a gene for flower color and a gene for seed shape. Each gene had several variations, or alleles. The gene for flower color had a white allele that produced white flowers and a purple allele that produced purple flowers.

Mendel's experiments revealed that organisms have two copies of any gene for a trait. Those two copies can be identical, two purple alleles of the flower color gene, for instance; or those two alleles can be different. A pea plant could have one purple allele of the flower color gene and one white allele of the flower color gene. When an organism has two identical alleles of a gene, it is homozygous for that gene. When an organism has two different alleles of a gene, it is heterozygous for that gene.

An organism inherits one allele, or copy of a gene, from one parent and one allele from the other parent. An organism, or cell, that has two copies of all of its genetic information is called diploid. In most sexually reproducing animals, the offspring are formed when a sperm cell from the male parent fertilizes an egg from the female parent. The sperm and the egg only contain half of all the genetic information. They are said to be haploid. However, the new organism they create is diploid because it gets one copy of the genetic information from the sperm and a second copy from the egg.

Mendel's first law, or the law of segregation, states that the two copies of each gene separate during the formation of gametes (eggs and sperm), and that fertilization of the egg by the sperm is a random event. Any sperm containing any allele of a gene can fertilize any egg of the same species, regardless of the allele carried by that egg.

Mendel noted that certain alleles seemed to dominate over others. For instance, when a plant had a purple allele for flower color and a white allele for flower color, the plant always had purple flowers. Mendel called the allele that was seen in the heterozygote, in this case the purple allele, the dominant allele. The allele that was hidden or masked, he called the recessive allele. In order to show a recessive allele, an organism has to have two identical copies of a gene, both containing the same recessive allele. This is known as the homozygous recessive condition. Garden peas that have white flowers are homozygous recessive for the white allele of the flower color gene.

Homozygous recessive describes the organism's genotype, or its genetic makeup. It has two copies of the recessive allele of the gene. The observable characteristic of the organism, having white flowers, is called its phenotype.

Mendel also demonstrated that the segregation of alleles of any one gene is not dependent on the segregation of alleles of any other gene. For instance, a gamete could receive a dominant allele for an eye color gene and a recessive allele for height, or that gamete could receive the recessive alleles for both genes or the dominant alleles of both genes. This is Mendel's second law, the law of independent assortment, and it applies to any genes that are located on separate chromosomes. Mendel's work was far ahead of its time. Although Mendel published his research in the 1800's, it was not until after his death that his work gained recognition in the scientific community. In 1900, three other scientists, each working separately on inheritance, came across Mendel's work in the course of their research. They gave him credit for his insights, and Mendel's research provided the foundation for the new discipline of genetics.

GENES AND CHROMOSOMES
Although Mendel described the gene as the factor that was responsible for a particular trait, nothing was known about the physical makeup of a gene. One of the first questions scientists needed to answer was where genes are found in cells. Early studies in frogs and sea urchins indicated that the nucleus of the sperm and the nucleus of the egg combined with each other during fertilization. This observation suggested that the genetic material that determined how the fertilized egg would develop might reside in the nucleus.

As microscopes improved, scientists were able to distinguish structures within the nuclei of cells. These long, threadlike structures stained blue and were called chromosomes (Greek *chroma*, "color"). Several scientists observed that when animal and plant cells divided, the chromosomes duplicated, then separated, and each daughter cell inherited a complete set of chromosomes. The one exception to this was the cell division that produced the gametes (eggs and sperm). When an egg or a sperm cell was produced, it only contained half the number of chromosomes as the cell that produced it. If genetic information was carried on chromosomes, scientists reasoned that a sperm and an egg could each contribute half of the genetic information to the new organism at fertilization.

Some of the first evidence that chromosomes were linked to observable traits came from the studies of American graduate student Walter S. Sutton. Sutton studied grasshoppers, and his observations indicated that male grasshoppers always had an X and a Y chromosome, whereas female grasshoppers contained two X chromosomes. Several other scientists observed similar things in other organisms, such as fruit flies, and concluded that the physical characteristic of sex was determined by the kind of chromosomes an organism possessed.

Since chromosomes determined the trait of sex, it was possible that chromosomes contained the genes that Mendel had shown to determine physical characteristics. The first scientist to demonstrate that genes were located on chromosomes was Thomas Hunt Morgan, who showed that an eye-color gene in the fruit fly, *Drosophila melanogaster*, was located on the X chromosome. Next, scientists wanted to know what kind of chemical molecule actually carried the genetic information.

Chromosomes contain two kinds of molecules, protein and a weak acid called deoxyribonucleic acid (DNA). Experiments in the early 1930's first demonstrated that DNA is the genetic material. Oswald Avery, Colin MacLeod, and Maclyn McCarty showed that adding DNA to these bacterial cells could change their physical traits. In their experiments, they mixed a harmless strain of bacteria with DNA from bacteria that caused disease in mice. When they did this, the previously harmless bacteria changed (or transformed) into disease-causing bacteria. Two other scientists, Alfred Hershey and Martha Chase, later obtained similar results by studying a virus that infects *E. coli*.

MOLECULAR GENETICS
By the 1940's, scientists knew that genetic information was carried by genes made of DNA molecules inside cell nuclei. However, scientists did not know how the genetic information was copied accurately from one generation to the next—from one cell division to the next. Nor did scientists know how the DNA could account for the appearance of inherited changes or mutations. In order to answer these questions, scientists needed to know the precise chemical structure of DNA.

Many scientists contributed to the understanding of the structure of DNA. Erwin Chargaff obtained data that indicated that specific molecular components of the DNA molecule were always present in equal parts. These components were nitrogen-containing molecules (or nitrogenous bases). Chargaff determined that the nitrogen-containing bases adenosine and thymine were always present in a one-to-one ratio, and the bases guanine and cytosine were always present in a one-to-one ratio, no matter what species' DNA was analyzed. Simultaneously, two scientists at Kings College in London, Rosalind Franklin and Maurice Wilkins, were attempting to make X-ray pictures of DNA molecules. Rosalind Franklin obtained an X-ray film that indicated that DNA was a helical molecule. Just previous to Franklin's work, an American chemist, Linus Pauling, had made a breakthrough in solving the structure of the protein alpha helix using a model-building approach.

Two scientists working at Cambridge University in England, James Watson and Francis Crick, decided to use Pauling's method of model building to attempt to solve the structure of the DNA molecule. Combining the data from a variety of sources including the data of Chargaff, Wilkins, and the crucial X-ray crystallography data of Rosalind Frankin, Watson and Crick solved the structure of the DNA molecule. Watson and Crick created a model of DNA: a double helix, like a twisted ladder. The DNA molecule was a long polymer of repeating nucleotides. Each nucleotide contained three chemical parts: a sugar, a phosphate group, and a nitrogen-containing base. The sides of the double helix ladder were formed by alternating sugars and phosphates, and the rungs were formed on the inside of the helix by specific pairings of the nitrogencontaining bases. Adenine paired with thymine to form one kind of rung. Guanine paired with cytosine to form a second kind of rung. The order of the bases provided the information within DNA. Certain combinations of bases could form "words" that stood for parts of protcins or other molecules encoded by the DNA. The double helix could unzip like a zipper, each strand serving as a template to guide the construction of a new strand. This provided an accurate means for copying the DNA molecules from a parent cell to a daughter cell.

Genetic Engineering

The details of how DNA is passed from one generation to the next, of how mutations arise, and of how the information of DNA is actually translated into the activities of cells forms the basis of genetic research at the beginning of the twenty-first century. One of the most important scientific discoveries that led to modern genetic technology was the discovery of a particular kind of protein, a restriction enzyme, from bacteria that cuts DNA molecules at specific sequences of bases. These restriction enzymes gave scientists the tool they needed to break DNA down into smaller pieces, eventually allowing the isolation of individual genes from the huge amount of DNA inside the nucleus of the cell.

Herbert Boyer and Stanley Cohen combined their knowledge of restriction enzymes and bacterial transformation (getting bacteria to take up DNA from the environment) to clone genes. Gene cloning involves isolating a gene of interest by using a restriction enzyme to cut it away from other DNA, and placing it in a piece of DNA called a vector that can be taken up by bacterial cells. One of the first applications of this technology was the production of human insulin. Scientists isolated the gene that encodes the information for making insulin from human DNA, cloned it into a bacterial vector, and placed the vector with the insulin gene in *E. coli*. The *E. coli* cells were able to produce large quantities of insulin. This new insulin was considerably cheaper and safer than insulin purified from human tissue.

Variations on this technique of taking a piece of DNA from one species and inserting it into the cells of another species are involved in genetic engineering of multicellular organisms. In multicellular organisms such as plants or monkeys, the DNA vector is usually a modified virus. These techniques are the basis of human gene therapy.

In the last decade of the twentieth century, entire organisms have been cloned. In Scotland, Ian Wilmut and colleagues reported the first mammalian cloning of a sheep named Dolly. In Wisconsin and Japan, scientists have cloned cattle. When an organism is cloned, all of its DNA, usually contained within an intact nucleus from a cell of the adult animal, is transferred to an egg cell from which all the genetic information has been removed. The egg is then allowed to develop into a new organism. Although the new organism is young, it has the same DNA as the parent from which the nucleus was obtained. Scientists have also developed techniques for sequencing DNA, determining the exact order and number of nitrogenous bases within the DNA of an organism's genome. In 2000, the Human Genome Project announced that the entire genome of the human had been sequenced. Many other genomes have been sequenced, including the roundworm, *C. elegans*, several plants, and even baker's yeast. The sequence of an organism gives scientists another tool in answering questions about how DNA regulates and determines the activities of cells.

The ethical consequences of genetic engineering are not clear. DNA forensic evidence is now used to convict or exonerate criminal suspects on a routine basis. The genetic engineering of food crops that are pest resistant or contain additional nutrients is fairly routine. With the cloning of entire organisms now possible, the cloning of a human is not science fiction. Parents can have an embryo tested for devastating genetic diseases before it is born. While many of these advances are clearly positive, many of them are double-edged swords, begging for informed public debate.

—*Michele Arduengo*

See also: Asexual reproduction; Cleavage, gastrulation, and neurulation; Copulation; Fertilization; Gametogenesis; Reproduction; Reproductive system of female mammals; Reproductive system of male mammals; Sexual development.

FURTHER READING

Hartwell, L., L. Hood, M. Goldberg, A. Reynolds, L. Silver, and R. Veres. *Genetics: From Genes to Genomes.* Boston: McGraw-Hill, 2000. Up-to-date, definitive genetic textbook covering all major fields of investigation in genetics: Mendelian, bacterial and viral, human, population, and genetic engineering.

Marshall, Elizabeth L. *The Human Genome Project: Cracking the Code Within Us.* New York: Franklin Watts. 1996. Discusses the goals and structure of the Human Genome Project and its implications. Contains a chapter on the contributions of animal model systems to this project.

Mousseau, T. A., B. Sinervo, J. A. Endler, eds. *Adaptive Genetic Variation in the Wild.* New York: Oxford University Press, 1999. Provides excellent background and current information for readers interested in population genetics and evolution in natural systems.

Sayre, Anne. *Rosalind Franklin and DNA.* New York: W.W. Norton, 1975. Details the invaluable contribution of Dr. Rosalind Franklin to the discovery of the structure of the DNA molecule.

Sponenberg, D. P. *Equine Color Genetics.* Ames: Iowa State University Press, 1996. A fascinating, in-depth look at the genetic mechanisms that cover coat color in horses and donkeys.

Watson, James D. *The Double Helix: A Personal Account of the Discovery of the Structure of DNA.* Edited by Gunther S. Stent. New York: W.W. Norton, 1980. Watson's personal story of the characters and events of the race to solve the structure of the double helix. This edition provides viewpoints and commentary from other players in the story as well as reprints of the original research articles.

ANIMAL MODEL GENETIC ORGANISMS

Roundworm (*Caenorhabditis elegans*): This millimeter-long worm allowed scientists to test the concepts of gene therapy, to develop methods for sequencing large amounts of DNA, and provided information about the biology of human diseases such as Alzheimer's disease and cancer. Research on this worm has also enabled scientists to develop effective control measures for plant and animal parasitic roundworms.

Fruit fly (*Drosophila melanogaster*): Studies in this organism allowed scientists to determine that genes reside on chromosomes, and gave insight into the nature of mutations. Studies of the development of complex structures such as the eye continue to provide insight into some of the ways cell specialization is regulated and directed by DNA.

Zebra fish (*Danio rerio*): The zebra fish, with its transparent embryos, provides an excellent system in which to study the genes that regulate vertebrate development. Additionally, zebra fish have been used in studies investigating bioaccumulation of organic compounds in the environment.

Common mouse (*Mus musculus*): Mice are useful for genetic study because of the availability of hundreds of single gene mutations. Studies in mice demonstrated that Gregor Mendel's laws of inheritance were as applicable to mammals as to plants. Transgenic genetic analysis of mice has allowed the creation of mouse strains that mimic human genetic diseases.

GERIATRICS AND GERONTOLOGY

FIELDS OF STUDY

Medicine; nursing; physical therapy; occupational therapy; psychology; sociology; political science; anthropology; public policy; education; statistics; dentistry; biology; pharmacy; social work.

SUMMARY

Examining life-span development, particularly the later years, has never been so important. Servicing an aging population requires diverse teams of medical doctors, biologists, psychologists, and other professionals, and training such individuals has produced dedicated schools and degrees focused exclusively on the aging process. Students interested in the study of the aged have never before had so many opportunities to expand their careers. Further, the need for teachers and specialists in geriatrics and gerontology is expanding at a rapid pace.

PRINCIPAL TERMS

- **activities of daily living:** basic chores of everyday life, such as eating and bathing, which are often used to determine level of independence.
- **Alzheimer's disease:** type of dementia that shrinks the brain, destroying memory and, in later stages, functional ability.
- **centenarian:** person who is one hundred years old or older.
- **chronic conditions:** ongoing diseases that account for the most deaths and disability in the united states, such as heart disease and cancer.
- **compression of morbidity:** belief that a lifetime of healthy activities produces a small period of disability at the end of life.
- **executive functions:** higher-order cognitive abilities including attention, decision making, planning, adapting to change, and inhibition.
- **preventable diseases:** leading causes of death in the united states—heart disease, stroke, cancer, rheumatoid arthritis, and diabetes—that are often the direct result of poor lifestyle choices.

BASIC PRINCIPLES

Before the first baby boomer turned sixty in 2006, geriatrics, the medical subspecialty of treating the aged, and gerontology, the comprehensive field of aging studies, both experienced a substantial growth of interest. Although old age might have been viewed as a period of disengagement, as of 2011 nothing is further from the truth. Aging is now often couched in terms of "successful" or "productive"; in fact, many seniors remain as busy after retirement with volunteering activities and the like as when they were employed. Reframing the negative language surrounding aging has changed the perception of aging from a period of consuming goods and services to a period of continued growth and productivity. With an estimated 69 million baby boomers turning sixty-five by 2029, the fields of geriatrics and gerontology have never been more in demand.

Maintaining the health and well-being of this graying section of society involves extensive education, research, and policy initiatives. Thus, the field of gerontology, by necessity, is interdisciplinary in nature. No one field encapsulates the varied systems of inquiry—especially when so many seniors now remain fit and active well into their nineties.

BACKGROUND

Gerontology as an official field of inquiry first began in 1903, when Russian biologist Élie Metchnikoff coined the term "gerontology" or the "study of old men." Six years later, Austrian physician Ignatz Nascher created the field of geriatrics and in 1914 published the first book on geriatrics. The Social Security Act, enacted in 1935, helped to pull millions of seniors, and others, out of poverty. The first organizations solely dedicated to the study of the aging process were formed in the 1940's: the American Geriatrics Society, founded in 1942, and the Gerontological Society of America, founded in 1945. In 1957, the National Institutes of Health formed the Center for Aging Research. In 1963, physician Sidney Katz and colleagues published the seminal work on gauging the independence of the elderly, the index of Activities of Daily Living—a concept still widely used. Irving Rosow published *Social Integration of the Aged* in 1967, which laid the groundwork for later theories on aging. The White House Conference on Aging, begun in 1961, spawned a formal division on aging at the National Institutes of Health. In 1976, Robert Butler was appointed the first director of that new National Institute on Aging. The 1980's through the 2000's saw an explosion in interest on aging as the baby boomers began to reach retirement age. Countless university centers on aging and degrees in geriatrics and gerontology have been created, and at many universities students may now declare gerontology as their major field of study.

THEORIES OF AGING

Scientific theories organize the how and why of empirical findings and provide a certain epistemology from which to examine new data. Using supported data-driven theory contextualizes known parameters, permitting a foundation of knowledge from which to build new projects and inquiries into the aging process. Previous data sets and explanations about interrelated phenomena streamline future interventions and research investigations. In geriatrics, biological theories of aging are the norm; however, the interdisciplinary nature of gerontology creates barriers to building comprehensive theories of aging. Although gerontology includes biological theories of aging, these two fields will be examined separately.

Geriatric theories on aging abound, and the more popular theories are briefly reviewed here. The free-eradical theory of aging posits that self-multiplying free radicals cause damage to deoxyribonucleic acid (DNA) and healthy cells. The hypothesized antidote is consumption of antioxidants, commonly found in vegetables. The programmed theory of aging proposes that every organism has an expected life span. Similarly, the wear-and-tear theory likens the human body to a machine where constant use degrades the machine, eventually leading to failure. Lastly, the immune-system theory holds that, with age, the human body becomes less able to fight off infection. In a world where health is increasingly viewed as more than the mere absence of disease, geriatricians use biological theories of aging to dispense health-promotion advice: Exercise both mind and body daily, consume the daily requirements of fruits and vegetables, and remain socially engaged in meaningful activities.

Gerontological theories of aging, often referred to as social theories of aging, are diverse in scope. Although largely refuted, disengagement theory suggests with increased age individuals slowly remove themselves from society. Alternatively and largely supported by early twenty-firstcentury research literature, activity theory holds a positive relationship between activity levels and happiness and health. Similarly, continuity theory maintains that new roles should be substituted for lost roles; for example, volunteering can replace retirement from a paying job. The life-course perspective on aging holds that aging is in fact a lifelong process and adaptation and change are continuous rather than enacted on by a specific age (sixty-five). These, and other social theories on aging, guide the development of new research endeavors and assimilate such new findings into the evolving language of aging research.

GERIATRICS IN THE FIELD

Geriatricians have advanced training and certifications in medicine and aging and work under the broad field of geriatrics. Such individuals actively develop health plans, treat comorbidities, promote health and wellness, focus on the prevention of disease and disability, and perform clinical evaluations. Unlike other disciplines, such as cardiology, geriatrics is not focused on a single organ or disease. Therefore other medical personnel with specialty training in aging often make up a treatment team; such a team can include osteopathic physicians, nurses, social workers, physical and occupational therapists, psychologists, and others. Such a team can create a complete wellness portfolio and assess a patient's activities of daily living.

Geriatricians have moved from a prevention of disease model to a health-promotion model that uses the best available research evidence, clinical knowledge, and patient feedback to create a holistic model of well-being. Such evidence-based approaches are now much in demand; with rising health care costs insurance companies demand proven treatments. Geriatrics as a subspecialty of medicine is especially important as older adults can often differ from younger persons in symptoms related to illness and react differently to treatment methods. The success of geriatrics is important for measures of public health as well, where reducing disability and disease would have far-reaching effects on the overburdened medical system.

Generally a person should consider seeing a geriatrician when he or she turns sixty-five, although individual health complications could necessitate an earlier visit. Often by the age of seventy-five, many older adults have multiple chronic conditions, such as sensory and cognitive impairments, that require the services of a geriatrician. Because the life span is seen as continuous, rather than demarcated by specific ages, it is important that individual decisions made throughout one's life are well-informed, and geriatricians are an important piece of the puzzle. Because the rate of aging is determined by the interaction of genetic and environmental conditions, which differ for every individual, it is important that aging seniors see a specialist. Only a qualified medical doctor should make medical decisions.

GERONTOLOGY IN THE FIELD

Applied gerontologists examine, study, and directly train the aging population and those who work with them in a variety of ways. Such areas include: learning to operate hearing aids, using assistive devices such as canes or walkers, maintaining proper nutrition, adjusting driving habits, proper use of corrective visual aids, and any other area affected by the aging process. Research by gerontologists suggests that change is the key to successful aging; static lives produce static minds and bodies.

Gerontologists also teach and promote preventative interventions to ensure successful aging. Such interventions can retrain mental acuity, strengthen ailing muscles and skeletal structure, teach positive behavioral methods to cope with loss and grieving. Gerontologists have led the aging revolution, where seniors are living longer, healthier, and more engaged lives. Accordingly, seniors are contributing to a level of human capital never before seen. For example, in 2010 Senior Corps, a federal governmental agency, saw its largest increase in senior volunteers since 2004. Part of the reason for this increase in productivity is because technology is changing the way productivity is perceived. No longer is physical stamina required for ongoing employment; technology has permitted older workers to stay in the workforce, thereby increasing social contribution and delaying age-related functional declines.

Gerontologists are quick to point out that productive aging includes activities outside of standard market contributions, such as volunteering. Such a revolution has caused many to rethink the very concept of old age. Often, age is a mixture of chronological age, biological age, psychological age, and sociological age. Functional age is a good marker of the aging process and aids in determining between three age conditions: normal aging, pathological aging, and successful aging. Another method of categorizing the diverse array of seniors is simply via chronological age. Gerontologists see three subcategories of seniors, young old (sixty-five through seventy-four), middle old (seventy-five through eighty-four) and old-old (eighty-five and older). Whatever the method of categorization, grouping the vast growing senior population is an arduous process given the inherent wide variability in human aging. Further, such grouping permits large-scale comparisons of health and wellness.

Unfortunately, the United States lags behind other countries, especially Great Britain, and the World Health Organization (WHO) recommendations on prevention and restraint of chronic diseases. The WHO's guide to *Global Age-Friendly Cities* provides eight guidelines for communities aiming for improvement: outdoor spaces, transportation, housing, social participation, respect, civic participation, communication and community support, and health services. In London and other European cities, free exercise playgrounds for the elderly are the norm. Costa Rica; Sardinia, Italy; and Ikaria, Greece, are examples of locales that possess the right mix of cultural and social factors that permit many seniors to live healthy lives into their nineties and beyond. Generally, such countries have a culture of respect for the aged. For example, those who study and treat the aging population, in Great Britain in particular, are held in high esteem. Conversely, the United States has historically stigmatized the aging and those who work with them. A slow tide of change is occurring as baby boomers prominently age in the American society, but changing preconceived notions is a slow process and stereotypes of the aged still abound.

Applications

Baby boomers possess a higher level of education than any previous generation; thus, the expectation is this cohort will be savvy consumers desiring the best proven treatments. Where daily life choices are more predictive of health status than genetic composition alone, the previous niche areas of Applications for the aging is now a rampant growth industry spanning every conceivable field.

Medicine. Medical implantation devices provide relief to ailing organs and prolong well-being. The left ventricle assist device aids the normal functioning of failing hearts while an implanted defibrillator prevents cardiac arrest by shocking the heart back into a normal rhythm. Cochlear implants are placed directly under the skin behind the ear and return the gift of sound to many hearing-impaired individuals. Cameras encapsulated in pill casings that the patient swallows take video of the intestinal track, eliminating the need for costly and invasive scoping procedures.

The increasing need for evidence-based medicine has produced some creative solutions to gathering information from patients. Wireless home-based transfer of medical information from accelerometers, glucose monitoring, and implanted devices, such as pacemakers, allows a patient to provide real-time health data while remaining independent. Cell phones with Global Positioning Systems (GPSs) track exercise regimens and allow for intermittent queries about self-perceived health and well-being. For example, patients newly released from the hospital can transmit responses to doctor-initiated questions eliminating the need and cost of in-person followups. This trend in distance-based medicine includes genetic-testing kits, available at local drug stores, that allow the user to mail in his or her sample for analysis. Although currently emerging technologies sound like science fiction, many of these products are closer to the marketplace than one might imagine. Thought-controlled mechanical limbs that receive feedback from the environment, for instance temperature and pressure, can closely mimic an individual's lost arm or leg. In development are microscopic cleaning robots, called blood bots, that can be guided to clean plaque-filled veins and take biopsies.

Noninvasive blood, saliva, and urine tests for Alzheimer's disease, cancer, and other difficult-to-detect diseases are currently in development. There is even an experimental Breathalyzer test in development that may replace expensive blood and urine analysis. Semipermanent prescription tattoos might be able to respond to glucose levels in a diabetic's bloodstream by changing color when placed under a handheld infrared light, eliminating the need for painful blood monitoring.

Common memory-storage cards, like those found in a digital camera, are being used as a portable patient medical archive. The cards fit into a wallet and facilitate communication and accuracy between the various health professionals many seniors visit. A computerized medical information system will likely soon replace inefficient paper-based records. Such an electronic system will permit comprehensive care while anywhere in the world and coordinate the spectrum of health care services seniors receive.

Pharmacology. Perhaps the field most engaged with the aging population is pharmacology. Clinical drug trials deliver numerous pharmaceuticals to the market each year—many designed to treat and extend wellness into advancing age. Drugs are being developed that may fight obesity and even change one's DNA. Drug encapsulation involves the coating of medicine either to delay activation or enter affected areas. For example, most oral medications are unable to pass the blood-brain barrier, which means they do not enter the brain. Encapsulating, or masking, the active drug compound could permit the body to pass the drug into the brain, eliminating the need for invasive surgeries.

Although, often with variable scientific evidence, herbs and supplements for the aged have expanded exponentially in the early twenty-first century. In 2011, a senior can take a pill that is purported to cure any ailment. However, geriatric researchers have found minimal scientific evidence for many of these claims. Ginkgo biloba was claimed to improve memory for many years; however, numerous large-scale clinical trials have found no such evidence. Large annual doses of vitamin D, thought to improve bone health, was found to increase fractures. Conversely, some herbs and supplements have proven effective. Omega-3 fatty acids have shown promise in improving heart health. Capsaicin, found in hot peppers, has recently been added to topical arthritis creams because of its analgesic properties. However, medicines work only when taken as directed. Medication non-adherence costs the health care industry millions of dollars per year. Patients frequently take the wrong dose or fail to fill the prescription, necessitating additional doctor visits. Accordingly, an industry has developed to correct this problem. Pill bottles with reminder alarms, automatic medication dispensers, Internet-based and cell-phone text reminders to assist with medication adherence are widely available.

Assistive Technology. Numerous home-based assistive devices are available to prolong independence: swing-down shelves in kitchens, easy-open door handles along with a bevy of structural changes to accommodate individuals with decreased strength, decreased stature, complications due to arthritis, and the like. Such assistive technologies have shown to decrease the need for outside personal assistance, further prolonging independence in the home. Motion-sensor systems eliminate wandering, which is often associated with later stages of dementia. Additional home-based applications for seniors include special bathtubs, mechanical chairs that climb stairs, motorized wheelchairs, and remote homemonitoring of health status.

Often advances in care for the aging have spillover benefits for the rest of the population. For example, the physical and occupational therapy fields have created user-friendly work environments for employees of all ages. Accordingly, ergonomics is now a household name with companies and therapists building optimal sitting and standing workstations that relieve pressure and support working and moving bodies. The automobile industry has responded with adjustable gas and brake pedals, backup cameras, audible turn-by-turn directions, and parallel-parking assistance.

Education. Lifelong learning colleges offer continuing education to seniors through a variety of formats. Online centers of learning, interactive CD and DVD training programs, and book and workbook training manuals abound that purport to increase memory, mental speed, and generally bolster one's brain power. Many traditional university and community colleges offer vacant classroom seats to seniors at a discount. Train the Trainer is an emerging public health program that trains seniors to educate other seniors, often in classroom-like settings. Select groups of seniors undergo an extensive program to learn the latest health-improvement information and how to deliver such information effectively. They each educate a room full of seniors, thereby increasing the scope of health-promotion programs.

Physical activity is the single best way to improve one's health at any age, and there are many products and services that promote an active lifestyle. The Nintendo Wii video game system has enjoyed popularity with seniors across the country. The Wii is a low-impact, hand-eye coordination system that is believed to increase balance, strength, and cognitive performance. Many senior centers have created Wii bowling leagues, and online goal-setting and health-improvement Web Sites permit seniors from across the globe to post their scores to encourage other seniors to maintain healthy lifestyles.

FUTURE PROSPECTS

In 2017, aging in America is viewed through the experience of the baby boomers—those individuals born between 1946 and 1964—who started turning sixty in 2006. The most educated of any senior cohort, baby boomers grew up during unprecedented economic growth and, accordingly, possess a unique view on the aging process. The boomers are not taking retirement lying down: This group is more redirected than retired. Traditional leisure activities, continuing education, volunteering, and often parttime employment now replace the full-time workweek. The baby boomers in particular expect an unprecedented retirement lifestyle. In response to this expectation, retirement communities now resemble theme parks with golf courses, activity centers, and staff solely dedicated to planning events. Future prospects have become reality. In central Florida, an entire city was developed for the retired. Specialized golf cart highways and parking spaces connect shopping malls and doctor offices to homes. Daily activities can include concerts, speeches, exercise facilities, and college classes. If such retirement cities become the norm, a complete redefining of aging will likely take place. Active and engaged theories of aging will continue to replace outmoded theories of aging such as the concept of disengagement. Whether the stigma of aging and corresponding stereotypes will also be replaced is yet to be determined. Biotechnology holds great promise for the future of aging research. Caloric-restrictive diets have extended longevity in mice, and such research has led investigators to explore the possibility of "turning off" mechanisms responsible for fat storage. Emerging research suggests B vitamins may decrease depression, phenolic compounds might kill certain cancer cells, a juice elixir may prevent the common cold, and capsaicin, mentioned earlier in connection with arthritis, may promote weight loss. Cortisone and prednisone, often found in anti-inflammatory medications, might have an unintended side effect of bone loss as recently reported in the journal *Cell Metabolism*. Perhaps the most visible biotechnology project involves using stem cells to regenerate aging cells.

Dana K. Bagwell, BS

FURTHER READING

Antonucci, Toni, and James Jackson, eds. *Annual Review of Gerontology and Geriatrics: Life-Course Perspectives on Late Life Health Inequalities*. New York: Springer, 2010. Yearly review of the vast field of aging studies. This volume's emphasis is on health disparities and aging.

Chodzko-Zajko, Wojtek, Arthur F. Kramer, and Leonard W. Poon, eds. *Enhancing Cognitive Functioning and Brain Plasticity*. Champaign, Ill.: Human Kinetics, 2009. A review of recent research supporting the notion that physical activity and brain exercises strengthen the brain. The fairly new concept of neural plasticity, that aging brains can grow and form new neural connections, is discussed extensively.

Halter, Jeffrey, et al. *Hazzard's Geriatric Medicine and Gerontology*. 6th ed. New York: McGraw-Hill, 2009. A compendium of evidence-based medicine and clinical applications for treating the aged. Includes 300 illustrations, numerous tables and figures, and additional online resources.

Palmore, Erdman B. *The Facts on Aging Quiz: A Handbook of Uses and Results*. New York: Springer, 1988. Contains several quizzes that test common misconceptions of the aging process.

Schaie, K. Warner, and Laura L. Carstensen, eds. *Social Structures, Aging, and Self-Regulation in the Elderly.* New York: Springer, 2006. Examines the evolution of personal and social roles in aging with particular emphasis on familial changes, immigration, and increased life span.

WEB SITES

AARP International http://www.aarpinternational.org
American Geriatrics Society http://www.americangeriatrics.org
American Psychological Association Division 20 (Aging) http://apadiv20.phhp.ufl.edu
Gerontological Society of America http://www.geron.org

FASCINATING FACTS ABOUT GERIATRICS AND GERONTOLOGY

- Globally, vast disparities exist in expected life span. The United States lags behind France, Sweden, Canada, and Japan in mean life expectancy. In some African countries people have a life expectancy of forty years.
- By 2050, 21 percent of the world's population will be sixty or older. Almost 1.5 million people are expected to be one hundred years old or older by 2050.
- In 2011, the first wave of baby boomers will reach age sixty-five and by this point boomers will be twice the population of the entire country of Canada.
- The aged do not suffer more crime than younger persons, although they are more fearful of crime.
- Many older adults enjoy a healthy sex life even into old age.
- Memory loss is not an inevitable part of the aging process. Most senior citizens do not get Alzheimer's disease.
- Men over fifty years old have the highest suicide rates; those eighty-five or older have the highest overall suicide rates.
- The majority of old people do not live alone.
- Religiosity does not increase with age.
- Although physical strength does decline, strengthtraining exercises maintain and improve flexibility and strength even into old age.

GROWTH

FIELDS OF STUDY

Cell biology, zoology

SUMMARY

Animal development begins with a single cell, the fertilized egg. Continued growth and formation through numerous cellular replications result in the formation of the complex animal body.

PRINCIPAL TERMS

- **differentiation:** the process during development by which cells obtain their unique structure and function
- **fertilization:** the union of two gametes (egg and sperm) to form a zygote
- **gamete:** a functional reproductive cell (egg or sperm) produced by the adult male or female
- **growth:** the increased body mass of an organism that results primarily from an increase in the number of body cells and secondarily from the increase in the size of individual cells
- **mitosis:** the process of cellular division in which the nuclear material, including the genes, is distributed equally to two identical daughter cells
- **zygote:** a fertilized egg

BASIC PRINCIPLES

Animal development has been a source of wonder for centuries. Development involves the slow, progressive changes that occur when a single cell—the zygote, or fertilized egg—undergoes mitosis. Mitosis is the process by which a cell divides into identical daughter cells. During development, mitosis occurs repeatedly, forming multiple generations of daughter cells. These cells increase in number and ultimately form all the cells in the body of a multicellular animal, such as a frog, mouse, or elephant. The simple experiment of opening fertile chicken eggs to observe the embryos on successive days of their three-week incubation period illustrates the process of embryonic development. A narrow band of cells can be seen increasing in number and complexity until the body of an entire, but immature, chick is seen.

Animal Growth and Development

An organism's growth occurs because of the increasing number of cells that form as well as because of the increasing size of individual cells. For example, a mouse increases from a single cell, the zygote, to about three billion cells during the period from fertilization to birth. Embryology is the study of the growth and development of an organism occurring before birth. Growth and development, however, continue after birth and throughout adulthood. Growth ceases only at death, when the life of the individual organism is ended. The bone marrow of human adults initiates the formation and development of millions of red blood cells every minute of life. About one gram of old skin cells is lost and replaced by new cells each day.

Development produces two major results: the formation of cellular diversity and the continuity of life. Cellular diversity, or differentiation, is the process that produces and organizes the numerous kinds of body cells. The first cell that determines an individual's unique identity, the zygote, ultimately gives rise to varying types of cells having diverse appearances and functions. Muscle cells, red blood cells, skin cells, neurons, osteocytes (bone cells), and liver cells are all examples of cells that have differentiated from a single zygote.

Reproduction

Morphogenesis is the process by which differentiated cells are organized into tissues and organs. The continued formation of new individual organisms is called reproduction. The major stages of animal development include fertilization, embryology, birth, youth, adulthood—when fertilization of the next generation occurs—and death. A new individual animal is begun by the process of fertilization, when the genetic material from the sperm, produced by the father, and the egg, produced by the mother, are merged into a single cell, the zygote. Fertilization may be external, occurring in freshwater or the sea, or internal, occurring within the female's reproductive tract. While fertilization marks the beginning of a new individual, it is not literally the beginning of life, since both the sperm and egg are already alive. Rather, fertilization ensures the continuation of life through the formation of new individuals. This guarantees that the species of the organism will continue to survive in the future.

Following fertilization, the newly formed zygote undergoes embryological development consisting of cleavage, gastrulation, and organogenesis. Cleavage is a period of rapid mitotic divisions with little individual cell growth. A ball of small cells, called the morula, forms. As mitosis continues, this ball of cells hollows in the middle, forming an internal cavity called the blastocoel. Gastrulation immediately follows cleavage. During gastrulation, individual cell growth as well as initial cell differentiation occur. During this time, three distinct types of cells are formed: an internal layer called the endoderm, a middle layer, the mesoderm, and an external layer, the ectoderm. These cell types, or germ-cell layers, are the parental cells of all future cells of the body.

Cells from the ectoderm form the cells of the nervous system and skin. The mesoderm forms the cells of muscle, bone, connective tissue, and blood. The endoderm forms cells that line the inside of the digestive tract as well as the liver, pancreas, lungs, and thyroid gland. The transformation of these single germ layers into functional organs is called organogenesis. Organogenesis is an extremely complex period of embryological development. During this time, specific cells interact and respond to one another to induce growth, movement, or further differentiation; this cell-to-cell interaction is called induction. Each induction event requires an inducing cell and a responding cell.

In the formation of the brain and spinal cord, selected cells from the ectoderm form a long, thickened plate at the midline of the developing embryo. Through changes in cell shape, the outer edges of this plate fold up and fuse with each other in the middle, forming a tubular structure (a neural tube). This tube-like structure then separates from the remaining ectoderm. At the head region of the embryo, the neural tube enlarges into pockets that ultimately form brain regions. For differentiation and development to occur, cells must be responsive to regulatory signals. Some of these signals originate within the responding cell; these signals are based in the genetic code found in the cell's own nucleus. Other signals originate outside the cell; they may include physical contact with overlying or underlying cells, specific signal molecules, such as hormones, from distant cells, or specialized structural molecules secreted by neighboring cells that map out the pathway along which a responding cell will migrate.

Postnatal Development

Embryological development climaxes in the formation of functional organs and body systems. This period is concluded by birth (or hatching, in the case of some animals). Following birth, development normally continues. In some animals, such as frogs, newly hatched individuals undergo metamorphosis during which their body structures are dramatically altered. Newly hatched frogs (tadpoles), for example, are transformed from aquatic, legless, fishlike creatures into mature adults with legs that allow them to move freely on land.

In mammals, development and growth occur primarily after birth, as the individual progresses through the stages of infancy, childhood, adoles- cence, and adulthood. Mature adulthood is attained when the individual can produce his or her own gametes and participate in mating behavior. Embryonic growth is especially impressive because the rate of cellular mitosis is so enormous. In the case of the mouse embryo, thirty-one cell generations occur during embryonic development. Thus, the zygote divides into two cells, then four, then eight, sixteen, thirty-two, and so on. This results in a newborn mouse consisting of billions of cells—produced in a period of only twenty-one days. When the newborn passes through its life stages to adulthood, its body cells may number more than sixty billion. One marine mammal, the blue whale, begins as a single zygote that is less than one millimeter in diameter and weighs only a small fraction of a gram. The resulting newborn whale (the calf) is about seven meters long and weighs two thousand kilograms: The embryonic growth represents a 200-millionfold increase in weight. Yet, for some animals, impressive growth periods also occur in the juvenile and adolescent stages of life.

In many cases, once an individual animal reaches its typical adult size, the rate of mitosis slows so that the number of new cells simply replaces the number of older, dying cells. At this maintenance stage, the individual no longer grows in overall size even though it continuously produces new cells. Since most of the cells in the mature adult have reached a final differentiated state, the function of mitosis is simply to replace the degenerating, aging cells. The slowing of the rate of cellular mitosis during this time may be attributable to the presence of specialized cell products called chalones. Chalones are thought to be local products of mature cells that inhibit further growth or mitosis.

Studying Growth

Historically, much study of animal development and growth was performed by simple observation. Aristotle, perhaps the first known embryologist, opened chick eggs during varying developmental periods. He observed and sketched what appeared to be the formation of the chick's body from a nondescript substance. With the invention of lenses and microscopes, growth and development could be studied on a cellular level. The concept of cellular differentiation arose, since investigators could see that embryonic muscle cells, for example, looked different from embryonic nerve cells. Again, much of the investigative information was descriptive in nature. Embryologists detailed the existence of the three germ-cell layers in gastrulation as well as the various tissues and primitive cells involved in organogenesis.

Experimentation as a method of investigating animal growth and development began during the nineteenth century. Lower animal species, such as the sea urchin and frog, were frequently investigated; their developmental patterns are simpler than those of mammals, their development occurs outside the maternal body, and they can be found in abundant numbers. Many of these experiments used separation or surgical techniques to isolate or regraft specific tissues or cells of interest. An attempt was made to determine how one tissue type would interact with and influence the development of another tissue type. Thus, the ideas of induction, in which some tissues affect other tissues, came into being. During this time, the descriptive and comparative observations resulting from these experimental manipulations were the major contributions of investigators.

The embryologists of the early twentieth century paid little attention to genetics. They believed that the major influences on development and growth were embryological mechanisms, although genes were thought to provide some nonessential peripheral functions. Chemical analyses of embryos attempted to establish the chemical basis for the cell-to-cell interactions that were seen during development and differentiation. During the middle portion of the twentieth century, geneticists began to investigate the role of the gene in cell function. The function of genes in the cellular manufacturing of specific proteins led to the hypothesis that each kind of cellular protein was the product of one gene. During this time, bacteria and fruit flies (*Drosophila*) were primary organisms of study because of their relatively simple genetic makeups.

In the latter part of the twentieth century, molecular biology techniques were applied to the study of development. Using techniques for transferring and replicating specific genes, researchers have greatly clarified the central importance of genes in development. Scientists came to believe that all the major developmental and differentiation influences that control cell growth are regulated through specific genes that are turned off or on.

Developmental Biology

The combination of molecular biology techniques with embryological investigations has led to a new field of study—developmental biology. New methods have been developed and used. Radioactive tracer technology has allowed the investigator to label particular genes or gene products and trace their movements and influences on cell growth through several generations. Recombinant deoxyribonucleic acid (DNA) technology has allowed the isolation and replication of significant genes that are important in development. Immunochemistry uses specific proteins (antibodies) to bind to differentiating cell products and quantify them. Cell-cell hybridization allows the introduction of specific genes into the nuclei of cells in alternate differentiation pathways.

Developmental biology, with its multidisciplinary approach, is solving many of the fundamental questions of development. As scientists become better able to understand the role of genetics and cell-to-cell interactions, they gain insight into the mechanisms that control cell growth and development. Consequently, the potential to control undesirable growth or to enhance underdeveloped growth is within reach.

The problem of cell aging is also under investigation. Questions about why mature cells stop dividing and growing and what the causes of aging are constitute important areas of developmental research. While various theories have been presented, the fundamental key to cellular aging remains to be discovered. One of the most challenging areas of continuing research is the determination of how developmental patterns guide evolutionary changes. Developmental principles may provide the answer to why evolution has given rise to animal diversity. In addition, developmental biology may give scientists the information needed to predict and determine future evolutionary trends. The individual animal is a growing organism that begins as a zygote and passes through the stages of embryonic development, birth, youth, adulthood, aging, and death. Preservation of the species depends on adult individuals' producing gametes that result in the formation of a future generation of zygotes and individuals. Remarkably, each zygote contains the necessary genetic instructions to regulate the orderly processes of growth and development. Thus, animal life continues from generation to generation.

—*Roman J. Miller*

See also: Cell types; Cleavage, gastrulation, and neurulation; Embryology; Fertilization; Gametogenesis; Multicellularity; Reproduction.

Further Reading

Alberts, Bruce, Dennis Bray, Julian Lewis, Martin Raff, Keith Roberts, and James D. Watson. *Molecular Biology of the Cell*. 3d ed. New York: Garland, 1994. This encyclopedic college text is greatly enhanced by its superior diagrams and illustrations. While all aspects of cell biology are covered in great detail, of special interest are the chapters on cellular mechanisms of development and differentiated cells and the maintenance of tissues. Highly recommended for the serious student who wishes to understand developmental processes on a cellular level.

Balinsky, B. I. *An Introduction to Embryology*. Philadelphia: Saunders College Publishing, 1981. While somewhat dated, this very well-written text covers the essentials of classical animal embryology while integrating with them the cellular and molecular mechanisms that regulate them. Of specific interest is the chapter on growth, which integrates this topic into the study of developmental biology.

Baserga, Renato. *The Biology of Cell Reproduction.* Cambridge, Mass.: Harvard University Press, 1985. This illustrated book can be read by the layperson who is interested in understanding more fully the role of cell division and reproduction in growth. The basics of the cell cycle are covered in detail, as are many of the influences that regulate it. In the latter portion of the book, the author describes the genetic mechanisms involved in growth and development.

Gilbert, Scott F. *Developmental Biology.* 6th ed. Sunderland, Mass.: Sinauer Associates, 2000. This is a college-level text written for the upper-level student who has a serious interest in development. The author's approach to development uses a historical experimental analysis of the progress of investigators in the research field. About half of the text deals with specific cell or genetic mechanisms that regulate growth and development. An excellent resource for advanced and accurate information regarding the field of development.

_____, ed. *A Conceptual History of Modern Embryology.* Baltimore: The Johns Hopkins University Press, 1994. A collection of essays concentrating on historical aspects of research on embryonic induction and the relationship between experimental embryology and genetics.

Gilbert, Scott F., and A. M. Raunio. *Embryology: Contructing the Organism.* Sunderland, Mass.: Sinauer Associates, 1997. A college-level textbook on comparative embryology. The first two chapters introduce general terms and concepts, while the following twenty chapters each concentrate on a different animal group.

Hartwell, Leland H., and Ted A. Weinert. "Checkpoints: Controls That Ensure the Order of Cell Cycle Events." *Science* 246 (November 3, 1989): 629-634. An excellent review article describing the biochemical controls of cell division and cell proliferation. Information is a bit technical and is written at the level of the more advanced reader who has some background in biology.

Wessels, Norman K., and Janet L. Hopson. *Biology.* New York: RandomHouse, 1988. The chapter on animal development gives an overview of classic embryology, including the role of growth. The chapter on developmental mechanisms and differentiation takes the reader through the major cellular and genetic regulatory mechanisms involved in development. This extremely well written introductory college text is easily read; it is well illustrated with color photographs and self-explanatory diagrams.

Hematology

FIELDS OF STUDY

Anatomy and physiology; biology; immunology; immunohematology; microscopy; chemistry; statistics; pathophysiology; microbiology; molecular biology.

SUMMARY

Hematology is the study of blood and blood-related disorders. It involves the diagnosis and treatment of a wide variety of diseases including anemia, leukemia, cancer, and clotting and bleeding disorders. As knowledge of the human genome has grown, many genetic mutations have been linked to hematological disorders. This has resulted in new methods of detection and novel treatments for diseases once thought incurable.

PRINCIPAL TERMS

- **anemia:** decrease in hemoglobin or the volume of red blood cells, resulting in less oxygen being delivered to body tissues.
- **automated cell counter:** instrument that uses electric impedance or optical light scatter to rapidly count blood cells.
- **flow cytometer:** instrument that can simultaneously measure multiple physical characteristics of a single cell as it flows in suspension through a measuring device.
- **hematopoiesis:** production, differentiation, and development of blood cells.
- **hematopoietic stem cell:** precursor cell that has the ability to replicate and proliferate into all the lymphoid and myeloid blood cell lines.
- **hemoglobin:** respiratory protein that is found in the cytoplasm of the red blood cells.
- **hemoglobinopathy:** genetic defect that results in either an amino acid substitution or diminished production of one of the protein chains of hemoglobin.
- **hemostasis:** process resulting in the balancing of bleeding and clotting within the circulatory system
- **leukemia:** disease characterized by the overproduction of immature or mature cells of various leukocyte types in the bone marrow or peripheral blood.
- **lymphoma:** disease characterized by malignant tumors of the lymph nodes and associated tissues or bone marrow.

BASIC PRINCIPLES

Hematology is the study of blood and its related disorders. The components of blood include the plasma (55 percent) and the formed elements, or cells (45 percent). Plasma is 91.5 percent water and 8.5 percent solutes, mainly the proteins, albumin, globulins, and fibrinogen. The cells can be subdivided into three types, erythrocytes (red blood cells), leukocytes (white blood cells), and platelets. The primary step in assessing hematologic function and disease processes is to examine the cellular elements, determining the percentage of each type of cell and its morphology.

Erythrocytes consist of a plasma membrane that surrounds a solution of proteins (mainly hemoglobin) and electrolytes. Mature red blood cells do not contain a nucleus. They are small, about 7-8 micrometers (μm) and shaped like a biconcave disk. Evaluation of red blood cells is important in the diagnosis and monitoring of anemia.

Platelets are small (1-4 μm) bits of cytoplasm that contain granules and no nucleus. Platelets are necessary for hemostasis. They have the ability to adhere, aggregate, and provide a surface for coagulation to occur. Platelets are important in bleeding and clotting disorders.

Leukocytes are white blood cells that are differentiated into five varieties, neutrophils, eosinophils, basophils, lymphocytes, and monocytes. Mature white blood cells contain a nucleus and often cytoplasmic granules. It is important to distinguish the different types of white blood cells and their relative amounts. Neutrophils have a segmented nucleus and are important in fighting infections. They also contain granules

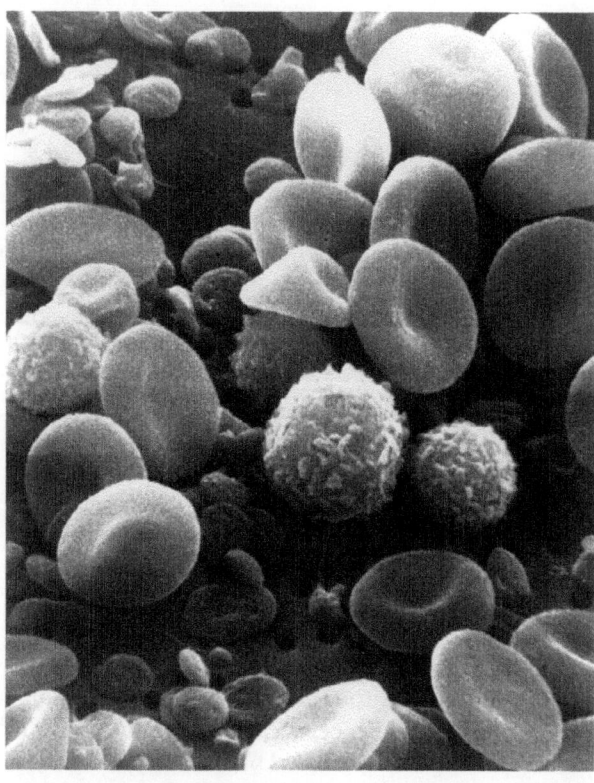

A scanning electron microscope image from normal circulating human blood. One can see red blood cells, several white blood cells including lymphocytes, a monocyte, a neutrophil, and many small disc-shaped platelets.

that secrete specific enzymes that aid in the killing of bacteria. Neutrophils play a key role in phagocytosis and inflammation. Eosinophils contain large granules that stain reddish orange. The number of eosinophils increases in cases of parasitic infections and some types of allergies. Basophils also contain granules; however, these stain a dark purple and contain the components heparin and histamine. Basophils play an important role in allergic reactions. Lymphocytes are the second most numerous white cells in the blood. They typically do not contain granules in the cytoplasm. Lymphocytes play an important role in immunity because of their ability to initiate an immune response and produce antibodies. Monocytes are the largest of the white blood cells. Often the monocyte's cytoplasm displays psuedopods. These cells are mobile and are important in fighting infections and removing foreign elements from the blood and tissues.

Hematopoiesis is the continual process of blood cell production and the development of the various cell lines. Blood cells are the progeny of a hematopoetic stem cell. Through the process of hematopoiesis, the body is supplied with ample blood cells of all lines. The hematopoetic stem cell can both replicate itself and differentiate into the mature cells found normally in the peripheral blood. The hematopoietic system includes the bone marrow, liver, spleen, lymph nodes, and thymus. Regulatory growth factors also play a role in hematopoiesis.

BACKGROUND

The field of hematology came into existence in 1642 when Antoni van Leeuwenhoek visualized cells using a microscope that he invented. It was not until 1842 that French microbiologist Alexandre Donne discovered platelets. In 1845, German physician Rudolf Virchow discovered an excess of white blood cells in a patient who had died from a condition he called leukemia (meaning white blood). A staining technique developed in 1877 by American scientist Paul Ehrlich allowed the visualization and differentiation of blood cells. French anatomist Louis-Charles Malassez invented the hemocytometer in 1874 and was able to quantitatively measure blood cells. The following year, French physician Georges Hayem developed a method for quantitatively measuring platelets using a hemocytometer. Many diseases of the blood and their treatments, cures, and causes were discovered in the twentieth century. In 1925, American pediatrician Thomas Cooley described Cooley's anemia (thalassemia major). In the 1970's, American geneticist Janet Davison Rowley demonstrated the translocation of chromosomes 8 and 21 in acute myelogenous leukemia and of chromosomes 9 and 22 in chronic myelogenous leukemia. In 1972, American immunologist and geneticist Leonard Herzenberg invented the fluorescence-activated cell sorter (FACS), which aided the study of cancer. Argentine immunologist César Milstein and German biologist Georges Köhler used hybridization to develop monoclonal antibodies in 1975. These were followed by many other discoveries and treatments, some of which were the result of knowledge about the human genome.

How It Works

Hematology diseases involve disorders of red blood cells, white blood cells, and platelets and hemostasis.

Disorders of Red Blood Cells. For the red blood cell to survive and function properly in the body, it must maintain a proper membrane, possess structurally correct and appropriately functioning hemoglobin, and have properly working metabolic pathways. Problems or defects in any of these areas will result in the red blood cell having a shortened life. Anemia results when the circulating red blood cells are unable to provide an adequate supply of oxygen to the tissues of the body. The many causes of anemia can be classified as nutritional deficiency (vitamin B12 or folic acid deficiency), blood loss, accelerated red cell destruction, hemoglobin defects (hereditary or acquired), and enzyme deficiencies.

Disorders of White Blood Cells. White blood cells perform a variety of functions in the body, including the destruction of bacteria, mediation of the inflammatory process, and production of antibodies or immunoglobulins.

White blood cell disorders include diseases that affect the number of cells (quantitative defects) and those that affect the functioning of the cells (qualitative defects). White blood cell disorders range from slight inflammation to acute leukemia. Leukemia is a malignant disease that involves the hematopoietic tissue. Abnormal cells can be found in both the bone marrow and the peripheral blood. Leukemia is classified as either chronic or acute and can affect any of the white blood cell types, red blood cells, or platelets. It is important to distinguish between the different types of leukemias to provide the proper and most effective treatment.

Disorders of Platelets and Hemostasis. Hemostasis is the process by which the body stops bleeding and maintains the fluid state of the blood. In other words, it is a balance between bleeding and clotting. It involves the platelets, blood vessels, and specialized coagulation proteins. Disorders of this system include qualitative and quantitative platelet disorders, such as idiopathic thrombocytopenic purpura (ITP) and von Willebrand's disease. Some disorders that involve problems with coagulation proteins include factor V Leiden mutation and hemophilia A and B. The interaction between the platelets, clotting factors, and blood vessels plays an immensely important role in the functioning of the cardiovascular system. The field of hematology has been integral in developing modern cardiovascular therapies, including heart catheterization and the use of stents.

Disseminated intravascular coagulation (DIC) is another hemostasis disorder. It is a complex disorder that involves the development of small clots within the blood vessels and the dissolution of these clots. Platelets and the clotting factors often are consumed during this disease process, and intensive therapy is necessary to resolve its occurrence.

APPLICATIONS

Automated Cell-Counting Instruments. Until the mid-1950's, cell counts were performed manually using a diluted fluid and a hemocytometer, and blood smears were viewed microscopically. Modern automated instruments can perform a complete blood count (CBC), which includes red blood cells, white blood cells, platelets, hemoglobin, hematocrit, and a five-part differential. This testing accounts for the primary hematology testing performed for almost every disorder. The instruments use the principles of impedance and optical light scattering to enumerate and determine the characteristics of each cell. Using the impedance principle, the cells are passed through an electrically charged aperture. Because blood cells are poor conductors of electricity, they create a resistance that can be measured as a pulse. The number of pulses is equivalent to the number of blood cells. The height of each pulse is equal to the cell's volume. Using the optical light scattering principle, each cell is passed through a beam of light (either optical or laser). Forward scatter and side scatter of the light is created by each cell. The forward scatter represents the size of the cell, and the side scatter represents the degree of complexity of the cells (cytoplasmic organelles can be assessed).

Flow Cytometry. Flow cytometry can detect molecules on the surface of a cell, making it possible to sort cells according to their surface composition. The molecule of interest is labeled using a fluorescent marker. Flow cytometry is used for immunophenotyping (identification of antigens on the cell surface), reticulocyte counting, and analysis of DNA. The information obtained about the subset of cells in flow cytometry is of critical use in the diagnosis, classification, and treatment of malignancies of mature lymphocytes, acute leukemia, and immunodeficiency disorders.

Cytogenic Analysis. In cytogenic analysis, cells are harvested and processed to visualize chromosomes in mitotically active cells. The cells are arrested in metaphase, and chromosome bands are visualized by various staining procedures. A similar technique, fluorescence in situ hybridization (FISH) may also be used. In this technique, fluorescent-labeled DNA probes are used to visualize specific chromosome centromeres (region on a chromosome joining two sister chromatids), whole arms, whole chromosomes, and individual genes. Chromosomes are visualized microscopically, and a karyotype can be constructed by using a video-computer-linked analysis system. Many chromosome aberrations are considered diagnostic or have significant prognostic implications for hematologic malignancies and solid tumors, such as chronic myelogenous leukemia, acute myelogenous leukemia, acute lymphoblastic leukemia, and lymphomas.

Chemotherapy. The goal of chemotherapy is to destroy all malignant cells within the bone marrow and allow the bone marrow to be repopulated with normal precursor cells. However, the drugs used for chemotherapy are not specific for leukemic cells, and many normal cells are also killed in the process. This results in the severe complications of bleeding, infections, and anemia.

Molecular-Targeted Therapy. Therapies that target specific genetic mutations can either silence the expression of a particular gene or reactivate a silenced gene. These types of therapies are better tolerated by patients than traditional chemotherapy and have been developed for chronic myelogenous leukemia and acute promyelocytic leukemia.

Bone Marrow Transplant. For a bone marrow transplant, drugs and radiation are used to first eradicate all leukemic cells that may be present. Then, bone marrow from a suitable closely matched donor is transplanted into the patient to provide a source of normal stem cells that can then repopulate the patient's bone marrow. Autologous bone marrow transplants are also an option. Some of the patient's bone marrow is removed while the patient is in remission. This specimen is treated to remove any residual leukemic cells and preserved. The patient is then treated to eradicate all leukemic cells and given back his or her own bone marrow. Many bone marrow transplants have been successful, and this procedure is being performed on an increasing basis.

Stem Cell Transplant. Stem cells can be found in either the bone marrow or the peripheral circulating blood. They also can be found in umbilical cord blood and fetal marrow and liver. Before stem cells can be harvested from a donor, mobilization of stem cells into the peripheral blood is stimulated by the use of cytokines. A process called apheresis is used to collect the stem cells from the donor's blood. These harvested stem cells are then injected into the patient. As with a bone marrow transplant, it is also possible to perform an autologous stem cell transplant. The transplanted patient will usually begin to produce new blood cells ten to twenty-one days after receiving the harvested stem cells.

FUTURE PROSPECTS

Never in the history of medical science have the opportunities for advancement in the field of hematology been greater. Many genes have been characterized as disease specific or disease related. This creates the possibility of finding cures or treatments for many diseases that have plagued humankind for centuries. With this newfound knowledge also comes the challenge of balancing the treatment of patients with new technologies and the ethical application of these treatments. Stem cell research has spurred some controversy as to how to ethically provide the stem cells needed for treatments. Another controversy arises when genetic testing for diseases or certain gene markers that are tested for as risk factors are used in the everyday practice of medicine.

Mary R. Muslow, MHS, MT(ASCP)SC

FURTHER READING

Carradice, Duncan, and Graham J. Lieschke. "Zebrafish in Hematology: Sushi or Science?" *Blood* 111, no. 7 (April, 2008): 3331-3342. Describes the potential of the use of zebrafish for the modeling of hematologic diseases.

Harmening, Denise M. *Clinical Hematology and Fundamentals of Hemostasis.* 5th ed. Philadelphia: F. A. Davis, 2009. Includes chapters on types of anemia, white blood cell disorders, hemostasis, and laboratory methods.

Herzenberg, Leonard A., and Leonore A. Herzenberg. "Genetics, FACS, Immunology, and Redox: A Tale of Two Lives Intertwined." *Annual Review of Immunology* 22 (2004) 1-31. The inventors of flow cytometry describe their lives and work.

McKenzie, Shirlyn B., and J. Lynne Williams. *Clinical Laboratory Hematology.* 2d ed. Upper Saddle River, N. J.: Pearson, 2010. Cover types of anemia, neoplastic hematologic disorders, hemostasis, and hematology procedures. Includes some excellent photomicrographs and illustrations.

Patlak, Margie. "Targeting Leukemia: From Bench to Bedside." *The Federation of American Societies for*

Experimental Biology 16, no. 3 (March, 2002): 273. An interesting review of leukemia diagnosis and treatment.

WEB SITES

American Society of Hematology http://www.hematology.org

American Society of Pediatric Hematology Oncology http://www.aspho.org

Hematology/Oncology Pharmacy Association http://www.hoparx.org

See also: Cardiology; Histology; Stem Cell Research and Technology.

FASCINATING FACTS ABOUT HEMATOLOGY

- A mature red blood cell does not contain a nucleus and therefore possess no DNA. Red blood cells serve the vital function of supplying oxygen to the rest of the body.
- The shortest-lived cell in the blood, the polymorphoneutrophil, circulates for only seventy-two hours. Other cells in the blood, such as the monocyte/ macrophages, can live for months or even years.
- Sickle cell anemia is a genetic defect that results in a single amino acid substitution in the protein chain of hemoglobin. Hemoglobin has four protein chains with more than 574 amino acids in a specific sequence.
- Iron-deficiency anemia is the most common anemia and affects an estimated 2 billion people worldwide, according to the World Heath Organization.
- Although leukemia can strike at any age, more than 50 percent of all cases of leukemia occur after the age of sixty-four.
- In 2010, an estimated 43,050 new cases of leukemia were diagnosed in the United States.
- Allogenic bone marrow transplants have been successful in curing chronic myelogenous leukemia. Long-term survival rates of as high as 78 percent have been reported. A majority of patients have successfully resumed their personal and professional lives.
- Hemophilia A is a bleeding disorder that affects men. This disorder was found in the Royal House of Stuart in Europe and Russia. Queen Victoria was later proved to be the carrier. The disorder, previously thought to be an absence of factor VIII, is actually a molecular defect in the factor that renders it unable to perform its clotting function.

HISTOLOGY

FIELDS OF STUDY

Microscopy; cell biology; embryology; anatomy and physiology; histopathology; histochemistry, immunohistochemistry; hematology, oncology, tissue transplantation.

SUMMARY

Histology is an interdisciplinary branch of science that focuses on the structure and function of normal and diseased tissues of the human body using microscopy and staining techniques. It studies comparative morphology of tissues, changes in tissues and organs during embryonic development, evolutionary changes of structure, and function of tissues in different species. In clinical medicine, histology is used as a diagnostic tool to understand and treat pathological developments in the body tissues. In forensics, histology is employed to understand the degenerative events in injured and dead tissues (autopsies) and to determine the cause of death in criminal investigations.

PRINCIPAL TERMS

- **autopsy:** medical test that involves the extraction of cells or tissues from a corpse to evaluate any disease or injury and determine a cause of death. it is usually performed by a specialized medical doctor called a pathologist.
- **biopsy:** medical test that involves the extraction of cells or living tissues for observation and analysis of structural and functional abnormalities by methods such as microscopy or histochemistry.

- **cytology:** science that focuses on the fine structure and function of cells.
- **embryology:** study of the development of an embryo from fertilization to the fetus stage.
- **hematology:** study of blood cell formation (hematopoiesis), the organs that form blood cells, and diseases of the blood.
- **histochemistry:** use of chemical reactions between specific laboratory reagents and components within tissue biopsy material for diagnostic purposes.
- **histology staining:** use of particular staining reagents to stain tissue sections for structural characterization of cells and tissues.
- **histopathology:** subdiscipline of histology that focuses on development of disease at a tissue level. it uses biopsies to diagnose and evaluate disease progression.
- **immunohistochemistry:** use of fluorescent-labeled antibodies on tissue sections to identify the location of particular structures (antigens) to which such antibodies would specifically bind.
- **microscopy:** study of objects too small to be seen with the naked eye, using a magnifying instrument called a microscope.
- **microtome:** device that slices ultra-thin sections of tissues that can be used for microscopic observation and analysis.
- **oncology:** branch of medicine that studies tumors (cancers).

BASIC PRINCIPLES

Histology studies the morphology of cells and tissues of the human body. It is sometimes called microscopic anatomy because it looks at the structure and function of the human body at the microscopic level. In the body, individual cells are organized in tissues, which form organs and organ systems to perform complex functions to maintain homeostasis. Although histology as a discipline is descriptive in many ways, it also involves a great deal of analysis because it focuses on structure and function relationships under normal and pathological conditions. The human body is made up of epithelial, connective, muscle, and nervous tissue types, but there are structural variations within each of these four groups. Changes in the body homeostasis are directly reflected in the changes—from temporary to irreversible— of the structure and function of tissues. Signs of infection, autoimmunity, aging, and malignant growths can be observed in various types of tissue by histological analysis of tissue probes. A trained histologist is capable of drawing a diagnostic conclusion based on such assessment alone or in combination with other methods.

Histology is based on the preparation of stained samples of tissue (from autopsies, biopsies, and cell cultures) and their microscopic analysis. The development of various staining techniques and tools for specimen preparation and microscopy have strengthened histology as a diagnostic approach in biomedical research and clinical medicine. The microscope evolved from a simple set of magnifying lenses to highly sophisticated optical (light), scanning, and electron microscopes that can use computer control to create high-resolution digital images for researchers, doctors, clinicians, and educators.

BACKGROUND

The word "histology" comes from the Greek *histos* (mast, or tissue) and *logie* (study) and means the study of tissues. Descriptions of tissues can be found in the works of the ancient Greek philosopher Aristotle, the eleventh-century Persian physician Avicenna, and the sixteenth-century German physician Andreas Vesalius, long before the microscope was invented. The English natural philosopher Robert Hooke was the first to introduce the term "cell" in 1665, while studying structures of a cork tissue using a magnifying device, a simple microscope. In 1674, Dutch scientist Antoni van Leeuwenhoek, using a microscope of his own invention, observed living organisms. He discovered that an animal cell is structurally organized and has a nucleus. The term "histology" was introduced by August Mayer in 1819, and in the nineteenth century, histology was established as an academic discipline. Studies of animal tissues were conducted by such prominent scientists as Marcello Malpighi, Camillo Golgi, Caspar Friedrich Wolff, and many others.

Histology developed much faster after around 1838, when microscopes with a greater magnification became available. Theodor Schwann developed the cell theory, which states that living organisms are made of similarly structured units, or cells, and that each individual cell exhibits characteristics of life. Rudolf Virchow, regarded as the father of modern pathology, contributed to the cell theory by stating that cells divide to produce daughter cells. In 1906, the Nobel Prize in Physiology or Medicine was awarded to Golgi for his outstanding work on the histology of the nervous system. Modern histology is a multidisciplinary branch of biomedical science. The results of histological observations contribute to the understanding of mechanisms of diseases and ways to treat diseases.

Preparation of Tissue Samples

Histology and histopathology examine tissue samples using microscopy; therefore, preparation of high-quality samples is an important step toward obtaining reliable results. For light microscopy, a sample of tissue is acquired, processed so that it can be thinly sliced or sectioned using a microtome, and stained with a specific dye. The tissue section is then analyzed using a microscope, and the results are interpreted, usually by comparison to normal tissue samples. A research scientist would examine how his experimental protocol affected the tissue, while a physician would analyze biopsied tissue to observe disease progression and determine the correct treatment. All steps of this process must be performed according to certain standards to avoid misinterpretation.

There are many methods of studying extracted tissues, and the choice of methodology depends on the type of tissue and the final objective. Because tissues degenerate quickly, it is important to preserve them as soon as possible, using a chemical process called fixation. The chemicals preserve the cells in the tissue without changing their structure. Individual protocols for fixation of different tissues vary and are the product of experimentation by many histologists. After fixation, tissue samples are treated to remove water so that they can be saturated with a waxy substance such as paraffin, which solidifies at room temperature. The paraffin-saturated tissue specimen is then sliced into sections using a microtome. The sections are mounted on a glass slide, treated to dissolve the paraffin, and stained with dye. The thickness of the tissue sections can vary from 0.002 to 0.02 millimeters (mm). Certain modifications to this process are made to prepare bone tissue, which is very hard because of its high mineral content, and adipose tissue, which has a high fat content. Staining is an important step, because stains have different chemical affinities for various cell and tissue components. For instance, the same kind of dye might stain the nucleus, cytoplasm, and membrane structures differently. These differing absorption rates make it possible to highlight particular structures of interest. The specimen must be thoroughly rinsed to remove any unbound stain and avoid nonspecific staining. Electron microscopy employs completely different techniques for preparation of specimens and allows observation of fine cell and tissue structures under very high magnification. A glass or diamond knife is used with an ultramicrotome to produce ultra-thin sections less than 0.001 mm thick. In some situations, extracted tissues can be frozen and sliced immediately, then stained and analyzed without lengthy tissue processing. Such tissue sections can deteriorate rapidly but are used to quickly analyze biopsies during surgery or in applications that do not require a detailed structural analysis.

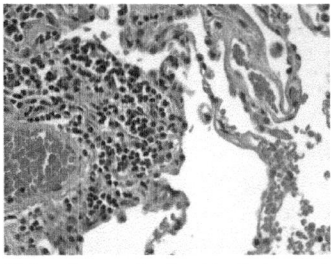

Light micrograph of a histologic specimen of human lung tissue stained with hematoxylin and eosin.

Microscopy

The most common tool in examining cells and cell structures in tissue samples is the microscope. Optical microscopy uses visible light, which passes thought the object (a tissue section mounted on a glass slide) and the lenses to create a magnified image. The maximum resolution that can be achieved with light microscopes is 0.2 micrometers (µm). Transmission electron microscopes send a beam of electrons through the sample and allow much higher resolution, around 0.05 nanometers (nm), but they require complex tissue processing and good technical skills to produce usable specimens on an ultramicrotome. Scanning electron microscopes produce a three-dimensional image of the cells and tissues. Electron microscopes are not suitable for routine or rapid observations of tissue samples.

Tissue Culture

Cell and tissue cultures can be grown on an artificial growth medium in a laboratory by providing adequate growing conditions in an incubator. Cultures are used to study living cells and tissues under experimental conditions. Tissue cultures are used to study tissue physiology, mechanisms of cell differentiation and development, cell-cell interaction, and gene regulation. Cell cultures are also used as a testing system for drug development. For example, fibroblasts, a cell type of connective tissue, grows well in the laboratory, providing a homogenous, live tissue material for drug testing.

Fluorescent Staining

Certain methods of tissue staining differentially detect the presence of particular chemical components in cells. One approach is to use a fluorescent label to trace components inside the cell. Fluorescent staining not only attests to the presence or absence of certain components in cells but also provides their precise localization in a particular cell or tissue type. For example, immunochemistry uses fluorescent antibodies that specifically bind with the target structures in the tissue so that the labeled antibodies will highlight the presence of the structures of interest in the analyzed specimen. Some applications of these techniques are staining of virus-infected cells in tissues and detection of immune-complex deposition in the kidneys and skin of patients with lupus.

Applications

Progress in histology has been dependent on the development of microscopy. Many companies are developing microscopes with various technical specifications because microscopes are being used not only in the biomedical field but also in fields such as engineering, optical physics, and biotechnology. Industry leaders such as Olympus and Carl Zeiss offer high-quality optics, superior construction, and highimage contrast to satisfy the growing demands of researchers. The companies develop and produce instruments, software, and accessories for microscope systems for use in industrial settings and clinical and research laboratories. Advances in technology have resulted in the creation of stereo microscopy, virtual microscopy, and total internal reflection fluorescence (TIRF) microscopy.

Fluorescence Microscopy. Improvements in fluorescence observation have resulted in images that are twice as bright as conventional fluorescence images. Fluorescence microscopy is used for staining and observation of tuberculosis bacteria and of tissues from pulmonary adenocarcinoma and breast cancer. The Carl Zeiss company, a member of the Stop Tuberculosis Initiative, provided the Primo Star iLED fluorescence microscope at a lower price to seventyfour countries with a high incidence of the disease. The microscope delivers up to four times faster detection of tuberculosis than can be achieved with traditional techniques.

NanoZoomer Digital Pathology is a system for scanning glass sides of tissue and rapidly converting them to high-resolution digital slides, known as virtual slides. By magnifying or shrinking a chosen area while changing the focus on a monitor screen, researchers can view intracellular structures in fine detail. This system of virtual microscopy can be employed for toxicology tests, gene-expression analysis, and protein localization. The great advantage of digital slides is that they eliminate the worry of the fluorescence fading and degrading the image quality.

Digital slide technology enables researchers to archive entire slides at high resolution, copy and edit them, and easily share them with others, allowing them to be used as the focal point for discussions. Stereo microscopy is used by cutting-edge biological and medical laboratories that require the most effective imaging and observation of a large quantity of live specimens. It achieves the world's highest zoom ratio of 1:16.4, enabling remarkably sharp three-dimensional imaging and considerably enhanced specimen manipulation.

Live microscopes, such as Olympus's VivaView fluorescence incubator microscope, achieve the dual objectives of culturing cells under monitored conditions and simultaneously observing the cultured cells under a microsope. These microscopes allows cell analysis in the most optimized environment.

Total internal reflection fluorescence (TIRF) microscopes are employed to study a diverse phenomena, including cell transport, signaling, replication, motility, adhesion, and migration; cell membranes and transport; the structure of ribonucleic acid (RNA); neurotransmitters; and virology. Industrial microscopy offers a wide range of both industrial microscopes and microimaging systems for metrology, semiconductor wafer manufacturing, quality assurance, advanced materials analysis, metallography, and other precision applications in various industrial sectors.

Sample Preparation. Isolation of single cells and cell groups, as well as biomolecules, from a heterogeneous tissue is crucial for preservation of the material. Therefore, the proper preparation and staining of high-purity samples is critical for production of reliable results. Laser microdissection with optical tweezers, a technology developed by Carl Zeiss, delivers absolute purity in sample isolation with maximum preservation of the material and without affecting the viability of live cells. The technology combines microdissection and advanced imaging and incorporates digital camera technology. Major achievements of this technology include instant viewing, editing, and sharing of images over a network from any remote location. It is employed for specimen capture and for isolation of

DNA, RNA, and proteins. Leica Microsystems was the first company to automate microtomy (microtome sectioning) and the first to introduce an integrated workstation for staining and cover slipping. The system does not require constant supervision, which provides an enormous time-saving advantage.

Clinical and Educational Applications. Histology is used in clinical, research, and educational settings. Biomedical research and clinical testing centers use fluorescent tags in staining to identify components of interest. For instance, flow cytometry analysis uses fluorescent-labeled antibodies as markers to perform rapid counts of blood cell types for clinical analysis. Hospitals, universities, and research centers provide training, education, and workshops on various protocols for specimen preparation as well as digital slide scanning for researchers and medical professionals. Histology is used in the manufacture of prepared tissue slides for biology, anatomy, and physiology courses at high schools and colleges.

Diagnosis of Diseases. Microscopic analysis of tissue samples is used to diagnose and classify many types of cancer, including cervical, breast, prostate, pancreatic, skin, and blood cancers, as well as noncancerous diseases. Histology tests can reveal if a tumor is cancerous, and if it is, they can determine the stage of progression, which leads to appropriate clinical treatment and an overall prognosis. Hematology analysis (of the blood cells) is usually the first step in the diagnostic process for cancer and noncancerous diseases.

FUTURE PROSPECTS

Histology is a field of continuous growth. Many areas of medicine—such as oncology and regenerative medicine—depend on and require knowledge of histology. Histological parameters such as tumor size and morphologic characteristics of tissues are the most important diagnostic factors for cancer. Tissue engineering, a subspecialty of regenerative medicine, deals with repairing and replacing tissues (such as skin, bone, cartilage, or blood vessels) in part or in their entirety. Examples of tissue engineering include artificial bladders and edible artificial animal muscle tissue (artificial meat). Many materials, both natural and synthetic (including carbon nanotubes), are being actively investigated for use in tissue engineering. Scientists are also pursuing stem cell development and manipulation for tissue regeneration. Interest in this area has expanded, leading to the establishment of the California Institute for Regenerative Medicine in 2004.

Elvira R. Eivazova, PhD

FURTHER READING

Croft, William J. *Under the Microscope: A Brief History of Microscopy.* Hackensack, N.J.: World Scientific, 2006. Traces the microscope from early beginnings to modern instruments, discussing how each works.

Hewitson, Tim D., and Ian A. Darby, eds. *Histology Protocols.* New York: Humana Press, 2010. This laboratory manual looks at tissue preparation and staining, with explanations of complex procedures.

Ovalle, William K., and Patrick C. Nahirney. *Netter's Essential Histology.* Philadelphia: Saunders/Elsevier, 2008. This atlas covers cells and tissues and the major bodily systems. Features a great collection of images by Frank H. Netter.

Ross, Michael H., and Pawlina Wojciech. *Histology: A Text and Atlas—With Correlated Cell and Molecular Biology.* 6th ed. Philadelphia: Wolters Kluwer/Lippincott Williams & Wilkins Health, 2011. Contains great illustrations, easy to follow diagrams. Recommended for medical, health professions, and undergraduate biology students

Tortora, Gerard J., and Bryan Derrickson. *Principles of Anatomy and Physiology.* 12th ed. Hoboken, N.J.: John Wiley & Sons, 2009. Excellent anatomy and physiology textbook with great illustrations and clinical references.

WEB SITES

American Society for Clinical Pathology http://www.ascp.org

California Institute for Regenerative Medicine http://www.cirm.ca.gov

Loyola University Medical Education Network Histology http://www.lumen.luc.edu/lumen/MedEd/Histo/frames/histo_frames.html

National Society for Histotechnology http://www.nsh.org

Olympus America Microscopy Resource Center http://www.olympusmicro.com

See also: Artificial Organs; Bionics and Biomedical Engineering; Biosynthetics; Cell and Tissue Engineering; Hematology; Pathology; Stem Cell Research and Technology.

FASCINATING FACTS ABOUT HISTOLOGY

- Antoni van Leeuwenhoek was the first person to observe human red blood cells. The preserved tissue samples that he used are part of the collections of the Royal Society of London. Leeuwenhoek was not a trained scientist but a merchant who made microscopes for a hobby.
- Olympus's VivaView fluorescence incubator microscope allows researchers to observe cultured cells growing in real time. This combination of a microscope and an incubator results in high-quality live-cell imaging with the freedom to control growth conditions of cells and use the cells for a significantly longer time than was previously possible.
- A transmission electron microscope allows researchers to see parts of individual human chromosomes, single proteins, and DNA.
- The life span of red blood cells is between 100 and 120 days, after which the cells undergo apoptosis (programmed cell death).
- There are about 100 trillion cells in the human body.
- A Pap smear (also known as a Papanicolaou test) involves microscopic examination of cells scraped from stratified squamous epithelium of the vagina and cervix of the uterus. It is performed for early detection of a precancerous condition or cancer. Collected cells are smeared on a microscope slide and sent to a laboratory for analysis.
- Tissue engineering is a technology in which living tissues are combined with synthetic materials to grow new tissues in the laboratory. Laboratory-grown versions of skin, cartilage, and bone are developed by using biodegradable synthetic materials or natural collagen fibers as a scaffolding system to immobilize and grow cells. The new tissue is then implanted into the patient.

HOMEOSTASIS

FIELD OF STUDY
Biology

SUMMARY

Homeostasis means the efforts of a physiological system in humans to maintain internal stability or metabolic equilibrium. All parts of the system coordinate to react appropriately to any situation or stimulus, whether coming from outside the body or from another body system, to keep the normal condition or function running smoothly by adjusting physiological processes. This keeps the body healthy and functioning despite any deviation in environment. The suffix "stasis" comes from the Greek "standing still."

PRINCIPAL TERM

- **metabolism:** Physical processes and chemical reactions within a living body that convert or use energy, including those associated with digestion, excretion, breathing, blood circulation, growing and using muscle tissues, communication through the nervous system, and body temperature. See also enzyme, protein.

BASIC PRINCIPLES

The human body actively tries to steadily maintain the conditions it needs to survive. All body systems, including circulatory, lymphatic, nervous, endocrine, respiratory, digestive, and urinary, participate in this process. This includes processes such as maintaining steady levels of elements needed in bodily fluids, such as blood, to keep the body running smoothly. Some of these elements are water, salts, sugars, proteins, fats, calcium, and oxygen. Without steady levels of these nutrients, the body cannot survive, so it uses its metabolic processes to maintain homeostasis.

Homeostasis is automatic if bodily systems are functioning correctly. The hypothalamus is involved in many of the processes used to maintain homeostasis because of its control over the feedback processes through the medulla oblongata, the pituitary gland, the thyroid, and the adrenal cortex.

It may take many bodily systems functioning together, feeding off the actions of one another, to maintain this equilibrium. This is called a negative feedback loop. In this sense, "positive" and "negative" do not mean "good" and "bad"; rather, they mean something more like "continuing" and "opposite." "Negative" in this sense means that a body system responds to the feedback by doing something "opposite" to correct the problem. "Positive" feedback loops respond to a change to take the body system even further down the road away from homeostasis. For example, if a person runs a fever of 107 Fahrenheit or above, the body stops responding to the negative feedback loop; it does not sweat to help cool down, and the person's skin remains hot and dry. Rather, a positive feedback loop kicks in, taking the body further and further away from homeostasis. The increased temperature speeds up body chemistry, which increases temperature, which speeds up body chemistry even more, creating a vicious cycle that must be stopped or must end in death.

The body uses both short-term and long-term ways to maintain homeostasis. For example, in attempting to keep body temperature stable, it uses short-term measures such as shivering or sweating. If the body temperature becomes unstable over the long term, the body will use measures such as increasing thyroxin to raise the metabolic rate, thus raising the body temperature consistently for a longer time. This may lead to other positive or negative feedback loops in other body systems.

EXAMPLES OF HOMEOSTASIS

Maintaining Body Temperature: In warm-blooded animals, the body attempts to maintain a steady temperature even in extreme heat or cold. In extreme cold, the body shivers to attempt to produce heat. In extreme heat, the body sweats to make more moisture on the skin available to evaporate, thus reducing temperature.

Stomach pH: The stomach must maintain a pH level that is different from the organs that surround it. It does so with the constant adjustment of acid and base levels through feedback loops.

Blood Sugar Levels: The body must maintain a balance between insulin, which decreases the concentration of glucose in the blood, and glucagon, which increases the concentration of glucose in the blood. If the constant push and pull (the balance between negative and positive feedback loops) between these substances is upset, the person develops diabetes.

Fluid Volume: The body must maintain the correct fluid level to keep its internal environment constant. If it doesn't have enough fluid, it signals the kidneys to decrease urine output. If it has too much fluid, the kidneys receive the signal to excrete more fluid through urine output.

Marianne Moss Madsen, MS

FURTHER READING

Chiras, Daniel. *Human Biology: Health, Homeostasis, and the Environment.* Jones and Bartlett, 2002. Discusses basic human biology using the perspective of homeostasis to explain behaviors in response to environment.

De Luca, Laurival Antonio, Jr., Jose Vanderlei Menani, Alan Kim Johnson, eds. *Neurobiology of Body Fluid Homeostasis: Transduction and Integration.* CRC Press, 2013. A more advanced discussion with examples of homeostasis from the fields of integrative neurobiology and regulatory physiology using animal and human models.

Janig, Wilfrid. *Integrative Action of the Autonomic Nervous System: Neurobiology of Homeostasis.* Cambridge University Press, 2006. Detailed descriptions of how the autonomic nervous system attempts to maintain homeostasis to keep the body running smoothly; covers neurobiological concepts from the cellular and integrative organization aspects.

Meeks, Robert. *Biology: Science of Life, Cell Theory, Evolution, Genetics, Homeostasis and Energy.* CreateSpace Independent Publishing Platform, 2016. A simple overview of many biological concepts, including homeostasis.

Princeton Review. *High School Biology Unlocked: Your Key to Understanding and Mastering Complex Biology Concepts.* Princeton Review, 2016. Straightforward and easy-to-understand lessons on biology concepts; includes practice questions with complete explanation of answers to enhance understanding of concepts such as homeostasis.

WEB SITES

Bright Hub, Science, Medical Science, Anatomy, 5 Common Examples of Homeostasis in the Human Body, http://www.brighthub.com/science/medical/articles/112024.aspx.

Khan Academy, Biology, Principles of Physiology, Homeostasis, https://www.khanacademy.org/science/biology/principles-of-physiology/body-structure-and-homeostasis/a/homeostasis.

FASCINATING FACTS ABOUT HOMEOSTASIS

- In 1930, Walter Cannon was attempting to find a word to convey the idea that the body maintains its temperature despite outside influences. He coined the term "homeostasis" in his book *The Wisdom of the Body.*

- "Homeostasis" is a term also used in social science fields to refer to how a person can maintain a stable psychological condition or how a society maintains stability, such as shown in the law of supply and demand.

McGraw Hill Higher Education, Body Systems and Homeostasis, http://www.mhhe.com/biosci/genbio/maderbiology/supp/homeo.html.

HOMINIDS

FIELDS OF STUDY

Anatomy, anthropology, evolutionary science, human origins, systematics (taxonomy)

SUMMARY

Though understanding of human ancestry is rudimentary at best, an astonishing series of discoveries since the nineteenth century has created a lively field of knowledge where there was none before.

PRINCIPAL TERMS

- **apes:** large, tailless, semierect anthropoid primates, including chimpanzees, gorillas, gibbons, and orangutans, and their direct ancestors—but excluding man and his ancestors
- **australopithecines:** nonhuman hominids, commonly regarded as ancestral to present-day humans
- **dryopithecines:** extinct Miocene-Pliocene apes (sometimes including *Proconsul,* from Africa) found in Europe and Asia; their evolutionary significance is unclear
- **humans:** hominids of the genus *Homo,* whether *Homo sapiens sapiens* (to which all varieties of modern man belong), earlier forms of *Homo sapiens,* or such presumably related types as *Homo erectus* and the still earlier (and more problematic) *Homo habilis*
- **primates:** placental mammals, primarily arboreal, whether anthropoid (humans, apes, and monkeys) or prosimian (lemurs, lorises, and tarsiers)

BASIC PRINCIPLES

The idea that humankind might be significantly older than the six thousand years previously allotted by biblical scholars, who tried to calculate the generations of man since Adam, was not widely maintained until 1859, when human stone tools and the bones of extinct animals were found lying close to each other in France. Charles Darwin's *On the Origin of Species* appeared on November 1 of the same year, but it suggested only that "light will be thrownon the origin ofmanand his history" by the theory of evolution he had just proposed.

A series of important and widely noticed books then followed, including J. Boucher de Perthes' *De l'homme antédiluvien et de ses oeuvres* (1860; of antediluvian man and his works); Thomas Henry Huxley's *Evidence as to Man's Place in Nature* (1863); and Charles Lyell's *Antiquity of Man* (1863), with others later by John Lubbock, James Geikie, and W. Boyd Dawkins. In *The Descent of Man* (1871), Darwin sagely hypothesized that the human line evolved in Africa (not Asia, as had previously been assumed) from a long-tailed, probably arboreal, ancestor. Yet, in Darwin's time only two fossil apes were known at all, together with some controversial bones of a creature known as Neanderthal man. Extinct species such as the australopithecines and *Homo erectus* had not been discovered.

HOW IT WORKS
Humans and Other Primates

The scientific name for mankind is *Homo sapiens sapiens* (wise man); the taxonomic family is the

Hominidae, which also includes chimpanzees and gorillas. This classification is reasonable because, bone for bone, their skeletons are almost identical to human skeletons. On other evidence as well, man and his cousins appear to be remarkably alike. The protein sequences in chimpanzee and human hemoglobin, for example, are identical; there are only two chemical differences between gorilla and human hemoglobin. Between humans and all other animals, there are more than two. Strands of deoxyribonucleic acid (DNA) from chimpanzees and humans, moreover, are 99 percent identical. Chimpanzees, finally, are second only to humans in intelligence; their brains closely resemble those of humans. These remarkable similarities attest a common ancestry for all the Hominidae, as the classification itself would imply, and a fairly recent differentiation among its members. Some biochemists have argued that man, the chimpanzee, and the gorilla shared a common ancestor nomore than six or eight million years ago.

Whatever the timing may have been, it is almost universally accepted that the link between humans and their protosimian ancestors was a now extinct genus of ape-men, the australopithecines (southern apes). The first of these, called, originally, the Taung child, was discovered in South Africa by Robert Dart in 1924. Its identity and significance remained controversial until 1936, when further discoveries by Robert Bloom convinced skeptical professionals and the public. Today, *Australopithecus* and *Homo* (man) are often grouped together as Hominini, as opposed to the Pongidae (apes), to which the chimpanzee and gorilla belong.

The Australophithecines

The australopithecines arose at least 4 million years ago, probably from dryopithecine ancestors. They lasted until two million years ago, evolving into a series of species. Of these, *Australopithecus afarensis* (found in the Afar region of Ethiopia) was the oldest and smallest. Males stood no taller than four feet, and the females were smaller. Most significantly, however, *afarensis* was fully bipedal; unlike the apes, it walked upright, with increasingly specialized hands, and with legs a bit longer than arms. Chimpanzee-like hips, together with curved toe and finger bones, suggest that it was essentially a tree-dweller living on fruits and seeds. A remarkably complete skeleton of *afarensis*, familiarly called Lucy, was found in 1974; it is about three million years old and is the oldest hominine skeleton yet found. *Afarensis* died out around 2.5 million years ago; it is thought to be ancestral to the later australopithecines and to modern humans.

Australopithecus africanus, deriving from Africa some three to one million years ago, probably evolved from *afarensis*. Most specimens come from Sterkfontein in South Africa, though others have been found in Ethiopia, Kenya, and Tanzania. This species was about the same size as *afarensis*, but it had a less apelike face. The arms were proportionately longer than a modern human's, yet shorter than those of *afarensis*; hands and teeth show similar "modernization." Dart's Taung child was the first (and is still the most famous) example of this species.

Australopithecus robustus (robust southern ape), found only in South African caves thus far, was once thought identical to *Australopithecus boisei*; in older literature, it was also known as *Paranthropus* (past man). Larger and more strongly built than *africanus*, *robustus* was more than a foot taller and had a larger brain. His teeth indicate that *robustus* was a plant eater. There is also a remarkable specimen (discovered by C. K. Brain at Swartkrans) of a child's skullcap in which the imprint of a leopard's lower canines can be seen. Since an exact-fit leopard's jaw was found nearby, it is assumed that the leopard killed the child and was then itself killed by an adult *robustus* armed with some kind of weapon. (A diorama at the Transvaal Museum, Pretoria, reconstructs this hypothetical incident.) *Australopithecus boisei*—a famous discovery by Mary Leakey, named for the Leakeys' sponsor, Charles Boise—called *Zinjanthropus*, was even bigger than *robustus* and lived at the same time, though in East Africa. A *boisei* skull discovered in 1985 in Kenya proved not only to be particularly massive but also considerably older (2.5 to 2.6 million years) than any known *robustus* specimen. *Australopithecus boisei* must therefore have been a separate and earlier species probably not descended from *africanus*. If so, then there was more than one australopithecine lineage, and the previously held idea that the australopithecines became increasingly robust through time must be reversed. As a result of this one find, there no longer are widely accepted ideas as to who gave rise to whom.

Throughout their history, the australopithecines manifest a regular progression from apelike characteristics to human ones. All the australopithecines walked upright, a fact that evidently encouraged increasing height, hand specialization, and brain development. Gradual changes in australopithecine dentition, moreover, suggest not only changing diet but revised habits as well. With advanced hands and evolving arms and shoulders, *Australopithecus* probably carried loads and used weapons, regardless of whether he was capable of making them. Though australopithecines may have done some hunting, they probably depended primarily upon foraging and scavenging—filching from leopard kills, for example.

The Evolution into Homo

At what point *Australopithecus* evolved into *Homo* is unclear, in part because the distinction between them is rather arbitrary. Though still regarded by some researchers as an advanced form of *Australopithecus* (that is, as more than one species) *Homo habilis* (handy man), another Leakey family discovery, is otherwise usually accepted as the earliest member of a distinctly human line. He was still only about five feet high (perhaps an optimum height for the environmental conditions), but had a larger brain, a rounder head, a less projecting face, advanced dentition, reduced jaws, and essentially modern feet. Stone artifacts have been found in close association with his remains; *habilis*, who lived between 2 and 1.5 million years ago, almost certainly made tools, hunted, built shelters, gathered plants, and scavenged. He is assumed to be ancestral to *Homo erectus*, and may have exterminated *Australopithecus*.

In 1859, when the prehistory of humankind was first broadly acknowledged, the only remains then known belonged to Neanderthal man (now called *Homo sapiens neanderthalensis*). Until 1924, when *Australopithecus* was discovered, all the intervening finds (excluding "Piltdown man," a deliberately planted fake) have since been classified as varieties of *Homo erectus* (erectman—a designation assuming that its predecessors stooped). Of these, the two best-known are Java man, discovered in 1891 by Eugène Dubois in Java, and Peking man, found in China by Davison Black in 1926. Only after a number of specimens had accumulated was it realized that Java man and Peking man were examples of the same species. An exceptionally complete *Homo erectus* skeleton was discovered in Kenya in 1984 and dated at 1.6 million years old. Overall, *Homo erectus* lived from some time before 1.6 million years ago to as recently as two hundred thousand years ago. He probably evolved in Africa but migrated from there (as no previous hominid had) to Europe and the Pacific shores of Asia. This species was as tall as modern humans, but more robust overall, with a noticeably thicker, somewhat "old-fashioned" skull that still included prominent brow ridges and a sloping forehead. He had large, projecting jaws, no chin, teeth that were larger than ours; he also had a bigger brain. *Homo erectus* was not only widespread in distribution, but also showed considerable regional variation. His success as a colonizer was attributable in large part to his intelligence, which was manifested in standardized but increasingly sophisticated toolkits, big-game hunting (almost certainly cooperative), the use of fire, and advanced housing. He lived during the Pleistocene, or glacial, epoch, and was probably stimulated to use his creative abilities by the deteriorating environments he sometimes encountered. Very late examples of *Homo erectus* are sometimes alternatively classified as *Homo sapiens*. Despite some continuing opposition, it is now usually accepted that *Homo erectus* gave rise to *Homo sapiens*.

Searching for the Fossil Remains

Considering the efforts that have been made to find them, hominid fossils are remarkably few. This is the case for three main reasons: First, the early hominids (unlike modern humans) did not exist in huge numbers; second, the majority of their bones were not preserved as fossils; and third, it is certain that only a small proportion of the hominid fossils that do exist have been found.

The hominid line evolved in Africa, and its earlier members (including the australopithecines and *Homo habilis*) have been found only there. *Homo erectus* was both more widespread and more numerous, but none of these types practiced ritual burials (Neanderthals were the first to do that), so the most usual agent of presentation was some sort of nonhuman carnivore. Predators, such as the large cats, might actually have hunted the early hominids; in any case, they certainly scavenged hominid carcasses. Hyenas and other cleanup animals then grabbed what they could, taking the leftover pieces to their dens in limestone caves. (There must be some truth to this scenario, because leopards and hyenas have left their toothmarks on australopithecine bones.) The gnawed bones, now thoroughly disarticulated, were scattered about the cave—and eventually solidified by limy deposits into a bone breccia.

When twentieth century investigators find such embedded hominid bones, or suspect their presence, they collect chunks of the breccia and dissolve the limy matrix with acetic acid, a procedure that does no harm to fossil bones. It is unlikely that any of the latter will be whole. Once the fragmented bones have been freed from the matrix, they are cleaned, preserved, sorted by type, and tallied. Whether big or small, routine or not, each must be identified. By far the great majority of the bones will belong to antelope of various kinds; less than one in a thousand, normally, proves to be hominid.

Such procedures are standard when dealing with cave deposits at the famous South African australopithecine sites of Sterkfontein, Kromdraai, and Swartkrans, all of which are adjacent to each other and to Pretoria. At Olduvai Gorge in Tanzania, where the Leakeys and others have found both australopithecine and *Homo habilis* remains, the geology is entirely different. Here, the erosive power of a now-defunct river has exposed primarily volcanic sediments that once bordered a shifting saline lake. The disadvantage of this site is that it lacked any obvious place for the location of bones; years of determined effort were required to locate productive sites. The advantages of the site were that the presence of early humans was virtually assured because stone tools were scattered about plentifully (whole campsites were eventually found); the bones involved had not been dragged about and disarticulated by animals; and the involved stratigraphy made fairly precise dating at least theoretically possible. In general, a specimen from Olduvai Gorge brings with it more useful information than does one found in the South African caves. Still other sites have provided additional unique information.

The collection, preservation, and interpretation of hominid fossil bones is very much a multidisciplinary effort. In particular, detailed geological understanding of the site is essential. Only through stratigraphical analysis, usually, can the age and situation of the discovered fossil be understood. Stratigraphy aside, certain rocks can also be dated according to the radioactive elements they contain. More often, bones can be dated approximately because they occur in association with a particular assemblage of animal bones, the animal species themselves being of reliable short-term ages. In some cases, pollen samples have been of use. All this additional information, together with comparative anatomical analysis, helps to give hominid fossils a defensible identity.

The Search for Ancestors and Origins
Thinking human beings have always been fascinated with the concept of origins, and of all origins, none has been of more interest than that of humans. Human groups have asserted deeply meaningful identities by attributing their present being to a particular origin. Most of these psychologically necessary genealogies relied upon some divine agency to explain human existence. From a surprisingly early time, however, civilized humans (such as the Greeks) recognized that there had been a time when humankind did not know the use of metals. Subsequently, many thinkers took the concept of cultural evolution for granted.

By the seventeenth century CE, anatomy had become a popular field of study. Comparisons soon established how like human anatomy that of the higher primates was. By the mid-eighteenth century, Carolus Linnaeus, the originator of modern biological classification, even ventured to place man and the apes within the same family. Yet this classification did not imply any necessary common ancestry. Linneaus and others of his time created the notion that individual species arose through a special, divine plan. The idea of special creation lost credibility when the fact of extinction became established at the end of the eighteenth century and as the diversity of species and varieties came increasingly to be appreciated. Nature, moreover, was no longer seen to be a benign reflection of its creator. As opinions of a distinctly human nature likewise declined, the realization that humans are animals encountered lessening resistance.

Before human evolution was generally accepted during the twentieth century, the evolution of cultures, language, law, institutions, and at least some animals had already been established. Though the idea was there, reliable evidence for the biological evolution of humans remained elusive. Before 1891, the only known prehistoric human bones belonged to Neanderthals; they were quickly and effectively dismissed as pathological freaks. Since Raymond Dart's Taung child of 1924 was likewise dismissed with ridicule, only a few specimens of what would later be recognized as *Homo erectus* (Javamanand Peking man) survived to satisfy the now fashionable quest for a "missing link." (There are still innumerable missing links, but the essential connection between ancestral apes and man was confirmed by the discovery of australopithecines.) It was Robert Broom who, during the 1930's, established the reality of the australopithecines and, by implication, of human evolution.

Though the evolution of the hominid line is certainly a worthwhile scientific topic, it has always been regarded as much more than that, because the claims to ancestry define humanity. Yet the formal constraints of science are limited. In a remarkable series of thirty-nine papers (published between 1949 and 1965), for example, Raymond Dart promulgated an interpretation of the australopithecines as aggressive, predatory, and cannibalistic hunters. Because of the recurrent wars in which civilization had engaged during this period and shortly before, this image of human (or almost human) nature appealed to the popular imagination—so much so that less technical restatements of the same views by Robert Ardrey were not only commercially successful but also politically influential. Interpretations of australopithecine and early hominid behavior have subsequently changed, however—many now see these species as abject scavengers disputing the possession of already picked-over animal corpses with hyenas. It is arguable whether such changing interpretations are attributable to scientific advances or are the result of changing philosophical views of humanity.

—Dennis R. Dean

See also: Apes to hominids; Evolution: Animal life; Evolution: Historical perspective; Genetics; *Homo sapiens* and human diversification; Human evolution analysis; Neanderthals.

Further Reading

Brain, C. K. *The Hunters or the Hunted? An Introduction to African Cave Taphonomy.* Chicago: University of Chicago Press, 1981. In a sophisticated but readable style, Brain reports the evidence about the australopithecines—what they hunted and what hunted them. (Taphonomy is the study of entire bone assemblages.) His study is intended primarily for graduate students and professionals in the field.

Campbell, Bernard. *Humankind Emerging.* 8th ed. Boston: Allyn & Bacon, 2000. A readable college-level text that is useful for its treatments of historical, biological, and anthropological topics. No other book presents so much information so well. Graphics and other illustrations are excellent. Campbell also includes topical bibliographies and a glossary. Chapters 7 through 10 are especially relevant.

Day, Michael H. *Guide to Fossil Man.* 4th ed. Chicago: University of Chicago Press, 1986. Despite some abstruse language (a glossary is included), this authoritative compilation can be of significant help. Organized geographically, it lists and often illustrates all known hominid fossils. Conflicting interpretations are reported but seldom reconciled. This is not, therefore, a book for those seeking the fast answer. Bibliographies accompany each section of the text. For college-level students and professionals, this is a standard work.

Johanson, Donald, and Maitland A. Edey. *Lucy: The Beginnings of Humankind.* Reprint. New York: Simon & Schuster, 1990. Intended for a broad audience, this popular account re-creates the excitement and explains the significance of the finding of an extraordinary specimen, known technically as *Australopithecus afarensis.* It is the oldest hominid whole skeleton extant.

Johanson, Donald, Lenora Johanson, and Blake Edgar. *Ancestors: In Search of Human Origins.* New York: Villard Books, 1994. A companion to the three-part *Nova* television series. Chronological coverage of *Australopithecus, Homo habilis, Homo erectus,* archaic *Homo sapiens,* the Neanderthals, and *Homo sapiens.* Focuses on the australopithecines and early *Homo.* Lavishly illustrated.

Jordan, Paul. *Neanderthal: Neanderthal Man and the Story of Human Origins.* Stroud, U.K.: Sutton, 1999. A clear exposition of a complex subject, highlighting the ideas involved in the study of human evolution and the relationship of Neanderthal Man to *Homo sapiens.*

Kurten, Bjorn. *Our Earliest Ancestors.* Translated by Erik J. Friss. New York: Columbia University Press, 1993. A succinct overview of hominid evolution. Focuses on the major changes in anatomy and culture that have marked the development of hominids from *Homo habilis* through *Homo sapiens,* including the emergence of bipedalism, stone toolmaking, and articulate speech. Numerous maps, charts, and drawings.

Leakey, Maeve, et al. "New Hominin Genus from Eastern Africa Shows Diverse Middle Pliocene Lineages." *Nature* 410 (March 22, 2001): 433-440. A new skeleton, dated to 3.5 million years ago and differing from the roughly contemporary *Autralopithecus afarensis,* throws the lineage of modern man into even further confusion.

Lewin, Roger. *Bones of Contention: Controversies in the Search for Human Origins.* New York: Simon & Schuster, 1987. This lucid but sophisticated treatment reviews scholarly debates regarding the Taung child (*Australopithecus africanus*), *Ramapithecus* (a fossil ape) and "Lucy" (*Australopithecus afarensis*). Too specialized for most high school students, it will appeal primarily to college graduates and professionals. Lewin has also written *Human Evolution: An Illustrated Introduction* (1984) and coauthored three popular books with Richard Leakey: *Origins* (1977), *People of the Lake* (1979), and *The Making of Mankind* (1981).

Mellars, Paul, ed. *The Emergence of Modern Humans: An Archaeological Perspective.* Ithaca, N.Y.: Cornell University Press, 1990. A volume of proceedings from a 1987 conference on the origins of modern humans and the adaptations of hominids. Concentrates on the archaeology of hominids in western and central Europe.

Reader, John. *Missing Links: The Hunt for Earliest Man.* Boston: Little, Brown, 1981. Organized chronologically, chapters cover the discoveries of the major hominids. Outstanding original photographs by the author are featured. Though other treatments have been more detailed (and more technical), Reader's agreeable combination of history and science makes his book irresistible. For readers at all levels.

Szalay, Frederick S., and Eric Delson. *Evolutionary History of the Primates.* New York: Academic Press, 1979. Though not intended for beginners, and now somewhat dated in places, this remains a standard reference.

Tattersall, Ian. *The Fossil Trail: How We Know What We Think We Know About Human Evolution.* New York: Oxford University Press, 1995. The reconstruction of human evolution on the basis of interpreting fossil remains, including the Laetoli footprints, Lucy, Turkana Boy, and others. Good coverage of the development of dating techniques such as fluorine analysis and mitochondrial DNA.

Theunissen, Berg. *Eugène Dubois and the Ape-Man from Java: The History of the First "Missing Link" and Its Discoverer.* Boston: Kluwer Academic, 1989. Java man, discovered by Dubois in 1891 (with one important discovery the previous year), was the first example of *Homo erectus* to be recovered. Theunissen's account of this historic addition to our knowledge of early man is detailed but nontechnical, suitable for readers at the high school level and beyond.

HOMO SAPIENS FACTS

Classification:
  Kingdom: Animalia
  Subkingdom: Metazoa
  Phylum: Chordata
  Subphylum: Vertebrata
  Superclass: Tetrapoda
  Class: Mammalia
  Subclass: Theria
  Infraclass: Eutheria
  Order: Primates
  Suborder: Anthropoidea
  Superfamily: Hominoidea
  Family: Hominidae

Genus and species: Homo sapiens

Geographical location: Originally Africa, now spread to all continents

Habitat: Originally savannas, now spread to all habitats

Gestational period: Nine months

Life span: Originally a maximum of twenty-five years; now over one hundred years, with averages from forty to eighty years, depending on habitat

Special anatomy: Large, well-developed brain; upright, bipedal posture; opposable thumbs; larynx, vocal chords, and tongue adapted to produce a wide variety of sounds for language

---

Homo sapiens AND HUMAN DIVERSIFICATION

FIELDS OF STUDY

Anthropology, evolutionary science, ecology, genetics, human origins, systematics (taxonomy)

SUMMARY

By studying fossil, cultural, and genetic evidence, scholars have attempted to trace the evolutionary development of the human species. It is believed that the earliest form of Homo sapiens appeared about 350,000 years ago and that the first modern humans (Homo sapiens sapiens) appeared somewhat before 100,000 years ago, with racial diversification following thereafter.

PRINCIPAL TERMS

- **gene pool:** the total collection of genes available to a species

- **generalized:** not specifically adapted to any given environment; used to describe one group of Neanderthal humans
- **hominid:** any living or fossil member of the taxonomic family Hominidae ("of man") possessing a human form hominoid: referring to members of the family Hominidae and Pongidae (apes) and to the taxonomic superfamily of Hominoidae
- **morphology:** the scientific study of body shape, form, and composition
- **natural selection:** any environmental force that promotes reproduction of particular members of the population that carry certain genes at the expense of other members
- **Pleistocene epoch:** the sixth of the geologic epochs of the Cenozoic era; it began about three million years ago and ended about ten thousand years ago
- **Würm glaciation:** the fourth and last European glacial period, extending from about seventy-five thousand years ago to twenty-five thousand years ago

BASIC PRINCIPLES

All human beings on the earth today are highly adaptive animals of the genus and species *Homo sapiens sapiens* (Latin for "wise, wise human"). In terms of physical structure and physiological function, *Homo sapiens sapiens*—modern humans—are classified taxonomically as members of the order Primates, which is part of the class Mammalia. Since humans and other members of Primates (monkeys and apes) are biologically related, scientists presume both groups to be the products of an evolutionary process similar to that which affected other divergent categories of animals. The evolutionary process that produced *Homo sapiens sapiens* from previously existing species is also believed to account for diversifications within the modern human population such as racial differentiation.

Modern humans and modern apes (the two most closely related of modern primate species) are believed to possess a common biological ancestry, or line, that diverged perhaps five or six million years ago. The scanty fossil record of this early period, in conjunction with modern genetic studies, seems to indicate that the branch of hominoid evolution which eventually led to *Homo sapiens sapiens* first gave rise to the earliest hominid type, called *Ramapithecus*. Next to appear, several million years ago, during the late Pliocene epoch, were the early forms of *Australopithecus*, which existed in Africa. They share certain characteristics with both humans and apes. Their brains are larger than those of apes but smaller than those of humans. There have been four species of *Australopithecus* identified.

HOW IT WORKS

The Emergence of the Genus Homo. Examples of the first undisputed members of the genus *Homo*—true human (though not *sapiens*)— appear in the fossil record about 1.5 million years ago. Samples of *Homo erectus* ("upright human") have been found in China, Africa, Java, and Europe. This creature habitually walked upright, made shelters, and used sophisticated tools. *Homo erectus* is also very important, since it is the first hominid to have used fire purposefully. It was suggested by John E. Pfeiffer, in a 1971 article entitled "When *Homo erectus* Tamed Fire, He Tamed Himself," that this first domestication of a natural force was a tremendous evolutionary step, changing the fundamental rhythms of life and human adaptability to environments. Most scholars accept the premise that *Homo erectus* was the hominid grade intermediate between the australopithecines and *Homo sapiens*.

Exactly when, where, and how advanced members of the species *Homo erectus* evolved into *Homo sapiens* are key questions in the study of human evolution, and they are questions that resist resolution. It might be thought that the closer one comes, in terms of time, to modern man, the easier it would be to find the answers. In actuality, such is not the case. The ancestral line or lines leading to modern man become hazy beginning approximately 500,000 years ago. Direct fossil evidence of the earliest members of the species *Homo sapiens* is scarce; moreover, finds of modern human fossils in the Middle East have intensified the debate about the immediate ancestry of *Homo sapiens sapiens*. All the evidence indicates, however, that the middle to upper Pleistocene epoch (beginning about 350,000 years ago), known as the Paleolithic or old stone age in archaeological terms, witnessed the emergence of early *Homo sapiens*.

The Earliest Homo Sapiens. In 1965, hominid fossil remains were found at a site named Vértesszöllös, near Budapest. They consisted of some teeth and an occipital bone (a bone at the back of the skull). The site also yielded stone tools and signs of the use of fire. Several features of the find recall *Homo erectus*, but the estimated cranial capacity of 1,400 cubic centimeters is well into the normal range for *Homo sapiens*. The age of the site was established at 350,000 years BP (before the present). These remains have been attributed to a *sapiens-erectus* intermediate type on the grounds that the remains, and the site, show a mixture of elements reflective of the transitional hominid evolutionary process. Such an assessment places Vértesszöllös man at the root of the *Homo sapiens* evolutionary line, some 100,000 years earlier than other specimens.

A better-known example of early *Homo sapiens* comes from a gravel deposit at Swanscombe near London, England. In 1935, 1936, and 1955, three related skull pieces were unearthed that fit together perfectly to form the back of a cranial vault with an advanced (over *Homo erectus*) cranial capacity of about 1,300 cubic centimeters. This has been dated to around 275,000 to 250,000 years BP. A more complete skull of approximately the same age (dated to the Mindel-Riss interglacial period about 250,000 years BP) was found at Steinheim, in southern Germany, in 1933. Swanscombe's and Steinheim's advanced morphological characteristics, in combination with relatively primitive ones, such as low braincase heights, suggest that they are primitive members of the species *sapiens* and are representatives of a population intermediate between *Homo erectus* and *Homo sapiens*.

The finds at Swanscombe and Steinheim have been augmented by others from France and Italy, and especially from the Omo River region in southern Ethiopia. One Omo skull displays more mixed features (between *erectus* and *sapiens*) including flattened frontal and occipital areas, a thick but rounded vault, large mastoid processes (pointed bony processes, or projections, at the base of the skull behind the ears), and a high cranial capacity. Another skull is more fully *sapiens*, or modern in appearance. Some paleoanthropologists assert that the Omo group of fossils also helps bridge the gap between advanced *Homo erectus* and *Homo sapiens*.

Neanderthal Man. The best-known examples of early *Homo sapiens* come from a group of fossils known collectively as Neanderthal man. Their name derives from the place where the first fossil type was discovered in 1865, the Neander Valley near Düsseldorf, Germany. Similar Neanderthal fossil types have been found at more than forty sites in France, Italy, Belgium, Greece, the Czech Republic Slovakia, the former Soviet Union, North Africa, and the Middle East.

Neanderthal fossils tend to show an aggregate of distinctive characteristics that at one time led to their being regarded as a separate human species, *Homo neanderthalensis*. They are generally regarded as a subspecies of humans, with the designation *Homo sapiens neanderthalensis*. The characteristic features of their morphology include large heads with prominent supraorbital tori (thick brow ridges), receding jaws, stout and often curved bones, and large joints.

Most important Neanderthal fossils disclose large brain capacities (1,500 cubic centimeters) and are found in sites revealing complex and sophisticated cultures. These two facts clearly separate Neanderthal humans from more primitive "presapiens" species that exhibit some of the same morphological features. Neanderthalers generally stood fully erect between 1.5 and 1.6 meters in height; they were not the stoop-shouldered brutes of early characterizations. They lived during the last glaciation (the Würm glacial stage) in Eurasia. The sites from which most examples of the Neanderthalers have been recovered have commonly yielded tools of the Mousterian complex, a stone-tool industry named for the kind found at Le Moustier, France, and dating from about 90,000 to about 40,000 years BP.

In fact two groups of Neanderthal humans seem to have existed. The first are referred to as classic Neanderthalers from such sites as Germany, France, Italy, Iraq (Shanidar man), and the former Soviet Union. The second group, known as either generalized or progressive Neanderthalers, lived contemporaneously with, as well as later than, classic Neanderthal humans. They display a combination of modern *sapiens* features and typical Neanderthal characteristics (especially the prominent supraorbital torus, the forehead ridge). Included in this category for the sake of simplification are those specimens termed neanderthaloid. Examples include Rhodesian man (from Zambia, formerly Northern Rhodesia) and Solo man (from Java), both unearthed at upper Pleistocene deposits (100,000 to 10,000 BP). Neanderthalers were cave dwellers and were well adapted to cold conditions (especially the classic Neanderthal variety). They used fire, manufactured stone flake tools, and buried their dead with care. They also seem to have practiced fairly complex religious rituals.

Cro-Magnon Man. Neanderthals were a successful group for many thousands of years, flourishing from about 127,000 BP to 37,000 BP, with a wide distribution geographically. Neanderthal traces suddenly and mysteriously disappear from the fossil record, however, and they seem to have been superseded around 37,000 BP by other *Homo sapiens* with a more advanced culture and different morphology. In Europe, these are known as the Cro-Magnon peoples, so named for the Cro-Magnon cave near Les Eyzies in southwestern France, where the first skeletons were found in 1868 and where more than one hundred skeletons have since been discovered. Indeed, Cro-Magnon skeletal anatomy is virtually the same as that of modern European and North African populations. The skull is relatively elongated, with a large cranial capacity of about 1,600 cubic centimeters; the brow ridges are only slightly projecting. The average height of Cro-Magnon man was between 1.75 and 1.8 meters.

Cro-Magnon humans produced a culture that, in variety and elegance, far exceeded anything created by their predecessors. They made weapons and tools of bone and stone, stitched hides for clothing, and lived in freestanding shelters as well as caves. Some Cro-Magnon people produced beautiful cave paintings (they have been found in southwestern France and northern Spain) and bone carvings, and they modeled in clay. Though Cro-Magnon samples are the best-known examples of early *Homo sapiens sapiens*, mounting fossil evidence from sites outside Europe as well as genetic research performed in the 1980's suggests a much older date of origin for the emergence of modern man.

At Qafzeh, a cave near Nazareth, Israel, anatomically modern fossils classified as *Homo sapiens sapiens* were discovered in 1988 and reliably dated to 92,000 years BP. In addition, newer fossil finds of progressive Neanderthalers from Kebara Cave in Israel, taken together with earlier Neanderthal finds from the caves of et-Tabun and es-Skhul, also in Israel, make it certain that progressive Neanderthalers and modern humans coexisted for many thousands of years.

Anthropologists have puzzled over the disappearance of the Neanderthals and, more important, over where they fit in the human family tree. It appears unlikely that classic Neanderthal humans were in the direct ancestral line of modern *Homo sapiens sapiens*. Reasons for their sudden disappearance are believed to include a combination of factors: extinction because of disease, lack of adaptation to the warmer climate following glaciation, and annihilation by the more advanced sapient groups.

Many scholars have considered the classic Neanderthals to be a cold-adapted, specialized side branch from the modern human line that became extinct as the climate became warmer. The generalized or progressive Neanderthals are considered by some to have avoided this specialization, perhaps continuing to exist through adaptation and ultimately being absorbed by flourishing modern human populations during the late Pleistocene epoch.

The Emergence of Modern Humans
Although the exact time place, and mode of the origin of the modern human species cannot yet be determined, genetic studies point to a date before 100,000 years BP. Examination of mitochondrial DNA (mtDNA) from a sampling of present-day humans representing five broad geographic regions has allowed researchers to propose a genetic family tree and calculate roughly (assuming a fairly constant mutation rate) a temporal origin for the modern human population. Further studies seem to indicate that the modern human ancestral line emerged between 280,000 years and 140,000 years BP. Genetic evidence, in concert with fossil finds, makes it plausible that a common ancestral population for *Homo sapiens sapiens* appeared in sub-Saharan Africa or the Levant (in the eastern Mediterranean region). Regional differentiation occurred, followed by radiation outward to other areas. The range of genetic and anatomical variability exhibited by fossil remains of modern humans is no greater than that known for the extant races of modern times.

During the late Pleistocene epoch (approximately 40,000 to 11,000 years BP) five different racial groups seem to have developed on the Eurasian and African landmasses. The last glaciation, approximately 30,000 to 10,000 years BP, absorbed enough water to lower the oceans ninety meters below present levels. Emerging land bridges allowed people to move from Asia into North America, Australia, and elsewhere. In time the major racial groups became subdivided into smaller ones that resulted in the major races seen in modern times.

This view of racial diversification emphasizes the effectiveness both of geographic barriers in reducing free gene flow among varied groups of *Homo sapiens* and of environmental pressure in selecting different adaptive responses from the gene pool. These are also key factors in the entire evolutionary process by which modern humans developed over epochs into their present taxonomic position in the animal kingdom.

The Study of Paleoanthroplogy and Physical Anthropology

The study of human evolution is primarily the concern of the physical anthropologist and the paleoanthropologist. Evolution maybe defined as change in the genetic composition of a population through time. Because evolution is thought to operate according to several principles and factors, modern human evolutionary theory is studied in the light of ideas and practices taken from different disciplines, including archaeology, biochemistry, biology, cultural anthropology, ecology, genetics, paleontology, and physics.

Early investigations into human evolution sought to establish the sequence of the human ancestral line through chronological and morphological analyses of hominid fossil remains (bones and teeth), thus placing them in their proper phylogenetic context (their natural evolutionary ordering). This remains the principal method of study, but it has been augmented by sophisticated techniques in fossil dating and new avenues of exploration into the evolutionary process, such as genetic research.

Determination of the accurate age of a fossil is most important, since it sets the fossil in a correct stratigraphic context that allows comparison with remains from the same geologic layer or level a great distance away. Accurate dating also has helped determine the order of succession for fossils that could not be established on morphological grounds alone.

The most valuable absolute dating methods are the radioactive carbon technique, which can effectively date specimens between 60,000 years BP and the present; the potassium-argon technique, which most easily dates material older than 350,000 years BP; and the fission-track method, which helps bridge the gaps between other methods. These methods are based on the constant or absolute rates at which radioactive isotopes of carbon, potassium, and argon decay. When absolute dating is impossible, investigators have ascribed a relative age to fossil remains by noting the contents of the layer of rock or the deposit in which the remains were found. A layer containing remains of extinct animals is likely to be older than one containing remains of present forms.

In conjunction with dating, anatomical studies of fossil remains and comparisons with the morphological features of known hominid types, as well as comparisons to primate skeletal structures, have been primary approaches to the study of the evolutionary path of *Homo sapiens*. The species *Homo sapiens* (of which the modern human races compose a number of geographical varieties) may be defined in terms of the anatomical characteristics shared by its members. In general, these include a mean cranial capacity of about 1,400 cubic centimeters, an approximately vertical forehead, a rounded occipital (back) part of the skull, jaws and teeth of reduced size, and limb bones adapted to fully erect posture and bipedalism. Scientists assume that any skeletal remains which conform to this pattern and cannot be classified in other groups of higher primates must belong to *Homo sapiens*. It is striking that the anatomical differences observed between *Homo erectus* and *Homo sapiens* have been confined to the skull and teeth. The limb bones thus far discovered for both are similar (though *erectus* appears more robust). Cranial capacity and morphology continue to be the dominant determining boundary separating sapient and presapient human species.

The Contribution of Other Sciences and Social Sciences

Human adaptability studies, using techniques from physiology, demographics, and population genetics, investigate all the biological characteristics of a population that are caused by such environmental stresses as altitude, temperature, and nutrition. It is believed that these normal stresses acted as genetic selectors in prehistoric times and continue to do so. Such racial variants as skin color and body hair are observable products of these stresses. The investigation of climatic changes during prehistoric epochs as revealed in the geologic record is important for understanding those pressures affecting the evolutionary history of man.

Genetic studies have become indispensable to the study of human evolution. Four forces have been identified as fundamental in the evolutionary process: mutation, natural selection, gene flow, and genetic drift. Since mtDNA is inherited through the female, it is possible to calculate how much time has elapsed since the mutations that gave rise to present variations originated in prehistoric populations.

Also important to the study of the evolution of *Homo sapiens* is the examination and classification of cultural remains preserved at hominid fossil sites. Not only can the relative date of a fossil be supported, but sometimes it is also possible to reconstruct the environmental situation that may have influenced the evolutionary process operating in a population. Cultural response is an integral part of hominid adaptation, and it in turn influences natural selection. Technology changes the physical and economic environment, and economic changes alter the demographic situation. Humans continue to promote or influence their own evolution by willingly or unwillingly altering the environment to which they must adapt.

The modern methods useful for investigating the evolutionary history of *Homo sapiens* are multidisciplinary. While each of them reveals an aspect of the emergence of modern man and complements the other methods of study, emphasis is placed on careful fieldwork, accurate dating, and comparative morphological analyses of hominid fossil remains. Increasing in importance, however, is the accumulating wealth of genetic data on human population relationships.

Human Evolution in the Context of Animal Evolution
Increasing attention is being given to the biological and behavioral changes that led to the emergence of *Homo sapiens sapiens*—the last major event in human evolution. Mounting evidence continues to push backward in time the point at which modern *Homo sapiens* made his appearance in the evolutionary scheme. The finds at Qafzeh, for example, indicated that modern man arose fifty thousand years earlier than had previously been thought. A clearer understanding of the evolutionary history of modern *Homo sapiens* has not only helped to define the place of the modern human species more accurately in relation to the rest of the animal kingdom but also helped to illuminate the pressures, adaptations, and changes that have made humans what they are.

The accumulating data on the evolutionary appearance of *Homo sapiens* have allowed biologists and anthropologists to see the rise of modern man as part of the evolutionary development of the animal kingdom in general and of primates in particular. The primates that exist today make up a remarkable gradational series that links *Homo sapiens sapiens* with small mammals of very primitive types.

Through the pressures and process of evolution (including adaptation and natural selection), *Homo sapiens* has become one of the most successful and adaptive animals that ever lived, because he came to possess an elaborate culture (culture is based on learned behavior). The key is *Homo sapiens*' superior mental capacity. Only human beings can assign arbitrary descriptions to objects, concepts, and feelings and can then communicate them unambiguously to others. In the late Middle Pleistocene, the hominid branch that gave rise to early *Homo sapiens* witnessed an increase in brain size, complex social organizations, continual use of fire, and perhaps even language. As to what initiated these changes, many have suggested tool use and, in turn, a hunting economy.

In a classic article published in 1960, entitled "Tools and Human Evolution," Sherwood Washburn argued that the anatomical structure of modern-man is the result of the change, in terms of natural selection, that came with the tool-using way of life. He stated that tools, hunting, fire, and an increasing brain evolved together. Washburn also argued that effective tool use led to effective bipedalism—another significant characteristic of *Homo sapiens*: Man is different from all other animals because he became a user of increasingly complex tools.

The other behavioral pattern that is seen to have been of utmost importance to sapient evolution is big-game hunting. Early *Homo sapiens* was undoubtedly a big-game hunter, as were all of his successors until approximately 8,000 years ago. It has been argued that human intellect, interests, emotions, and basic social life are the evolutionary products of the success of hunting adaptations. Success in hunting adaptation dominated the course of human evolution for hundreds of thousands of years. The agricultural revolution and the industrial and scientific revolutions are only now releasing human beings from conditions characteristic of 99 percent of their evolutionary history.

Scholars have suggested that research into human origins and development is much more relevant than is often realized. It has been argued that, although man no longer lives as a hunter, he is still physically a hunter-gatherer. Some investigators in the field of stress biology, the study of how the human body reacts to stressful situations, feel that man is biologically equipped for one mode of life (hunting) but lives another. Thus, there would be some link between an emotional reaction, such as explosive aggression, and human evolutionary history. Tools and more efficient hunting helped produce great change in hominid evolution and made man what he is. Humans continue to be users of increasingly complex tools, such as computers, and perhaps this continued development of technology may determine the future evolutionary path of *Homo sapiens*.

—*Andrew C. Skinner*

See also: Apes to hominids; Evolution: Animal life; Evolution: Historical perspective; Gene flow; Genetics; Hominids; Neanderthals; Natural selection.

FURTHER READING

Brauer, Gunter, and Fred H. Smith, eds. *Continuity or Replacement: Controversies in "Homo Sapiens" Evolution*. Brookfield, Vt.: A. A. Balkema, 1992. A collection of technical papers from a 1988 international symposium on the biological and cultural evolution of hominids.

Cann, R. L., M. Stoneking, and C. A. Wilson. "Mitochondrial DNA and Human Evolution." *Nature* 325 (January 1, 1987): 31-36. Though written for the knowledgeable student of human evolution, this important article presents the results of a major study using DNA to trace genetic differences in widely varied samples of the present population and, by constructing an evolutionary family tree, to propose a time and place for the emergence of modern man. It shows the increasing importance of genetics for the study of the modern human evolution.

Eldredge, Niles, and Ian Tattersall. *The Myths of Human Evolution*. New York: Columbia University Press, 1982. In fewer than two hundred pages the authors give the nonspecialist a brief, comprehensive look at human evolutionary history, while attempting to show that the once-standard expectations of evolution—slow, steady, gradual progress—are not supported by the evidence. According to the authors, fossil evidence shows human evolution to be the result of long periods of stability interrupted by abrupt change, occurring in smaller populations. The thesis is an important consideration presented in well-written form.

Johanson, Donald, Lenora Johanson, and Blake Edgar. *Ancestors: In Search of Human Origins*. New York: Villard Books, 1994. A companion to the three-part *Nova* television series. Chronological coverage of *Australopithecus*, *Homo habilis*, *Homo erectus*, archaic *Homo sapiens*, the Neanderthals, and *Homo sapiens*. Focuses on the australopithecines and early *Homo*. Lavishly illustrated.

Leakey, Richard E. *The Making of Mankind*. New York: E. P. Dutton, 1981. This book, by the son of Louis and Mary Leakey, is written from the perspective of one who grew up with famous physical anthropologists and scientists who were tracing human origins. The text is complemented with many color photographs. Chapters discussing the development of language and early *Homo sapiens* culture are important. It also distills and synthesizes important ideas of others.

Lewin, Roger. *The Origin of Modern Humans*. New York: Scientific American Library, 1998. Overview of the key concepts in the origin of modern humans, including the molecular biology and archaeology supporting the out-of-Africa hypothesis versus the multiregional hypothesis of human evolution.

Mellars, Paul, ed. *The Emergence of Modern Humans: An Archaeological Perspective*. Ithaca, N.Y.: Cornell University Press, 1990. A volume of proceedings from a 1987 conference on the origins of modern humans and the adaptations of hominids. Concentrates on the archaeology of hominids in western and central Europe.

Phenice, Terrell W. *Hominid Fossils*. Dubuque, Iowa: Wm. C. Brown, 1973. One of the best introductory illustrated keys of hominid fossil remains of the Pleistocene epoch. The drawings are all oriented the same way, and each fossil illustration is listed with its measurements and with information concerning tools found at the respective sites. This book (and similar types of atlases) is very helpful to the nonspecialist.

Tattersall, Ian. *The Fossil Trail: How We Know What We Think We Know About Human Evolution*. New York: Oxford University Press, 1995. The reconstruction of human evolution on the basis of interpreting fossil remains, including the Laetoli footprints, Lucy, Turkana Boy, and others. Good coverage of the development of dating techniques such as fluorine analysis and mitochondrial DNA.

Waechter, John. *Man Before History*. Oxford, England: Elsevier-Phaidon, 1976. This is a concise, readily accessible volume on the evolutionary history of humans. It contains some of the finest visual and graphic representations (photographs and illustrations) on the subject found in any single work on the evolutionary journey of the human species. It also contains an extensive glossary of terms with numerous illustrations. Particularly strong in its discussion and representation of late prehistoric culture (especially art), it is an excellent reference work geared toward the nonspecialist.

HUMAN EVOLUTION ANALYSIS

FIELDS OF STUDY

Anthropology, evolutionary science, genetics, human origins, systematics (taxonomy)

SUMMARY

The study of human evolution traces the descent of the hominids from their primate ancestors and focuses particularly on the most recent stages that led to modern *Homo sapiens*.

PRINCIPAL TERMS

- **deoxyribonucleic acid (DNA):** the large molecular chains of nucleic acid that make up genetic material
- **gracile:** slender and light-framed, as opposed to robust
- **hominid:** belonging to the taxonomic family Hominidae, which includes all humans and their evolutionary ancestors as far back as the split from the great apes
- **mitochondria:** self-replicating units in a cell that are responsible for the metabolic generation of energy for cell processes

BASIC PRINCIPLES

About fifteen million years ago, there was a great flowering of diversity among the primates. One branch of this evolutionary spurt gave rise to the family Pongidae (gorillas, orangutans, and chimpanzees). A second branch led to the family Hominidae. Fossil evidence suggests that the hominids diverged from the common ancestral primate at about the same time as the pongids, but according to molecular analysis of the genes of modern humans and gorillas, the divergence may have been much later—perhaps as recently as two million or three million years ago. The earliest known examples of the hominid family have been named *Ramapithecus*. Fossils of this creature, more ape than human, have been found in India, Greece, Africa, Turkey, and Pakistan. It is likely that this genus evolved into another hominid group, *Australopithecus*. Many australopithicine fossils have been found, mostly in Africa, and the genus is divided into two types, the large, heavyboned robust species and the smaller, gracile types exemplified by "Lucy," the nearly complete skeleton discovered by Donald Johanson in 1974. The current view is that the robust australopithicines became extinct but that the smaller forms gave rise to the genus *Homo*, leading to modern humans.

The earliest discovery of a fossil of the *Homo* genus was given the species name *erectus* by its discoverer, Eugène Dubois, in 1891. It was found not in Africa but in Java. Dubois believed that Java man, as he called his find, was the first hominid to walk erect. Later discoveries showed that the more primitive *Australopithecus* had already achieved the erect stance of modern humans, but the name *Homo erectus* persists for the link between *Australopithecus* and more modern hominids.

TOOL USE AND CULTURE

The genus *Homo* was the first to leave clear evidence of toolmaking. Pebble tools of *Homo erectus* were described by Louis Leakey, and he named the collection of artifacts "Oldowan" culture, from the Olduvai Gorge where some of his most famous excavations were done. On the basis of his collection of African fossils dated to between one million and two million years ago, Leakey proposed a new species, *Homo habilis*, which would be intermediate between *Homo erectus* and *Homo sapiens*, but there were too few fossils from this critical period to make the sequence clear. In any case, the simple pebble tools of the earliest members of the human genus were gradually supplanted by somewhat more sophisticated implements made by chipping both sides, along with a growing preference for flint over softer stones. There are many more examples of flint tools than there are of actual bones of human ancestors from this period. The techniques used for making these tools were evidently handed down from generation to generation, and the patterns were quite conservative, so tools from a given culture can be identified wherever they are found. The tools and artifacts of the later members of *Homo erectus* are known as the Acheulian culture. The most characteristic tool of this culture is the hand axe, used for chopping, cutting, scraping, and possibly even as a weapon.

The species to which modern humans belong first appeared between 200,000 and 300,000 years ago. In 1856, when the first example of *Homo sapiens* was discovered, it was named Neanderthal man because the skeleton was found in a cave by the Neander Valley near Düsseldorf, Germany. Neanderthals were a robust species with a somewhat larger brain than modern humans. They wandered widely, leaving their remains over much of Europe, Asia, and Africa. The name Mousterian culture was coined to describe their artifacts, which were much more sophisticated than Acheulian tools.

Finally, perhaps as recently as fifty thousand years ago, Neanderthals were replaced by fully modern humans. The two groups can be separated into the subspecies *Homo sapiens neanderthalensis* and *Homo sapiens sapiens*. At about the same time the Neanderthals disappeared, the Mousterian artifacts were replaced by much more complex and widely varied implements of the Aurignacian culture. The explosion of cultural innovation—including cave paintings, carvings, and other artistic and technological inventions—had a remarkable effect on life.

The first fossils of the *sapiens* subspecies were found in a limestone cliff at Cro-Magnon in southern France in 1868. They and numerous later finds from the same period were given the name Cro-Magnon man, and from the very beginning, they were recognized as being fully modern in form.

EVOLUTIONARY CONTROVERSIES

Controversy over the relationship between the modern human subspecies and the Neanderthals centers around the fact that the two groups overlapped in time. Evidently some subpopulation of Neanderthal evolved into Cro-Magnon while the more primitive type was still flourishing. For some unknown period, both types existed, but eventually—quite abruptly on a geological time scale—Neanderthals vanished from the face of the earth, and *Homo sapiens sapiens* reigned alone.

Some anthropologists put the divergence between Neanderthals and the *sapiens* subspecies quite early and suggest that most of the known fossils of Neanderthals represent a dead end of evolution and not human ancestors. A few fossils that appear to be of the modern human type are tentatively dated much earlier than fifty thousand years ago, which is the time most of the fossil evidence would indicate that modern humans first appeared. Whether humans evolved from early or late Neanderthals, there is certainly no other candidate for an immediate ancestor nor any evidence of an alternative link between *Homo erectus* and modern humans.

The Neanderthal question is one of three major controversies about the details of human evolution. Another is over the site of major stages in human evolution. The "out-of-Africa" view is that Africa, with its rich store of fossil evidence, had to be the place where, successively, *Australopithecus*, *Homo erectus*, and modern humans emerged. Countering this is the fact that fossils and artifacts of each stage of human evolution are found in many regions of the globe. Some authorities feel that it is more probable that this mobile, adaptive, wide-ranging species was already dispersed during the final million years or so of human evolution rather than exclusively in Africa.

The third major disagreement among anthropologists is about how to interpret evidence from molecular biology. The random mutation of deoxyribonucleic acid (DNA) chains provides a molecular clock because the mutations cause a genetic drift. In an interbreeding population, the genetic changes are shared by all members, but if a population divides, the genetic changes vary between the two groups, resulting in more and more differences as the time since the separation increases. Studies of DNA differences among humans and between humans and the great apes suggest that the genetic divergence between humans and the gorillas has taken only about three million years. This is a much shorter period than fossil evidence would indicate. However, the time scale for DNA changes is not well enough established to persuade many of the geologists and fossil hunters to give up their chronology. In 1987, the even more disturbing claim was made that all living humans had descended from one woman and that this "Eve" had lived in Africa as recently as 200,000 years ago. The work that led to this announcement was based on the study of mitochondrial DNA, which is much less variable than the DNA in the nucleus of the cell.

Each tiny mitochondrion has its own set of genes, and they divide independently of the cell nucleus. Because the sperm does not contribute any mitochondria to the fertilized egg, mitochondrial DNA is inherited only through the female line of descent. In one study, mitochondrial DNA from 147 individuals, representing populations in Africa, Europe, Asia, Australia, and New Guinea, was compared. The samples seemed to fall into two major groups when analyzed by a computer program: a group containing the African samples and one holding all the rest. The group containing the African individuals appeared to be the more primitive. From the rate of mutations in mitochondrial DNA, it was calculated that the observed variations in the samples would have occurred over a time span of about 200,000 years.

The convergence of molecular evidence and fossil interpretation has given a boost to the outof- Africa hypothesis for human origins. Africa is certainly a rich source of fossils of the earliest progenitors of the human species. It is also the home of most of the great apes. The controversy is mainly about when the migration of humans or proto-humans into the rest of the world occurred. Fossils of every stage of human evolution have been found in many parts of Europe and Asia and even Australia. It is possible, of course, that sample populations of each stage of human evolution emigrated from Africa, leaving their bones and their tools scattered all over Eurasia, only to be replaced by successive waves of emigrants from later stages. There is no hypothesis to suggest why the further evolution of these intermediate species should not have taken place outside Africa.

Fossils and Paleoanthropology

The major technique for the study of human evolution is the analysis of fossils and the artifacts associated with human activities. Fossil hunting is more an art than a science, and many of the major finds have been accidental. However, knowing where to look and the use of systematic excavation have been very fruitful. Dating the fossils and artifacts is the most critical part. When fossils are excavated from undisturbed layers of sediment, the dating can be determined by knowledge of the geological strata. Layers of lava from volcanic eruptions, sediment from floods, the presence of other fossils of plants or animals of known periods all contribute to the determination of the age of the deposit. Radioactive decay provides another method. Radioactive carbon-14 is continuously produced in the upper atmosphere and is absorbed by plants. The carbonate in fresh bone contains about one atom of radioactive carbon for every one hundred forty atoms of the inactive isotope. The radioactive isotope decays at a constant rate and is reduced to half of its original concentration in about five thousand years. The radioactivity of bones (or charcoal or other organic debris) can be measured, and the older the sample, the smaller the proportion of radioactive to nonradioactive carbon atoms will be. Unfortunately, the sensitivity of the method limits its usefulness to materials less than about thirty thousand years old. Also, it is often hard to rule out the possibility of contamination by groundwater or organic sources of "fresh" carbon. Older materials can sometimes be dated by other radioactive isotopes such as argon. Another dating method is to use the accumulation of atomic dislocations caused by cosmic ray bombardment in hard materials, especially stone and pottery. Such dislocations "heal" when the material is heated, so measuring the amount of cosmic ray damage in a material can reveal its age since the last heating. This method is good for much longer time periods than radiocarbon dating.

Molecular biology is relatively new, and its use as a method for studying human evolution is still experimental. The basic premise on which it operates is that mutations of DNA molecules are random events, relatively independent of environmental factors, and therefore relatively constant in time. Although exposure to natural or artificial radiation or certain chemical mutagens can increase the rate of DNA change, it is assumed that such factors would not be likely to affect a whole population; therefore, the assumption of a uniform slow rate of genetic drift is probably justified statistically.

DNA molecules are very large, and with modern techniques, it is possible to detect small changes in the molecule that may not produce any observable mutation in the organism. This means that chemical comparison of DNA samples from two individuals is a much more sensitive measure of their degree of relatedness than simple comparison of visible features. Analysis of DNA differences (and similarities) correlates well with older techniques of classification. Human DNA, for example, is much more similar to sheep DNA than to earthworm DNA. Monkey DNA is much closer to human DNA than is sheep DNA. By assuming that DNA changes at a uniform rate, these differences can be interpreted in terms of evolutionary time since humans and monkeys or humans and sheep shared a common ancestor.

The Implications of Human Origins

The study of human origins has social, intellectual, religious, and philosophical implications. The fact that humans are all one subspecies, with a remarkably homogeneous genetic makeup, should minimize the importance of race, culture, or language differences. Anthropology is the discipline that studies the origins of humans. Anthropologists also study living human cultures, and there is a two-way transfer of information between cultural anthropologists and physical anthropologists. Knowledge of human descent deepens understanding of the diversity of cultures and capabilities among living humans. Studies of contemporary and recent cultures give great insight into how Stone Age ancestors might have lived and worked. Even the study of developmental psychology provides insights into the ways the modern mind might have evolved.

Naturalist Charles Darwin's theory of human evolution met with powerful resistance not only because it was contrary to a literal interpretation of the Christian scriptural story of human creation but also because it seemed to make humans helpless products of random processes. Humans may be powerless to control their biological evolution, but they are fully in control of the spectacular cultural evolution that provided people with language, knowledge, and civilization.

Although many people still see a conflict between religious beliefs about human origins and the scientific view, others have been able to reconcile the two doctrines and are comfortable with both their religious faith and their scientific knowledge.

Philosophers have always discussed what it means to be human and the origins of humankind. Scientific knowledge of human origins has had immeasurable impact on philosophy. Modern philosophy emphasizes the importance of language to humans, and anthropological studies of human evolution also point to the critical importance of language development in the final stages of human evolution. There is a growing conviction that the acquisition of language was the single most important step in the final evolutionary jump from Neanderthal to modern human. Language made it possible for humans to communicate with others and to pass along the accumulated wisdom and experience of one generation to the next. Language helped the pace of cultural development accelerate to a point where further biological evolution became irrelevant. Clothing and fire substituted for fur in a cold climate. Weapons were better than sharp claws and long teeth. Human evolution thus became cultural evolution.

—*Curtis G. Smith*

See also: Apes to hominids; Evolution: Historical perspective; Gene flow; Hominids; *Homo sapiens* and human diversification; Neanderthals.

Further Reading

Cann, R. L., M. Stoneking, and A. C. Wilson. "Mitochondrial DNA and Human Evolution." *Nature* 325 (1987): 31-36. This brief article presents a computer analysis of the mitochondrial DNA studies and makes the conclusion that all living humans descended from one African woman who lived about two hundred thousand years ago.

Johanson, D., and M. Edey. *Lucy: The Beginnings of Humankind.* New York: Simon & Schuster, 1981. This is the exciting story, told by a world-famous anthropologist, of his discovery of "Lucy," the diminutive three-and-a-half-million-year-old skeleton of one of humankind's australopithecine ancestors. Also gives a readable account of the state of modern research into human evolution, with careful explanations, diagrams of dating methods, and chronologies. Contains many pictures and diagrams, an appendix, and a good Bibliography of books and scientific papers.

Jolly, Alison. *Lucy's Legacy: Sex and Intelligence in Human Evolution.* Cambridge, Mass.: Harvard University Press, 1999. Traces four major transitions in human evolution, all based in cooperation, and posits that we are in the process of undergoing a fifth, incorporating species–wide, global communication.

Leakey, Mary. *Disclosing the Past.* New York: Doubleday, 1984. This is Mary Leakey's autobiography, but it serves as an unsurpassed history of field research into human origins over the past fifty years, told by an individual who was at the forefront of that story. The book contains photographs, maps, diagrams, and a completely captivating narrative of human evolution. Contains a list of twelve books for Bibliography.

McKee, Jeffrey K. *The Riddled Chain: Chance, Coincidence, and Chaos in Human Evolution.* New Brunswick, N.J.: Rutgers University Press, 2000. In contrast to the traditional top-down approach, McKee takes a bottom-up view of human evolution, in which each genetic variation must be tested through successive levels within a species. He argues that the large size of the current human population poises us for a rapid rate of evolution.

McKie, Robin. *Dawn of Man: The Story of Human Evolution.* New York: Dorling Kindersley, 2000. Published in conjunction with the Learning Channel series. Clearly written for a general audience, with many photographs and illustrations.

Shreeve, J. *The Neanderthal Enigma.* New York: William Morrow, 1995. The author wrote this book after spending many months interviewing leading anthropologists. The book presents the controversies over Neanderthal interpretation in an exciting and readable fashion. One may disagree with the author's hypothesis, but he tells a truly fascinating story. Contains an extensive Bibliography of both research articles and books.

Smith, C. G. *Ancestral Voices: Language and the Evolution of Human Consciousness.* Englewood Cliffs, N.J.: Prentice-Hall, 1985. This book is written for the general reader and presents a fusion of anthropology, linguistics, and neuroscience to describe the final emergence of the human species from its animal background. It emphasizes the importance of cultural evolution in shaping modern humans. Contains many illustrations, a list of references, and a list of suggested readings.

Solecki, R. S. *Shanidar, the First Flower People.* New York: Alfred A. Knopf, 1971. Solecki was the anthropologist who excavated a Neanderthal skeleton that pollen analysis revealed to have been laid on a bed of flowers. This is a poetic book, with interesting speculation about the "humanness" of the Neanderthals, and a very good story about the process of excavation. Solecki has been an authority in the field of late human evolution and gives a good account of the state of knowledge in the 1960's. Bibliography, pictures, and diagrams.

HUMAN GENETIC ENGINEERING

FIELDS OF STUDY

Biotechnology; genetics; cytogenetics; molecular genetics; biochemical genetics; genomics; population genetics; developmental genetics; clinical genetics; genetic counseling.

SUMMARY

Human genetic engineering is a branch of genetic engineering focusing on the understanding of human genes to produce applications that can improve human life. Genes, formulated by DNA (deoxyribonucleic acid), determine genotype—the complete genetic information carried by an individual, even if not expressed. Visible human characteristics, by contrast, are formed as the result of human genes interacting with the environment and are called the phenotype. Human genetic engineering aims to alter genotypes to cause changes in phenotypes; also, and more often, the knowledge of human genetics is used to engineer products, such as medications, that can cure or improve the quality of human life by addressing genetic disorders. Many of the applications of what is now known as human genetic engineering arose out of the mapping of the human genome during the Human Genome Project, completed in 2003.

PRINCIPAL TERMS

- **bioinformatics:** science of compiling and managing genetic and other biology data using computers, requisite in human genome research.
- **biologics:** medicines produced using genes and genetic manipulations.
- **clones:** genetically identical living organisms produced via genetic engineering.
- **DNA (deoxyribonucleic acid):** molecule, found in all living organisms, that by reproducing itself allows for the inheritance of characteristics from one generation to the next.
- **dysmorphology:** abnormal physical development resulting from a genetic disorder.
- **forensic genetics:** application of genetics, particularly DNA technology, to the analysis of evidence used in criminal cases and paternity testing.

Genetics

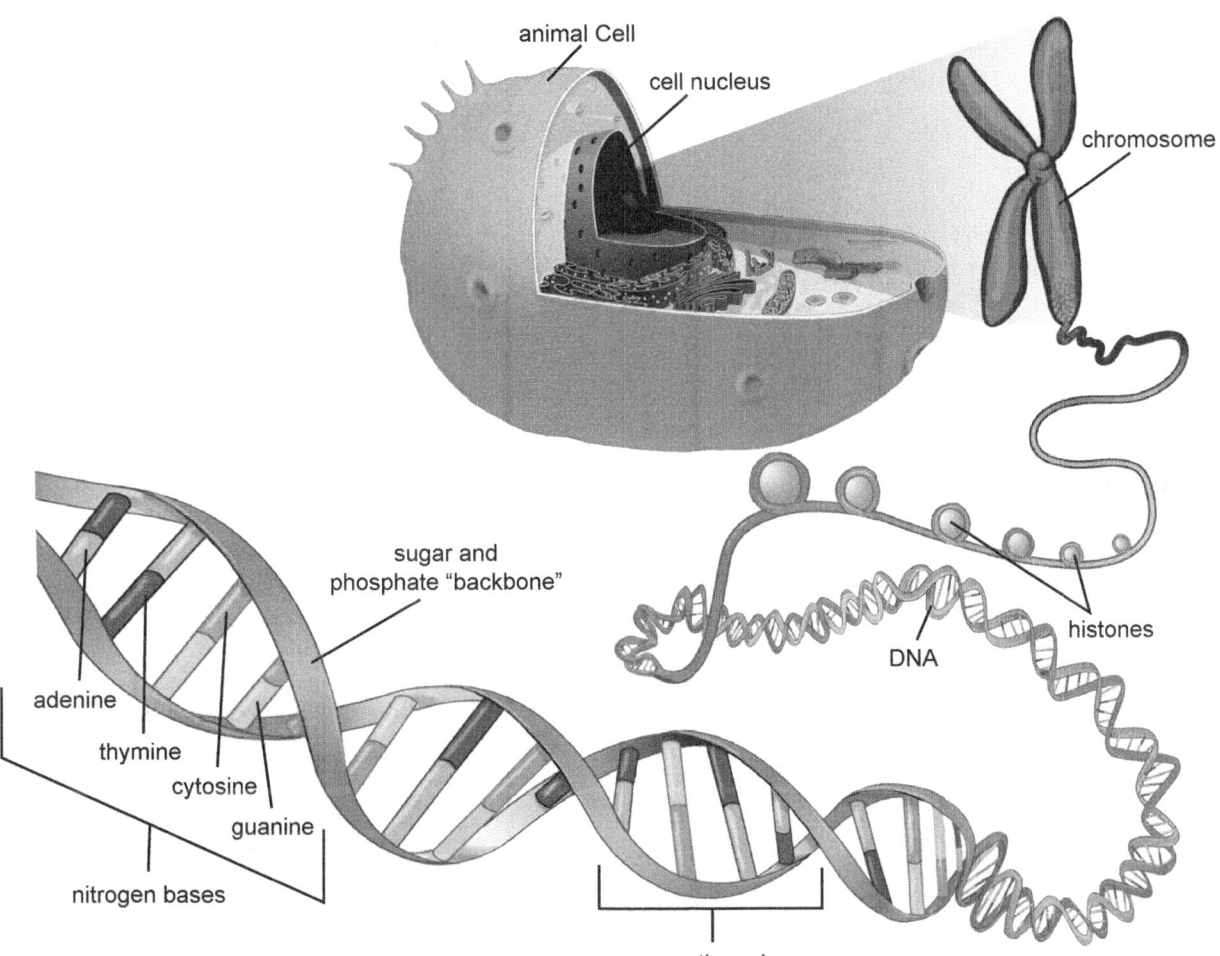

- **Gene:** specific DNA sequence that codes for a specific protein.
- **gene therapy:** use of a viral or other vector to incorporate new DNA into a person's cells with the objective of alleviating or treating the symptoms of a disease or condition.
- **genetic screening:** use of the techniques of genetics research to determine a person's risk of developing, or his or her status as a carrier of, a disease or other disorder.
- **gene transfer:** using a viral or other vector to incorporate new DNA into a person's cells; used in gene therapy.
- **genetic testing:** process of investigating a specific individual or population of people to detect the presence of genetic defects.
- **genomics:** branch of genetics dealing with the study of the genetic sequences of organisms, including the human being.
- **pharmacogenomics:** branch of human medical genetics that evaluates how an individual's genetic makeup influences his or her response to drugs.
- **proteomics:** study of how proteins are expressed in different types of cells, tissues, and organs.
- **recombinant DNA:** DNA that has been transferred from one cell to another. Genes are recombined from a human chromosome to another cell, usually from bacteria; if the transferred human genes code for insulin, the bacteria accepting the transferred genes will now produce human insulin.
- **stem cell:** progenitor cell that has the capability to become a more specialized cell, such as a kidney,

liver, or heart cell. once a cell becomes a specific kind of cell, it cannot change to another type of cell.
- **toxicogenomics:** science of evaluating ways in which genomes respond to chemical and other pollutants in the environment.

BASIC PRINCIPLES

Human genetic engineering is the science and technology of manipulating or changing human genes to alter or control visible characteristics of a human newborn or adult. Genes, formulated by DNA (deoxyribonucleic acid), are called the genotype. Visible human traits or characteristics, formed from the interaction of genes and the environment, are called the phenotype. Human genetic engineering aims to change genotypes to cause change in the phenotype. To understand human genetic engineering capabilities, it is important to understand basic genetic principles.

BACKGROUND

Human genetic engineering is a scientific endeavor, and as such this field builds on the information and knowledge gained from the decades of experimentation accomplished in years past. Without this foundation, human genetic engineering would not exist. This foundation brings the prospect of human cloning and the use of human genetics for therapeutic purposes.

A keystone event in modern genetics occurred in 1953, when American biologist James D. Watson and English physicist Francis Crick deduced the double-helical structure of DNA. This structural information enabled effective study of how genetic material codes for life. In 1968, the DNA genetic code was deciphered. Armed with this important genetic information, geneticists undertook the first recombinant DNA experiments on bacteria in 1973. The ambitious Human Genome Project started in 1990, with the goal of mapping out the entire human genetic sequence. In June, 2000, the first working draft of the human genetic sequence was produced from the efforts of this project. April, 2003, saw the announcement of the first complete human genetic sequence—breakthrough information for human genetic engineering.

Cloning, a subdiscipline of bioengineering, is the reproduction of genetically identical living organisms. In 1996, the first mammal was cloned, a sheep named Dolly. Other animals have been cloned since this pioneering event, including a bull in 1999 and a pig in 2000. The year 2003 saw the cloning of a mule, a horse, and a rat, followed by the cloning of a dog in 2005. Attempts at pet cloning have occurred: John Sperling, a wealthy and influential American educator, has funded pet-cloning projects, and researchers at Texas A&M University successfully cloned a cat in 2002. Commercial attempts at pet cloning started in April, 2004, with a company called Genetics Savings and Clone (now defunct) offering pet gene banking and cloning. Korean researchers published claims of successful human embryonic cloning in 2004, but these claims were later retracted because of fabricated data and other problems with the research. In May, 2010, the journal *Science* reported that scientists J. Craig Venter, Clyde Hutchison III, and Hamilton Smith had created a living creature in the laboratory. This new life-form, a bacterium, was artificially produced using genetic-engineering techniques. It had no ancestor and it reproduced, a key ability of living organisms.

Milestones in other areas of human genetic engineering—with more practicable and practical results—occurred in the areas of medicine, pharmaceuticals, forensics, and even psychology, as identified below in Applications. Many, if not most, of these blossomed shortly after the mapping of the human genome was completed in 2003. Many more will be developed as scientists and researchers continue to investigate the data that were gathered through that monumental accomplishment.

HOW IT WORKS

Until the middle of the eighteenth century, most biologists believed in spontaneous generation—that life arises from combinations of decaying matter, as if flies arose from garbage. It is now known that DNA genetically codes for many physical characteristics. The story of genetics starts with DNA and ends with protein. DNA, the genetic material found in the nucleus of every cell, codes for (that is, creates instructions for the building of) various proteins by means of components of the DNA molecule called nucleotides, which form the building blocks of DNA. The nucleotides establish the code.

Proteins make up the structural elements of the body, including collagen, ligaments, tendons, and muscles; some hormones, such as insulin, are made of protein as well. Perhaps most important, however, are the protein enzymes. All the enzymes in the body are made up of protein and protein alone. Enzymes are key because they accelerate chemical reactions. Thousands of chemical reactions occur in human bodies all the time. Protein enzymes catalyze all these reactions.

DNA, by dictating the production of enzymes, controls these chemical reactions. DNA dictates which proteins are produced by living and staying in the cell nucleus during the entire protein-making process. Much like a general in a command center, the DNA sends out orders but does not leave the nucleus. DNA is made up of nucleotide bases, and the first step in protein production involves the reading of the nucleotide code in the DNA. When the nucleotides are read, a strand of messenger ribonucleic acid (mRNA) is produced and sent from the nucleus into the cytoplasm of the cell. The DNA stays in the nucleus, while the mRNA leaves the nucleus. This process of reading the DNA nucleotide code and producing mRNA is called transcription.

The next step involves reading the nucleotide code on the mRNA in the cytoplasm of the cell. The cytoplasm is the liquid environment inside the cell where all the cellular organelles float. Transfer RNA (tRNA) reads the code on the mRNA, and this process is called translation. tRNA is called transfer because it transfers a specific amino acid when it reads the appropriate code on the mRNA.

The basic building blocks of protein are called amino acids. About twenty-two different amino acids build all the various proteins the body uses. It is like the English alphabet: Twenty-six letters and vowels comprise the English alphabet, and combining these various letters and vowels results in tens of thousands of different combinations and all the various words in the English language. Likewise, the body uses the different amino acids to form the tens of thousands of different proteins in the body. During translation, a specific amino acid is coded for and carried by the transfer RNA to a ribosome. Ribosomes (along with the rough endoplasmic reticulum) are where the cell's proteins are produced. Along ribosomes, different amino acids are transported by tRNA and linked, forming a protein molecule.

Some proteins may be only eighty or ninety amino acids long, whereas others, such as hemoglobin, may have more than 300 amino acids as their amino acid backbone. The way DNA codes for all this involves the nucleotide bases that make up DNA. Four nucleotide bases make up DNA: adenine, cytosine, guanine, and thymine. Adenine will chemically bind with thymine, and cytosine always chemically binds with guanine. When DNA is transcribed to form mRNA, if the nucleotide sequence in the DNA reads cytosine-cytosine-guanine, these nucleotide bases will code for guanine-guanine-cytosine in the mRNA. Then when mRNA is translated by tRNA, the code goes back to the original DNA code, cytosine-cytosine-guanine. Cytosine-cytosine-guanine can code for a specific amino acid, and in that fashion DNA codes for the amino acid sequence of all protein molecules.

The nucleotide base sequence in the mRNA is called the codon and the complimentary base sequence found in tRNA is called the anticodon. The example of how Dolly the sheep was cloned demonstrates how genetic engineering in mammals works, and, hence, how human cloning could work. Cloning is an ultimate example of genetic engineering because cloning produces an entire living organism via genetic engineering.

Dolly was a Finn Dorset sheep, which is all white. A Blackface ewe, named because of the distinctive black face these sheep have, was used as an egg donor and as a surrogate mother. Cells taken from a Finn Dorset ewe were grown in a tissue culture. An egg cell, from the Blackface ewe, had the nucleus removed. The nucleus contained the genes and DNA. The nucleus and genetic information from the Finn Dorset ewe were placed in the enucleated Blackface ewe egg cell. The Blackface ewe egg cell, now containing genetic information from the Finn Dorset ewe, was placed in the uterus of the Blackface ewe after an electric pulse is applied to stimulate growth and duplication of the cells. The Blackface ewe gave birth to Dolly, the all-white Finn Dorset ewe.

The newborn Finn Dorset ewe was an identical genetic copy of the Finn Dorset ewe originally used to harvest the genetic information found in the nucleus. Recombinant DNA refers to DNA transfer from one cell to another. In human genetic engineering, genes transfer from a human chromosome to another cell, usually bacteria. If the transferred human genes code for insulin, the bacteria accepting the transferred genes will now produce human insulin. In this process, the desired genes are isolated and removed from the human cell.

Bacterial cells have small, circular strips of DNA called plasmids. These circular plasmids are removed from the bacterial cells and opened up. Various enzymes are used to cut the human DNA and bacterial

DNA sequences at specific points. Restriction enzymes cut the original DNA in specific locations. DNA ligase pastes strips of DNA together. Scientists mix isolated human genes with the opened bacterial plasmids, along with DNA ligase. The human genes are spliced into the bacterial plasmid and the circle of genetic information in the bacterial plasmid closes. The bacterial plasmid with the spliced human genes is now called a vector. The plasmid vectors are taken up by the bacterial cells. Once inside the bacterial cells, the bacteria multiply and reproduce the spliced human genes. Whatever specific human genes were selected for splicing, for example, human insulin genes, are now functioning in the reproduced bacteria, and human insulin is harvested from the bacterial clones.

APPLICATIONS

Early Medical Applications. Recombinant DNA techniques are remarkable biological life adaptations, and many medicines based on this technology are used. Medications generated through human genetic engineering techniques have been in use since 1982, with the production of human insulin using recombinant DNA techniques. Human genes are inserted into a bacterial host that then makes human insulin. Prior to this type of genetic manipulation, diabetics needing insulin had to rely on insulin harvested from pigs or cows. Genetic techniques produce human growth hormone, previously only available from human cadavers. A genetically engineered hepatitis B vaccine has been in use since 1987. Since these first human genetic medicines and vaccines, many types of biological products have been introduced or are under current investigation and development. These new medicinal products are called biologics to distinguish them from chemically synthesized medicines. Genes and genetic manipulations produce biologics. Major types of biologics include hormones, antibodies, and cell-receptor proteins. Insulin and human growth hormone, discussed above, are classic protein hormones produced with recombinant DNA technology. The immune system produces protein antibodies that attack disease causing-agents such as bacteria and viruses. Genetic antibody production interferes with or attacks entities associated with diseases such as psoriatic arthritis and Crohn's disease. Recombinant DNA technology produces proteins binding with specialized white blood cells to reduce inflammation associated with rheumatoid arthritis.

Bioinformatics. The purpose of bioinformatics is to help organize, store, and analyze genetic biological information in a rapid and precise manner, dictated by the need to be able to access genetic information quickly. In the United States the online database that provides access to these gene sequences is called GenBank, which is under the purview of the National Center for Biotechnology Information (NCBI) and has been made available on the Internet. In addition to human genome sequence records, GenBank provides genome information about plants, bacteria, and animals other than humans.

Proteomics. Bioinformatics provides the basis for all modern studies of human genetics, including analyzing genes and gene sequences, determining gene functions, and detecting faulty genes. The study of genes and their functions is called proteomics, which involves the comparative study of protein expression. That is, it studies the metabolic and morphological relationship between the protein encoded within the genome and how that protein works. Geneticists are now classifying proteins into families, superfamilies, and folds according to their configuration, enzymatic activity, and sequence. Ultimately proteomics will complete the picture of the genetic structure and functioning of all human genes.

Toxicogenomics. Another newly developing field that relies on bioinformatics is the study of toxicogenomics, which is concerned with how human genes respond to toxins. As of 2011, this field is specifically concerned with evaluating how environmental factors negatively interact with mRNA translation, resulting in disease or dysfunction.

Gene Testing. In a gene-testing protocol, a sample of blood or body fluids is examined to detect a genetic anomaly such as the transposition of part of a chromosome or an altered sequence of the bases that comprise a specific gene, either of which can lead to a genetically based disorder or disease. As of 2011, more than 600 tests are available to detect malfunctioning or nonfunctioning genes. Most gene tests have focused on various types of human cancers, but other tests are being developed to detect genetic deficiencies that cause or exacerbate infectious and vascular diseases.

The emphasis on the relationship between genetics and cancer lies in the fact that all human cancers are genetically triggered or have a genetic basis. Some cancers are inherited as mutations, but most result from random genetic mutations that occur in specific cells, often precipitated by viral infections or environmental factors not yet well understood. At least four types of genetic problems have been identified in human cancers. The normal function of oncogenes, for example, is to signal the start of cell division. However, when mutations occur or oncogenes are overexpressed, the cells keep on dividing, leading to rapid growth of cell masses. The genetic inheritance of certain kinds of breast and ovarian cancers results from the nonfunctioning tumor-suppressor genes that normally stop cell division. When genetically altered tumor-suppressor genes are unable to stop cell division, cancer results. Conversely, the genes that cause inheritance of colon cancer result from the failure of DNA repair genes to correct mutations properly. The accumulation of mutations in these "proofreading" genes makes them inefficient or less efficient, and cells continue to replicate, producing a tumor mass.

If a gene screening reveals a genetic problem several options may be available, including gene therapy and genetic counseling. If the detected genetic anomaly results in disease, then pharmacogenomics holds promise of patient-specific drug treatment.

Gene Therapy. The science of gene therapy uses recombinant DNA technology to cure diseases or disorders that have a genetic basis. Still in its experimental stages, gene therapy may include procedures to replace a defective gene, repair a defective gene, or introduce healthy genes to supplement, complement, or augment the function of nonfunctional or malfunctioning genes. Several hundred protocols are being used in gene-therapy trials, and many more are under development. As of 2011, trials are focusing on two major types of gene therapy, somatic cell gene therapy and germ-line gene therapy.

Somatic cell gene therapy concentrates on altering a defective gene or genes in human body cells in an attempt to prevent or lessen the debilitating impact of a disease or other genetic disorder. Some examples of somatic cell gene therapy protocols now being tested include ones for adenosine deaminase (ADA) deficiency, cystic fibrosis, lung cancer, brain tumors, ovarian cancer, and AIDS. In somatic cell gene therapy a sample of the patient's cells may be removed and treated and then reintegrated into body tissue carrying the corrected gene. An alternative somatic cell therapy is called gene replacement, which typically involves insertion of a normally functioning gene. Some experimental delivery methods for gene insertion include use of retroviral vectors and adenovirus vectors. These viral vectors are used because they are readily able to insert their genomes into host cells. Hence, adding the needed (or corrective) gene segment to the viral genome guarantees delivery into the cell's nuclear interior. Nonviral delivery vectors that are being investigated for gene replacement include liposome fat bodies, human artificial chromosomes, and naked DNA (free DNA, or DNA that is not enclosed in a viral particle or any other "package"). Another type of somatic gene therapy involves blocking gene activity, whereby potentially harmful genes such as those that cause Marfan syndrome and Huntington's disease are disabled or destroyed.

Two types of gene-blocking therapies being investigated include the use of antisense molecules that target and bind to the mRNA produced by the gene, thereby preventing its translation, and the use of specially developed ribozymes that can target and cleave gene sequences that contain the unwanted mutation. Germline therapy is concerned with altering the genetics of male and female reproductive cells (gametes) as well as other body cells. Because germline therapy will alter the individual's genes as well as those of his or her offspring, both concepts and protocols are still very controversial. Some aspects of germ-line therapy now being explored include human cloning and genetic enhancement.

Clinical Genetics. Clinical genetics is that branch of medical genetics involved in the direct clinical care of people afflicted with diseases caused by genetic disorders. Clinical genetics involves diagnosis, counseling, management, and support. Genetic counseling is a part of clinical genetics directly concerned with medical management, risk determination and options, and decisions regarding reproduction of afflicted individuals. Support services are an integral feature of all genetic counseling themes. Clinical genetics begins with an accurate diagnosis that recognizes a specific, underlying genetic cause of a physical or biochemical defect following guidelines outlined by the National Institutes of Health (NIH) Counseling Development Conference. Clinical practice includes several hundred genetic tests that are able to detect mutations such as those associated with breast and colon cancers, muscular dystrophy, cystic fibrosis, sickle-cell disease, and Huntington's disease.

Genetic counseling follows clinical diagnosis and focuses initially on explaining the risk factors and human problems associated with the genetic disorder. Both the afflicted individual and family members are involved in all counseling procedures. Important components include a frank discussion of risks, of options such as preventive operations, and of options involved with regard to reproduction. All reproductive options are described along with their potential consequences, but genetic counseling is a support service rather than a directive mode. That is, it does not include recommendations. Instead, its ultimate mission is to help both the afflicted individuals and their families recognize and cope with the immediate and future implications of the genetic disorder.

Pharmacogenomics. That branch of human medical genetics dealing with the correlation of specific drugs to fit specific diseases in individuals is called pharmacogenomics. This field recognizes that individuals may metabolically respond differentially to therapeutic medicines based on their genetic makeup. It is anticipated that testing human genome data will greatly speed the development of new drugs that not only target specific diseases but also will be tailored to the specific genetics of patients.

Forensic Genetics. Forensic genetics is the use of human genetics in criminal or paternity cases. For example, DNA testing on blood, saliva, or other tissue can be used to determine the source of evidence, such as blood stains or semen, left at a crime scene. Forensic DNA analysis is also used to determine paternity and other kinship. Finally, with the increasing use of forensic genetics since the 1990's, some incarcerated prisoners have been released after it was clearly determined that they could not possibly have been guilty of crimes they were convicted of, as DNA evidence eliminated them from suspicion.

Potential for Human Cloning. Human therapeutic cloning involves the production of cloned human embryos, with the idea of harvesting embryonic stem cells. The hope is that the stem cells can be grown into a wide variety of cells to replace or repair organs, such as liver, kidney, or heart cells. Although human cloning has not yet reached this potential, future applications could offer identically matched kidneys for people with failing kidneys or even a genetically duplicate heart for someone in severe heart failure.

FUTURE PROSPECTS

The Human Genome Project painstakingly mapped out the human DNA sequence in 2003, after a decade and a half of meticulous multicenter collaboration. Genetic databases are now rapidly filling with genetic detail because of technological advances in the speed of analyzing DNA sequences. While the speed of this analysis has increased considerably, the price of such investigations has dropped significantly. Genetic databases currently hold information on a wide variety of life-forms, and significant amounts of new generic information is added frequently. The DNA sequencing found in genetic databases provides the burgeoning field of synthetic biology with important basic information needed for human genetic engineering. This information can be used for modeling and as supply depots for the mixing and matching of genes. As the speed of genetic analysis has increased significantly and the price of genetic investigations has dropped considerably, the process of DNA synthesis is much less expensive and faster than it was in the beginning of the twenty-first century. More genetic information, faster artificial DNA synthesis, and significant technological cost savings result in more feasible human genetic engineering projects. Genes are the stuff of life, and the field is on the verge of changing life and even making new life-forms, via genetic engineering. How and what changes are made will present significant bioethical and societal challenges, along with potentially fantastic and beneficial results.

Richard P. Capriccioso, MD

FURTHER READING

Andrews, Lori B. *The Clone Age: Adventures in the New World of Reproductive Technology.* New York: Henry Holt, 1999. A lawyer specializing in reproductive technology, Andrews examines the legal ramifications of human cloning, from privacy to property rights.

Baudrillard, Jean. *The Vital Illusion.* Edited by Julia Witwer. New York: Columbia University Press, 2000. A sociological perspective on what human cloning means to the idea of what it means to be human.

Capriccioso, Richard P. "Genetic Testing." In *Salem Health: Cancer*, edited by Jeffrey A. Knight. Pasadena, Calif.: Salem Press, 2009. A comprehensive overview of genetic testing covering different types of genetic tests with a review of the science behind the testing. *The Economist.* "Artificial Lifeforms: Genesis Redux." 395, no. 8683 (May 20, 2010): 81-83. Informative article on synthetic biology and the creation of a new form of life in the laboratory.

Hartwell, Leland, et al. *Genetics: From Genes to Genomes.* 4th ed. New York: McGraw-Hill, 2011. A comprehensive textbook on genetics, including human genetics discussed in a comparative context.

Hekimi, Siegfried, ed. *The Molecular Genetics of Aging: Results and Problems in Cell Differentiation.* Berlin: Springer, 2000. Examines various genetic aspects of the aging process. Illustrated.

Jorde, Lynn B., John C. Carey, and Michael J. Bamshad. *Medical Genetics.* 4th ed. Philadelphia: Mosby, 2010. Provides both an introduction to the field of human genetics with chapters on clinical aspects of human genetics, such as gene therapy, genetic screening, and genetic counseling.

Lewis, Ricki. *Human Genetics: Concepts and Applications.* 9th ed. New York: McGraw-Hill, 2010. This textbook provides a broad overview of human genetics and genomics.

Pasternak, Jack J. *An Introduction to Human Molecular Genetics: Mechanisms of Inherited Diseases.* 2d ed. Hoboken, N.J.: Wiley-Liss, 2005. Discusses treatment advances, fundamental molecular mechanisms that govern human inherited diseases, the interactions of genes and their products, and the consequences of these mechanisms on disease states in major organ systems such as muscles, the nervous system, and the eyes. Also addresses cancer and mitochondrial disorders.

Rudin, Norah, and Keith Inman. *An Introduction to Forensic DNA Analysis.* 2d ed. Boca Raton, Fla.: CRC Press, 2002. An overview of many DNA typing techniques, along with numerous examples and a discussion of legal implications.

Shostak, Stanley. *Becoming Immortal: Combining Cloning and Stem-Cell Therapy.* Albany: State University of New York Press, 2002. Examines the question of whether human beings are equipped for potential immortality.

Wilson, Edward O. *On Human Nature.* 1978. Cambridge, Mass.: Harvard University Press, 2004. A look at the significance of biology and genetics for the way people understand human behaviors, including aggression, sex, and altruism, and the institution of religion.

WEB SITES

American Society of Human Genetics (ASHG) http://www.ashg.org

Center for Genetics and Society http://www.geneticsandsociety.org

Genetics Education Center University of Kansas Medical Center http://www.kumc.edu/gec

Human Genome Project http://www.ornl.gov/sci/techresources/Human_Genome/home.shtml

National Institutes of Health Stem Cell Information http://stemcells.nih.gov/info/basics

National Society of Genetic Counselors http://www.nsgc.org

See also: Bioengineering; Cell and Tissue Engineering; Cloning; DNA Analysis.

HUMAN-COMPUTER INTERACTION

FIELDS OF STUDY

Computer science; computer graphics; software engineering; systems analysis; graphic design; industrial design; ergonomics; mechanical engineering; information science; information architecture; robotics; artificial intelligence; cognitive science; psychology; social psychology; linguistics; neurobiology; psychophysics; social neuroscience; anthropology; scientific computing; data visualization; typography; anthropometrics.

SUMMARY

Human-computer interaction (HCI) is a field concerned with the study, design, implementation, evaluation, and improvement of the ways in which human beings use or interact with computer systems. The importance of human-computer interaction within the field of computer science has grown in tandem with technology's potential to help people accomplish an increasing number and variety of personal, professional, and social goals. For example, the development of user-friendly interactive computer interfaces, Web Sites, games, home appliances, office equipment, art installations, and information distribution systems such as advertising and public awareness campaigns are all applications that fall within the realm of HCI.

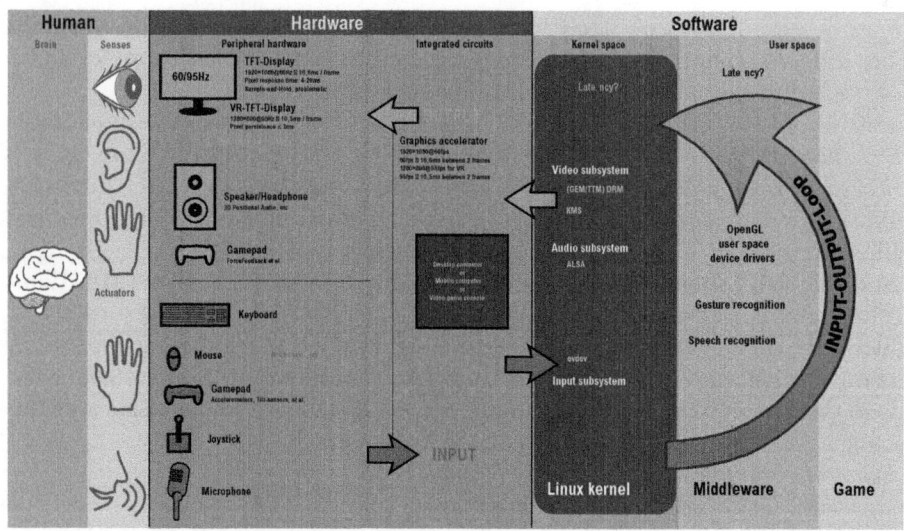

The user interacts directly with hardware for the human input and output such as displays, e.g. through a graphical user interface. The user interacts with the computer over this software interface using the given input and output (I/O) hardware. Software and hardware must be matched, so that the processing of the user input is fast enough, the latency of the computer output is not disruptive to the workflow.

PRINCIPAL TERMS

- **accessibility:** extent to which an interface can be used by people with visual, auditory, cognitive, or physical impairments.
- **direct manipulation:** interacting with a graphic representation of an object to accomplish a task.
- **ethnography:** process of observing, interviewing, and analyzing the activities of people in their everyday environments in order to gain the perspective of a user.
- **graphical user interface (GUI):** means of interacting with an electronic device based on images rather than text.
- **heuristic:** guideline or rule of thumb used to quickly evaluate an interface or product.
- **information architecture:** study and practice of organizing data so that they can be found and used efficiently.
- **Likert scale:** method of ascribing quantitative value to qualitative data, used in questionnaires; typically asks respondents to rate how much they agree or disagree with a statement.
- **ubiquitous computing:** model of computing that sees technology as being fully integrated into every aspect of daily life rather than limited to the functionality of a machine on a desktop.
- **usability:** extent to which a product can easily, efficiently, and effectively be used to achieve a certain goal.
- **wayfinding:** ways in which users orient themselves and navigate from place to place within an interface.
- **widget:** interactive component on a Web site, such as one that allows the user to click through a list of options and choose one.
- **wire frame:** skeleton version of a Web site that leaves out visual design elements and focuses on how pages will be linked.

BASIC PRINCIPLES

Human-computer interaction is an interdisciplinary science with the primary goal of harnessing the full potential of computer and communication systems for the benefit of individuals and groups. HCI researchers design and implement innovative interactive technologies that are not only useful but also easy and pleasurable to use and anticipate and satisfy the specific needs of the user. The study of HCI has applications throughout every realm of modern life, including work, education, communications, health care, and recreation.

The fundamental philosophy that guides HCI is the principle of user-centered design. This philosophy proposes that the development of any product or interface should be driven by the needs of the person or people who will ultimately use it, rather than by any design considerations that center around the object itself. A key element of usability is affordance, the notion that the appearance of any interactive element should suggest the ways in which it can be manipulated. For example, the use of shadowing around a button on a Web site might help make it look three-dimensional, thus suggesting that it can be pushed or clicked. Visibility is closely related to affordance; it is the notion that the function of all the controls with which a user interacts should be clearly mapped to their effects. For example, a label such as "Volume Up" beneath a button might indicate exactly what it does. Various protocols facilitate the creation of highly usable applications. A cornerstone of HCI is iterative design, a method of development that uses repeated cycles of feedback and analysis to improve each prototype version of a product, instead of simply creating a single design and launching it immediately. To learn more about the people who will eventually use a product and how they will use it, designers also make use of ethnographic field studies and usability tests.

Background

Before the advent of the personal computer, those who interacted with computers were largely technology specialists. In the 1980's, however, more and more individual users began making use of software such as word-processing programs, computer games, and spreadsheets. HCI as a field emerged from the growing need to redesign such tools to make them practical and useful to ordinary people with no technical training. The first HCI researchers came from a variety of related fields: cognitive science, psychology, computer graphics, human factors (the study of how human capabilities affect the design of mechanical systems), and technology. Among the thinkers and researchers whose ideas have shaped the formation of HCI as a science are John M. Carroll, best known for his theory of minimalism (an approach to instruction that emphasizes real-life applications and the chunking of new material into logical parts), and Adele Goldberg, whose work on early software interfaces at the Palo Alto Research Center (PARC) was instrumental in the development of the modern graphical user interface.

In the early days of HCI, the notion of usability was simply defined as the degree to which a computer system was easy and effective to use. However, usability has come to encompass a number of other qualities, including whether an interface is enjoyable, encourages creativity, relieves tension, anticipates points of confusion, and facilitates the combined efforts of multiple users. In addition, there has been a shift in HCI away from a reliance on theoretical findings from cognitive science and toward a more hands-on approach that prioritizes field studies and usability testing by real participants.

How It Works

Input and Output Devices. The essential goal of HCI is to improve the ways in which information is transferred between a user and the machine he or she is using. Input and output devices are the basic tools HCI researchers and professionals use for this purpose. The more sophisticated the interaction between input and output devices—the more complex the feedback loop between the two directions of information flow—the more the human user will be able to accomplish with the machine.

An input device is any tool that delivers data of some kind from a human to a machine. The most familiar input devices are the ones associated with personal computers: keyboards and mice. Other commonly used devices include joysticks, trackballs, pen styluses, and tablets. Still more unconventional or elaborate input devices might take the shape of head gear designed to track the movements of a user's head and neck, video cameras that track the movements of a user's eyes, skin sensors that detect changes in body temperature or heart rate, wearable gloves that precisely track hand gestures, or automatic speech recognition devices that translate spoken commands into instructions that a machine can understand. Some input devices, such as the sensors that open automatic doors at the fronts of banks or supermarkets, are designed to record information passively, without the user having to take any action.

An output device is any tool that delivers information from a machine to a human. Again, the most familiar output devices are those associated with personal computers: monitors, flat-panel displays, and audio speakers. Other output devices include wearable head-mounted displays or goggles that provide visual feedback directly in front of the user's field of vision and full-body suits that provide tactile feedback to the user in the form of pressure.

Perceptual-Motor Interaction. When HCI theorists speak about perceptual-motor interaction, what they are referring to is the notion that users' perceptions—the information they gather from the machine—are inextricably linked to their physical actions, or how they relate to the machine. Computer systems can take advantage of this by using both input and output devices to provide feedback about the user's actions that will help him or her make the next move. For example, a word on a Web site may change in color when a user hovers the mouse over it, indicating that it is a functional link. A joystick being used in a racing game may exert what feels like muscular tension or pressure against the user's hand in response to the device being steered to the left or right. Ideally, any feedback a system gives a user should be aligned to the physical direction in which he or she is moving an input device. For example, the direction in which a cursor moves on screen should be the same as the direction in which the user is moving the mouse. This is known as kinesthetic correspondence. Another technique HCI researchers have devised to facilitate the feedback loop between a user's perceptions and actions is known as augmented reality.

With this approach, rather than providing the user with data from a single source, the output device projects digital information, such as labels, descriptions, charts, and outlines, on the physical world. When an engineer is looking at a complex mechanical system, for example, the display might show what each part in the system is called and enable him or her to call up additional troubleshooting or repair information.

APPLICATIONS

Computers. At one time, interacting with a personal computer required knowing how to use a command-line interface in which the user typed in instructions—often worded in abstract technical language—for a computer to execute. A graphical user interface, based on HCI principles, supplements or replaces text-based commands with visual elements such as icons, labels, windows, widgets, menus, and control buttons. These elements are controlled using a physical pointing device such as a mouse. For instance, a user may use a mouse to open, close, or resize a window or to pull down a list of options in a menu in order to select one. The major advantage graphical user interfaces have over text-based interfaces is that they make completing tasks far simpler and more intuitive. Using graphic images rather than text reduces the amount of time it takes to interpret and use a control, even for a novice user. This enables users to focus on the task at hand rather than to spend time figuring out how to manipulate the technology itself. For instance, rather than having to recall and then correctly type in a complicated command, a user can print a particular file by selecting its name in a window, opening it, and clicking on an icon designed to look like a printer. Similarly, rather than choosing options from a menu in order to open a certain file within an application, a user might drag and drop the icon for the file onto the icon for the application.

Besides helping individuals navigate through and execute commands in operating systems, software engineers also use HCI principles to increase the usability of specific computer programs. One example is the way pop-up windows appear in the word-processing program Microsoft Word when a user types in the salutation in a letter or the beginning item in a list. The program is designed to recognize the user's task, anticipate the needs of that task, and offer assistance with formatting customized to that particular kind of writing.

Consumer Appliances. Besides computers, a host of consumer appliances use aspects of HCI design to improve usability. Graphic icons are ubiquitous parts of the interfaces commonly found on cameras, stereos, microwave ovens, refrigerators, and televisions. Smartphones such as Apple's iPhone rely on the same graphic displays and direct manipulation techniques as used in full-sized computers. Many also add extra tactile, or haptic, dimensions of usability such as touch-screen keyboards and the ability to rotate windows on the device by physically rotating the device itself in space. Entertainment products such as video game consoles have moved away from keyboard and joystick interfaces, which may not have kinesthetic correspondence, toward far more sophisticated controls. The hand-held device that accompanies the Nintendo Wii, for instance, allows players to control the motions of avatars within a game through the natural movements of their own bodies. Finally, HCI research influences the physical design of many household devices. For example, a plug for an appliance designed with the user in mind might be deliberately shaped so that it can be inserted into an outlet in any orientation, based on the understanding that a user may have to fit several plugs into a limited amount of space, and many appliances have bulky plugs that take up a lot of room.

Increasingly, HCI research is helping appliance designers move toward multimodal user interfaces. These are systems that engage the whole array of human senses and physical capabilities, match particular tasks to the modalities that are the easiest and most effective for people to use, and respond in tangible ways to the actions and behaviors of users. Multimodal interfaces combine input devices for collecting data from the human user (such as video cameras, sound recording devices, and pressure sensors) with software tools that use statistical analysis or artificial intelligence to interpret these data (such as natural language processing programs and computer vision applications). For example, a multimodal interface for a GPS system installed in an automobile might allow the user to simply speak the name of a destination aloud rather than having to type it in while driving. The system might use auditory processing of the user's voice as well as visual processing of his or her lip movements to more accurately interpret speech. It might also use a camera to closely follow the movements of the user's eyes, tracking his or her gaze from one part of the screen to another and using this information to helpfully zoom in on particular parts of the map or automatically select a particular item in a menu.

Workplace Information Systems. HCI research plays an important role in many products that enable people to perform workplace tasks more effectively. For example, experimental computer systems are being designed for air traffic control that will increase safety and efficiency. Such systems work by collecting data about the operator's pupil size, facial expression, heart rate, and the forward momentum and intensity of his or her mouse movements and clicks. This information helps the computer interpret the operator's behavior and state of mind and respond accordingly. When an airplane drifts slightly off its course, the system analyzes the operator's physical modalities. If his or her gaze travels quickly over the relevant area of the screen, with no change in pupil size or mouse click intensity, the computer might conclude that the operator has missed the anomaly and attempt to draw attention to it by using a flashing light or an alarm.

Other common workplace applications of HCI include products that are designed to facilitate communication and collaboration between team members, such as instant messaging programs, wikis (collaboratively edited Web Sites), and video-conferencing tools. In addition, HCI principles have contributed to many project management tools that enable groups to schedule and track the progress they are making on a shared task or to make changes to common documents without overriding someone else's work.

Education and Training. Schools, museums, and businesses all make use of HCI principles when designing educational and training curricula for students, visitors, and staff. For example, many school districts are moving away from printed textbooks and toward interactive electronic programs that target a variety of information-processing modalities through multimedia. Unlike paper and pencil worksheets, such programs also provide instant feedback, making it easier for students to learn and understand new concepts. Businesses use similar programs to train employees in such areas as the use of new software and the company's policies on issues of workplace ethics. Many art and science museums have installed electronic kiosks with touchscreens that visitors can use to learn more about a particular exhibit. HCI principles underlie the design of such kiosks. For example, rather than using a text-heavy interface, the screen on an interactive kiosk at a science museum might display video of a museum staff member talking to the visitor about each available option.

FUTURE PROSPECTS

As HCI moves forward with research into multimodal interfaces and ubiquitous computing, notion of the computer as an object separate from the user may eventually be relegated to the archives of technological history, to be replaced by wearable machine interfaces that can be worn like clothing on the user's head, arm, or torso. Virtual reality interfaces have been developed that are capable of immersing the user in a 360-degree space that looks, sounds, feels, and perhaps even smells like a real environment—and with which they can interact naturally and intuitively, using their whole bodies. As the capacity to measure the physical properties of human beings becomes ever more sophisticated, input devices may grow more and more sensitive; it is possible to envision a future, for instance, in which a machine might "listen in" to the synaptic firings of the neurons in a user's brain and respond accordingly. Indeed, it is not beyond the realm of possibility that a means could be found of stimulating a user's neurons to produce direct visual or auditory sensations. The future of HCI research may be wide open, but its essential place in the workplace, home, recreational spaces, and the broader human culture is assured.

M. Lee, MA

FURTHER READING

Bainbridge, William Sims, ed. *Berkshire Encyclopedia of Human-Computer Interaction: When Science Fiction Becomes Fact.* 2 vols. Great Barrington, Mass.: Berkshire, 2004. Contains more than a hundred HCI topics, including gesture recognition, natural language processing, and education. Each article includes Bibliography listings and may contain sidebars, figures, tables, or photographs.

Helander, Martin. *A Guide to Human Factors and Ergonomics.* 2d ed. Boca Raton, Fla.: CRC Press/Taylor & Francis, 2006. Discusses the cognitive and physical aspects of human capabilities that inform product development, and includes an appendix examining the use of checklists to improve safety and effectiveness in human factors design.

Sears, Andrew, and Julie A. Jacko, eds. *The Human-Computer Interaction Handbook: Fundamentals, Evolving Technologies, and Emerging Applications.* 2d ed. New York: Lawrence Erlbaum Associates, 2008. Prominent researchers in the field address issues in design, development, testing, and evaluation. Contains hundreds of explanatory figures and tables.

Sharp, Heken, Yvonne Rogers, and Jenny Preece. *Interaction Design: Beyond Human-Computer Interaction.* 2d ed. Hoboken, N.J.: John Wiley & Sons, 2007. Each chapter on interaction design contains an outline, Summary, subsections, and text boxes highlighting case studies and other points of focus. Includes a comprehensive Bibliography.

Thatcher, Jim, et al. *Web Accessibility: Web Standards and Regulatory Compliance.* New York: Springer, 2006. Presents an overview of laws, policies, and technical standards for creating accessible Web Sites.

Tufte, Edward R. *The Visual Display of Quantitative Information.* 2d ed. Cheshire, Conn.: Graphics Press, 2007. A seminal work on ways in which complex information can be clearly and simply displayed in visual forms such as graphics, maps, charts, and tables.

WEB SITES

Human Factors and Ergonomics Society Internet Technical Group http://www.internettg.org

Special Interest Group on Human Computer Interaction http://www.sigchi.org

---

FASCINATING FACTS ABOUT HUMAN-COMPUTER INTERACTION

- Children working with scientists at a University of Maryland HCI laboratory created their own toy: a set of animal blocks that each plays a recorded factoid about an animal and reveal its name. When the blocks are jumbled up, they create a nonsensical animal using syllables from each animal's name.
- Roomba, a robotic vacuum cleaner, appeals so well to users' emotions (through features such as "cute" chirping alert noises when the robot finds itself stuck in a corner) that many people find themselves naming and talking to their robots.
- To print a file using a command-line interface rather than a graphical user interface, a user would have to type in the instruction "print [/d:Printer] [Drive:][Path] FileName [. . .]."
- HCI researchers have coined the term critical incident to describe times when a piece of technology makes a person feel frustrated, angry, or confused. Some 80 percent of computer users say frustrating experiences with their machines have led them to shout expletives out loud.
- One common application of HCI research in Japan is smart toilets with automatic seat warmers and speakers that emit flushing sounds to mask bathroom noises.
- The mouse was invented in 1964 and began its life as a simple wooden box with rolling wheels.

HYDROPHILIC AND HYDROPHOBIC

FIELDS OF STUDY

Biochemistry; Chemical Engineering; Physical Chemistry

SUMMARY

The characteristic properties of hydrophilic and hydrophobic materials are discussed. Hydrophilic materials readily interact with liquid water, while hydrophobic materials do not. These properties are important in biochemical systems and chemical engineering processes.

PRINCIPAL TERMS

- **cohesion:** the tendency for like molecules of a substance to stick together due to their shape and electronic structure.
- **hydrogen bond:** a weak type of chemical bond formed by the attraction of a hydrogen atom to an electro-

HYDROPHILIC AND HYDROPHOBIC

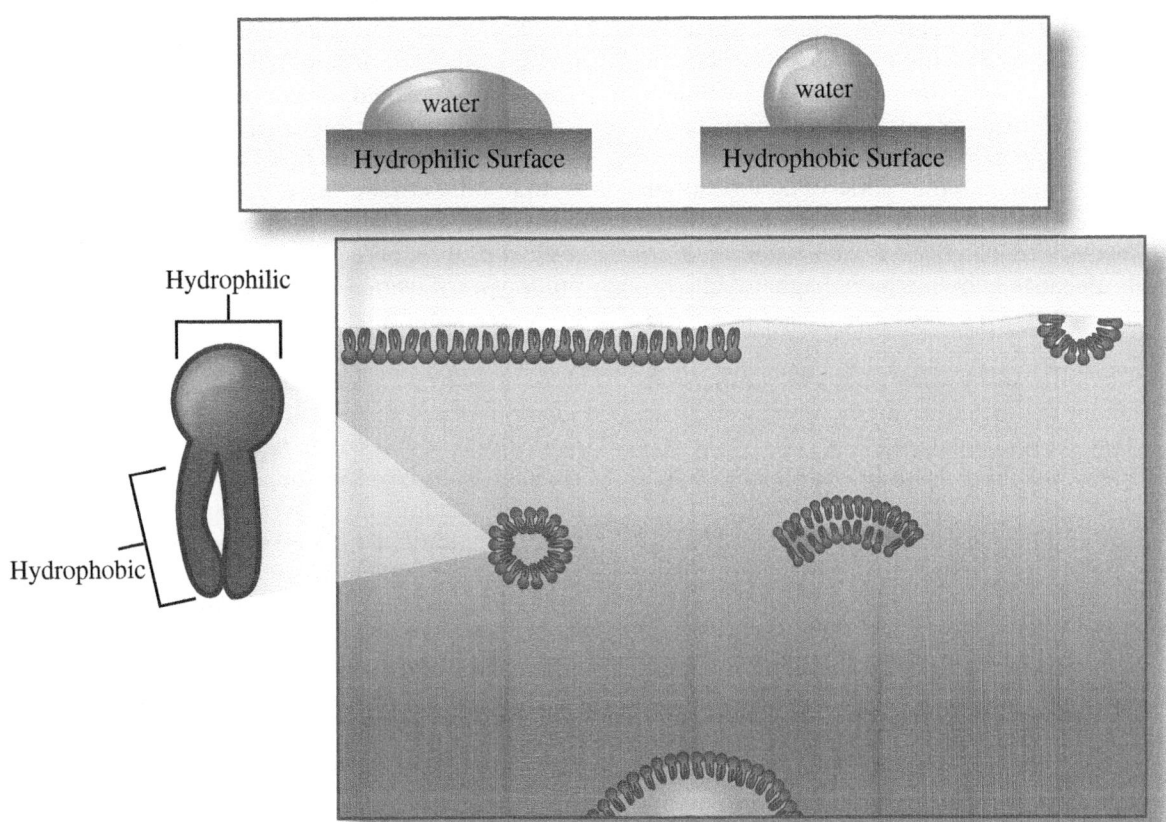

Water with hydrophilic and hydrophobic molecules.

negative atom—an atom with a strong tendency to attract electrons—in the same or another molecule.
- **polarity:** a characteristic of a molecule or functional group in which there is a difference in the distribution of electronic charge, causing one part of the molecule or group to be relatively electrically positive and another part to be relatively electrically negative.
- **repulsion:** an oppositional force that pushes two entities apart, such as the electrostatic repulsion between particles of like electrical charge.
- **solubility:** the ability of a particular substance, or solute, to dissolve in a particular solvent at a given temperature and pressure.

The Nature of Hydrophilic and Hydrophobic Materials

The terms "hydrophilic" and "hydrophobic" refer specifically to the interaction of a particular material with liquid water. The two words literally mean "water loving" and "water fearing," respectively, and they aptly describe the behavior of the corresponding materials: hydrophilic materials readily interact with water, while hydrophobic materials do not.

To understand this, one must understand the structure and electronic properties of the water molecule. A water molecule consists of two hydrogen atoms covalently bonded to a single oxygen atom. Each hydrogen atom contributes its single electron to the formation of the covalent bond. The electronegativity of the oxygen atom tends to keep the electron density of the oxygen-hydrogen bond locked between those two atoms, leaving the other side of each hydrogen atom bare of electron density and effectively exposing the positive charge of the hydrogen nuclei. At the same time, the oxygen has two lone pairs of electrons essentially bulging out from the other side of the molecule, creating a high electron density and negative charge. There is therefore a high degree of charge separation between the exposed nuclei of the hydrogen atoms and the lone electron pairs of the oxygen atom, giving the molecule as a whole a pronounced electrical polarity and creating what is known as a "dipole moment."

This dipole moment is the source of water's rather unique physical properties. The polarity of the molecules effectively enables them to stick together, positive end to negative end, like little magnets. The resulting cohesion is due to the formation of hydrogen bonds between water molecules. This gives water very high melting and boiling points in comparison to other molecules of similar mass, as well as a great ability to dissolve other materials.

Hydrophilic materials are able to interact with the structural and electronic properties of water to different degrees. They may undergo solvation (that is, dissolve), ionic dissociation, or surface wetting (that is, adhere to the water molecules, thus becoming wet). Hydrophilic materials also tend to have polar molecular structures or similar electronic properties. Hydrophobic materials, on the other hand, are generally nonpolar covalent materials that do not dissolve in water and do not exhibit surface wetting.

SOLVENT PARTITIONING AND PHASE-TRANSFER CATALYSIS

Compounds have different solubilities in different solvents. Ionic compounds dissolve well in water but not in organic solvents such as diethyl ether or dichloromethane, while covalently bonded compounds, such as organic molecules, generally do not dissolve well in water. For any particular compound, solubility is a matter of degree, and manipulating the form of the molecule greatly affects its solubility. For example, a neutral amine will dissolve well in acidic water, but if the amine solution is neutralized or made basic by the addition of a stronger base, the amine will precipitate out of the solution as an undissolved solid. Organic acids behave in a similar manner with respect to stronger acids. By adjusting the acidity or basicity of a solution containing different components, it is a simple exercise to isolate the different compounds from each other with a second solvent that is immiscible (that is, does not mix) with the first. If one of the solvents is water or some other hydrophilic substance, the other must be a hydrophobic solvent that will not become dissolved in the existing solution. When the second solvent is added and mixed together with the solution, a material will migrate from one solution into the other, according to its solubility in each one. The process is called solvent partitioning.

A related process, phase-transfer catalysis, uses two immiscible solvents such as water and the organic compound benzene (C_6H_6). Each solution in the process is considered a "phase"—that is, matter with distinct properties that touches but does not mix with other types of matter—and the surface between them is known as the "phase boundary." This technique is used when two reactant materials will not dissolve in the same solvent. Such reactants are typically an ionic compound and a nonionic compound and may even be in different states of matter. In the procedure, each reactant is dissolved in the appropriate solvent, and the two solutions are stirred together vigorously. A third compound, the phase-transfer catalyst, is added and binds reversibly to one of the reactants to form an addition product, or adduct, that will dissolve well in the other solvent. The adduct transfers from one liquid phase into the other, where the two reactant species can contact each other and undergo the desired reaction. As in solvent partitioning, one solvent must be hydrophobic and the other must be hydrophilic.

BIOLOGICAL SYSTEMS

Hydrophilic and hydrophobic properties are essential in biological systems and are the functional basis of such fundamentally important aspects as cell structure. Every animal cell, and the various organelles that it contains, is enclosed within a cell membrane composed of phospholipid molecules. One common type of phospholipid consists of a glycerol molecule backbone to which are attached two long-chain fatty acids and a phosphate ion (PO_4^{3-}) bonded to an alcohol group. The phosphate ion is bonded to one of the hydroxyl (–OH) substituents of the glycerol molecule, while the carboxyl groups (–COOH) of the fatty acids are bonded to the other two hydroxyl substituents of the glycerol. Overall, a phospholipid molecule has the general structure:

Hydrophilic section   **Hydrophobic section**

A phospholipid molecule thus has two distinct regions. The phosphate end of the molecule is strongly hydrophilic, while the long-chain fatty acids, which are nonpolar and cannot form hydrogen bonds, are decidedly hydrophobic and are often referred to as "greasy." The fatty-acid section of the phospholipid molecule is actually much larger than the phosphate portion (the diagram is not to scale). There is a simple rule of solubility that "like dissolves like," meaning that polar solvents such as water generally dissolve polar materials, and nonpolar solvents such as hydrocarbons generally dissolve nonpolar materials. The two different regions of the phospholipid molecule therefore do not interact well with each other but are instead attracted to the corresponding sections of other molecules.

In large quantity in an aqueous environment, the long-chain fatty-acid portions of the phospholipid molecules are forced to aggregate and effectively blend into each other by the surrounding water molecules. The fatty acids cannot dissolve into each other completely, however, due to the repulsion between the phosphate groups to which they are attached. As a result, the molecules automatically form a sandwich-like structure known as a "bilayer." The two outer surfaces of the bilayer consist of the hydrophilic phosphate portions of the phospholipid molecules, often called the "heads." Between them is a thick, greasy hydrophobic layer of intertwined long-chain hydrocarbon groups, often called the "tails." This bilayer fully encloses the interior contents of animal cells, forming an integral membrane that can permit the passage of materials from one side to the other by various mechanisms. Being hydrophilic, the inner and outer surfaces can interact freely with the water-based fluids on either side of the membrane.

Richard M. Renneboog, MSc

Further Reading

Askeland, Donald R., Wendelin J. Wright, D. K. Bhattacharya, and Raj P. Chhabra. *The Science and Engineering of Materials*. Boston: Cengage Learning, 2016. Print.

Berg, Jeremy M., John L. Tymoczko, Gregory J. Gatto, and Lubert Strye. *Biochemistry*. 8th ed. New York: W. H. Freeman, 2015. Print.

Fenichell, Stephen. *Plastic: The Making of a Synthetic Century*. New York: Harper, 2009. Print.

Lehninger, Albert L. *Biochemistry: The Molecular Basis of Cell Structure and Function*. 2nd ed. New York: Worth, 1975. Print. Lodish, Harvey, et al. *Molecular Cell Biology*. 7th ed. New York: Freeman, 2013. Print.

Morrison, Robert Thornton, and Robert Neilson. Boyd. *Organic Chemistry*. 7th ed. Englewood Cliffs, N.J.: Prentice Hall, 2003. Print.

Myers, Richard. *The Basics of Chemistry*. Westport: Greenwood, 2003. Print. Reece, Jane B., et al. *Campbell Biology*. 10th ed. San Francisco: Cummings, 2013. Print.

HYDROPHILIC AND HYDROPHOBIC SAMPLE PROBLEM

Examine the structures of the following compounds and determine whether the compound is hydrophilic or hydrophobic:
- benzene
- ethylamine
- benzoic acid
- *N*-ethylbenzamide
- stearic acid

Answer:

Benzene (C_6H_6) is a nonpolar hydrocarbon and therefore hydrophobic.

Ethylamine ($H_3C-CH_2-NH_2$) has a lone pair of electrons on the nitrogen atom and polarized N−H single bonds. It is therefore hydrophilic.

Benzoic acid (C_6H_6-COOH) is a carboxylic acid derived from benzene, with a polar C=O double bond and an O−H single bond in the carboxyl functional group. The benzene ring in the molecule is hydrophobic, but the carboxyl functional group is hydrophilic.

N-Ethylbenzamide is formed from benzoic acid and ethylamine. The ethyl group and the benzene group in the molecule are hydrophobic, and only the C=O double bond in the carbonyl group and the nitrogen atom in the amide have any polarity. Overall, the compound is more hydrophobic than hydrophilic.

Stearic acid ($H_3C-(CH_2)_{16}-COOH$) is a long-chain fatty acid. The carboxylic acid functional group (−COOH) is hydrophilic, but the long-chain alkyl substituent is hydrophobic. The compound has both hydrophobic and hydrophilic characteristics.

Hypnosis

FIELDS OF STUDY

Psychology; humanistic psychology; psychoanalysis; neuroscience; philosophy of mind; placebo research; altered states of consciousness; complementary medicine; alternative healing; ethno-psychotherapy; shamanism.

SUMMARY

Hypnosis, a subfield of psychology, studies the influence of hypnotherapy (therapy undertaken in hypnosis) and hypnosis (as an altered state, social role, or response expectancy) on people, especially its impact on psychological and physical health. Hypnosis is believed to give access to the unconscious mind, and there is empirical evidence that in highly susceptible subjects, hypnosis is successful as a therapeutic treatment (or as adjunct treatment) against a variety of psychological and psychosomatic disorders, such as post-traumatic stress disorder and depression. Hypnotherapy has also proven to be helpful as a substitute for anesthetics, for pain reduction in general, and for relaxation. Finally, hypnosis is used in the context of sports, education, advertising, and marketing.

PRINCIPAL TERMS

- **age regression and progression:** attempts to focus the mental attention of a subject on an earlier or later age, which can be therapeutically constructive in the context of experiencing roles, elucidating expectations, and testing consequences but does not necessarily produce real-life memories.
- **altered state of consciousness (ASC):** mental state different from wakefulness and sleeping, includes drug-induced and trance states, as well as those brought about by mental, physical, and holistic techniques; also known as altered state of awareness or altered state of mind.
- **hypnotherapy:** form of psychotherapy or counseling that employs hypnosis.
- **induction:** first set of suggestions in the hypnotic process, inducing relaxation, concentration, and the establishment of hypnotic rapport, or the definition of the roles of the hypnotist and subject; also known as hypnotic induction.
- **nocebo effect:** unpleasant or harmful effect brought about by a placebo or a medical intervention that simulates the actual procedures.
- **placebo effect:** positive therapeutic effect caused by a simulated medical intervention.
- **rapport:** subjective experience of consciously or unconsciously feeling trust or an emotional affinity with another person; also known as hypnotic rapport.
- **suggestion:** guidance of thoughts, imaginations, feelings, and behaviors of one person by a different person (hetero-suggestion) or by the same person (auto-suggestion, self-suggestion); also known as hypnotic suggestion.
- **susceptibility:** ability to be hypnotized or to respond to hypnotic suggestion, as measured by such scales as the Stanford Hypnotic Susceptibility Scale and the Harvard Group Scale of Hypnotic Suggestibility.
- **trance:** altered state of consciousness achievable through hypnosis, meditation, prayer, shamanistic rituals, or drug use; also known as hypnotic trance.
- **unconscious:** domain of the mind that is not conscious, of which people are unaware during the normal waking state.

BASIC PRINCIPLES

Hypnosis is a wakeful state in which the attention is focused on one or several issues by diminished peripheral awareness and heightened suggestibility, usually induced by suggestions (hypnotic induction). Although the word "hypnosis" is derived from the Greek word for sleep (*hypnos*) and a hypnotized person might at times appear to be asleep, neurological research has revealed that brain waves during hypnosis do not resemble those of sleep. Hypnosis is an altered state of consciousness and a specific interactive situation with voluntarily assumed, defined roles in which a subject follows the suggestions of a hypnotist (in hetero-hypnosis) or the subject's own suggestions (in self-hypnosis). Some researchers think that a hypnotic state can occur without suggestion, as part of everyday life when people become extremely focused on a particular issue, for example, during concentrated learning or the creative process. The depth and the success of hypnosis are determined by psychological factors such as positive motivation, an appropriate attitude, expectations, susceptibility, and an active imagination. Hypnotherapy is hypnosis in a psychotherapeutic or counseling setting, with goals such as stress management, pain reduction, and the modification of attitudes, habits, or behavior.

Background

Hypnosis as psychosocial phenomenon is as old as human culture; images from several cultures show trances in the context of what are probably religious rituals, in some cases, possibly induced for medical reasons. Hypnosis is similar to some forms of trance brought about by eastern meditative techniques and religious or shamanistic rituals in diverse traditional cultures.

In the eighteenth century, the German physician Franz Anton Mesmer (from whose name comes the word "mesmerize") invented a treatment dubbed "animal magnetism." The Scottish physician and surgeon James Braid developed a treatment known as neuro-hypnotism, or hypnotism, which shared some features with Mesmer's technique. Braid is the first advocate of hypnosis and hypnotherapy to gain scientific acceptance. In France, neurologists Jean-Martin Charcot and Hippolyte Bernheim conducted research and developed clinical forms of hypnosis. Austrian neurologist and psychiatrist Sigmund Freud was trained by Charcot and initially was enthusiastic about hypnotherapy, but he later abandoned it in favor of his own psychoanalytic approaches.

In the twentieth century, psychiatrist Milton H. Erickson, who founded the American Society of Clinical Hypnosis, was an advocate of hypnotherapy. Erickson's approaches were both innovative and controversial, according to his collaborator, André Muller Weitzenhoffer, one of the most prolific hypnosis researchers of the second half of the twentieth century.

How It Works

In the first hypnosis or hypnotherapy session, the subject is informed about the basics of hypnosis. Each session usually begins with an introduction in which the subject, in most cases, will be asked to recline and to relax. This is followed by the first inductive suggestions (impressions of gravity, feelings of

Braid reserved the term "hypnotism" for cases in which subjects entered a state of amnesia resembling sleep

heaviness, and the like), followed by further suggestions that guide the subject toward becoming more relaxed but also more alert and focused toward his or her inner impressions, images, and imagination. The methods vary considerably depending on the therapist's training and philosophical worldview, the subject's aims or problems, and the respective circumstances. For example, induction can use the eye-fixation method, whereby the subject is told to keep his or her eyes fixed on a certain object such as the hypnotist's finger. Various other induction methods employ one or more senses and the imagination. According to the altered state theory, induction helps the subject transfer into an altered state of awareness or consciousness, and social role and response expectancy theories view the induction as a means of defining the roles of the client and hypnotist, increasing expectations, focusing attention, and increasing concentration. Posthypnotic suggestions given during the session are intended to trigger or support the subject's therapy goals, usually to change a behavior or alter an attitude in daily life. Hypnosis can be conducted with individuals and groups. Before the end of the session, the hypnotist conducts an exit procedure, in which any suggestions that are not posthypnotic are taken back and the subject is gradually brought back to a normal condition. The session can end with a review.

Altered State and Dissociation Theories. Braid, Erickson, and Weitzenhoffer, among many others, believed that hypnosis is an altered state of consciousness or an altered state of awareness. American psychologist Ernest Hilgard was of the opinion that consciousness is dissociated and that parallel streams of consciousness coexist and have certain degrees of autonomy (for example, one feeling pain and the other not). Therefore, the coordination and the emphasis of such streams of consciousness can be altered by suggestions. For example, a feeling of pain can be suggested to have less gravity, while a pleasant feeling can be emphasized. Additionally, many experts believe that hypnosis can give access to more remote, subordinated, or covered streams of the consciousness.

Social Role and Response Expectancy Theories. Social role theories (like those of American psychologist Theodore R. Sarbin), sociocognitive theories, and response expectancy theories emphasize the similarity of hypnosis and a placebo. Empirical evidence and meta-analyses suggest that the two main parameters that contribute to the effect of hypnosis in a significant way are the willingness to act socially compliant and an imaginative suggestibility (about one-third of subjects respond to imaginative suggestions and social pressure). Hypnotic suggestibility is not correlated with intelligence, social position, willpower, motivation, gender, introversion, extroversion, or credulity. Researchers such as psychology professor Irving Kirsch think that the effect of hypnosis, as well as that of placebos, is grounded in a kind of self-fulfilling prophecy, namely that largely subjects experience what they expect to experience. In patients with depression, the difference between a placebo and an antidepressant drug is not clinically significant. Research in this area attempts to prove that the altered state theories are wrong, but that the effects of hypnosis are nevertheless real and that subjects do not fake the effects of hypnosis, as social role and response expectancy theorists believe. A number of researchers hold that both altered state and social role/response expectancy theories are correct and responsible for the effect of hypnosis. Additionally, it can be argued that a hypnotic state is a deeper form of hypnosis, following a nonaltered or less altered state in which suggestions are given and taken according to the social role/response expectancy theory.

APPLICATIONS

Hypnotherapy, Psychotherapy, and Counseling. In clinical psychology, hypnosis, as an adjunct method, and hypnotherapy, as a stand-alone treatment, are successfully used to deal with pain reduction, psychosomatic symptoms, obsessive-compulsive disorder, post-traumatic stress disorder, anxiety disorders, and depression. Less successful is the use of hypnosis or hypnotherapy to treat problem habits such as excessive drinking, eating, and smoking, which are not manageable by self-control. Hypnosis is also used for pain and stress management, for self-improvement, and to change behavior and attitudes.

Hypnotic Analgesia. Hypnosis and shamanistic trance rituals have been used as analgesia. Hypnosis has been employed successfully to achieve relaxation and reduce anxiety, fear, discomfort, and pain before and during childbirth, in dental settings, and also during minor surgery. Hypnosis does not reduce the physical reception of pain, but its perception can be manipulated by hypnotic suggestions, whether administered by a hypnotist or the self.

Nonclinical Applications. Hypnosis and hypnotic suggestions (self- and hetero-) have been successfully used to cope with stage fright, to reduce stress levels, and to increase the degree of concentration and focus. They can also intensify relaxation and concentration in the context of creative arts, education, and sports. Research has been conducted on applications for military intelligence, investigations, and forensics, but there is no scientific evidence that such applications are of value. A number of business applications, for advertising, marketing, and improving sales, have been created; however, such applications are ethically questionable. Interest in raising athletic performance levels, losing weight, or quitting smoking has resulted in the proliferation of self-hypnosis products, usually in the form of CDs, DVDs, or books. However, self-hypnosis is best learned from a qualified practitioner. Stage hypnosis is usually considered to be neither of therapeutic interest nor an overly important issue in academic research. Leaving aside stage hypnosis, most of the applications still take place in the medical or clinical field.

Future Prospects

Hypnosis still receives a lot of attention in the discourse concerning memory recovery. Debate exists as to in which circumstances, under which conditions, and how reliably forgotten memories of past events, especially traumatic experiences, can be "recovered" through hypnosis. The more research conducted on hypnosis and more empirical data collected, the more effectively and appropriately hypnotherapy can be used and the more it will gain acceptance in mainstream medicine. The question of whether hypnosis is a social role/expected response or a true altered state will not be conclusively answered until neuroscience advances and anthropologists use neurological tools to study shamanist-cultic and religious trance rituals. Anthropology and sociology have made it evident that the use of hypnosis as therapy is not a European invention but rather a phenomenon that can be traced back to therapeutic shamanistic rituals and religious trances in various parts of the world. Another issue concerns the role of the subject's imagination in the curative powers of hypnosis. A subject in a hypnosis show is similar to the subject of a traditional cultic healing ritual in that the involvement and participation of the public is taken for granted. If both subjects experience a curative effect, then hypnosis is acting like a placebo, and its actual therapeutic benefits are questionable. The subject's own power of imagination and its neurological, biochemical, physiological, psychological, and holistic effects may be what is producing the curative effect. Certain scholars hold that quite a number of everyday settings—such as intensive educational settings, artistic performances, and mass political events—have a hypnotic character. Research in this field will bring to light how business applications of hypnosis are possible. The ethical problem in such contexts is that subjects should not be hypnotized against their own will. Critics hold that this has already been done in the sphere of marketing.

From a feminist perspective, it can be argued that hypnosis cements and even perpetuates patriarchal structures, since most of the well-known hypnotherapists were men and the hypnotherapist exerts a kind of dominance over the patient or client, unlike as in a guided imagery setting in which the relationship is less hierarchical and less suggestive. Therefore, a unique feminist approach to hypnosis is also the subject of research.

Roman Meinhold, MA, PhD

Further Reading

Erickson, Milton H. *The Wisdom of Milton H. Erickson.* Compiled by Ronald A. Havens. 1985. Reprint. Williston, Vt.: Crown House, 2003. Erickson's thoughts on hypnosis and psychotherapy.

Kirsch, Irving, Steven J. Lynn, and Judith W. Rhue, eds. *The Handbook of Clinical Hypnosis.* 2d ed. Washington, D.C.: American Psychological Association, 2010. Broad introduction to hypnosis covering how it is used clinically to treat conditions and disorders.

Nash, Michael R., and Amanda J. Barnier, eds. *The Oxford Handbook of Hypnosis: Theory, Research and Practice.* New York: Oxford University Press, 2008. Presents both academic theory and practical approaches. Contains name and subject indexes.

Pattie, Frank A. *Mesmer and Animal Magnetism: A Chapter in the History of Medicine.* Hamilton, N.Y.: Edmonston, 1994. Mesmer specialist Pattie provides a thoughtful biography of Mesmer that includes the probable source of his thought.

Pintar, Judith, and Steven Jay Lynn. *Hypnosis: A Brief History.* Malden, Mass.: Wiley-Blackwell, 2008. Offers a compact but critical overview of the origins and the history of hypnosis, including accounts on debates within psychology, references and index.

Temes, Roberta. *The Complete Idiot's Guide to Hypnosis.* 2d ed. Indianapolis, Ind.: Alpha Books, 2004. Provides the basics of hypnosis. Contains a glossary, an index, and suggestions for Bibliography.

Web Sites

American Psychological Association Division 30: Society of Psychological Hypnosis http://www.apa.org/about/division/div30.aspx

American Society of Clinical Hypnosis http://www.asch.net

European Society of Hypnosis http://www.esh-hypnosis.eu

International Society of Hypnosis http://www.ish-hypnosis.org

Society for Clinical and Experimental Hypnosis https://netforum.avectra.com/eWeb/StartPage.aspx?Site=SCEH

FASCINATING FACTS ABOUT HYPNOSIS

- In Greek mythology Hypnos, the god of sleep, is the brother of Thanatos, the god of death. Morpheus, the god of dreams, is Hypnos's son.
- The famous German writer Thomas Mann, in his 1930 novel *Mario und der Zauberer* (*Mario and the Magician*, 1930), explores how hypnosis relates to the mesmerizing power of Fascist political leaders.
- In a 1956 article, professor Frank A. Pattie claimed that Franz Anton Mesmer, one of the fathers of hypnosis, plagiarized his dissertation from a work by the English physician Richard Mead, an acquaintance of Sir Isaac Newton.
- American psychiatrist Milton H. Erickson (1901–1980), sometimes referred to as Mr. Hypnosis, was born tone-deaf and color-blind, and he attributed much of his heightened sensitivity to altered modes of sensory-perceptual functioning, body dynamics, and kinesthetic cues to his innate infirmities.
- Despite a lack of scientific evidence, many subjects claim to have past-life experiences under hypnosis; experiments undertaken by psychology professor Nicholas Spanos in the 1980's suggest that such past-life memories reflect social constructions.
- According to Guinness World Records, in 1987, the German stage hypnotist Manfred Knoke was able to hypnotize 1,811 people in the city of Bochum in a six-day period.
- In 1995, the German magazine *Der Spiegel* reported that International Society of Hypnosis president Walter Bongartz, appearing on a television show featuring Manfred Knoke and his feats, pleaded for a law to prohibit such stage-hypnosis spectacles, given the medical risks that they could pose.
- Dream Theater, a progressive metal band, released a concept album, *Metropolis Pt. 2: Scenes from a Memory* (1999), which tells the story of a man who undergoes hypnosis to experience the mystery of his past lives.

Immune system

FIELDS OF STUDY

Cell biology, immunology, pathology

SUMMARY

The immune system distinguishes "self" from "nonself" in the body, fighting off foreign invaders such as bacteria, viruses, and parasites. It works through the production of proteins called antibodies, and of cells that recognize and kill foreign pathogens.

PRINCIPAL TERMS

- **antibody:** protein produced by lymphocytes, with specificity for a particular antigen antigen: chemical that stimulates the immune system to respond in a very specific manner
- **cell-mediated immunity:** production of lymphocytes that specifically kill cells with foreign antigens on their surfaces
- **humoral immunity:** production of antibodies specifically reactive against foreign antigens carried in body fluids (humors)
- **lymphocyte:** white blood cell that produces either cell-mediated or humoral immunity in response to foreign antigens
- **macrophage:** mature phagocytic cell that works with lymphocytes in destroying foreign antigens

Basic Principles

An animal must keep itself distinct from its environment, recognizing its own tissues and keeping them from being invaded or mixed with tissues of other organisms. There are two types of protection used by animals in keeping out invaders and resisting foreign substances, nonspecific and specific defenses. In both types, the body distinguishes between cells that belong to the animal, which are "self," and anything that does not belong, or "nonself."

How It Works

Nonspecific Defenses. Even animals as primitive as sponges have the ability to recognize and maintain self-integrity. Scientists have broken apart two different sponges of the same species in a blender, intermixing the separated cells in a dish. Cells crawled away from nonself cells and toward self cells, reaggregating into clusters of organized tissues containing cells of only one particular individual. Phagocytic cells that engulf and destroy foreign invaders were first identified by a scientist who had impaled a starfish larva on a thorn. He observed that, over time, large cells moved to surround the thorn, apparently trying to engulf and destroy it, recognizing it as nonself. Even earthworms have the ability to recognize and reject skin grafts from other individuals. If the graft comes from another worm of the same population, the skin is rejected in about eight months, but rejection of skin from a worm of a different population occurs in two weeks. Phagocytic cells in earthworms have immunological memory, enabling a worm to reject a second transplant from the same foreign source in only a few days.

Barriers, chemicals, and phagocytic cells are nonspecific protective mechanisms, which do not distinguish among different kinds of invaders. Tough outer coverings such as skin, hide, scales, feathers, or fur provide surface barriers. Nonspecific defenses also include secretions of mucus, sweat, tears, saliva, stomach acid, and urine, as well as body-fluid molecules, such as complement and interferon. Damaged tissues or bacterial invaders signal other cells to produce inflammation, a nonspecific response characterized by heat, redness, swelling, and pain. Cellular defenses associated with inflammation include phagocytes such as neutrophils and macrophages, which engulf and digest bacteria and debris, and natural killer cells, which destroy cancer cells or virally infected cells by poking holes in them.

Immunity. The only specific defense in vertebrates is provided by the immune system, in which the component parts react against particular antigens on invaders, such as individual strains of bacteria or types of viruses. This more sophisticated protection is produced by lymphocytes that provide either cell-mediated or humoral immunity against particular antigens. Cell-mediated immunity depends on T lymphocytes (T cells) that become mature as they pass through the thymus, from which they get the "T" of their name. Humoral immunity is the function of antibodies, proteins released by B lymphocytes (B cells) that have matured and developed into plasma cells. The B lymphocytes reach maturity in the bursa of Fabricius in birds, where they were first recognized and from which they were named. Other vertebrates lack the bursa of Fabricius, and B cells mature in the bone marrow instead, so the name "B lymphocyte" still applies.

Antigen molecules are usually proteins or glycoproteins (proteins with sugars attached) that generate either an antibody response or a cellular immune response when they are foreign to the responding animal. So-called self antigens are molecules on cell membranes that identify the cells as belonging to the animal itself. An animal would not normally produce an immune response against its own antigens, but the same antigens would generate an immune response if placed in another animal to whom they were foreign. These antigens are the means by which self and nonself distinctions are made by the immune system, so the system can determine whether to ignore cells or attack them. Occasionally, self antigens, for some reason, are no longer recognized by the animal's immune system, and are attacked as if they were foreign. This causes an autoimmune disease, where the immune system destroys the body's own tissues.

Scientific understanding of how the immune system functions is largely dependent on work done using laboratory animals, including rabbits, mice, and hamsters. Laboratory mice have been highly inbred into strains where all the animals are genetically identical and their genes and antigens are well known. Studies on these mice have been essential in determining how the immune system normally works, and how it fails to work in autoimmune diseases and the inability to prevent cancer cells from proliferating.

Antigen Presentation and Receptors. Central to the functioning of the immune system in mammals is a system of genes called the major histocompatibility complex (MHC). These genes encode a collection of cell-surface glycoproteins that are the self antigens by which the immune system recognizes its own body cells. Class I MHC molecules are expressed on the surfaces of all nucleated cells, while Class II MHC markers are produced only by specialized cells, including cells of the thymus, B lymphocytes, macrophages, and activated T lymphocytes. Both Class I and Class II MHC molecules identify the cells bearing them as self, and these also serve as the context in which the immune system recognizes foreign antigens that are presented on the cell surface. Cells with self antigens are tolerated by the immune response of that individual animal, while cells that show foreign antigens are attacked and destroyed. Rejection of a graft or transplanted organ is reduced with more closely matched tissues, which are better tolerated by the immune system.

When bacteria evade protective barriers and chemicals to enter an animal's body, the animal's macrophages attack, engulfing and digesting the invaders. One bacterium may have thousands of different antigenic segments that can be recognized on its surface or inside the cell. Small parts of these digested cells, the individual antigens, are joined to the macrophage's newly formed MHC Class I and Class II before they are exposed on the cell surface. The foreign antigens fit into a space or pocket within the MHC molecule and are recognized by T cells that have the same MHC molecules and can respond specifically to the foreign antigen. Cytotoxic T cells (T) react to antigens held in the pocket of a Class I molecule, while helper T cells (T) respond to those presented by Class II molecules. T cells are the agents of cellular immunity, producing perforin molecules that puncture and kill cells bearing the foreign antigen against which the T cells are specific. T cells, when activated by encountering their specific antigens presented with Class II molecules, release cytokines that help to activate both TC cells and B lymphocytes. Activated B cells divide to produce memory B cells and lymphocytes that mature into plasma cells, which secrete about two thousand antibody molecules per second over their active lifespan of four or five days.

Both T and B lymphocytes can react with their specific foreign antigens because the antigen-MHC complex binds to receptor molecules on the lymphocyte surfaces. Each clone of lymphocytes has the genetic ability to respond to a particular shape that fits its receptors. There may be millions of different receptors among the lymphocytes of a single animal, capable of binding millions of different antigens, even artificial chemicals not existing in nature. This enormous variability in response capability makes the immune system of each animal protective against many kinds of foreign invaders. Since each individual has its own set of immune responses, a population is less likely to have all its members die in an epidemic. Certain animals will be more resistant to the pathogens, so some will survive to reproduce and keep the population from extinction.

Primary and Secondary Immune Responses. When a foreign antigen is encountered by an animal for the first time, both T and B cells that can bind the antigen are activated, but not immediately. In a series of reactions, macrophages first break down the antigen-bearing cell, processing and presenting the antigen on its surface with MHC. The T cell specific to that antigen then encounters the antigen-MHC complex on the macrophage and divides to produce a clone of memory T cells and a clone of effector (activated) T cells. Activated T cells release cytokines that activate T and B cells so that they can attack the same foreign antigen. The first encounter with antigen produces a slow primary response, taking more than a week to reach peak effectiveness. During the time needed to generate this response, pathogenic bacteria or viruses can produce disease in the animal under attack. The memory T and memory B cells remain alive but inactive until the same foreign antigen is encountered again, even years later. The secondary response that results immediately when these memory cells are activated occurs so quickly that the disease process does not recur.

The importance of the immune system is seen in humans who lack its function, those with acquired immunodeficiency syndrome (AIDS). Human immunodeficiency virus (HIV) is the causative agent of AIDS, and is similar to viruses that attack other species in the same way. Most who die with AIDS really succumb to one of many opportunistic infections that cause diseases in HIV positive individuals, but which are eradicated by the immune system in normal individuals.

—*Jean S. Helgeson*

See also: Anatomy; Bone and cartilage; Brain; Circulatory systems of vertebrates; Digestive tract; Diseases; Eyes; Noses; Reproductive system of female mammals; Reproductive system of male mammals; Respiratory system; Skin.

FURTHER READING

Campbell, Neil A., Jane B. Reece, and Lawrence G. Mitchell. *Biology*. 5th ed. Menlo Park, Calif.: Benjamin/Cummings, 1999. Chapter 43 on "The Body's Defenses" is a clear presentation of how the immune system works, in a college textbook for science majors. Color diagrams explain more difficult concepts.

Paul, William E. *Fundamental Immunology*. 4th ed. Philadelphia: Lippincott-Raven, 1999. Despite the

---

FIVE CLASSES OF ANTIBODIES

IgM is produced first in a response to foreign antigen, but its concentration declines rapidly. With five Y-shaped monomer subunits forming a pentamer structure, IgM is very effective in binding many copies of the same antigen and agglutinating them, but is too big to cross the placenta.

IgG is the most abundant class of antibodies in circulating blood, a monomer capable of passing through vessel walls to protect cells and tissues. In some species, including humans, it crosses the placenta to pass on the mother's immune protection to the fetus. Produced after IgM in an immune response, it is much more effective against bacteria, viruses, and toxins.

IgA is secreted as a dimer (two subunits) into milk, sweat, saliva, and tears. It is especially important in colostrum, the secretion before milk production begins that is the only way some newborn animals receive their mother's antibodies. IgA prevents bacteria and viruses from binding to epithelial cell surfaces, especially in the digestive tract.

IgE antibodies bind to the surfaces of mast cells and basophils with the arms of the Y-shaped monomer extended. Foreign antigens bind to the ends of the Y arms and trigger these cells to release histamine and other chemicals that cause the inflammation of allergy. IgE is also the antibody that attacks parasites inside the body, such as worms.

IgD molecules are monomers located mainly on the surfaces of B cells, apparently acting as receptors for the antigen that is recognized by each B cell and triggers its activation.

name, this medical text covers all aspects of human immune function in over 1,500 pages. Regulation of the immune system and its functions in fighting disease are discussed extensively.

Roitt, Ivan, Jonathan Brostoff, and David Male. *Immunology*. 5th ed. Philadelphia: C. V. Mosby, 1998. Both vertebrate and invertebrate immune functions are discussed in Chapter 15, "Evolution of Immunity." Good color photos and diagrams help explain cellular and molecular aspects of immunology.

Staines, Norman, Jonathan Brostoff, and Keith James. *Introducing Immunology*. 2d ed. St. Louis: C. V. Mosby, 1993. For readers with no previous knowledge of immunology, this primer covers the complex language without using jargon and explains the information clearly.

IMMUNOLOGY AND VACCINATION

FIELDS OF STUDY

Microbiology; general biology; cell biology; chemistry; molecular biology; biochemistry; biophysics; microbial genetics; immunology; genetic engineering.

SUMMARY

The function of vaccination is to induce immunity in humans and other animals for the purpose of providing protection against disease-causing organisms. Vaccination is generally carried out through the injection of attenuated or inactivated microorganisms such as bacteria or viruses, or the inactivated toxins produced by bacteria. The first vaccinations were directed against smallpox during the eighteenth century and involved the use of cowpox virus, similar but not identical to the virus that caused smallpox. Modern vaccinations often use purified components of the organism rather than the entire bacterium or virus, producing similar immunity without the danger of side effects or illness.

PRINCIPAL TERMS

- **acellular vaccine:** vaccine prepared from subunits or genes from the agent rather than the whole bacterium or virus.
- **antibody:** protein produced by B lymphocytes in response to foreign molecules.
- **antigen:** commonly a protein, but any molecule perceived as being foreign to the body.
- **B lymphocyte:** type of white blood cell that produces and secretes antibodies.
- **cellular immunity:** immunity based on activation of macrophages, natural killer cells, and antigen-specific cytotoxic T-lymphocytes.
- **complement:** pathway triggered by antigen/antibody complexes that consists of proteins that augment the immune response.
- **humoral immunity:** immunity based on soluble proteins such as antibodies found in blood and body fluids.
- **monoclonal antibodies (mAB):** antibodies produced by a clone of B lymphocytes, all of which are identical.
- **phagocytosis:** ingestion and digestion of material, including bacteria, by white blood cells.
- **T lymphocyte:** white cell that matures in the thymus gland (indicated by the T) and that regulates the immune response.

BASIC PRINCIPLES

Immunology is the science that studies the reactions of immune cells within the body to foreign molecules referred to as antigens. The majority of immune cells are represented by populations of white blood cells, or leukocytes, found circulating within the bloodstream and lymphatic system. Although all immune cells originate and mature largely within the bone marrow, they undergo differentiation into highly specialized categories. Monocytes and neutrophiles are phagocytic cells circulating in the bloodstream and lymphatic system that function to ingest and digest both foreign antigens from outside the body and old or dying cells within the body. Monocytes mature within tissues and organs such as the spleen and lymph nodes into a class of cells called macrophage, transport the ingested antigens, to the cell surface, and present the digested molecules to a second class of white cells called B and T lymphocytes. Only those lymphocytes that express a receptor specific for the presented antigen will respond, with

B cells differentiating into plasma cells that secrete antibodies directed against the original antigen. The principle underlying vaccination is that administration of a killed or attenuated form of bacterium, bacterial toxin, or virus will result in production of antibodies against the target, producing immunity in the individual against the bacterial or viral antigens. If the person is later exposed to the same microorganism or toxin, the presence of preexisting antibodies will protect against infection or poisoning.

Background

Immunitas originally referred to the freedom from taxes among ancient Romans. The medical concept of freedom from disease—immunity—was described by the Greek historian Thucydides, who in his description of the plague (actually probably a typhoid fever epidemic) in Athens in 430 BCE noted that individuals who survived "were never attacked twice." An understanding of the cellular basis for immunity was not reached until late in the nineteenth century. However, the principle of prior exposure to a disease resulting in immunity had been known since about 1000 CE, when the Chinese carried out a practice called variolation. In this practice, a person inhaled dried crust from the pocks that developed on smallpox victims. If he or she developed only a mild form of the disease, the most common outcome, the person was immune for life. The practice traveled to Eastern Europe and then England by the early eighteenth century. Although variolation was generally successful, sometimes it resulted in the person contracting smallpox.

The first successful active immunization is ascribed to Edward Jenner, an English country physician who in the 1790's tested a belief common among local dairymaids—that prior exposure to a mild cowpox infection of the udder on a cow provided immunization against smallpox. Beginning in 1796, Jenner carried out tests in which he intentionally infected people by applying cowpox "lymph" obtained from a lesion to small slits cut in the arms of volunteers. During a subsequent epidemic, none of the inoculated individuals developed smallpox. Jenner called the practice "vaccination," from *vacca*, Latin for cow.

How It Works

An understanding of the cellular mechanism underlying successful vaccinations did not begin until the late nineteenth century and was the outcome of both a scientific and a nationalistic rivalry between French and German scientists. The major proponent of a cellular theory of immunity was the Russian scientist Élie Metchnikoff. While studying the differentiation of cells in animals such as starfish larvae, Metchnikoff observed that insertion of a wooden splinter into the larvae resulted in the infiltration of both large and small white blood cells. He called these macrophage, "large eaters," and microphage, "small eaters." Microphage later became known as neutrophils. Metchnikoff subsequently joined the laboratory of French scientist Louis Pasteur, where he became a proponent of the cellular theory of immunity. The competing theory was defined by the German school, and became known as humoral immunity. Robert Koch, Emil von Behring, and their associates noted that blood plasma obtained from animals previously exposed to etiological agents of disease or to bacterial toxins could directly kill bacteria or neutralize these toxins. Behring and Paul Ehrlich applied their discovery in the development of the first vaccines against diphtheria. Soluble proteins, including antibodies, became the basis for humoral immunity. It was not until the mid-twentieth century that the basis for immunization was established as a combination of both cellular and humoral immunity. Phagocytosis is indeed carried out by several classes of white blood cells, while antibodies are produced by a class of white cells called B lymphocytes. The actual immune mechanisms involve a complex interaction between these classes of cells and their soluble products in which the phagocytic cell presents the digested antigen on its surface to the appropriate lymphocyte. The end result is that B lymphocytes mature and differentiate into an end-stage antibody-producing factory called a plasma cell. Each plasma cell produces a single type of antibody, selected on the basis of possessing a receptor specific for the antigen presented by the phagocyte.

Vaccine Production

Vaccine production is based on activation of lymphocytes through exposure to bacterial or viral antigens (proteins), the result of which is the production of antibodies or the stimulation of phagocytic cells. Vaccines have historically been produced by three major mechanisms: use of inactivated or killed microorganisms, use of attenuated or cross-reacting organisms, or use of purified portions of microorganisms in recombinant vaccines. The smallpox vaccine is an example of a cross-reacting organism; the cowpox virus is similar enough to smallpox that the immune response is protective against both.

Most viral vaccines have used attenuated strains of the original virus, selected either by passage through nonhuman animals or cells or by artificial selection on the basis of avirulence, an inability to cause disease. The strains of poliovirus vaccines developed by Albert Bruce Sabin, as well as vaccines against rabies, chickenpox, measles, and mumps all consist of attenuated viruses. The polio vaccine developed by Jonas Salk is a formalin-killed virus. A later generation of viral vaccines, those directed against viruses such as hepatitis B and human papillomavirus (which causes warts and cervical cancer), are subunit types consisting of surface proteins obtained from the virus, which through DNA recombination are linked to harmless carrier proteins. Vaccines directed against tetanus toxin are similar to those originally developed by Behring and Ehrlich against diphtheria toxin. The toxin is chemically modified and injected. The principle behind all vaccinations is the same. Exposure to the agent results in an immune response within the individual. Antibodies are produced, and cellular immunity is activated. The response is already in place in the event of future exposure to the same organism or toxin. In most cases, immunity is long-lasting, though periodic boosters are recommended to ensure a proper level of immunity.

AUTOIMMUNE DISEASE

In principle the immune response is directed only against foreign agents that could potentially cause disease. However, in certain circumstances, alterations in immune regulation take place, and antibodies are produced against the person's own tissues. The precise molecular mechanism that triggers autoimmune function is unclear. Some diseases run in families or are gender specific (women are more likely to contract certain autoimmune diseases), and other illnesses may be triggered by crossreaction with viral or bacterial antigens. The tissue involved depends on the type of autoantibody produced, but the mechanisms for damage are similar.

Autoimmune diseases are placed into two major categories, organ specific or systemic, reflecting the sites or systems involved. Examples of organ-specific diseases include type 1 diabetes, in which the B cells of the pancreatic islets of Langerhans are targeted; Crohn's disease, a form of inflammatory bowel disease; and multiple sclerosis, characterized by inflammation of tissue in the central nervous system. Systemic autoimmune diseases include systemic lupus erythematosus, in which antigen/antibody complexes lodge in different organs, and rheumatoid arthritis, characterized by immune complexes that lodge in joints or bone. Although the type of antibody may differ in autoimmune diseases, pathologies are similar in that each activates the complement pathway, components of which include degradative enzymes that contribute to inflammation and tissue destruction.

APPLICATIONS

Vaccine Production. Historically, vaccines fell into two categories: live vaccines in which the agent was altered so as to be unable to cause disease but still able to replicate in the human host, triggering the immune response, and killed vaccines in which the organism was identical to the parent strain but unable to replicate. Each had advantages. Live vaccines produced a greater response and often a lifelong immunity, and killed vaccines would not result in reversion to the wild strain, causing disease. For example, before they were discontinued in 1990, the Sabin strains of attenuated poliovirus had a reversion rate of about one in one million persons inoculated in the United States, resulting in about ten vaccine-associated cases of polio per year.

Live, or attenuated, vaccines were originally created by passage in nonhuman animals or in cell cultures in a laboratory. This was particularly true for vaccines for viruses, including those against rabies, polio, measles, mumps, rubella, and chickenpox. Because viruses develop random mutations, variant strains were selected on the basis of sensitivity to pH (acidity-alkalinity) or to elevated temperatures (fever), or for their inability to infect certain tissues. The Sabin poliovirus strains represent a prototype of attenuated viruses, being both temperature sensitive and incapable of infecting tissue in the central nervous system. Later methods of developing attenuated strains have involved active modification of viral genetic material or creation of recombinant viruses in which those genes necessary for replication have been deleted.

Most viruses can be grown in cell culture for vaccine production. Animal cells are easy to maintain in the laboratory, and viruses for vaccines can be grown to necessary concentrations. Influenza viruses are exceptions, which is one reason quick production of yearly influenza vaccines has been difficult. The influenza virus genome consists of eight individual segments; coinfection of cells with two different strains, often involving viruses from two different species such as humans and birds, routinely creates a new recombinant strain that is not recognized by the human immune system. Influenza viruses do not grow well in cell culture, so vaccines must be produced using viruses grown in eggs. The lead time necessary to produce sufficient quantities of vaccine for the influenza season, which begins in the fall, is about six months. Therefore, health agencies such as the World Health Organization and the Centers for Disease Control and Prevention must decide in late winter which strains are most likely to produce an outbreak later that year.

Monoclonal Antibodies (mAB). Exposure to antigens such as those found on bacteria or viruses triggers the production of a large number of different antibodies, each of which is specific for a particular molecular determinant on an organism. In the 1960's, it was discovered that persons with multiple myeloma, a cancer of plasma cells, produced large quantities of homogeneous antibodies. British scientists Georges J. F. Köhler and César Milstein found that because myeloma cells, like those of most cancers, are immortalized, they could artificially fuse myeloma cells with immune cells of known specificity to produce a clone of "immortal" cells producing identical antibodies; because these cells represented a clone, the product became known as monoclonal antibodies.

Because antibodies, in theory, can be generated in the laboratory against any target, monoclonal antibodies can be used as a probe for detection of any cellular molecule. Initial applications used monoclonal antibodies for detection of cell surface proteins for identification of cell types or the maturation stage of cells during differentiation. Because these surface proteins exhibited clustering, they became known as cluster of differentiation (CD) proteins. Nearly two hundred cluster of differentiation proteins are now known. The ability of monoclonal antibodies to bind surfaces on specific cells has led to their use in the diagnosis or treatment of certain types of cancers. Immuno-conjugates are prepared by chemically attaching a toxin or radioisotope to a monoclonal antibody and injecting the molecule into a patient. Binding of the conjugated monoclonal antibody to the tumor cell results in killing of the target. Although in theory immunotherapy could be applied to many forms of cancers, most tumors do not express proteins unique to that type of cancer.

FUTURE PROSPECTS

The effective control of most childhood infectious diseases by the end of the twentieth century has caused the fear of such diseases to all but disappear among most modern populations. As some segments of European, British, and American populations have grown up in a time in which childhood infectious diseases appeared to be a thing of the past, many of these people do not fully understand the devastating nature of these diseases and question the value of vaccines. Also, the sheer number of recommended vaccinations has created concern among parents, some of whom are afraid their children's immune systems could be overwhelmed, perhaps resulting in autism. Although no evidence for a link between autism and vaccination has been found, some parents still believe that such a link exists. Future immunizations are likely to rely less on whole virus or bacterial vaccines and more on acellular or subunit vaccines. Side effects resulting from the pertussis vaccine, generally mild fever or inflammation but occasionally a more serious problem, led to the development of an acellular pertussis vaccine using only bacterial proteins. Similar vaccines, some containing only genetic information for production of viral proteins, are likely to be used against other diseases in the future. Immunization against some agents such as influenza viruses, which undergo yearly changes, will probably involve some form of combination vaccines, incorporating proteins that are common to most major strains of the virus. The simplicity of world travel in the twenty-first century means scientists must take a worldview of new strains, as an outbreak in a few countries can rapidly develop into a worldwide pandemic. The ability of the human immunodeficiency virus (HIV) to undergo rapid mutations, even within the same individual, means a vaccine against the acquired immunodeficiency syndrome (AIDS) remains unlikely in the foreseeable future.

Richard Adler, PhD

Further Reading

Allen, Arthur. *Vaccine.* New York: W. W. Norton, 2007. Discusses the history of vaccine development, from eighteenth century variation to modern times, and the controversies that have surrounded vaccination.

Heller, Jacob. *The Vaccine Narrative.* Nashville, Tenn.: Vanderbilt University Press, 2008. Tells how four of the major vaccines of the twentieth century were developed.

Link, Kurt. *The Vaccine Controversy: The History, Use, and Safety of Vaccinations.* Westport, Conn.: Praeger, 2005. Short synopses of the history behind most major vaccines, from smallpox to acellular vaccines.

Plotkin, Stanley A., Walter A. Orenstein, and Paul A. Offit, eds. *Vaccines.* 5th ed. Philadelphia: Saunders/ Elsevier, 2008. An excellent source for understanding the history of vaccine development against most major agents.

Tauber, Alfred. "Metchnikoff and the Phagocytosis Theory." *Molecular Cell Biology* 4 (November, 2003): 897-901. Discussion of Metchnikoff's discovery of phagocytosis. Includes original illustrations.

Williams, Tony. *The Pox and the Covenant.* Naperville, Ill.: Sourcebooks, 2010. Story of Cotton Mather and variolation in Boston during the 1720's.

Web Sites

American Association of Immunologists http://www.aai.org

Centers for Disease Control and Prevention Vaccines and Immunization http://www.cdc.gov/vaccines

World Health Organization Immunization Surveillance, Assessment, and Monitoring http://www.who.int/immunization_monitoring/en

See also: Epidemiology; Human Genetic Engineering; Pathology; Virology.

FASCINATING FACTS ABOUT IMMUNOLOGY AND VACCINATION

- Edward Jenner, the English physician who developed the first vaccine against smallpox, refused an offer by Captain James Cook of the HMS *Endeavour* to accompany the captain on his next trip to Pacific islands during the 1770's.
- Bishop Cotton Mather of Boston is notorious for his condemnation of witchcraft, but in the 1720's, he was the primary proponent of variolation.
- During the 1918 influenza epidemic, the bacterium *Haemophilus influenza*e was mistakenly thought to be the etiological agent of the disease. A vaccine was produced against it, which had no effect on the epidemic.
- Louis Pasteur's rabies vaccine was first tested on nine-year-old Joseph Meister in 1885. In 1940, Meister, then a caretaker at the Pasteur Institute, committed suicide rather than open Pasteur's tomb for the Nazis.
- The attenuated Edmonston strain of measles virus developed by physician John Enders was isolated in 1954 from an eleven-year-old boy, David Edmonston, son of a Bethesda mathematician. As an adult, Edmonston participated in the 1963 civil rights march on Washington and later became a schoolteacher in Mississippi.
- The JL strains of the mumps virus vaccine were developed by physician Maurice Hilleman from an isolate obtained in 1963 from his six-year-old daughter, Jeryl Lynn.

Intelligence

FIELDS OF STUDY

Developmental biology, ethology, evolutionary science

SUMMARY

Animals are guided by more than instinct when interacting with their environment, yet the exact measurement of intelligence in various species remains problematic. While scientists devise numerous problem-solving tasks to assess intelligence, anthropomorphism leads to exaggerated claims of intelligence through anecdotal evidence.

PRINCIPAL TERMS

- **anthropomorphism:** attributing human characteristics or states of mind to animals cognition: transformation and elaboration of sensory input
- **cognitive ethology:** scientific study of animal intelligence

- **lexigrams:** symbols associated with objects or places in keyboard communication experiments with primates
- **protogrammar:** word coined to signify the early foundation for grammar development found in primates
- **recapitulation:** stages of human development reappearing in different animal species

Basic Principles

Both the general public and scientific community have long been intrigued with questions about how animals think and what are they thinking. Published reports of animal cognition increased dramatically in the last half of the twentieth century. Chimpanzees in the Ivory Coast have demonstrated extensive use of rocks as tools in cracking nuts. These primates have also been reported to hide undesirable expressions from their faces and act as if blind or deaf. Vervets have been found to use an elaborate system of alarm calls that seem to function as words. Parrots can demonstrate the ability to count, and birds exhibit the capacity to make and use tools to gather food. Dolphins apparently understand and follow simple commands. Primates have been trained to use signs in a symbolic fashion, communicating their needs, desires, and thoughts.

Theories of Cognitive Ethology

Cognitive ethology is a relatively new discipline that studies animal intelligence. Donald Griffin is considered to have founded this branch of study through the publication of *Animal Thinking* (1984) and *Animal Minds* (1992). Since the appearance of his books, numerous instances of animal intelligence have been gathered from observation and experimentation.

Traditionally, attitudes about animal intelligence can be sorted into those that place animals on a continuum with humans and those that see animals as distinct from humans. From the former perspective, animal behavior is readily interpreted as a definite sign of various cognitive skills and special abilities along a continuum of development. From a discontinuity perspective, only humans are considered to possess the higher cognitive skill of reasoning. The higher cognitive abilities are considered to be a uniquely human capacity that sets them apart from the lower animals, who are controlled by instinct.

Charles Darwin, in *The Descent of Man* (1871), defended the idea of the intelligence of animals existing on a continuum with humans. Since animals and humans have a common ancestry, animals would have the fundamental capacities for rational choice, reflection, and insight. Darwin concluded that the differences between the minds of humans and animals were of degree rather than of kind. Following Darwin's proclamation, a number of anecdotal studies concerning animal intelligence appeared that suggested extensive cognitive ability in animals. Unfortunately, many of the examples illustrated anthropomorphism. This is the process whereby humanlike characteristics are attributed to animal behavior.

Some interpretations of Darwin's statement created a distorted view about evolution that persisted long into the twentieth century. The idea that life on earth represents a chain of progress from inferior to superior forms began to influence the view of animal intelligence. The theory that ontogeny recapitulates phylogeny also became popular in the early years of the twentieth century. This theory, which does not have any scientific support, suggested that the advancement of life forms corresponded to the stages of development for humans. This stepladder approach to animal intelligence led to a ranking of animals compared to the developmental stages of human infants and children. This approach to animal intelligence is flawed because it relies on the notion that some animals are more highly evolved than others are. Evolution does not have a single point of greatest evolution. The branches of the evolutionary tree have culminated with many different species occupying special niches. Thus, the "degree" of a species' evolution depends on the extent to which it successfully occupies its niche.

Animals Who Might Think

In addition to this tendency to attribute states of mind to animals that are found in humans, there were a number of cases of labeling trained behavior in animals as signs of reasoning skills. One of the most famous examples was the case of the horse, Clever Hans, in the early 1900's. Wilhelm von Osten owned a horse that demonstrated extensive arithmetic skills. When von Osten presented a written arithmetic problem to Hans, the horse would tap out the answer with his forefoot. Clever Hans also appeared adept at telling time, and answered questions about sociopolitical events by nodding or shaking his head yes or no. The horse's abilities suggested to many individuals the similarity between animal and human minds. Eventually, the Prussian Academy of Sciences discovered that Hans was not answering the questions by means of any reasoning skills, but was an astute observer of the behavior of his owner and those around him. When questions were posed to Hans, cues were provided unconsciously to the horse about the correct answer. Since horses have evolved to ascertain subtle visual cues from others in their herd, Hans was able to form a number of cued associations which led to a reward. The owner of Clever Hans was not attempting to perpetrate fraud. He believed in the possibility that a horse could have reasoning ability, but von Osten was not sophisticated in how he tested for the skills. The inadvertent cueing of an animal to respond in a certain fashion is one of the major confounding factors found in the investigation of animal intelligence.

The case of Clever Hans illustrates two other problems that confound reports concerning the level of intelligence in animals. First is the problem of anthropomorphism. People develop an emotional bond with animals and interpret behavior in order to enhance the closeness they feel to them. The second problem concerns the methods used to measure intelligence. The classic case of Köhler's chimpanzees illustrates this problem.

In the early part of the twentieth century, Wolfgang Köhler assessed the reasoning ability of chimpanzees to obtain food outside of an enclosure. After a rake was left in the enclosure, food was placed out of reach of the caged chimpanzees. The chimpanzees were able to use the rake to bring food to the cage. Köhler concluded that the animals had insight into the nature of the problem and used reasoning to achieve a solution. A further study, requiring the fitting together of two sticks in order to reach the food, also supported Köhler's conclusions. However, later experimentation has revealed that chimpanzees without a history of playing with sticks could not solve the problem. Apparently, in order to solve the problem, the chimpanzees needed an extensive history of playing with sticks, which enabled them to learn how sticks could be used at a later time. In solving the problem, they were using an instinctual tendency to play with sticks and scraping them over the ground.

PRIMATES AND SIGN LANGUAGE

A contemporary example of the problem of measurement can be provided with the case of Washoe, the first chimpanzee to be taught sign language. Because of physical inability to vocalize human speech, chimpanzees were taught sign language as a mode of communication with humans. Soon Washoe and another signing chimpanzee, Nim Chimsky, were reported to have spontaneously created novel sentences through their signing. For example, Washoe was reported to have signed the combination water and bird after seeing a swan. Being a novel combination of signs, the trainers of Washoe explained the behavior as creative insight. Unfortunately, Washoe had also shown repeated signing of meaningless combinations, leading to the conclusion that a significant pairing of signs would eventually appear not because of the primate's cognitive reasoning but as a result of chance. Inevitably, these early attempts to demonstrate animal intelligence were widely discredited as exaggeration or self-delusion on the part of the animal's trainers, and this animal language research from the late 1970's fell into disrepute.

In order avoid the ambiguities of sign language, later researchers used keyboards that related symbols to a variety of objects, people, and places. Much of this research has taken place at the Language Research Center at Georgia State University in Atlanta under the guidance of Dr. Sue Savage-Rumbaugh. In the first experiments, two chimpanzees, Austin and Sherman, were familiarized with a system of symbols or lexigrams. Each was abstract and arbitrarily associated with an object, person, place, or situation. Eventually Austin and Sherman learned to communicate with symbols illustrated on a keyboard. For example, an experiment was devised where one chimpanzee was shown where food was being deposited in a certain container while the other had control of a tool to open the container. With the keyboard present, the chimpanzees were able to communicate with one another to use the tool on the correct container.

Soon a bonobo chimpanzee, Kanzi, became the star pupil of this technique and learned a vocabu- lary of two hundred symbols. Kanzi eventually showed the capacity to construct rudimentary sentences that were generated spontaneously. The chimpanzees trained using the keyboards appear to be exhibiting a protogrammar. This is a term to indicate the beginnings of grammar, roughly equivalent to the verbal skills seen in a human child about two to three years old.

In the late 1990's, another bonobo chimpanzee, Panbanisha, surpassed the capacities evidenced by Kanzi. Panbanisha has been reported to understand complex sentences and use the keyboard to communicate spontaneously with the outside world. Although the results have been impressive, critics of the Center's activities remain. The question remains whether the chimpanzees are demonstrating extremely effective training or some level of abstract reasoning.

—Frank J. Prerost

See also: Brain; Emotions; Tool use.

FURTHER READING

Budiansky, Stephen. *If a Lion Could Talk.* New York: Free Press, 1998. A good collection of contemporary and historical cases of animal intelligence. The stories cover a wide range of examples seen in various animal species.

Moss, Cynthia. *Elephant Memories.* New York: William Morrow, 1988. An interesting account of thirteen years of field observations concerning the behavior of elephants in the Amboseli National Park in Kenya.

Page, George. *Inside the Animal Mind.* New York: Doubleday, 1999. The author begins with a historical account of the popular and scientific views about animal intelligence. He provides good details about the various attempts to communicate with primates by teaching them sign language or through the use of keyboards.

Savage-Rumbaugh, Sue. *Kanzi: The Ape at the Brink of the Human Mind.* New York: John Wiley&Sons, 1994. This book presents an apparent breakthrough in the communication with chimpanzees using symbols and a keyboard. It includes a number of incidences of the spontaneous construction of sentences by the primates.

Savage-Rumbaugh, Sue, Stuart G. Shanker, and Talbot J. Taylor. *Apes, Language, and the Human Mind.* New York: Oxford University Press, 1998. A book written from an academic and scientific perspective about the ability of chimpanzees to communicate by means of symbols and a keyboard display.

MEASURING ARITHMETIC SKILLS IN PRIMATES

The chimpanzee Sheba was involved in a number of widely reported experiments that purported to demonstrate arithmetic ability in primates. Sheba was taught to associate a tray containing one, two, or three pieces of candy with cards containing the corresponding number of marks. If two candies were present on the tray, Sheba learned to pick the card with two marks. If the correct card was selected, Sheba would be rewarded by being allowed to eat the candy. In the next stage of the experiment, the marks on the cards were replaced by the numerals 1, 2, and 3. Then Sheba had to match the number of candy pieces with the correct number. After Sheba showed success in this phase of the experiment, the candy pieces were replaced with other inedible objects. If Sheba chose the number that matched the number of objects, she was rewarded with the corresponding number of candies. Eventually the numbers were expanded to include 0 and 4. Next Sheba was given the challenge of counting the total number of candies presented on a series of trays. Sheba demonstrated a rudimentary ability for addition showing correct responses 75 percent of the time. Initially the sums were restricted to one, two, or three. Sheba was eventually able to exhibit the skill to add numbers when the candies were replaced with the numerals 1, 2, and 3. Although Sheba was able to show mathematical ability, even the experimenter acknowledged some important limitations. Sheba required extensive training and "heroic" effort on the part of the trainer to accomplish the counting performances. A true mathematical ability should generalize to other situations, yet animals such as Sheba do not demonstrate easy or automatic transfer of numerical performance from one realm to another.

Kinesiology

FIELDS OF STUDY

Anatomy; physics; biomechanics; physiology; exercise physiology; chemistry; biochemistry; motor behavior; sport psychology; sport sociology; coaching; ergonomics.

SUMMARY

Kinesiology is a multidisciplinary field that specializes in the science of human movement. It can focus on improving health or performance or on preventing injuries. Traditionally, kinesiology concentrated on the structural anatomy and the mechanics of movement. Later, the field expanded to include the physiological and mental aspects of movement. Some common applications of kinesiology include proper running and jumping mechanics, correct weight-lifting techniques, and perfecting the execution of sports skills. Kinesiology also can encompass the mechanics of work-related activities such as lifting, repetitive movements, and sitting at a desk.

PRINCIPAL TERMS

- **action:** type of movement made by a muscle contraction.
- **adenosine triphosphate:** form of stored energy that can be used directly by muscle cells for contraction.
- **aerobic exercise:** physical activity that is vigorous, continuous, and rhythmical.
- **alveoli:** lung structure in which oxygen is transferred from air to the blood.
- **antagonist:** muscle that has an opposite action from its paired muscle and limits its action.
- **applied kinesiology:** diagnostic technique used in chiropractic medicine.
- **hemoglobin:** component in blood that attaches to and transfers oxygen.
- **insertion:** point of muscle attachment on the bone that moves.
- **isokinetic machine:** instrument that measures the strength of muscles through the joint's full range of motion.
- **metabolism:** sum of all chemical reactions in a living organism.
- **motor:** relating to or involving movements of a muscle.
- **origin:** point of muscle attachment on the bone that remains stationary.

Basic Principles

Kinesiology is the study of human movement. Although it technically is not limited to humans, as an applied science, it almost always is. Kinesiology is primarily concerned with all kinds of physical activity, including competitive sports and activities designed to maintain and improve health. It is a major part of physical rehabilitation and injury prevention, and it can be used to design workstations that are ergonomic and minimize hazards.

A key component of kinesiology is to understand which muscles are involved in specific movements. Each major joint in the human body has identifiable planes of movement. Each movement in the plane is named according to its direction, and each movement has certain muscles that contribute to it. Muscles can be primary movers at the joint or can assist in the movement. Using this information, kinesiologists can observe movement patterns, identify which joints are involved, determine the movement at each joint, and ascertain which muscles contribute to the movement. Furthermore, they can develop training programs to work the appropriate muscles to improve the strength, movement, and muscle balance at the joint. The field of kinesiology should not be confused with applied kinesiology, which is a diagnostic technique used in chiropractic medicine. The clinician applies a force to a muscle or muscle group, and the patient resists the force. Based on the patient's response to the force, the clinician makes a diagnosis. This is not a generally accepted practice in medicine and not related to the field of kinesiology.

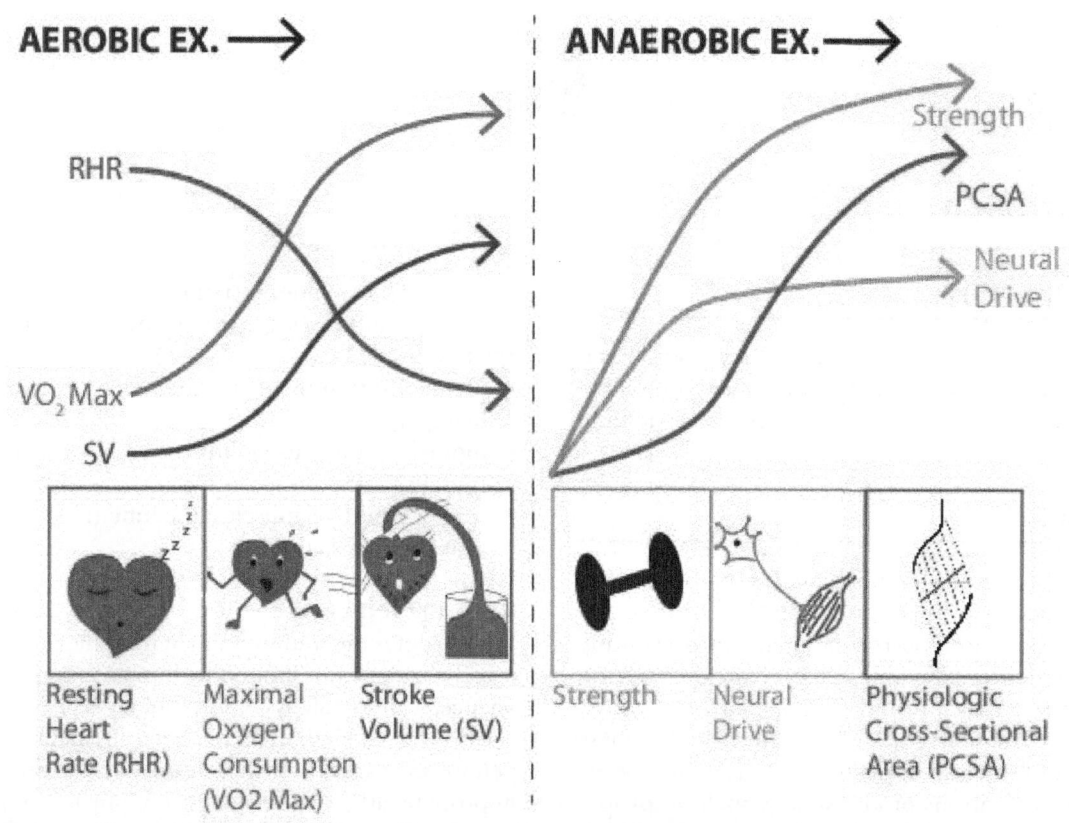

Summary of long-term adaptations to regular aerobic and anaerobic exercise. Adaptations include an increase in stroke volume (SV) and maximal aerobic capacity (VO2 max), as well as a decrease in resting heart rate (RHR). Long-term adaptations to resistance training include muscular hypertrophy, an increase in the physiological cross-sectional area (PCSA) of muscle(s), and an increase in neural drive, both of which lead to increased muscular strength. Neural adaptations begin more quickly and plateau prior to the hypertrophic response.

BACKGROUND

The field of kinesiology is believed to have begun in ancient Greece. The term "kinesiology" comes from the Greek words *kinein* meaning "to move" and *logos* meaning "discourse." Aristotle is considered the father of kinesiology for his work using geometry to describe the movement of humans. It was not until the fifteenth century that Leonardo da Vinci helped expand the knowledge of human movement by studying the mechanics of standing, walking, and jumping.

One of the greatest contributors to kinesiology, Sir Isaac Newton, did not actually study human movement. In the late 1600's and early 1700's, Newton developed three laws of rest and movement that laid the foundation for the analysis of human movement in the following years. The development of photography in the 1800's enabled researchers to study how animals moved by taking a number of pictures in rapid succession. They first studied horses, then humans. Cinematography further enabled researchers to understand human movement.

In 1990, the American Academy of Physical Education (later the American Academy of Kinesiology and Physical Education) recommended that programs of study involving human movement be called kinesiology. This idea gained wide acceptance, and kinesiology as a field came to include many other specialized fields beyond the traditional anatomy and biomechanics. Kinesiology can include any area that relates to human movement, such as history, sociology, psychology, physiology, philosophy, and motor behavior.

HOW IT WORKS

Movement Analysis. Traditionally, courses in kinesiology focused on the anatomy and mechanics of human movement. Although more advanced courses sometimes take the same approach, introductory courses tend to be an overview of the broader field of kinesiology.

A thorough understanding of muscle and skeletal anatomy is required to understand movement. This includes the names of the bones and the names, origins, insertions, and actions of the major muscles of the human body. The origin and insertion of a muscle are the locations where it attaches to bones. These locations determine the action of the muscle, based on the angle of pull on the bones. This knowledge is important in understanding which muscles are involved in specific movements.

To describe movements of the body, planes and rotations are defined at the various joints. Some of these movements include flexion, extension, adduction, abduction, internal rotation, and external rotation. Kinesiologists watch a specific movement and determine the actions at the joint or joints being evaluated. They can also evaluate movements to determine if they are completed properly. Slow-motion cinematography is helpful when analyzing the very fast movements often found in sports.

After determining the movement at a joint, kinesiologists can use their knowledge of anatomy to determine which muscles are involved. Additionally, they can use this information to develop strength training programs to develop the specific muscles used in the activity. It is important to note that muscles that generate a movement (agonist muscles) have antagonist muscles that stop the movement. Therefore, strengthening an antagonist muscle is just as important as strengthening the muscle that initiates the movement.

Physiological Function. Human movement requires oxygen and energy beyond the levels needed to simply survive. Exercise physiology includes the study of how the body gets food and oxygen from the environment to the working muscles. Oxygen is very important for sustaining activity during intense exercise, during which the cardiovascular, respiratory, and muscular systems are primarily involved. The respiratory system transfers oxygen from the atmosphere into the blood. Air, which is about 21 percent oxygen, is inhaled into the lungs, and much of it enters the alveoli at the ends of the airways. Oxygen diffuses from the alveoli into the blood, where it binds with the hemoglobin. The pumping action of the heart carries the blood to the muscles. When the oxygenated blood gets near muscle cells that are low in oxygen, the hemoglobin releases the oxygen, which goes into the cell. A strong, healthy heart and blood with normal amounts of hemoglobin are capable of delivering sufficient amounts of oxygen to support high levels of muscle movement and activity. When oxygen enters the muscle cell, it is metabolized. It goes through a series of chemical reactions in which oxygen and energy are converted into adenosine triphosphate (ATP). Only ATP can be used by the muscle to make the fibers contract and the body move. When highly trained people exercise at a very high intensity, the amount of oxygen consumed can increase more than twenty times. It is the ability of the muscle cells to use oxygen to produce ATP that limits high-intensity human movement. The oxygen consumed and several other measures can be determined with a metabolic cart, which is an important type of testing equipment used in exercise physiology.

Behavioral Control. The areas of kinesiology that involve the brain and nervous system are motor development, sport psychology, and sport sociology. Motor development is concerned with skills that take stored movement patterns in the brain and communicates them to the muscles. Most skilled movements are an organized, synchronous set of smaller movements. Therefore, the series of muscle contractions needed to perform the skill must be stored and retrieved often. Practicing the movement patterns on a regular basis is required to refine the skills. Sport psychology and sociology are the segments of kinesiology that relate to the mental aspects of human movement and performance. One of the largest fields is clinical sport psychology, in which psychologists assist athletes with aggression, stress, motivation, mood, adherence, and leadership as well as a number of other related issues. Sport sociology is a smaller element of kinesiology that focuses on social relationships in sport and how sports affect different segments of society and organizational structures.

Applications

Rehabilitation. Kinesiology techniques, especially those regarding the muscular and skeletal systems, are used in physical rehabilitation every day. Physicians, chiropractors, physical therapists, and athletic trainers use their knowledge of muscle origins, insertions, and actions to diagnose injuries and determine exercises that will help people recover from them. Isokinetic machines can be used to determine the muscular strength at any joint through the entire range of motion. Identifying points of weakness during movement helps determine which muscles need to be strengthened. A muscle and its antagonist can be tested to see if one is weaker than the other. Good muscle balance is needed to prevent injuries and reinjury. Limbs (arms and legs) can be tested to see if the left and right sides are equally strong or if one side, usually the injured side, is weaker. Kinesiology is a very important component of muscular and skeletal rehabilitation.

Cardiovascular rehabilitation focuses on the area of kinesiology that includes the cardiovascular and respiratory systems. In this area, exercise is used to strengthen the heart muscle. When a person has a heart attack, some of the heart tissue dies and is replaced by connective tissue. With some of the muscle gone and not able to pump blood, the heart is weaker. In cardiac rehabilitation, patients perform aerobic exercises to increase the strength of the heart muscle that is left. Through kinesiology applications, patients can strengthen their hearts and improve their ability to engage in daily living activities.

Health Promotion and Injury Prevention. The principles of kinesiology are used to maintain good health and prevent injuries. For many years, exercise has been recognized as an important component of health promotion. Kinesiology studies have involved developing and researching the best types of exercise to improve health. Exercise specialists such as physical education instructors, fitness trainers, and sports coaches rely on kinesiology to help people exercise safely and efficiently. They create exercise programs or develop exercise sessions based on kinesiology principles. Exercise can take the form of group classes or individual instruction (personal training). Cardiovascular exercise is very important for health and fitness. Exercise specialists are often charged with helping a person develop the endurance to engage in physical activity for extended periods. Based on an assessment of the person's health and fitness, exercise specialists use exercise physiology principles to write a prescription for activity. The prescription, or exercise plan, is often based on heart rate so that the individual can use his or her pulse to gauge whether the right level of effort is being attained. Another important type of exercise for health and fitness is strength training. Exercise specialists use kinesiology principles to demonstrate proper lifting techniques and help clients get stronger. Attention is paid to balancing muscle strength across the body. Movement mechanics are also used to improve flexibility. A stretching program is designed to improve or maintain flexibility throughout the body and help clients move more freely. A good training program will help clients stay healthier throughout their lives and enable them to perform the activities of daily living more easily and longer.

An overriding factor in exercise for health is adherence to the program. Health benefits are obtained only with regular participation. The psychological area of kinesiology studies provides information about getting and keeping exercisers motivated.

Coaching. Coaches use kinesiology in many of their activities. Sports skill development requires regular evaluation of movement to determine if the sports skill is being performed properly, which maximizes performance and reduces the chances of injuries. Coaches must consider all the involved joints, the type of muscle contractions, and the planes of movements. Of great importance is the synchronization of the movements around the involved joints. Energy transfer from one joint to the next is critical for superior performance. Coaches use their knowledge of kinesiology to teach proper sport skills.

Coaches also must use kinesiology for strength and conditioning. Weight-lifting and other exercises must be performed with proper mechanics. The variety of available equipment makes a basic understanding of kinesiology imperative for teaching athletes effectively. Additionally, coaches must determine which exercises should be performed and which muscles are to be strengthened.

Coaches also use sport psychology to motivate athletes to perform their best and to develop the leadership skills that are important for success. Athletic competition can be stressful, especially as most athletes must deal with demands on their time and concentration that stem from their social and academic or work-related obligation. Coaches often teach athletes stress management techniques to help them cope.

Ergonomics. Kinesiology can be applied at the workplace in the area of ergonomics, which science uses body mechanics concepts to design workstations that are more comfortable and minimize overuse injuries from repetitive movements. Any workstation can be analyzed and appropriately modified. Ergonomic solutions for people in desk jobs are relatively simple, but finding answers for people in jobs that require lifting and carrying require the use of more kinesiology principles. After the proper techniques for lifting and carrying have been determined, the worker must be trained to perform the movements properly.

Future Prospects

Programs that use kinesiology professionals have traditionally been paid for by the participants. Kinesiology benefited those who could pay for good sport trainers, health and fitness clubs, and sports medicine. More programs, including youth sports programs, are emerging to provide access to exercise facilities and sports training to those who cannot afford to pay. Some insurance plans provide free or discounted gym membership, and many companies offer employee gyms or discounted gym memberships. Kinesiology continues to be a growing field. Professional sports remain popular around the world, and the emergence of new competitive sports such as extreme sports results in more research and more need for people trained in kinesiology. Also, more people are interested in maintaining good health, and positions in fitness training are likely to increase to meet the demands of these exercisers. With more people exercising and participating in sports, the number of sports-related injuries is likely to grow. Kinesiology education will be needed to train rehabilitation professionals to research injuries, educate the public about injuries, and rehabilitate those injured.

Bradley R. A. Wilson, PhD

Further Reading

Floyd, R. T. *Manual of Structural Kinesiology.* New York: McGraw-Hill, 2009. An introductory text that focuses on movements of the major joints.

Hoffman, Shirl J., ed. *Introduction to Kinesiology: Studying Physical Activity.* 3d ed. Champaign, Ill.: Human Kinetics, 2009. Includes an overview of the major areas of kinesiology and a discussion of professions.

Klavora, P. *Foundations of Kinesiology: Studying Human Movement and Health.* Toronto: Sport Books, 2007. An overview of basic anatomy and physiology, human performance, and the major components of kinesiology.

_____. *Introduction to Kinesiology: A Biophysical Perspective.* Toronto: Sport Books, 2009. A succinct text on basic kinesiology and human movement.

Kornspan, Alan S. *Fundamentals of Sport and Exercise Psychology.* Champaign, Ill.: Human Kinetics, 2009. Discusses basic opportunities and goals in the field.

Oatis, Carol A. *Kinesiology: The Mechanics of Human Movement.* Baltimore: Lippincott Williams & Wilkins, 2009. Covers biomechanics, movement at the major joints, and posture.

Web Sites

American Academy of Kinesiology and Physical Education http://www.aakpe.org

American Kinesiology Association http://www.americankinesiology.org

Energy Kinesiology Association http://www.energykinesiology.com

See also: Biomechanics.

FASCINATING FACTS ABOUT KINESIOLOGY

- In 1972, University of Oregon track coach Bill Bowerman invented the Nike Waffle Racer, a shoe with extra traction, when he poured a liquid rubber compound into his wife's waffle iron.
- Exercise is believed to boost mental function, increase energy, reduce stress, build relationships, prevent disease, strengthen the heart, and allow a person to eat more. However, on an average day, only about 16 percent of Americans over the age of fifteen participated in sports or exercised.
- In 2007, there were 9.9 million health club members over the age of fifty-five, more than four times the number in 1990.
- In 2010, the Washington, D.C., area was ranked as the fittest metropolitan area in the United States for the third time in a row.
- A leading professional organization in the kinesiology field, the American College of Sports Medicine, was founded in 1954.
- There are 639 skeletal muscles and 206 bones in the human body. Numbers vary slightly depending on how the muscles and bones are counted.
- Indiana University psychologist Norman Triplett is considered the first sport psychologist for his work that found exercisers performed better in a group than when exercising alone. He concluded that people became more competitive in a group setting.
- Exercise physiology in the United States is believed to have originated at the Harvard Fatigue Laboratory in 1927.

L

LACTATION

FIELDS OF STUDY
Biochemistry, developmental biology, embryology, physiology

SUMMARY
Lactation is the process by which female mammals produce milk for the nourishment of their offspring. For all species except humans, successful lactation is essential for the survival of the young.

PRINCIPAL TERMS

- **alveoli:** the milk-producing areas within the mammary glands
- **colostrum:** the precursor to milk that is formed in the mammary gland during pregnancy and immediately after birth of the young
- **ducts:** the tubular structures that carry milk from the alveoli to the outside through the nipple or teat
- **lactation:** the process of producing and delivering milk to the young; also, the time period during which milk is produced
- **mammary glands:** the milk-producing glands found in all mammals; for example, the cow's udder contains the mammary glands
- **milk ejection:** also known as milk letdown, this is the reflex response of the mammary gland to suckling of the nipple; the hormone oxytocin mediates this reflex
- **myoepithelial cells:** the specialized cells within the mammary gland that surround the alveoli and contract to force milk into the ducts during milk ejection
- **nipple:** the raised area on the surface of the skin over the mammary gland that contains the duct openings
- **teat:** an elongated form of nipple that contains one duct opening

BASIC PRINCIPLES
Lactation is the process by which female mammals produce milk to feed their young. The ability to produce milk is one of the defining characteristics of the class Mammalia: All mammals, but no other animals, possess the highly specialized glands necessary for lactation. In evolutionary terms, the appearance of lactation coincides with the tendency of mammals to produce only a few offspring at a time; the provision of milk for these offspring helps to ensure their survival while removing competition between the adults and the young for food.

HOW IT WORKS

The Mammary Glands. The mammary glands are the milk-producing organs. The number varies among species from two to about twenty, with a rough correlation between the number of young born and the number of glands present. The glands are located on the ventral surface of the body, either in the thoracic (in humans, for example) or abdominal region (in horses and cows) or in two lines extending almost the length of the body (in dogs and rodents). Both male and female mammals have mammary glands, because in early mammalian development the basic body plan of male and female embryos is identical. The mammary glands of males are nonfunctional, however, since they lack the hormonal stimulation necessary for lactation.

Internally, the mammary glands of all mammals follow the same basic plan, consisting of alveoli that produce milk and ducts that carry the milk to openings on the surface of the skin. The alveoli are surrounded by myoepithelial cells that contract to squeeze the milk into the ducts during suckling by the young. Externally, considerable variation exists among the mammals in the appearance of the mammary glands and their associated openings. In the spiny anteater and platypus, the many lobes of the mammary glands each open directly to the surface of the abdominal skin through individual ducts, and the young suck the hair-covered skin to obtain the milk. In other mammals, the mammary glands are more obvious as swellings beneath the skin, with a raised area, the nipple or teat, that contains the duct openings. In some four-legged animals (cows, horses, and goats) the mammary glands are located in a baglike structure called the udder, from which are suspended the elongated teats. In humans, the nipple, which contains the openings of fifteen to twenty-five ducts, is surrounded by pigmented skin, the areola. The areola contains glands (tubercles of Montgomery) that secrete a lubricating fluid.

Milk Production. Lactogenesis (milk production) does not begin until a female has produced young. During pregnancy, a complex of hormones prepares the mammary glands for milk production by promoting their growth and internal development. These hormones include prolactin from the mother's anterior pituitary gland, placental lactogen from the placenta within the uterus, and estrogen and progesterone, which are produced in the corpus luteum of the mother's ovary and in the placenta. Other hormones, including cortisol from the adrenal gland, thyroxine from the thyroid gland, and insulin from the pancreatic islets, may also be involved. Progesterone appears to participate in the induction of mammary development, but, paradoxically, it also prevents milk secretion during pregnancy.

Although true milk is not produced during pregnancy, a precursor to milk, colostrum, can be produced in small amounts by the mammary glands of most species. Colostrum is a sticky, yellowish, transparent liquid. Colostrum secretion continues in the first few days after birth of the young; there is then a gradual transition to production of true milk.

Milk contains water, proteins, fats, vitamins, minerals, and a unique sugar, lactose. The exact concentration of the various components varies greatly between species according to the nutritional demands of the young. The milk of seals is high in fat and other solids that contribute to rapid weight gain in the pups, a strategy that appears to be essential for their survival.

Noteworthy among the constituents of milk are antibodies produced by the mother. These antibodies help protect the newborn from disease in the period when the newborn's own immune system is immature and incapable of providing significant defense. The antibody concentration of colostrum is higher than that of true milk, and for this reason the first few days of nursing are considered the most important for immunological protection of the newborn.

The transition in production from colostrum to true milk is brought about by a change in the hormonal status of the mother. At the time of birth, the placenta is expelled from the mother's body, thus removing the source of progesterone, estrogen, placental lactogen, and other hormones. The decrease in progesterone levels is thought to be essential for the onset of lactogenesis. In addition, at the time of birth, there are changes in prolactin secretion that may play a role in initiating milk secretion.

Suckling. Once lactogenesis is established, a set of hormonal reflexes act to match milk production and delivery to the needs of the newborn. Suckling of the nipple involves motions similar to chewing as the infant takes the nipple between the tongue and the palate. This suckling motion stimulates nerve endings in the mother's nipple that relay signals about the stimulation back to the mother's brain. Within thirty to sixty seconds, these signals result in the release of prolactin from the mother's anterior pituitary gland and oxytocin from her posterior pituitary gland. Prolactin causes continued production of milk by the alveolar cells of the mammary glands. Oxytocin acts immediately on the myoepithelial cells of the mammary gland, causing them to contract and push milk from the alveoli into the ducts and thence through the nipple into the infant's mouth. Thus, the infant does not actually remove milk from the mammary gland by suction, but instead is responsible for promoting a hormonal reflex that results in active milk ejection, or letdown, from the mammary gland.

Because of the operation of the prolactin and oxytocin reflexes during suckling, lactation is a biological example of the principle of supply meeting demand. All that is necessary to increase milk production is to increase the suckling stimulus by nursing the young more often. Once established, lactation in some species can be sustained in this manner for years, assuming the nutritional needs of the mother are met. On the other hand, if the mother fails to nurse her offspring, the absence of the suckling stimulus will cause the mammary glands gradually to cease milk production.

The exact composition of the milk is altered as lactation continues to meet the changing nutritional needs of the growing offspring. The most extreme example of the ability of the mammary gland to change the composition of milk is seen in the kangaroo. In this animal, the newborn attaches to a teat in the mother's pouch shortly after birth and remains there for a month or more. A mother kangaroo may nurse offspring of different ages from separate teats, and each teat supplies a milk with the appropriate nutritional composition for that young.

Species vary in time spent suckling the young. The rabbit nurses her litter for only about five minutes once a day, while the rat nurses for about half an hour at a time, at intervals throughout the day. Lactation lasts about ten days in rodents, but it may persist for months in large species such as horses and cows. Continued lactation has a suppressive effect on ovulation that is thought to be attributable to interference by prolactin with the normal hormonal mechanisms that cause ovulation.

THE STUDY OF LACTATION
Although it has always been clear that milk is expelled from the mammary glands, the realization that the glands themselves actually produce the milk is a relatively recent one. Early anatomists erroneously assumed that milk must be a product of the uterus, since the uterus is involved in support and nourishment of the fetus. Thus, much of the early anatomical work attempted to show some sort of connection between the uterus and the mammary glands. It was not until the late 1800's that the light microscope clearly demonstrated that milk is formed within the mammary glands. In the twentieth century, electron microscopy showed that during pregnancy the intracellular organization of the alveoli becomes increasingly more complex as the cells become capable of milk secretion.

Various techniques for labeling compounds with radioactive or fluorescent markers have been used in conjunction with electron microscopy to examine how milk is synthesized in the alveoli. The alveoli cells extract necessary precursors from blood flowing through the mammary gland, assemble the precursors into milk components, and then secrete the constituents of milk into the mammary gland ducts. Specific routes of secretion have been identified for the major components of milk.

More recently, researchers have used cell-free systems to study the biochemical pathways involved in milk synthesis. These systems use isolated fragments of deoxyribonucleic acid (DNA), ribonucleic acid (RNA), and perhaps some cell organelles to examine the intermediate chemical steps in the synthesis of milk components. Using these techniques, researchers have been able to "watch" as complex milk proteins and constituents are assembled step by step. The knowledge of how the components of milk are assembled is leading to a fuller understanding of how the amounts of these substances in milk are hormonally regulated.

Knowledge of the hormones involved in inducing and maintaining lactation has come about through systematic assembly of information from several lines of research. Test animals can be treated with a specific hormone to determine if that hormone causes or suppresses lactation. The test animals may be males or immature females, with the goal being the duplication of the specific mix of hormones that cause lactation in the adult female. The opposite approach may also be taken: An endocrine gland can be removed from a lactating female to determine if the hormonal products of that gland are necessary for lactation. Another approach is to make careful measurements of the levels of hormones circulating in the blood as the lactational state changes; any hormone that shows a correlated change may be a good candidate for further investigation by treatment or removal from test animals. These methods have led to an understanding of the importance of prolactin, placental lactogen, oxytocin, estrogen, and progesterone in promoting lactation, but researchers still do not understand how the system is fine-tuned. For example, considerable variation exists in the volume and quality of milk produced by different individuals—or by the same individual at different times—but these differences cannot currently be explained by any known change in hormone levels. Research is focusing not only on describing changes in circulating levels of hormones but also on elucidating the exact effects of these hormones on the biosynthetic pathways within the mammary gland.

—*Marcia Watson-Whitmyre*

See also: Birth; Digestion; Endocrine systems of vertebrates; Nutrient requirements; Reproductive system of female mammals; Sexual development.

FURTHER READING

Larson, Bruce L., and Vearl R. Smith, eds. *Lactation: A Comprehensive Treatise.* 4 vols. New York: Academic Press, 1974. Every aspect of lactation is covered: development, structure, and diseases of the mammary gland; hormonal control; lactogenesis; milk com- ponents; and maintenance of lactation. Some of the chapters are specific to humans, and others cover domestic and laboratory animals.

Mepham, T. B., ed. *The Biochemistry of Lactation.* Amsterdam: Elsevier, 1985. Concentrates on ruminant lactation.

Peaker, M., R. G. Vernon, and C. H. Knight, eds. *Physiological Strategies in Lactation.* London: Academic Press, 1984. The purpose of this book is not only to explore mechanisms of lactation but also to analyze the impact of evolution on lactational patterns in different mammal species.

LACTIC ACID FERMENTATION

FIELDS OF STUDY

Bioengineering; Industrial Fermentation; Metabolic Engineering; Nutrition and Dietetics; Pharmacology; Zymurgy and Zymology

SUMMARY

Energy captured during the formation of glucose during photosynthesis is released during the processes involved in respiration in the human body. The glycolytic pathway breaks down the glucose molecule into carbon dioxide and water, and uses the energy released by the chemical changes to drive the processes of metabolism. When oxygen is not available, pyruvic acid from glucose is broken down to lactic acid rather than degraded further. Buildup of lactic acid in overworked muscles causes a burning sensation. Prokaryotic cells derive metabolic energy from glucose anaerobically via the lactic acid fermentation process, and are unable to grow and produce adenosine triphosphate (ATP) in the presence of oxygen. Eukaryotic cells require oxygen for growth and the production of ATP.

PRINCIPAL TERMS

- **enzyme:** a biological catalyst, any protein molecule within a living organism that speeds up biochemical reactions to a rate that will sustain life. The effect may speed up metabolic reactions by a factor of one million, compared with what would occur chemically outside the body. Names and classification of enzymes are regulated by the International Commission on Enzymes. Most enzymes are named by adding -ase to the root of a corresponding substrate, the molecule an enzyme acts upon. Sucrase catalyzes the hydrolysis of sucrose into glucose and fructose. A living cell has a unique set of 3,000 enzymes, each defined by the cell's DNA.
- **fermentation:** a biological process for breaking down complex organic compounds into simpler compounds. One of the most familiar in human history is the conversion by yeast of sugar to carbon dioxide, alcohol, and water. Fermentation also occurs in cells, including animal muscle cells, breaking down glucose to produce lactic acid, lactate, carbon dioxide, and water, as well as adenosine triphosphate, a source of energy. It is less efficient than cellular respiration but occurs when muscles are short of oxygen. Many anaerobic bacteria ferment sugars: Lactobacillus ferment milk to produce yogurt. Fermentation also produces lactic acid in a variety of foods, such as sauerkraut and sourdough bread.
- **metabolism:** physical processes and chemical reactions within a living body that convert or use energy, including those associated with digestion, excretion, breathing, blood circulation, growing and using muscle tissues, communication through the nervous system, and body temperature. *See also* enzyme, protein.
- **microbe:** any microscopic form of life, also called a microorganism, particularly bacteria, protozoa, fungi, or virus. Most commonly, this term refers to pathogenic microscopic life—those that cause infection, disease, decay, sepsis, or gangrene. However, biologists are identifying an increasing number of microbes that are beneficial, even essential to life, including a variety of those found in the human intestine.

Energy and Metabolism

Living organisms, including humans, derive their energy from the solar energy released by the sun. That energy is captured by green plants in the process of photosynthesis, which combines carbon dioxide and water in the presence of chlorophyll and sunlight to produce the simple hexose sugar glucose. The glucose so formed is the fundamental food source of biological systems. When consumed as food by animals, the glucose is broken down again by the processes of glycolysis to provide the energy needed to drive the processes of metabolism. In glycolysis, the glucose molecule is disassembled in stages mediated by enzymes.

At each stage, the energy that has been captured in the glucose molecule in chemical bonds is released and transferred for the formation of the main energy carriers of metabolism, which are adenosine triphosphate and diphosphate (ATP and ADP) and nicotinamide adenine dinucleotide (NAD) and its phosphate derivative NADPH. Through the process of "respiration" consisting of glycolysis and the "citric acid cycle", or Krebs' cycle, the glucose molecule is broken down, in the presence of oxygen, again ultimately to carbon dioxide and water. This is the carbon dioxide that is exhaled in the breath during respiration.

Enzymes in the mitochondria of the cells carry out the citric acid cycle, finally releasing carbon dioxide into the cell protoplasm. From there the carbon dioxide passes through the cell membranes and into the bloodstream where an enzyme called carbonic anhydrase catalyzes its conversion to the more soluble bicarbonate ion. In the lungs, bicarbonate is converted back to carbon dioxide and passes through the lung lining to be exhaled into the atmosphere again, ready for another round of photosynthesis and glycolysis. Throughout the process of respiration, many molecules of ATP are produced, providing the energy required for metabolism.

In the Absence of Oxygen

Respiration requires the presence of oxygen, and is termed an aerobic process. When the system is deprived of available oxygen, however, as when prolonged muscle contraction exceeds the rate at which oxygen can be replenished in the muscle tissue, the anaerobic process of respiration becomes the anaerobic process of fermentation. In the initial stage of glycolysis, the glucose molecule is split into two molecules of pyruvic acid. Normally, these then go on into the citric acid cycle to be converted sequentially through several reaction steps into carbon dioxide and water. When oxygen is not available to aerobic cells, the pyruvic acid molecule is reduced to lactic acid rather than being oxidized. This is the essential process of lactic acid fermentation. Most eukaryotic cells are "obligate aerobes," meaning that they require the presence of oxygen for their growth and the metabolism of glucose to carbon dioxide with the concomitant production of ATP. Some eukaryotic cells can grow either aerobically or anaerobically, and are termed "facultative anaerobes." Almost all prokaryotic cells, all of which are microbes, are obligate anaerobes, and do not grow or function in the presence of oxygen.

Lactic Acid Fermentation Versus Respiration

Both the processes of respiration and lactic acid fermentation have as their principal logic the production of ATP to provide energy for the processes of metabolism. Respiration, through glycolysis and the citric acid cycle, is more efficient at the production of ATP than is the anaerobic process of lactic acid fermentation. Aerobic metabolism releases more energy from glucose, and at a faster rate, than does its anaerobic counterpart. Accordingly, anaerobic organisms are simple prokaryotic organisms, while all higher life forms are composed of eukaryotic cells having higher metabolic function.

Richard M. Renneboog MSc

See also: Cell Organelles, Eukaryotes and Prokaryotes.

Further Reading

Lahtinen, Sampo, Ouwehand, Arthur C., Salminen, Seppo and Von Wright, Atte *Lactic Acid Bacteria. Microbiological and Functional Aspects* 4th ed., Boca Raton, FL: CRC Press, 2012.

Holzapfel, William H. and Wood, Brian J.B., eds. *Lactic Acid Bacteria. Biodiversity and Taxonomy* New York, NY: John Wiley & Sons, 2014.

Ewing, W.N. And Tucker, L.A. *The Living Gut* 2nd ed., Nottingham, UK: Nittingham University Press, 2008.

Kang, Jie *Bioenergetics Primer for Exercise Science* Champagne, IL: Human Kinetics, 2008.

Patton, Kevin T. and Thibodeau, Gary A. *Anatomy & Physiology* 9th ed., St. Louis, MO: Elsevier, 2016.

FASCINATING FACTS ABOUT LACTIC ACID FERMENTATION

- In anaerobes such as yeast, pyruvic acid is decarboxylated to acetaldehyde, which is subsequently reduced to ethanol. This, of course, is the fundamental process of the beer, wine and spirits industry.
- Lactic acid fermentation is an essential process for the production of many food products, including yoghurt, cheeses, sourdough baked goods, fermented vegetable preparations.
- It is now believed that most digestion in the human gut is carried out by the trillions of probiotic bacteria living there. The bacteria break down food materials to their molecular components, often altering their molecular structure in the process, and these nutrients are then transported out of the digestive tract for use by the many different cells in the body.
- The same chemical reaction of the lactic acid fermentation process that causes the feeling of 'muscle burn' is responsible for the taste of sour cream and yogurt.

LIFE SPANS

FIELDS OF STUDY

Biochemistry, cell biology, developmental biology, genetics, population biology

SUMMARY

An animal's life span, the time lapsing between its birth and death, differs greatly from species to species because of differing environmental pressures, chance variations in physical conditions, and heredity.

PRINCIPAL TERMS

- **life cycle:** the sequence of development beginning with a certain event in an organism's life (such as the fertilization of a gamete), and ending with the same event in the next generation
- **life expectancy:** the probable length of life remaining to an organism based upon the average life span of the population to which it belongs
- **life span:** the maximum time between birth and death for the members of a species metabolism: the biochemical action by which energy is stored and used in the body to maintain life
- **mortality rate:** the percentage of a population dying in a year

BASIC PRINCIPLES

Life span has two common meanings, often confused. Popularly, the term can refer to the longevity of an individual, but in biology it is more abstract, a characteristic of the entire species rather than individual members. In this sense, life span is the maximum time an individual can live, given its environment and heredity, and life expectancy is the amount of time remaining it at any point during its life span.

The great variety within the animal kingdom complicates the definition of life span. For some species, life span is essentially the same as its life cycle, the return to the same developmental stage from one generation to the next. Salmon, for instance, hatch from eggs in small streams, migrate to the sea where they reach maturity, then struggle up rivers to return to the site of their hatching in order to produce more eggs. After completing the reproduction cycle, both males and females die. For most animal species, however, an individual may produce several generations of offspring before dying. In the case of humans, individuals can live long after their fertility ends.

It is often difficult to measure the life span for specific species, and in fact, a time period is seldom definitive for all species members. Rather, scientists recognize that mortality, the percentage of individuals that die each year (or each day in some cases), increases for a population of organisms until at some age it reaches 100 percent, and all individuals in a generation are dead. Finding the life span of laboratory animals is fairly simple: A population is given the best possible living conditions, and observers wait for the last individual to die. That, presumably, is the optimal life span for the species.

Much the same procedure can determine the life spans of pets and animals in zoos, except that the population to be observed is much smaller, sometimes only a single individual. Likewise, the records of thoroughbred domestic animals, born and raised in captivity, provide evidence for the maximum species life span. Most animals live in the wild, however, where investigations face a great variety of conditions and anecdotal evidence can be misleading. For example, biologists long thought that bowhead whales lived only about fifty or sixty years, but in the late 1990's, various new kinds of historical and biochemical evidence identified bowheads that lived well into their second century. Moreover, species whose members can, if need be, go into dormancy show a large variation in individuals' apparent longevity, even when all members pass through a single life cycle.

In general, however, the life span variation among species falls into a fairly narrow range of time. The shortest life spans, which last a single life cycle, can be a matter of days, while the longest last more than two hundred years. Humans enjoy the greatest longevity among primates; scientists estimate the theoretical maximum human life span to be from 130 to 150 years, more than double that of the species with the next greatest endurance, gorillas and chimpanzees. However, several kinds of invertebrates live more than two centuries.

How It Works

Life Span Limiters and Extenders. While each species has a theoretical maximum life span, few if any individual animals reach it. Three general influences limit longevity: environmental pressures, variations in physiological processes, and heredity.

Most domesticated species have longer life spans, frequently two times longer, than their wild relatives, and wild animals in captivity often live longer than in their natural habitat. The gray squirrel, for example, lives for three to six years in the wild, but from fifteen to twenty years in a zoo. The reason is clearly the safer, healthier, less stressful environment. Predators are one of the biggest threats in the wild, as are fluctuations in climate that affect the availability of food and shelter. Natural calamities such as hurricanes, wild fires, and earthquakes also take their toll.

Disease and chance injury also kill off many organisms, but even if disease is absent, many physiological processes in the body appear to degenerate or stop with age. Biochemists find that individual cells age and die. Cross-connections among connective tissue, such as ligaments and cartilage, gradually reduce the body's flexibility and inhibit motion. Chemical plaques build up in brain tissue, hindering the electrochemical connection among neurons. Highly reactive oxidants, the ionized molecular byproducts of cellular metabolism, also build up with age and degrade the operation of cells' mitochondria, the generators of chemical energy, as well as other cellular organelles. Moreover, laboratory tests reveal that cells can divide only a certain number of times—about fifty for some human cells—and then they die, a limit called the Hayflick finite doubling potential phenomenon. In connection with this finding, geneticists discovered that the lengths of deoxyribonucleic acid (DNA) at the tips of chromosomes, called telomeres, shorten with age and appear to play a role in cell dysfunction and death.

Other genetic factors help determine the age limit of cells and the entire organism. Research in the late 1990's with mice and roundworms uncovered genes that regulate cell life, including one kind that causes cells to suicide if their DNA or in- ternal structure is too damaged to function properly or if the cells turn cancerous. Loss of function by genes damaged during cell division (mitosis) or by environmental toxins can impair a cell's ability to maintain metabolism, or set loose functionless or even outrightly harmful proteins and enzymes into the blood stream and lymph system— all of which can bring on illness. Organisms that escape such effects still may face genetic disease. Most human populations, for instance, now have a longer life expectancy than ever before; because of it, neurodegenerative diseases (such as Alzheimer's disease), cardiovascular dysfunction, and immunological disorders, caused or made possible by genes, grow ever more common. Because many of these life-shortening maladies were rare or nonexistent before the twentieth century, natural selection has not had a chance to cull them from the human genome. Similar diseases crop up in domestic animals and wild animals in captivity.

Late twentieth century research also identified several ways to extend life. A sharply reduced diet extends the lives of mice and fruit flies beyond their normal life span (although for the brown trout, a richer diet is life-extending), and lower than normal temperature has the same effect on fruit flies, fish, and lizards. Discovery of the relation of telomere length to cell death and isolation of genes that order cells to suicide allowed scientists to bioengineer animals with cells whose DNA self-repaired the telomeres and genes that failed to trigger cell-suicide; the result was individuals that in some cases had life spans twice the normal length or more. However, the most pervasive life-extending method is the development of social life—colonies, herds, packs—in which individuals work together to protect, shelter, and feed themselves and rear their offspring. In the case of humans (and perhaps some whales), culture and intelligence permit the species to pass on recently acquired information from one generation to the next and even to alter the environment to lengthen collective and individual life spans.

Size and Life Span. Biologists noted long ago that large animals live longer than small animals, but why this should be true was a mystery. In 1883, Max Rubner proposed that the relation had to do with metabolism. Large animals have a smaller skin surface to body mass ratio than smaller animals; accordingly, large animals lose heat more slowly and so can maintain body functions with less energy burned per unit of mass; in other words, a slower metabolism. Scientists since found that individuals of different animal species all use about the same amount of chemical energy during their lives, twenty-five to forty million calories per pound per lifetime. (There are significant exceptions: Humans consume about eighty million calories per pound.) Because small animals must burn energy at a higher rate, the argument holds, they physically wear out faster.

In 1932, Max Kleiber derived a mathematical relationship for Rubner's proposal. According to Kleiber's law, also known as the quarter-power scaling law, as mass rises, pulse rate decreases by the one-fourth power. So elephants, which have 104 times the mass of chickens, have a pulse rate one tenth as fast. Scientists suggest that the relation results from the geometry of circulatory systems and point out that the quarter-power scaling law is pervasive in nature, but the underlying reason for it remains unknown. In any case, plenty of exceptions to the mass-life span correlation exist. With a life span of about one hundred years, box turtles outlive fellow reptiles, for example, and humans outlive all mammals (with the possible exception of some whale species) regardless of size. Exceptions also occur among domestic species living sheltered lives: Cats have longer life spans than dogs.

Theories of Life Span. A 1995 review of data from earlier animal studies suggested that heredity accounts for about 35 percent of the variation in life spans among invertebrates and mammals; 65 percent comes from unshared environmental influences. Nonetheless, theorists in the life sciences continue to debate the relative influence of genetic and other biochemical factors on the one hand and environment fac- tors on the other hand. The debate derives from the premise that life spans are the product of the natural selection that ensured species' reproductive success. The proposals fall into three categories: random damage (stochastic) theory, programmed self-destruction theory, and ecological theory.

Random damage theories emphasize the wear and tear on the body that accumulates with metabolic action. It is the source of damage that differs from one theory to another. One holds that the buildup of metabolically produced antioxidants is the key factor, a spinoff of the long-standing conjecture that the faster an animal's metabolism is, the shorter its life span. A second theory focuses on proteins that change over time until their effect on the body alters for the worse, especially when the proteins are involved in cellular repair. There is, for example, the altered connective tissue that causes the cross-connections stiffening tendons and ligaments. Another such change is the glycosylation of proteins or nucleic acids, in which a carbohydrate is added. Glycosylation is involved in such age-related disorders as cataracts, vascular degeneration among diabetics, and possibly atherosclerosis. A third theory points to the buildup of toxins inside cells, and a fourth concerns the potential problems that come from errors in metabolism or viral infection which slowly impair or kill cells. Fifth, the somatic mutations theory proposes that chance mutations accumulate in a person's nuclear or mitochondrial genome and induce cell death or produce proteins and enzymes that have aging effects.

Programmed death theories hold, as the name suggests, that a species' genetic heritage includes a built-in timer or damage sensor. Telomeres shorten as DNA ages until the genes at the end of chromosomes are unprotected and subject to deterioration during the splitting and gene crossover of mitosis. The genes then lose their ability to produce essential biochemicals, whose absence harms the body or leaves it defenseless against damage from infection or injury. Damage sensors can include the genes that instruct cancerous or malfunctioning cells to die. Although such genes clearly are a means to check the spread of disease, their cumulative effect may be harmful. Furthermore, scientists discovered genes that produce much more of, or less of, their metabolic products as cells age, which also contributes to the overall aging of the body.

The ecological theory draws conclusions about life span from a species' role in its environment. Small animals have faster metabolisms and live shorter lives, it is argued, because they are not likely to escape predators for very long. Therefore, they evolved to mature and reproduce rapidly. Large animals typically have more defenses against predators and can afford to take life slowly. Moreover, animals that evolve defensive armor, spines, or poison also avoid predation and live longer than related species that do not. Finally, species that evolve mechanisms to withstand environmental stress, as from extreme temperatures or food scarcity, also have long life spans.

The theories assume that the life span for individuals within a species serves the survival of the entire species. Yet even a species' days on earth are numbered. Environmental change can slowly squeeze them from their habitats, a catastrophe may wipe them out indiscriminately, or they may evolve into a new species. Scientists estimate that the average life span for a multicellular species lasts from one to fifteen million years. That average is stretched by several notable exceptions in the animal kingdom—such living fossils as crocodiles (140 million years old), horseshoe crabs (200 million), cockroaches (250 million), coelacanths (a type of fish, 400 million), and certain mollusks of the genus *Neopilina* (500 million).

—*Roger Smith*

See also: Aging; Birth; Death and dying; Demographics; Genetics; Growth; Reproduction; Zoology.

FURTHER READING

Austad, Steven. *Why We Age*. New York: John Wiley & Sons, 1997. Written by a leading investigator of aging, the book is primarily devoted to human concerns, but there is much discussion of theories of aging, as well as specific biological and environmental influences, pertinent to animals, especially mammals.

Bova, Ben. *Immortality*. New York: Avon Books, 1998. Although human life span is his primary topic, Bova discusses animal life spans and the biochemistry of aging in clear, engaging prose for general readers.

Finch, Caleb B., and Rudolph E. Tanzi. "The Genetics of Aging." *Science* 278, no. 5337 (October 17, 1997): 104-109. A concise, well-explained survey of the genetic influences on life span, drawing from studies conducted on animals. Best for readers who understand basic biology and genetics.

Furlow, Bryant, and Tara Armijo-Prewitt. "Fly Now, Die Later." *New Scientist* 164, no. 2209 (October 23, 1999): 32-35. The authors describe the age differences between closely related creatures of the same size to support their theory that environmental factors, such as predation, influence the evolution of life spans for species.

MacKenzie, Dana. "New Clues to Why Size Equals Destiny." *Science* 284, no. 5420 (April 6, 1999): 1607-1608. Discusses theories about the relation of size to life span in a style accessible to general readers.

Margulis, Lynn, and Karlene V. Schwartz. *The Five Kingdoms: An Illustrated Guide to the Phyla of Life on Earth*. 3d ed. New York: W. H. Freeman, 1998. A richly illustrated explanation of taxonomy with detailed information about the biology of the thirty-seven phyla in the kingdom Animalia, including discussion of life cycles and development.

Walford, Roy L. *Maximum Life Span*. New York: W.W. Norton, 1983. Delightfully written and lucid, this book, written by a pioneering researcher in aging, is somewhat out of date and focuses primarily upon human aging, but it still supplies solid information about the biological and ecological limits on animal life spans.

THE BIG, THE OLD, AND THE IMMORTAL

In some species, the older an animal gets, the larger it grows. Big sharks are old sharks, as are big crocodiles and snakes. In 1912, a reticulated python was found that measured thirty-two feet, ten inches in length, the longest ever, but no one knows how old it was. The age record for snakes goes to Popeye, a boa constrictor at the Philadelphia Zoo that died in 1977 at the great age of forty years and three months. Even for reptiles, however, Popeye's life span was modest. One Madagascar radiated turtle lived 188 years.

Most species, including humans, grow to a maximum size and then stop. The longest attested life span for a person is 122 years, achieved by France's Jeanne Louise Calment, who died in 1997, followed by Japan's Shigechiyo Izumi, who died just short of 121 years in 1986. Despite theories relating large body mass to long life span, some big and small animals in the wild have about the same life expectancy at birth: fifty years for both golden eagles and most whales, for instance, or twenty-four years for rhesus monkeys compared with twenty-five years for lions. There is some disagreement over which animal species has the longest life span. The tubeworms growing on the ocean floor near hydrocarbon-seep sites in the Gulf of Mexico are thought to be as much as 250 years old, and so have been called the longest-lived noncolonial animals without backbones. However, some mollusks of the class Lamellibranchia, such as quahogs, a variety of clam, have also been called the longest-lived animal. They, too, can lie snug in their ocean bed for 250 years.

Whether classified as animals or not, bacteria hold the record for life span. If they are not eaten by predators or killed by environmental change, these single-cell creatures are theoretically immortal. Furthermore, they do not get larger with age: They divide.

M

Metabolic engineering

FIELDS OF STUDY

Biology; biochemistry; molecular biology; organic chemistry; analytical chemistry; microbiology; genetic engineering; biotechnology.

SUMMARY

Metabolic engineering is a new science that appeared in the 1990's. It is associated with biology and chemistry. Metabolic engineering allows the designing of biochemical pathways that do not exist in the natural world, as well as the redesign of existing biochemical pathways often with the use of genetic engineering. Metabolic engineers often modify biochemical pathways by reducing cellular energy use or waste production, by changing the nutrient flow to the cells, or improving the productivity and yield of a particular pathway. In addition, metabolic engineers may potentially design new organisms that are tailormade for the desired chemicals and production processes. Many novel compounds of industrial and medical interest can be produced by metabolic engineering. In the twenty-first century, the main efforts of metabolic engineers are concentrated on biofuels and pharmaceuticals.

PRINCIPAL TERMS

- **bioreactor:** apparatus for cell growth with practical purpose under controlled conditions.
- **DNA sequencing:** determining of the precise order of nucleotides (such as adenine, guanine, cytosine, and thymine) in a DNA.
- **enzymes:** biological catalysts made of proteins.
- **genetic engineering:** modification of genetic material to achieve specific goals.
- **metabolism:** sum of biochemical reactions within an organism.
- **substrate:** substance that is acted on (as by an enzyme).

BASIC PRINCIPLES

Metabolic engineering is a relatively new field that deals with the modification and optimization of metabolic pathways, mainly in microorganisms, by altering genes, nutrient uptake, or metabolic flow to allow production of novel compounds that are of industrial and medical interest. Metabolic pathways of living organisms are not optimal for specific practical applications, but they can be modified using the tools of modern biotechnology such as genetic engineering.

The redesign of existing, natural metabolic pathways for useful purposes is a main objective of metabolic engineering. Metabolic engineering usually includes two phases: careful analysis of the metabolic pathway and genes involved in the pathway (analytical phase) and its modification (synthesis phase). Pathway analysis often includes the metabolic control analysis: determining which compounds can control the productivity and yield of particular pathway. Different tasks of metabolic engineering are as follows: improvements of productivity and yield of particular pathway; expansion of substrate range; elimination of waste; improvement of process performance; improvements of cellular activities; and extension of product array. Metabolic engineering is becoming one of the principal fields of biotechnology. Production of many chemicals and fuels uses nonrenewable resources or limited natural resources.

Metabolic engineering creates many alternatives to replace dangerous chemicals and petroleum-based transportation fuels with clean, green, and renewable chemicals and biofuels.

BACKGROUND

The term "metabolic engineering" first appeared in the early 1990's. Since that time, the range of products that can be generated has increased significantly, partly because of remarkable advances in other fields related to metabolic engineering, such as DNA sequencing and genetic engineering. With

DNA sequencing, scientists were able to identify the majority of metabolic genes and enzymes in many organisms. In the post-sequencing era, the obtained information is used for practical construction of biochemical pathways or whole organisms with optimized functions through metabolic engineering.

In the 1990's, scientists developed new genetic tools that gave metabolic engineers more precise control over metabolic pathways. They also created analytical tools that allowed the metabolic engineer to track metabolites in a cell to identify new biochemical pathways more precisely. Earlier in the twenty-first century, metabolic engineers joined other scientists in their quest for alternative fuels, which are in high demand because of increasing oil prices and concern about climate change.

How It Works

Metabolic engineering is based mainly on microbial metabolism. Microbes produce different kinds of substances that they use for the growth and maintenance of their cells. These substances can be useful for humans. The goal of metabolic engineering is to enhance the microbial production of useful substances. To achieve this goal, metabolic engineers must follow a particular route. They need to choose a friendly organism (host) for their metabolic manipulations. They need to find cheap and available substrates to use for modified metabolic pathways. Finally, metabolic engineers must be able to perform genetic manipulations of metabolic routes. Metabolic engineers can also alter nutrient uptake or metabolic flow. All these steps are dependent on each other. For example, genes cannot be manipulated in every organism; products or metabolic intermediates may be toxic to its host.

Host and Host Design. Generation of products by metabolic engineering has been achieved by transferring product-specific enzymes or entire metabolic pathways into so-called user-friendly microorganism hosts, which were used traditionally in industry. These industrial microorganisms grow rapidly on inexpensive culture media available in bulk quantities, are open to genetic manipulation (and genetic manipulation tools are available), and are nonpathogenic (do not cause disease). In addition, it is important that the host can survive (and thrive) under the desired process conditions (ambient versus extremes of temperature, pH). It is essential that the host is genetically stable (with the introduced pathway) and not susceptible to virus or another microbe's attack. Among the host microorganisms most widely used are *Saccharomyces cerevisiae* and *Escherichia coli*. *Saccharomyces cerevisiae*, or baker's yeast, has been used for making bread and alcohol for thousands of years. It is one of the earliest domesticated organisms. This organism has come to be used in a large number of different processes within the biotechnological and pharmaceutical industries. Comprehensive knowledge of *S. cerevisiae* has been accumulated over a long period of time. In addition, the complete genome sequence of yeast is available, and yeast is nonpathogenic. The well-established fermentation and process technology for large-scale production with *S. cerevisiae* in bioreactors makes this organism very attractive for several industrial purposes.

Escherichia coli, commonly known is *E. coli*, is a bacterium that is widely used as a research (model) organism. It is easy to grow and genetically manipulate this bacterium, and its genome sequence is available. Several important products such as interferon (flufighting drug), insulin, and growth hormone are manufactured by genetically modified *E. coli*. In addition to *E. coli* and *S. cerevisiae*, several other microorganisms are widely used as hosts for metabolic engineering manipulations, including bacteria *Bacillus subtilis* and *Streptomyces coelicolor*. Finally, in addition to redesigning particular metabolic processes, metabolic engineers may also design de novo artificial cells that will produce desired products.

Substrates. To make metabolically engineered products, chemical substrates are needed. To make these products economically viable, inexpensive sources of substrates are required. Substrates must contain different chemical components, such as carbon, nitrogen, oxygen, and hydrogen. For example, metabolic engineers are looking at sugars from cellulosic biomass as potential substrates for biofuel production. Cellulosic biomass is a very attractive biofuel feedstock because of its abundant supply. On a global scale, plants produce almost 100 billion tons of cellulose per year, making it the most abundant organic compound on Earth.

Genetic Manipulation of Metabolic Routes. Genetic manipulation of metabolic pathways by adding or deleting genes or modifying the expression of existing genes in the host can serve several useful purposes. It can extend the existing pathways or shifting metabolic route into a desired pathway or increase the rate-determined step of the particular metabolic route. Adding genes into the host consists of the following steps.

- The gene the for desired pathway is obtained from the non-host organism.
- The gene is inserted into the host cell.
- Host cells are induced to express (to cause the gene to manifest its effects) this "foreign" gene in order to produce the desired product.

One example of how gene manipulation is used in areas relevant to metabolic engineering is as follows: In the mold *Aspergillus terreus*, the producer of cholesterol-lowering drug lovastatin, genes were modified to increase their expression levels in order to change its metabolism in terms of drug production. Another example is the introduction of bovine lactic acid pathway into *S. cerevisiae*. As a part of this, a gene responsible for speeding up removal of hydrogen, which participates in lactic acid production, was expressed in *S. cerevisiae*, and lactic acid was produced at rate of eleven grams per liter per hour. Because it tolerates acid, yeast may serve as an alternative to bacteria, which is usually used in industry for lactic acid production. Lactic acid is widely used as a food preservative.

Altering Nutrient Uptake or Metabolic Flow. Alteration of nutrient uptake or metabolic flow can be done not only by genetic manipulation but also by using inhibitors—simple chemicals or physical factors such as light or temperature.

The alteration of molecular hydrogen (H_2) production in green algae using high-intensity light is an example of metabolic flow modification by physical factors. H_2 is one of the possible energy carriers of the future. Microscopic green algae produce H_2 in photosynthetic reactions from water using sunlight as an energy source, usually in anoxic (without oxygen) conditions. Oxygen (O_2) produced by photosynthesis in green algae is an inhibitor of H_2 production. Brief illumination of algal cells by high-intensity light was accompanied by rapid suppression of photosynthetic O_2 evolution. The decline in the rate of O_2 evolution was accompanied by stimulation of H_2 production in algal cells.

Production Systems. All of the above-mentioned considerations are very important in metabolic engineering, although it is also important to ensure that the production of desired compounds by modified cells can be reproduced. This can be achieved by using bioreactors, in which the important parameters such as pH, temperature, substrate supply, and other variables are controlled. It is even possible to modify cell metabolism by using bioreactors.

APPLICATIONS

There are a wide range of metabolic engineering products and applications. Undoubtedly, a number of novel Applications will arise in the future.

Pharmaceuticals. Metabolic engineering is most promising in the production of pharmaceuticals. These include pharmaceuticals from different classes of natural products: alkaloids, isoprenoids, and flavonoids. Biosynthesis of natural products is an emerging area of metabolic engineering that offers significant advantages over conventional chemical methods. Some pharmaceutical compounds are too complex to be chemically synthesized or extracted from biomass organisms inexpensively.

Alkaloids are mainly plant-derived compounds that have been used as drugs such as morphine. Alkaloids are produced by simple extraction from plants. Studies show that alkaloids can be synthesized from amino acids by metabolic engineering in *E. coli* and *S. cerevisiae*.

Isoprenoids, organic compounds composed of two or more hydrocarbons, have a range of functions: pigments, fragrances, and vitamins. Isoprenoids are also the precursors to sex hormones. Many isoprenoids have been produced using microorganisms, including carotenoids and various plant-derived terpenes. Metabolic engineers are using *S. cerevisiae* as a cell factory for the biosynthesis of isoprenoids.

One metabolic-engineering success is the production of Taxol, which is used to treat breast cancer. It is an isoprenoid that was first isolated in the bark of the Pacific yew (*Taxus brevifolia*). The demand for Taxol greatly exceeds the supply that can be obtained from its natural source. A partial Taxol biosynthetic pathway has been engineered in *S. cerevisiae*. Another metabolic engineering success is the production of isoprenoids-carotenoids. Carotenoids are naturally occurring yellow, orange, and red pigments commonly found in plants such as carrots as well as in bacteria, algae, and fungi and play an important role in fighting disease. Metabolic engineers have successfully introduced carotenoid biochemical pathways into nonproducing carotenoid microbes such as *E. coli* and *S. cerevisiae*.

Flavonoids are a group of secondary plant metabolites. These compounds can be used as antioxidants or antiviral, antibacterial, and anticancer drugs. Many flavonoid biosynthetic pathways are known, and a wide array of flavonoid compounds from *S. cerevisiae* are expected to be produced by metabolic engineering in the near future.

Chemicals. Numerous chemicals, such as amino acids, organic acids, vitamins, flavors, fragrances, and nutraceuticals can be manufactured by metabolic engineering. Glycerol (or glycerin) is a chemical produced by metabolic engineering. Glycerol is used to synthesize many products, ranging from cosmetics to lubricants. It is a by-product of soap or biodiesel manufacturing and its production is 1.2 billion liters annually. It can be also used a fuel. Metabolically engineered *S. cerevisiae* strain produced more than 200 grams of glycerol per liter of liquid medium.

Another example of chemicals produced with help of metabolic engineering are sterols. The most well-known sterol is cholesterol. Sterols are important for living organisms as they are a part of the cellular membrane, participate in the synthesis of several hormones, and are also nutrient supplements. Several sterols are being produced from metabolically engineered *S. cerevisiae*.

Fuels. Metabolic engineering can be used in the production of biofuels. Several scientific laboratories have demonstrated the feasibility of manipulating microorganisms to produce molecules similar to oilderived products, although the yield is very low. Adjusting metabolic pathways of microbes to produce fuels similar to gasoline has the potential to save an enormous amount of money. These fuels can be used in existing engines, unlike other biofuels that require modified engines or fueling stations.

Several research groups are trying to metabolically engineer microorganisms to produce ethanol fuel using cellulose as substrate. Another example of the work of metabolic engineers is biodiesel production. Biodiesel is a diesel substitute primarily obtained from vegetable oils such as soybean. However, the production of this fuel is limited by the absence of sufficient vegetable oil feedstocks. Another problem is that in order to produce biodiesel, oils should be modified by transesterification, a chemical reaction with methanol, catalyzed by acids or bases (such as sodium hydroxide). *E. coli* has been metabolically engineered to produce biodiesel directly, using low-cost materials.

FUTURE PROSPECTS

Though the redesign of life forms for the benefit of mankind is definitely an exciting career, metabolic engineers are paying particular attention to ethical, legal, and political issues. To continue in this work, the field as a whole will need sustained support from the public and government. At present, metabolic engineering is more a collection of successful experiments than an established science. In the future, metabolic engineering may play a significant role in production of chemicals and fuels from inexpensive and renewable starting materials. Continued development of the techniques of metabolic engineering will be necessary to expand the range of products. The role of metabolic engineering in science is likely to expand in the future as a result of increasing needs for pharmaceuticals and biofuels.

Sergei A. Markov, PhD

FURTHER READING

Bailey, James E., and David F. Ollis. *Biochemical Engineering Fundamentals*. 2d ed. New York: McGraw-Hill, 1986. Classic textbook on biochemical engineering.

Bourgaize, David, Thomas R. Jewell, and Rodolfo G. Buiser. *Biotechnology: Demystifying the Concepts*. San Francisco: Benjamin Cummings, 2000. Excellent introduction to biotechnology.

Lewin, Benjamin. *Genes VIII*. San Francisco: Benjamin Cummings, 2003. In-depth look at genes and molecular biology. Madigan, Michael T., et al. *Brock Biology of Microorganisms*. 12th ed. San Francisco: Benjamin Cummings, 2008. Several chapters of this popular textbook describe microbial metabolism and the application of microorganisms in industry.

Marguet, Philippe, et al. "Biology by Design: Reduction and Synthesis of Cellular Components and Behavior." *Journal of the Royal Society Interface* 4, no. 15 (2007): 607-623. Review on metabolic engineering and synthetic biology written for the general public.

Ostergaard, Simon, Lisbeth Olsson, and Jens Nielsen. "Metabolic Engineering of *Saccharomyces cerevisiae*." *Microbiology and Molecular Biology Reviews* 64, no. 1 (2000): 34-50. Describes metabolic engineering techniques using *S. cerevisiae* as an example.

Stephanopoulos, Gregory N., Aristos A. Aristidou, and Jens Nielsen. *Metabolic Engineering: Principles and Methodologies.* San Diego: Academic Press, 1998. Classic text on metabolic engineering.

WEB SITES

Biotech Career Center http://www.biotechcareercenter.com/biotech.html

Biotechnology Industry Organization http://www.bio.org

Nature Technology Corporation http://www.natx.com

See also: Biochemical Engineering; Cloning.

FASCINATING FACTS ABOUT METABOLIC ENGINEERING

- In 2010, a team of scientists led by J. Craig Venter created the first synthetic cell. The team synthesized the artificial chromosome, which was then transplanted into the recipient cell. The artificial chromosome was able to take over the recipient cell. This research opens the door for creation of useful artificial cells to make products such as vaccines and biofuels.
- Antimalarial drug artemisinin has been produced from metabolically engineered laboratory yeast. The antimalarial comes from the *Artemisia annua* plant, which grows in Southeast Asia. Artemisinin could also possibly be used in cancer treatment.
- Scientists are able to synthesize large DNA molecules in the laboratory. Researchers have made artificial DNA containing all twenty-one genes encoding the small ribosomal subunit (cell protein factory) from *Escherichia coli* (*E. coli*).
- Using metabolic-engineering methods, researchers were able to modify *E. coli* bacterium to produce butanol. Among the types of biofuels that are on the road to commercialization, butanol has been the most promising. It is another alcohol fuel, but when compared with ethanol, it has higher energy content (roughly 80 percent of gasoline energy content). Butanol can also be stored and transported using existing infrastructure—and it does not occur naturally in *E. coli*.

MULTICELLULARITY

FIELDS OF STUDY

Cell biology, evolutionary science, paleontology

SUMMARY

By studying the fossil record, scientists have found that the first multicellular life appeared on the earth about 1 billion years ago. Before that time, only single-celled organisms existed. The appearance of multicellularity paved the way for the evolution of all higher organisms.

PRINCIPAL TERMS

- **Ediacarian (Ediacaran) fauna:** a diverse assemblage of fossils of soft-bodied animals that represents the oldest record of
- **multicellular animal life on the earth eukaryotic cell:** a cell that has a nucleus with chromosomes and other complex internal structures; this is the type of cell which makes up all organisms except bacteria
- **fossils:** the remains of ancient life preserved in sediment or rock
- **multicellular organisms:** organisms consisting of more than one cell; there are diverse types of cells, specialized for different functions and generally organized into tissues and organs
- **Precambrian eon:** the earliest chapter of the earth's history, covering the time interval between the formation of the earth, about 4.6 billion years ago, and the beginning of the Cambrian period, about 570 million years ago
- **prokaryotic cell:** a primitive cell that lacks a nucleus, chromosomes, and other well-defined internal cellular structures; only members of the kingdom Monera (such as bacteria) are prokaryotic cells—all higher organisms have eukaryotic cells

BASIC PRINCIPLES

Multicellular organisms are those consisting of more than one cell. Three of the five kingdoms of living organisms are multicellular: the plants (Plantae), the animals (Animalia), and the fungi (Fungi). The other two kingdoms consist of single-celled organisms: the bacteria (Monera), which have primitive prokaryotic cells, and the protists (Protista), which have complex eukaryotic cells. Prokaryotic cells lack a nucleus and other internal cell structures and are found today only among the bacteria. Eukaryotic cells, on the other hand, contain a nucleus, other complex internal cell structures called organelles (such as mitochondria, which perform respiratory functions), and sometimes chloroplasts (which contain chlorophyll and perform photosynthesis). All multicellular organisms are composed of eukaryotic cells, so the eukaryotic cell must have evolved before multicellular organisms could develop. It is generally accepted that simpler types of organisms evolved first, followed by more complex organisms.

The Multicellular Kingdoms

Plants are multicellular organisms that have chlorophyll (a green pigment used for photosynthesis), plastids (internal structures on the cell that contain chlorophyll), and a cell wall that contains cellulose. Plants are sometimes called "primary producers" because they can manufacture their own food from carbon dioxide and water through a process called photosynthesis, using sunlight for energy and producing oxygen and organic matter (carbohydrates) as by-products. Animals are multicellular organisms that cannot produce their own food and must feed on other organisms. They are "consumers." Metazoans have many types of cells, which are organized into tissues, and groups of tissues, which form organs. There are two primary embryonic tissue layers present in all metazoans (except the sponges). These are the ectoderm (outer layer) and the endoderm (inner layer). More advanced metazoans also have a third embryonic cell layer, the mesoderm, which lies between the other two layers. Fungi (mushrooms and their relatives) possess cell walls like plants, but unlike plants, they lack chlorophyll. Although fungi appear plantlike, they cannot produce their own food because of the absence of chlorophyll, so they must feed by ingesting organic material and therefore are consumers. Because they are neither plants nor animals, the fungi are placed in a separate kingdom (Fungi).

Advantages and Origins of Multicellularity

Multicellularity probably evolved because it gave organisms some sort of advantage, assuring them of a greater chance of survival. Multicellularity allows organisms to become larger (which helps them to outcompete other organisms and provides a greater internal physiological stability), to have a longer life (because individual cells are replaceable), to produce more offspring (because many cells can be dedicated to reproduction), and to have a variety of body plans (which permits adaptation to various modes of life or environmental conditions). Specialization of cells for particular functions allows organisms to become more efficient.

Evidence for the origin of multicellularity comes from thefossil record, studies of the organization and biochemistry of living cells and organisms, and from studies of the embryonic and larval stages of animals. During the Archean eon (between 3.8 and 2.5 billion years ago), only singlecelled, prokaryotic life (bacteria-like organisms) existed on the earth. Some of the prokaryotes were photosynthetic, including the cyanobacteria (or so-called blue-green algae). Some of these prokaryotes were colonial, with cells organized into structures such as chains, or filaments, or algal mats. These colonies differ fromtrue multicellular organisms because they generally consist of only one type of cell rather than many types of cells. Colonies of blue-green algae formed mound-like structures called stromatolites, which were quite common during the Precambrian but are present only in a few areas today. Stromatolites appeared about 3 billion years ago but did not become abundant until about 2.3 billion years ago.

The Rise of Eukaryotic Organisms

Eukaryotic organisms appeared during the Proterozoic era. The eukaryotic cell probably evolved from prokaryotic ancestors some time before about 1.4 billion years ago. The oldest convincing fossils of eukaryotic cells are generally considered to be those from the 1.3 billion-year-old Beck Spring dolomite of California. Eukaryotic fossil cells have also been found in chert from the approximately 850 million-year-old Bitter Springs formation of Australia. The earliest eukaryotes were animal-like protozoans. This evolution occurred when photosynthetic prokaryotic cyanobacteria were ingested by protozoans and then developed a symbiotic (mutually beneficial) relationship with them. The evolution of the plantlike eukaryotes probably occurred by at least 1.4 billion years ago. (This date has been suggested because primitive multicellular algae fossils are present in rocks 1.3 billion years old.)

The eukaryotic cell was a prerequisite for the development of multicellular organisms. The plantlike eukaryotes are considered to be ancestral to the multicellular algae and higher plants. The protozoans are considered to be the ancestors of the metazoans (animals). The first multicellular organisms may have been algae. Fossils that appear to be primitive multicellular algae are known from the 1.3 billion-year-old sedimentary rocks of the Belt supergroup of Montana, and the 800 million- to 900 million-year-old Little Dal group of northwestern Canada. Multicellular algae can be found living today in both freshwater and marine environments.

Fossil fungi first appear in the fossil record in the 790 million- to 1,370 million-year-old Bitter Springs formation cherts of Australia. The fossil record of fungi is poor and not well known.

The Rise of Multicellular Animals

Multicellular animals evolved independently of the multicellular plants, probably arising from protozoan ancestors. The oldest evidence of metazoans (multicellular animals) in the geologic record is in the form of trace fossils. Trace fossils are imprints such as tracks, trails, or burrows made in sediment by moving animals. Over time, the sediment hardened into sedimentary rock as a result of compaction and cementation. The earliest trace fossils consist of simple trails and tubelike burrows. In some places, there is a succession of types of trace fossils from simple tubelike burrows in older rocks to more complex structures in younger rocks. This change suggests that the evolution and diversification of increasingly complex burrowing organisms occurred during the latter part of the Precambrian. The oldest trace fossils are less than one billion years old, and many scientists believe that it is unlikely that any trace fossils exist in rocks much older than about 700 million years old. The trace fossils appear in the geologic record just before the first appearance of soft-bodied metazoan fossils. Structures that resemble trace fossils, however, have been reported from much older rocks. Among these questionable traces are the one-billion-year-old Brooksella, which resembles a jellyfish. In addition, tubelike structures from the upper Medicine Peak quartzite in Wyoming have been dated at 2 billion to 2.5 billion years, at least 1 billion years older than the oldest known metazoans. The origin of these older traces is uncertain and may be a result of inorganic processes (such as dewatering of sediment), rather than of organisms.

One possible fossil metazoan that appears to be more than 850 million years old has been reported from the Tindir group of Alaska. This fossil is less than one millimeter long and appears to be a flatworm (phylum Platyhelminthes). Both the age and identification of this fossil have been disputed, but if valid, it is a very important find because some biologists theorize that the earliest metazoans would have been primitive flatworms.

The oldest unquestioned metazoan fossils are the imprints of a diverse assemblage of relatively well-developed, soft-bodied marine animals. More than half of the organisms appear to be some type of cnidarian or coelenterate (related to jellyfishes), about 25 percent appear to be segmented worms (related to annelids), and a small percentage appear to be arthropods (related to insects, crabs, and lobsters). Trace fossils are also present. This assemblage of soft-bodied fossils is called the Ediacaran fauna. It was discovered in 1946 in sandstones of the Pound subgroup in the Ediacara Hills of the Flinders Range in South Australia. The exact age of the Ediacaran fauna is uncertain because there are no nearby rocks of the proper type for radiometric dating. The Ediacaran fauna is clearly Precambrian, however, judging from its position in the geologic sequence. The soft-bodied Ediacaran fossils are separated from the younger fossil shells of the Cambrian (570 million years old) by a thick section of unfossiliferous rock (up to several hundred meters thick). Since the 1950's, fossils similar to the Ediacaran fauna have been found in rocks of approximately the same age on virtually every continent on the earth (with the possible exception of Antarctica). In some of these other areas, it is possible to date radiometrically the rocks associated with the fauna. These radiometric dates indicate that the early metazoan soft-bodied fossils range from about 620 million to 700 million years old. It is likely that the metazoans evolved some time prior to 700 million years ago because these fossils represent well-developed, complex animals.

Skeletonized faunas (animals with shells or other hard parts) did not appear until approximately 580 million years ago. The skeletonized faunas are represented by microscopic scraps, cones, tubes, and plates made of calcium phosphate or a hard organic material called chitin. It is not known exactly what types of organisms produced these skeletal remains, but the tiny fossils are so diverse and complex that it is assumed that the organisms must have had a long history of evolution during the Precambrian. The origin of skeletons was advantageous to marine organisms because hard parts provide protection against predators as well as the mechanical functions of support and muscle attachment.

Theories of the Origin of Multicellular Life

There are a number of theories to explain the origin of multicellular life. Most of the theories are derived from studies of various types of cells and living organisms, including advanced protozoans, early developmental stages (embryos), and larval stages. Four types of cells are central to these theories, and they are grouped into two categories: motile (capable of movement) and nonmotile (not capable of movement). The motile protists include flagellate cells (those with a whiplike "tail," or flagellum) and amoeboid cells (those such as Amoeba, which move by pseudopodia, or fingerlike extensions of the cell membrane). The nonmotile stages include coccine cells (those with many nuclei, sometimes called multinucleate cells) and sporine cells (those that divide and stick together to form multicellular aggregates).

There are many theories that have been proposed to explain the origin of plants, fungi, and metazoans (animals). Formerly, it was thought that plants evolved from prokaryotic algae (cyanobacteria, or blue-green algae), but it is more likely that plants arose from a eukaryotic ancestor, such as a flagellate cell. Flagellate algae are similar to flagellate protozoans, but it is not certain whether the algae evolved from theprotozoan or vice versa. The presence of plastids (such as the chloroplasts that contain chlorophyll used for photosynthesis) may be the key feature separating plants from protozoans, fungi, and animals. According to a theory proposed by Lynn Margulis, plastids evolved from prokaryotic blue-green algae that were captured by eukaryotic cells. The sporine cell is another possible ancestor of the plants. Sporine cells appear to have had the capacity to evolve beyond the colony level and to produce complex tissue-level green algae and higher plants.

Fungi used to be considered as plants that had lost (or never evolved) chlorophyll. The discovery of a single-celled stage with flagellae among the more primitive fungi, however, suggests that fungi probably evolved from protozoans. The ancestral multicellular organisms, which gave rise to all the more-complex living animals, are all extinct. The simplest multicellular animal living today is the sponge (phylum Porifera). The sponges are not considered to be ancestors of the more complex animals because their body organization and developmental history are very different. Sponges have no tissues, mouth, or internal organs. Instead, they consist of an aggregate of flagellate and amoeboid cells (and a few other types) roughly arranged in layers. The sponges may have evolved independently from the other metazoans. Sponges are classified as a distinct side branch of the animal kingdom (Parazoa), with a primitive multicellular grade of organization (no tissues). The remaining multicellular animals are grouped into the Eumetazoa.

Theories of Metazoan Origin

Several theories have been proposed to explain the origin of the metazoans. These theories can be placed into the following categories: evolution from single-celled protozoans; evolution from colonial protozoans; evolution from multinucleate coccine cells as a result of development of internal cell boundaries; and evolution from sporine cells. There are several versions of each of these theories, and there is no general agreement on which theory is best. Some researchers promote the colonial theory as the most widely accepted theory, whereas others claim no longer to take it seriously. Most experts agree that evolution of metazoans from colonial protozoans would seem to be easier than evolution directly from a single cell. Multicellularity may have arisen independently several times, in several different ways.

The colonial theory suggests that the metazoans evolved from flagellate or amoeboid protozoans that lived together in colonies, much like the modern green alga Volvox, which is shaped like a hollow sphere. From an original hollow spherical form, the shape of the ancestral metazoan changed as an indentation or invagination formed in the side. The indentation became larger, producing a double-walled "cup" (envision pushing one's thumb into the side of a deflated ball until that side becomes nested into the other side of the ball, forming a cuplike shape). The double-walled cup shape is referred to as a diploblastic body plan, meaning two layers of body tissue. These two layers are the ectoderm (outer layer) and the endoderm (inner layer). This process of indentation to produce a diploblastic (double-walled) form occurs in the embryos of many animals. The jellyfishes are a good example of animals with a diploblastic body plan. Nearly all groups of animals have ectoderm and endoderm (except the sponges), suggesting that nearly all groups of animals are related. Because the jellyfishes (phylum Cnidaria) have the simplest body plan, they are believed to be the most primitive. The diploblastic ancestral form has been called a gastrea. Ernst H. Haeckel, a prominent nineteenth century German biologist who studied animal embryos, believed that all bilaterally symmetrical animals evolved from a gastrea.

A second theory for the origin of metazoans suggests that the ancestral form was a bilaterally symmetrical animal resembling a flatworm. Some scientists believe that the complex organs and organ systems of metazoans are beyond the evolutionary potential of flagellate and amoeboid cells.

The flatworm may have evolved from "cellularization" of a multinucleate coccine cell (formation of cell membranes around each of the nuclei) or from clumping of sporine cells. Most of the cells in the metazoans are sporine cells that stick together to form multicellular aggregates. Sporine protozoans do not exist, so it is hypothesized that sporine ancestors of the metazoans must have evolved from "pre-protozoans." These hypothetical ancestors may have been solid balls of cells resembling the early stages of many embryos. At some point, the exterior cells may have developed (or redeveloped) flagellae and become specialized for locomotion, and the interior cells may have become specialized for digestion and reproduction. Such colonies of cells would have resembled the larval (immature) form of cnidarians, called a planula larva, and, hence, they are called planuloids. Planuloids are believed to have given rise to two groups of metazoans, the cnidarians (jellyfishes and their kin) and the flatworms. The primitive flatworms are believed to have been ancestral to all other bilaterally symmetrical metazoans.

Evidence from Rocks

Theories to explain the origin of multicellular life have been developed by biologists as a result of studies of various types of cells and living organisms, including advanced protozoans, early developmental stages (embryos), and larval stages. Geologists (scientists who study rocks) and paleontologists (scientists who study fossils) have a variety of techniques that they use to search for the evidence of life in Precambrian rocks (older than 570 million years). These include searching for fossil remains and chemical analysis of organic residues that are probably the breakdown products of once-living organisms.

The first step in the search for Precambrian life is to locate rocks of the proper age. Geologic maps exist for virtually all parts of the world. From an examination of these maps, it is possible to identify areas that contain rocks of the proper age. (The age of a rock is determined by radiometric dating.) Age, however, is not the only consideration. For fossil remains to be preserved, the rocks must also remain little altered from the way they were originally deposited. Metamorphism (geologic alteration caused by heat and/or pressure) has deformed many Precambrian rocks to the extent that any fossils that may have been present can no longer be recognized.

Assuming that undeformed rocks of the proper age can be located, the search begins for fossil remains. Unfortunately, most Precambrian rocks are not fossiliferous. Precambrian multicellular fossils are found in only a few places in the world. In Australia, soft-bodied Precambrian metazoan fossils are restricted to a few thin layers of sandstone in a sequence of Precambrian rock more than one thousand meters thick. In most places in the world, however, there is a thick section of unfossiliferous rock separating the Precambrian metazoan fossils from the shelly faunas in the Cambrian rocks. This unfossiliferous sequence of rock is an interval for which there is little or no information on the types of life that existed.

Before multicellular organisms appeared (prior to perhaps one billion years ago), only microscopic, single-celled organisms existed on the earth. Microscopic fossils of single-celled organisms are found by careful examination of fine-grained, dark-colored rocks such as black cherts. The black color of the rocks commonly indicates the presence of carbon, which is present in all living organisms and which may be preserved in some fossils. Very thin slices of rock are prepared and mounted on glass slides so that the organic matter can be studied.

These slices of rock, called thin sections, are so thin that light can pass through them, and they are examined with a microscope. Much of the carbon in these rocks is present as amorphous (indistinct or shapeless) patches, but in some places, microscopic structures are present that appear to be the fossilized remains of single-celled organisms. Pieces of rock can also be prepared for examination using a scanning electron microscope. The search for microfossils is difficult and painstaking. Among the problems involved are the possibility of contamination by modern-day organic matter in the laboratory and the possibility that the microscopic structures may really be inorganic in origin.

Chemical tests are used to search for the products of biological activity, which may be preserved in rocks. In principle, rocks that have been influenced by biological activity should contain certain characteristic isotopic ratios. There are a number of problems inherent in searching for organic residues. Organic material may have been preserved in the rock, but it could easily have been altered subsequently by heat and pressure or by circulating fluids. In addition, circulating fluids can contaminate the rocks by introducing organic material from much younger rocks.

Multicellularity in the Evolutionary Process

Studying the origin of multicellularity helps one to understand the conditions that led to the evolution of plant and animal life on the earth. As one begins to understand how multicellular life evolved, one may begin to wonder about why it was such a slow process. It is known that the earth formed about 4.6 billion years ago and that the first cells appeared about 3.5 billion years ago, but that the first multicellular life did not appear until approximately 1 billion years ago. In other words, it took more than 3.5 billion years for multicellular life to develop. More than three quarters of the earth's history had passed before multicellular life ever appeared.

One may also begin to wonder about the conditions that promoted the origin of multicellular life. Of all the planets in the solar system, the earth seems uniquely suited to life. Two of the most important factors involved are the presence of liquid water (which requires a specific temperature range) and the presence of an oxygen-rich atmosphere. None of the other planets in this solar system has either of these two characteristics. Interestingly enough, the earth originally did not have liquid water or an oxygenated atmosphere. Geologic evidence suggests that the earth's early atmosphere was the result of volcanic outgassing and that it consisted of gases such as carbon dioxide, carbon monoxide, ammonia, methane, hydrogen sulfide, nitrogen, and water vapor. As the planet cooled from its original molten state, the water vapor in the atmosphere condensed to form liquid water, which fell to the earth as rain and accumulated to form the oceans, rivers, and lakes.

There is abundant geologic evidence that the earth's early atmosphere lacked the free oxygen that is breathed today. In the absence of free oxygen, chemical evolution in the oceans or lakes led to the formation of organic compounds, or what has been called the "primordial soup." The first living cells, the prokaryotes, evolved in this organics-rich water. As time passed, some of the early prokaryotic cells became photosynthetic, which allowed them not only to produce their own food from water and carbon dioxide but also to produce oxygen as a waste product. Oxygen was toxic to these early organisms. In order to survive, the cells had to develop a mechanism to adapt to the presence of increasing levels of oxygen. The buildup of oxygen led to the development of the ozone layer in the atmosphere and to the appearance of the eukaryotic cell. As the percentage of oxygen in the atmosphere increased, it is believed that some threshold level was reached, and it became possible for the environment to support multicellular organisms. That allowed a rapid diversification of life on the earth.

Hence, it appears that multicellular life on the earth appeared as a result of some prehistoric accident that resulted in global atmospheric change—the buildup of a toxic waste product (oxygen) as a result of photosynthesis by early life-forms. One might also speculate on the possible global effects of the increasing waste products that humans are now producing. The thinning of the atmospheric ozone layer is but one manifestation of the way that life is presently changing the earth's fragile environment. Knowing that the formation of the ozone layer was probably essential to the appearance of multicellular life on the earth, it is alarming to speculate on the consequences of its destruction. Life as humans know it depends on an earth with environmental conditions in a precarious balance.

—*Pamela J. W. Gore*

See also: Cell types; Cleavage, gastrulation, and neurulation; Evolution: Animal life; Evolution: Historical perspective.

Further Reading

Ayala, Francisco J., and James W. Valentine. *Evolving: The Theory and Processes of Organic Evolution.* Menlo Park, Calif.: Benjamin/Cummings, 1979. This book covers the basic principles of evolution and contains a section on the geologic record, which is extremely important in studies of the origin of early life-forms. It is designed for a college-level audience but is clearly written and easy to read.

Bittar, E. Edward, and Neville Bittar, eds. *Evolutionary Biology.* Greenwich, Conn.: JAI Press, 1994. A collection of essays on the structure and dynamics of DNA, RNA, and protein, and more general topics on the origin and cellular basis of life.

Boardman, Richard S., Alan H. Cheetham, and Albert J. Rowell, eds. *Fossil Invertebrates.* London: Blackwell Scientific Publications, 1987. A textbook that is designed for use in college courses in invertebrate paleontology. It begins with a number of chapters on basic principles (such as ecology, paleoecology, evolution, classification, and fossil preservation), followed by a run-through of kingdom Protista and twelve invertebrate phyla of kingdom Animalia that are commonly preserved as fossils. Well illustrated with photographs and line drawings. Will be most useful for those who have had courses in high school biology and historical geology.

Cloud, Preston. *Oasis in Space: Earth History from the Beginning.* New York: W.W. Norton, 1988. This book is part of a series designed to make science more accessible to the general public. It is clearly written, well illustrated, easily understood, and up-to-date. Unfamiliar scientific terms are explained on first mention in the text, and the reader can easily locate definitions by using the index at the back of the book. A two-page time chart indicating major events in geologic history is in the front of the book. Earth history is clearly covered, in addition to many of the basic principles of geology that contribute to the interpretations.

Futuyma, Douglas J. *Evolutionary Biology.* 3d ed. Sunderland, Mass.: Sinauer Associates, 1998. A textbook aimed at advanced undergraduates and graduate students, emphasizing the history of life and microevolutionary processes. Comprehensive glossary, index, Bibliography, abundant illustrations.

Minkoff, Eli C. *Evolutionary Biology.* Reading, Mass.: Addison-Wesley, 1983. A college textbook that deals with evolution in an interdisciplinary way. It is a good reference book that will be most useful to those who have had an introductory course in college-level biology, but it can also be understood by beginners because it clearly explains new terms and has a good index and a glossary.

Selander, Robert K., Andrew G. Clark, and Thomas S. Whittam, eds. *Evolution at the Molecular Level.* Sunderland, Mass.: Sinauer Associates, 1991. A collection of essays focused on four themes: bacteria and viruses, organelles, "selfish" genes, and nuclear multigene families.

Stanley, Steven M. *Earth and Life Through Time.* 2d ed. New York: W. H. Freeman, 1989. A textbook for introductory college historical geology classes that can be readily understood by those without previous training in geology. It clearly covers the origin of life and rise of multicellular organisms during the Precambrian. It also presents a clear picture of the interrelationship between the physical and biological history of the earth. It is amply illustrated and has a glossary, comprehensive index, and Summary of the classification of major fossil groups in the back.

Mutations

FIELDS OF STUDY

Biochemistry, cell biology, developmental biology, embryology, evolutionary science, genetics

SUMMARY

Mutations have been used to work out metabolic pathways, define genes and their controlling sites, understand how multicellular organisms develop, and study how organisms evolve.

PRINCIPAL TERMS

- **allele:** one of many possible sequences of a gene
- **controlling site:** a sequence of nucleotides generally fifteen to sixty nucleotides long, to which a transcriptional activator or repressor binds
- **gene:** a sequence of one thousand to ten thousand nucleotides, which usually specifies a protein
- **mutation:** a change in the nucleotide sequence of a gene or of a controlling site; changes in genes alter the protein, whereas changes in controlling sites determine where and how much of a protein is produced

BASIC PRINCIPLES

In all living organisms, the hereditary information consists of two complementary strands of deoxyribonucleic acid, known as DNA. DNA strands are constructed of subunits called nucleotides that consist of a nitrogenous base, a deoxyribose sugar, and a phosphate. Generally, DNA strands consist of millions of nucleotides attached to each other like the rings of a chain. There are four different nucleotides. The two complementary strands are held together by hydrogen bonds between the bases. If there is an adenine (A) in one strand, it hydrogen bonds to a thymine (T) in the complementary strand. Similarly, if there is a guanine (G) in one strand, it hydrogen bonds to a cytosine (C) in the other strand. Thus, the amount of A is equal to T and the amount of G is equal to C. The order of nucleotides in a strand specifies the order of amino acids in proteins.

Genetics is the study of how the information in DNA molecules is expressed and how DNA molecules account for the heredity of an organism. Changes in the sequence of nucleotides in DNA may alter an organism's proteins, which in turn may change one or more of an organism's traits. DNA changes are called mutations, and the organisms that harbor mutations are known as mutants. The commonly encountered trait or organism is referred to as the wild-type. The characterization of mutations and mutants has been and still is one of the best ways of discovering the function of genes and determining how organisms maintain themselves, evolve, and develop. A study of mutations and mutants also has shed light on numerous genetic diseases.

HEAT: THE CAUSE OF MOST SPONTANEOUS MUTATIONS

Most mutations are caused by the instability of the nucleotide bases. Sometimes bases hit by rapidly moving water molecules briefly alter their chemistry. These chemical changes are known as tautomeric shifts. Tautomeric shifts alter the distribution of electrons and protons in the bases so that bases in the complementary strands no longer pair normally. The redistribution of electrons and protons in the bases causes abnormal pairings to occur. For example, an abnormal adenine (A*) pairs with C and an abnormal guanine (G*) pairs with T.

When a DNA molecule is being replicated, spontaneous tautomer shifts can result in permanent mutations. Spontaneous mutations occur, for example, when an A in the template strand undergoes a tautomer shift (A *) just as the DNA polymerase reaches it. A cytosine pairs with the A* and becomes part of the new strand being synthesized by the DNA polymerase. When this new strand, with a C in it instead of a T, functions as a template, the complementary strand will have a G in it rather than an A. This type of tautomeric shift during DNA replication converts what normally would have been an A = T base pair in "granddaughter" DNA to a G = C base pair.

CHEMICALS THAT CAUSE MUTATIONS

Mutations are induced by many chemical and physical agents that are called mutagens. Many chemicals act as mutagens. Nitrous acid, for example, diffuses into cells and removes amino groups from DNA bases. These chemically altered bases no longer base pair normally. When DNA is replicated or repaired, incorrect nucleotides are inserted opposite the chemically altered bases. Nitrous acid changes adenine to hypoxanthine, which pairs with cytosine. It also changes guanine to xanthine, which pairs with T.

Base analogues are molecules that closely resemble normal nucleotides and consequently are incorporated into DNA that is being repaired or replicated. A base analogue to thymine, such as 5-bromouracil (5BU), is efficiently incorporated into DNA. 5BU spontaneously undergoes tautomeric shifts at a high rate. The abnormal form of 5BU pairs with G rather than A. Thus, 5BU introduces many base pair transitions in newly synthesized DNA molecules.

The most potent mutagens are alkylating agents, such as nitrosamines, methyl bromide, and ethylene oxide. These mutagens attach methyl or ethyl groups (alkyl groups) to A and G. This causes A and G to undergo tautomeric shifts at a higher than normal rate.

High-Energy Electromagnetic Radiation and Particles

Ultraviolet (UV) light is a powerful mutagen. It generally penetrates cells but is readily absorbed by thymine and cytosine bases in DNA. When two thymines or two cytosines next to each other in a strand absorb UV light, they often react chemically with each other to form thymine dimers or cytosine dimers that distort the DNA. These distorted regions stimulate a repair system that cuts out the dimers and some DNA on either side and replaces the DNA with normal nucleotides. Excessive repair leads to an increased occurrence of spontaneous mutations. Sometimes a distortion in the template allows the DNA polymerase to add or to leave out nucleotides as it moves along the template during strand synthesis. This may explain how some additions and some deletions occur. Very energetic electromagnetic radiation, such as X rays and gamma rays, as well as high-energy particles released from radioactive atoms, also induce mutations. These energetic mutagens easily penetrate cells and chemically alter many molecules in their path by stripping away electrons. Ions and radicals formed by these mutagens react with the DNA, causing bases to be released and DNA to break. DNA deletions, DNA transpositions, and DNA inversions may be promoted by DNA breakage.

When a gene is mutated, the protein the gene specifies generally becomes nonfunctional. In bacteria that have only one copy of each gene, traits are immediately altered by a mutation. On the other hand, in animals and plants that may have more than one copy of a gene, a mutation in only one gene may not produce a new trait because the wild-type (normal) gene often provides enough of the essential protein. When developing animals and plants are missing both genes, however, they may fail to develop or they may develop, but in a different way.

A few mutations are beneficial to the organism that acquires them and may make the organism better adapted to its environment. These beneficial mutations may make a protein work a little better or in a different way. Some mutations are also beneficial because they create diversity in a population. Diversity promotes the survival of a population by ensuring that some organisms survive if the environment drastically changes. A population that is too well adapted to a particular environment will not survive if there are significant changes in the environment. There have been at least five major mass extinctions during the history of life on earth, in some cases eliminating more than 85 percent of all species. The organisms that survived these mass extinctions were much less specialized than the organisms that did not.

Usefulness of Mutations

Mutations have been extremely useful in the study of organisms. Mutations allow scientists to understand what a particular gene and its product do. If the mutation eliminates the gene (and product), scientists can guess what the gene does by looking at the affected organism. For example, if a mutation changes eye color (red to white), the affected gene most likely has something to do with pigment synthesis or deposition of the pigment in the eye. The study of mutations and mutant organisms has helped scientists unravel anabolic (synthetic) and catabolic (degrading) pathways, determine how parental genes combine to produce new characteristics in progeny, clarify what genes are and what they do, establish how genes are regulated, and even decipher how multicellular organisms develop and evolve.

Mutations in Development and Evolution

The study of mutations and mutant organisms at the end of the twentieth century led to an understanding of how multicellular organisms develop and evolve. One of the most useful organisms in unraveling the development problem has been the small fruit fly *Drosophila*. Thousands of mutations that affect development of this organism have been characterized. Scientist found that a hierarchy of genes are involved in development. First, maternal genes are expressed. These genes activate gap genes and these, in turn, activate pair-rule genes. All of these gene catagories are known to be involved in regulating the expression of homeotic genes. Maternal, gap, pair-rule, and homeotic gene products all function as transcriptional activators and repressors. For example, the maternal gene product called bicoid stimulates its own synthesis, whereas it inhibits the synthesis of another maternal gene product called nanos.

This gene hierarchy is responsible for the anterior-posterior segmentation seen in *Drosophila*. Edward B. Lewis, Christiane Nüsslein-Volhard, and Eric Wieschaus shared the 1995 Nobel Prize in Physiology or Medicine for their studies of the genes that control *Drosophila* development.

Homeotic genes are found in all multicellular organisms. Homeotic genes similar to those found in *Drosophila* control the development of segments most visibly exemplified by the vertebrae and the bones in animals' appendages. Mutations in homeotic genes or their controlling sites affect the development of segments. Segments can be eliminated or modified by homeotic gene controlling site mutations.

One well-studied homeotic gene in *Drosophila* is the gene antennapedia, *antp*. Certain mutations in the controlling sites for the antennapedia gene result in legs developing rather than head antennae. Another homeotic gene is ultrabithorax, *ubx*. Some mutations in the controlling sites for ultrabithorax gene result in a second pair of wings developing where the pair of halteres normally develop. Halteres are tiny, winglike appendages that all flies have, which promote stable flight. Other mutations in the controlling sites for *ubx* produce a second pair of winglike structures that are half haltere (anterior portion), half wing (posterior portion). By studying mutations and the altered traits, scientists have discovered that controlling site mutations change when and where proteins are synthesized. For example, if a protein is to be produced in seven segments along the anterior-posterior axis of an animal, there must be at least seven different controlling sites that can respond to the different activators and repressors produced in each segment.

Numerous studies suggest that antennapedia and ultrabithorax are transcriptional repressoractivators that not only repress the development of legs and wings, but also stimulate the development of antennae and halteres, respectively. The study of *Drosophila* mutants is beginning to clarify how antennae and mouth parts evolved from leglike appendages and how halteres evolved from wings. The study of genes and controlling sites has led to the understanding of their role in the maintenance, development, and evolution of every organism.

—*Jaime Stanley Colomé*

See also: Asexual reproduction; Cleavage, gastrulation, and neurulation; Copulation; Fertilization; Gametogenesis; Reproduction; Reproductive system of female mammals; Reproductive system of male mammals; Sexual development.

FURTHER READING

Brennessel, Barbara. "Inborn Error of Metabolism." In *Encyclopedia of Genetics*, edited by Jeffrey A. Knight and Robert McClenaghan. Pasadena, Calif.: Salem Press, 1999. A short history of Sir Archibald Garrod's discoveries and a short discussion of other inborn errors of metabolism in humans.

Colomé, Jaime S. "Gene Regulation: Bacteria." In *Encyclopedia of Genetics*, edited by Jeffrey A. Knight and Robert McClenaghan. Pasadena, Calif.: Salem Press, 1999. Regulation of the lactose operon, arabinose operon, tryptophan operon and a flagellin operon is discussed.

Foran, John M. "Thomas Hunt Morgan: 1933." In *The Nobel Prize Winners: Physiology or Medicine*, edited by Frank N. Magill. Pasadena, Calif.: Salem Press, 1991.A history of Morgan's discoveries, in particular his work showing that one of the genes that controls eye color is linked to the X chromosome.

Fornari, Chet S. "Homeotic Genes." In *Encyclopedia of Genetics*, edited by Jeffrey A. Knight and Robert McClenaghan. Pasadena, Calif.: Salem Press, 1999. Discusses the function of homeotic genes.

Gliboff, Sander. "Gregor Mendel and Mendelism." In *Encyclopedia of Genetics*, edited by Jeffrey A. Knight and Robert McClenaghan. Pasadena, Calif.: Salem Press, 1999. A well-done history and Summary of Mendel's research with wild-type and mutant pea plants.

Kalumuck, Karen E. "*Drosophila melanogaster*." In *Encyclopedia of Genetics*, edited by Jeffrey A. Knight and Robert McClenaghan. Pasadena, Calif.: Salem Press, 1999. A brief review of T. H. Morgan and Sturtevant's work with *Drosophila* followed by a discussion of the use of *Drosophila* to study development.

Kang, Manjit S. "One Gene-One Enzyme Hypothesis." In *Encyclopedia of Genetics*, edited by Jeffrey A. Knight and Robert McClenaghan. Pasadena, Calif.: Salem Press, 1999. A short history of the concept that each gene specifies a polypeptide.

Morgan, T. H. "Sex Limited Inheritance in *Drosophila*." *Science* 32 (1910): 120-122. Morgan demonstrates how a spontaneous mutation (sport) affecting eye color in *Drosophila* is passed to succeeding generations. He showed that a mutated form of the gene is recessive to the normal form of the gene and that the mutated and normal gene are linked to the X chromosome. Genes that are linked to the X chromosome are now known as sex-linked genes.

Sturtevant, A. H. "The Linear Arrangement of Six Sex-Linked Factors in *Drosophila*, as Shown by Their Mode of Association." *Journal of Experimental Zoology* 14 (1913): 43- 59. Sturtevant provides evidence that genes are arranged in a linear sequence along the X chromosome. He was able to map the genes relative to each other and determine a distance between them.

Thompson, James N., and R. C. Woodruff. "Mutation and Mutagenesis." In *Encyclopedia of Genetics*, edited by Jeffrey A. Knight and Robert McClenaghan. Pasadena, Calif.: Salem Press, 1999. Discusses the processes of mutation and mutagenesis.

THE USE OF A MUTATION TO DETERMINE A CATABOLIC PATHWAY

Sir Archibald Garrod (1857-1936) was an English physician who studied the effects of a mutation in humans and discovered a catabolic pathway. Garrod's research into the cause of alkaptonuria, a condition where the urine turns black upon exposure to the air, led to the idea that alkaptonuria occurred in persons with two defective genes for the enzyme that eliminates homogentisic acid in the urine. Homogentisic acid turns black when oxidized by oxygen.

In a paper published in 1902, Garrod analyzed a number of families where alkaptonuria occurred, establishing that the trait was recessive and followed simple Mendelian genetics. This was the first account of recessive inheritance in humans. When Garrod fed alkaptonuric patients homogentisic acid, their urine blackened upon exposure to air and contained nearly the same amount of homogentisic acid that they consumed. Normal individuals given homogentisic acid contained no detectable homogentisic acid in their urine. By feeding alkaptonuric patients various compounds that might give rise to homogentisic acid, Garrod discovered the metabolic pathway that led from the amino acids phenylalanine and tyrosine to homogentisic acid. In his book published in 1908, *Inborn Errors of Metabolism*, Garrod concluded that alkaptonuria is caused by a block in the catabolic pathway that eliminates homogentisic acid. He reasoned that the catabolic block was caused by the lack of a specific enzyme. The fact that the trait followed Mendelian genetics suggested that a functional gene normally specified the enzyme involved in the breakdown of homogentisic acid. This was the beginning of the idea that genes specify enzymes.

MUTATIONS USED TO DETERMINE AN ANABOLIC PATHWAY

Between 1937 and 1941, George W. Beadle and Edward L. Tatum isolated fungal mutants that were unable to synthesize various vitamins and amino acids. They isolated these mutants to try to understand how genes control specific reactions in various metabolic (synthetic) pathways. The mutants would not grow on a simple medium unless supplied with the nutrient they were unable to synthesize. Beadle and Tatum induced large numbers of mutations in the fungus they were studying and isolated hundreds of mutants. Seven mutants were isolated, each carrying a defective form of a gene involved in the synthesis of the amino acid arginine. Using these mutants, Beadle and Tatum were able to order the chemical reactions by feeding the fungal mutants different metabolic intermediates. If the fungal mutants grew when a particular metabolic intermediate was provided, then the gene mutation affected a step leading to the synthesis of the intermediate. If the mutant did not grow, however, then the gene mutation affected a step that converted the intermediate into arginine. They concluded from their experiments that each gene controlled a different chemical reaction. They confirmed that a specific chemical reaction fails to take place in diploid organisms if both representatives of a given gene were defective. Beadle and Tatum's research strengthened the idea that genes specify enzymes. Their idea became known as the "one gene-one enzyme" hypothesis.

THOMAS HUNT MORGAN

Born: September 25, 1866; Lexington, Kentucky

Died: December 4, 1945; Pasadena, California

Fields of study: Genetics

Contribution: Morgan's studies popularized the use of the fruit fly *Drosophila* for the study of animal genetics. He is credited with discovering the first sex-linked trait in *Drosophila* and with demonstrating how new characteristics could be passed on to successive generations. Morgan and his students showed that chromosomes exchanged genes, a process known as crossing over. In 1933, Morgan was awarded the Nobel Prize for Physiology or Medicine for his work on *Drosophila* genetics.

In his classic paper of 1910, Thomas Hunt Morgan described very rare white-eyed flies that appeared spontaneously in a red-eyed population. Because these mutants were always males, Morgan suspected that the gene controlling eye color was linked to the X chromosome that influenced the development of sex. The observation that white-eyed females could be produced from certain matings, however, indicated that the white-eyed trait was not limited to males.

Morgan's experiments demonstrated for the first time that a gene controlling eye color was linked (or limited) to the X chromosome. Red and white eyes are caused by different alleles of the same gene. These alleles on the X chromosome are represented in the following manner: XR and Xr. Any gene linked to the X chromosome is called a sex-linked gene. In *Drosophila*, two X chromosomes generally result in a female fly, whereas one X chromosome results in a male. Morgan found that the R allele inducing red eyes is dominant over the r allele that allows white eyes to develop when it is the only allele a fly has.

The Y chromosome that pairs with the X chromosome in males (XY) lacks sex-linked genes. Thus, mating red-eyed females (XRXR) to white-eyed males (XrY) results in all first filial (F1) generation flies, XRXr females and XRY males, having red eyes. Matings between the F1 flies demonstrated that the red-eye-inducing allele and the white-eye-promoting allele always remained associated with the X chromosome. This suggested that the alleles were linked to the X chromosome.

In 1913, A. H. Sturtevant, working in Morgan's laboratory, reported on mutations linked together on a fruit fly's X chromosome. Sturtevant demonstrated that recombination between two X chromosomes could separate genes controlling different traits. In addition to using the alleles that determined eye color, Sturtevant used alleles that influenced wing formation. A normal wing forms under the influence of the L gene, but a miniature wing is associated with the l allele of the L gene. Genes linked on the X chromosome may be shown as follows: XRL. Sturtevant observed recombination when he characterized the offspring from certain crosses.

A female fly with red eyes and normal wings, XRlXrL, usually produces two types of eggs. One type of egg has the XRl chromosome, whereas the other type of egg has the XrL chromosome. Very infrequently, when there is a crossover between the X chromosomes, rare eggs are produced with recombinant X chromosomes, one type of egg has the Xrl chromosome, whereas the other type of egg has the XRL chromosome. When these recombinant eggs fuse with a sperm carrying only a Y chromosome, recombinant male flies result, those that have normal eyes and normal wings (XRLY) and those that have white eyes and miniature wings (XrlY). By using various mutant flies, Morgan and Sturtevant discovered that they could both order a number of different genes on the X chromosome and determine how far they were from each other. The farther a gene is from another gene, the greater the number of recombinant offspring. The pattern of offspring was used to determine the sequence of genes on the X chromosome. Finding flies with mutations in different genes was essential for determining the sequence of genes and the distances between them.

—Jaime Stanley Colomé

Natural selection

FIELDS OF STUDY
Ecology, evolutionary science, genetics

SUMMARY
Natural selection is the process of differential survival and reproduction of individuals resulting in long-term changes in the characteristics of species. This process is central to evolution.

PRINCIPAL TERMS

- **adaptation:** the process of becoming better able to live and reproduce in a given set of environments
- **evolution:** any cumulative change in the characteristics of organisms or populations over many generations
- **fitness:** the relative ability of individuals to pass on genes to subsequent generations heritability: the extent to which variation in some trait among individuals in a population is a result of genetic differences
- **population:** a group of individuals that occupy a common area and share a common gene pool
- **species:** the group of all individuals or populations that interbreed or potentially interbreed with one another under natural conditions

BASIC PRINCIPLES

Natural selection is a three part process. First, there must exist differences among individuals in some trait. Second, the trait differences must lead to differences in survival and reproduction. Third, the trait differences must have a genetic basis. Natural selection results in long-term changes in the characteristics of the population.

As one of the central processes responsible for evolution, natural selection results in both fine-tuning adaptations of populations and species to their environments and creating differences among species. The importance of natural selection was first recognized by Charles Darwin, who provided the first widely accepted mechanism for evolutionary change. Natural selection is one of several processes responsible for changing the characteristics of populations and leading to an increase in adaptiveness. Other processes include genetic drift and migration. These processes interact with the processes responsible for producing variation (mutation and development) and those responsible for determining the rate and direction of evolution (mating system, population size, and long-term ecological changes) to establish the path of evolution of a species.

THE PROCESS OF NATURAL SELECTION

Natural selection occurs through the interaction of three factors: variation among individuals in a population in some trait, fitness differences among individuals as a result of that trait, and heritable variation in that trait. If those three conditions are met, then the characteristics of the population with respect to that trait will change from one generation to the next until equilibrium with other processes is reached. An example that demonstrates this process involves the peppered moth. It has two forms in the United Kingdom, a light-colored form and a dark-colored form; there is variation in color among individuals. Genetic analysis has shown that this difference in color is caused by a single gene; the variation has a heritable basis. The moth is eaten by birds that find their food by sight. The light-colored form cannot be seen when sitting on lichen-covered trees, while the dark-colored form can be seen easily. Air pollution kills the lichen, however, and turns the trees dark in color. Then, the dark-colored form is hidden and the light-colored form visible. Thus, differences in color lead to fitness differences. In the early nineteenth century, the dark-colored form was very rare. In the last half of the nineteenth century, however, air pollution increased, and the dark-colored form became much more frequent as a result of natural selection.

The characteristics of a population can be changed by natural selection in several ways. If individuals in a population with an extreme value for a trait have the greatest fitness on average, then the mean value of the trait will change in a consistent direction, which is called directional selection. For example, the soil in the vicinity of mines contains heavy metals that are toxic to plants. Individuals with the greatest resistance to heavy metals have the highest survivorship. Evolution leads to an increase in resistance. If individuals in a population with intermediate values for a trait have the greatest fitness on average, then the variation in the trait will be reduced, which is called stabilizing selection. For example, in many species of birds, individuals with intermediate numbers of offspring have the greatest fitness. If an individual has a small number of offspring, that parent has reduced reproduction and a low fitness. If the number of offspring is large, the parent will not be able to provide enough food for all the young, and most, or all, will starve, again resulting in reduced reproduction and a low fitness. Evolution leads to all birds producing the same, intermediate number of offspring. If individuals in a population with different values for a trait have the greatest fitnesses on average and intermediates have low fitness, then the variation in the trait will be increased. This is called disruptive selection. For example, for Darwin's finches, individuals with long, thin bills are able to probe into rotting cactus to find insects. Individuals with short, thick bills are able to crack hard seeds. Individuals with intermediate-shaped bills are not able to do either well and have reduced fitness relative to the more extreme types. Evolution leads to two different species of finch with different bills.

A Slow and Holistic Process

Natural selection is a slow process. The rate of evolution—that is, response to selection—is determined by the magnitude of fitness differences among individuals and the heritability of traits. Fitness differences tend to be small so that more fit individuals on average may have only a few more offspring than less fit individuals. Heritabilities of most traits are low to intermediate, meaning that most differences among individuals are not a result of genetic differences. So, even if one individual has many offspring and another has few offspring, they may not differ genetically and no change will occur. For example, if all the beetles in a population were between one and two centimeters in length and there was selection for larger beetles, it could take five hundred generations before all beetles were larger than two centimeters. Also, the direction of selection may change from one generation to the next, so that no net change occurs.

Natural selection does not act on traits in isolation. How a trait affects fitness in combination with other traits, called correlational selection, is important. For example, fruit flies lay their eggs in rotting fruit. Considered in isolation, a female should always lay as many eggs as possible. One fruit is not big enough for all the eggs she might lay, however, so she must fly from fruit to fruit. Flying requires energy, and the more energy that is used in flight the less that can be used to make eggs. So natural selection results in the division of energy between eggs and flight that yields the greatest overall number of offspring. This example demonstrates that the result of natural selection is often a trade-off among different traits.

By acting differently on males and females, natural selection results in sexual selection. This form of selection can explain differences in the forms of males and females of a species. In general, because male gametes, sperm, are much smaller and "cheaper" to produce than female gametes, eggs, more sperm than eggs are produced. As a result, it is possible for one male to fertilize many eggs, while other males fertilize few or no eggs. For example, a lion pride usually consists of one or a few males and many females. Other males are excluded, and they live separately; larger males are able to chase away smaller males. The thick mane on male lions helps to protect their throats when they fight other males. Thus, larger males with thicker manes father more cubs than other males, leading to selection on these traits. Only males are under natural selection since all females, regardless of size, will mate. The result is that males are larger than females and have manes.

Group Selection

Natural selection can occur not only among individuals but also among groups. This process is generally known as group selection; when the groups are composed of related individuals, it is called kin selection. Group selection operates the same way as individual selection. The same three conditions are necessary: variation among groups in some trait, fitness differences among groups because of that trait, and a heritable basis for that trait. For example, in Australia, rabbits introduced from Europe in 1859 spread rapidly during the next sixty years. In order to control the rabbits, a virus was introduced in 1950. At first, the virus was very virulent, killing almost all infected animals within a few days. After ten years, however, the virus had evolved to become more benign, with infected rabbits living longer or not becoming sick at all. Virulent strains of the virus grow and reproduce faster than benign strains. So, within a single rabbit, virulent strains have a higher fitness than benign strains. The longer a rabbit lives, however, the more opportunity there is for the virus to be passed to other rabbits. Thus, a group of benign viruses infecting a rabbit are more likely to be passed on than a group of virulent viruses. In this example, group selection among rabbits resulted in evolution opposite to individual selection within rabbits; however, group selection and individual selection can result in evolution in the same direction. In general, natural selection can act at many levels: the gene, the chromosome, the individual, a group of individuals, the population, or the species.

Natural selection is the primary process leading to adaptation of individuals. It involves many traits acting together, differences among males and females, and differences among levels. The interaction of all these processes of natural selection determines the path of evolution.

MEASURING NATURAL SELECTION

Natural selection is investigated in two ways: by use of indirect measurements and direct measurements. The indirect methods involve observing the outcome of natural selection and inferring its presence. The direct methods involve measuring the three parts of the process and following the course of evolution. Although the direct methods are preferred, as they provide direct proof of natural selection, in most instances, only indirect methods can be used.

Indirect methods involve three kinds of observations. First, comparisons are made of trait similarities or differences among populations or species living in the same or different areas. For example, many species of animals living in colder climates have larger bodies than those living in warmer climates. It is inferred, therefore, that colder climates result in natural selection for larger bodies. Second, long-term studies are done of traits, in particular changes in a group in the fossil record. For example, during the evolution of horses, their food, grasses, became tougher and horses' teeth became thicker. It is inferred, therefore, that tough grass resulted in natural selection for thicker teeth. Third, comparisons are made of gene frequencies of natural populations, with predictions from mathematical models. Gene frequencies are measured using various techniques, including scoring differences in appearance, as with light-colored and dark-colored moths; using electrophoresis to observe differences in proteins; and determining the sequence of base pairs of deoxyribonucleic acid (DNA). The models make predictions about expected frequencies in the presence or absence of selection. Indirect methods are best at revealing long-term responses to evolution and general processes of natural selection that affect many species. The indirect methods suffer from the problem that often many processes will result in similar patterns. So, it must be assumed that other processes were not operating, or other predictions must be made to separate the processes.

Direct methods involve two kinds of observation. First, there is observation of changes in a population following some change in the environment. There are many types of environmental changes, including man-made changes, natural disasters, seasonal changes, and introductions of species into new environments. For example, from the changes in the peppered moth following a change in pollution levels, one can measure the effects of natural selection. The second type of observation is the direct measurement of fitness differences among individuals with trait differences. For example, individual animals are tagged at an early age and survival and reproduction are monitored. Then, statistical techniques are used to find a relationship between fitness and variation among individuals in some trait. Alternatively, comparisons of traits are made between groups of individuals, such as breeding and nonbreeding, adults and juveniles, or live and dead individuals, again using statistical techniques. For example, lions that breed are larger than lions that do not breed. Direct methods are best at revealing the relative importance to natural selection of the three factors (variation, fitness differences, and heritability). The direct methods suffer from two limitations. It takes a long time for evolution to occur. So, although one can measure natural selection, it is often not known if it results in evolution. Also, for many species, it is impossible or impractical to mark individuals and follow them through their lives.

Many methods can be used to study natural selection and evolution. Each method provides information about different parts of the process. Only through the integration of these methods can the entire process of evolution be revealed.

ADAPTATION AND EVOLUTION

Natural selection is the central process in adaptation and evolution. By understanding how the process operates and where its limits lie, scientists hope to determine why evolution has proceeded in the fashion that it has. Historically, it was only after Darwin presented his theory of natural selection that the idea of evolution became widely accepted in the nineteenth century. In the twentieth century, much of the work of evolutionary biologists during the 1930's, 1940's, and 1950's was to integrate the fields of genetics, ecology, paleontology, and systematics, using natural selection and evolution as the unifying concepts.

Knowledge of natural selection is still growing; many questions proposed by Darwin and others are yet to be answered. It is still not known to what extent organisms are well adapted to their environments or whether the evolution of the parts of the chromosome that are not translated into proteins are a result of processes that do not involve natural selection. Of the many theories of how natural selection works, it is still unknown which ones are the most important in nature and to what extent evolution is caused by natural selection at the level of the individual, the group, and the species.

Genetic engineering requires knowledge of natural selection. The addition of a new gene into an organism will result in natural selection on that gene and change selection on other genes. Efficiency will be gained if successful and unsuccessful outcomes can be predicted beforehand. If genetically engineered organisms are to be released into nature, scientists need to be able to forecast their fates, such as whether the organism will remain benign or will become a pest. Genes added to one organism could possibly spread to other, native species. The solutions to these dilemmas involve predictions of the outcome of natural selection.

An understanding of natural selection is critical for conservation biology. During the twentieth century, the rate at which natural areas are being destroyed and species are becoming extinct has accelerated tremendously. Conservation biology attempts to stop that destruction and preserve species diversity. For extinction of endangered species to be halted, it must be understood how natural selection will affect these species given massive environmental changes. By discovering how evolution is occurring under natural conditions, researchers will learn how to design nature preserves to maintain species.

—*Samuel M. Scheiner*

See also: Development: Evolution: Animal life; Evolution: Historical perspective; Gene flow; Genetics; Human evolution analysis; Sex differences: Evolutionary origin.

FURTHER READING

Avers, Charlotte J. *Process and Pattern in Evolution.* New York: Oxford University Press, 1989. An intermediate-level college text that lays out the evidence for evolution and the processes that cause it. Chapters 6 and 7 present natural selection and its mechanisms and are a good introduction to the mathematical theories.

Bell, Graham. *Selection: The Mechanism of Evolution.* New York: Chapman and Hall, 1997. Clearly explains the processes of natural selection and explores its possible consequences. Offers many examples and gives an extensive review of the literature. Written for nonspecialists with some science background.

Brandon, Robert N. *Adaptation and Environment.* Princeton, N.J.: Princeton University Press, 1990. Explores the varying roles of environment at different levels of natural selection, the external, ecological, and selective environments.

Darwin, Charles. *On the Origin of Species by Means of Natural Selection.* London: J. Murray, 1859. The most important book written on natural selection. The basic premises are laid out, and data from natural and domestic species are presented. The ideas presented here are still being explored and tested.

Endler, John A. *Natural Selection in the Wild.* Princeton, N.J.: Princeton University Press, 1986. This book, although a somewhat more technical presentation, presents the best Summary of the process of natural selection. Chapters 1 and 2 are a nontechnical description of the process and a discussion of the relationship between natural selection and evolution.

Futuyma, Douglas J. *Evolutionary Biology.* 3d ed. Sunderland, Mass.: Sinauer Associates, 1998. A textbook for an intermediate-level college course in evolution. Chapters 6 and 7 provide an overview of natural selection, its mechanisms, and its consequences. It is well illustrated and contains many examples.

Gould, Stephen J. *The Panda's Thumb.* New York: W.W. Norton, 1980. A compilation of columns from *Natural History* magazine, this book provides a very entertaining and accessible view of evolution and natural selection. This and Gould's other books are the nonscientist's best introduction to the subject.

Provine, William B. *Sewall Wright and Evolutionary Biology.* Chicago: University of Chicago Press, 1986. Sewall Wright was one of the most important figures in the development of evolutionary biology in the twentieth century. Besides describing his life, this book places his work within the context of the development of evolutionary theory. Chapters 7, 8, and 9 present an excellent overview of theories of evolution and natural selection and are understandable by a general audience.

Suzuki, David T., Antony Griffiths, Jeffrey Miller, and Richard Lewontin. *An Introduction to Genetic Analysis.* 4th ed. New York: W. H. Freeman, 1989. An intermediate-level college text that presents the experimental basis of the understanding of genetic processes. Chapter 23 focuses on natural selection and evolution and is useful for gaining a more mathematical understanding of the phenomenon.

CHARLES DARWIN

Born: February 12, 1809; Shrewsbury, England

Died: April 19, 1882; Downe, England

Fields of study: Entomology, evolutionary science, human origins, invertebrate biology

Contribution: Darwin was not the first philosopher or scientist to posit a theory of evolution, but his theories of natural and sexual selection provided much of the foundation for later scientific evolutionary theory.

Charles Robert Darwin had briefly studied medicine at the University of Edinburgh and attended Cambridge University, intending to prepare for the ministry, when he was offered a chance to sail on the HMS *Beagle* as a naturalist and companion to Captain Robert FitzRoy. The fifty-seven month voyage, from December 7, 1831, to October 2, 1836, allowed Darwin unique opportunities to explore fossils, fish and sea mammals, and coral reefs. Lengthy land excursions allowed him to examine land animals and fossils, primarily in South America.

Returning to England, he first published his findings as *Journal and Remarks*, volume 3 in the series *Narrative of the Surveying Voyages of H.MS "Adventure" and "Beagle" Between 1826 and 1836* (1839). This work was revised and published the same year as *Journal of Researches into the Geology and Natural History of the Various Counties Visited by H.MS "Beagle"* (1839). His findings caused him to question generally accepted assumptions about animal creation and to posit evolutionary change as occurring mainly through natural selection. In later works, he increasingly stressed the importance of sexual selection. Challenged by peers to examine individual species before generalizing about life as a whole, he began lengthy examinations of such life-forms as beetles and barnacles, which he published.

By the 1850's, despite his aversion to public controversy, he accepted the need to publish his general theories. They appeared in 1859 as *On the Origin of Species by Means of Natural Selection: Or, The Preservation of Favoured Races in the Struggle for Life.* Although he avoided discussion of human origin in this work, the controversy he dreaded was forthcoming. His ideas, however, were adopted by young scientists, most notably Thomas Henry Huxley, who sought to establish the natural sciences as disciplines separate from the natural theology that then prevailed in universities. These scientists became his spokespersons, as he continued his experiments at his estate; his theories gained widespread acceptance. In 1871, he dealt directly with the origin of human life in *The Descent of Man*; he followed this, in 1872, with *The Expression of Emotions in Man and Animals.* These works clearly placed man within the animal kingdom, not the product of a separate creation.

Darwin also published on narrower topics involving animal life and fossils, and extensively revised *On the Origin of Species*, ultimately producing six revised editions in the quarter-century after its initial publication. He wrote a brief autobiography, published posthumously in 1887. He was awarded numerous honors in England and on the Continent. At his death, his work was so widely respected that, despite his religious skepticism, England honored him with burial in Westminster Abbey.

—Betty Richardson

ALFRED RUSSEL WALLACE

Born: January 8, 1823; Usk, Monmouthshire, Wales

Died: November 7, 1913; Broadstone, Dorset, England

Fields of study: Entomology, evolutionary science, population biology, zoology

Contribution: Wallace, a pioneer of the science of zoogeography, proposed a theory of evolution by natural selection in 1855 that predated and stimulated the publication of Charles Darwin's *On the Origin of Species* (1859).

Alfred Russel Wallace grew up in rural Wales and then in Hertford, England. His formal education was limited to six years at the Hertford Grammar School. From 1837 to 1844 Wallace worked in his brother William's surveying business. In 1844, Wallace taught at the Collegiate School in Leicester, England. In 1848, Wallace and the entomologist Henry Walter Bates embarked on an expedition to Brazil. Wallace and Bates planned to collect and identify biological specimens and then pay for their trip by selling their collections. Wallace spent a total of four years exploring the Amazon River basin, collecting birds, butterflies and other insects.

Unfortunately, on the return voyage Wallace lost his precious collections when his ship caught fire and sank. Nevertheless, the expedition led to the publication of several articles and two books (*Palm Trees of the Amazon and Their Uses* and *Narrative of Travels on the Amazon and Rio Negro*, 1853). These reports attracted the attention of the Royal Geographical Society, which helped to fund his next expedition. For eight years (1854-1862) Wallace continued his research in the Malay Archipelago (Indonesia). Wallace's research on the geographic distribution of animals among the islands of the Malay Archipelago provided crucial evidence for his evolutionary theories and led him to devise what became known as Wallace's Line, the boundary that separates the fauna of Australia from that of Asia. By the time Wallace returned to England in 1862 he had collected over 125,000 animal specimens.

During an attack of a tropical fever, Wallace experienced a flash of insight in which he realized that natural selection could serve as the mechanism of evolution. Within a few days he completed his essay "On the Tendency of Varieties to Depart Indefinitely from the Original Type" and sent it to Charles Darwin for review and possible publication. Darwin was shocked to find that Wallace had developed a theory of evolution identical to that outlined in his own unpublished 1842 essay. Darwin's friends Charles Lyell and Joseph Hooker arranged for a joint presentation of the papers written by Wallace and Darwin and simultaneous publication in the August, 1858, *Proceedings of the Linnean Society*.

Graciously allowing Darwin to claim priority for the discovery of evolution by means of natural selection, Wallace continued to publish works on natural history and travel, including *The Malay Archipelago* (1869), *Contributions to the Theory of Natural Selection* (1870), *Geographical Distribution of Animals* (1876), and *Island Life* (1880). It was Wallace who called evolution by means of natural selection "Darwinism" in order to distinguish this theory from its predecessors. Unlike Darwin, however, Wallace continued to believe that natural selection could not account for the higher faculties of human beings.

—*Lois N. Magner*

NEANDERTHALS

FIELDS OF STUDY

Anthropology, archaeology, evolutionary science, human origins, physiology, systematics (taxonomy)

SUMMARY

Neanderthals are the best-known extinct members of the human lineage. It is generally agreed that they were close relatives of modern humans, but the nature of the relationship is vigorously debated. They have been assumed to be a direct ancestor, a diseased member of the species, or an extinct side branch of the family tree.

PRINCIPAL TERMS

- **deoxyribonucleic acid (DNA):** the chemical that carries the instructions for all living things; closely related organisms have very similar DNA
- **genus:** the first part of the scientific name of an organism; members of the same genus but different

- species are closely related, but cannot mate and produce fertile offspring
- **mitochondria:** subcellular structures containing DNA used to estimate the relationships between groups of organisms; the more similar the DNA, the more closely related the groups
- **species:** the second part of the scientific name of an organism; members of the same species can mate and produce fertile offspring
- **subspecies:** the third part of a scientific trinomial, assigned to one of two groups that can mate and produce fertile offspring, but that have some strikingly different characteristics
- **taxonomy:** the science of classifying and naming living and fossil organisms, or the classification and scientific name of a living or fossil group

Basic Principles

The fossil that gave the Neanderthals their name was found in a cave being quarried for limestone in Germany's Neander Valley in 1856. At least two Neanderthal fossils were discovered before the Neander Valley individual; however, neither was recognized as a member of an extinct human group until after the name "Neanderthal" was assigned. Many similar fossils have been found in scattered locations all over Europe and the Middle East since the Neander Valley discovery. Dates assigned to the various fossils indicate that the Neanderthals originated late in the Ice Age and became extinct a few thousand years before the last glacial retreat (from about 200,000 years ago to about 30,000 years ago). Thus the Neander Valley specimen lent its name to a fossil relative of modern humans that occupied Europe and the Middle East late in the Ice Age.

Though the Neanderthals were very similar to modern humans, they had several distinctive characteristics. Neanderthals were short and exceptionally stout-bodied with broad supportive bones and joints. This body form suggests a life filled with intense physical effort. Perhaps the compact body also helped them cope with cold stress under Ice Age conditions. Their brains were somewhat larger than modern human brains. That size may have compensated for the more massive total body size of the Neanderthals, since large-bodied organisms generally have larger brains. Their foreheads sloped up from their exceptionally heavy eyebrow ridges, their jaws extended forward beyond the plane of the face, and their chins were weakly developed. These and several other characteristics are used to define a fossil find as a Neanderthal.

Structure and Behavior

Rudolph Virchow's initial interpretation of the Neander Valley fossil as a diseased human was popular for a time. Virchow held that the fossil was a modern human whose unique features were the result of disease. However, as more fossils with the same characteristics were discovered all around Europe and the Middle East, this explanation became untenable. Later, misinterpretation of the characteristics of Neanderthal fossils led Marcellin Boule and others to interpret Neanderthals as stooped, bent-kneed, ape-like subhumans with an animal nature to match.

Additional fossil discoveries, including evidence for toolmaking and burials, sometimes with flowers placed in the grave, caused anthropologists to rethink the presumed animal nature of the Neanderthals. Although the evidence for flowers has been challenged, the evidence for burials, presumably accompanied by mourning, is accepted by many anthropologists. In addition, fossils showed that some Neanderthals lived much of their lives with deformed limbs and other disabilities, which would have made it difficult or impossible for them to fend for themselves. Yet they apparently lived many years in that condition, suggesting the support of other members of a social group. Such behavior was not in keeping with Boule's picture of the Neanderthals as nonhuman animals.

Reinterpretation of the anatomic evidence also suggested that, instead of a bent-kneed, stooped posture, the Neanderthals walked on two rather straight legs and had hands capable of manipulating materials and making tools, much as modern humans do. All this indicated that the Neanderthals were more like modern humans than Boule's interpretation, and they came to be thought of in that light.

Taxonomic Relationship to Modern Humans

Neanderthals have always been recognized as close relatives of modern humans, but the specific taxonomy of the relationship is still a point of contention. They are placed in the same genus (*Homo*) as modern humans by almost all anthropologists, but researchers debate whether they were members of our species, *Homo sapiens* or belonged in their own species, *Homo neanderthalensis*. The discussion of the structure and behavior of Neanderthals bears directly on this question. If the Neanderthal characteristics were the result of disfigurement caused by disease, Neanderthals were simply aberrant humans and not especially interesting from the perspective of human evolution. However, if they were stooped, bent-kneed, and animal-like in behavior, they were probably a separate species, perhaps ancestral to modern humans, and therefore more interesting from the evolutionary perspective. On the other hand, if they were upright in stature, were skilled toolmakers, were supportive of their handicapped and elderly, and buried their dead with mementos such as flowers, they might earn the designation *Homo sapiens* and take on an even greater interest to the more modern members of that species.

Such arguments are part of the practical taxonomy of the Neanderthals, but the real key to species identification and species separation is (at least theoretically) interbreeding. If the members of two groups can mate with each other and produce fertile offspring, and if these offspring can produce fertile offspring, the two groups are generally considered to be members of the same species. Therefore, the real taxonomic question becomes: Could Neanderthals and early modern humans interbreed?

Because it is difficult to determine whether fossil groups interbred with one another, Neanderthal taxonomy has been primarily determined by anatomic and presumed behavioral characteristics, such as those already discussed. That taxonomy has vacillated with changing interpretations of those characteristics. Neanderthals have been placed in their own species (*Homo neanderthalensis*) for much of their history, but they have been identified as a human subspecies (*Homo sapiens neanderthalensis*) at other times. The latter designation implies that the Neanderthals and modern humans (*Homo sapiens sapiens*) were members of the same species and therefore could interbreed.

Determination of the Neanderthals' taxonomic position is an integral part of arguments over the mechanism of the origin of modern humans. There are two main hypotheses for that origin: the replacement hypothesis of Christopher Stringer and the multiregional hypothesis vigorously supported by Milford Wolpoff. The replacement hypothesis is also designated "out-of-Africa" because it assumes that a population of African origin expanded throughout Africa, Europe, and Asia and rapidly replaced the more primitive humanlike species living there, including the Neanderthals. Whether this replacement was by competition or by more direct and violent means is undetermined. The multiregional hypothesis suggests that the widespread, more primitive humanlike populations evolved into modern humans rather than being replaced by new immigrants. Both hypotheses hold that the more primitive populations also originated in Africa and spread to Europe and Asia at a much earlier date.

Because the Neanderthals are the best known and best understood early human group, an understanding of the Neanderthal relationship is critical to an understanding of the evolutionary history of humanity. A Neanderthal contribution to modern human ancestry would support the multiregional hypothesis, and the lack of such a contribution would be consistent with the replacement hypothesis.

Advances of the Late Twentieth Century

By the 1990's, the Neanderthals were well established as a group related to modern humans, but the questions remained: How close was the relationship? Did the two groups interbreed? Were Neanderthals a part of the evolutionary heritage of modern humans? During the 1990's, improved techniques and additional fossil discoveries led to greater understanding of the Neanderthals but little consensus on these questions. A few examples will illustrate the situation.

In a 1996 study, Jean-Jacques Hublin and several coworkers determined that Neanderthals found at an archeological site in France made bone tools and wore decorative emblems on their bodies, behaviors not uncovered with older Neanderthal fossils. They concluded that the Neanderthals were influenced by early modern humans who lived in the same area at the same time and that a reasonably elaborate cultural exchange must have occurred between the two groups. However, based on the strikingly different anatomy of the two groups' inner ears, they also concluded that the Neanderthals and modern humans did not interbreed. The investigators reasoned that if interbreeding had occurred, the two groups would have shared a common ear structure.

In 1997, Matthias Krings, Svante Paabo, and their colleagues isolated and engineered deoxyribonucleic acid (DNA) from the mitochondria of Neanderthal bones and compared it to DNA from modern human mitochondria. They found the Neanderthal DNA to be quite different from that of modern humans and concluded not only that the two groups were different species but also that Neanderthals were not ancestral to modern humans.

In 1998, Daniel Lieberman proposed that a reduction in the length of the sphenoid bone during embryology can explain most differences between the two groups' skulls. The sphenoid is a bone in the skull of both Neanderthals and modern humans, and Lieberman showed it to be shortened in modern humans but not in Neanderthals. He hypothesized that the impact of shortening the sphenoid resulted in the modern human skull characteristics, while the longer sphenoid resulted in the Neanderthal skull. Based on the fundamental nature of the change, he concluded that Neanderthals do not belong to the same species as modern humans and were probably not ancestral to modern humans.

In 1999, Cidàlia Duarte, Erik Trinkaus, and several colleagues discovered the buried remains of a four-year-old child in southern Spain. The skeleton was estimated to be about 24,500 years old, and they interpreted its anatomy to be a mixture of modern human and Neanderthal characteristics. Most anthropologists agree that southern Spain supported Neanderthal populations longer than other parts of the world, perhaps as late as 27,000 years ago, and that modern humans and Neanderthals coexisted in the region. Duarte, Trinkaus, and their group suggested that the skeleton they found demonstrated that the two groups did interbreed and that Neanderthals were part of the ancestry of *Homo sapiens*.

SIGNIFICANCE

Consideration of this short list of studies in the 1990's demonstrates the state of knowledge about the Neanderthals' place in human evolution. Viewed alone, each study seems to clinch the position of its authors. In fact, the first three reinforce one another so well that Neanderthals would seem to be eliminated from direct participation in the evolution of modern humans. However, Duarte and Trinkaus's study would seem to clinch the opposite position, that Neanderthals were direct participants in the evolution of modern humans. This situation symbolizes the absence of consensus in the field. There are also established scientists with alternative viewpoints for each of these studies. Lieberman himself is a coauthor of a letter that criticizes his own conclusions about the sphenoid and points to the need for a better understanding of the development of primate skulls to help clarify the situation.

A number of anthropologists have pointed out that Krings and Paabo's conclusions are extrapolated from a single, short segment of the mitochondrial DNA and that more extensive studies, including studies of DNA from the nucleus, are necessary before definitive conclusions can be drawn. In fact, nuclear DNA studies of modern humans have suggested that modern human DNA comes from a number of sources rather than a single African source as in the out-of-Africa hypothesis. Clearly, extensive DNA comparisons would be helpful; however, DNA from fossils is difficult to find and difficult to work with, so an extensive collection of such studies is not likely to accumulate.

Ian Tattersall, who rejects the Neanderthals as direct contributors to modern human evolution, has criticized Duarte and Trinkaus's data and their interpretation of the data. The verbal exchange has been bitter, not an unusual circumstance for disagreements in this field.

Although anthropologists have learned an enormous amount about the Neanderthals, their relationship to modern humans continues to escape consensus. This is, without question, a result of the difficulty of the problem and the tentative nature of the evidence. Most agree that the Neanderthals were a successful group closely related to modern humans. Everyone's hope is that more fossils, improved technology, and fresh insight will clarify the question because understanding the Neanderthals is likely to contribute to an understanding of humanity.

—*Carl W. Hoagstrom*

See also: Apes to hominids; Cannibalism; Communication; Communities; Evolution: Animal life; Evolution: Historical perspective; Hominids; Homo sapiens and human diversification; Human evolution analysis.

FURTHER READING

Akazawa, Takeru, Kenichi Aoki, and Ofer Bar-Yosef, eds. *Neandertals and Modern Humans in Western Asia*. New York: Plenum, 1998. A scholarly but understandable group of papers with contributions from many of the major workers in Neanderthal evolution and biology.

Ciochon, Russell L., and John G. Fleagle, eds. *The Human Evolution Source Book.* Englewood Cliffs, N.J.: Prentice Hall, 1993. The broad spectrum of human evolution is covered in this book, but a large section titled "The Neanderthal Question and the Emer- gence of Modern Humans" covers Neanderthals. Several of the most fundamental questions are dealt with in a clear and interesting fashion.

Fox, Richard G. "Agonistic Science and the Neanderthal Problem." *Current Anthropology*, supp. 39 (June, 1998). All articles in the supplement concern "The Neanderthal Problem and the Evolution of Human Behavior," the supplement's title. The interaction between Neanderthals and modern humans in Europe and in the Middle East and an archaeological consideration of the out-of-Africa model are the subjects of three other articles.

Shreeve, James. *The Neanderthal Enigma.* New York: William Morrow, 1995. Written by a science writer rather than a scientist, this is an interesting account of the history of Neanderthals and Neanderthal studies. Stringer, Christopher, and Robin McKie. *African Exodus.* New York: Henry Holt, 1996. A small, interesting book on the out-of-Africa hypothesis written by an anthropologist intimately involved with the development of the hypothesis (Stringer) and a science writer.

Tattersall, Ian. *The Last Neanderthal.* New York: Macmillan, 1995. An extensively illustrated account of what is known about Neanderthals and how it was learned. The prologue outlines two interesting, if imaginative, characterizations of the "last Neanderthal."

Tattersall, Ian, and Jeffrey H. Schwartz. "Hominids and Hybrids: The Place of Neanderthals in Human Evolution." *Proceedings of the National Academy of Science* 96 (June, 1999): 7. This commentary is a reply to the paper by Duarte, Trinkaus, and their colleagues (pages 7604-7609 of this same issue) in which they report the description of a 24,500-year-old fossil with characteristics of both Neanderthals and early modern humans. The two positions are clearly laid out in the article and the commentary.

Trinkaus, Erik, and Pat Shipman. *The Neanderthals: Changing the Image of Mankind.* New York: Alfred A. Knopf, 1992. An interesting and well-written account of the Neanderthals and the scientists who study them.

NEPHROLOGY

FIELDS OF STUDY

Dialysis; transplantation; pediatric nephrology; proteomics; genetics; electrolyte physiology; hypertension; plasmapheresis; mineral metabolism; pharmacology; internal medicine; nephrolithiasis.

SUMMARY

Nephrology is a division of medical science associated with internal medicine, concentrated on the kidneys and the associated anatomy, physiology, diseases, and disorders. The word "nephrology" is of Greek origin: nephros, meaning kidney, and logos, meaning word, reason, thought, or discourse. Kidney diseases may include electrolyte-balance disorders and hypertension. Therapies associated with kidney disease manage limitations or failure of the renal system and may include dialysis or kidney transplant. Disorders of the kidney are often a result of systemic or congenital disorders, affecting more than one system or organ in the body, which can make treatment complicated and often challenging. A medical doctor who specializes in the diagnoses and treatment of the kidneys is a nephrologist.

PRINCIPAL TERMS

- **calyx:** cuplike urine-collection cavity located at the tip of each pyramid within the kidney.
- **cortex:** outer layer of the kidney that contains millions of microscopic nephrons.
- **dialysis:** process of removing toxic materials from the blood and maintaining fluid, electrolyte, and acid balance often using automated equipment when kidneys have become damaged or have been removed.
- **kidney:** one of a pair of reddish brown, bean-shaped organs located on either side of vertebral column

in the retroperitoneal area. their function is to form urine from blood plasma by regulating water, electrolyte, and acid balance of the blood.
- **nephrolithiasis:** presence of calculi in the kidney, commonly called a kidney stone.
- **nephron:** functional unit of the kidney, consisting of a glomerulus, bowman's capsule, renal tubule, and peritubular capillaries. urine is formed by nephrons by a process of filtration, reabsorption, and secretion.
- **pyramid:** triangular tissue in the medulla (the inner region) of the kidney, which contains the loops and collecting tubules of the nephrons.
- **renal artery:** blood vessel that carries blood from the aorta into the kidney. inside the kidney, the renal artery branches into kidney tissue until the smallest artery (arteriole) leads to a glomerulus, where filtration can begin.
- **renal vein:** blood vessel that carries blood from the kidney, back to the heart.
- **ureter:** muscular tube with a mucosal lining that leads from each kidney to the urinary bladder.
- **urine:** fluid created by the kidney that consists of 95 percent water and 5 percent dissolved solids (salts and nitrogen-containing wastes), which is eliminated by the body.

BASIC PRINCIPLES

Nephrology is a branch of medical science, a specialty of internal medicine, concerned with the structure and function of the kidneys. Any pathology within this system and the management of many systemic diseases affecting the kidneys are key responsibilities of a nephrologist. A nephrologist will determine the stage or degree of kidney disease, treat any associated complications (hypertension, bone disorders, vitamin imbalances), manage anemia, educate the patient about nutrition, risk factors, treatments, and transplantation. The nephrologist also commonly provides the vascular access placement for dialysis and coordinates treatment.

The purpose of the kidneys is to regulate water, electrolytes, and acid-base content of the blood and, indirectly, all other body fluids. Filtration by the kidneys is a continuous process and the rate is affected by blood flow through the kidneys and daily fluid intake. Blood enters the kidney and passes through the glomerulus, where water and dissolved substances are filtered through capillary membranes, and the

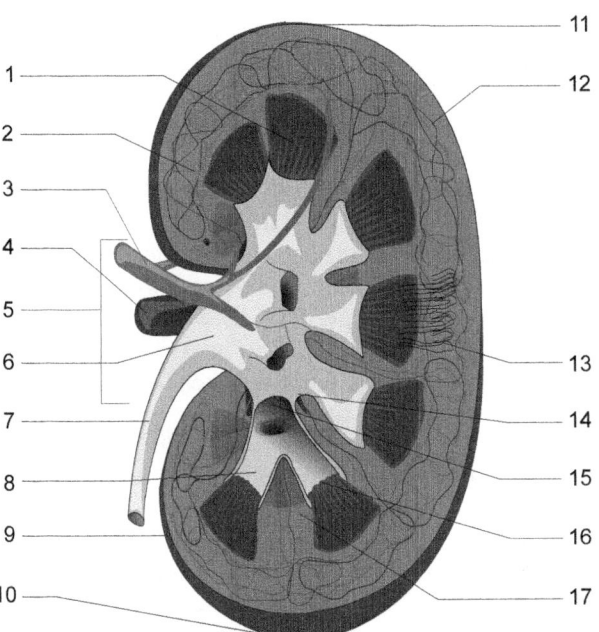

Structures of the kidney: 1.Renal pyramid 2.Interlobular artery 3.Renal artery 4.Renal vein 5.Renal hilum 6.Renal pelvis 7.Ureter 8.Minor calyx 9.Renal capsule 10.Inferior renal capsule 11.Superior renal capsule 12.Interlobar vein 13.Nephron 14.Minor calyx 15.Major calyx 16.Renal papilla 17.Renal column (no distinction for red/blue (oxygenated or not) blood, arteriole is between capilaries and larger vessels

inner layer of Bowman's capsule, where it becomes glomerular filtrate. The blood cells and larger protein cells stay in the capillaries during this time. The filtrate travels through a series of renal tubules, where useful substances such as water, glucose, amino acid, minerals, and vitamins are reabsorbed into the capillaries to be used in the body. The amount of water that is reabsorbed is regulated by antidiuretic hormone and indirectly by aldosterone. The products that are not reabsorbed, such as metabolic products of medications, are considered waste and remain in the filtrate and become part of urine. The collecting tubules join together to form papillary ducts that empty urine into the calyxes and eventually into the ureter and bladder. The process from the ingestion of a large amount of liquid to the production of urine takes about forty-five minutes for most well-hydrated people. The average volume of urine produced each day is about 1,500 milliliters, but it depends on many factors—age, climate, activity, diet, and blood pressure.

Patient pathology or disease may affect the urine production; however, people who have had part of a kidney removed or have only one kidney can have normal renal function. There are five recognized stages of kidney disease.

- Stage 1: Slightly diminished function. Kidney damage with normal or relatively high glomerular filtration rate: >90 milliliters per minute per 1.73 m_2.
- Stage 2: Mild reduction in kidney function. Glomerular filtration rate: 60-89 milliliters per minute per 1.73 m_2.
- Stage 3: Moderate reduction in kidney function. Glomerular filtration rate: 30-59 milliliters per minute per 1.73 m_2.
- Stage 4: Severe reduction in kidney function. Glomerular filtration rate: 15-29 milliliters per minute per 1.73 m_2.
- Stage 5: Established kidney failure. Glomerular filtration rate: <15 milliliters per minute per 1.73 m_2.

Background

English physician Richard Bright first established the relationship between the symptoms and pathology of renal failure in 1827. It was not until 1854 when Scottish chemist Thomas Graham described osmotic force. In 1861, Graham went on to explain the process of dialysis using a hoop form dialyzer. In the late 1800's and early 1900's, several scientists began performing dialysis and kidney transplants on animals. Austrian surgeon Emerich Ullmann performed a kidney autotransplant on a dog, and in 1914, pharmacologist John Jacob Abel and his colleagues Leonard Rowntree and Benjamin Turner discovered that salicylic acid could be removed from the blood of rabbits using dialysis. In 1924, German physician George Haas performed the very first dialysis procedure on a human. Dutch physician Willem Kolff treated sixteen patients with acute kidney failure between 1943 and 1944 but with limited success. The first success came in 1945 with the seventeenth patient, a sixty-seven-year-old woman in uremic coma due to acute renal failure from gram-negative sepsis. After eleven hours of hemodialysis, the patient regained consciousness and began to produce urine. She went on to live seven more years.

The Scribner shunt, a U-shaped Teflon tube, is the creation of Chicago-born Belding Scribner. The shunt, inserted between an artery and vein in a patient's forearm, could be opened and connected to the artificial kidney machine during dialysis. Teflon was relatively new to the biomedical community at the time, and its nonstick properties made it less likely to clot. Before Scribner's shunt, a patient could receive only a few dialysis treatments before doctors would run out of places to connect the machine to the patient.

The shunt was first used on March 9, 1960, on Clyde Shields, who was dialyzed repeatedly for eleven years. Another patient was dialyzed for thirty-six years, undergoing 5,700 cycles of hemodialysis, before his death. In 1962, Scribner and American physician James Haviland developed the first free-standing dialysis center in the world, the three-bed Seattle Artificial Kidney Center.

How It Works

Nephrology is the science that concentrates on the kidneys and the associated anatomy, physiology, diseases, and disorders. The improper functioning of the kidney may disrupt electrolyte balance or lead to hypertension (high blood pressure) and is often related to other systemic or congenital conditions.

Diagnosing Kidney Disease. Treating the underlying condition may be complicated by reduced kidney function, and treatment of the kidney may be a challenge because of the underlying condition. Deteriorating kidney function has very unspecific symptoms, which makes kidney disease difficult to diagnose. Patients may feel generally unwell or have a reduced appetite. Quite often, kidney disease is not recognized until a major complication such as anemia, pericarditis, or cardiovascular disease has been detected. People diagnosed with high blood pressure or diabetes often have their kidney function assessed as part of normal screening procedure.

Chronic Kidney Disease Detection. Chronic kidney disease is detected by a blood analysis for levels of creatinine. Higher levels of creatinine indicate a decreased glomerular filtration rate resulting in a decline in normal kidney function. A glomerular filtration rate of less than 60 milliliters per minute per 1.73 m_2, for a period of three months, is classified as having chronic kidney disease. Red blood cells or excess protein detected in urinalysis may cause a physician to investigate more thoroughly.

Kidney Stone Detection. Patients may develop nephrolithiasis or renal calculus, also known as a kidney stone, which is a solid mass normally composed of mineral salts. In the kidney, a calculus can block the ureter and urine flow. Symptoms may include a severe, sudden pain, chills, fever, and appearance of blood in the urine. If the blockage cannot be passed on its own by relaxation of surrounding smooth muscle, it should be removed surgically or disintegrated ultrasonically.

APPLICATIONS

Dialysis, the common name for hemodialysis, is the procedure used to treat end-stage kidney failure, transient kidney failure, and some poisoning or drug overdose situations. Other indications for dialysis include: hyperkalemia, uremia, uremic pericarditis, acidosis, and fluid overload. It is used to act as an artificial kidney, outside the body when a disorder is causing fluid, acids, electrolytes, and some drugs from being effectively eliminated.

Dialysis. Dialysis involves a series of five steps:

- Establishment of access to the patient's circulatory system via an arteriovenous fistula, graft, or catheter.
- Anticoagulating the patient's blood in order to prevent clotting during its circulation outside the body.
- Pumping of the patient's blood to a dialysis membrane.
- Adjusting the diffusion of solutes from the blood into a buffered dialysis solution.
- Returning the cleaned and buffered blood to the patient's circulation.

Typically, the entire process takes between three and four hours to complete and must occur several times per week. Even with regular dialysis, mortality rates remain quite high for end-stage renal disease patients. Most deaths are attributed to stroke, heart disease, or complications from diabetes. Complications of dialysis include hypotension (low blood pressure), infection at access site, sepsis, air embolism, bleeding, anemia, and muscle cramping.

Dialysis is the only treatment option for most renal disease patients, and the procedure that has not changed much since it was first performed. Unfortunately, during the days off of treatment, patients experience toxin buildup leaving them feeling bloated, tired, and uncomfortable.

There are two common types of dialysis, the most common being hemodialysis. Hemodialysis removes wastes and water by circulating blood outside the body through an external dialyzer that acts as a filter and contains a semipermeable membrane. Peritoneal dialysis uses a peritoneal membrane inside the body to filter wastes and water from the blood.

Home Dialysis Machines. In an effort to meet the growing need for dialysis treatment and provide some convenience to patients, home-treatment machines were created. The first home-treatment units were very large, very expensive, and cleaning and maintenance were difficult, but things are improving. Modern home dialysis machines are portable, and after rigorous training sessions, patients can treat themselves at home in six shorter sessions per week, rather than in several three- or four-hour sessions at a facility. Only about 1 percent of dialysis patients are being treated at home. Machines cost almost $25,000, are the size of a microwave, and perform the sophisticated filtration using a disposable cartridge.

Anemia Treatments. Anemia is an independent predictor of mortality in chronic kidney disease patients and is also associated with worsening of cardiovascular morbidity and accelerated rate of kidney damage. The administration of recombinant human erythropoietin (rHuEpo) has greatly reduced anemia in patients with chronic kidney disease. Unfortunately, almost 15 percent of patients show limited or no response to rHuEpo.

Genomic and Proteomic Research. Genomics (the study of genomes or DNA sequence of organisms) and proteomics (the study of structure and function of proteins) in nephrology are still in the early phases of research. The application of these approaches has recently produced promising new urinary biomarkers for kidney injury and chronic kidney disease, which may provide better understanding of renal physiology and assist in the development of new therapeutic strategies.

FUTURE PROSPECTS

Patients requiring nephrology treatment are increasing at an alarming rate. In 2008, there were more than 1.64 million people on dialysis. This growth in end-stage renal disease patients is five times the world population growth (1.3 percent). Diabetes alone is expected to grow by 165 percent by 2050; therefore it is estimated that one out of every eight United States residents will have some level of kidney disease. Improving the facilitation of dialysis treatment to patients is a key concern as well as managing the growing number of people requiring treatment. The average cost to treat each patient with end-stage renal disease is about $85,000 annually. Research has found that daily treatment provides better outcomes for patients. These findings are not accommodated by Medicare coverage, which, on average, pays for three treatments per week. Homedialysis programs are being established to meet the growing need for treatment and allow patients the flexibility to accommodate treatment into their lives. Patients receive training on how to operate the machines as well as basic dialysis knowledge. About one-third of patients who have a friend or family member willing to become a living kidney donor will not be able to receive the donor kidney because of an incompatible blood type or cross match. Kidney exchange programs are becoming more popular and are being facilitated by major medical centers and health providers. A kidney exchange increases the pool of available donors through an exchange between incompatible donorrecipient pairs, or a donor chain, managed by a specialized computer program that matches donors and recipients.

April D. Ingram, B.Sc.

Further Reading

Goldsmith, David, Satish Jayawardene, and Penny Ackland, eds. *ABC of Kidney Disease*. Malden, Mass.: Blackwell Publishing, 2007. This book contains excellent illustrations and clearly written information to provide a greater understanding of renal disease.

Greenberg, Arthur, ed. *Primer on Kidney Diseases*. Philadelphia: Saunders Elsevier, 2009. This book is endorsed by the National Kidney Foundation and provides very good background and basics of renal anatomy and physiology. Defines and classifies common kidney disorders and outlines treatment protocols and modalities.

Jörres, Achim, Claudio Ronco, and John A. Kellum, eds. *Management of Acute Kidney Problems*. Berlin: Springer-Verlag, 2010. This excellent reference contains information regarding the definition, epidemiology, pathophysiology, and clinical causes of acute kidney failure.

Lai, Kar Neng, ed. *Practical Manual of Renal Medicine: Nephrology, Dialysis and Transplantation*. Hackensack, N.J.: World Scientific, 2009. This manual provides practical information about dialysis, transplantation, and general nephrology in straightforward language. It describes treatment rationale and kidney disease management.

Schrier, Robert W. *Manual of Nephrology*. 7th ed. Philadelphia: Lippincott, Williams & Wilkins, 2009. An excellent clinical reference guide for the advanced student.

Web Sites

American Board of Internal Medicine http://www.abim.org
American Society of Nephrology http://www.asn-online.org
International Society of Nephrology http://www.isn-online.org
National Institute of Diabetes and Digestive and Kidney Diseases http://www2.niddk.nih.gov

See also: Artificial Organs.

FASCINATING FACTS ABOUT NEPHROLOGY

- The World Health Organization (WHO) reports that more than 68,300 kidney transplants are performed every year worldwide.
- Kidney disease affects more than 600 million people globally.
- The 1990 Nobel Prize in Physiology or Medicine was awarded to Joseph E. Murray, the surgeon who performed the first-ever kidney transplant between identical twins, which demonstrated that previous failures were due to immunologic incompatibilities rather than surgical methods.
- The average kidney is about 4.5 inches long and weighs between four and six ounces.
- The kidneys have a higher blood flow than the heart, liver, or brain.
- World Kidney Day started in 2006 and is now an annual event held every March in more than one hundred countries.
- Most people are born with two kidneys but can survive with a single healthy kidney.
- The Voluntary Health Association of India estimates that about 2,000 Indians sell a kidney every year.
- The largest reported kidney stone weighed just more than thirty-one pounds.
- The United States has one of the highest incidences of end-stage renal disease in the world, 363 per million people, compared with Iceland, which has less than 60 per million.
- Chinese medicine believes that kidney stones may be caused by blockage or imbalance of chi (vital energy) in the kidney and acupuncture can restore positive energy flow.

Neural engineering

FIELDS OF STUDY

Biomedical engineering; computational neuroscience; medicine; neural prostheses; neuroscience; chemistry; neural implants; neural interfacing; bioelectrical engineering; biology; biomaterials; neurosurgery; electrical engineering; materials science; nanotechnology; neural imaging; neural networks; tissue engineering.

SUMMARY

Neural engineering is an emerging discipline that translates research discoveries into neurotechnologies. These technologies provide new tools for neuroscience research, while leading to enhanced care for patients with nervous-system disorders. Neural engineers aim to understand, represent, repair, replace, and augment nervous-system function. They accomplish this by incorporating principles and solutions derived from neuroscience, computer science, electrochemistry, materials science, robotics, and other fields. Much of the work focuses on the delicate interface between living neural tissue and nonliving constructs. Efforts focus on elucidating the coding and processing of information in the sensory and motor systems, understanding disease states, and manipulating neural function through interactions with artificial devices such as brain-computer interfaces and neuroprosthetics.

PRINCIPAL TERMS

- **cochlea:** coiled part of the inner ear where the hearing receptors reside.
- **electrode:** solid conductor through which electrical current enters or leaves a medium.
- **motor cortex:** area of cerebral cortex (outer brain layer) that processes motor information and control movement.
- **photodiode:** semiconductor component with light-sensitive electrical characteristics.
- **retina:** light-sensitive layer lining the inner eyeball.
- **thalamus:** mass of neural tissue situated deep in the brain.
- **vagus nerve:** tenth and longest cranial nerve, which passes through the neck and thorax into the abdomen.
- **visual cortex:** area of cerebral cortex that processes visual information.

BASIC PRINCIPLES

Neural engineering (or neuroengineering, NE) is an emerging interdisciplinary research area within biomedical engineering that employs neuroscientific and engineering methods to elucidate neuronal function and design solutions for neurological dysfunction. Restoring sensory, motor, and cognitive function in the nervous system is a priority. The strong emphasis on engineering and quantitative methods separates NE from the "traditional" fields of neuroscience and neurophysiology. The strong neuroscientific approach distinguishes NE from other engineering disciplines such as artificial neural networks. Despite being a distinct discipline, NE draws heavily from basic neuroscience and neurology and brings together engineers, physicians, biologists, psychologists, physicists, and mathematicians.

At present, neural engineering can be viewed as the driving technology behind several overlapping fields: functional electrical stimulation, stereotactic and functional neurosurgery, neuroprosthetics and neuromodulation. The broad scope of NE also encompasses neurodiagnostics, neuroimaging, neural tissue regeneration, and computational approaches. By using mathematical models of neural function (computational neuroscience), researchers can perform robust testing of therapeutic strategies before they are used on patients.

The human brain, arguably the most complex system known to humankind, contains about 10^{11} neurons and several times more glial cells. Understanding the functional neuroanatomy of this exquisite device is a sine qua non for anyone aiming to manipulate and repair it. The "neuron doctrine," pioneered by Spanish neuroscientist Santiago Ramón y Cajal, considers the neuron to be a distinct anatomical and functional unit. The extension introduced by American neuroscientist Warren S. McCullogh and American logician Walter Pitts asserts that the neuron is the basic information-processing unit of the brain. For neuroengineers, this means that a particular goal can be reached just by manipulating a cell or group of cells. One argument in favor of this view is that stimulating groups of neurons produces a regular effect. Motor activity, for example, can be induced by stimulating the motor cortex with electrodes. In addition, lesions to specific brain areas due to neurodegenerative disorders or stroke lead to more or less predictable clinical manifestation patterns.

BACKGROUND

Electricity (in the form of electric fish) was used by ancient Egyptians and Romans for therapeutic purposes. In the eighteenth century, the work of Swiss anatomist Albrecht von Haller, Italian physician Luigi Galvani, and Benjamin Franklin set the stage for the use of electrical stimulation to restore movement to paralyzed limbs. The basis of modern NE is early neuroscience research demonstrating that neural function can be recorded, manipulated, and mathematically modeled. In the mid-twentieth century, electrical recordings became popular as a window into neuronal function. Metal wire electrodes recorded extracellularly, while glass pipettes probed individual cells. Functional electrical stimulation (FES) emerged with a distinct engineering orientation and the aim to use controlled electrical stimulation to restore function. Modern neuromodulation has developed since the 1970's, driven mainly by clinical professionals. The first peripheral nerve, then spinal cord and deep brain stimulators were introduced in the 1960's. In 1997, the Food and Drug Administration (FDA) approved deep brain stimulation (DBS) for the treatment of Parkinson's disease. An FES-based device that restored grasp was approved the same year.

In the 1970's, researchers developed primitive systems controlled by electrical activity recorded from the head. The U.S. Pentagon's Advanced Research Projects Agency (ARPA) supported research aimed at developing bionic systems for soldiers. Scientists demonstrated that recorded brain signals can communicate a user's intent in a reliable manner and found cells in the motor cortex the firing rates of which correlate with hand movements in two-dimensional space. Since the 1960's, engineers, neuroscientists, and physicists have constructed mathematical models of the retina that describe various aspects of its function, including light-stimulus processing and transduction. In addition, scientists have made

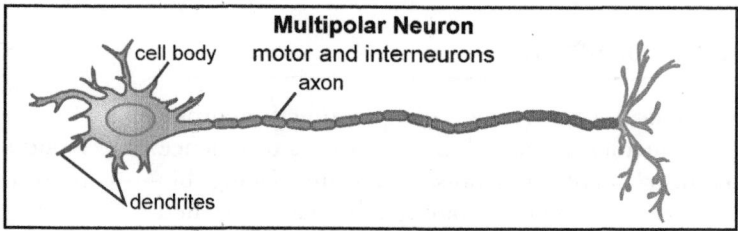

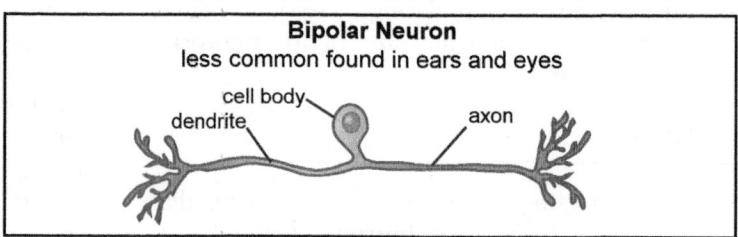

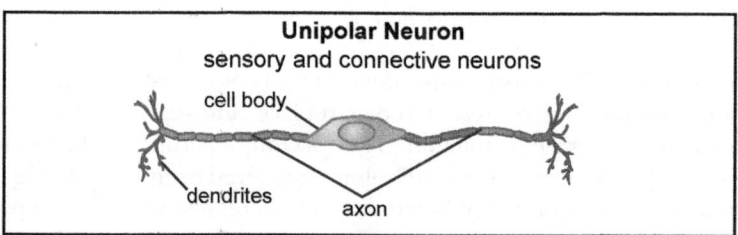

attempts to treat blindness using engineering solutions, such as nonbiological "visual prostheses." In 1975, the first multichannel cochlear implant (CI) was developed and implanted two years later.

NEUROMODULATION AND NEUROAUGMENTATION

Neural engineering applications have two broad (and sometimes overlapping) goals: neuromodulation and neuroaugmentation. Neuromodulation (altering nervous system function) employs stimulators and infusion devices, among other techniques. It can be applied at multiple levels: cortical, subcortical, spinal, or peripheral. Neural augmentation aims to amplify neural function and uses sensory (auditory, visual) and motor prostheses.

NEUROMUSCULAR STIMULATION

Based on a method that has remained unchanged for decades, electrodes are placed within the excitable tissue that provide current to activate certain pathways. This supplements or replaces lost motor or autonomic functions in patients with paralysis. An example is application of electrical pulses to peripheral motor nerves in patients with spinal cord injuries.

These pulses lead to action potentials that propagate across neuromuscular junctions and lead to muscle contraction. Coordinating the elicited muscle contractions ultimately reconstitutes function.

Neural Prosthetics

Neural prostheses (NP) aim to restore sensory or motor function—lost because of disease or trauma—by linking machines to the nervous system. By artificially manipulating the biological system using external electrical currents, neuroengineers try to mimic normal sensorimotor function. Electrodes act as transducers that excite neurons through electrical stimulation, or record (read) neural signals. In the first approach, stimulation is used for its therapeutic efficacy, for example, to alleviate the symptoms of Parkinson's disease, or to provide input to the nervous system, such as converting sound to neural input with a cochlear implant. The second paradigm uses recordings of neural activity to detect motor intention and provide input signal to an external device. This forms the basis of a subset of neural prosthetics called brain-controlled interfaces (BCI).

Microsystems

Miniaturization is a crucial part of designing instruments that interface efficiently with neural tissue and provide adequate resolution with minimal invasiveness. Microsystems technology integrates devices and systems at the microscopic and submicroscopic levels. It is derived from microelectronic batch-processing fabrication techniques. A "neural microsystem" is a hybrid system consisting of a microsystem and its interfacing neurons (be they cultured, part of brain slices, or in the intact nervous system). Technologies such as microelectrodes, microdialysis probes, fiber optic, and advanced magnetic materials are used. The properties of these systems render them suitable for simultaneous measurements of neuronal signals in different locations (to analyze neural network properties) as well as for implantation within the body.

Applications

Some of the most common applications of NE methods are described below.

Cochlear Implants. Cochlear implants (CI), by far the most successful sensory neural prostheses to date, have penetrated the mainstream therapeutic arsenal. Their popularity is rivaled only by the cardiac pacemakers and deep brain stimulation (DBS) systems. Implanted in patients with sensorineural deafness, these devices process sounds electronically and transmit stimuli to the cochlea. A CI includes several components: a microphone, a small speech processor that transforms sounds into a signal suitable for auditory neurons, a transmitter to relay the signal to the cochlea, a receiver that picks up the transmitted signal, and an electrode array implanted in the cochlea. Individual results vary, but achieving a high degree of accuracy in speech perception is possible, as is the development of language skills.

Retinal Bioengineering. Retinal photoreceptor cells contain visual pigment, which absorbs light and initiates the process of transducing it into electrical signals. They synapse onto other types of cells, which in turn carry the signals forward, eventually through the optic nerve and into the brain, where they are interpreted. Every neuron in the visual system has a "receptive field," a particular portion of the visual space within which light will influence that neuron's behavior. This is directly related to (and represented by) a specific region of the retina. Inherited retinal degenerations such as retinitis pigmentosa (RP) or age-related macular degeneration (AMD) are responsible for the compromised or nonexistent vision of millions of people. In these disorders, the retinal photoreceptor cells lose function and die, but the secondary neurons are spared.

Using an electronic prosthetic device, a signal is sent to these secondary neurons that ultimately causes an external visual image. A miniature video camera is mounted on the patient's eyeglasses that captures images and feeds them to a microprocessor, which converts them to an electronic signal. Then the signal is sent to an array of electrodes located on the retina's surface. The electrodes transmit the signal to the viable secondary neurons. The neurons process the signal and pass it down the optic nerve to the brain to establish the visual image.

Several different versions of this device exist and are implanted either into the retina or brain. Cortical visual prostheses could entirely bypass the retina, especially when this structure is damaged from diseases such as diabetes or glaucoma. Retinal prostheses, or artificial retinas (AR), could take advantage of any remaining functional cells and would target photoreceptor disorders such as RP. Two distinct retinal placements are used for AR. The first type slides under the retina (subretinal implant) and consists of small silicon-based disks bearing microphotodiodes. The second type would be an epiretinal system, which involves placing the camera or sensor outside the eye, sending signals to an intraocular receiver. In addition to challenges related to miniaturization and power supply, developing these systems faces obstacles pertaining to biocompatibility, such as retinal health and implant damage, and vascularization.

Functional Electrical Stimulation (FES). Some FES devices are commercialized, and others belong to clinical research settings. A typical unit includes an electronic stimulator, a feedback or control unit, leads, and electrodes. Electrical stimulators bear one or multiple channels (outputs) that are activated simultaneously or in sequence to produce the desired movement. Applications of FES include standing, ambulation, cycling, grasping, bowel and bladder control, male sexual assistance, and respiratory control. Although not curative, the method has numerous benefits, such as improved cardiovascular health, muscle-mass retention, and enhanced psychological well-being through increased functionality and independence.

Brain-Controlled Interfaces. A two-electrode device was implanted into a 1998 stroke victim who could communicate only by blinking his eyes. The device read from only a few neurons and allowed him to select letters and icons with his brain. A team of researchers helped a young patient with a spinal cord injury by implanting electrodes into his motor cortex that were connected to an interface. The patient was able to use the system to control a computer cursor and move objects using a robotic arm.

Brain-controlled interfaces (BCIs), a subset of NP, represent a new method of communication based on brain-generated neural activity. Still in an experimental phase, they offer hope to patients with severe motor dysfunction. These interfaces capture neural activity mediating a subject's intention to act and translate it into command signals transmitted to a computer (brain-computer interface) or robotic limb. Independent of peripheral nerves and muscles, BCI have the ability to restore communication and movement. This exciting technological advance is not only poised to help patients, but it also provides insight into the way neurons interact.

Every BCI has four main components: recording of electrical activity, extraction of the planned action from this activity, execution of the desired action using the prosthetic effector (actuator), and delivery of feedback (via sensation or prosthetic device). Brain-controlled interfaces rely on four main recording modalities: electroencephalography, electrocorticography, local field potentials, and singe-neuron action potentials. The methods are noninvasive, semiinvasive, or invasive, depending on where the transducer is placed: scalp, brain surface, or cortical tissue. The field is still in its infancy; however, several basic principles have emerged from these and other early experiments. A crucial requirement in BCI function, for example, is for the reading device to obtain sufficient information for a particular task. Another observation refers to the "transparency of action" in brain-machine interface (BMI) systems: Upon reaching proficiency, the action follows the thought, with no awareness of intermediate neural events.

Deep Brain Stimulation (DBS) and Other Modulation Methods. Deep brain stimulation of thalamic nuclei decreases tremors in patients with Parkinson's disease. It may alleviate depression, epilepsy, and other brain disorders. One or more thin electrodes, about 1 millimeter in diameter, are placed in the brain. An external signal generator with a power supply is also implanted somewhere in the body, typically in the chest cavity. An external remote control sends signals to the generator, varying the parameters of the stimulation, including the amount and frequency of the current and the duration and frequency of the pulses. The exact mechanism by which this method works is still unclear. It appears to exert its effect on axons and act in an inhibitory manner, by inducing an effect akin to ablation of target area, much like early Parkinson's treatment. One major advantage of DBS over other previously employed methods is its reversibility and absence of structural damage. Another valuable neuromodulatory approach, the electrical stimulation of the vagus nerve, can reduce seizure frequency in patients with epilepsy and alleviate treatment-resistant depression. Transcutaneous electrical nerve stimulation (TENS) represents the most common form of electrotherapy and is still in use for pain relief. Cranial electrotherapy stimulation involves passing small currents across the skull. The approach shows good results in depression, anxiety, and sleep disorders.

Transcranial magnetic stimulation uses the magnetic field produced by a current passing through a coil and can be applied for diagnostic (multiple sclerosis, stroke), therapeutic (depression), or research purposes.

FUTURE PROSPECTS

Bioelectrodes for neural recording and neurostimulation are an essential part of neuroprosthetic devices. Designing an optimal, stable electrode that records long-term and interacts adequately with neural tissue remains a priority for neural engineers. The implementation of microsystem technology opens new perspectives in the field. More than 200 million people around the world suffer from hearing loss, mainly because sensory hair cells in the cochlea have degenerated. The only efficient therapy for patients with profound hearing loss is the CI. Improvements in CI performance have increased the average sentence recognition with multichannel devices. An exciting new development, auditory brainstem implants, show improved performance in patients with impaired cochlear nerves. Millions of Americans have vision loss. The need for a reliable prosthetic retina is significant, and rivals the one for CI. Technological progress makes it quite likely that a functioning implant with a more sophisticated design and higher number of electrodes will be on the market soon. The epiretinal approach is promising, but providing interpretable visual information to the brain represents a challenge. In addition, even if they prove to be successful, retinal prostheses under development address only a limited number of visual disorders. Much is left to be discovered and tested in this field. The coming years will also see rapid gains in the area of BCI. Whether they achieve widespread use will depend on several factors, including performance, safety, cost, and improved quality of life. The advent of gene therapy, stem cell therapy, and other regenerative approaches offers new hope for patients and may complement prosthetic devices. However, many ethical and scientific issues still have to be solved. Implanted devices are changing the way neurological disorders are treated. An unprecedented transition of NE discoveries from the research to the commercial realm is taking place. At the same time, new discoveries constantly challenge the basic tenets of neuroscience and may alter the face of NE in the coming decades.

People's understanding of the nervous system, especially of the brain, changes, and so do the strategies designed to enhance and restore its function.

Mihaela Avramut, MD, PhD

FURTHER READING

Blume, Stuart. *The Artificial Ear: Cochlear Implants and the Culture of Deafness.* New Brunswick, N.J.: Rutgers University Press, 2010. Historical study of implant development and implementation.

DiLorenzo, Daniel J., and Joseph D. Bronzino, eds. *Neuroengineering.* Boca Raton, Fla.: CRC Press, 2008. Essential review of neuroengineering developments written by leaders in the field.

Durand, Dominique M. "What Is Neural Engineering?" *Journal of Neural Engineering* 4, no. 4 (September, 2006). Written by the editor in chief of the journal, who defines NE and its scope.

He, Bin, ed. *Neural Engineering.* New York: Kluwer Academic/Plenum Publishers, 2005. Introductory overview of research in neural engineering.

Katz, Bruce F. *Neuroengineering the Future: Virtual Minds and the Creation of Immortality.* Hingham, Mass.: Infinity Science Press, 2008. Fascinating introduction to this field, describing the state of the art and speculating on long-term developments.

Montaigne, Fen. *Medicine By Design: The Practice and Promise of Biomedical Engineering.* Baltimore: The Johns Hopkins University Press, 2006. Bioengineering (including neuroengineering) applications made accessible to the nonspecialist through vignettes and portraits of researchers.

WEB SITES

Engineering in Medicine and Biology Society http://www.embs.org/index.html

International Functional Electrical Stimulation Society (IFESS) http://ifess.org

National Institute of Biomedical Imaging and Bioengineering http://www.nibib.nih.gov

Whitaker International Fellows and Scholars Program http://www.whitaker.org

See also: Biomechanical Engineering; Bionics and Biomedical Engineering; Cell and Tissue Engineering; Neurology.

> **FASCINATING FACTS ABOUT NEURAL ENGINEERING**
>
> - Even though neuroengineering is still in its infancy, ethical questions are already arising. Will it affect human identity? Could it be used in the future to control thought processes? This is just the beginning.
> - Cochlear implants are a great achievement of modern medicine and represent the most successful of all neural prostheses developed to date.
> - Cell-containing polymer implants that release therapeutic factors hold promise for treating retinal disorders.
> - An exciting new development in antiepilepsy therapy, "closed-loop" devices record electroencephalograph (EEG) signals, process them to detect imminent seizures, and deliver stimuli to stop them.
> - The limb prostheses of the future will be equipped with multichanneled sensors that send tactile and proprioceptive feedback to the brain, continuously informing it about the effector's function. This approach will improve the patient's "sense of ownership" of the artificial limb.
> - Scientists developed neuroprostheses that restore urinary bladder function by stimulating the spinal cord or nerves controlling the lower urinary tract.
> - Advances in miniaturization and biosensors are expected to facilitate noninvasive monitoring of neuronal signaling and intracellular environment, thus greatly improving the diagnosis and treatment of nervous-system disorders.
> - In a quest to replace the conventional, inadequate brain stimulation methods, scientists developed neural cells that become active when exposed to light and implemented carbon nanotube-based stimulators.

NEUROLOGY

FIELDS OF STUDY

Internal medicine; neurosurgery; neuroscience; biology; molecular biology; biochemistry; pharmacology; neurophysiology; electrophysiology; neurobiology; neuroanatomy; neuroendocrinology; geriatric neurology; pediatric neurology; interventional neurology; veterinary neurology; psychology; behavioral neurology; cognitive neurology; neuroimaging; sleep medicine; movement disorders; neuromuscular disorders; cerebrovascular disorders; neurodegenerative disorders.

SUMMARY

Neurology is a rapidly developing branch of medicine dedicated to the diagnosis and treatment of disorders involving the brain, spinal cord, nerves, and muscles. The nervous system is crucial to human life. Understanding its operation under normal and pathological conditions is the focus of a group of disciplines that includes neurology, neuroscience, neurosurgery, and psychiatry. Considerable overlap and interaction exist between these specialties. Among the most important and challenging neurological disorders are neurodegenerative conditions (such as Parkinson's disease, Alzheimer's disease, and amyotrophic lateral sclerosis), stroke, brain tumors, epilepsy, migraine, and multiple sclerosis.

PRINCIPAL TERMS

- **anticholinergic:** agent that blocks the neurotransmitter acetylcholine.
- **ataxia:** lack of muscle coordination during voluntary movements.
- **basal ganglia:** group of nuclei deep within the cerebral hemispheres, functioning in movement coordination.
- **blood-brain barrier:** protective physical and functional barrier formed by a specialized brain capillary blood vessel structure; serves to keep the brain environment stable.
- **brainstem:** lower part of the brain, structurally continuous with the spinal cord, that plays important roles in consciousness, arousal, breathing, cardiovascular function, and digestion.
- **central nervous system:** main processing center of the nervous system, consisting of the brain and spinal cord.
- **cerebrospinal fluid:** clear liquid that surrounds the brain and spinal cord and fills their cavities.

The Lobes of the Brain

(Labels: frontal lobe, parietal lobe, occipital lobe, cerebellum, temporal lobe, brain stem)

The Inner Brain

(Labels: cerebral cortex, caudate nucleus, thalamus, amygdala, hypothalamus, pituitary gland, medula, hippocampus)

- **cortical:** pertaining to the cerebral cortex, the outer part of the cerebrum, which is responsible for high brain functions.
- **dopaminergic:** related to the neurotransmitter dopamine.
- **levodopa (L-dopa):** metabolic precursor of dopamine used in the treatment of parkinson's disease.
- **meninges:** membranes that cover the central nervous system.
- **myelin:** insulating lipid and protein sheath covering nerve fibers that increases the speed of impulse transmission.
- **peripheral nervous system:** part of the nervous system outside the brain and spinal cord, consisting of nerves that connect with target organs, muscles, blood vessels, and glands.
- **synapse:** specialized point of connection between two neurons or a neuron and an effector cell.

BASIC PRINCIPLES

Neurology is the medical discipline dedicated to the diagnosis, treatment, and management of diseases affecting the central nervous system, peripheral nerves, muscle effectors, and corresponding blood vessels. Its mission is thus distinct from that of other disciplines that focus on nervous system disorders, although some degree of overlap exists. Neuroscience studies the nervous system from a structural, functional, and molecular point of view, while neurosurgery is the surgical specialty involved in the treatment of nervous system disorders. Psychiatry deals with the diagnosis and treatment of mental illness. It is often taught that neurological diagnosis is based on answering two distinct questions: Where the lesion is located and what the nature of the disease is. The concept of localization is central to the practice of neurology. A disorder can affect the cerebral hemispheres, the brainstem, the spinal cord, the peripheral nerves, and the muscles. Malfunctions in these territories fall into several categories: motility disturbances, pain, disordered sensorium, seizures, altered consciousness, disturbed intellect or behavior, and changed speech and language. Alterations in mood or hormonal function are sometimes encountered. How the disorder manifests depends on the localization of the pathological process. Brainstem disease, for example, is suggested by cranial nerve palsies, ataxia of gait or limbs, or tremor. Involuntary movements often indicate that the basal ganglia might be affected. Identifying the precise nature of the disease process (for example, tumor, vascular malformation, neurodegeneration, infection) is facilitated by a variety of diagnostic tests.

BACKGROUND

The structure and function of the human brain have preoccupied physicians and philosophers alike since the dawn of history. Prominent Greek physicians such as Alcmaeon of Croton, Hippocrates, and Galen correctly considered the nervous system to be the source of sensations, emotions, and cognitive faculties. In the sixteenth and seventeenth centuries, the anatomy of the brain was described in detail by Flemish physician

Andreas Vesalius and English physician Thomas Willis. The term "neurology" was first used by Willis to describe the study of the brain. Despite these efforts, the function of the nervous system remained obscure until the early 1800's, when studies of functional localization were performed. Using a microscope, Czech anatomist Jan Evangelista Purkyně described brain cells that he called "neurons." French physician Pierre Paul Broca's studies of speech localization and French anatomist Jean Cruveilhier's analysis of stroke-induced lesions laid the foundations of modern neuroscience and neurology. English physician James Parkinson described the "shaking palsy" later known as Parkinson's disease. In the second half of the nineteenth century, clinicians with exceptional observational abilities such as French neurologists Jean-Martin Charcot and Joseph Babinski made crucial contributions to the advancement of clinical neurology.

In the early twentieth century, Spanish histologist Santiago Ramón y Cajal and Italian physician Camillo Golgi refined the description of neurons as separate cells, which led to an improved understanding of synapses and interneuronal communication. German psychiatrist Alois Alzheimer gave a lecture in 1906 in which he first described what became known as Alzheimer's disease. The complex processes occurring within a single neuron became accessible. Imaging techniques first introduced in the 1970's have been continuously improved and constitute essential diagnostic and research tools.

All these advances, and countless others, have shaped neurology into a discipline that is strongly anchored in scientific reasoning. Neurology remains a fast-growing medical field, rapidly incorporating advances in basic neuroscience, molecular genetics, and imaging technologies.

How It Works

Understanding of the pathogenetic mechanisms in neurological disorders has improved tremendously, mainly because of significant progress in imaging, electrophysiology, and molecular genetic techniques. The diagnostic, however, still relies heavily on clinical history and neurological examination. Ancillary tests are frequently employed.

Clinical History. It is important for neurologists to know as much as possible about their patients' backgrounds and socioeconomic status. The tempo and duration of the disease are essential. Some conditions, such as stroke, occur suddenly, while others, such as tumors or dementia, have a gradual onset. Many disorders manifest continuously; others are characterized by remissions and exacerbations (multiple sclerosis, myasthenia gravis) or bouts (migraine headaches).

Neurological Examination. Neurological exams target multiple areas and are performed in an organized, step-wise manner. First, the patient's vital signs and general appearance are evaluated, with special focus on posture, motor activity, and potential signs of meningeal involvement. Alertness, speech, and language are assessed, usually in conjunction with a Mini Mental Status Exam. The integrity of cranial nerves 1 through 12 is evaluated. Motor system examination includes an evaluation of muscle tone and strength and possible muscular atrophy. Reflexes, coordination, and gait are also tested. Sensory function analysis involves testing the senses of touch, pain, temperature, vibration, and position.

Imaging Studies. Computed tomography (CT) and magnetic resonance imaging (MRI) are the core neurological imaging methods. In CT, the image is reconstituted with high speed from sets of X-ray measurements. The method proves especially useful in evaluating acute strokes, head injuries, and acute infections, as well as in analyzing blood flow and identifying hemorrhages. It is also used for medically unstable or uncooperative patients and for patients with pacemakers or metallic implants, who cannot undergo an MRI. Magnetic resonance imaging is based on the ability of protons in water molecules to align to the direction of a strong magnetic field; on subsequent exposure to radio waves, the protons spin and emit signals that are detected by a receiver and processed by a computer. The technique plays an important role in detecting and delineating cerebral and spinal lesions. One advantage is the fact that it can be used when details on tissue physiology and biochemistry are needed.

Electrophysiological Studies. The arsenal of electrophysiological methods includes measures of brain electrical activity (such as electroencephalography and evoked potentials), electromyography, and nerve conduction studies. Electroencephalograms (EEGs) are noninvasive tests that display the electrical activity of the outer cortical layer neurons and serve as sensitive indicators of focal or generalized disturbances in neuronal activity. They can also reflect paroxysmal disturbances in neuronal function associated with seizure disorders. Evoked potentials (EPs) test the functional integrity of the sensory system and reflect the activity of the central nervous system in response to various stimuli. These measures of nervous system function are often important for differential diagnosis, especially in disorders that are not accompanied by morphological changes detectable with a CT or MRI scan.

Electromyograms (EMs) evaluate the electrical properties of the muscle. Nerve conduction studies (NCSs) assess the speed and strength of the electric impulse conduction along a peripheral nerve. The information derived from these studies allows the physician to establish if the patient's condition is of peripheral origin and whether the muscle or nerve is involved.

Lumbar Puncture and Cerebrospinal Fluid Studies. Studies of cerebrospinal fluid (CSF) for pressure, cellularity, pigments, protein, or glucose are undertaken after a thorough clinical evaluation and consideration of the value and risks of the lumbar puncture. They are especially valuable in the differential diagnosis of infections (such as meningitis and encephalitis), bleeding, stroke, malignancies, and demyelinating diseases. It is not undertaken when intracranial hypertension and brain herniation are suspected.

Other Diagnostic Tools. Additional techniques that can be employed for neurodiagnosis include ultrasonography, muscle and nerve biopsy, and DNA methods.

APPLICATIONS

Neurological assessment aims to render an accurate diagnosis, provide therapy (including pain management), and lead rehabilitative efforts. At present, several categories of nervous system disorders are at the forefront of clinical and research efforts worldwide.

Movement Disorders. Many diseases with various pathological features and manifestations are included in this category. Patients diagnosed with this type of disorder often have difficulty walking, speaking, and performing basic daily tasks. They may exhibit involuntary movements, tremors, rigidity, and muscle spasms. The cause of these manifestations is incompletely understood but often involves the degeneration of neurons in central nervous system areas responsible for motor function. Examples of diagnostic entities commonly encountered in movement disorder clinics are Parkinson's disease, Huntington's disease, dystonia, myoclonus, and Gilles de la Tourette's syndrome. Therapeutic choices vary for different conditions and include medications (such as dopaminergic and anticholinergic drugs) and surgery that severs neural pathways, restores neuronal population, or stimulates certain brain areas. Physiotherapy represents an important component of disease management. Parkinson's disease treatment in particular is the object of an enormous basic and translational research effort, stemming from the inadequacy of long-term levodopa (L-dopa) administration and the absence of a treatment that addresses the degenerative process. Stem cell and gene therapy are examples of novel therapeutic approaches under investigation for Parkinson's disease.

Vascular Diseases. Stroke ranks in the top four causes of death in many countries and accounts for a large proportion of the neurological disease burden. It is caused by a reduction in cerebral blood flow (ischemia, usually induced by a blood clot, or thrombus) or bleeding (hemorrhage). The main clinical characteristic is a sudden onset of clinical signs of abnormal cerebral function, such as weakness, difficulty walking, sensory loss, vision and speech disturbances, and cognitive impairment. Treatment is directed at minimizing and preventing blood vessel occlusion, using antithrombotic drugs and surgical procedures. Extensive research studies focus on possible neuroprotective agents that can minimize the effect of nutrient deprivation on brain cells. Rehabilitation is essential and requires a team that includes neurologists and physical therapists.

Dementias. These disorders are diagnosed based on the development of multiple cognitive deficits that cause impairment in occupational and social functioning. Various forms of dementia have been identified. They can be grouped into four major categories, according to their pathogenesis: degenerative, vascular, metabolic, and infectious disorders. The most frequently encountered degenerative dementia is named after Alzheimer, the German psychiatrist who first described its clinical features. Alzheimer's disease evolves progressively, with memory impairment (initially, for new information) and gradual cognitive loss, terminating in incapacitation and death. Neurological examination may show rigidity and postural changes, but primary motor and sensory functions are usually spared. Pathological changes in the brain include a striking atrophy, with loss of neurons; senile plaque microscopic lesions outside the cells, consisting of beta amyloid peptide; and tangle structures inside the neurons composed of filaments of a protein called tau. The exact causes and mechanisms of neuronal death need to be clarified and are under intense scrutiny by neuroscientists and neurologists all over the world. Medications that decrease the degradation of the neurotransmitter acetylcholine, thus increasing its levels, have been approved for treatment of this disorder. A drug that protects cells from the detrimental effects of the chemical messenger glutamate is used in intermediate to late stages. None of these agents, however, has disease-modifying effects. Alzheimer's disease remains a large part of the neuroscience and neurology research environment. Several strategies aimed at inhibiting disease progression have advanced to clinical trials. Among these, approaches targeting the production and clearance of the beta amyloid peptide are the most advanced. Attempts to modulate the abnormal aggregation of tau filaments and alleviate neuronal metabolic dysfunction are also being evaluated at the clinical level.

Demyelinating Diseases. Multiple sclerosis is a chronic disease characterized by multiple areas of inflammation in the white matter of the nervous system, associated with the loss of myelin sheath (demyelination) and scarring. It usually begins in young adults and has a relapsing and remitting course. Clinical manifestations are as variable as the distribution of the lesions; they often include unusual sensations, impaired vision, fatigue, muscle weakness, spasticity, ataxia, tremor, and bladder dysfunction. Cerebrospinal fluid examination, MRI, and evoked potentials are of great diagnostic value. The management of the condition is extremely challenging, and no cure has been developed. Treatment modalities include corticosteroids, beta interferons, glatimer acetate, mitoxantrone, and natalizumab. Symptomatic treatment and physical therapy are an important part of disease management. The use of transplanted stem cells that release immunomodulatory and neuroprotective factors is being explored in clinical trials.

Spinal Cord Diseases. Amyotrophic lateral sclerosis (ALS), a disease of unknown pathogenesis, is characterized by progressive muscle weakness and atrophy because of the selective degeneration of motor neurons and pathways in the brain and spinal cord. Muscle cramps and weight loss are characteristic symptoms. Breathing is compromised because of paresis of respiratory muscles. Treatment is symptomatic, as no effective drug treatment exists. In the twenty-first century, ALS research has seen a dramatic expansion, comparable to that of Parkinson's disease and dementia. Studies that hold significant promise are investigating mechanisms of disease and testing growth factors and chemical messenger receptor antagonists, as well as gene and stem cell therapies.

Paroxysmal Disorders. Migraines are paroxystic, intense hemicranial pain episodes, accompanied by vomiting and light sensitivity. The cause of these episodes is still unclear. Acute treatment of migraine employs drugs such as sumatriptan and ergotamine, while prophylactic therapy uses various agents such as propranolol, verapamil, amitriptyline, and valproate. Epilepsy is a chronic condition characterized by recurrent seizures that are typically unprovoked and seldom predictable. It affects more than 40 million people worldwide. Seizures are the result of a temporary perturbation in the function of cortical neurons that causes a self-limited, hypersynchronous electrical discharge. The electrophysiological disturbance can be detected on a scalp EEG recoding. Several distinct forms of epilepsy (such as partial, generalized tonic-clonic, myoclonic, and absence) have been identified, each of which has its own behavioral changes, EEG activity, natural history, and response to treatment. Therapy is aimed at eliminating or reducing seizures, avoiding side effects of long-term treatment, and restoring the patient's psychosocial function. Many antiepileptic drugs are in use; phenytoin, carbamazepine, phenobarbital, and valproate are some of the mainstays. Surgical treatment becomes necessary when seizures cannot be controlled pharmacologically, and they affect the patient's quality of life.

Other Disease Categories. Tumors, infections, genetic and developmental diseases, birth injuries, trauma, peripheral neuropathies, neuromuscular diseases, and myopathies are also the focus of neurological assessment.

FUTURE PROSPECTS

Neurology is expected to be one of the most rapidly advancing disciplines of the twenty-first century. Its realm is expanding to encompass new areas of genetics, metabolism, sleep disorders, vascular diseases, and neuroimmunology. With an improved understanding of the nervous system's function in health and disease, the pressure to correctly diagnose neurologic disease increases. The neurologist may function as a consultant or as principal physician for patients with primary nervous system disorders. These diseases represent an important area of medicine because of their high prevalence, disabling outcomes, and high costs for health systems. The World Health Organization's Global Burden of Disease report shows that while neurological and mental disorders are responsible for about 1 percent of deaths, they account for almost 11 percent of disease burden the world over. As life expectancy increases and the general population ages, it is likely that the incidence of age-related neurological disorders will increase. The World Health Organization is implementing programs aimed at prevention, diagnosis, and treatment of neurological disorders that are of public health importance. These include epilepsy, headache, dementia, multiple sclerosis, Parkinson's disease, stroke, pain syndromes, and brain injury. It is vital to ensure that an appropriate range of care is made available to all people with neurological disorders in every country of the world. Identifying biochemical markers of disease for early detection implementing preventative measures are fast becoming central goals of worldwide research efforts. Scientists are investigating the use of agents that open the blood-brain barrier to allow drug penetration, as well as various alternative routes of drug delivery. They are also testing novel therapies based on gene replacement, growth factors, and stem cells that have the potential to promote neuronal repair and regeneration.

Mihaela Avramut, MD, PhD

Further Reading

Evans, Randolph W., ed. *Common Neurological Disorders.* Philadelphia: Saunders, 2009. Describes the most common nervous system disorders and their treatment.

Martin, Joseph B. "The Integration of Neurology, Psychiatry, and Neuroscience in the Twenty-first Century." *American Journal of Psychiatry* 159 (May, 2002): 695-704. Examines the historical basis for the divergence of neurology and psychiatry and discusses prospects for a potential convergence.

Mumenthaler, Marco, and Heinrich Mattle. *Fundamentals of Neurology: An Illustrated Guide.* New York: Thieme, 2006. A well-written, logically ordered textbook for the novice in neuroscience and neurology.

Rowland, Lewis P., Thomas A. Pedley, and H. Houston Merritt, eds. *Merritt's Neurology.* 12th ed. Philadelphia: Lippincott Williams & Wilkins, 2010. Provides the essential facts about common and rare diseases. Includes chapters on neurological symptoms, diagnostic tests, and rehabilitation.

Sacks, Oliver. *Awakenings.* 1973. Reprint. New York: Vintage Books, 1999. Exceptional case history and literary work, describing the outcome of treating encephalitis lethargica patients institutionalized since World War I.

_____. *The Man Who Mistook His Wife for a Hat.* 1970. Reprint. New York: Simon and Schuster, 2007. Sacks explores fascinating cases of excesses and losses in neurological conditions.

Web Sites

American Academy of Neurology http://www.aan.com

FASCINATING FACTS ABOUT NEUROLOGY

- The human brain consists of about 200 billion neurons and many more supportive cells.

- The adult brain weighs around 1,500 grams and represents only 2 percent of total body weight; however, its oxygen supply needs (72 liters per day) account for 20 percent of the body's total oxygen consumption.

- Synesthesia is a neurological phenomenon in which a stimulus directed at a certain sense modality triggers sensations in a different sense territory. Individuals with this synesthesia can, for example, see yellow or smell cinnamon when they hear a G-sharp note.

- Locked-in syndrome sufferers have intact cognitive function and can see and hear; however, they cannot move, speak, or otherwise communicate except through coded eye movements.

- Prions are nonbacterial, nonviral agents that do not contain genetic material. They are responsible for some of the most intriguing disease entities in neurology, such as mad cow disease, Creutzfeldt-Jakob disease, fatal insomnia, and kuru, which affects tribe members in Papua New Guinea who eat the tissue of affected people during cannibalistic funeral rituals.

- In 1969, neurologist Oliver Sacks administered L-dopa (then a new miracle drug) to his institutionalized, lethargic patients suffering from Parkinson's disease symptoms after the mysterious World War I encephalitis epidemic. The drug gave most of these patients a spectacular, sudden awakening and allowed them to temporarily experience active life.

- Interventional neurology, also known as neuroendovascular therapy, is a neurology subspecialty that deals with blood vessel lesions of the brain and spine in a minimally invasive manner. A microscopic catheter is inserted through a nick in the leg, then advanced via the femoral artery to the site of the lesion.

National Institute of Mental Health http://www.nimh.nih.gov

National Institute of Neurological Disorders and Stroke http://www.ninds.nih.gov

Neurotechnology Industry Organization http://www.neurotechindustry.org

World Health Organization Neurology http://www.who.int/topics/neurology/en

See also: Epidemiology; Geriatrics and Gerontology; Pathology; Stem Cell Research and Technology

NOSES

FIELDS OF STUDY

Anatomy, biochemistry, biophysics, cell biology, developmental biology, genetics, herpetology, histology, immunology, neurobiology, ornithology, pathology, physiology, zoology

SUMMARY

The nose is the major access channel for the sense of smell. The nose signals danger by detecting the smell of spoiled food, smoke, or natural gas. The nose also warms and humidifies the air as it moves onto lungs. It filters out particles and bacteria found in the inspired air and so prevents these particles from entering the lungs.

PRINCIPAL TERMS

- **mucus:** the watery material covering the internal nasal structures that aids in humidification, warming, and particle filtration
- **olfaction:** the sense of smell
- **septum:** the bony structure that divides the nose into two sections
- **turbinates:** bony structures that define the internal nasal anatomy

BASIC PRINCIPLES

In vertebrates, the back of the throat is connected to the outside air through a passageway called the nose. The outside opening of the nose is referred to as the external naris, whereas the opening from the nose to the back of the throat is called the internal naris. In reptiles and amphibians, air drawn into the lungs passes through the external naris and into a tubelike structure which is the simplest form of a nose. In animals like the salamander, the nasal air passageway is a straight tube. Air exits this tube via the internal naris where it proceeds into the back of the throat. In some amphibians, such as the bullfrog, the channel between the external and internal naris has a bony bump on the floor of the nasal passageway. This bump, called the ementia, apparently functions like a baffle plate, so that the incoming air stream is deflected. As a result of this deflection, air moving through the nose is turbulent. This turbulence likely increases the ability to detect smells as well as improves the efficiency of the other nasal functions.

In mammals, the internal anatomy of the nose is defined by bony structures called turbinates. These turbinates produce multiple and convoluted air flow paths for the inspired air. As in the frog, these convoluted flow paths facilitate the various nasal functions.

In mammals the nose is divided into two halves along the midline by a bony structure called the nasal septum. In some mammals, such as the rat, the septum is incomplete, and so there is some mixing in the nose of air that comes into the left and right nostrils.

Air entering the nose travels in the airspace between the turbinates and the nasal septum. The surface of these nasal structures has a very rich blood supply, and so blood flow to these areas can be quickly changed. By changing the amount of blood flow to the nasal structure, the diameter of the nasal airspace itself can be quickly changed. Because of the speed and amount of diameter change that can occur in these areas they are called nasal "swell spaces."

HEAT, HUMIDITY, AND PURIFICATION

When cold air enters the nose, it causes the blood flow to the swell spaces to increase dramatically. This causes a swelling of the nasal tissue and so reduces the size of the airspace through which the incoming air must travel. Because the air passageways are now more narrow, more heat can be transferred from the blood stream to the incoming air. Thus, the cold air is effectively warmed before it enters the lungs. When air that is warmer than the body temperature enters the nose the reverse happens, and the nasal air passageways are made wider.

The material covering the turbinates and nasal septum is called the mucus. This mucus layer is mostly composed of water and serves to humidify the incoming air. When the air is dry, water is evaporated from the mucus into the inspired air. When the air is dry, a considerable quantity of water can be lost through the nose. For animals living in dry, desert conditions, the nasal humidification process is critical because it is necessary to save every drop of water while still humidifying the incoming air. When these animals take air into the lungs, the inspired air passes through the extensive turbinate structures of the nose where it is humidified by the evaporation of water from the nasal mucus. As evaporation takes place, the surface of the mucus is cooled. When it is time to discharge the air from the lungs, the expired air passes over the cooled surface of the mucus. As a result, water in the inspired air is now picked up by the cooled surface and so it is not lost in the expired air. This type of water conservation method has been seen in kangaroo rats living in the deserts of Australia and in certain birds, such as the cactus wren, which inhabit warm, dry areas.

In addition to supplying water for humidification purposes, the nasal mucus serves as a trap for particulate matter. Smoke and dust particles and even airborne bacteria are trapped in the mucus of the nose. Beneath the nasal mucus is a layer of cells that contain hairlike protrusions called cilia. As the cilia of these cells beat, they create a wavelike action in the mucus. The mucus is thus moved through the nose and to the back of the throat. Once in the throat the mucus is swallowed and the particles and bacteria are dealt with in the stomach. The movement of the nasal mucus is an ongoing process and so trapped particles are continually being removed from the nose. There are also white blood cells and enzymes found in the mucus that destroy bacteria.

When particles (and sometimes even air) touch parts of the nose a sneeze occurs. In this case, breathing is stopped and air is forcibly expelled from thelungs at a high flow rate. The sneeze is an attempt to expel the particles frominside the nose. However, when the odorant is very foul or stings the nose, breathing may be temporarily stopped. This protects the lungs from potentially damaging chemicals. The detection of potentially harmful chemicals occurs both through the smell receptors as well as through pain receptors that are found in the nose.

—*David E. Hornung*

See also: Anatomy; Bone and cartilage; Brain; Circulatory systems of vertebrates; Digestive tract; Eyes; Immune system; Reproductive system of female mammals; Reproductive system of male mammals; Respiratory system; Skin; Smell.

FURTHER READING

Association for Chemoreception Sciences. http://www.achems.org. ThisWeb site contains a description of current work in the field as well as a discussion of smell disorders. Getchell, T. V., R. L. Doty, L.M. Bartoshuck, and J. B. Snow, eds. *Smell and Taste in Health and Disease.* New York: Raven Press, 1991. Discussion of the clinical aspects of smell problems.

Gibbons, Byron. "The Intimate Sense of Smell." *National Geographic* 170, no. 3 (1986): 321-361. An excellent overview of the anatomy, physiology, and psychology of the sense of smell. The making of perfumes, use of dogs for tracking, and the history of smell are all well covered.

Vroon, Piet. *Smell: The Secret Seducer.* New York: Farrar, Straus and Giroux, 1997. Acultural history and compendium of odd facts about noses.

NUTRIENT REQUIREMENTS

FIELDS OF STUDY

Biochemistry, physiology

SUMMARY

All animals depend on external sources of nutritional raw materials for energy, growth, maintenance, and functioning. Food, used to provide material for production of new tissue and the repair of old tissue as well as used as an energy source, is obtained from a variety of plant, animal, and inorganic sources. Regardless of the source, food must provide its consumer with a sufficient amount of the essential nutrients. A nutrient is any substance that serves as a source of metabolic energy, raw material for growth and repair of tissues, or general maintenance of body functions.

PRINCIPAL TERMS

- **carbohydrate:** an organic molecule containing only carbon, hydrogen, and oxygen in a 1:2:1 ratio; often defined as a simple sugar or any substance yielding a simple sugar upon hydrolysis
- **lipid:** an organic molecule, such as a fat or oil, composed of carbon, hydrogen, oxygen, and sometimes phosphorus, that is nonpolar and insoluble in water
- **mineral:** one of the many inorganic elements other than carbon, hydrogen, oxygen, and nitrogen that an organism requires for proper body function
- **protein:** an organic molecule containing carbon, hydrogen, oxygen, nitrogen, and sulfur and composed of large polypeptides in which over a hundred amino acids are linked together
- **vitamin:** an organic nutrient that an organism requires in very small amounts and which generally functions as a coenzyme

GENERAL NUTRITIONAL REQUIREMENTS

Animals differ widely in their specific nutritional needs, depending on the species. Within any given species, those needs may vary according to variations in body size and composition, age, sex, activity, genetic makeup, and reproductive functions.

A small animal requires more food for energy per gram of body weight than does a larger animal, because the metabolic rate per unit of body weight is higher in the smaller animal. Likewise, an animal with a cool body temperature will have less energy needs and require less food than an animal with a high body temperature. An egg-producing or pregnant female

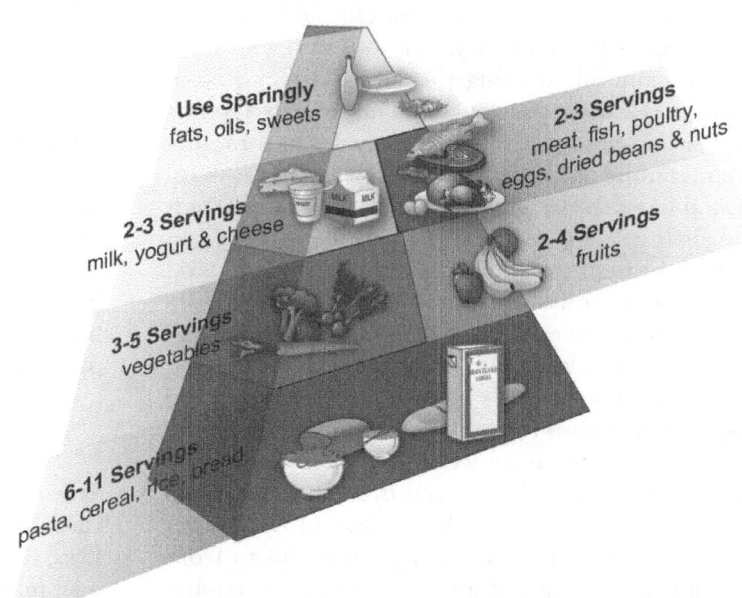

Daily Food Pyramid

will require more nutrients than a male. In order for an animal to be in a balanced nutritional state, it must consume food that will supply enough energy to supply power to all body processes, sufficient protein and amino acids to maintain a positive nitrogen balance and avoid a net loss of body protein, enough water and minerals to compensate for losses or incorporation, and those essential vitamins that are not synthesized within the body.

Activities such as walking, swimming, digesting food, or any other activity performed by an animal require fuel in the form of chemical energy. Adenosine triphosphate (ATP), the body's energy currency, is produced by the cellular oxidation of small molecules, such as sugars obtained from food. Cells usually metabolize carbohydrates or fats as fuel sources; however, when these carbon sources are in short supply, cells will utilize proteins. The energy content of food is usually measured in kilocalories, and it should be noted that the term "calories" listed on food labels is actually kilocalories (1 kilocalorie = 1000 calories). Cellular metabolism must continually produce energy to maintain the processes required for an animal to remain alive. Processes such as the circulation of blood, breathing, removing waste products from the blood, and in birds and mammals, the maintenance of body temperature, all require energy. The calories required to fuel these essential processes for a given amount of time in an animal at rest is called the basal metabolic rate (BMR). For a resting human adult, the BMR averages from thirteen hundred to eighteen hundred kilocalories per day. As physical activity increases, the BMR increases.

Energy balance requires that the number of calories consumed for body maintenance and repair and for work (metabolic and otherwise) plus the production of body heat in birds and mammals be equal to the caloric intake over a period of time. An insufficient intake of calories can be temporarily balanced by the utilization of storage fats, carbohydrates, or even protein, and will result in a loss of body weight. On the other hand, an excessive intake of calories can lead to the storage of energy sources. Animals normally store glycogen, but when the glycogen stores are full, food molecules, such as carbohydrates and protein, will be converted to fats.

NUTRIENT MOLECULES

Proteins are composed of long chains of amino acids and serve a number of important functions in all living organisms, but they are primarily used as structural components of soft tissues and as enzymes. Proteins can also be utilized as energy sources if they are broken down into amino acids. Animal tissues are composed of about twenty different amino acids. The ability to synthesize amino acids from other carbon sources, such as carbohydrates, varies among species, but few, if any, animal species can synthesize all twenty required amino acids. Those amino acids that cannot be synthesized by an animal, but are required for the synthesis of essential amino acids, are the so-called essential amino acids, and must be included in the diet. Humans, for example, require nine essential amino acids. Both plant and animal tissues can serve as protein sources, but animal protein generally contains larger quantities of the essential amino acids.

Carbohydrates are primarily used as immediate sources of chemical energy, but they can also be converted to metabolic intermediates or fats. Some carbohydrates are also structural components of larger molecules. For example, the nucleic acids deoxyribonucleic acid (DNA) and ribonucleic acid (RNA) contain the sugars deoxyribose and ribose, respectively, as an integral component of their structure. Most animals can also convert proteins and fats into carbohy- drates. The principle sources of carbohydrates are the sugars, starches, and cellulose in plants and the glycogen stored in animal tissue.

Lipids are an important and essential component of all biological membranes. In addition, several animal hormones, such as the sex hormones, are lipoidal in nature. Fats and lipids are also especially suitable as concentrated energy reserves, because each gram of fat supplies twice as much energy as a gram of carbohydrate or protein and does not have to be dissolved in water. Hence, animals commonly store fat for times of caloric deficit when energy expenditure exceeds energy uptake. Some animals, such as migratory birds and hibernating mammals, store large quantities of fat to offset the times that they are not actively feeding. Lipid molecules include fatty acids, monoglycerides, triglycerides, sterols, and phospholipids.

All animals require an adequate supply of essential inorganic minerals. Carbonate salts of the metals calcium, potassium, sodium, and magnesium as well as some chloride, sulfate, and phosphate are important constituents of intra- and extracellular fluids. Calcium phosphate is present as hydroxyapatite, a crystalline material that gives hardness and rigidity to the bones of vertebrates and the shells of mollusks. Certain metals, such as copper and iron, are required for oxidation-reduction reactions and for oxygen binding and transport. The catalytic function of many enzymes requires the presence of certain metal atoms. Animals require moderate amounts of some minerals and only trace quantities of others.

Animals require a variety of vitamins, diverse and chemically unrelated organic substances. Vitamins primarily function as coenzymes for the proper catalytic activity of essential enzymes. As with amino acids, the ability to synthesize different vitamins from other carbon sources varies among species. Those essential vitamins that cannot be synthesized by the animal itself must be obtained from other sources, primarily from plants but also from dietary animal flesh or from intestinal microorganisms. Vitamin C (ascorbic acid) can be synthesized by many animals, but not by humans. Vitamins K and B_1 are produced by intestinal bacteria in humans. Vitamins such as A, D, E, and K are fat soluble and can be stored in fat deposits within the body; however, water soluble vitamins such as vitamin C are not stored and are excreted through the urine. Hence, the water soluble vitamins must be consumed or produced continually in order to maintain adequate levels.

Although not commonly thought of as a nutrient, water is tremendously important and comprises up to 95 percent or more of the weight of some animal tissue. Water is replaced in most animals by drinking, ingestion with food, and to some extent, by the metabolism of carbohydrates and lipids.

—D. R. Gossett

See also: Biology; Cannibalism; Digestion; Digestive tract; Osmoregulation; pH maintenance; Thermoregulation.

FURTHER READING
Campbell, N. A., L. G. Mitchell, and J. B. Reece. *Biology: Concepts and Connections.* 3d ed. San Francisco: Benjamin/Cummings, 2000. An outstanding introductory biology text that gives a clear concise description of nutrient requirements.

FEEDING MECHANISMS

Since most animals cannot absorb nutrients directly from their environments, they must exert energy in order to obtain food. Although animals show tremendous diversity in their methods of obtaining food, most feeding habits can be classified in one of three different types. Many animals, particularly those living in aquatic environments, feed on particulate matter, the free-floating material made up of plankton, microscopic aquatic life-forms, and organic remains of dead and decaying plants and animals. Some of the animals utilize a technique referred to as suspension feeding, in which the food material is drawn into the digestive tract by currents created by external structures, such as cilia or setae. Other particulate feeders feed on deposits of detritus (decaying organic matter) that accumulate at the bottom of lakes or oceans. A second feeding method involves the consumption of food masses. Most of these animals utilize specialized adaptions that allow them to capture and manipulate solid food. Herbivores have special adaptions for cutting and crushing or grinding plant tissues, while carnivores such as predators must have the ability to capture, hold, and either swallow the prey whole or shred it into smaller pieces. Feeding on liquids is a third type of feeding habit. This method, which involves the sucking of fluids from a host plant or animal, is utilized primarily by parasites, but it can also be observed in certain insects.

FUNCTIONS OF ESSENTIAL MINERALS AND VITAMINS

Minerals

Calcium: nerve and muscle function, blood clotting, bone and tooth formation

Chlorine: acid-base balance, gastric juice

Chromium: associated with glucose and energy metabolism

Cobalt: component of Vitamin B_{12}

Fluorine: maintenance of tooth structure

Iodine: component of thyroxine, a thyroid hormone

Iron: component of essential enzymes and electron carriers in energy metabolism

Phosphorus: transfer of chemical energy, bone and tooth formation, nucleic acid synthesis

Potassium: proper nerve function, acid-base balance, water balance

Selenium: component of essential enzymes and functions in association with vitamin E

Sodium: proper nerve function, water balance, acid-base balance

Sulfur: component of certain amino acids

Copper, Magnesium, Manganese, Molybdenum, Zinc: components of essential enzymes

Vitamins

Vitamin A (Carotene): formation of visual pigments, maintenance of certain membranes

Vitamin B_1 (Thiamine), Vitamin B_2 (Riboflavin), Niacin: coenzymes for certain enzymes in energy metabolism

Vitamin B_6 (Pyridoxine): coenzyme for amino acid synthesis and fatty acid metabolism

Vitamin B_{12} (Cyanocobalamin): required for nucleoprotein synthesis Vitamin C (Ascorbic acid): vital to collagen formation, serves as an important antioxidant

Vitamin D (Calciferol): increases calcium absorption from gut, bone and tooth formation

Vitamin E (Tocopherol): maintains pregnancy in mammals; serves as an important antioxidant

Vitamin K (Naphthoquinone): required for synthesis of a protein necessary for blood clotting

Biotin: coenzyme for enzymes associated with protein synthesis

Folic acid: required for nucleoprotein synthesis and formation of red blood cells Pantothenic acid: forms part of coenzyme A associated with energy metabolism

Carr, D. E. *The Deadly Feast of Life*. Garden City, N.Y.: Doubleday, 1971. A keen insight on what and how animals eat.

Fox, I. S. *Human Physiology*. 6th ed. Boston: WCB/McGraw-Hill, 1999. An excellent treatment of the physiology of nutrition in humans, but applicable to other mammals.

Jennings, J. B. *Feeding, Digestion, and Assimilation in Animals*. 2d ed. New York: St. Martin's Press, 1972. An excellent comparative approach to nutrition among animals. Randall, David, Warren Burggren, Kathleen French, and Russell Fernald. *Animal Physiology: Mechanisms and Adaptations*. 4th ed. New York: W. H. Freeman, 1997. An advanced text that gives an excellent description of the nutrient requirement in animals.

Weindrach, R. "Caloric Restriction and Aging." *Scientific American* 274 (1996): 46-52. This article gives a very good discussion of how organisms from protists to mammals live longer on well-balanced but low-calorie diets.

Obstetrics and Gynecology

FIELDS OF STUDY

Women's health care; maternal-fetal medicine; reproductive endocrinology; genetic counseling; genetics; infertility; perinatology; postpartum care; prenatal care; urogynecology.

SUMMARY

Obstetrics and gynecology is a medical specialty focused on women's health care. Obstetrics entails the care of a woman and her developing fetus from the moment of conception, through delivery, and following delivery (postpartum care). Gynecology encompasses the evaluation and treatment (both medical and surgical) of the female reproductive system (uterus, Fallopian tubes, ovaries, vagina, and external genitalia).

PRINCIPAL TERMS

- **Cesarean section:** surgical removal of a full-term infant from the uterus.
- **hysterectomy:** surgical removal of the uterus (womb).
- **hysteroscopy:** insertion of viewing device (hysteroscope) via the cervix (opening of the uterus) for visualizing and treating problems within the uterus.
- **in vitro fertilization:** fertilization of an ovum (egg) in a laboratory for later introduction into the uterus for development.
- **laparoscopy:** surgical procedure that involves introduction of surgical instruments and a viewing device (laparoscope) into the abdomen to accomplish surgical procedures.
- **maternal-fetal medicine:** subspecialty of obstetrics and gynecology that focuses on high-risk pregnancies.
- **Pap smear:** scraping of cells from the cervix to check for abnormalities such as cancerous and precancerous conditions; also known as a Papanicolaou smear.
- **reproductive endocrinology:** subspecialty of obstetrics and gynecology that focuses on the surgical and medical treatment of women with fertility problems.
- **vaginal delivery:** delivery of an infant via the vaginal route; a normal delivery.

BASIC PRINCIPLES

Obstetrics and gynecology is a medical specialty limited to women's health care. Many obstetriciangynecologists are generalists and provide both obstetrical and gynecologic care; however, some limit their practice to either obstetrics or gynecology. Some focus on problems of the menopause, and some focus on gynecologic care of children and adolescents. Some obstetrician-gynecologists receive additional training in a specific area of the field. The following subspecialties are recognized: gynecologic oncology, the medical and surgical treatment of cancers of the female genital tract (the ovaries, uterus, cervix, vagina, and external genitalia); reproductive endocrinology, the treatment of infertility in women; maternal-fetal medicine, the treatment of high-risk pregnancies); and urogynecology, the medical and surgical treatment of problems of the female urinary tract. Generalists often refer their more complicated and challenging cases to these subspecialists and, in many cases, comanage the patients' care.

As with other medical specialties, obstetrics and gynecology has become more sophisticated and technologically advanced. For example, infertility specialists are able to help women with certain conditions have children, although fifty years ago, women with those same conditions would not have been able to give birth. Endoscopy, the visualization of internal structures with a small viewing device, has enabled physicians to better understand their patients' conditions and provide appropriate treatment. Although a breast examination is a common part of the gynecologic exam, treatment of breast disease is usually referred to surgeons who specialize in breast disease.

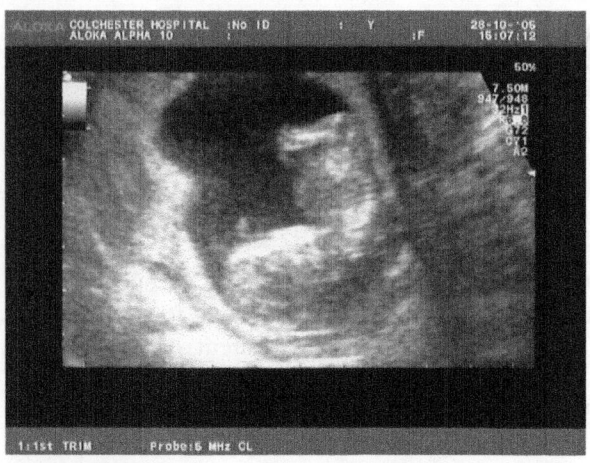
A dating scan at 12 weeks.

BACKGROUND
In Western civilization, obstetrics was first practiced by female midwives in the seventeenth century. In colonial America, about one in eight women would die in childbirth, and of every ten infants born, between one and three would die before reaching the age of five. Midwives would call male surgeons to the birthing room if they determined that the infant had died while in the womb. The surgeon's task was to use instruments to reduce the size of the baby's skull and extract the body. Toward the end of the seventeenth century, doctors began to perform cesarean sections. In the eighteenth century, forceps were developed to assist in the delivery process. In the nineteenth century, painkillers began to be used during childbirth, and in the 1940's and 1950's, the use of anesthesia increased to the point that women were often rendered semiconscious during the labor process. This excessive use of sedation produced a backlash, and in the 1970's, natural childbirth became popular. Advocates argued that childbirth was a natural process, not a disease needing treatment. Women began to use a variety of techniques to avoid using medication during labor and sought alternatives to hospital births, including using midwives rather than obstetricians.

In the nineteenth century, the direct viewing of a woman's genitals by a physician was regarded as immoral and immodest, and physicians had to rely on palpitation of the area. This made it difficult to diagnose and treat gynecologic problems. Gynecologic surgery, as well as surgery in general, did not develop until after the introduction of anesthesia in the nineteenth century. In the mid-1800's, J. Marion Sims developed a surgical treatment for vesicovaginal fistula, a complication of childbirth that caused urine leakage and discomfort, and he became known as the father of gynecology. Because Sims worked on slave women, he was able to directly observe their genitals. In 1869, Commander D. C. Pantaleoni performed the first operative hysteroscopy, paving the way for modern endoscopy (hysteroscopy and laparoscopy). In 1978, English gynecologist Patrick Steptoe announced the birth of a child through in vitro fertilization.

HOW IT WORKS
Some family practice doctors, typically in rural areas with a low population density, offer obstetrical services, but in urban areas, obstetricians generally handle pregnancy care. An obstetrician cares for a woman and her fetus/infant throughout pregnancy (antepartum), during labor (intrapartum), and following delivery (postpartum). An obstetrician (or obstetrical group) may also be associated with nurse practitioners and sometimes nurse midwives. Obstetricians confirm a woman's pregnancy, then set up a schedule of visits, ranging from once a week to once a month, depending on the stage of pregnancy and whether complications are present.

During a prenatal visit, the obstetrician listens for fetal heart tones, records the woman's weight and blood pressure, and checks her urine for abnormalities such as the presence of protein or sugar. Protein in the urine, accompanied by elevated blood pressure and increased reflexes, may indicate that the woman has toxemia (preeclampsia), which is life-threatening to both the woman and her fetus. The presence of sugar may indicate that the woman has developed gestational diabetes, which poses a threat to her fetus. Both conditions are treatable but require extra monitoring of the woman and fetus.

Blood is drawn to check for anemia, screen for birth defects, and determine the blood type and Rh, or rhesus, factor. If a woman is Rh negative and her fetus is Rh positive, the risk of an immune reaction is present. This incompatibility may not cause problems during the woman's first pregnancy, but if a subsequent fetus is Rh positive, it could develop Rh disease, a life-threatening condition that could cause anemia or other serious problems. An obstetrician can also provide invaluable information on many aspects of pregnancy, such as nutrition, foods or medication to avoid, and exercise guidelines. Some family practitioners and internists, especially those in rural areas, offer gynecologic services, although in urban areas, women often visit gynecologists for these services.

Many women turn to gynecologists not only for problems with their reproductive system but also for other medical conditions; thus, their gynecologists function as primary care or family physicians. Many gynecologists also function as obstetricians and will continue to care for their patients if they become pregnant.

A gynecologic visit entails a breast examination and a pelvic examination. During the pelvic examination, a gynecologist inserts a speculum (a twobladed instrument) into the vagina to visualize the cervix and takes a Pap smear, which involves scraping some superficial cells from the cervix to check for cancerous or precancerous conditions. The gynecologist next places one hand in the vagina and the other on the lower abdomen to palpate (feel) the internal pelvic organs (uterus, Fallopian tubes, and ovaries). Unless a woman is unusually tense, the examination is relatively painless. The visit also includes a gynecologic history, which includes asking about any problems or concerns the patient may have in regard to her reproductive system. A woman may also be advised regarding birth control (contraception) or pregnancy.

APPLICATIONS

Obstetrics and gynecology have developed markedly since the 1960's. Technological advances such as ultrasound and endoscopy have dramatically changed patient care. New medications have been developed to both prevent and increase the likelihood of pregnancy, as well as to treat numerous diseases and conditions. The understanding of disease processes has also grown. Old techniques have been replaced by more appropriate and beneficial ones.

Obstetrics. Most pregnancies progress uneventfully through delivery; however, complications can arise suddenly. The majority of women deliver vaginally; some of these deliveries are assisted by the use of forceps or a vacuum extractor (a suction device applied to the fetal head). Some deliveries are conducted via a cesarean section (operative delivery) because of problems that arise during labor or the intrapartum period. Common reasons for cesarean sections include lack of progress during labor or the development of fetal distress.

Modern-day obstetrical care involves a great deal of technology. Obstetricians visualize the development of the fetus via ultrasound; this modality does not involve X rays and is virtually harmless to the fetus and mother. Ultrasound can be used to evaluate fetal growth and to detect genetic abnormalities and the presence of multiple fetuses. About ten to twelve weeks into the pregnancy, chorionic villus sampling can be done to test for genetic defects in the fetus. In this procedure, ultrasound is used to guide a needle or catheter into the placenta and remove a small piece of tissue adjacent to the fetus. Amniocentesis is used for genetic testing at about fourteen to eighteen weeks into the pregnancy. In amniocentesis, ultrasound is used to guide a needle into the fetal sac (amnion) to withdraw fluid for genetic analysis.

Fetal monitoring involves recording the fetal heart rate on a continuous roll of graph paper and can be conducted during both pregnancy and labor. If a problem is noted, appropriate treatment or intervention can be initiated. For example, an obstetrician can induce (start) or augment (stimulate) labor with a variety of medications. A medication commonly used to promote labor is Pitocin (synthetic oxytocin), which acts directly on the uterus to stimulate contractions. The medication is administered by an intravenous drip, and the amount of medication administered is titrated (adjusted) so that contractions are strong enough for labor to progress but not excessively strong. Excessively strong contractions could cause harm to the fetus (even death) and possible rupture of the uterus. Instead of inducing labor, an obstetrician can also perform a cesarean section.

Gynecology. In an office setting, gynecologists offer contraceptive (birth control) advice and modalities. Oral contraceptives (birth control pills) and intrauterine devices (IUDs) are popular methods for preventing pregnancy. Gynecologists can also advise patients in regard to other contraceptive methods. As women age, their ovarian function slows and stops. This condition is known as the menopause. Menstrual flow may change and then cease. Annoying symptoms, such as hot flashes, dry skin, and mood changes, may accompany the menopause and cause women to consult with gynecologists. Hormonal therapy or other medication may be prescribed to alleviate these symptoms.

Gynecologists often perform colposcopies when the results of a Pap smear are abnormal. The colposcope is a microscope that magnifies the cervix. If cervical abnormalities are found, a gynecologist can remove them by a cone biopsy, cryocautery (freezing), or the loop electrosurgical excision procedure (LEEP).

Gynecologists use ultrasound to aid in making diagnoses. Ultrasound can detect fibroid growths on the uterus and ovarian cysts. Cancerous cysts are more likely to have multiple cavities; therefore, ultrasound may detect the likelihood of a malignancy before surgery. Ultrasound can also measure the thickness of the endometrium (uterine lining); a thickened endometrium might indicate a cancerous condition.

Fertility specialists commonly use ultrasound for in vitro fertilization. With ultrasound guidance, the specialist can insert a needle through the vaginal wall and into an ovarian follicle, which contains an ovum (egg). The ovum is aspirated into the needle and then fertilized with sperm. After an embryo develops, it is inserted into the uterine cavity to grow. Gynecologic surgery is limited to the female reproductive system. A common gynecologic procedure is dilatation and curettage (D&C), which involves dilating the cervix (opening the womb) and scraping the uterine lining (endometrium); this procedure is done for diagnostic purposes and to control bleeding.

Some problems for which a D&C was previously performed are now diagnosed or treated by hysteroscopy (the insertion of a hysteroscope into the uterine cervix for viewing the interior of the uterus). A small piece of tissue (biopsy) can be removed for evaluation of an abnormality such as endometrial cancer (cancer of the uterine lining). Growths such as fibroid tumors (benign growths, which can cause pain or bleeding) can be removed, and abnormalities of the cavity can be corrected.

A hysterectomy involves removal of the uterus; sometimes this can be performed via the vagina (vaginal hysterectomy). Some hysterectomies are done because of fibroids that can cause pain or bleeding. Sometimes, a hysterectomy can be avoided by removing only the fibroids (myomectomy). Other common procedures are a salpingo-oophorectomy (removal of a tube and ovary) and vaginal repair surgery (vaginal damage is usually the result of childbirth). A number of gynecologic procedures, including a hysterectomy, are done with a laparoscope. Laparoscopy is often done to coagulate a portion of the Fallopian tubes for women who do not wish to have children.

Future Prospects

Any physician who deals with birth control and pregnancy may also be called on to deal with abortion. Abortion is a highly debated topic in many areas of the globe and an accepted procedure in others. Opinions range from the belief that life begins at the moment of conception and should not be disrupted under any circumstances to the view that it is a woman's right to terminate a pregnancy at any point up to full term. Some physicians will not perform an abortion under any circumstances, and some limit their practice to abortion and sterilization procedures. In the United States, this topic is likely to remain controversial for the foreseeable future. Surgical techniques will continue to evolve, particularly in endoscopy (laparoscopy and endoscopy), where the use of robotics is likely to increase. As new equipment is developed, techniques will evolve. For example, during in vitro fertilization, traditionally, several embryos were implanted into the uterus because some were not expected to survive. However, this practice increases the likelihood of multiple births, which raises the risk of premature birth and miscarriage. Because of improvements in technology, a single embryo now has a good chance of survival, and many physicians have begun implanting only one embryo. Ultrasound and other imaging methods allow many birth defects to be diagnosed before birth, and intrauterine surgery is evolving to correct these problems.

Robin L. Wulffson, MD, FACOG

Further Reading

Beckmann, Charles R. B. *Obstetrics and Gynecology*. 6th ed. Baltimore: Lippincott Williams & Wilkins, 2010. Covers all aspects of the specialty and is endorsed by the American College of Obstetricians and Gynecologists for its compliance with the organization's standards and procedures.

Carlson, Karen J., Stephanie A. Eisenstat, and Terra Diane Ziporyn. *The New Harvard Guide to Women's Health*. Cambridge, Mass.: Harvard University Press, 2004. Brings together doctors from Harvard Medical School, Massachusetts General Hospital, and Brigham and Women's Hospital to provide complete information on women's health.

Elit, Laurie, and Jean Chamberlain Froese, eds. *Women's Health in the Majority World: Issues and Initiatives*. New York: Nova Science Publishers, 2006. This first part of the book focuses on health issues that specifically affect women. The second part discusses how agencies such as governments, nongovernmental organizations, and professional societies can work together and improve standards for women.

Norwitz, Errol R., et al., eds. *Oxford American Handbook of Obstetrics and Gynecology*. New York: Oxford University Press, 2007. Covers all aspects of obstetrics and gynecology as well as diagnosis and treatment options.

Warsh, Cheryl Lynn Krasnick. *Prescribed Norms: Women and Health in Canada and the United States Since 1800*. Toronto: University of Toronto Press, 2010. Examines the history of medical health treatment of women in the United States and Canada, examining menstruation, the menopause, childbirth, and medical education, including nursing.

WEB SITES

American Congress of Obstetrics and Gynecology http://www.acog.org

National Institutes of Health Women's Health Initiative http://www.nhlbi.nih.gov/whi

U.S. Department of Health and Human Services National Women's Health Information Center http://www.womenshealth.gov

See also: Geriatrics and Gerontology; Reproductive Science and Engineering.

FASCINATING FACTS ABOUT OBSTETRICS AND GYNECOLOGY

- In 1900, infant mortality in the United States was 1,650 per 100,000 births and maternal mortality was 900 per 100,000 births. In 2000, infant mortality had declined to 70 per 100,000 births and maternal mortality to 10 per 100,000 births.
- An estimated 350,000 to 500,000 women worldwide die in childbirth each year.
- Charles Clay performed the first abdominal hysterectomy in 1843. Although the operation was a success, the patient died fifteen days later of other causes. The first successful abdominal hysterectomy in which the woman lived was performed ten years later, by Ellis Burnham.
- Quintuplets are an obstetrical challenge even with modern technology. In 1934, the Dionne quintuplets were delivered by a country doctor with the assistance of two midwives. The infants were placed in a wicker basket containing heated blankets. They were brought into the kitchen and placed next to the open door of the oven to keep warm. All five infants survived into adulthood.
- Puerperal sepsis (severe infection occurring at the time of childbirth) became widespread in the nineteenth century when home delivery was replaced by hospital delivery. In 1843, Boston physician Oliver Wendell Holmes determined that the fever was carried from bed to bed on the unwashed hands of the physician. In 1847, Hungarian doctor Ignaz Semmelweis came to the same conclusion.
- In 1894, the use of rubber gloves became popular in surgery. The use was encouraged by Hunter Robb, a gynecologist at The Johns Hopkins University who had a special interest in wound contamination and sterile surgical technique.

OPTOMETRY

FIELDS OF STUDY

Eye examination; low-vision rehabilitation; visual development; refractive correction; vision therapy; behavioral optometry; biomedical ocular research; electrodiagnostics/physiology.

SUMMARY

Optometry is a regulated and recognized branch of primary health care focused on the evaluation and management of eye health and vision. It is a key part of the eye health care team, which may also include ophthalmologists, orthoptists, and opticians. Doctors of optometry, called optometrists, are trained to provide comprehensive assessment of a patient's visual system, including refractive care (prescriptions for glasses and contact lenses) as well as disease detection, rehabilitation, and management. Optometrists do not typically provide surgical management of eye conditions but in most states are permitted to prescribe drugs for the treatment or management of eye diseases.

PRINCIPAL TERMS

- **binocular vision:** ability to use both eyes together to enhance vision.

- **cornea:** transparent front of the eye, covering the iris and pupil, the first layer of the eye to refract light.
- **low vision:** condition in which an individual has reduced vision even with the use of glasses or contact lenses.
- **ophthalmologist:** physician/surgeon who specializes in treatment of disorders of the eye.
- **ophthalmoscope:** instrument used to examine the interior of the eye, particularly the retina.
- **optician:** person who specializes in filling prescriptions for corrective lenses, such as glasses or contact lenses.
- **orthoptist:** person who specializes in assessing and treating defects in binocular vision resulting from abnormal optic musculature.
- **pupil:** contractile opening at the front of the eye, at the center of the iris.
- **refraction:** determination of the amount of ocular refractive error, used to determine a prescription for corrective glasses or contact lenses.
- **retina:** innermost layer of the eye. receives images transmitted through the lens and contains receptors for vision, the rods and cones.
- **tonometer:** instrument used to measure pressure or tension in the eye, often in the detection of glaucoma.
- **vision therapy:** nonsurgical techniques aimed at correcting and improving binocular, oculomotor, visual-processing, and perceptual disorders.
- **visual acuity:** measure of the ability to read various sizes of letters at a standard distance. normal visual acuity is 20/20, which means that at a distance of 20 feet, the eye can see what an eye would normally see at that distance.

Basic Principles

Optometry is an applied science of diagnosing, managing, and treating diseases and conditions of the eye. The word optometry comes from the Greek *optos*, which means seen or visible, and *metria*, meaning measurement. This measurement of vision is based on principles of optics (the relationship between light and vision) and is conducted using specialized devices such as refractors, lenses, and ophthalmoscopes. Optometrists provide primary-level health care focused on the structure and function of eyes. An optometric eye exam usually involves testing a patient's visual acuity, color vision, depth perception, and binocular vision, and evaluating the structures of the eye to check for eye diseases such as cataracts, glaucoma, or retinopathy. The examination leads to a treatment plan for the patient and may involve referral to other eye care professionals, such as opticians, ophthalmologists, or orthoptists. In most states, optometrists can prescribe and administer drugs for the treatment of specific conditions; however, if more specialized or surgical intervention is required, they will refer the patient to an ophthalmologist. The optometrist will often remain involved in the patient's care after referral and provide low vision rehabilitation and preoperative and postoperative care for cataract or laser refractive surgery. Optometrists can diagnose potentially serious systemic diseases such as high blood pressure and diabetes, which would then necessitate referral to other health care professionals.

Background

The origins of optometry can be traced back to early studies of optics and image formation; however, the term "optometry" was not used until centuries later. The basic principles of optics, reflection, and light angles appear in writings by the Greek mathematician Euclid that date to 280 BCE. Around 150 CE, Ptolemy described his observations of angles of incidence and reflection as light traveled through air to water. In the eleventh century, the Arab physicist Alhazen wrote *Kitāb al-Manāzir*, commonly known as *Optics*. His work was published in Latin in 1572 and partially translated into English in 1989. The use of a spectacle, although a rather primitive version, for magnification was first noted in 1260 in Italy. Johannes Kepler published *Dioptrice* (1611; partial translation of the preface, 1880) describing mathematical concepts associated with mirrors, prisms, and lenses. In *Uso de los antojos y comentarios a proposito del mismo* (1623; *The Use of Eyeglasses*, 2004), Benito Daza de Valdés discusses eyes and refraction, spectacles, and clinical conditions. In the nineteenth century, professional opticians, either dispensing or refracting, provided vision correction. Refracting opticians were later termed optometrists.

Schools of optometry began to appear in the 1870's. The American Association of Opticians was formed in 1898, and by 1919, the group had changed its name to the American Optometric Association. In 1922, the American Academy of Optometry was organized and began to disseminate information through its journal, *Optometry and Vision Science*. By 1924, all states in the United States had passed laws regarding optometry licensing.

How It Works

Optometry encompasses eye physiology and the measurement of vision, based on principles of optics. Examinations that measure how well a person can see include tests of visual acuity, visual field, and refraction. Those that focus on the physiology of the eye include tonometry, and slit lamp, ocular motility, and fundus examinations.

Visual Acuity Examination. Visual acuity is a measurement of sharpness or clarity of vision. It is often tested using a series of black letters or symbols on a white background to provide optimal contrast. The most well-known way to test distance visual acuity is to use the Snellen chart, which has the large "E" on the top and a series of progressively smaller letters below. Normal visual acuity is 20/20. This value indicates that 20 feet from the chart, the eye can clearly see what can be normally read by an eye with no refractive error at this distance. A visual acuity of 20/40 would reflect that the eye being tested can clearly see at 20 feet what a normal eye could see at 40 feet away, meaning a visual acuity of 20/40 is only half as good as 20/20. Testing can be done on each eye separately, both eyes together, or with corrective lenses. Near vision is tested similarly but using a modified Snellen chart, such as a Rosembaum chart, held at 15.7 inches from the eyes.

Ocular Motility Examination. There are six extraocular muscles that control eye movements. An ocular motility examination begins with an observation of the eyes together, as the patient is asked to fixate on an object. Some patients have eyes that are obviously misaligned. In these cases, one eye is typically able to fixate and the other eye is deviated. The deviations can be outward (exotropia), inward (esotropia), or upward (hypertropia). Cover/uncover testing determines if one eye dominates fixation or if fixation switches between the eyes. Misalignment of the eyes can be benign or a sign of something more serious, such as cranial nerve palsy or orbital floor fracture.

Refraction. The principles of optics are used in this clinical test to determine the eye's refractive error and prescribe corrective lenses to achieve the best possible visual acuity. Refractive errors can be spherical or cylindrical. Spherical errors mean that the optical power of the eye is either too large or too small to focus light on the retina, which causes blurry vision. Spherical errors can be further classified into myopia (nearsightedness) or hyperopia (farsightedness). Cylindrical errors occur when the optical power of the eye is too

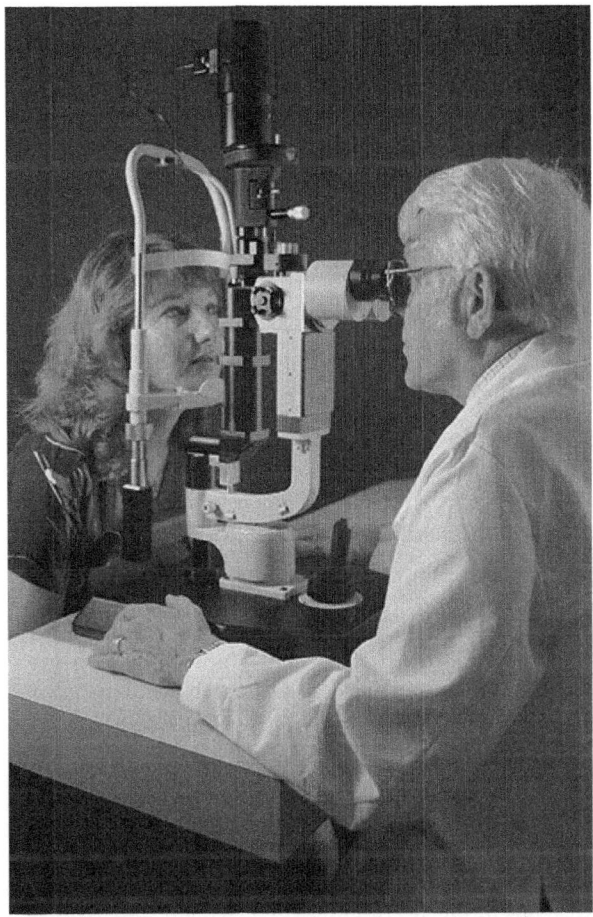

An Optometrist examining the eyes of a patient with slit lamp biomicroscope

powerful or too weak across one meridian, known as astigmatism. A phoropter, retinoscope, or automated refractor can be used to measure refractive error.

Visual Field Examination. Visual fields are the areas in which objects can be seen with (peripheral) vision while the eyes are focused on a central point in front of the person. Visual field defects can be detected by an optometrist moving his or her hand or a target object to the side of a patient's head or with specialized devices such as Goldmann field examinations or automated perimetry.

Slit Lamp Examination. Also known as biomicroscopy, a slit lamp examination allows the optometrist to closely inspect the anatomical structures of the eye. If a biomicroscope is used, the optometrist can obtain a magnified and detailed view of the eye. The slit lamp is an instrument that can focus an intense, thin sheet of light into the eye. During this examination, the patient is asked to place his or her chin and forehead against a support or to keep his or her head and eyes steady.

Fundus Examination. Also known as ophthalmoscopy, a fundus examination is done to see the anatomical structures at the back of the eye, such as the fundus, macula, and retina. For the best view, the patient's pupils must be dilated with specialized eye drops that may cause light sensitivity and blurry vision for a few hours after the exam. When the pupil has sufficiently dilated, the optometrist, usually wearing a head-mounted lighting device, asks the patient to fixate on a distant point on a wall or ceiling. The optometrist then shines the light source into the patient's eye while looking through a handheld condensing lens that permits him or her to inspect the back of the eye.

Tonometry. Tonometry measures the pressure inside the eye, which is a common way to screen for eye diseases such as glaucoma. The two tonometry methods are contact and noncontact. In contact tonometry, a drop of anesthetic is placed in the eye, gentle contact is made with the cornea, and a measurement is determined. Normal eye pressure ranges from 10 to 21 millimeters of mercury. In the noncontact method, the patient feels a brief puff of air at the eye. The machine measures eye pressure by looking at how the light reflections change as the air hits the eye.

APPLICATIONS

Specializations in optometry deal with the study of eye examination techniques, contact lenses, low vision rehabilitation, visual development, refractive correction, vision therapy, biomedical ocular research, and electrodiagnostics/physiology. Diagnostic equipment in optometry has undergone major changes since the 1970's. Numerous automated and computerized testing devices—topographers, pupillometers, autorefractors, tonometers, pachymeters, and computerized acuity charts—have emerged, taking the place of numerous charts, rulers, and screens. Technology has increased the capacity to refine results and create an electronic record that can be transferred between practitioners. Computerization and enhanced communication has allowed for consultation with other practitioners and specialists to become routine and to take place in real time, even if the clinicians are practicing in different parts of the country or even the world. Digital fundus photography provides a digital record of the patient's eye, replacing hand drawings that were difficult to interpret and compare. These improvements in detailed diagnostics and record keeping also assist in billing, coding, and medical-legal issues. New contact lens materials have been developed, and new diagnostic technology, such as corneal topography (mapping the surface of the cornea), has become integral to delivering care to contact lens patients. Modern consumers would find early contact lenses to be rather primitive. Those early lenses could be worn only for limited periods of time because they did not allow oxygen to permeate through to the cornea, and they were rigid and very expensive. The silicone hydrogel materials used in modern contact lenses allow oxygen to reach the cornea. Contact lenses can be worn for prolonged periods and are so inexpensive that patients can wear a new pair of disposable lenses every day. Further advances have allowed for the development of special toric contact lenses for people with astigmatism and contact lenses with multiple focuses for people who require both near and distance correction. Color and costume contact lenses are available with or without refractive correction for those who wish to temporarily change their eye color or to create a special effect. Despite their name, eyeglasses rarely contain glass. The traditional glass lens has largely been replaced by plastic polymers. High-index, polycarbonate, and aspheric lenses are thinner and lighter than the traditional glass lenses. Photochromic lenses darken and lighten according to light conditions. Scratch-resistant and ultraviolet-blocking coatings are also available. The thick, heavy glasses of the 1950's have been replaced with thin functional lenses in stylish frames. Low vision rehabilitation is still a significant part of optometric practice. The devices and aids to daily living that have been developed assist patients with low vision and allow them to lead productive and functional lives with greater independence than ever before. Vision therapy and behavioral optometry use nonsurgical methods to treat common visual problems such as strabismus, double vision, convergence insufficiency, and some learning disabilities. These methods are best described as physical therapy for the eyes and brain. Under the direction of an optometrist, the patient performs a series of exercises, some incorporating the use of specialized equipment such as patches, computer software, prisms, and optical filters. In ophthalmic and medical circles, controversy surrounds certain vision therapy practices because of doubts concerning their efficacy. Biomedical ocular research encompasses a vast area of optometry. This work, which often takes place in a laboratory setting, examines the basic physiology of eye diseases or the application of technological advances in eye care, covering both structure and function of the eye as well as visual processing and vision assessment. These laboratories are found within academic institutions or as part of large eye and vision care corporations. Studies are ongoing in all areas of eye anatomy, physiology, microbiology, biomechanics, and immunology. Research into the treatment of conditions such as refractive error and amblyopia are spotlighted at annual academic meetings.

Future Prospects

Optometry has gone far beyond its beginnings as the fitting of spectacles, becoming, through continuing research and rigorous specialized training, a facet of primary health care. Optometry still provides patients with prescriptions for vision correction; however, the examinations are far more extensive. This broadened knowledge has allowed optometrists to work more closely with ophthalmologists, neurologists, and other health care practitioners. Some optometrists are specializing, focusing on areas such as contact lenses, vision therapy, pediatrics, low vision, sports vision, head trauma, learning disabilities, and occupational vision. Optometry is expected to experience a high level of growth to meet the demand of the aging population. In 2008, there were more than 34,000 optometrists in the United States, about one-quarter of whom were self-employed in a private practice or in a group practice, while the remainder were employed by a larger facility, organization, or retail business. Optometry is considered to be very rewarding and interesting work, and according to the American Optometric Association, the average annual income for self-employed optometrists was $175,329 in 2007. The regular hours and good income make optometry an attractive and satisfying career choice. Globally, expertise in optometry is lacking in many underserved areas. The World Health Organization (WHO) estimates that 500 million people who have refractive error do not have access to eye care services. In highly developed countries such as the United States, there is one optometrist for every 10,000 people. In contrast, the ratio in developing countries is one optometrist for every 600,000 people or more. The WHO's Vision 20/20 program is striving to provide more optometric expertise to combat the underserved areas of the world.

April D. Ingram, B.Sc.

Further Reading

Griffin, John R., and J. David Grisham. *Binocular Anomalies: Diagnosis and Vision Therapy*. 4th ed. Boston: Butterworth-Heinemann, 2002. Deals with vision therapy as a way to manage some disorders.

Keirl, Andrew William, and Caroline Christie. *Clinical Optics and Refraction: A Guide for Optometrists, Contact Lens Opticians, and Dispensing Opticians*. Oxford, England: Butterworth-Heinemann, 2007. A guide to visual optics, providing straightforward information on clinical optics and refraction.

Kitchen, Clyde K. *Fact and Fiction of Healthy Vision: Eye Care for Adults and Children*. Westport, Conn.: Praeger, 2007. Starts with anatomy and proceeds to discuss how to maintain healthy eyes. Includes information on many conditions and the types of refractive surgery.

Millodot, Michel. *Dictionary of Optometry and Visual Science*. Oxford, England: Butterworth-Heinemann, 2008. Provides understandable definitions, tables, and illustrations.

Phillips, Anthony. *The Optometrist's Practitioner-Patient Manual*. Oxford, England: Butterworth-Heinemann, 2008. Information and illustrations of procedures and conditions pertaining to eye health.

FASCINATING FACTS ABOUT OPTOMETRY

- The average eye blinks more than 10 million times per year.
- Sailors once believed that wearing a gold earring would help improve their eyesight.
- Prescriptions for glasses and contact lenses may be similar but are not interchangeable.
- In 1620, Peter Brown, a pilgrim, became the first person known to wear glasses on the North American continent.
- The cornea is the only living tissue in the human body that does not contain any blood vessels.
- The eyeball of a giant squid, at 18 inches in diameter, is the largest eyeball on the face of the Earth.
- Newborn babies may cry, but they do not produce any tears until they are one to three months old.
- Sam Foster sold the first pair of Foster Grant sunglasses at Woolworth's store on the Atlantic City Boardwalk in 1929.

WEB SITES

American Optometric Association http://www.aoanet.org

Association of Schools and Colleges of Optometry http://www.opted.org

National Institutes of Health National Eye Institute http://www.nei.nih.gov

See also: Geriatrics and Gerontology.

ORTHOPEDICS

FIELDS OF STUDY

Physiology; kinesiology; human anatomy; biology; cellular biology; chemistry; biochemistry; physics; mathematics; pathology; psychology.

SUMMARY

Orthopedics is a branch of medicine focusing on treating the skeleton, muscles, joints, ligaments, tendons, and nerves of the human body, collectively referred to as the musculoskeletal system. Orthopedic conditions may be treated by a family practice physician or a physician who specializes in treating disorders of the musculoskeletal system. Other medical specialists and health care providers such as physical and occupational therapists also treat orthopedic disorders and often play a part in the plan of treatment. A multidisciplinary team approach is important in managing the symptoms of an orthopedic condition, especially when symptoms are chronic and ultimately will change in severity.

PRINCIPAL TERMS

- **arthritis:** inflammation of one or more joints.
- **arthroscope:** type of surgical device inserted into the joint through a small incision.
- **arthroscopy:** examination or treatment of the inside of a joint using an arthroscope.
- **fracture:** break, rupture, or crack of a bone.
- **joint prosthesis:** artificial joint that replaces structural elements within a joint to improve and enhance the function of the joint.
- **magnetic resonance imaging (MRI):** test that uses a magnetic field to create pictures of organs and structures inside the body.
- **musculoskeletal system:** system encompassing the joints, ligaments, tendons, muscles, and nerves.
- **occupational therapist:** allied health care professional specializing in assisting patients with various disabilities, from decreased motor skills to short-term memory loss.
- **orthopedic surgeon:** physician educated in the workings of the musculoskeletal system and specializing in the surgical treatment of this system.
- **physical rehabilitation:** process of returning function to an individual afflicted by conditions or disorders such as injury, chronic disease, or genetic dysfunction.
- **physical therapy:** allied health care profession specializing in returning function to an individual.
- **X ray:** test that produces images of structures such as bones within the body.

BASIC PRINCIPLES

Orthopedics is the science that studies the bones, muscles, and joints in the human body, collectively referred to as the musculoskeletal system. Medical professionals who specialize in this branch of medicine treat a wide variety of conditions. Some of the many conditions that are treated by orthopedic physicians are arthritis, sports injuries, back pain, and leg and foot disorders. Orthopedic surgeons are specifically trained to deal with conditions and disorders associated with the musculoskeletal system and to perform surgical repair when necessary.

The care of bone fractures and the reconstruction or replacement of joints are common surgical procedures performed by orthopedic surgeons. Some orthopedic conditions do not require surgery and can be managed by medication, injections, or rehabilitation by medical professionals such as physical and occupational therapists. Some orthopedic surgeons specialize in a particular part of the body, such as the spine, hands, or feet. Some further specialize in orthopedics by exclusively treating children and are referred to as pediatric orthopedic surgeons. Pediatric orthopedic problems—such as curvature of the spine, hip joint disorders, and limb length discrepancies—can occur as a child grows and develops. In addition, active children often experience bone fractures, severe sprains or strains, and dislocated joints, which are treated by pediatric orthopedists. Orthopedics is a specialty that can be called on in emergency situations as well.

Background

Orthopedics can be traced as far back as prehistory, when humans underwent crude orthopedic procedures such as amputations of limbs and fingers in order to survive. Paintings and drawings on walls have suggested that early humans used forms of assistive walking aids such as crutches. The Greek historian Herodotus writes of a soldier who escapes from chains by cutting off his foot and later creates a prosthesis for himself. Other early writings mention wooden legs, iron hands, and artificial feet. The early writings of the Greek Hippocrates and his understanding of fractures and fracture care had an impact on orthopedics.

In the twelfth century and after the Dark Ages, universities and hospitals were beginning to be established in Europe. In many of these institutions, researchers were performing human dissections and gaining a better understanding of human anatomy. In addition, the ancient Greek texts were being translated from Arabic to Latin.

Orthopedics began to come into its own in the early 1900's. The discovery of the X ray was a major advancement in medicine, especially for orthopedics. Orthopedics was being seen as a true medical specialty. The British dominated orthopedic developments during this time; however, later in the century, Americans would make progressively more contributions.

Orthopedic Assessment

An individual with an orthopedic concern is often referred to a specialist through a primary care provider, often a family practice physician. An orthopedic assessment begins with the physician taking a thorough history. For example, he or she may ask where the problem is located, what (if any) event brought on the problem, and what type of activities increase or decrease the discomfort. The physician may order tests, such as X rays, magnetic resonance imaging (MRI) scans, or computed tomography (CT) scans. Once the problem has been identified, a plan of care

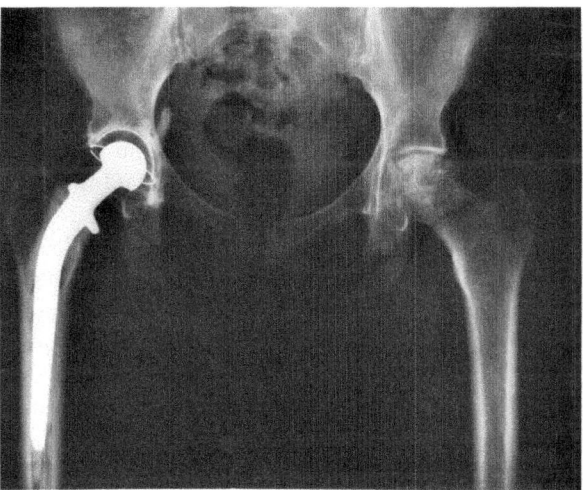

X-ray of a hip replacement.

is developed. This may consist of physical or occupational therapy, medication (including injections), and possibly surgery.

Fracture Care

Specialists in orthopedics are trained in the early management of fractures. This typically consists of the realignment of a broken limb, followed by the immobilization of the fractured extremity in a cast or splint. Nerve and blood vessels are assessed for possible injury associated with the fracture and documented before and after realignment.

Joint Pain and Dysfunction

Medical professionals involved in orthopedics treat joint pain and dysfunction on a regular basis. Orthopedic specialists often treat osteoarthritis (degenerative joint disease), with the primary goal of controlling pain. Physical therapy is used to decrease pain and swelling and increase muscle strength and joint motion. The orthopedic physician may also prescribe medications, including nonsteroidal anti-inflammatory drugs (NSAIDs). Surgery may be necessary if symptoms have not responded to conservative therapy such as medication and physical therapy. Surgical procedures include arthroscopy, which involves the removal of torn cartilage and the roughened joint surface (debridement), usually from the knee; hip resurfacing; and arthroplasty, replacement of the hip or knee. In arthrodesis, bones in a damaged joint (usually the spine, hand, ankle, or foot) are fused.

Developmental Disorders

Orthopedic physicians can become more specialized and treat a variety of developmental dysfunctions acquired at or before birth, such as cerebral palsy, a condition that describes a group of brain disorders affecting the communication between the brain and the muscles. The orthopedic team, which often includes physical and occupational therapists, addresses cerebral palsy and other developmental dysfunctions causing a permanent state of uncoordinated movement and posturing. Other developmental conditions seen and treated by the orthopedic physician are upper extremity misalignment and contractures, and joint deformities of the spine, hip, knee, ankle and foot.

Sports Medicine

Orthopedic specialists are trained to perform a comprehensive assessment, to treat, and to provide follow-up care to children, adolescents, and adults with sports-related orthopedic injuries. This is often accomplished with a team approach that includes sports medicine physicians and a physical therapy staff and results in the development of long-term treatment and activity plans.

Emergency Medicine

Orthopedic specialists serve as members of hospital-based emergency care teams, treating trauma-related injuries. In the emergency care setting, the physician assesses and treats acute illnesses and injuries that require immediate medical attention. Although emergency medicine physicians do not usually provide long-term or continuing care, those who specialize in orthopedics diagnose a variety of fractures and soft-tissue injuries and undertake acute interventions to stabilize patients.

Orthopedic Prosthesis

Arthroplasty (joint replacement) consists of the surgical removal of diseased or worn surfaces or parts of a joint and replacing them with a metal and plastic prosthesis. The prosthesis can be cemented in place or it can be attached by means of a porous coating designed to allow the bones to grow and adhere to the artificial joint. The decision made by the orthopedic surgeon to use bone growth or cement can depend on the age of the patient. The use of cement relieves pain more quickly, and the patient can bear weight much sooner, usually immediately after surgery. Although healing is slower with a porous coating and weight bearing is done in a much more progressive manner, attachment achieved through bone growth may last longer; therefore, a porous coating is used in younger patients. Joint replacements are available for joints such as the knee, hip, shoulder, elbow, wrist, ankle, and fingers. One problem that can arise with joint replacements is that the surfaces of the two parts of the implant can wear out, which leads to failure of the implant. Instead of traditional metal-on-plastic implants, orthopedic surgeons sometimes use ceramic-on-ceramic or metal-on-metal implants. Companies that design joint replacement implants are trying to improve wear characteristics.

Joint resurfacing is becoming a feasible alternative to complete joint replacement and is more prevalent in the hip. It is becoming more popular among younger and more active patients. This type of operation delays the need for the more traditional and less bone-conserving total hip replacement. However, this procedure does have a significant risk of early failure from fracture and bone death.

Spinal Stabilization

Since the early 1900's, orthopedic surgeons have performed spinal stabilization of the neck, truck, and lower back regions. Through trauma or progressive wear of each segment of the spine, tissues and nerves become aggravated and even compressed. The involved levels of the spine also become unstable, causing additional problems. This can result in pain in the arms or legs, reduced motion of the spine, and significant limitations in function. Orthopedic surgeons may use fixation devices, often a screw-and-rod plate system, to fuse the two levels of the spine. Orthopedic surgeons may also use a bone graft from another area of the body, such as the pelvis, rather than a fixation device. The bone graft is inserted between the unstable segments. Once the bone grafts heals and becomes solid, it fuses the unstable and worn segments. In both cases, surgery is followed by a relatively lengthy period of physical therapy before the patient is able to return to adequate function.

Arthroscopy is a procedure commonly used in orthopedic surgery, either to examine or treat a joint. By using a device called an arthroscope, the surgeon is able not only to see the inside the joint and examine the surfaces for damage but also to repair or remove floating debris and torn surface cartilage or perform reconstruction of ligaments. Arthroscopy has several advantages over traditional surgery because the joint does not have to be opened and exposed. Instead, several small incisions are made, one for the arthroscope and one or two for the surgical instruments used within the joint. This reduces recovery time and can improve the outcome because of reduced trauma to the surrounding tissues. It is a useful application in sports medicine, especially with college and professional athletes, who may require minimal healing time. Knee arthroscopy is one of the most common operations performed by orthopedic surgeons.

PHARMACOTHERAPY AND MEDICAL DEVICES
Orthopedic surgeons prescribe low to moderate doses of simple analgesics and nonsteroidal anti-inflammatory medications such as acetaminophen, aspirin, ibuprofen, and naproxen. When these medications fail to relieve a patient's pain, alternative or additional pharmacologic agents are considered. Such pain and anti-inflammatory medications are carefully selected after the patient's risk factors are considered. Orthopedic products are designed to improve surgical and rehabilitative outcomes, decrease pain, promote motion and strength, or simply improve overall comfort. Products in this medical field can range from an improved joint replacement device to strength training equipment for the physical therapist to an orthopedically designed pillow to help improve overall comfort while sitting or sleeping. Products made for the orthopedic industry are constantly changing in an effort to improve overall quality of life.

FUTURE PROSPECTS
As the number of elderly people in the United States continues to rise, the incidence of age-related musculoskeletal conditions will increase, creating a greater demand for all health care professionals associated with orthopedics, including orthopedic surgeons. In addition, many soldiers in the conflicts in Afghanistan and Iraq need the services of orthopedic medical professionals to deal with musculoskeletal injuries that include amputations. Another area of growth is sports medicine, with orthopedic surgeons becoming trained in sports medicine or teaming with physicians specializing in the area to treat professional and amateur athletes. As new technologies and advances in robotics and bone substitutes guide the way to less-invasive surgical interventions, health care experts predict that many more orthopedic procedures will be performed in outpatient settings or result in overnight hospital stays and that fewer procedures will involve extended stays in the hospital. As demand for orthopedic surgeons increases, hospitals are likely to form alliances and partnerships with other hospitals so that they may continue to provide orthopedic care at their facilities. As with health care in general, among the challenges facing orthopedics are the cost of health care and the changes brought about by health care reform. Modern prostheses incorporate advanced technology so that they function more like the limbs they replace, but these devices are very expensive to produce and maintain. Adaptive devices, which range from grab bars for bathtubs to step-in bathtubs to leg prostheses designed for running to sports wheelchairs, are increasingly available, but they are not always covered by insurance and can be prohibitively expensive.

Jeffrey Larson, PT, ATC

FURTHER READING
Brotzman, S. Brent, and Kevin E. Wilk. *Clinical Orthopaedic Rehabilitation.* 2d ed. Philadelphia: Mosby, 2003. Includes chapters on rehabilitation of patients who have had a total knee replacement, lumbar fusion, and knee arthroscopy.

Magee, D. J. *Orthopedic Physical Assessment.* 5th ed. St. Louis, Mo.: Saunders 2008. Includes chapters on orthopedic assessments and rehabilitation of the cervical, thoracic, and lumbar spine.

Maxey, Lisa, and Jim Magnusson, eds. *Rehabilitation for the Postsurgical Orthopedic Patient.* 2d ed. St. Louis, Mo.: Mosby, 2007. Provides detailed descriptions of each orthopedic surgery and addresses patient rehabilitation and how to adapt therapy to geriatric, athletic, and pediatric populations.

Pagliarulo, Michael A., ed. *Introduction to Physical Therapy.* 3d ed. St. Louis, Mo.: Mosby, 2007. Examines the history of physical therapy, discusses financial and legal aspects, and contains a chapter on physical therapy for musculoskeletal conditions.

Scuderi, Giles R., and Peter D. McCann, eds. *Sports Medicine: A Comprehensive Approach.* 2d ed. Philadelphia: Mosby-Elsevier, 2005. Provides informative chapters on each body part with a focus on sports medicine injuries and rehabilitation.

Skinner, Harry B., ed. *Current Diagnosis and Treatment in Orthopedics*. 4th ed. New York: Lange Medical Books/McGraw-Hill Medical Publishing Division, 2006. An accessible resource for diagnosis of orthopedic conditions, covering trauma, sports medicine, oncology, surgery, amputations, and rehabilitation

WEB SITES

American Academy of Orthopaedic Surgeons http://www.aaos.org

American Association of Hip and Knee Surgeons http://www.aahks.org

American Orthopaedic Association http://www.aoassn.org

American Orthopaedic Foot and Ankle Society http://www.aofas.org

American Orthopaedic Society for Sports Medicine http://www.sportsmed.org

American Physical Therapy Association Orthopaedic Section http://www.orthopt.org

Orthopedic Surgical Manufacturers Association http://www.osma.net/index.htm

See also: Bionics and Biomedical Engineering; Geriatrics and Gerontology; Kinesiology.

FASCINATING FACTS ABOUT ORTHOPEDICS

- The human body has more then 206 bones.
- In 2006, according to the American Academy of Orthopaedic Surgeons, musculoskeletal conditions were the number-two reason for visits to physicians.
- Exercising at least thirty minutes per day will reduce the risk of bone and joint injury.
- About 300,000 hip replacements and 500,000 knee replacements are performed in the United States each year.
- Every pound of weight gained places 3 pounds of additional stress on one's knees and six times the pressure on one's hips.
- Arthritis is the leading chronic condition reported by the elderly; however, more than half of those with arthritis are under the age of sixty-five.
- Many modern leg prostheses have microprocessors that allow the knee and ankle to adapt to changes in terrain or walking speed.
- The myoelectric arm is an electric prosthesis whose movements are controlled by electric signals produced by muscle contractions in the amputee's body

---

OSMOREGULATION

FIELDS OF STUDY

Biophysics, cell biology

SUMMARY

Osmoregulation is the ability an organism must have to adjust its internal concentrations of solutes and of water so that it can maintain an osmotic pressure appropriate to its functioning.

PRINCIPAL TERMS

- **euryhaline:** the ability of an organism to tolerate wide ranges of salinity
- **hyperosmotic:** describes a solution with a higher osmotic pressure, one containing more osmotically active particles relative to the same volume, than the solution to which it is being compared
- **hypoosmotic:** a solution with a lower osmotic pressure, fewer osmotically active particles relative to the same volume, than the solution to which it is being compared
- **isosmotic:** a solution having the same osmotic pressure, the same number of osmotically active particles relative to the same volume, as the solution to which it is being compared
- **osmoconformer:** an organism whose internal osmotic pressure approximates the osmotic pressure of its environment; such an organism is also

referred to as "poikilosmotic"
- **osmoregulator:** an organism that maintains its internal osmotic pressure despite changes in environmental osmotic pressure; such an organism is also referred to as "euryosmotic"
- **stenohaline:** the inability of an organism to tolerate wide ranges of salinity

BASIC PRINCIPLES

Osmotic pressure of a solution is the measure of the tendency of water to enter a solution from pure water. Osmoregulation, the regulation of osmotic pressure, is vital to every organism. The phenomenon collectively is called "osmosis." It is the difference of hydrostatic pressure that must be created between that solution and pure water to prevent any net osmotic movement of particles in the water when the solution and pure water are separated by a semipermeable membrane. Hydrostatic pressure is a measuring device: It is a means of assessing the tendency of a solution to take on water osmotically.

Only dissolved solutes contribute to osmotic pressure. The number of individual particles determines the strength of osmotic pressure. Each particle makes a roughly equal contribution to osmotic pressure. The same number of molecules of a substance such as sodium chloride (table salt), which ionizes in water to release two ions (one sodium ion and one chlorine ion) display twice as much osmotic pressure as the same number of molecules of glucose, which retains its molecular form in water. Cells and other suspended materials do not contribute to osmotic pressure.

Several characteristics of solutions depend upon the number of particles in the solution. These are called "colligative properties." Increasing the number of particles of solute impairs the ability of the solvent to change state. The colligative properties are the freezing point, the boiling point, the osmotic pressure, and the vapor pressure. Only freezing point depression and vapor pressure are used to determine osmotic pressure.

Osmoticity refers to the osmotic pressure of solutions. Isosmotic solutions have equal osmotic pressures. A hypoosmotic solution has an osmotic pressure lower than the solution to which it is being compared; a hyperosmotic solution is one with a greater osmotic pressure than the solution to which it is being compared. These solutions can be body fluids, environmental liquids, or laboratory solutions.

The terms used to describe the changes in volume of cells exposed to solutions of differing concentrations are often confused with those comparing osmotic pressure. Changes of cell volume are described by the term "tonicity." Solutions isotonic to a cell cause no change in cell volume. Hypotonic solutions will cause the cell to swell as water diffuses into the cell; the cell may even burst. Hypertonic solutions will cause a cell to shrink as water diffuses across the cell membrane into the solution.

THE CHALLENGES OF OSMOREGULATION

This discussion reveals some of the problems that an organism encounters in the environment as concentrations of water and salts vary. Most organisms attempt to regulate both their volume and their ion content. If volume is not regulated, the chemicals within the cell will become too dilute to react or too concentrated to interact. If ions are not regulated, chemical reactions will be affected by inappropriate levels of ions, which may change the electrochemical properties of the cellular solution. Thus, there are independent challenges to volume regulation, to ion regulation, and to osmotic regulation. The homeostatic physiological responses to all three types of challenges are interconnected but distinct.

Osmoregulation is the regulation of the ratio between all dissolved particles, regardless of their chemical nature as ions or molecules, and water. All organisms are exposed to osmotic stress. Any organism incurs obligatory water losses. These occur during respiration, urination, and defecation. The organs most often thought of as participating in osmoregulation are the kidneys. They are intimately concerned with the elimination or conservation of water. Some salts are also found in urine. The proportions of the salts excreted in urine may be different from those in the body fluids because the kidney can retain required ions while eliminating less desirable ones.

Freshwater organisms are in danger of dilution, and they excrete great quantities of dilute urine. Saltwater organisms usually produce small quantities of isosmotic urine, which preferentially excretes divalent ions such as magnesium and sulfate. Other surfaces lose water, causing desiccation. The composition of the diet also influences the need for excretion of urine. Nitrogenous wastes from protein metabolism must be eliminated in urine, often as urea, which requires water for its excretion.

Carbon dioxide released during metabolism of carbohydrates and fats is eliminated by the respiratory organs. In terrestrial organisms, air leaving the lungs is usually saturated and some water is lost on expiration. The respiratory organs of aquatic organisms are gills. Their surfaces must be permeable to water. Freshwater organisms gain water through them and hypoosmotic marine organisms lose water through them.

The metabolism of carbohydrates produces what is known as metabolic water, which can be used to prevent desiccation. Metabolic water produced from the metabolism of fats is lost because of the higher rate of respiration required to supply the oxygen needed in fat oxidation.

Salt Loss

Preformed water is present in any food. Even the driest seeds contain a small amount of water. The nutrients also always include salts. The presence of great quantities of salts may require urinary loss of water in excess of the preformed water found in the food.

Although feces may appear to be solid, they contain some water that was not absorbed in the gut. The presence of salts and other solutes in the digesta may also draw water from the hypoosmotic body fluids into the gut. One of the reasons that humans cannot drink seawater, in fact, is that the magnesium ions in ocean water increase the permeability of the gut and increase water loss, because the seawater is hyperosmotic to body fluids. More water is lost than can be gained.

Salts can be lost from the body by means other than urine formation and defecation. Marine reptiles and birds have salt glands located on the head. Since neither can produce hyperosmotic urine, these glands allow the elimination of salt with a minimum loss of water as the secretion may be four to five times as concentrated as body fluids. The cloaca of birds and the rectal glands of sharks also have the capacity to excrete salts.

One of the most fascinating mechanisms of osmoregulation is found in elasmobranch fish— the sharks, skates, and rays. Their body fluids are hyperosmotic but hypoionic to seawater. Blood salt concentrations are below those of seawater. Excess osmotic pressure is supplied by two molecules: urea and trimethylaminoxide (TMAO). Urea is toxic to most organs, but some organs resist its deleterious effects. Others would be harmed but are apparently protected by the TMAO. Retention of both urea and TMAO minimizes enzymatic disturbances by urea and allows the elasmobranchs to avoid the salt gain associated with hypoosmotic body fluids.

Organisms exposed to environmental variations have two choices: they can maintain internal constancy or homeostasis at the expense of metabolic energy, or they can allow their internal conditions to follow that of the environment. Organisms that maintain their internal osmotic pressure despite changes in external osmotic pressure are called osmoregulators. These euryosmotic organisms are protected from environmental changes. Their metabolism can continue to function, but much of the energy will be used to maintain their body fluids at the appropriate osmotic pressure.

Organisms that allow their osmotic pressure to follow that of the environment are called osmoconformers. These poikilosmotic organisms often have a limited tolerance for such changes. They are stenohaline. They may be less vigorous at salinities other than their optimal levels. The adults of such groups (for example, mollusks such as oysters and mussels) may be found in salinity extremes not tolerated by their young. These populations must be maintained by immigration of young spawned in more favorable salinity conditions.

Hormonal Regulation of Osmotic Pressure

The internal osmotic pressure is affected by the hormones present in the body fluids. In invertebrates such as annelids, mollusks, and arthropods, neuroendocrine changes are seen upon changing the osmotic pressure of the environment. These changes indicate that nervous and endocrine systems are at work regulating the osmotic pressure of the organism. In most invertebrates, the biochemical nature of these hormones is unknown. Some freshwater pulmonate snails, however, produce an antidiuretic hormone and a neurosecretory factor associated with electrolyte balance. Depending upon the demands placed on them, insects such as grasshoppers and cockroaches can synthesize diuretic or antidiuretic hormones.

The best-known hormonal factors in ion regulation are studied in vertebrates. The pituitary gland produces antidiuretic hormone (ADH), which promotes water retention in terrestrial vertebrates. In fish and amphibians, ADH may induce urine formation and increase water loss through diuresis.

The adrenal gland also produces hormones that influence ion retention. In mammals, aldosterone increases reabsorption of sodium in the kidney and promotes the excretion of potassium. In nonmammalian vertebrates, extrarenal glands maintain salt and water balance by affecting the gills and intestines of fishes, the urinary bladder and skin of amphibians, and the salt glands of elasmobranchs, reptiles, and birds.

Measuring Osmoregulation

Osmoregulation involves the balancing of water and solutes in the body so that the animal can continue to function. Because the presence of particles influences certain physical characteristics of the solution, these colligative properties can be used to determine osmotic pressure of solutions. Colligative properties change with increasing numbers of particles in solution: The osmotic pressure increases; the boiling point increases; the freezing point decreases; the vapor pressure decreases. Freezing point depression and vapor pressure can be used to measure a solution's osmotic pressure.

Freezing point is used most often. It works for the same reason that salt is spread on ice on sidewalks in winter. The salt lowers the freezing temperature of the water. Body fluids are much more dilute than the salt and water mixtures that melt ice, but the salinity of the ocean (approximately thirty-five parts salt for each thousand parts of solution) causes it to freeze as much as 1.6 degrees Celsius lower than pure water. Only marine organisms have body fluid osmotic pressures in that range. Terrestrial organisms have much less salt and therefore much lower osmotic pressures in their body fluids. Because most body fluids are so dilute, a large sample maybe required to determine freezing point depression. When only small volumes of body fluid exist, the determination becomes more difficult.

Vapor pressure determinations are also used in osmometry. Usually, a small amount of the fluid being studied is tested in a capillary tube. In one ingenious method, the capillary tube is placed in a solution more concentrated than the experimental fluid. The higher osmotic pressure of the reference solution pushes a meniscus up the tube. The rate of movement of the meniscus depends upon the difference in concentrations between the experimental and the reference solutions.

Another ingenious method of using vapor pressure to determine the osmotic pressure of an experimental solution requires enough fluid to fill a depression. A glass plate with capillary tubes filled with reference solutions of known osmotic concentrations is mounted over the experimental fluid. The reference tube that exhibits no movement is at equilibrium with the experimental fluid. One of the rigors of this method is that all movement must stop.

Another method involves capturing a precise volume of experimental fluid in a capillary tube. The shape of drops of the same volume of reference solutions is compared to that of the experimental solution. Those of the same concentration will have the same shape because their vapor pressures are exerting equal force on the drop. Thermocouples are also used in vapor pressure determinations. This procedure is delicate and costly and is used infrequently. All these procedures are difficult and require patience. Now, electronic instruments analyze the constituents of solutions and allow easier calculation of the osmotic pressures of solutions than ever before.

Freshwater Versus Saltwater Environments

All organisms experience osmotic stress. There is no environment in which the osmotic pressure and the ion composition exactly match the requirements of the cells. Every organism must expend metabolic energy to maintain appropriate water and ion concentrations.

Freshwater organisms are hyperosmotic to their environment. They risk losing scarce ions through their permeable gills and in their urine and also tend to take up water through their gills or other surfaces and in their food. They face the problem of dilution of their body fluids by the environment.

Marine organisms are often isosmotic to the salt water they inhabit. They must change the concentrations of some ions, however, in order to attain this state. Magnesium is present in greater amounts in seawater than is desirable in their body fluids and must be eliminated. These organisms ion regulate even though they are not in danger of volume changes.

Marine organisms that evolved from freshwater or terrestrial ancestors are often hypoosmotic to seawater. They are in danger of desiccation as water from body fluids diffuses into the hyperosmotic ocean water. They also must regulate the types of ions that are retained and eliminated from their bodies.

Marine organisms may also be exposed to freshwater when they enter rivers, which dilute the salt content of the incoming tidal water. Under these conditions, the water is brackish—not as salty as the sea, but not as pure as freshwater. The criticality of this situation depends upon whether the organism is tolerant or intolerant of salinity changes. Organisms that can live in only a narrow range of salinities are called stenohaline. Organisms that are tolerant of wide ranges of salinities are called euryhaline.

Terrestrial organisms are always hyperosmotic to their environment, so they continually face desiccation in the air. They also must adjust the ion composition of their body fluids, because the foods that they eat may not have inorganic ions in the desired ratios and because some ions are always lost in urine.

One example of the influence of these effects concerns the interaction of oysters and the protistan parasite known as MSX. (MSX stands for "multinucleate sphere unknown," which refers to the protista *Haplosporidium nelsoni*.) The MSX organism survives in osmotic pressures greater than 0.4 osmolar. Oysters are osmoconformers which grow in saline, brackish, and nearly fresh water. At osmotic concentrations less than 0.4 osmole, oysters can survive and are unaffected by MSX. When rainfall is abnormally low, however, the salinity of brackish water increases, and oysters which were protected in low-salinity water are exposed to higher-salinity water, which allows the MSX organism to infect them.

Organisms exposed to tides may protect themselves from exposure to variations in osmotic pressure by sealing themselves off, the way snails and bivalve mollusks do. Others may move offshore to more saline waters or onshore, away from the increasing salinity. Worms that burrow in the sediments of salt water are protected from transient changes in salinity because there is little exchange of solutes with the overlying salt water.

The vertebrates adapted to their various environments by using hormones to regulate salt and water balance. Because of the differing demands of aquatic and terrestrial environments, in different groups, the same hormone may have opposite effects, but that effect is always to maintain the optimal osmotic pressure to ensure survival.

—*Judith O. Rebach*

See also: Nutrient requirements; pH Maintenance; Thermoregulation.

FURTHER READING

Brown, A. D. *Microbial Water Stress Physiology: Principles and Perspectives.* New York: John Wiley & Sons, 1990. Explores the means by which organisms adapt to survive acute and chronic water stress.

Gilles, R., E. K. Hoffmann, and L. Bolis, eds. *Volume and Osmolality Control in Animal Cells.* New York: Springer-Verlag, 1991. A comprehensive review and Summary of current literature in the field of osmolality and volume control in animals in terms of both organic and inorganic ions.

Hadley, Neil F. *Water Relations of Terrestrial Arthropods.* San Diego, Calif.: Academic Press, 1994. Brings together data on the physiology and anatomy of arthropods in the context of their adaptations to terrestrial environments. Written for advanced students and researchers in the field.

Krogh, August. *Osmotic Regulation in Aquatic Animals.* Reprint. New York: Dover, 1965. Reprint of the 1939 edition, a classic work in the field. It is clearly written and covers all aspects including techniques used to determine osmotic pressure.

Prosser, C. Ladd. *Adaptational Biology: Molecules to Organisms.* New York: JohnWiley & Sons, 1986. This text, by the leader in the field of adaptational physiology, is one of the most comprehensive in the field.

Smith, HomerW. "The Kidney." *Scientific American* 188 (January, 1973): 40-48. Aclassic exploration of the structure and function of the kidney and its role in osmoregulation. Strange, Kevin, ed. *Cellular and Molecular Physiology of Cell Volume Regulation.* Boca Raton, Fla.: CRC Press, 1994. Atextbook on cellular osmoregulation and membrane transport physiology.

OSMOSIS

FIELDS OF STUDY

Biochemistry; Genetics; Molecular Biology

SUMMARY

The process of osmosis is defined, and its importance in chemistry-related fields is elaborated. Osmosis is a fundamental and necessary process for living systems, as well as the basis of a valuable technology used to produce potable water and to control water-based fluid systems.

PRINCIPAL TERMS

- **concentration gradient:** the gradual change in the concentration of solutes in a solution across a specifi c distance.

OSMOSIS

SALT WATER
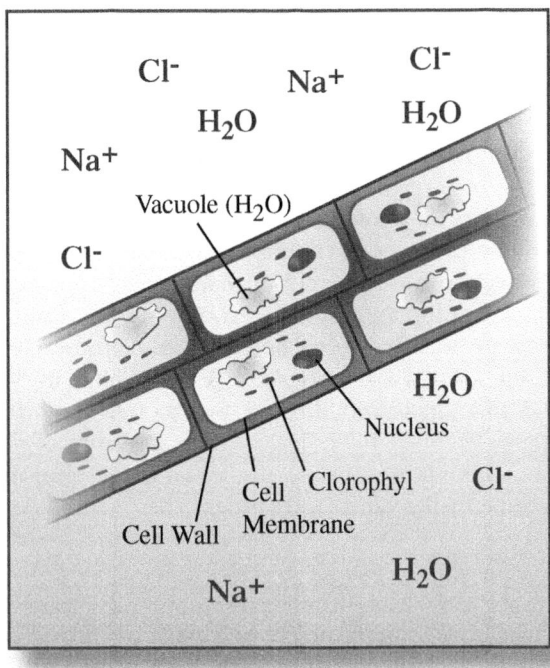

DISTILLED WATER
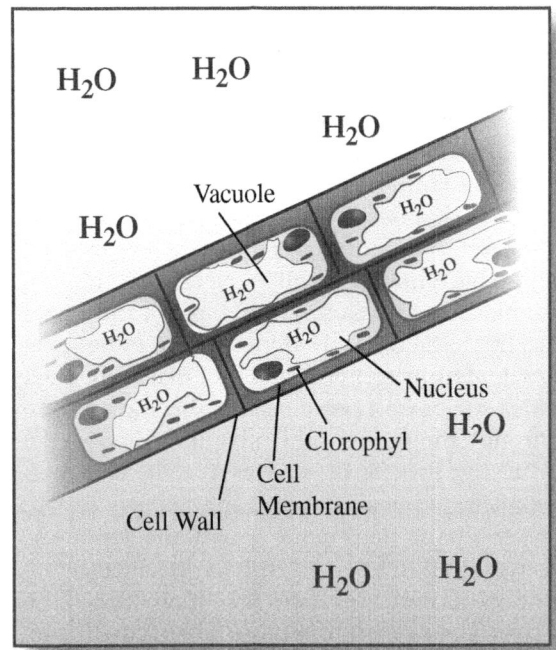

- **diffusion:** the process by which different particles, such as atoms and molecules, gradually become intermingled due to random motion caused by thermal energy.
- **equilibrium:** the state that exists when the forward activity of a process is exactly equal to the reverse activity of that process.
- **hypertonic:** describes a solution with a greater concentration of solutes than the solution to which it is being compared; in biology, a solution with a greater solute concentration than the cytoplasm of a cell.
- **hypotonic:** describes a solution with a lower concentration of solutes than the solution to which it is being compared; in biology, a solution with a lower solute concentration than the cytoplasm of a cell.
- **isotonic:** describes a solution with the same concentration of solutes as the solution to which it is being compared; in biology, a solution with the same solute concentration as the cytoplasm
- **osmotic pressure:** the pressure that would have to be applied to a solution to prevent the flow of solvent through a semipermeable membrane.
- **reverse osmosis:** the application of pressure to a solution in order to overcome the osmotic pressure of a semipermeable membrane and force water to pass through it in the direction opposite to normal osmotic flow.
- **semipermeable membrane:** a membrane that allows the passage of a material, such as water or another solvent, from one side to the other while preventing the passage of other materials, such as dissolved salts or another solute.
- **solute:** any material that is dissolved in a liquid or fluid medium, usually water.
- **solvent:** any fluid, most commonly water, that dissolves other materials.

VISUALIZING THE CONCEPT OF OSMOSIS
Osmosis is the process by which molecules of a solvent pass through a semipermeable membrane that is separating two solutions with differing concentrations of solute. The solvent moves from the solution with the lower concentration (the hypotonic solution) toward the one with the higher concentration (the hypertonic solution). The process will continue

until the concentrations of both solutions are equal and equilibrium is achieved.

A semipermeable membrane is necessary for osmosis to occur. Such a membrane acts as a porous barrier that will allow the passage of solvent molecules but not dissolved materials, such as various mineral salts. Water molecules, though polar in nature, are electrically neutral and very small. When salts such as sodium chloride are dissolved in water, they dissociate into ions, which are both electrically charged and significantly larger in size than the surrounding water molecules. A porous membrane that has pores big enough to allow electrically neutral water molecules to pass through, but not the larger electrically charged dissolved ions, is said to be semipermeable. Other types of dissolved materials, such as various sugars and proteins, are also too large to pass through the pores of the membrane and so are subject to the process of osmosis as well. The presence of a semipermeable membrane can produce an osmotic system. In an osmotic system, the hypotonic solution is confined to one side of the membrane and the hypertonic solution is contained on the other side. The process of osmosis will occur spontaneously and continue until the two solutions become isotonic, meaning that both solutions have the same concentration of solutes.

Diffusion and Osmotic Pressure

Diffusion can be demonstrated simply by adding a few drops of food coloring to a container of water, being careful not to mix them together, and then letting the water stand undisturbed. At first, the food coloring will remain where it was placed, but over time it will become evenly distributed throughout the water. The water molecules are in constant motion, and as they continually bump into the molecules of food coloring, they eventually spread them throughout so that the two become mixed together. During this process, the distribution of the food coloring in the water follows a concentration gradient, which is the difference in concentration of a solute when the concentration is not constant throughout the solution. Once the food coloring is evenly distributed, the result of this mixing is the same as if the solution had been stirred or agitated, but it requires a much longer time.

Diffusion is the mechanism that drives water molecules through the pores of a semipermeable membrane. Once they are through the membrane, the water molecules interact with the dissolved salts in that solution and remain. The process is reversible, so some water molecules are driven through the membrane in the opposite direction at the same time. However, the difference in concentration ensures that the net flow of water molecules is toward the hypertonic solution until equilibrium is achieved.

Osmosis can be prevented by applying pressure to the hypertonic solution. The amount of pressure that must be applied to stop osmosis is termed the osmotic pressure of the membrane, and it is dependent on both the temperature and the difference in concentration between the two solutions. Osmotic pressure was first described by Jean-Antoine Nollet (1700–1770), also known as Abbé Nollet, in 1748 and first measured directly by Jacobus van't Hoff (1852–1911) in 1877. The osmotic pressure is given the symbol ϖ and, in the case of an ideal solution, is defined by the van't Hoff equation as:

$$\varpi = RT(C_B - C_A)$$

where T is temperature in kelvins and C is the concentration in moles per liter (mol/L), or molars (M). R is the gas constant and can be written as:

$$R = 0.0821 \frac{\text{L atm}}{\text{mol K}}$$

where L is liters, atm is atmosphere (a unit of pressure), mol is moles, and K is kelvins. An ideal solution is a solution in which the molecules of solute and solvent interact with each other in the same way they interact with themselves. If the solution is not ideal, then an osmotic coefficient, must be included in the equation.

By applying pressure in excess of the osmotic pressure, the process can be driven in reverse. Reverse osmosis forces water molecules to pass through a semipermeable membrane from the hypertonic solution into the hypotonic solution. It is through this process that salt-free potable water can be produced from salty seawater or other non-potable sources.

A Demonstration of Osmosis

The process of osmosis and osmotic pressure can be readily demonstrated and observed. The essential feature of the demonstration is that two solutions are separated by a semipermeable membrane and so are not able to mix. This can be done by using the membrane as a partition to separate one half of the inside of a beaker from the other half. The solution on one side of the membrane is a solution of salt in water, while on the other side is just plain water. As osmosis takes place, the level of the saltwater solution will increase as the level of the unadulterated water decreases. The rate at which the levels change depends on the area of the membrane that is exposed to both solutions.

If both solutions contain dissolved salt but in different amounts, the same effect will be observed, but it will cease when the concentrations of the two solutions become equal. As water molecules pass through the membrane, the concentration of dissolved salt in the hypertonic solution decreases and the concentration of dissolved salt in the hypotonic solution increases. At the equilibrium point, when the two solutions become isotonic, water molecules pass through the membrane in both directions at the same rate.

Osmosis in Biological Systems

Cell membranes function as semipermeable membranes in living systems, allowing water, oxygen, carbon dioxide, sugars, enzymes, ions, hormones, metabolites, and various other cellular components to pass through as necessary. In order to maintain the proper amount of water in cells and prevent dehydration, living systems use a complex mechanism of osmoregulation that actively brings water into the cells to replace water that is lost through osmosis. Anything that interferes with this mechanism, such as the consumption of alcohol, use of drugs, smoking, or lack of sufficient water in the diet, adversely affects the viability of the system. Hangovers, for example, are partly the result of dehydration of cell fluids as alcohol is metabolized and often persist until the osmotic balance of the cells is restored.

Richard M. Renneboog, MSc

Further Reading

Costanzo, Linda S. *Physiology: Cases and Problems*. 4th ed. Baltimore: Lippincott, 2012. Print.

Kucera, Jane. Reverse Osmosis: Design, Processes, and Applications for Engineers. Hoboken: Wiley, 2010. Print.

Lafferty, Peter, and Julian Rowe, eds. *The Hutchinson Dictionary of Science*. 2nd ed. Oxford: Helicon, 1998. Print.

Lodish, Harvey, et al. *Molecular Cell Biology*. 7th ed. New York: Freeman, 2013. Print.

Pelczar, Michael J., Jr., E.C.S. Chan, and Noel R. Krieg. *Microbiology: Concepts and Applications*. New York: McGraw, 1993. Print.

Reece, Jane B., et al. *Campbell Biology*. 10th ed. San Francisco: Cummings, 2013. Print.

OSMOSIS SAMPLE PROBLEM

Use the van't Hoff equation to determine the osmotic pressure (π) between two ideal solutions of sodium chloride, one with a concentration of 0.01 M and one with a concentration of 0.02 M, in water at a temperature of 20°C. Use $R = 0.0821 \frac{\text{L atm}}{\text{mol K}}$.

Answer:

Convert the temperature from degrees Celsius (°C) to kelvins (K):

$$K = °C + 273.15$$
$$K = 20 + 273.15$$
$$K = 293.15$$

The van't Hoff equation is

$$\pi = RT(C_B - C_A)$$

Substitute in the values of R ($0.0821 \frac{\text{L atm}}{\text{mol K}}$), T (temperature), C_B (concentration of solution B), and C_A (concentration of solution A) and calculate, paying attention to the units throughout:

$$\pi = RT(C_B - C_A)$$
$$\pi = (0.0821 \frac{\text{L atm}}{\text{mol K}})(293.15 \text{ K})(0.02 \frac{\text{mol}}{\text{L}} - 0.01 \frac{\text{mol}}{\text{L}})$$
$$\pi = 0.24067615 \text{ atm}$$

The osmotic pressure between 0.01 M and 0.02 M solutions of salt in water is 0.24067615 atm, or 24386.51 pascals (Pa).

Parasitology

FIELDS OF STUDY

Biology; internal medicine; ecology; infectious disease; public health; evolutionary biology; microbiology; pharmacology; zoology; bacteriology.

SUMMARY

Parasitology is the study of parasites and the relationship between parasites and host organisms. The primary interest in parasitology is to investigate the role of parasites in diseases that affect humans, livestock, and pets. Parasitologists work closely with medical professionals and pharmacologists to develop drugs that combat parasitic infections. Parasitology became a distinct branch of medicine and biology in the mid-nineteenth century after the development of microscope technology. Although medical parasitology dominates the field, some ecologists and zoologists are studying parasites to learn more about the role parasitism plays in evolution and ecology.

PRINCIPAL TERMS

- **bacterium:** single-celled organism that lacks a defined nucleus within its cell body; also called prokaryote.
- **ectoparasite:** parasitic organism that lives outside or on the surface of the host organism.
- **endoparasite:** parasitic organism that lives within the host organism.
- **helminth:** parasitic worm, such as a roundworm or tapeworm, in the intestines of vertebrates.
- **host:** organism that carries a parasite or organism and is the target organism for a parasite.
- **protozoan:** any of a diverse group of single-celled organisms from the eukaryotic group that has a nucleus contained within a discrete chamber inside the cell.
- **vector:** intermediary organism that delivers a parasite or parasitic reproductive stage to the ultimate host.

BASIC PRINCIPLES

Parasitology is the study of parasite organisms, their relationship with host organisms, and their role in disease. Parasitism is defined as a prolonged and intimate association between two organisms in which one organism (the parasite) benefits at the expense of the other organism (the host). Scientists have identified thousands of parasites from a variety of groups, including bacteria, protozoa, animals, plants, and fungi. The study of parasitic bacteria is covered under bacteriology, while parasitologists concentrate on parasites from the eukaryotic group, which includes protozoa, animals, fungi, and plants.

Parasites infect their hosts for the purpose of obtaining food or completing a portion of their reproductive cycle. They may infect their host by either attaching to the outside of the host's body, entering an existing opening such as the mouth or anus, or tunneling through the host's tissues. Some parasites are specialized to infect a single type of host organism, and others can parasitize organisms from a variety of species. Some parasites display complex life cycles that may include infecting hosts from different species during specific parts of their life cycles. Humans are vulnerable to infection by at least three hundred types of parasitic worms and more than seventy species of protozoa.

A key feature of parasitism is that it causes damage to the host. Some parasites reduce the host's ability to obtain or absorb nutrients, while others damage tissues directly. Some parasites are known to cause disease, such as malaria and Lyme disease.

BACKGROUND

Early civilizations, including the ancient Greeks, kept records of patients suffering from infection by large, easily visible parasites such as the tapeworm. However, scientists were not able to examine parasites in detail until the invention of the microscope in the mid-seventeenth century. The next major advance was the discovery of bacteria in the late nineteenth century,

The Italian Francesco Redi, considered to be the father of modern parasitology, was the first to recognize and correctly describe details of many important parasites

which precipitated the discovery of the link between parasites and disease.

In the early twentieth century, scientists conducted the first detailed studies into the nature of malaria, giardiasis, and many other diseases caused by parasites. Although scientists have been unable to develop vaccines for malaria, sleeping sickness, and many other types of parasite-related diseases, researchers have been able to drastically reduce instances of infection because of a greater understanding of the life cycles of the parasites and the vectors (dispersal organisms) they use. In addition, cooperation between parasitologists and pharmacologists led to the development of medications that can effectively reduce the intensity of parasitic infections.

Although the medical study of parasitism remains the most active facet of the field, ecologists and evolutionary biologists began studying the role of parasites in nature in the 1980's, ushering in a new age of parasitology research. Studying the role of parasites in nature has also helped further the study of medical and agricultural parasitology.

Parasitology is divided into three major branches of study: helminthology (parasitic worms), entomology (insect parasites), and protozoology (parasitic protozoa). Each field is further divided into specialties, of which the most important are medical, agricultural, and ecological parasitology.

MEDICAL PARASITOLOGY

Medical parasitologists focus on parasites that cause disease and infection in humans. They work closely with pharmacologists in the development of vaccines and antiparasite drugs and with physicians to develop new therapeutic techniques. Parasitologists also collect information that is used to create guidelines for diagnosis. This includes documenting symptoms common to each type of parasitic infection and cataloging the life stages of parasitic species. Intestinal parasites may cause diarrhea, intestinal pain, and the development of granulomas, or tumorlike masses. Other parasites can cause muscle and joint pain, fatigue, and various skin lesions and rashes.

In most cases, patients with parasitic infections are treated with drugs that function by differential toxicity, meaning that the chemicals in question are more toxic to the parasites than to the host. However, most antiparasite medications are also somewhat toxic to the host and can cause a variety of side effects.

AGRICULTURAL AND VETERINARY PARASITOLOGY

The subfields of agricultural and veterinary parasitology are extremely important from an economic perspective. A 2009 study from India estimated that crop parasites cause losses amounting to $200 billion annually worldwide, while an additional $3.5 billion is spent combating animal parasites. Some parasites, such as the parasitic roundworm, are capable of infecting plants, animals, and humans.

In many cases, parasites lead to the death of livestock or crop plants. In addition, research has shown that even minor infections can affect an animal's metabolism in such a way that the animal will never obtain full growth or development. Parasites therefore lead to lower yields from both crops and livestock. Agricultural parasitologists perform a variety of laboratory experiments to develop antiparasite drugs and vaccines. Medications developed in agricultural laboratories are sometimes co-opted by medical parasitologists to create medications used to treat human patients.

Ecological Parasitology

Ecological parasitologists investigate the role of parasites in nature. This includes making field observations of the various methods parasites use to survive in their environments and their ultimate effect on the evolution of species.

Many evolutionary biologists believe that parasitism has had an important influence on the evolution of species. In a 2009 study of a species of snail from New Zealand, researchers found that in populations with high levels of parasites, the snails will switch from asexual reproduction (producing clones) to sexual reproduction, which produces genetically diverse offspring that are more resistant to infection. This discovery led the researchers to speculate that parasites may have been instrumental in the evolution of sexual reproduction.

Applications

The development of antiparasite drugs involves research from all branches of parasitology as well as organic chemistry, pharmacology, and infectious disease medicine. Some antiparasite medicines have been found to have effects on nonparasitic illnesses, leading to greater integration between parasitology and other branches of medicine.

Malaria. Malaria is one of the most common and widespread diseases caused by parasite infection. According to the Centers for Disease Control and Prevention (CDC), between 350,000 and 500,000 cases of malaria are reported annually, and malaria has one of the highest fatality rates of all diseases. Malaria is caused by plasmodium, a parasite that is typically spread through mosquito bites.

The typical strategy for treating malaria involves using an array of antimicrobial drugs designed to kill plasmodium as it develops. Treatment is difficult because plasmodia have tremendous genetic diversity and many strains are resistant to existing treatments. In 2009, a team at the Monash University ARC Center announced the results of a set of experiments indicating that it may soon be possible to use a drug to deactivate an enzyme essential for the malaria parasite to digest nutrients. Deprived of this key enzyme, the malaria parasite will starve inside its host.

Parasite Prevention. In addition to treating parasites with drugs, researchers also focus on preventing infection through better hygiene and behavioral modification. In the case of malaria, controlling mosquito populations is an essential step toward preventing the spread of malaria. Parasitologists also contribute to the development of antibug sprays, mosquito netting, and other types of insect repellents in an effort to prevent parasite infection.

Another area of research in parasite prevention involves finding ways to prevent parasites from spreading through food or water. Ultraviolet (UV) light treatments have been effective in disinfecting food and water and killing parasites in various stages of life. Although UV radiation has often been used to disinfect foods, water treatment using UV radiation has just begun to become widespread.

In 2009, New York City became the first city in the United States to mandate UV light treatment for water processing in an effort to eliminate cryptosporidium, a parasitic microbe that causes a diarrheal infection known as cryptosporidiosis. Before the process was introduced in the United States, it had been used effectively in Britain to reduce instances of cryptosporidium infection.

Impact on Other Medical Fields

Discoveries from parasitology have filtered into other areas of research and development. For example, the physiological reaction to parasite infection gives important clues about the function of the immune system and the development of resistance. In some cases, medications developed to combat parasite infection have been found to have beneficial therapeutic effects for other types of diseases. In 2008, for instance, it was discovered that miltefosine, a drug used to treat patients suffering from protozoan infections, could potentially be effective in treating people infected with the human immunodeficiency virus (HIV). Miltefosine works by preventing the development of macrophage reservoirs, cells that, because of their long life, allow the virus to proliferate in secret before it becomes detectable.

Future Prospects

With millions of people worldwide suffering from parasitic diseases and many more living with the threat of infection, the study of parasitism is vital to public health. Given the tremendous economic and social effect of parasitic diseases, research in parasitology has the potential to create major changes around the world. Many parasitologists believe that the future of parasitology depends on developments in genomics and genetic medicine. Scientists have been examining the genetic components of parasitic organisms with the goal of finding more effective treatments for parasite-related diseases. The next generation of antiparasite medications may be genetically tailored to combat parasite organisms, thereby reducing the unpleasant side effects caused by differential toxicity.

Micah L. Issitt, MA

FURTHER READING

De Kruif, Paul. *Microbe Hunters*. 1926. Reprint. New York: Harcourt, Brace, 2006. A classic work that introduces some aspects of microbiology and bacteriology and contains some information about parasitology and eukaryotic microbes that cause disease as well as interesting coverage of malaria.

Esch, Gerald W. *Parasites, People and Places: Essays on Field Parasitology*. New York: Cambridge University Press, 2004. Anecdotal accounts of parasitology illustrate many aspects of the field and cover elements of the history of parasitology as well as medical and evolutionary investigations of parasite behavior.

John, David T., et al. *Markell and Voge's Medical Parasitology*. 9th ed. St. Louis, Mo.: Saunders Elsevier, 2006. Covers the basics of medical parasitology from treatments and pharmacology to genomic research. Written for readers with a strong medical or biological background.

Moore, Janice. *Parasites and the Behavior of Animals*. 2002. Reprint. New York: Oxford University Press, 2005. An interesting look at a variety of parasites and other animals, written for the general reader; contains information about medical, agricultural, and ecological parasitology.

Zimmer, Carl. *Parasite Rex: Inside the Bizarre World of Nature's Most Dangerous Creatures*. New York: Simon and Schuster, 2001. An account of interesting parasites, their ecology, evolution, and effect on culture and society; covers medical issues, parasite ecology, and other facets of the field.

Zuk, Marlene. *Riddled with Life: Friendly Worms, Ladybug Sex, and the Parasites That Make Us Who We Are*. Fort Washington, Pa.: Harvest Books, 2008. This introduction to parasites that affect humans in both negative and negligible ways focuses on the evolutionary aspect of parasitism and also contains information on medical and agricultural parasitology.

WEB SITES

Centers for Disease Control and Prevention A-Z Index of Parasitic Diseases http://www.cdc.gov/ncidod/dpd/parasites/index.htm

U.S. Department of Agriculture, Food Safety and Inspection Service Parasites and Foodborne Illness http://www.fsis.usda.gov/factsheets/parasites_and_foodborne_illness/index.asp

See also: Immunology and Vaccination; Virology.

FASCINATING FACTS ABOUT PARASITOLOGY

- There have been four Nobel Prizes awarded for work pertaining to the treatment of malaria.
- The parasitic isopod *Cymothoa exigua* infects fish by eating the fish's tongue. The worm then attaches to the fish's mouth and functions as a replacement tongue, living off scraps from the fish's meals.
- One of the methods being tested for fighting malaria involves using chocolate, which sticks to the blood fats on which the parasites feed and thereby weakens them.
- The filarial worm, a parasite that burrows into an animal's organ systems and causes damage, is recognized as the second leading cause of permanent and long-term disability in the world.
- In a 2006 study from the Imperial College of London, researchers found a potential link between toxoplasmosis, a parasite found in cat feces, and the family of mental disorders known as schizophrenia.
- A species of parasitic hairworm infects the brain of a grasshopper and influences the insect to commit suicide by leaping into open water. Once in the water, the parasite emerges from its host and morphs into its aquatic adult form.

PATHOLOGY

FIELDS OF STUDY

Biology; chemistry; biochemistry; immunology; chemical pathology; clinical pathology; forensic pathology; anatomic pathology; surgical pathology; blood banking; hematology; histology; cytology; genetics; microbiology; toxicology.

SUMMARY

Pathology is the scientific study of the nature of disease and its causes, processes, development, consequences, and resolution. Pathologists examine tissue and body fluids to determine if disease is present and, if so, its nature and extent. Many decisions on whether and how to treat disease are based on results of tests delivered by pathologists as laboratory reports. Pathologists also work to determine the cause of death in questionable or puzzling circumstances. This field is changing rapidly as new technologies that allow disease to be found in earlier stages are developed and implemented. Genetic mutations and diseases caused by them are also an emerging pathology field.

PRINCIPAL TERMS

- **atrophy:** wasting away of an organ or tissue because of disease or injury.
- **biopsy:** removal and examination of tissue.
- **degeneration:** deterioration or loss of function.
- **hyperplasia:** abnormal increase in number of cells.
- **hypertrophy:** abnormal increase in size of an organ or tissue.
- **inflammation:** response of tissue to injury or infection.
- **positive predictive value:** likelihood that the positive results of a test are reliably positive; a high positive predictive value means the test is reliable, while a low positive predictive value means further testing should be pursued.
- **reagent:** substance used to create a chemical reaction that detects, measures, or produces another substance.
- **sensitivity:** proportion of individuals who will correctly test positive when testing for a particular disease; if a test has high sensitivity, it will most likely recognize all those tested who truly have the disease.
- **specificity:** probability that a test will correctly identify those who do not have a specific disease; a test with high specificity can be used to rule out that disease.

BASIC PRINCIPLES

Pathology is the study of disease and involves examining biopsied tissues to determine whether atrophy, degeneration, hyperplasia, hypertrophy, or inflammation has occurred and, if so, to what extent. Pathologists also examine body fluids for signs of disease. They interpret the results of biopsies and other tests and send them to the practicing clinician who requested the test and will pass those results on to the patient. There are many subspecialties of pathology that focus on specific types of testing or specific body tissues or fluids. Genetic testing is another area in which pathologists are involved. They examine a patient's genetic materials to determine whether there are any genetic mutations that are likely to be passed on to the patient's children or that could eventually result in a genetic disease. For example, examining a woman's genetic material for the presence of a gene associated with breast cancer could help that woman determine whether her personal risk of breast cancer is increased and decide whether she will make lifestyle changes that may help her avoid breast cancer.

Pathologists are often involved in research studies. Volunteers donate tissues or body fluids to help these scientists study the effects of long-term behaviors, such as smoking or exercise, on body tissues to help understand diseases and the disease process. Through research and development, pathologists may create better laboratory tests.

Pathologists are medical doctors and are very involved in determining how a patient's disease should be treated, although they are usually not involved in seeing patients in a clinical environment. They are much more likely to be involved in patient care through testing and research.

BACKGROUND

Disease and its causes have been poorly understood throughout history. In many Western societies,

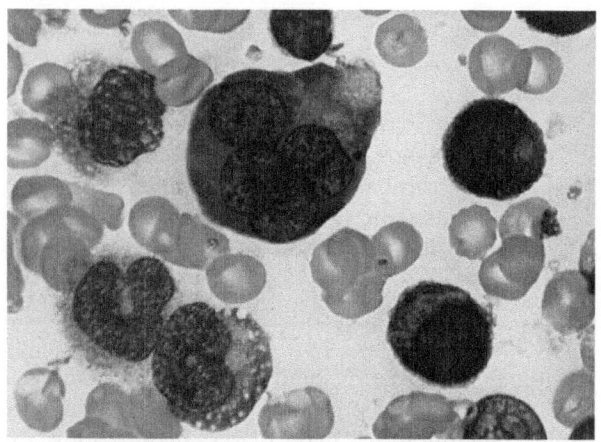

A bone marrow smear from a case of erythroleukemia showing a multinucleated erythroblast with megaloblastoid nuclear chromatin

autopsies were prohibited for religious reasons until the late Middle Ages, and scientists and doctors knew next to nothing about how death occurred and how diseases could affect body tissues and fluids at the cell level. In 1761, Italian anatomist Giovanni Battista Morgagni published the first book to discuss diseases in individual organs. Not until the mid-nineteenth century did German physician Rudolf Virchow's theories of cell-based disease replace the archaic notion that humors caused infections. That same century, German physician Robert Koch and French scientist Louis Pasteur theorized that bacteria caused some diseases.

Because pathology is based on the study of tissues and fluids, its history is closely tied to that of the microscope. Until the microscope was improved in the nineteenth century, it was impossible for scientists to closely examine tissues to determine whether disease was present and, if it was, what effect it had on the body. Advances in microscopic technology, including the electron microscope; new fields of study such as immunohistochemistry and molecular biology; and methods of preparing tissues such as staining and culturing have greatly improved pathologists' ability to study disease and disease processes. Scientists have come to understand that all diseases are reflected by changes that extend down to the molecular level, and testing methods such as polymerase chain reaction (PCR), fluorescence in situ hybridization (FISH), and mass spectrometry have changed the way laboratory tests are conducted.

How It Works

The process of analyzing tissue or a body fluid depends on the type of tissue or fluid being examined and the reason for testing it. Usually a sample is taken in a clinical setting and prepared for delivery to a laboratory. For example, if the sample is blood, it may be drawn at a clinic, then placed in special tubes with a substance that keeps the blood from clotting. When the sample arrives at a laboratory, it is prepared for the selected test. For example, tissue may be sliced very thinly, placed on a microscope slide, and stained so that a technician can examine it, or blood may be spun in a centrifuge to separate it into its components, then different reagents may be added to the blood to cause a chemical reaction that will help the technician determine whether the substance for which he or she is looking is in the blood sample. Another type of processing is that used for genetic material: A small sample of the genetic material is replicated many times to make it easier to determine if the sample contains errors. The sample is then processed, either automatically with specialized machines that determine the test results or by trained workers who examine the specimen and determine the results. The test results are compared with values that have been determined to be normal, and if the results fall out of the normal range, they are further interpreted. For example, if a pregnant woman takes a standard blood test and some test values fall outside of the normal range for a woman who is not pregnant, the test values must be compared to what is normal for a woman in that particular stage of pregnancy. After the results are interpreted, the laboratory sends a report containing the test values and an interpretation of the results to the clinician who ordered the test so that the results can be shared with the patient. The results often play a central role in determining the course of treatment.

Applications

Although pathology generally refers to laboratory testing, the science is usually broken down into the following specialties: Anatomic pathology, chemical pathology, forensic pathology, genetic pathology, hematology, immunology, and microbiology.

Anatomic Pathology. Sometimes called surgical pathology, anatomic pathology uses biopsied tissue from a person (living or dead) to diagnose disease (or a possible cause of death). A common example of this type of test is a Pap smear, which tests for cervical cancer. Another example is the examination of tissues for the presence of disease during surgery. A surgeon removes a tumor, and a pathologist examines its edges, or margins, to determine if the surgeon removed the entire tumor. If the pathologist does not see healthy tissue in all the margins, he or she informs the surgeon, who removes additional tissue. This process continues until the pathologist and surgeon are convinced that the tumor is fully excised. Subcategories of this specialty include histology (preparing tissues for examination), cytology (performing tests on tissues to determine if cancer exists), and forensic pathology (performing autopsies and analyzing tissues to determine the cause of death).

Chemical Pathology. Also called biochemistry, chemical pathology detects substances and changes in blood and bodily fluids. An example of this type of testing would be examining blood glucose levels to determine risk factors for diabetes or its progression. This type of testing also detects enzymes and proteins in blood that may change with the progression of illness or measures cancer tumor markers that show whether a tumor is increasing or in remission. A subspecialty is toxicology, which involves testing to look for poisons, drugs, or other toxins in the body.

Forensic Pathology. A subspecialty of anatomic pathology, forensic pathology focuses on investigating cases of sudden or unexpected death. Forensic pathologists perform autopsies to identify the cause of death and examine tissues and body fluids to reconstruct how the death occurred. This type of pathologist may be required to visit crime scenes or help law enforcement personnel in other ways.

Genetic Pathology. In genetic pathology, chromosomes and DNA are tested to diagnose diseases that may have genetic components. Examples include a chromosomal test to determine whether a fetus has Down syndrome and an evaluation of a DNA specimen to determine whether a woman has a gene associated with breast cancer. Subcategories include biochemical genetics (identifying specific genetic markers with biochemical testing), cytogenetics (performing an analysis of chromosomal abnormalities using a microscope), and molecular genetics (analyzing mutations in the DNA in genes).

Hematology. Pathological hematology focuses on diseases that affect blood and the organs that create blood. For example, blood could be tested to see if the patient has a clotting disorder. This specialty includes transfusion medicine, which involves performing blood typing and compatibility testing and managing the supply of blood products to ensure that blood is safe for transfusion into patients.

Immunology. Pathological immunology focuses on allergies, inflammations, and autoimmune diseases. For example, a patient's blood could be tested to determine if he or she is allergic to something and to identify the allergen. These tests can also determine if the immune system is malfunctioning, as in diseases such as lupus or multiple sclerosis, where the immune system targets normal systems as allergens and attempts to destroy the normal system's tissues.

Microbiology. Testing in pathological microbiology involves diseases caused by infectious agents. For example, a patient's urine sample could be tested to see if the individual has a urinary tract infection. Tests look for the presence of bacteria, fungi, parasites, and viruses. Pathologists may work with public health officials and be involved in efforts to control disease outbreaks, or they may attempt to solve problems related to drug-resistant bacteria.

Other Applications. Some pathologists practice as clinical pathologists. They may work in a rural area or with a community hospital and are usually trained in chemical pathology, hematology (including blood typing), and microbiology but usually do not practice anatomic pathology. General pathologists are trained in all areas of pathology (including anatomic) but at a lesser depth of knowledge, so if necessary, they refer cases to specialized pathologists. Another type of pathologist is a specialist in a body system, such as a dermatopathologist, who specializes in diseases of the skin, or a nephropathologist, who focuses on the renal system.

Pathologists are involved in research and development. They may develop laboratory tests that have more specificity, more sensitivity, better positive predictive value, or a faster turnaround time, or work on reducing the cost of testing. They also may conduct fundamental research, such as evaluating tissues or body fluid samples from patients with a specific disease and patients from a control group to determine similarities and differences.

FUTURE PROSPECTS

The aging population in the United States is creating an increased need for laboratory testing to diagnose disease and determine an appropriate course of treatment. Pathologists can help reduce the rising cost of health care by identifying which tests are the most cost-effective to perform and which are unlikely to help determine a clinical path of treatment. For example, in men over the age of eighty-five, many studies have shown that the negatives (loss of continence and function) outweigh the positives of treatment for prostate cancer and, therefore, recommend no treatment for these men. Pathologists can help write and publicize guidelines for clinicians so that testing is not performed when the treatment does not vary depending on the outcome of the test. Pathologists can thus shape health care policies and practices that are cost-effective and improve patient care outcomes.

Genetic testing is another area in which pathologists can help guide informed decision making. This type of testing can help a couple with a history of family genetic problems determine their risk of having a child with genetic mutations. For example, a man and woman with a family history of cystic fibrosis can be tested to determine if they carry the gene that causes that disease. The results allow pathologists to calculate the couple's risk of conceiving a child who will carry that gene or who will develop the disease. Knowing the calculated risk, the couple can determine whether they want to have a child naturally or to investigate other options. Genetic testing on a fetus can show whether it has genes or genetic mutations that are likely to cause diseases, thus allowing the parents to make an informed decision as to whether to continue or terminate the pregnancy. If the decision is made to continue the pregnancy, the parents and health care workers can be better prepared to deal with the altered circumstances and health issues.

As health care becomes more personalized, a pathologist can interpret test data as it pertains to a particular patient and that patient's family history. For example, patients who have a family history of alpha-1-antitrypsin deficiency, a disease that has genetic components, could be tested to see whether they have inherited those genes, and if so, which level of lung disability they might be likely to have. This could affect lifestyle decisions, such as not smoking, not working in an industry that might negatively affect their lungs, or avoiding being treated for asthmalike symptoms with drugs that do not help this disease. Another application of pathology to personalized health care is that of obtaining detailed information about the disease or condition that a patient has to determine the best course of treatment. For example, breast cancer is thought to be the result of several different disease processes, and the most effective treatment depends on the process involved. Identifying the type of breast cancer allows physicians to prescribe the most appropriate type of treatment. For example, if the patient's type of cancer does not respond well to chemotherapy, the doctor might recommend surgery.

Marianne M. Madsen, MS

FURTHER READING

Damjanov, Ivan. *Pathology for the Health Professions.* 4th ed. Maryland Heights, Mo.: Elsevier/Saunders, 2012. A basic overview of pathology with clear pictures and review questions; discusses normal pathology and diseases in various systems.

_____. *Pathology Secrets.* 3d ed. Philadelphia: Mosby/Elsevier, 2009. A basic review in question-and-answer format, focusing on practical knowledge.

Hayes, A. Wallace. *Principles and Methods of Toxicology.* 5th ed. Boca Raton, Fla.: CRC Press, 2008. Includes history of toxicology and discusses interpretation of data and problems that may arise. Comprehensive glossary.

Kemp, William L., Dennis K. Burns, and Travis G. Brown. *Pathology: The Big Picture.* New York: McGraw-Hill Medical, 2008. Focuses on broad pathology concepts. Contains full-color illustrations, summary tables and figures, and questions and answers.

Kumar, Vinay, et al. *Robbins and Cotran Pathologic Basis of Disease.* 8th ed. Philadelphia: Saunders/Elsevier, 2010. Regarded by some as the keystone book and used as the primary text in pathology at many medical schools. Includes illustrations and case studies.

Richards, Ira S. *Principles and Practice of Toxicology in Public Health.* Sudbury, Mass.: Jones & Bartlett, 2008. An overview of toxicology in public health practice, written in an easy-to-read style. Includes glossary and index.

Rubin, Raphael, and David S. Strayer, eds. *Rubin's Pathology: Clinicopathologic Foundations of Medicine.* 6th ed. Philadelphia: Lippincott Williams & Wilkins, 2011. Award-winning book taking a clinical approach to pathology. Includes full-color, enhanced illustrations.

Zaher, Aiman. *Pathology Made Ridiculously Simple.* Miami: MedMaster, 2007. A book of mnemonics and cartoons to aid in memory and retention.

WEB SITES

American Pathology Association https://www.apfconnect.org

American Society for Clinical Pathology http://www.ascp.org

Association for Molecular Pathology http://www.amp.org
College of American Pathologists Lab Tests Online http://www.labtestsonline.org

See also: DNA Analysis; Hematology; Immunology and Vaccination; Parasitology; Virology.

FASCINATING FACTS ABOUT PATHOLOGY

- Forensic pathologists are said to speak for the dead. Their investigations tell the story of a person's death after that individual can no longer speak.
- The popular television series *CSI: Crime Scene Investigations*, which features the work of forensic evidence investigators, including pathologists, premiered in 2000 and spawned several spinoffs and imitators.
- Laboratories in the United States perform more than 10 billion tests each year.
- More than 2,000 laboratory tests can be performed on blood and body fluids alone.
- More than 99.9 percent of the samples that pathologists analyze are from living people, not from autopsies of corpses.
- In its 2006 annual report, the Los Angeles County Department of Coroner reported performing 4,401 complete and 450 partial autopsies. It also conducted 5,499 toxicology studies.
- Veterinary pathologists may participate in drug development as they monitor and describe the effects of drugs on laboratory animals.
- Environmental pathology deals with diseases resulting from environmental factors, such as chemicals.
- Phytopathology is the study of plant diseases.

Placental mammals

FIELDS OF STUDY

Anatomy, biochemistry, cell biology, embryology, physiology, reproductive science

SUMMARY

Placental reproduction is the most common form of mammalian reproduction. All mammals except the marsupials (such as the kangaroo and opossum) and the monotremes (platypus and echidna) form placentas.

PRINCIPAL TERMS

- **chorion:** the outer cellular layer of the embryo sac of reptiles, birds, and mammals; the term was coined by Aristotle
- **embryo:** a young animal that is developing from a fertilized or activated ovum and that is contained within egg membranes or within the maternal body
- **endometrium:** an inner, thin layer of cells overlying the muscle layer of the uterus
- **fetus:** a mammalian embryo from the stage of its development where its main adult features can be recognized, until birth maternal: referring to the female parent
- **ovum:** an unfertilized egg cell
- **uterus:** in female mammals, the organ in which the embryo develops
- **viviparous:** producing young that are active upon birth (often referred to as live birth); the embryo is nurtured within the uterus

BASIC PRINCIPLES

Monotremes are oviparous, or egg-laying mammals. Marsupials are ovoviviparous, meaning that the egg is large and has a yolk adequate to nourish the embryo during its early development, but it remains unattached to the wall of the uterus. Gestation in marsupials is necessarily short, therefore, and the young are born in an immature, fetal stage. They make their way to the mother's pouch and continue to grow, nourished by the mother's milk. All other mammals, termed placental, are viviparous. The small egg, lacking food substance, becomes attached to the uterine wall, and the developing embryo is nourished by the mother's blood passing through a placenta. This process allows longer gestation, and as a result the young are born in a more advanced (precocial) state of development.

The placental mammals form a diverse and successful group that includes the insectivores (such as shrews, hedgehogs, and moles), bats, sloths, anteaters, armadillos, primates (to which humans belong), rodents, rabbits, whales, dolphins and porpoises, carnivores (such as cats, dogs, and bears), seals, aardvarks, elephants, hyraxes, manatees, uneven-toed mammals (such as tapirs, horses, and rhinoceroses), and even-toed (cloven-hoofed) mammals such as pigs, camels, deer, sheep, cattle, and goats.

PLACENTA STRUCTURE

The term "placenta" comes from the Latin meaning "flat cake," and is used to describe the flat structure (in most animals) which attaches the developing embryo and fetus to the wall of the uterus. The term was first used in 1559 CE, although knowledge of the placenta, at least in humans, goes far back into antiquity. Reference to it may be found in many ancient texts and drawings, including the Old Testament books of the Bible. Early Egyptians considered the placenta to be the seat of the external soul.

The placenta is the organ responsible for the transmission of materials between mother and fetus prior to birth—the only bridge between them. In many species, it has important endocrine functions, producing hormones necessary for devel- opment of the fetus or for maintenance of the pregnant state. It is essentially a product of both the developing ovum and the mother. The placenta is commonly referred to as the afterbirth, extruded from the mother's uterus at the end of the birth process.

When the ovum is released by the ovary and fertilized by sperm, it eventually comes to rest in the hollow cavity of the uterine horn (one of two chambers) or the uterus (single chamber). While it moves toward that area it begins to divide, forming a ball of cells known as a morula. As development continues, the ball becomes hollow, and it is then referred to as a blastocyst. Within this hollow blastocyst a few cells protrude into the cavity, forming a knob. These are the only cells that will eventually develop into the embryo and fetus. The rest of the cells are responsible for forming the supportive structures, including the chorion, the placenta, and the umbilical cord (which attaches the embryo to the placenta).

Among the eutherian (placental) mammals, a variety of placental configurations occurs. In many species the entire chorionic sac becomes connected to the uterine wall, and transfer of materials between the maternal and fetal compartments occurs over the whole surface. In other species, a much more specialized system develops. Here parts of the chorion (a membrane equivalent to the one that lines the shell of reptile and bird eggs) become highly specialized, establishing an intimate relationship with the uterine tissues. Thus, transfer of materials occurs only in one select region of the chorion, referred to as the placenta. It is these tissues, with their flattened, cakelike appearance, from which the name derives.

The most important feature of the placenta is the close contact between the fetal blood vessels in the placenta and the maternal blood vessels in the uterine wall. While it is a common misconception that the fetal and maternal blood mix or flow together, this is not a correct picture. What actually occurs is that the two blood systems come close to each other, at which point materials that can diffuse out of one vessel may diffuse into another. Thus, transfer of materials from mother to fetus, and vice versa, can occur. The closer the two blood pools come to each other, the better it is for transfer. Nutritional, respiratory, and excretory products are transferred.

PLACENTA TYPES

In the epithelio-chorial placenta, as found in the hoofed mammals (such as Artiodactyla and Perissodactyla), whales, and lemurs, the wall of the uterus retains its surface epithelium. The minimum separation of bloods is four cells thick (two epithelial cell layers and the endothelium of the blood vessels). Thus, it is vital to have a very large surface area to allow for adequate movement of materials. While this is a large improvement over the marsupial system, it is nonetheless 250 times less efficient at salt transfer than the placenta used by humans.

The separation between mother and fetus is reduced in carnivores and sloths. Here the chorion invades the uterine epithelium and comes in direct contact with the epithelium of the maternal blood vessels, allowing a more uniform transfer of materials. The most advanced form, showing minimal separation, is the hemo-chorial placenta, found in humans, rodents, bats, and most insectivores. Here the maternal blood vessel walls are chemically broken down and the invading chorion is now in direct contact with the maternal blood stream. Because this is so much more efficient for materials exchange, the size of the interacting surfaces can be much reduced.

In addition to the exchange of materials, the placenta plays an important role in immunology, and without the placenta, the mother's body would reject the developing embryo like any other foreign body. It is this tolerance of the embryo that separates the placental mammals from the marsupials, and allows for gestation periods to be extended. Fetuses in placental mammals receive antibodies from their mothers, thus enhancing their early immunity to disease.

The epithelio-chorial placenta, being only in contact with the uterine wall and not being invasive, is readily shed by the uterus when the fetus is born. There is no damage to the maternal tissue. The more invasive types of placenta, including the hemo-chorial type, can only be lost by separation through the uterine tissues. Thus, birth in species with hemo-chorial placentas is of necessity associated with some degree of maternal bleeding. In fact, in many hemo-chorial placentas, the blastocyst actually digests (chemically) the endometrial lining of the uterus and comes to lie completely within it. The endometrium then heals over the blastocyst, which then grows, fully surrounded by endometrium. The true placenta forms on the deep pocket of the endometrium in which the blastocyst lies. When the fetus is expelled from the uterus (that is, at birth), it ruptures the now very thin layer of stretched endometrium that covers the chorion. The placenta separates from the uterus as a result of rupture of the uterine blood vessels and tissues when the uterus contracts down after expulsion of the fetus. Bleeding between uterus and placenta produces a clot which eventually seals off the broken blood vessels and forms the basis for endometrial repair.

Most mammalian placentas, regardless of type, have some type of endocrine (hormonal) function. While the specific hormones may vary from species to species, two in particular are found in most placental mammals. Chorionic gonadotropin is a hormone secreted by the placenta which acts upon the ovary to increase progesterone synthesis. Progesterone, in return, is responsible for maintenance of pregnancy during the early phases. Placental lactogen, another hormone secreted by most placentas, acts on the mother to stimulate mammary gland development. This occurs throughout gestation, so that the mammary glands are ready for suckling by the time the offspring is born.

Gestation Periods

The length of gestation varies tremendously among placental mammals. In elephants the gestation period is as long as twenty-two months. However, size alone is not the determining factor. The giant among all mammals, the blue whale, has a gestation period of only eleven months, not appreciably longer than the human (nine to ten months).

Many bats and other mammals have a delayed implantation, in which the fertilized ovum remains dormant or its development is retarded at first, thus considerably extending the gestation period and delaying birth until the optimal season of warm weather or abundant food is present. Often a placenta is not found during this period of delay, with arrest of development occurring in the blastula stage. Thus, the gestation period of the fisher, a small North American carnivore with delayed implantation, is forty-eight to fifty-one weeks, or about the same as that of the blue whale. Gestation varies from twenty-two to forty-five days in squirrels, twenty to forty days in rats and mice, two to seven months in porcupines, six months in bears, and fourteen to fifteen months in giraffes.

Gestation length is ultimately constrained by the size of skull which will fit through the maternal pelvis. Where agility, speed, or long distances of travel put a premium on the mother's athleticism, gestation length is often shortened, and the birth weight of the offspring will be low.

Animals having long gestation periods, or whose young mature slowly and are suckled for a long period of time, generally do not breed as often as others. Many species of mice breed repeatedly throughout the spring, summer, and fall, having a gestation period of about twenty days, and being mature and ready to breed by twenty-one days of age. Many others, such as bears, coyotes, and weasels, breed only once a year. Environmental conditions, and the adaptability of various species to these conditions, play a large role in breeding cycles. It is clearly advantageous for young to be born during the season of least severe weather and to be weaned when food is most abundant. Many tropical mammals breed and give birth throughout the year, whereas in temperate or cold climates young are usually born in the spring or summer.

Similar factors also influence the number of young born in each litter among different species. Their rate of growth until weaned, mortality rates, adult activity cycles, and other factors no doubt help determine the litter size as well. Many rodents have three to six young per litter; a few species of mice can have as many as eighteen. Seals, whales, and most species of bats and primates bear only a single young at one time.

—*Kerry L. Cheesman*

See also: Asexual reproduction; Birth; Cleavage, gastrulation, and neurulation; Copulation; Gametogenesis; Reproduction; Reproductive system of female mammals; Reproductive system of male mammals; Sexual development.

Further Reading

Gilbert, Scott. *Developmental Biology*. 6th ed. Sunderland, Mass.: Sinauer Associates, 2000. Beautiful artwork showing placental formation and the process of implantation. Easy-to-read text.

Harris, C. Leon. *Concepts in Zoology*. 2d ed. New York: HarperCollins, 1996. A college text. Excellent chapters on reproduction in mammals, including photographs of developing blastocysts and placentas.

Macdonald, David, ed. *The Encyclopedia of Mammals*. New York: Facts on File, 1984. An excellent chapter on "what is a mammal," along with in-depth reviews of each family or group of mammals. A lot of attention to social aspects of mammals.

Randall, David, Warren Burggren, and Kathleen French. *Eckert Animal Physiology*. 4th ed. New York:W. H. Freeman, 1997. An in-depth college text that is not easy to read, but well written for the advanced student.

Silverthorn, Dee Unglaub. *Human Physiology: An Integrated Approach*. 2d ed. Upper Saddle River, N.J.: Prentice Hall, 2001. Chapters on reproduction illustrate the placenta well, but the book covers only the human.

Polymers and monomers

FIELDS OF STUDY

Biochemical Engineering; Biosynthetics; Chemical Engineering; Enzyme Engineering; Epoxies and Resin Technology; Pharmacology; Plastics; Polymer Science.

SUMMARY

Polymers are macromolecules produced by repetitive bonding of small molecules that are termed 'monomers', as the individual units from which polymers are formed. Celluilose is the most common biopolymer, formed from glucose molecules. Amino acids are polymerized through the formation of peptide bonds to produce an incredible variety of proteins. All proteins are produced from just twenty amino acids. The DNA molecule is the most complex polymer known, and is built upon a sugar-based backbone, rather like cellulose.

PRINCIPAL TERMS

- **amino acids:** biological molecules that serve as the building blocks of proteins and enzymes. Amino acids are incorporated into proteins by transfer RNA, according to the genetic code contained in DNA. The majority of amino acids have names ending with -ine, and are complex arrangements of atoms of carbon, nitrogen, hydrogen, and oxygen. *See also* enzyme, protein.
- **enzyme:** a biological catalyst, any protein molecule within a living organism that speeds up biochemical reactions to a rate that will sustain life. The effect may speed up metabolic reactions by a factor of one million, compared with what would occur chemically outside the body. Names and classification of enzymes are regulated by the International Commission on Enzymes. Most enzymes are named by adding -ase to the root of a corresponding substrate, the molecule an enzyme acts upon. Sucrase catalyzes the hydrolysis of sucrose into glucose and fructose. A living cell has a unique set of 3,000 enzymes, each defined by the cell's DNA.
- **photosynthesis:** a chemical process in which carbon dioxide and water are converted into carbohydrates and oxygen by the energy from a light source, generally the Sun. This reaction, which all plants and many bacteria rely on, is the source of the unnatural presence of oxygen in Earth's atmosphere and

supplies the entire food chain upon which animal life depends for existence.
- **protein:** a long chain of amino acids. There are thousands of different proteins in each cell of the human body, and since each species has slightly different proteins in its cells, there are millions of different proteins in the biosphere. A balance of all necessary proteins is essential to the continued life of any organism. Food consumption must either supply each complete protein that the human body cannot manufacture for itself or a wide variety of incomplete proteins that can be assembled into complete proteins.

BASIC PRINCIPLES

A polymer is a large molecule that typically consists of and is formed by the repetitive connection of a large number of smaller, identical molecules. Each of these smaller molecules is a monomer. The terms are taken from the Greek root words 'poly', meaning 'many', and 'mono', meaning 'one.' Polymers are either synthetic or natural, and the two types can be combined to form a new 'semisynthetic' material. Synthetic polymers are well known as plastics, such as polyethylene, which is formed by the repeated head-to-tail bonding of ethylene molecules to produce very long hydrocarbon molecules, according to

$$n\ H_2C=CH_2 \rightarrow -(CH_2-CH_2)_n-$$

in which 'n' is some whole number greater than one and ranging into the hundreds of thousands. This type of polymerization produces linear polymers. If a monomer has an additional functional group that will allow a polymerization reaction to occur from the same molecule, then cross-linking can occur in which different polymer chains become bonded together to form a three-dimensional network. Synthetic polymers are typically produced from just one monomer to produce a material that is suited to a specific application or technology. Biopolymers obey the same basic principles as synthetic polymers, but are based on biomolecules such as sugars and amino acids, and their formation within biological systems is mediated by enzymes. The most common biopolymer in the world, cellulose, is produced in green plants by the process of photosynthesis, in which carbon dioxide and water are combined and converted into the sugar glucose. This requires the presence of sunlight, and the mediator complex called chlorophyll, and is the foundation of almost all life on this planet.

BIOPOLYMERS

The most common biopolymer is produced by the head-to-tail bonding of individual glucose molecules in plants. This produces more complex sugars, the starches, and finally the celluloses. In combination with a three-dimensional biopolymer known as lignin, cellulose is found as wood. The variety of sugar and sugar-like molecules that are produced from the basic material glucose is important in the formation of other biopolymers, and it is not a stretch of reality to say that all carbon-based life as it is known depends on this ability. But glucose is not the only compound that is used to produce biopolymers. Amino acids are also capable of undergoing polymerization by forming a peptide bond between the carboxylic acid function on one amino acid molecule and the amine function on another amino acid molecule. This is the essential feature in the formation of proteins. All of the myriad of proteins in human biology, and in biology in general, are produced by the enzyme-directed polymerization of just twenty 'essential' amino acids. Depending on the chain length of the biopolymeric structure of a protein, the number of possible combinations of this limited number of amino acids is staggering, and it has been estimated that the number of proteins involved in human biology is between 10^7 and 10^9, and each individual protein has its own unique role in the human body. Proteins provide the structural material of muscles and other organs. As enzymes, proteins moderate and control essentially all of the biochemical processes that characterize life, from the relatively simple task of exchanging carbon dioxide for oxygen in breathing to the manufacturing of every other material in the body. The most fascinating aspect of protein biochemistry is its role as the mediator of processes involving DNA.

DNA AND PROTEINS

DNA is without doubt the most complex biopolymer known. It is not a protein itself, but contains the structural blueprint for each and every protein in the human body. The DNA molecule more closely resembles cellulose than it does a protein, as it is essentially a structure of many millions of repeating units of the sugar deoxyribose (which is closely related to glucose) bonded through phosphate groups. Each of these deoxyribophosphate groups also bears one of four purine and pyrimidine bases, forming a series of nucleotides. The order of the nucleotides in a segment of DNA determines the identity of the corresponding protein that will be produced through interaction with proteins that moderate the opening of the DNA molecule at that location. The pattern is transcribed by the formation of a complimentary strand of RNA, which then is used by enzyme complexes to construct the desired protein from free amino acids in the cytoplasm. Enzyme proteins control the formation of DNA, while DNA determines the identity of the proteins that carry out that function. It is a fascinating cyclic relationship.

Richard M. Renneboog MSc

See also: Deoxyribose Nucleic Acid; RNAase

FURTHER READING

Blake, R.D. Informational Biopolymers of Genes and Gene Expression. Properties and Evolution Sausalito, CA: University Science Books, 2005.

Malacinski, George M. Essentials of Molecular Biology 4th ed., Sufbury, MA: Jones and Bartlett Education, 2005.

Whitford, David Proteins: Structure and Function New York, NY: John Wiley & Sons, 2013.

Tanford, Charles and Reynolds, Jacqueline Nature's Robots: A History of Proteins New York, NY: Oxford University Press, 2003.

FASCINATING FACTS ABOUT MONOMERS AND POLYMERS

- Cross-linking during polymer formation can potentially turn a large volume of monomer material, such as a railway tank car, into a single molecule.
- Biopolymer formation is almost exclusively due to the ability of the carbon atom to form four equivalent bonds to other atoms in the same period, particularly with nitrogen and oxygen.
- While physical forces do provide functional limitations as the size of a polymer molecule increases, in theory a polymer molecule is limited in size only by the quantity of monomer molecules that are available for its formation.
- Diamond is a polymer consisting of pure carbon. Each carbon atom is bonded to four other carbon atoms in a three-dimensional array that produces that hardest known substance in the world.

PROTEIN SYNTHESIS

FIELDS OF STUDY

Biochemistry; Genetics; Molecular Biology

SUMMARY

The process of protein synthesis is described, and its importance in the biochemistry of living systems is elaborated. Protein synthesis is a vital process for the production of enzymes and other polypeptides, which are in turn responsible for carrying out many functions.

PRINCIPAL TERMS

- **amino acid:** an organic compound that contains both an amine and a carboxylic acid functional group and can, in some cases, combine with other amino acids to form proteins and other polypeptides.
- **anticodon:** a sequence of three nucleotide bases in transfer RNA (tRNA) that bonds to a complementary codon in messenger RNA (mRNA).
- **codon:** a sequence of three nucleotide bases that specifies a particular amino acid or control point in the process of protein synthesis.
- **ribosome complex:** a structure consisting of ribosomal RNA (rRNA) and enzymes that decodes messenger RNA (mRNA) and coordinates the assembly of proteins from amino acids carried by transfer RNA (tRNA).
- **translation:** the overall process of RNA-mediated protein synthesis, entailing transcription of genetic information from the DNA molecule by

PROTEIN SYNTHESIS

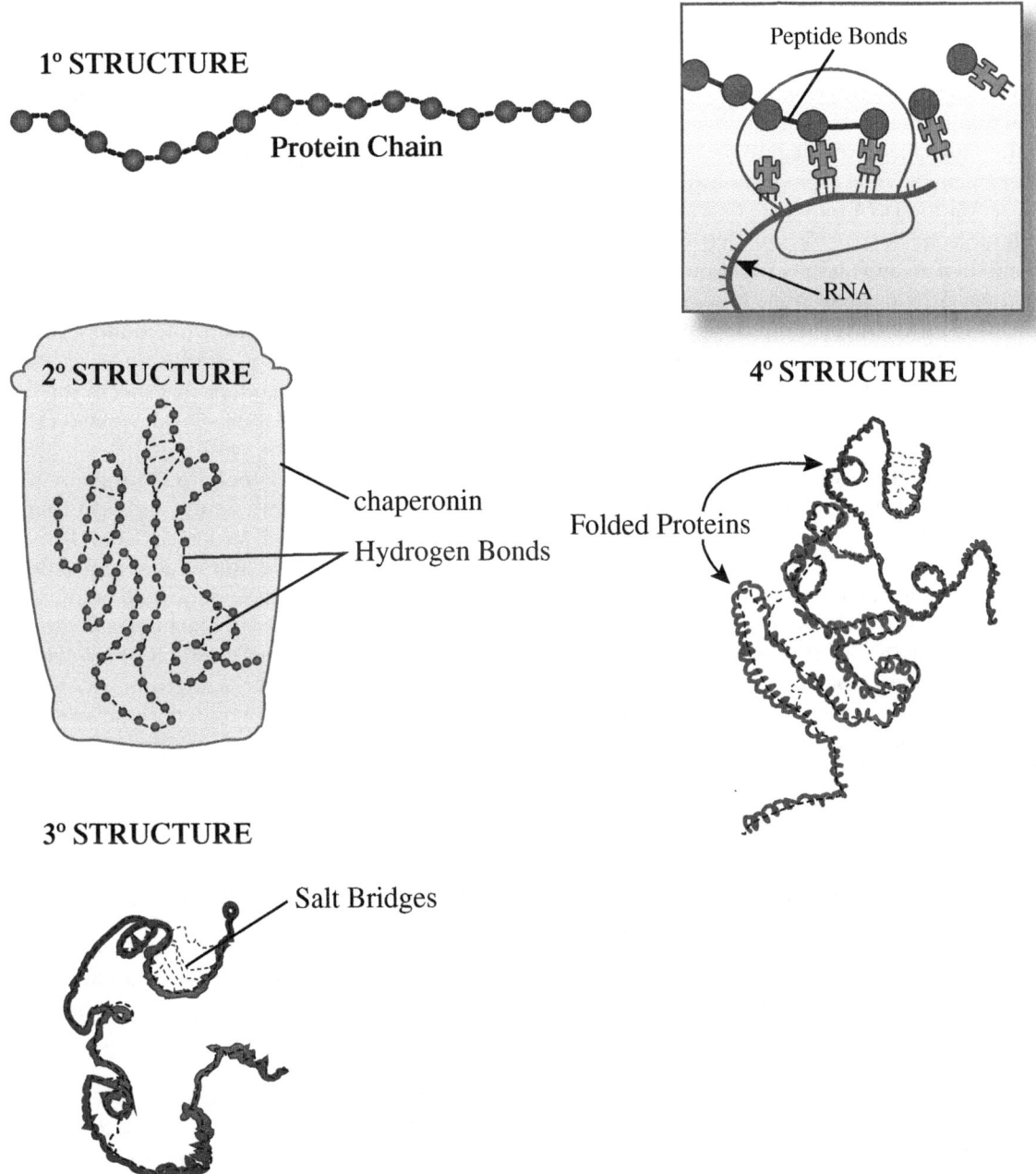

messenger RNA (mRNA), assembly of the corresponding amino acids by transfer RNA (tRNA), and formation of the protein molecule in ribosomes by ribosomal RNA (rRNA).

Visualizing Protein Synthesis

The process of protein synthesis is similar to any modern manufacturing facility. DNA is analogous to the design department, where all the product specifications are kept. The communications department is the role of mRNA, transmitting the product specifications to the production line. The production line is found in the ribosome, where rRNA operates the protein assembly line. All of the components needed to keep the assembly line running are delivered by tRNA, the just-in-time suppliers.

The actual process of protein synthesis takes place at the molecular level, beginning with the transcription of genetic code from a molecule of DNA. Enzymes act on the DNA molecule first to straighten out the double helix structure of the duplex DNA strand and then to initiate a separation of the two DNA strands. At the separation, mRNA assembles along one of the strands, forming a fragment of RNA that is complementary to the nucleotide sequence of the DNA strand. The nucleotides pair up in the process, the cytosine nucleotides pairing with guanine nucleotides and the adenine nucleotides pairing with thymine nucleotides. However, when the adenine nucleotide is in the DNA strand, the complementary nucleotide in the RNA strand is the uracil nucleotide, instead of its thymine counterpart.

The genetic information contained in DNA is "written" as a series of nucleotides called genes. The genes differ in length, but each gene spans several thousand nucleotides. Specific sequences indicate both the starting and end points of a gene. The overall sequence of nucleotides is transcribed into the mRNA strand by this process of gene expression, and when completed, the RNA strand separates as a molecule of mRNA. Enzymes then act on the DNA molecule again to seal up the location at which it had been separated for transcription and gene expression.

mRNA, tRNA, and rRNA

Three forms of RNA are involved in the translation of the genetic code from DNA to protein synthesis. The mRNA carries a transcribed copy of genetic code from DNA. Once freed from its association with the DNA molecule, mRNA makes its way to the nearest ribosome complex, a structure consisting of about 60 percent ribosomal RNA and 40 percent various enzyme proteins. The function of the ribosome is to direct the assembly of new proteins according to the code carried by the mRNA. At the ribosome, the mRNA coordinates with the rRNA in a manner that reveals the nucleotide sequence as codons, which are then used for recognition by corresponding units of tRNA.

A codon is a series of three nucleotides in mRNA that correspond to a complementary anticodon on a tRNA molecule and, thus, to the specifi c amino acid carried by that tRNA molecule. Because RNA, like DNA, is assembled from only four different nucleotides and proteins are assembled from twenty different amino acids, it is impossible for a single nucleotide to specify a single amino acid. Thus, there must be combinations of nucleotides, with three being the minimum number capable of specifying all twenty essential amino acids. In fact, the system of three-nucleotide codons is capable of specifying more than twenty amino acids, which has resulted in a number of redundancies in the codon-anticodon series. Some amino acids can correspond with up to six different codons, and there are codons that specify the starting and ending points of amino acid sequences in the protein molecules being synthesized.

"Amino acid" is a general term for any organic compound that has both amine and acid functional groups in its molecular structure. Therefore, there are millions of possible amino acid molecules. The genetic code specifying the sequence of amino acids in proteins in essentially all biochemical systems uses only a limited set of twenty essential amino acids. Each one has both a carboxylic acid group (–COOH) and an amine group (–NH$_2$) bonded to the same carbon atom. The third substituent, or side chain bonded to the carbon atom, is different in each of the twenty amino acids, and ranges from a simple hydrogen atom in glycine to more complex cyclic, bicyclic, and electrically charged substituents in others. The substituents in nine of the amino acids are neutral and nonpolar. Six others are also neutral but polar, and the remaining five are electrically charged and thus highly polar. The sequence of the amino acid residues in a protein molecule determines the structure and function of that protein.

Protein Structure and Function

Protein molecules have four aspects of their structure and function. The first aspect, known as the "primary structure," is the sequence of amino acid residues in the molecule. The shapes of molecules are guided by strict rules with regard to their geometric shape and conformation, corresponding to the geometry of the elecelectronic orbitals in the atoms. Therefore, the bonds between component atoms of a molecule must form specific angles with each other, determined by the geometry of the atomic orbitals. As a result, protein molecules (and effectively all others) are not straight-line structures, but rather twist and turn as bond angles demand. This motion produces the secondary structure of protein molecules, forming various segments into coil and sheetlike conformations. The tertiary structure of protein molecules is produced by the interaction of amino acid substituents at different locations, and especially by the formation of disulfide bridges between methionine residues in different parts of the molecule. A disulfide bridge bonds the two residues together chemically, locking the protein molecule into a certain conformation. The fourth aspect of protein structure, termed the "quaternary structure," arises when two or more separate protein molecules intertwine and form a single functional entity.

Taken altogether, the four levels of structure determine the protein's function and specificity. Many proteins serve structural purposes, helping to create muscle, cartilage, hair, and fingernails. Others function as enzymes to facilitate and mediate biochemical processes such as digestion, respiration, and metabolism. In enzymes, the overall shape of the protein molecule forms an active site into which only materials having the proper corresponding structure can fit.

APPLICATIONS OF PROTEIN SYNTHESIS

Many medical conditions are caused by improper protein synthesis in individuals. The genetic code in DNA may be lacking the proper trigger for the transcription and synthesis of certain proteins, and as a result, the corresponding protein or enzyme is lacking. Many people suffer from lactose intolerance, for example. Those that do lack the ability to produce the enzyme lactase, which is responsible for the digestion of lactose, the sugar found in milk and other dairy products. The lactose affects such individual's biochemistry in much the same way that a chemical poison would, producing symptoms of discomfort. Avoiding milk and dairy products can diminish the negative effects, but developing a method by which the body naturally produces lactase provides an absolute cure for lactose intolerance.

Richard M. Renneboog, MSc

FURTHER READING

Berg, Jeremy M., John L. Tymoczko, Gregory J. Gatto, and Lubert Strye. *Biochemistry*. 8th ed. New York: W. H. Freeman, 2015. Print.

Lafferty, Peter, and Julian Rowe, eds. *The Hutchinson Dictionary of Science*. 2nd ed. Oxford: Helicon, 1998. Print.

Lehninger, Albert L. *Biochemistry: The Molecular Basis of Cell Structure and Function*. 2nd ed. New York: Worth, 1975. Print.

Lodish, Harvey, et al. *Molecular Cell Biology*. 7th ed. New York: Freeman, 2013. Print.

Pelczar, Michael J., Jr., E. C. S. Chan, and Noel R. Krieg. *Microbiology: Concepts and Applications*. New York: McGraw, 1993. Print.

Reece, Jane B., et al. *Campbell Biology*. 10th ed. San Francisco: Cummings, 2013. Print.

PROTEIN SYNTHESIS SAMPLE PROBLEM

Describe the four levels of structure in proteins with respect to the interactions that determine each aspect.

Answer:

The primary level of structure in a protein is the sequence of amino acids. This is due to the formation of peptide, or amide, bonds between the carboxylic-acid functional group of one amino acid and the neighboring amine-functional group.

The secondary level of structure is the gross shape of the parts of the protein molecule as coils and sheets. These formations arise from the restrictions of the bond angles between the atoms.

The third level of structure results from the interaction of certain functional groups and atoms with other functional groups and atoms in other parts of the molecule. Both the electronic affects of polar and charged side chains and the formation of disulfide linkages play roles in this process.

The fourth level of structure is the overall conformation of a protein complex involving two or more separate protein molecules, which results from both the manner in which the different proteins fill out the space available in the other molecules and the interactions between atoms and functional groups in the different molecules

Proteins, Enzymes, Carbohydrates, Lipids, and Nucleic Acids

FIELD OF STUDY

Organic Chemistry

SUMMARY

The basic characteristics of proteins, enzymes, carbohydrates, lipids, and nucleic acids are presented, and their polymeric nature is described. These macromolecules are the building blocks of biological systems and have a variety of biochemical functions.

PRINCIPAL TERMS

- **biochemistry:** the chemistry of living organisms and the processes incidental to and characteristic of life.
- **functional group:** a specific group of atoms with a characteristic structure and corresponding chemical behavior within a molecule.
- **macromolecule:** a very large molecule; most often refers to polymers but can also refer to single molecules with extended, non-polymeric structures.
- **monomer:** a molecule capable of bonding to other molecules to form a polymer.
- **organic compound:** generally, a compound containing one or more carbon atoms, although some carbon-containing compounds are considered inorganic.

Basic Principals

Biomolecules are the organic molecules that make up living organisms. Five particular classes of biomolecules are particularly significant in the study of biochemistry: proteins, enzymes, carbohydrates, lipids, and nucleic acids, all of which can be described as macromolecules. A macromolecule is, quite simply, a molecule with a large mass; macromolecules are often polymers, but this is not always the case.

All biochemical macromolecules are organic compounds, their essential molecular structure being composed of carbon atoms. The particular chemical and biochemical behaviors of the macromolecules derive from the various functional groups that are present in their molecular structures.

Proteins and Enzymes. The genetic code in deoxyribonucleic acid (DNA) carries the blueprint for more than tens of thousands of different polypeptide compounds that combine in polymeric chains to form proteins and enzymes. A polypeptide consists of four or more amino-acid subunits chemically bonded together in a linear head-to-tail fashion by peptide bonds, which are simply amide bonds formed between a carboxyl functional group ($-COOH$) and an amine, or amino, functional group ($-NH_2$).

A protein is a polypeptide that performs a specific role in biochemical processes, typically either as an enzyme or as part of the many types of tissue. Fibrous proteins provide structural support to tissues such as muscle, hair, and cartilage, while globular proteins transport and store nutrients and can act as catalysts for a number of biochemical reactions necessary to maintaining life. A globular protein that catalyzes or mediates specific biochemical reactions is considered to be an enzyme. Essentially all biochemical processes are enzyme mediated, with each particular enzyme serving to catalyze or facilitate a specific biochemical transformation. Enzymes control the transcription and translation of genetic information by which proteins are formed.

The broad versatility of proteins and enzymes in carrying out biochemical functions is a result of the nearly infinite possible combinations of their component amino acids, which number approximately twenty and can be repeated all but indefinitely. The linear sequence of the amino acids in a protein chain determines the protein's primary structure, and even slight differences in the order of the component amino acids can create an entirely different protein. The secondary structure of a protein is largely determined by hydrogen bonding between the carboxyl and amino groups of the various amino acids, which can cause various segments of the macromolecule to assume either sheetlike or coiled shapes, giving the protein flexibility and strength. The tertiary structure of a protein derives from the three-dimensional shape of the protein molecule, which can be determined by interactions between specific side-chain functional groups. These interactions may determine the shape of the active site of an enzyme or the specific compound or part of a compound that is amenable to enzyme catalysis. The quaternary structure of an enzyme results when two separate enzyme

PROTEINS/ENZYMES, CARBOHYDRATES, LIPIDS, NUCLEIC ACIDS

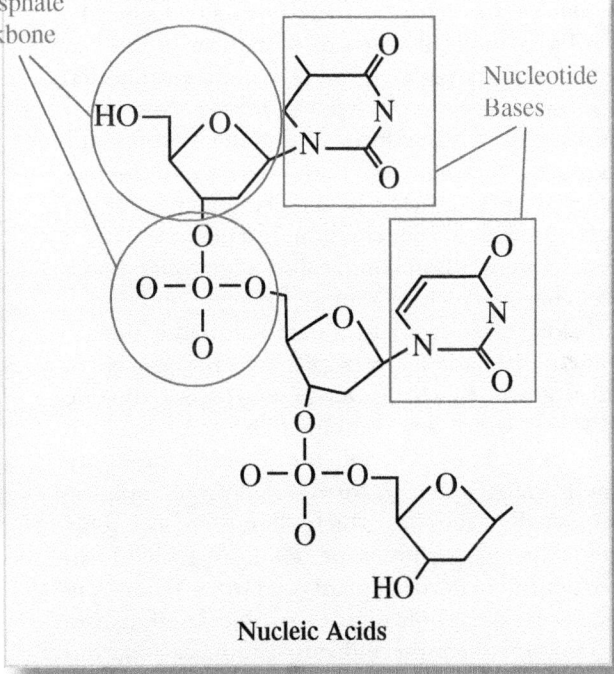

Proteins and enzymes have a characteristic bond formation of a carbon double-bonded to oxygen and single-bonded to the nitrogen of another amino acid. Lipids are composed of hydrocarbon tails and polar heads, often with a carbon double-bonded to an oxygen and hydrogen group. Carbohydrates have a characteristic molecular formula of oxygen and hydrogen in a 1:2 ratio. Nucleic acids have a characteristic sugar-phosphate backbone and nucleotide bases.

proteins interact without bonding chemically to each other to form a protein complex with specific enzymic activity.

Carbohydrates. Carbohydrates, including sugars, starches, and cellulose, serve as a food source for most organisms and provide structural support to plants. Most carbohydrates have the empirical formula CH_2O, meaning that there are typically two hydrogen atoms for every carbon and oxygen atom. Generally, however, the term "carbohydrate" is used to refer to the sugars, or saccharides, that form the most basic units of carbohydrates.

Simple sugars such as glucose and fructose are called monosaccharides because they cannot be broken down further through hydrolysis, a chemical reaction that splits bonds in the presence of water. Monosaccharides are characterized by the presence of one carbonyl group (C=O) and a hydroxyl group (–OH) on each of the non-carbonyl carbon atoms. They may be either ketoses (polyhydroxy ketones), in which the carbonyl group is attached to two carbons atoms, or aldoses (polyhydroxy aldehydes), in which it is attached to a hydrogen atom. When two monosaccharides are chemically bonded to each other, they form a disaccharide, which has a molecular structure that can be broken down by hydrolysis to form two monosaccharides. The disaccharide sucrose (CHO), 122211, commonly known as table sugar, is formed from the monosaccharides glucose and fructose (both isomers with the molecular formula $C_6H_{12}O_6$) by the elimination of water (H_2O) through a condensation reaction to form a carbon-oxygen-carbon bond, called a glycosidic bond. Polysaccharides, such as cellulose, glycogen, and starch, feature long chains or rings of monosaccharide units.

Glucose is made in green plants by the process of photosynthesis, in which atmospheric carbon dioxide (CO_2) and water are combined through the heat energy of sunlight. The glucose molecules that are formed link together to form much larger polysaccharides called starches, which often serve as energy stores for living organisms, or celluloses, which form the structural material of plants and trees. When consumed as food, carbohydrates are subjected to hydrolysis that separates the individual monosaccharides from the polymeric chain. In respiration, each glucose molecule is oxidized back into carbon dioxide and water, resulting in the release of energy stored in the glucose's chemical bonds, which can then be used by the organism to fuel other biochemical processes.

Lipids. The term "lipids" describes a wide variety of compounds, including fats, phospholipids, waxes, and steroids. It is most often used in reference to fats, which are formed as esters of glycerol and fatty acids. A fatty acid is a long-chain carboxylic acid with the basic formula R–COOH, where R is a hydrocarbon chain of variable length, most often containing between twelve and twenty carbon atoms. The glycerol end of the esters may become bonded to a phosphate group, making the lipid into a phospholipid. The phosphate group enhances the hydrophilic (literally "water loving," meaning it is attracted to water and other polar substances) character of that end of the molecular chain. The long hydrocarbon chains, on the other hand, are hydrophobic ("water fearing," repelled by water and other polar substances). This leads to the natural formation of a structure called a lipid bilayer. The phospholipid bilayer is a major component of cell membranes, as well as the membranes surrounding many of the organelles and other components within cells.

Nucleic Acids. The nucleic acids DNA and ribonucleic acid (RNA) are complex polymeric molecules constructed by the sequential addition of hundreds of thousands of similar structural units called nucleotides. In DNA, each nucleotide contains a molecule of deoxyribose sugar ($C_5H_{10}O_4$), while in RNA, each nucleotide contains a ribose sugar ($C_5H_{10}O_5$); in both, the sugar is bonded to either a purine or pyrimidine nitrogenous base at one end and an inorganic phosphate group at the other. The structure of both DNA and RNA thus consists of a long chain of sugar molecules alternating with phosphate groups. In the case of DNA, the base on each nucleotide matches up with a complementary base on a nucleotide in a second DNA strand, and the two strands twist together to form the double-helix structure of a DNA molecule. Only four nitrogenous bases are used to construct either DNA or RNA. In DNA, the component bases are adenine, guanine, thymine, and cytosine; in RNA, the base uracil is used instead of thymine, but all the others are the same. The order of nucleotides in the DNA molecule specifies the order of amino acids in all of the proteins and enzymes in an organism.

Polymeric and Non-Polymeric Biomolecules. Polymers are formed by the sequential addition of individual units called monomers, which are small molecules that easily bond together to form a chain. Proteins and enzymes are polymers that are formed from amino acids that link together via peptide bonds. Nucleic acids, including DNA and RNA, are also a form of polymer, constructed from hundreds of thousands of individual units called nucleotides, which themselves are composed of a nucleoside (a five-carbon sugar and a nitrogen-containing base) and a phosphoric acid. Carbohydrates, particularly the starches, celluloses, and glycogen, are formed by the sequential addition of thousands of molecules of monosaccharides such as glucose and fructose.

Lipids are the smallest of the biochemical macromolecules and are insoluble in water. There are various classes of lipids, the most well known of which are the triglycerides, or fats. Triglycerides generally consist of a molecule of glycerol, an alcohol containing three hydroxyl groups, that has been esterified by long-chain carboxylic acids called fatty acids. Because the molecules that make up a lipid are not structured in the form of a repetitive chain, lipids are non-polymeric macromolecules.

Richard M. Renneboog, MSc

FURTHER READING

Berg, Jeremy M., John L. Tymoczko, Gregory J. Gatto, and Lubert Strye. *Biochemistry*. 8th ed. New York: W. H. Freeman, 2015. Print.

Lehninger, Albert L. *Biochemistry: The Molecular Basis of Cell Structure and Function*. 2nd ed. New York: Worth, 1975. Print.

Lodish, Harvey, et al. *Molecular Cell Biology*. 7th ed. New York: Freeman, 2013. Print.

Morrison, Robert T, Robert N. Boyd, and Saibal K. Bhattacharjee. *Organic Chemistry*. Noida: Dorling Kindersley, 2013. Print.

PROTEINS, ENZYMES, CARBOHYDRATES, LIPIDS, AND NUCLEIC ACIDS SAMPLE PROBLEM

Classify the following compounds as protein, nucleic acid, lipid, or carbohydrate:
  fructose
  glyceryl tripalmitate
  amylase
  mRNA

Answer:

Fructose is a simple sugar with the chemical formula $C_6H_{12}O_6$. The formula shows that there are two hydrogen atoms and one oxygen atom for each carbon atom, so fructose is a carbohydrate. Carbohydrate names typically end in "-ose."

Glyceryl palmitate is the triester of glycerol and palmitic acid. It is a lipid. Ester names typically end in "-ate."

Amylase is the starch-cleaving enzyme secreted in saliva. It is a protein. Enzyme names typically end in "-ase."

mRNA is the abbreviation for messenger ribonucleic acid. It is a nucleic acid.

Pulmonary medicine

FIELDS OF STUDY

Biology; organic chemistry; statistics and methodology; physiology; epidemiology; internal medicine.

SUMMARY

Pulmonary medicine is a medical specialty concerned with the diagnosis and treatment of diseases of the respiratory system. Pulmonologists complete several years of postgraduate medical training before they begin caring for patients who have respiratory diseases. Pulmonologists often specialize in a specific area of pulmonary medicine, such as asthma management or lung transplantation. Pulmonary medicine has also spawned areas of focus that have become specialties, notably critical care medicine and sleep medicine.

PRINCIPAL TERMS

- **auscultation:** evaluative technique in which the physician places a stethoscope on different areas of the chest and back and asks the patient to take deep breaths to identify irregularities in the lungs.
- **crackles:** discontinuous, interrupted explosive sounds in the lung that are indicative of pulmonary disease; also known as rales.

- **critical care medicine:** medical subspecialty concerned with the management and treatment of acutely ill patients in an intensive care unit.
- **interstitial lung disease:** lung disease occurring in the small areas between the lungs or parts of the lung.
- **lungs:** primary organs of respiration, found in all land mammals, including humans.
- **percussion:** evaluative technique in which the physician places a hand over certain parts of the chest or back directly over the lungs and, with the other hand, taps with the fingertips to identify irregularities in the lungs.
- **pulmonary:** of, relating to, or affecting the lungs.
- **pulmonology:** study of the basic anatomy, physiology, and function of the respiratory system.
- **respiratory system:** system of organs used to take in and exchange oxygen for carbon dioxide in an organism.
- **ventilator:** device that helps pulmonary patients breathe when they cannot adequately respirate on their own.

Basic Principles

Pulmonary medicine is a branch of medicine that is concerned with the maintenance and function of the respiratory system. It deals with causes, diagnoses, and treatments of diseases that affect the lungs and related systems, such as sleep, that are strongly supported by an efficiently working respiratory system. Pulmonary medicine is often confused with pulmonology, which is the scientific study of the basic anatomy, physiology, and function of the respiratory system. Practitioners of pulmonary medicine are called pulmonologists.

The practice of pulmonary medicine has evolved to include a number of distinct areas of focus. Examples include obstructive airway diseases (such as asthma and chronic obstructive pulmonary disease, or COPD), occupational and environmentally caused diseases (such as asbestosis), congenital lung diseases (such as cystic fibrosis), interstitial lung diseases (such as sarcoidosis), neoplasms (such as small cell lung cancer, or SCLC), diseases of the pleural space (such as pneumothorax), vascular diseases (such as deep venous thrombosis), lung transplants, sleep disorders (such as obstructive sleep apnea), and critical care medicine. The last two areas of focus—sleep disorders and critical care medicine—have become distinct areas of subspecialization that have significant overlap with the general practice of pulmonary medicine.

Background

Pulmonary medicine came into being as a subspecialty of internal medicine in the early part of the twentieth century. However, interest in treating the underlying cause of diseases of the lungs can be traced back to Hippocrates, who described the symptoms, physical findings, and treatment of thoracic empyema. Interest in developing treatments for people suffering from ailments of the respiratory system increased in the nineteenth century with the introduction of resorts—often situated in mountainous areas—that claimed to offer a healing climate for sufferers of afflictions of the lungs. Some early pioneers in pulmonary medicine were scientist-practitioners who had a particular interest in the physiological structure of the lungs and the respiratory tract. The scientific study of pulmonology and its application to the diagnosis and treatment of lung diseases began systematically and earnestly in the early twentieth century and has progressed steadily since that time. During the late twentieth century, advances in the treatment of respiratory diseases were numerous and dramatic. Some of the most significant advances include lung volume reduction surgery, in which the diseased parts of the lungs of emphysema or COPD patients are surgically removed to allow the remaining, healthy parts of the lung to expand and perform better, and lung transplantation, in which badly diseased lungs are partially or completely replaced by donor organs. Advances in mechanical ventilation technology have allowed critically ill patients to be stabilized for survival and patients with chronic respiratory conditions (such as sleep apnea) to achieve a much higher quality of life.

The Pulmonary Examination

Pulmonologists practice medicine in both outpatient (ambulatory) and inpatient (hospital) settings. In the United States, most pulmonologists see patients only after they are referred from primary care physicians because of the complexity of the patient's respiratory condition. When a patient sees a pulmonologist, the physician's initial examination will include questioning about pulmonary symptoms. The physician may ask about breathing problems that the patient may be having and whether the problems change

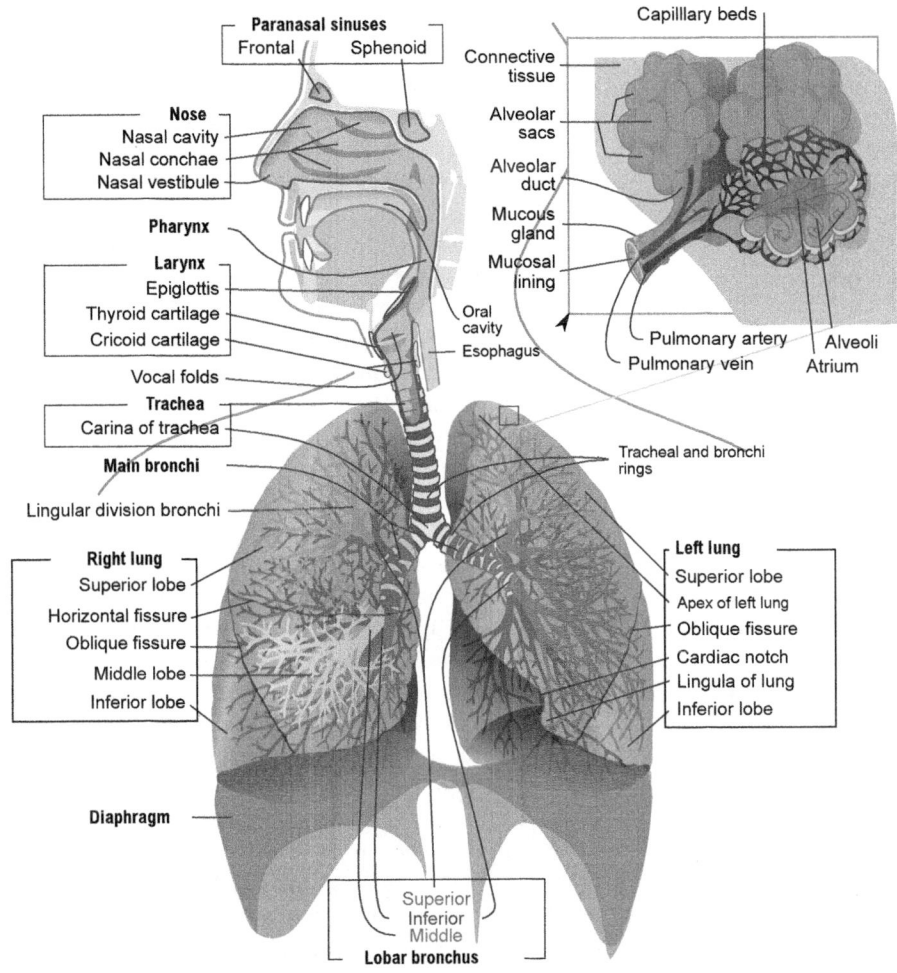

The respiratory system consists of the airways, the lungs, and the respiratory muscles that mediate the movement of air into and out of the body.

after exertion. The pulmonologist may also ask about the quality of the patient's sleep, because certain pulmonary issues may be particularly evident at rest.

Auscultation and Percussion

After questioning the patient about symptoms, the pulmonologist will perform the initial pulmonary physical examination. The complete pulmonary examination takes about ten minutes to complete. The primary component of this examination requires the physician to listen to the patient's lungs for the distinct sounds of different respiratory diseases. These lung sounds have been identified empirically by a computerized examination of differences in the lung sounds of normal patients versus patients who have various confirmed lung diseases. Two primary components of the pulmonary physical examination are auscultation and percussion.

Auscultation is the act of listening to the lungs in an effort to diagnose potential respiratory disorders. Successful auscultation requires substantial clinical experience and listening skill. The physician places the stethoscope on different areas of the chest and back and asks the patient to take deep breaths. The pulmonologist is listening for several potential indicators of pulmonary dysfunction, including crackles, rhonchi, wheezes, and rubs in both lungs. Crackles (also known as rales) are defined as discontinuous, interrupted explosive sounds in the lung.

Crackles may sound low-pitched (coarse crackles) or high-pitched (fine crackles). The area of the lung in which crackles are heard may be indicative of different pulmonary conditions. Rhonchi are low-pitched rumbling or gurgling sounds that suggest that air is being forced through fluid obstructions in the airways. Wheezes are continuous and high-pitched hissing sounds. The physician may also listen for rubs—sounds that have been traditionally described as being similar to two pieces of leather being rubbed together. Successfully identifying the type and intensity of different lung sounds through listening with a stethoscope is a skill that requires several years of practice. Having a naturally strong listening ability may be one factor in choosing pulmonary medicine as a subspecialty.

Percussion is somewhat different from auscultation in the pulmonary examination, but the two techniques ultimately serve the same purpose—to identify some dysfunction of the respiratory system through audible observation. Percussion is a technique in which the physician places a hand over parts of the chest or back directly over the lungs and, with the other hand, taps with his or her fingertips. The physician then notices the pitch of the resonant sound from the tap (called percussed sounds). Percussion is particularly useful for the identification of pneumothorax, a pulmonary condition in which air or gas is trapped in the pleural cavity. In modern pulmonary medicine, percussion is almost always used in conjunction with auscultation.

Imaging Studies

Auscultation and percussion are integral to the initial examination of the pulmonary patient. The next step in evaluation is often imaging studies. The most common imaging studies in pulmonary medicine are radiographs of the chest (X rays), computed tomography (CT) scans, high-resolution CT scans, pulmonary angiography, and ventilation/perfusion lung scans. Pulmonologists will often start with less invasive and more economical chest radiography to look for clues about the cause of the patient's symptoms. They may be looking for many different signs depending on the results of the patient's physical examination. One of the most common findings on radiography of the chest or chest CT is opacities on the film of the lungs that indicate the presence of some kind of lung infiltrate. The clarity of the radiograph, the availability of other imaging studies, and the physician's clinical suspicion about the etiology of the patient's condition will determine whether any of the other imaging studies are used. Some practices may immediately image the patient's respiratory system with a chest CT and bypass the radiograph altogether, but getting the chest radiograph first is still the most common practice.

Pulmonary Function Tests

Beyond imaging techniques, pulmonologists often need to test the functional performance of the lungs. Pulmonary function tests measure the functional performance of the lungs by measuring how much air they can hold, how quickly a person can move air into his or her lungs, and how well a person's lungs can distribute oxygen to the bloodstream and remove carbon dioxide from it. Lung function is most commonly measured with a process called spirometry. Some of the most common spirometric measures of lung function are FEV1 (forced expiratory volume), FVC (forced vital capacity), RV (residual volume), and TLC (total lung capacity). FEV1 is a measure of the amount of air that the patient can exhale forcefully in one second. FVC is a measure of the amount of air that the patient can exhale after inhaling deeply. RV is a measure of how much air is left in the lungs after the patient has completely exhaled. TLC is a measure of the amount of air deposited in the patient's lungs after he or she has inhaled deeply. Other common pulmonary function studies include arterial blood gas studies, which measure the amount of oxygen and carbon dioxide in the patient's bloodstream. These and other pulmonary function tests are often employed after the pulmonologist has developed some clinical suspicion of the patient's diagnosis.

Applications

Bronchodilators. Some of the most common problems of pulmonary patients are obstructive airway diseases, in which something is preventing the normal flow of air in and out of the lungs and, thus, affecting the normal function of the respiratory system. Two of the most common obstructive airway diseases are asthma and chronic obstructive pulmonary disease (COPD). Physicians have a variety of methods at their disposal to treat obstructive airway diseases. One of the most common treatments is bronchodilator therapy. Bronchodilators are medications that work by relaxing the muscles that surround the airways in the lungs. This allows more air to flow easier into the lungs and prevents attacks of labored breathing in obstructive airway disease patients. Short-acting bronchodilators help with rapid, acute problems such as asthma attacks and acute COPD exacerbations, while long-acting bronchodilators work to prevent underlying symptoms that can lead to acute attacks.

Ventilators. Pulmonary patients who have severe conditions often must be placed on machines that aid their lungs in the respiratory process. These machines are called ventilators. The proper use of ventilators for various lung conditions is a primary part of the knowledge base of pulmonologists. Ventilators may be used for management of acute conditions in the emergency department or, more commonly, in the intensive care unit (ICU) for conditions such as acute lung injury. They may also be used for more chronic conditions such as COPD. Ventilators are either negative-pressure or positive-pressure. Negative-pressure ventilation involves directing air directly into the lungs, and positive-pressure ventilation involves directing air into the trachea. Some ventilators require intubation, the placement of a tube into the trachea from the nose or mouth. Ventilation requiring intubation is typically used for patients who will require ventilation for a protracted period. Other ventilators work with a breathing mask that can be placed over the mouth and nose. With the increase in respiratory-related sleep disorders (such as obstructive sleep apnea), use of two positive airway pressure systems—continuous positive airway pressure (CPAP) and bilevel positive pressure ventilators (BiPAP)—has become very common.

Lung Volume Reduction Surgery. Lung volume reduction surgery (LVRS) is a procedure that has been successful for patients who have emphysema and severe COPD. The procedure involves surgically removing the disease-damaged part of the lung to allow the other (healthy) part of the lung to expand and compensate.

Lung Cancer Therapy. Many types of respiratory cancers can affect pulmonary patients. Common neoplasms are small cell lung or nonsmall cell lung cancers. Small cell lung cancers are less common and include subtypes such as sacromatoid carcinoma, salivary gland tumors, and carcinoid tumors. More common nonsmall cell cancers include lung carcinoma, mediastinal lymph node cancer, and pleural cancers. Lung cancers are primarily treated with chemotherapy, radiation therapy, or combination therapy. Less common lung cancer treatments include adjuvant chemotherapy and radiotherapy. Patients who have lung cancer may be treated by an oncologist, a pulmonologist who specializes in lung cancer, or a multidisciplinary team.

Lung Transplant. In some very severe cases of lung disease, the patient may require lung transplantation, in which a diseased lung is surgically replaced by a live donor organ. Lung transplantation may be performed by a skilled pulmonologist specializing in transplant or, more commonly, by a thoracic surgeon. A national registry lists people who are waiting for lung transplants. Many factors, including the severity of disease and lifestyle factors (for example, continuing to smoke cigarettes), can determine a person's eligibility and position on the list.

Future Prospects

A functioning respiratory system is vital to life, and the goal of pulmonary medicine is to ensure that the respiratory system is functioning properly and adequately. Pulmonary medicine is a vital medical specialty that offers many interesting potential research areas and a variety of respiratory-related subspecialties that are at the cutting edge of modern medicine (such as critical care and sleep medicine). Training in pulmonary medicine is long and laborious but can ultimately be very rewarding. As demand for pulmonary and critical care medicine practitioners is at an all-time high and expected to increase, the prospects for a successful career in pulmonary medicine are very good.

Jeremy Dugosh, MS, PhD

Further Reading

Fishman, Alfred P., et al. *Fishman's Pulmonary Diseases and Disorders.* 4th ed. New York: McGraw-Hill Professional, 2008. A basic pulmonary textbook and a good source for basic information about procedures for maintaining respiratory health and the history of the science of pulmonary medicine.

Henschke, Claudia I., Peggy Mcarthy, and Sarah Wernick. *Lung Cancer: Myths, Facts, Choices—and Hope.* New York: W. W. Norton, 2002. This book is a nontechnical examination of the prognosis and treatment of lung cancer for patients and their caregivers.

Kovitz, Kevin L. "Pulmonary and Critical Care: The Unattractive Specialty." *Chest* 127, no. 4 (April, 2005): 1085-1087. A frank discussion of the workforce issues surrounding pulmonary and critical care medicine and an accurate description of the day-to-day activities of a pulmonologist.

Macnee, William, and Stephen I. Rennard. *Chronic Obstructive Pulmonary Disease.* London: Health Press, 2009. Spends equal time on the science and clinical treatment of COPD. It is a useful reference to understand the disease and its manifestations at a deeper level.

Mason, Robert J., Jay A. Nadel, and John F. Murray. *Murray and Nadel's Textbook of Respiratory Medicine.* 5th ed. Philadelphia: Elsevier Saunders, 2010. The quintessential textbook on pulmonary medicine. Includes chapters on every aspect of pulmonary disease management from basic anatomy of the lungs to lung transplantation.

Wilkins, Robert L., and James R. Dexter. *Respiratory Disease: A Case Study Approach to Patient Care.* Philadelphia: F. A. Davis, 2006. Describes different pulmonary conditions by examining specific case studies of different pulmonary patients.

WEB SITES

American Board of Internal Medicine http://www.abim.org

American College of Chest Physicians http://www.chestnet.org

American Thoracic Society http://www.thoracic.org

Lung Cancer Alliance http://www.lungcanceralliance.org

Society of Critical Care Medicine http://www.sccm.org

See also: Artificial Organs; Bionics and Biomedical Engineering; Geriatrics and Gerontology; Otorhinolaryngology; Pathology.

FASCINATING FACTS ABOUT PULMONARY MEDICINE

- In 1928, the first rudimentary hyperbaric chamber, the Timken tank, was constructed in Cleveland, Ohio, at a cost of $1 million. The tank was 64 feet in diameter and five stories tall.
- According to a national health survey, as many as 24 million Americans are affected by chronic obstructive pulmonary disease (COPD).
- Lung cancer is the leading cause of cancer death in both men and women for all ethnic groups in the United States.
- According to the Asthma and Allergy Foundation of America, about 20 million Americans (1 in 15) suffer from asthma, and the prevalence of asthma has been steadily increasing since 1980 for both men and women in all ethnic groups in the United States.
- Human lungs are filled with tiny air-filled structures called alveoli. The average adult lungs contain 600 million of these alveoli—enough to cover an area the size of a tennis court.
- Human lungs inhale more than two million liters of air everyday.

Reproduction

FIELDS OF STUDY
Ethology, genetics, reproduction science

SUMMARY
Reproduction is a prerequisite for life. Sexual organisms reproduce when the gametes of two adults unite. Similarities and differences in how these adults maximize their own lifetime offspring production influence the diversity of mating behaviors observed in nature.

PRINCIPAL TERMS

- **gamete:** a sexual reproductive cell that must fuse with another cell to produce an offspring: a sperm or egg mate choice: the tendency of members of one sex to mate with particular members of the other sex
- **mate competition:** competition among members of one sex for mating opportunities with members of the opposite sex
- **natural selection:** the process that occurs when inherited physical or behavioral differences among individuals cause some individuals to leave more offspring than others
- **reproductive success:** the number of offspring produced by one individual relative to other individuals in the same population
- **sexual dimorphism:** an observable difference between males and females in morphology, physiology, and behavior
- **sexual selection:** the process that occurs when inherited physical or behavioral differences among individuals cause some individuals to obtain more matings than others

BASIC PRINCIPLES
The ability to reproduce is central to the existence of any organism. Simple one-celled organisms reproduce asexually by duplicating their genetic material and dividing in half. For reproduction in sexual species the participation of two individuals is essential. Mating partners have a common interest: production of successful offspring. They also have conflicting interests. Appreciating the diversity of reproductive behaviors that occur in different organisms requires understanding of the reproductive conflicts that exist between mates even more than their mutual reproductive interests.

Early sexual species most likely consisted of individuals that produced gametes of similar size. Except that one set of individuals produced smaller gametes that would be called sperm and another set produced larger gametes called eggs, males and females did not exist. Yet, reproduction was sexual. Through time it is thought that distinct sexes evolved because some individuals obtained reproductive advantages by producing smaller than average gametes whose greater motility increased their likelihood of fertilizing other gametes, while other individuals obtained reproductive advantages by producing larger than average gametes whose greater stores of nutrients increased the survival prospects of their young. Such evolution of gamete dimorphism would then lead to the many specializations in appearance and behavior of present-day males and females.

REPRODUCTIVE SUCCESS
For species whose parents do not provide care for their young, the maximum number of offspring that an individual female can produce is determined by the number of eggs that she can manufacture. For species in which parents do provide care, the number of young a female can produce might be limited more by the number of young she can raise than by the number of eggs she can make. For both these types of species, females that copulate with more than one male produce roughly the same number of young each season as females that mate with only one male. Thus, the quality of an individual mate, or the resources of paternal care that he can provide,

should affect the reproductive success of females more than the number of mates they can obtain. As a result, females of most species are expected to be selective about which male fertilizes their eggs.

In contrast, the reproductive success of males of most species is not limited by the number of sperm they can produce. In species lacking parental care, males that mate with the most females usually leave the most offspring. In parental species, the reproductive consequences of mating with more than one female are slightly more complex. If one parent can provide sufficient care to raise young to independence, males that mate with multiple females usually leave more offspring than males that mate with only one female. If both parents are necessary to raise young, males that mate with only one female should leave more offspring than males with multiple mates. Given the reproductive advantages of mating with multiple females and a tendency for breeding males to outnumber breeding females, competition for mating opportunities is more pronounced in males than in females for many species. Furthermore, males should usually be less selective in mate choice than females.

Mate Competition and Mate Choice

Charles Darwin realized that the struggle to obtain mates could be an important process affecting the evolution of organisms. He deemed this process "sexual selection" to distinguish it from the struggle for existence, or "natural selection," and suggested two components of sexual selection: mate competition and mate choice. The importance of mate competition is rarely questioned by biologists because it often involves frequent and conspicuous aggressive interactions and distinctive weapons (for example, antlers of various ungulate species, horns of bovids, and enlarged canine teeth of primates). Mate choice is another matter. This exceedingly subtle behavior is difficult to document in any species. Although females of most species are demonstrably more reluctant to mate than males and are remarkably good at rejecting the mating attempts of males of other species as well as those of closely related males of their own species, few studies can convincingly show that females actively choose particular types of males as mates. The lack of direct evidence for the pervasiveness of mate choice in nature has caused manybiologists to doubt its significance ever since Darwin first suggested it. Mate choice is also often misunderstood. The decision process it embodies is not thought to result from conscious deliberation; rather, individuals might simply react more favorably to some members of the opposite sex than to others.

Fighting for Mates

Similarities and differences in the details of mate competition and mate choice of different species can be exemplified by comparing three North American vertebrates: bullfrogs, sage grouse, and elephant seals. Male bullfrogs fight each other for small areas in ponds that females use as egg deposition sites. The wrestling matches that occur between males do not involve weapons as such, but being larger than an opponent almost always confers success. During nighttime choruses, females move among the territorial males, apparently assessing features of the male and his territory. Pairing begins when a female approaches a particular calling male and touches him. The male clasps the female, and within an hour, the female releases up to twenty thousand eggs, which the male fertilizes externally. Neither the male nor the female provides parental care for their young. Each year, roughly half of the males in a population obtain mates, and the most successful male may mate with six or seven different females.

Male elephant seals can weigh as much as three thousand kilograms and are highly aggressive. Rather than fight for territories, they fight directly for groups of females, called harems, that haul out on land to give birth. Males that monopolize large groups of females might mate with ninety or more females each year. One study revealed that less than 10 percent of the males sire all the pups in a breeding colony. Success in male competition not only involves being large but also involves having formidable canine teeth to use as weapons. During a fight, males rear up to half their length, slam themselves against the opponent, and bite him on the neck. Skin on the chests of males is highly cornified; these "shields" provide some protection against such onslaughts, but injury is still common. Mate choice by females is limited to vocalizing before and during copulation. If the male attempting to mate with her is a subordinate individual, the dominant male quickly responds to the call and attacks the copulating male.

Male sage grouse often congregate, or "lek," in traditional areas where they display and fight to control small territories. The territories function as courtship sites and places to copulate. Males provide neither resources nor parental care for their young; yet, females initiate mating and appear to be highly selective in mate choice. Near unanimity in preferred mates by the females in the population results in only a few males obtaining all the matings.

In all three of these species, many males compete for mates by fighting for territories; however, some males employ different tactics to obtain mates. Small, young bullfrog males remain silent near large, calling males and attempt to intercept any female attracted by the calling male. Small, young elephant seal males lurk about on the surf and grab females as they leave land to feed in the ocean. These males force females to copulate. Some sage grouse males attempt copulation with females away from the display arena. In another lekking bird species, the ruff, two types of males exist: territorial males and satellite males. The latter males are nonaggressive and appear to capitalize on the ability of territorial males to attract females. Unlike the case of bullfrog males, however, genetic differences of male ruffs produce the striking differences in plumage and behavior, rather than their ages.

Mate choice by females is well developed in bullfrogs and sage grouse. The benefit that female bullfrogs obtain by choosing particular males is relatively straightforward: Chosen males tend to control superior egg deposition sites that increase offspring survival. Benefits that female sage grouse obtain from mate choice are unknown. Mate choice in such lekking species continues to pose a significant question for biologists.

COMPETITION AND REPRODUCTIVE SUCCESS

Differential success of males in mate competition and mate attraction may translate to large differences in reproductive success. In contrast, variation in reproductive success among females is usually low because most females mate and produce at least some offspring. The relative amount of variation in reproductive success within each sex can influence the evolution of sexual traits. When only a few adults produce most of the offspring in a population, genes affecting the traits that underscore their success will be passed on to their offspring and quickly become the predominant characteristics of future generations. In contrast, if the most successful individual produces only slightly more offspring than other individuals, genes from all these parents will be present in roughly similar numbers in subsequent generations.

A consequence of greater variation in reproductive success of males relative to females is the evolution of elaborate sexual characteristics that are expressed only in males. These traits can be morphological, physiological, or behavioral. The extent of sexual dimorphism is predicted to be related to the relative variation in reproductive success in the sexes. Thus, species in which one or a few males sire most of the offspring produced in the population would tend to be species with considerable phenotypic differences between the sexes.

FIELD RESEARCH AND LABORATORY STUDIES

Studies of the reproductive behavior of organisms usually involve observation and experimentation of male and female interactions in nature or the laboratory. Early studies were mostly observational and cataloged the most typical behavior patterns observed in each sex. These studies ignored differences in behaviors among individuals. Because such differences can have significant consequences in terms of reproductive success, more recent studies usually involve marking males and females for individual recognition and recording various features of their morphology, behavior, and reproductive success.

Quantifying reproductive success in nature is a difficult task, and various methods have been used for different organisms. For all studies, the identity of each individual must be known. For some species, researchers can assess only the number of copulations that individuals obtain; for other species, they can count the number of young born; in yet others, they can determine the number of young that survive to independence or even sexual maturation.

Laboratory and field experimentation has been used to study a plethora of questions concerning the acquisition, function, and evolutionary significance of a variety of reproductive behaviors. Early studies of bird song investigated not only the role of song in attracting mates but also how individuals acquired their species-typical song. Choice experiments on female frogs using either naturally calling males or playbacks of recorded male calls revealed the call characteristics that females use in species and mate recognition. Crossing different species that varied in reproductive behaviors and noting the characteristics of their hybrid offspring provided some insights into the genetic basis of various behaviors. Staging aggressive interactions between males that differ physically in some regard demonstrated the significance of various male characteristics.

Researchers investigate general trends in reproductive behaviors by using the comparative method. This method usually works best when fairly closely related taxa (for example, species within the same genera or family) are considered. Using the comparative method, researchers can look for the relationships between the degree of sexual dimorphism in some characteristic and sex-specific differences in reproductive success variation, or the ecological and social conditions that affect male behaviors such as territoriality or female behaviors such as the amount of maternal care provided to young.

A thorough understanding of reproductive behaviors requires the creative use of all three methods of investigation: observations of individuals in nature to document normal behavioral patterns, precise experimentation on behaviors under controlled conditions to understand mechanisms of behavior and the stimuli that produce them, and comparison of trends in closely related species to gain insights into the evolutionary history of behavioral traits.

Reproduction is one of the defining attributes of life itself, and for most organisms, reproduction is sexual. Yet, despite the universality of sexual reproduction, the behavior patterns associated with it in different organisms are exceedingly diverse. For solitary organisms, sexual reproduction may be the only form of social behavior. For highly social species, individual interactions can be much more complex but still usually influenced by sex in some manner. Biologists seek to find some order to the variety of reproductive behaviors observed in different organisms. A unifying theme for this diversity is an evolutionary one: How do the sexes differ in maximizing the number of offspring they can produce? More specifically: How do males or females maximize the number of offspring they produce given the behavior patterns of the other sex and the various ecological factors that affect them? Thus, current research on reproductive behaviors has gone well beyond the point of merely describing what animals do to reproduce to determining why they do what they do.

—*Richard D. Howard*

See also: Asexual reproduction; Birth; Copulation; Embryology; Evolution: Historical perspective; Fertilization;; Lactation; Reproductive systems of female mammals; Reproductive systems of male mammals; Sex differences: Evolutionary origin.

FURTHER READING

Alcock, John. *Animal Behavior: An Evolutionary Approach.* 7th ed. Sunderland, Mass.: Sinauer Associates, 2001. An extremely up-to-date textbook, utilizing case studies to illustrate topics. Covers mating and mating behavior in the context of natural selection.

Clutton-Brock, Tim H., F. E. Guinness, and S. D. Albon. *Red Deer: Behavior and Ecology of Two Sexes.* Chicago: University of Chicago Press, 1982. Provides an in-depth analysis of a fifteen-year study of the reproductive behavior of male and female red deer. This study has produced some of the very best information on the reproductive biology of a long-lived vertebrate.

Ferraris, Joan D., and Stephen R. Palumbi, eds. *Molecular Zoology: Advances, Strategies, and Protocols.* New York: Wiley-Liss, 1996. Discusses the uses and methods of molecular biology as related to a wide variety of zoological topics, including reproduction.

Gould, James L., and Carol Grant Gould. *Sexual Selection.* 2d ed. New York: Scientific American Library, 1997. Summarizes some of the best research to date on the reproductive behaviors of a variety of animals. It is well written and well illustrated. Provides a small but select Bibliography for more detailed reading.

Halliday, Tim. *Sexual Strategy.* Chicago: University of Chicago Press, 1980. Covers a range of topics related to sexual behavior in a clear, informative manner.

Krebs, John R., and Nicholas B. Davis. *An Introduction to Behavioural Ecology.* 4th ed. Sunderland, Mass.: Sinauer Associates, 1997. Covers various aspects of reproductive behavior and develops some of the theoretical approaches used by biologists.

Short, R. V., and E. Balaban, eds. *The Differences Between the Sexes.* New York: Cambridge University Press, 1994. A collection of symposium papers covering a wide variety of topics related to sex differences in animals.

Wiley, R. Haven. "Lek Mating System of the Sage Grouse." *Scientific American* 238 (May, 1978): 114-125. Gives additional information on the reproductive behavior of this fascinating species and on the general phenomenon of lekking behavior.

Reproductive science and engineering

FIELDS OF STUDY

Cell biology; molecular biology; biochemistry; developmental biology; medicine; pediatrics; physiology; obstetrics; gynecology; andrology; embryology; endocrinology; surgery; genetics; biomedical engineering; chemical engineering; reproductive technology; pharmacology; neurobiology; urology; pathology; immunology.

SUMMARY

Reproductive science and engineering is concerned with the examination and regulation of the physiological mechanisms involved in human reproduction, such as conception and birth, and with diagnosing and treating disorders of reproduction, such as male and female infertility. Because the ability to successfully bear offspring is the core objective driving the success of all animal species, reproductive research is of deep importance on a purely scientific level. In addition, by providing methodologies that enable infertile couples to conceive, such as artificial insemination and in vitro fertilization, this discipline has profound effects on both individual human lives and the population trends of societies, nations, and the world as a whole.

PRINCIPAL TERMS

- **artificial insemination:** any procedure, other than sexual intercourse, by which sperm is placed directly into a woman's cervix.
- **assisted reproductive technology (ART):** any technique, including in vitro fertilization, in which ova (also known as oocytes or eggs) are removed from a woman before being fertilized with sperm.
- **cloning:** process by which an embryo, either human or animal, can be generated from the genetic information of another individual, creating an identical copy, or clone.
- **cryopreservation:** process by which gametes (sex cells) and embryos are preserved by freezing.
- **follicle-stimulating hormone (FSH):** hormone produced by the pituitary gland that stimulates the production of eggs; it can be produced in pill form and prescribed to improve fertility.
- **human chorionic gonadotrophin (hCG):** hormone produced by the placenta; it helps preserve the uterine lining and its presence is used to confirm pregnancy.
- **intracytoplasmic sperm injection (ICSI):** procedure by which a single spermatozoa is inserted directly into an individual ovum in a laboratory.
- **micromanipulation:** procedure in which sperm, ova, or embryos are manipulated under a microscope; also known as microinsemination.
- **preimplantation genetic diagnosis (PGD):** procedure by which DNA is removed from a fertilized egg or embryo and examined for the presence of genetic abnormalities.
- **sperm bank:** facility that collects and stores (usually via cryopreservation) human sperm from donors.

BASIC PRINCIPLES

Reproductive science and engineering is the study of the physical and chemical processes that underlie human reproduction. It involves the application of scientific technologies to treat disorders of reproduction, such as infertility. It also includes the development and use of methods to interfere with or prevent impregnation, such as contraception and sterilization. Among the many approaches reproductive scientists take to these issues, three of the most significant involve mechanical, chemical, and genetic strategies for engaging with reproduction. Micromanipulation is an umbrella term for any reproductive assistance technique that involves the physical handling of sperm, oocytes, or embryos on a microscopic scale, using specialized tools. Another of the key tools that is widely used by reproductive scientists is a set of pharmaceutical products that mimic or interfere with the chemical signals naturally produced by the body. These artificial hormones can be used to manipulate the human reproductive system in a variety of ways. Reproductive genetics is a rapidly expanding subfield of reproductive science that applies the tools of genetic research and DNA-based technologies to issues of conception, childbirth, and inheritance.

BACKGROUND

Written records indicate that humans have struggled with infertility since ancient times. For example, infertility is mentioned in texts from ancient Greece and Persia. For much of European history, infertility was generally attributed to women, not men, and was considered a sign of impiety because bearing children was considered a blessing from God. The first experiments with artificial insemination were carried out in the middle of the nineteenth century by the American gynecologist J. Marion Sims, who also carried out surgical procedures designed to widen the cervix and thus facilitate the entry of sperm. In the late nineteenth century, scientists began to acknowledge the potential role of male infertility, using microscopes to test the potency of sperm.

The first half of the twentieth century witnessed the discovery of the three most important hormones involved in reproduction, the female hormones estrogen and progesterone, and the male hormone testosterone. Soon after, companies began to manufacture the first synthetic hormones for the treatment of infertility. In 1944, the first laboratory test showing that human oocytes could be fertilized in vitro was carried out. Thirty-four years later, the first so-called test-tube baby was born in England, and sperm banks became more common. The late twentieth century saw two more milestones in infertility treatment: the first successful implantation and pregnancy with an egg that had been cryopreserved and the development of intracytoplasmic sperm injection technology.

MICROMANIPULATION

The basic setup of a micromanipulator is a microscope connected to robotic arms that are powered by electric motors and moved by hydraulic or pneumatic controls that may require foot pedals or joysticks, or both. The robotic arms are in turn connected to incredibly tiny glass tools. By looking through the microscope, which magnifies the cells hundreds of times, and manipulating the controls, the operator is able to tinker with the gametes and embryos with great precision—in essence performing a kind of microsurgery. Some micromanipulation techniques involve lasers, which can move segments of a cell from place to place and slice open the thick membrane around an oocyte. This membrane, known as the zona pellucida, is often slit open to facilitate the entry of sperm. Other techniques involve the use of electric currents that can cause the membranes of two different cells to join together or transfer genetic material from one cell to another. Intracytoplasmic sperm injection is the most common micromanipulation procedure.

REPRODUCTIVE PHARMACOLOGY

Reproductive pharmacology makes use of synthetic hormones to produce a desired effect—whether contraceptive in nature or intended to increase fertility. For example, most birth control pills use some combination of artificial forms of the female hormones estrogen and progesterone. These substances interfere with the normal cycle of ovulation and menstruation, thus suppressing a woman's ability to conceive. For example, the chemical signals sent by the pill may prevent a woman's pituitary gland from releasing a chemical signal that induces ovulation, so that the ovaries do not release any oocytes. Or it may prevent the lining of the uterus from thickening, thus inhibiting the implantation of a fertilized egg. Other ways in which artificial hormones may prevent pregnancy include thickening the mucus found in the cervix so that it is more difficult for sperm to travel through it and reducing the rate at which oocytes migrate from the ovaries toward the uterus. Artificial hormones can also be used to treat infertility. For example, injections of follicle-stimulating hormone (FSH) and human chorionic gonadotropin (hCG) stimulate the process of ovulation, while gonadotropin-releasing hormone alters the timing of ovulation, making a woman's fertile period more regular and ensuring that an oocyte is not released into the uterus until it has developed properly. These and other pharmaceutical tools can help physicians correct problems associated with common female fertility disorders. The most common disorder leading to infertility in women is polycystic ovary syndrome (PCOS), which results in excessively high levels of androgen (a male hormone) and ovulation that is irregular or entirely absent.

GENETICS

Some infertility disorders are associated with a genetic defect of one kind or another. Male infertility has been linked with microdeletions, or tiny missing parts, in the Y chromosome, and with mutations in particular genes. To identify these genetic components of infertility, scientists often conduct what is known as a genome-wide association study, or whole-genome association study. This is a technique by which the entire set of genetic material belonging to each member of a group of subjects is scanned and compared in order to pinpoint specific genetic variations that are more prevalent in people with a certain trait, such as infertility.

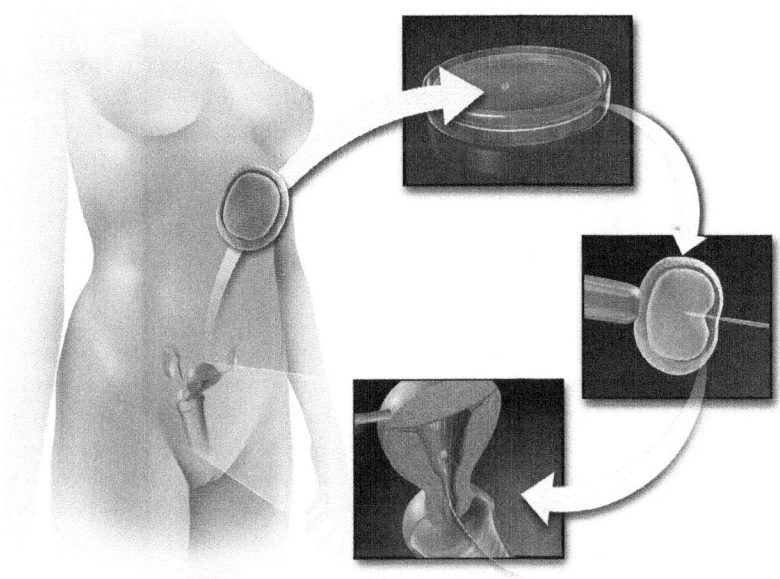

Assisted Reproductive Technology.

Screening Tests

Another common tool of reproductive genetics is the use of screening tests to identify genetic traits in embryos produced by in vitro fertilization before they are implanted in a woman's uterus. After a cell sample has been retrieved from the blastocyst or embryo, various techniques can be used to screen its DNA for possible genetic abnormalities associated with diseases such as Down syndrome or cystic fibrosis. For example, short pieces of DNA can be artificially produced that are specially designed to bind to and mark mutated DNA in the sample, if it exists. Alternatively, the DNA in the sample can be directly examined to look for known mutations. Tests can also be carried out that reveal enzymes and proteins produced by specific genes.

Applications

In Vitro Fertilization. In vitro fertilization (IVF) is one of the most common applications of assisted reproduction technology. The Latin term *in vitro* literally means "in glass." In vitro fertilization is a medical procedure in which egg cells are fertilized not inside the body (in vivo) but within an artificial laboratory environment. Because this procedure is both complex and expensive, it is often employed to help couples for whom other infertility treatments have already failed.

The first step in an IVF cycle involves the use of artificial hormones, such as follicle-stimulating hormone and human chorionic gonadotropin, in drug form. During normal ovulation, a woman's ovaries produce a single egg; this treatment, known as superovulation, causes the ovaries to produce multiple oocytes. Next, the oocytes, along with some follicular fluid, are retrieved from the patient's ovaries via a needle. They are allowed to incubate under controlled laboratory conditions, with sperm collected from the patient's partner or with donor sperm. After a day or two, the eggs are examined to determine which, if any, have been successfully fertilized. If necessary, intracytoplasmic sperm injection may be used to inseminate the eggs. This is a micromanipulation procedure that can be used to artificially induce fertilization when sperm have low motility (do not move well). In this process, a micromanipulator with a thin glass pipette on the end is used to pick up and inject a single sperm directly into an oocyte. Next, the fertilized eggs—now known as embryos—are either frozen for later use or transferred into the uterus using a speculum and a catheter. Typically, multiple embryos are inserted into the uterus so as to increase the chances of at least one successful implantation. This also increases the possibility of a multiple birth. Because the success of any given cycle of IVF treatments is by no means guaranteed, many couples choose to use cryopreservation techniques to freeze embryos produced in one cycle for future use. This streamlines the process a couple must go through if treatment does, in fact, have to be repeated. It also reduces the need for performing invasive procedures on the female patient.

Intrafallopian Transfer. Gamete intrafallopian transfer (GIFT) is a procedure that resembles in vitro fertilization. The main difference between the two applications is that with GIFT, fertilization takes place not within a laboratory setting but rather inside the female patient's body. GIFT is a minimally invasive surgical procedure in which a catheter is placed through a small keyhole incision. Oocytes and semen are inserted through the catheter into the Fallopian tubes. At this point, fertilization and implantation of the embryo may or may not occur.

Zygote intrafallopian transfer (ZIFT) is a procedure that combines elements of both in vitro fertilization and GIFT. First, gametes are extracted and fertilized under controlled laboratory conditions. Next, they are inserted into the Fallopian tubes using the same method as in GIFT. Because GIFT and ZIFT ization, they are much less commonly performed. They may be recommended for couples whose struggles with infertility are more severe or who have not responded positively to previous cycles of IVF treatment.

Artificial Insemination. Artificial insemination is a widely used, minimally invasive procedure used to help couples with a variety of infertility problems, such as a male partner with low sperm count or motility, the existence of natural antibodies in either the male or female partner that attack sperm, or characteristics of the cervix shape that make fertilization difficult. Artificial insemination can be carried out using either the intended father's sperm or that of a donor obtained from a sperm bank. In either case, once the sperm has been collected, it is physically inserted, via a catheter, into the woman's cervix. Though the technique is simple, timing is extremely important—artificial insemination must take place either just before or on the day of ovulation to be successful. In practice, it is often carried out on two consecutive days to increase the chances of fertilization.

Surrogacy. Surrogacy is an attractive option for women who are infertile because their uterus is abnormal or has been removed or who are believed to be at high risk for miscarriage or other complications of pregnancy. In gestational surrogacy, an embryo fertilized though in vitro fertilization is implanted into the uterus of a woman who is healthy and fertile, and has agreed to act as a surrogate. In many cases, the surrogate donates her own egg to be fertilized via in vitro fertilization with the biological father's sperm or that of a donor, then carries the baby to term. Some surrogates perform this service out of pure altruism or generosity; these women are usually close friends or relatives of the intended mother. Others are unrelated strangers who are compensated financially by the parents.

Contraception and Sterilization. Some applications of reproductive science are intended to prevent, not facilitate, impregnation. Most of the available contraceptive methods are designed for use by the female partner alone, although a few are meant for shared use or use by the male partner alone. Barrier methods, such as the sponge, the diaphragm, the male and female condoms, and the cervical cap, prevent sperm from reaching the egg inside the Fallopian tube. Hormonal methods, such as the pill, the vaginal ring, injected or implanted devices, and certain intrauterine devices, release artificial hormones into a woman's body to interfere with the process of ovulation and prevent eggs from being released into the Fallopian tubes. Emergency contraception, which can be used up to three days after intercourse, uses high doses of estrogen to prevent a fertilized egg from being implanted in the uterus.

Other contraceptive methods are less reliable. These include the rhythm method, in which partners carefully monitor the woman's body temperature and menstrual cycle to determine the date of ovulation and avoid intercourse during this time. Sterilization is the most decisive method of preventing impregnation. A vasectomy is a reversible surgical procedure that results in sterility for the male partner. It involves cutting or otherwise sealing both the right and the left vas deferens, the tubes through which sperm travel into the penis. A tubal ligation is a nonreversible surgical procedure that results in sterility for the female partner. It involves sealing the Fallopian tubes so that eggs are unable to pass from them into the uterus.

FUTURE PROSPECTS

Perhaps the most significant social impact of reproductive science and engineering is the way it has transformed the opportunities available to women in the workplace and, more broadly, the shape of families themselves. Because technologies such as in vitro fertilization enable women to bear children successfully later in their lives, many choose to delay parenthood until they have established themselves fully within their careers. Artificial insemination not only has enabled infertile couples to fulfill their desire to bear children but also—because sperm can be acquired through donor banks—has facilitated the rise of the modern phenomenon of single parenthood by choice. Also, assisted reproductive technologies have provided a means for same-sex couples to become biological parents. Some very useful technologies created by reproductive science and engineering have the potential to be turned into what some observers fear are unethical applications. For example, preimplantation genetic diagnosis has profound benefits because it enables couples to raise their chances of having healthy babies. However, it has also generated a certain amount of controversy because the same techniques could, in theory, be used to allow couples to select embryos with certain very specific traits. For example, embryos could be chosen or engineered to have genes encoding for eye or hair color or perhaps traits such as intelligence or physical beauty—the so-called designer baby concept. Other ethical questions provoked by assistive reproductive technologies include the question of what to do with leftover frozen embryos and whether and how much women should be compensated for egg or embryo donation. Finally, cloning is a highly controversial and often misunderstood area in biomedical research. Reproductive cloning, which involves creating a precise genetic copy of an existing organism through a process known as somatic cell nuclear transfer, is banned from being done with humans in most countries.

M. Lee, MA

FURTHER READING

Elder, Kay, Doris Baker, and Julie Ribes. *Infections, Infertility, and Assisted Reproduction*. 2004. Reprint. New York: Cambridge University Press, 2010. A detailed, illustrated examination of the microbiology of assisted reproductive technologies. Each chapter includes references, Bibliography suggestions, and frequently appendixes outlining procedures and protocols.

Green, Ronald Michael. *Babies by Design: The Ethics of Genetic Choice*. 2007. Reprint. New Haven, Conn.: Yale University Press, 2009. A bioethicist tackles moral dilemmas provoked by emerging genetic engineering technologies. Contains a glossary of relevant technical terms.

Jones, Richard E., and Kristin H. Lopez. *Human Reproductive Biology*. Rev. ed. London: Academic Press, 2010. A comprehensive introductory textbook, heavily illustrated with diagrams and photographs. Each chapter includes a Summary, Bibliography, and advanced reading list.

Romundstad, Liv Bente, et al. "Effects of Technology or Maternal Factors on Perinatal Outcome After Assisted Fertilisation: A Population-based Cohort Study." *The Lancet* 372, no. 9640 (August, 2008): 737-743. Finds that adverse outcomes associated with births following assisted reproduction are not caused by technological factors. Includes several tables.

Spar, Debora. *The Baby Business: How Money, Science and Politics Drive the Commerce of Conception*. Boston: Harvard Business School Press, 2006. A critical overview of the fertility industry and how reproductive technologies are used in the marketplace. Includes numerous tables and extensive end notes listing sources.

WEB SITES

Centers for Disease Control and Prevention Assisted Reproductive Technology http://www.cdc.gov/art

National Institute of Child Health and Human Development Reproductive Sciences Branch http://www.nichd.nih.gov/about/org/cpr/rs

Society for Assisted Reproductive Technology http://www.sart.org

See also: Bioengineering; Cloning; DNA Analysis; Human Genetic Engineering; Obstetrics and Gynecology; Stem Cell Research and Technology.

FASCINATING FACTS ABOUT REPRODUCTIVE SCIENCE AND ENGINEERING

- In 1996, the first mammal, a sheep, was successfully cloned. Dolly had the exact same DNA as the animal from which she was cloned.
- Louise Joy Brown was known as the first test-tube baby, conceived in a laboratory rather than in her mother's body. Brown was conceived in 1978 by in vitro fertilization, which was then a new technology.
- Through a process known as gestational surrogacy, one woman's fertilized egg can be carried to term in the uterus of a surrogate. The resulting baby will not be related to the woman who gestated and gave birth to it.
- The longest a human embryo created through in vitro fertilization has been known to survive outside of the womb is twenty-nine days.
- Frequently using saunas or hot tubs can lead to infertility in men because elevated temperatures are known to impair the body's ability to produce sperm.
- In 2009, artificial insemination helped a man who had become sterile as a result of radiation treatments father a baby girl. He had frozen his sperm twenty-two years earlier.
- In 2009, American Nadya Suleman made the headlines when she gave birth to what was only the second set of octuplets to be born in the United States and the first to have all the babies survive.

Reproductive system of female mammals

FIELDS OF STUDY

Developmental biology, embryology, physiology

SUMMARY

The mammalian female reproductive system is a complex system of organs and hormones that functions to produce offspring. The female also provides a protective environment in which the offspring develop until birth.

PRINCIPAL TERMS

- **anterior pituitary gland:** the front portion of the pituitary gland, which is attached to the base of the brain; the source of luteinizing hormone (LH) and folliclestimulating hormone (FSH)
- **estrus cycle:** hormonally controlled changes that make up the female reproductive cycle in most mammals; ovulation occurs during the estrus (heat) period
- **external genitals:** the external reproductive parts of the female
- **gonad:** the primary reproductive organ (the ovary in females and the testes in males), which produces sex cells (gametes) and sex hormones
- **menstrual cycle:** a series of regularly occurring changes in the uterine lining of a nonpregnant primate female that prepares the lining for pregnancy
- **ovary:** the female gonad, which produces ova and the hormones estrogen and progesterone
- **ovum (pl. ova):** the female reproductive cell (gamete); a mature egg cell
- **uterus:** the hollow, thick-walled organ in the pelvic region of females that is the site of menstruation, implantation, development of the fetus, and labor

BASIC PRINCIPLES

The function of the mammalian female reproductive system, in cooperation with the male reproductive system, is to produce offspring. The role of the female is very complex: She must produce gametes (sex cells) called ova (singular, ovum) or eggs, provide the site for the combination of ova with sperm from the male (fertilization), and nourish and protect a developing fetus during pregnancy. She also must provide for the delivery of offspring from her body to the outside. These functions are carried out by a group of organs or structures and a number of chemicals called hormones. The major organs of the system include ovaries, which produce ova and hormones; uterine (Fallopian) tubes that transport the ovum and provide the site of fertilization; the uterus, which houses the developing offspring; the vagina, which receives the male penis and sperm during sexual intercourse and also functions as a birth canal; and the external genitals. Hormones important to the function of the female reproductive system include estrogen and progesterone from the ovaries, and follicle-stimulating hormone (FSH) and luteinizing hormone (LH) from the anterior pituitary gland. Release of FSH and LH is under control of chemicals called releasing factors, which are produced by a small region of the brain called the hypothalamus.

THE OVARIES

The paired ovaries are the primary sex organ, or gonad, of the female. They are analogous to the testes in the male reproductive system and actually develop from the same tissue. The size and shape of the ovary depend on the age and size of the female and whether the female usually has a single offspring or several at one time. Before birth, small groups of cells called follicles are formed in each ovary. In the center of each follicle is a single large cell called an oocyte, which is able to mature into an ovum. In other words, all the oocytes a female will ever produce were already in place before she was born. The ova are special cells; they are formed by a process (meiosis) that results in a cell with only half of the chromosomes found in other body cells. The only other cells that divide by this process are those that form sperm in males. When an ovum and sperm unite, then, each cell contributes half the necessary chromosomes to make a new complete cell. This new cell, the first cell of an offspring, will have characteristics of each parent. The follicles develop in response to the hormone FSH from the anterior pituitary gland. A mature follicle releases its mature ovum through the wall of the ovary into the pelvic cavity. This process is called ovulation and is controlled by LH and FSH.

The ovary also produces the female sex hormones—estrogen and progesterone. Estrogen is produced by the maturing follicle cells. In addition to causing growth of the sex organs at puberty and stimulating growth of the uterine lining each month, estrogen is responsible for the appearance of female secondary sex characteristics. The follicle cells that are left behind following ovulation form a structure called the corpus luteum, which produces both estrogen and progesterone. The most important function of progesterone is to stimulate the lining of the uterus to complete its preparation for pregnancy.

THE ACCESSORY ORGANS

The rest of the internal structures of the reproductive system are called accessory organs. The first of these is a pair of uterine (Fallopian) tubes, or oviducts, which extend from each ovary into the uterus. They are frequently shaped like funnels, with fingerlike ends, called fimbria, that partially surround each ovary. Movements of the fimbria sweep the ovum and some attached cells into the uterine tube following ovulation. If fertilization is to take place, it will be in the uterine tube.

The uterus in most mammals consists of two horns and a body, although much variation occurs. Marsupials, mammals that have pouches, such as the opossum, have two completely separate uteri, each opening to the outside through a separate vagina. Rats, mice, and rabbits have uteri with two horns. Primates have simple uteri with no horns. The uterus has an amazing ability to expand during pregnancy. In all cases, the wall of the uterus is thick and muscular. This muscle layer, the myometrium, is able to contract rhythmically and powerfully to move the young down the birth canal and out of the mother's body during the birth process. The lining of the uterus is called the endometrium.

The uterus narrows down into a muscular, necklike region called the cervix. This structure acts like a valve to keep the opening into the uterus closed most of the time. This prevents bacteria and other harmful objects from entering. The final internal accessory structure is the vagina, a thin-walled muscular tube. The vagina surrounds the cervix of the uterus at its anterior end and extends to its opening to the outside of the body. It allows for childbirth, sexual intercourse, and, in primates, menstrual flow. The walls of the vagina normally touch one another and have deep folds that allow for stretching without damage. A thin fold of tissue called the hymen partially covers the external opening of the vagina. This structure has no function and varies considerably in different mammals.

The external structures of the female reproductive system are called the external genitals. These include the labia majora, labia minora, and clitoris. Two thick, hair-covered folds of skin, the labia majora, protect and enclose other structures. In some mammals, two smaller hair-free folds of skin are located within the labia majora. These folds, the labia minora, are very prominent in primates but small in most other mammals and completely lacking in some. They enclose a region called the vestibule. Within the vestibule are located the clitoris, the external opening from the urinary system (the urethra), and the external opening of the vagina. The clitoris is a small structure almost covered by the anterior ends of the labia minora. It is very sensitive, being richly supplied with nerve endings and blood vessels.

SEXUAL MATURITY

Reproduction can occur only after females reach sexual maturity. In mammals, this requires the full development of the reproductive structures. The point at which maturity is attained is ultimately under the control of the hypothalamus, as it controls the release of FSH and LH. Many factors, such as attainment of a particular body weight, temperature, day length, and climate may influence the release of hormones.

After the female reaches maturity, reproductive activities are cyclic. In mammals, there are two different kinds of reproductive cycles. Most mammals have an estrus cycle in which females will mate with a male only if they are "in heat," which happens at certain restricted times. An estrus cycle is divided into stages: an inactive phase, called anestrus, which may last for days, weeks, months, or years; proestrus, during which the follicles are developing; estrus, when ovulation occurs; and metestrus, when the ova are moving into the oviduct. Females mate, and may become very aggressive about finding a mate, during estrus only. Usually ovulation is triggered by LH from the pituitary gland. In some mammals, including cats and rabbits, ovulation does not occur until the animal mates. Many females signal that they are in estrus. The signals may be chemical—a special scent which carries for a long distance, for example—or visual. Chimpanzees, for example, develop pink swollen skin on the external genitals during estrus.

The Menstrual Cycle

Primates have a menstrual cycle instead of an estrus cycle. The menstrual cycle is coordinated by estrogen and progesterone from the ovary. These hormones, in turn, are controlled by FSH and LH from the anterior pituitary gland, so all the functions of the reproductive structure are coordinated and synchronized. The three stages of the menstrual cycle are the menses, proliferation, and secretion stages.

In menses, the thick endometrium is sloughed off and flows out of the uterus and out of the body through the vagina. This is also called the menstrual flow or menstrual period. The menstrual fluid consists of roughly equal parts of blood and other accumulated bodily fluids. In the proliferation phase, the endometrium again grows thick. Ovulation occurs in the ovary at the end of this stage, following a sudden increase in the release of LH from the anterior pituitary gland. In the secretion stage, the endometrium becomes very thick and cushiony and prepared to nourish a developing embryo if fertilization has occurred. If fertilization did not occur, the endometrial cells die and the cycle begins again. These stages are controlled by estrogen and progesterone from the ovary. Female primates will mate throughout the entire menstrual cycle.

Studying the Female Reproductive System

Detailed examination of individual reproductive tissues is performed using a variety of very thin tissue slices, various dyes and stains, and microscopes. Frequently, preserved tissue is used. Electron microscopes have made it possible to magnify single cells, or parts of cells, several thousand times to observe minute details of structure. Fresh tissue is also examined. It is possible to freeze a small tissue sample quickly, slice it very thin, and then expose the tissue to chemicals, which can add to researchers' understanding of the function of particular cells.

The study of reproductive hormones and the understanding of their function demand the use of many different methods. Again, much information comes from nonhuman studies. The procedures vary widely but a typical laboratory experiment may involve removing the ovaries from a female rat and then injecting small amounts of estrogen or progesterone to observe the response of the endometrium. It is also possible to use chemicals that block, or inhibit, one or more specific hormones. By creating an abnormal, controlled situation and observing the results, an understanding about the role of individual hormones within a complex interrelated system can be obtained.

It is also frequently necessary to measure how much hormone is present in some bodily fluid, either for research to gain understanding of normal function or for medical diagnosis. This is very difficult, as most hormones occur in very minute concentrations. Procedures called radioimmunoassay (RIA) techniques, introduced during the late 1950's and early 1960's, represent a very important advance in the study of hormone concentrations. These procedures, which use special recognition molecules for each hormone, plus certain hormones that have been purified and made radioactive, make it possible to measure levels of hormones as low as one trillionth of a gram (a picogram).

In a system as complex in its function as the female reproductive system, it is not surprising that information is obtained from a variety of sources. Each technique has made a contribution to an understanding of the whole system.

Understanding the Female Reproductive System

The reproductive system is unique among all body systems. It is the only system not called upon to function continuously for the well-being of the individual. It is nonfunctional during the early part of the female's life, then is activated by chemical messages from the anterior pituitary gland. Its primary function is not, after all, the well-being of one individual but rather the continued existence of the species. It is also unique in that it must interact with another individual, a male, in order to fulfill this function. Throughout the reproductive years, all the functions of the female reproductive systems are directed toward pregnancy.

—*Frances C. Garb*

See also: Copulation; Endocrine systems of vertebrates; Fertilization; Gametogenesis; Lactation; Reproductive systems of male mammals; Sex differences: Evolutionary origins; Sexual development.

Further Reading

Banks, William J. *Applied Veterinary Histology*. 3d ed. St. Louis: Mosby-Year Book, 1993. A good general reference on animal physiology and organ systems.

Berne, Robert M., and Matthew N. Levy. *Principles of Physiology*. 3d ed. St. Louis: C. V. Mosby, 2000. A clearly written textbook at the college level. Emphasis is on cellular mechanisms and regulation. The overview of reproductive function includes an excellent description of interactions between brain, pituitary gland, and gonads. The female system is covered separately. Excellent Summary charts and figures. References provided for each chapter. Index.

Hayssen, Virginia, Ari van Tienhoven, and Ans van Tienhoven, eds. *Asdell's Patterns of Mammalian Reproduction: A Compendium of Species-Specific Data.* Ithaca, N.Y.: Comstock, 1993. An encyclopedic reference work listing information on the reproductive characteristics of mammalian species, including size and mass of newborns, litter size, age at sexual maturity, length of estrus, gestation, and lactation, and details of frequency and seasons of reproduction.

Hickman, Cleveland P., Jr., Larry Roberts, and Frances Hickman. *Integrated Principles of Zoology.* 11th ed. St. Louis: Mosby College Publishing, 2001. A comprehensive textbook, richly illustrated. Chapters on mammals, the reproductive process, and principles of development survey reproductive strategies. A very readable work. A historical perspective and references are included with each chapter. The book has an index, a glossary, and appendices.

Rijnberk, A., and H.W. De Vries. *Medical History and Physical Examination in Companion Animals.* Translated by B. E. Belshaw. Boston: Kluwer, 1995. A handbook for veterinarians, includes coverage of the reproductive systems of animals that are commonly found as pets.

REPRODUCTIVE SYSTEM OF MALE MAMMALS

FIELD OF STUDY

Physiology

SUMMARY

The male reproductive system is a group of organs that function together to produce sperm and carry them to the outside of the male body. Sperm carry the male's genetic information to his offspring. Continued survival of the species is assured by the proper functioning of the male and female reproductive systems.

PRINCIPAL TERMS

- **chromosome:** a molecule of deoxyribonucleic acid (DNA) that contains a string of genes, which consist of coded information essential for all cell functions, including the creation of new life
- **ejaculation:** the process of expelling semen from the male body endocrine glands: glands that produce hormones and secrete them into the blood
- **erection:** the process of enlargement and stiffening of the penis because of increased blood volume within it
- **fertilization:** the union of a sperm with an ovum; fertilization is the first step in the creation of a new individual
- **gamete:** a reproductive cell—sperm in the male, ovum in the female; produced in the gonads, gametes contain a set of chromosomes from the adult male or female
- **gonad:** the organ responsible for production of gametes—the testis in the male, the ovary in the female
- **gonadotropin:** a hormone that stimulates the gonads to produce gametes and to secrete other hormones
- **semen:** the sperm-containing liquid that is expelled from the male body

BASIC PRINCIPLES

The reproductive systems of all male mammals have the same basic design. The reproductive organs produce sperm and deliver it to the outside of the body. The sperm can be regarded as packages of chromosomes that the animal passes on to his offspring. Hormones and nerves control and coordinate the functions of the reproductive organs.

THE BRAIN AND REPRODUCTION

Although the brain is not usually considered to be a component of the reproductive system, part of the brain is, in fact, essential to the function of the reproductive organs because of the hormones produced there. This part of the brain is the hypothalamus, a relatively small area that acts without conscious control. The hypothalamus is located in the lower middle of the brain; it contains centers that control eating, drinking, body temperature, and other essential functions.

Hypothalamic control over reproduction in the male is primarily by way of the hormone called gonadotropin-releasing hormone (GnRH). GnRH is released from the hypothalamus to enter blood vessels that carry it to the pituitary gland, a small gland suspended just below the hypothalamus. When GnRH arrives at the pituitary, it stimulates the pituitary to produce and release two more hormones, follicle-stimulating hormone (FSH) and luteinizing hormone (LH). FSH and LH in the male are identical to hormones of the same names in the female. The names of these hormones describe their functions in the female. Like other hor- mones, FSH and LH are released into the blood and circulate throughout the body. FSH and LH are called gonadotropin hormones: gonadotropin means "gonad stimulating." These are the hormones that stimulate the gonads (testes in the male, ovaries in the female) to produce sperm or eggs and to secrete gonadal hormones. In the male, the gonadal hormones are primarily testosterone and related hormones. There is a chain of hormonal commands, with GnRH from the hypothalamus at the top of the chain. GnRH stimulates the pituitary to secrete FSH and LH, which in turn stimulate the testes to produce sperm and testosterone.

In addition to the chain leading from the brain to the pituitary to the testes, information is sent back to the brain from the testes, a checks-and-balances system using principles of negative feedback to ensure that the hormones are produced in the appropriate quantities. If, for example, the hormone system gets slightly out of balance, leading to too much testosterone being produced, this excess of testosterone will be sensed by the hypothalamus. It will cause a temporary shutdown of GnRH production, leading to the system's correcting itself, because then a little less testosterone will be produced. If testosterone levels fall too low, the opposite will happen: GnRH, and then FSH and LH, and then finally testosterone, will all increase, again resulting in a correction of the original aberration. The hormonal system is a delicately balanced network that ensures the proper functioning of the testes.

THE TESTES

The testes are the sites of sperm production. Within the testes are hundreds of tiny tubes, the seminiferous tubules, that are responsible for sperm production. The sperm develop gradually from round cells called spermatogonia, which are located in the walls of the seminiferous tubules. As a sperm matures, it develops a long, whiplike tail attached to an oval head. The head of the sperm contains chromosomes, the genetic information of the male that will be passed on to his offspring. The sperm of some mammals can be distinguished under the microscope by characteristic differences in their appearance.

Between the seminiferous tubules are clusters of hormone-producing cells, the interstitial or Leydig cells. The Leydig cells produce testosterone and related hormones. Testosterone is essential for proper sperm development. In addition, testosterone is responsible for the development of male body features, including, in most species, a large muscle mass, and for the growth of the reproductive organs during puberty. In some animals, testosterone is also linked to aggressive and reproductive behaviors.

The testes of most mammals are located in the scrotum, a pouch of skin and muscle that is suspended outside the abdomen. In some animals, the testes may be withdrawn into the abdomen when the animal is startled or when it is not in the breeding condition.

The function of the scrotum is to maintain the temperature of the testes at a few degrees lower than average body temperature. The capability to maintain this temperature of the scrotum is rooted in the fact that the muscles within the scrotum are responsive to temperature. Under warm conditions, the scrotum relaxes, allowing the testes to move away from the body and lose heat. In cool temperatures, the opposite occurs: The scrotum wrinkles, pulling the testes closer to the body and allowing them to stay warmer. The reduced temperature maintained by the scrotum is mandatory for the production of normal fertile sperm. Fever or other situations that raise the temperature of the scrotum can interfere with sperm production, even resulting in temporary infertility. In a few large mammals (such as elephants, whales, and dolphins), the testes are not located within a scrotum, but instead occupy a position in the abdomen. It is not known why these species apparently do not require a temperature lower than that of the body for sperm production.

THE EPIDIDYMUS AND VAS DEFERENS

Sperm are removed from the testes by a system of tubes that lead out of the body. Located next to each testis within the scrotum is the epididymis, a highly coiled tube that is directly connected to the seminiferous tubules of the testes. The epididymis serves two functions: sperm maturation and sperm storage. The epididymis is drained by a long, thin tube called the vas deferens, which carries sperm out of the scrotum through the inguinal canal into the abdomen. The inner end of the vas deferens is a widened area that may serve as a site of storage for mature sperm.

The vas deferens passes in a loop next to and under the bladder, the sac that stores urine until it can be removed from the body. Immediately beneath the bladder, the vas deferens is connected by a short tube, the ejaculatory duct, to the urethra. The urethra is the long, fairly straight tube that carries either urine from the bladder or sperm from the reproductive system. A valve located in the urethra below the bladder opens and closes to prevent sperm and urine frommixing, so that only one type of fluid is in the urethra at a time.

From their site of production in the testes, sperm pass through the epididymis, the vas deferens, the ejaculatory duct, then finally the urethra to the outside of the body. As sperm are expelled from the body along this route, they are mixed with seminal fluid to produce semen. Seminal fluid is secreted into the tubes by three sets of glands: the seminal vesicles, the prostate, and the bulbourethral (Cowper's) glands. The sperm never enter these glands; fluid is squeezed out of them into the tubes where the sperm are located.

THE PENIS

The penis is designed to deliver sperm to the female system. The penis consists of a long shaft with an enlarged head, the glans. The skin of the penis, especially the glans, is extremely sensitive to touch. In some species, the penis is withdrawn into a sheath of skin except during sexual arousal.

Internally, the penis contains the outer segment of the urethra, as well as erectile tissue. This erectile tissue is designed like a sponge. The many blood vessels in the erectile tissue are capable of greatly expanding and increasing the quantity of blood that they contain. When this happens, the erectile tissues swell, and the entire penis increases in length and width and becomes stiff. This process, called erection, is an involuntary reflex: It cannot be consciously prevented or caused. Erection can result from direct stimulation of the penis, as during sexual contact, or from erotic sights or sounds. In some animals, a bone within the penis, the baculum, assists in maintenance of the erection.

Continued sexual stimulation will eventually result in an ejaculation, with semen being forced out of the body by contractions of muscles in the fluid-producing glands and along the tube system. Ejaculation is coordinated by nerves that arise in the spinal cord. The normal volume of fluid ejaculated varies from species to species. In man, it is usually two to six milliliters; it may be up to one hundred milliliters in pigs. The ejaculate of most animals contains many millions of sperm per milliliter of fluid.

STUDYING MALE REPRODUCTION

The hormonal system that controls the male reproductive system is the subject of much research. The most straightforward type of hormonal research is simply descriptive: The scientist seeks to describe the levels of the reproductive hormones when the animal of interest is in different physiological states. The hormones can be measured in blood samples taken from the animals. Obtaining a blood sample from an experimental animal may pose difficulties: Some large animals may be difficult to restrain, and some small animals may not have veins large enough for an easy puncture. Another consideration is how often blood samples should be taken. Endocrinologists have become increasingly aware of the importance of the pattern of hormone release over time. In particular, it now appears that fluctuations in hormone levels within a time frame of minutes or hours may be critical in regulating the responses of hormone target sites. To obtain blood samples with such a high frequency, researchers usually implant a cannula into a vein of the animal; the cannula can be left in place for repeated blood sampling with very little stress to the animal.

Scientists interested in hormonal feedback may examine the roles of specific hormones by removing one of the endocrine glands from the system, and then examining the effects on the remaining hormones. For example, the testes (as the site of testosterone production) can be removed from an experimental animal. Blood samples after the surgery can then be assayed to determine the circulating levels of LH, FSH, and GnRH. The endocrine glands may be left in place, but the researcher may administer hormones either by injection or by implanting timed-release capsules containing the hormone under the skin. Then, blood samples taken from the animal will reveal how levels of hormones produced by the animal's own endocrine glands have changed as a result of the exposure to the added hormone.

A technique that is widely used to study males of seasonally breeding species is to subject the animals to carefully controlled environmental conditions. Length of exposure to light, temperature, rainfall, nutrients in the diet, and other factors can be controlled in the laboratory to determine which acts as the cue for seasonal reproduction. The status of the reproductive system can be determined by various methods. The testes can be measured: Inactive testes are usually smaller and lighter in weight. Hormone levels in the blood can be measured: Testosterone and other hormones may decrease when the animal is reproductively quiescent, or the male can be exposed to a female to determine whether he will show mating behavior.

For some types of research, the most revealing experiments may not use the entire animal (referred to as in vivo research), but will instead focus on specific organs. Living samples of organs can be maintained in the laboratory for such in vitro experimentation. For the in vitro approach, a small piece of living tissue can be removed from an animal and the cells suspended in a liquid that contains the nutrients necessary for their life. Under these isolated conditions, the scientists can investigate a number of areas such as which hormones tissues produce and the hormones that make the tissue itself respond. Organs respond optimally to a particular pattern of hormonal stimulation, and this is another important area of research. By combining the results of in vivo and in vitro experiments, scientists are able to piece together a complete picture of how the reproductive system functions.

CONTROLLING REPRODUCTION

Knowledge of how the male reproductive system functions has allowed scientists to develop technologies for controlling reproduction to enhance or curtail fertility in domestic animals. Knowledge of male reproductive physiology has been applied to the management of domestic breeding populations. Hormone measurements and sperm counts can be used to determine the optimum age at which to begin breeding young stock. Techniques for collecting and storing semen can be combined with artificial insemination of females to increase the number of offspring produced by valuable males, thus resulting in improvement of the population. These methods are particularly valuable to breeders of large animals because maintaining large numbers of males of these species (such as stallions and bulls) can be costly and difficult because of the aggressive behaviors that these males may exhibit.

The study of seasonal breeding has also been of value in agriculture. Scientists now know much about the environmental conditions that are responsible for promoting reproductive activity in many domestic species. Farmers can apply this knowledge to their breeding stock to increase production throughout the year. Another area in which reproductive studies are of vital importance is the enhancement of the breeding of captive animals that are endangered in the wild. Zoos, once considered merely spectacles for entertainment, are now seen by many as the last hope of saving many species on the verge of extinction. Knowledge of the conditions necessary for successful breeding of exotic animals will help to increase their numbers and, perhaps, to return them to the wild.

—*Marcia Watson-Whitmyre*

See also: Copulation; Endocrine systems of vertebrates; Fertilization; Gametogenesis; Lactation; Reproductive systems of female mammals; Sex differences: Evolutionary origins; Sexual development.

FURTHER READING

Carter, Carol Sue, I. Izja Lederhendler, and Brian Kirkpatrick, eds. *The Integrative Neurobiology of Affiliation.* Cambridge, Mass.: MIT Press, 1999. Ac ollection of papers from a symposium, with coverage of the neurobiology of affiliation associated with reproductive behaviors.

Knobil, Ernst, and Jimmy D. Neill, eds. *The Physiology of Reproduction.* 2 vols. 2d ed. New York: Raven

Press, 1994. Covers the entire scope of mammalian reproduction, looking at both human and nonhuman species.

Marshall Graves, J. A., R. M. Hope, and D.W. Cooper, eds. *Mammals From Pouches and Eggs: Genetics, Breeding, and Evolution of Marsupials and Monotremes.* Canberra, Australia: CSIRO, 1990. A collection of papers on the often-overlooked topic of marsupial and monotreme reproduction.

Nalbandov, A. V. *Reproductive Physiology of Mammals and Birds: The Comparative Physiology of Domestic and Laboratory Animals and Man.* San Francisco: W. H. Freeman, 1976. This book is written for beginning students of reproductive physiology, and it focuses on the essentials. As the title implies, the author takes a comparative perspective, surveying reproduction in different species.

Setchell, B. P. *The Mammalian Testis.* Ithaca, N.Y.: Cornell University Press, 1978. Although written for a scientific audience, the work is noteworthy because of its completeness. Contains detailed information about all aspects of the testis, scrotum, spermatogenesis, and hormonal control, including historical perspectives of the research in male reproduction. The book is also useful because it covers all mammals, not only humans, so the reader is able to appreciate similarities and differences among the mammals.

Van Tienhoven, Ari. *Reproductive Physiology of Vertebrates.* 2d ed. Ithaca, N.Y.: Cornell University Press, 1983. This is a good reference for anyone interested in reproduction among nonmammalian vertebrates. The text covers many subjects not often found in one volume, including effects of the environment, immunological aspects of reproduction, puberty, and animals with intersex characteristics.

RESPIRATION AND LOW OXYGEN

FIELDS OF STUDY

Anatomy, biochemistry, physiology

SUMMARY

Nearly all animals require free oxygen in order to convert food to living tissue and energy, yet many animals face periods when oxygen supply is diminished. Adaptation to low oxygen involves developing mechanisms to compensate for the reduced oxygen supply.

PRINCIPAL TERMS

- **hyperventilation:** an increase in the flow of air or water past the site of gas exchange (lung, gill, or skin)
- **hypoxia:** from two Latin words, *hypo* and *oxia*, meaning "low oxygen"
- **metabolism:** the sum of all of the reactions that take place in an animal allowing it to move, grow, and carry out body functions
- **respiratory pigment:** a protein that "supercharges" the body fluid (blood) with oxygen; the oxygen can bind to the pigment and then be released
- **respiratory surface:** the gill, lung, or skin site at which oxygen is taken up from the air or water into the animal, with the release of carbon dioxide at the same time and site
- **systemic:** referring to a group of organs that function in a coordinated and controlled manner to accomplish some end, such as respiration
- **ventilation:** the movement, often by pumping, of air or water to the site of gas exchange; commonly thought of as breathing

BASIC PRINCIPLES

Adaptation to low oxygen refers to a number of different changes in metabolism or body function, or both, that animals use to survive low-oxygen conditions. Low-oxygen conditions mean a reduction in the amount of oxygen available in relation to the need or demand for oxygen by the cells, or tissues. Low oxygen, or hypoxia, therefore, can result from either a decrease in the supply at constant demand, or an increase in demand at a constant supply. The former, a reduction in oxygen supply, is the focus of the present discussion. Low oxygen resulting from increased oxygen demand usually is referred to as tissue hypoxia and is discussed only briefly here.

Oxygen is required by animal cells in order to produce energy used for growing, moving around, or simply maintaining normal body functions. At times when less oxygen is available, animals must either move to some place where there is sufficient oxygen, or change some internal function or process. A change in internal function or process is an adaptation that allows the animal to live with less oxygen or that will be a means of keeping the supply of oxygen to the tissues great enough to meet the needs of the cell.

External or environmental hypoxia results from one of two conditions: a greater utilization of oxygen by plants and animals than can be renewed by natural processes, or a lower density of air (at high elevations). It is necessary that animals cope with a decrease in oxygen in some way, in part because of the consequences of a decrease in internal oxygen supply. If oxygen supply to the tissues and cells falls, then the functions that require oxygen will fail or at least be reduced. The functions that fail include the "maintenance functions" of a cell, apart from growing or producing specialized chemicals. The types of low-oxygen conditions that have received the most attention from researchers are high altitude, diving by air breathers, and oxygen depletion in water. Some of these are temporary; others, such as high altitude, can last for a lifetime for animals that do not migrate.

INCREASE IN VENTILATION

One common adaptation to hypoxia is an increase in ventilation—the amount of air or water that the animal breathes. This increase is referred to as hyperventilation, and it makes up for the reduced amount of oxygen in the air or water by breathing a greater volume. This response is especially common in mammals that move to high altitude, fish and crabs, and some other water-breathing animals. A simple mathematical example will show how the response is effective. If an animal normally breathes one liter of air per minute and removes half the oxygen (air contains 209 milliliters of oxygen per liter), then it is taking in 104.5 milliliters of oxygen per minute. If the amount of oxygen in the air falls to only 104.5 milliliters per liter, and the animal still uses half of that (52.25 milliliters), then it must breathe two liters of air per minute to keep taking up 104.5 milliliters of oxygen per minute. Such an increase in ventilation can be accomplished by an increase in the frequency of the respiratory pump, or by increasing the volume of water or air that is moved with each "breath." For this response to occur, the nervous system must sense the reduction in oxygen and provide a nerve impulse to the brain, which then stimulates the ventilatory pump(s) to increase activity.

It may seem that this adaptation is all that would be needed for animals to survive hypoxia, but there are some limitations to this adaptation. First, hyperventilation causes increased muscular activity and an increase in oxygen used to move the respiratory muscles. The greater ventilation volume is a benefit, but the cost is a greater demand for oxygen. For animals that breathe air, the increase is rather small, but for animals that breathe water, the increase in the muscular activity causes a substantial increase in the oxygen used to pump the water, so that the "cost" may be greater than the "benefit" when the oxygen falls to low levels. Another problem exists for air breathers. Air breathers are in danger of losing water in the air that is exhaled (desert animals, such as camels, have elaborate mechanisms to conserve this respiratory water loss). Hyperventilation increases the water loss and requires the animal to drink more water. A final problem for both air and water breathers is that carbon dioxide is lost from the same respiratory surface where oxygen is taken up. Hyperventilation thus increases the loss of carbon dioxide, changing the chemical balance of the body as a whole.

BLOOD FLOW AND OXYGEN DELIVERY

In response to an internal hypoxia, many animals also exhibit an increase in the flow of blood to the tissues. This response is similar to that described for the ventilation system. An increased rate of flow compensates for a smaller amount of oxygen delivered for a given volume of blood (or respiratory medium in the case of ventilation). As with ventilation, blood flow can be elevated by increasing heart rate or by increasing the volume of blood pumped with each beat. There are numerous limitations to the effectiveness of this response, and it is only short term. The limitations center on the critical role of blood flow and blood pressure in the function of other systemic body functions. An excellent example is how kidney filtration rate increases with blood pressure.

Long-term adaptations to low oxygen often increase the ability of systemic respiratory functions to maintain oxygen delivery to the tissues. In the case of internal oxygen transport, this can be accomplished by increasing the amount of oxygen carried by the blood. A higher concentration of respiratory pigment accomplishes this, increasing either the number of red blood cells or the concentration of respiratory pigment in the blood. This adaptation requires the synthesis of new proteins and possibly new cells. Not surprisingly, many days or even weeks may be needed to increase respiratory pigment levels. Another way in which oxygen transport by the respiratory pigment may be improved is by increasing the concentration of a chemical that affects oxygen binding. This adaptation requires a change in metabolism and is discussed below.

METABOLIC ALTERATION

One final type of adaptation to low oxygen is an alteration in the basic metabolism of the animal. Metabolic changes can take one of several forms. First, a simple reduction in metabolism will lower the need and demand for oxygen by the cells. To be effective, this must occur before the oxygen has been exhausted, so as not to impair normal functions. A few animals show this type of adaptation, which is thought to result from the metabolic reactions being limited by the availability of oxygen. Second, the chemical reactions involved in metabolic pathways (a series of chemical reactions) may be altered in low-oxygen conditions so that different reactions take place to maintain energy production. The nature of these adaptations is that an alternative metabolic pathway requires different enzymes and perhaps different chemicals in the reactions. Last, a metabolic adaptation may yield a product that has an enhancing effect on oxygen transport. An enhancement of oxygen transport occurs when certain chemicals increase the ability of the respiratory pigment to bind oxygen or cause the respiratory pigment to bind oxygen at lower oxygen levels; this is called an increase in oxygen affinity. The change in metabolism at low oxygen thus results in an improvement in the supply of oxygen to the tissues. This response is seen in both vertebrates and invertebrates.

An excellent example of an animal that shows nearly all of the adaptations to low oxygen is the blue crab, the common commercial crab found throughout the Gulf Coast of North America and on the East Coast from Florida to New York. To compensate for low oxygen, the blue crab increases the flow of water over the gills, thereby keeping the amount of oxygen that actually passes over the gills nearly constant. Blue crabs also increase the heart rate, thereby increasing the rate of blood flow in the gills and to the muscles and organs of the body. This increase helps maintain the oxygen supply. If the period of hypoxia is brief, only a few hours, then these reactions may be all that is required for the animal to survive. If, however, the hypoxia continues for days or even weeks, then other responses come into play. There are changes in metabolism and in the way in which oxygen is transported in the body. Metabolism actually decreases so that less oxygen is needed by the animal. When that happens, then there must be some activity, such as swimming, that the animal gives up for lack of energy. The other change is an improvement in the way oxygen is transported to the tissues by the respiratory pigment, hemocyanin—a certain kind of protein, dissolved in the blood, that binds oxygen at the gills and can release the oxygen at the tissues where it is used by the cells. This improvement takes the form of increasing the level of a chemical in the blood that changes the binding of oxygen to hemocyanin. The hemocyanin then works as well when the animal is in hypoxic water. In addition, the crab can, and does, make the hemocyanin in a new form, so that it works better in the hypoxic conditions.

UNDERSTANDING THE RESPIRATORY AND CIRCULATORY SYSTEMS

Adaptation to low oxygen (either high demand or reduced supply) has been studied with the idea of understanding the functional capabilities of the respiratory and circulatory systems that supply oxygen to the tissues. Many different experimental protocols and procedures are used to assess the balance between oxygen uptake and oxygen demand when the external supply of oxygen is limited and demand remains constant. One approach to the study of low-oxygen conditions has been to compare animals that live at sea level in high-oxygen habitats with those living in habitats in which oxygen levels are low. A comparison of water breathers and air breathers is, strictly speaking, within the realm of consideration. Water holds much less oxygen than air, and is therefore a low-oxygen condition. Freshwater at room temperature contains about 0.8 milliliter of oxygen per 100 milliliters of water; there are 20.9 milliliters of oxygen in the same volume of air. To obtain the same amount of oxygen, an animal must thus take all the oxygen from either 2,600 milliliters of water or 100 milliliters of air. Consequently, air breathers have much lower ventilation rates, at the same temperature, than do water breathers of the same size. The lower ventilation rate of air breathers is considered functional by reducing the loss of water from the respiratory surface.

Such adaptation is principally evolutionary and involves the transition from water breathing to air breathing in the evolutionary transition to land. There are a great many morphological as well as physiological consequences of this transition.

Adaptation to short-term hypoxia has been studied under controlled conditions in the laboratory in a variety of animals. Short-term conditions may mean anything from a few hours to weeks or even months. The length of the low-oxygen exposure generally depends on the animal used, its ability to withstand low oxygen, and the nature of the inquiry into the responses. Some clams, for example, are able to live in the absence of oxygen for several weeks. These experiments require careful monitoring of the animal and the conditions to ensure that oxygen neither rises nor falls too low and that the animals will survive.

A method that has been used to study adaptation to low oxygen in mammals is to conduct field studies in which the subjects are temporarily moved to high elevations. Mountain-climbing expeditions have been involved in some of these experiments in areas throughout the world. Additionally, experimental stations have been established at certain locations for the purpose of conducting these research projects. In this way, medical researchers are able to bring in appropriate equipment and supplies necessary to make complex and precise measurements.

All approaches to the study of low-oxygen adaptation require measurements of respiratory function or metabolic processes, or both. These measurements assess the uptake and transport of oxygen and the transport and excretion of carbon dioxide. The specific measurements are of the rate of oxygen uptake, ventilation volume and rate, blood flow, heart rate, oxygen transport properties of the blood, and oxygen uptake. In long-term monitoring studies of free-ranging animals, the animals are frequently fitted with implanted electrodes and blood sampling tubes. In this way, measurements can be made routinely over long periods without disturbing the animals.

A common measure of respiratory function is the total amount of oxygen used by an animal in a given period of time. The rate of excretion of carbon dioxide is another measure of overall function. The ratio of carbon dioxide loss to oxygen uptake is used to determine the nature of the metabolic pathways at a given time. Different metabolic pathways have characteristic oxygen uptake and carbon dioxide excretion ratios, and these are used in a predictive or diagnostic fashion. Some of the methods do not impair normal activity and can be used in low-oxygen experiments. The technique of placing small animals in respiratory chambers is used in these types of experimentation. Measurement of single organ function is used more often with larger animals, such as humans.

EVOLUTION, METABOLISM, AND ECOLOGY

Adaptation to low oxygen has been studied to understand three concepts better: evolutionary changes associated with oxygen availability, cell metabolism, and ecology of hypoxic habitats. Literally every aspect of oxygen uptake, transport, and utilization has received some attention.

Some of the evolutionary changes during the transition from water to land and from low to high altitudes have been studied as problems related to low oxygen. Results indicate that air breathers have lower ventilation rates than do water breathers of the same size. The lower ventilation rates are possible because of the higher oxygen levels, but they result in higher internal carbon dioxide levels. Animals such as insects, reptiles, and mammals have respiratory structures that are internalized and are inpocketings of the body wall. This arrangement aids in water conservation and helps keep the respiratory surfaces moist.

Just as important as research on the transition to land has been the information gained about the evolution of life in high-oxygen environments as compared to the low-oxygen conditions that are believed to have occurred in the ancient oceans. From this research, it is clear that the major advances in respiratory systems are present in invertebrates and probably evolved quite early in the history of life on earth. Marine worms possess closed circulatory systems, respiratory pigments, red blood cells, special gas exchange structures, gills, and alternate metabolic pathways.

Biologists interested in metabolism and the factors that cause metabolic rate to change have examined the relationship between metabolic rate and other physiological functions. Specifically, oxygen supply, carbon dioxide removal, and glucose supply have been examined because all three are directly involved in aerobic metabolism. Imposing a limitation on external oxygen supply has therefore been used as an experimental tool to probe the limits and capabilities of cellular metabolism.

One of the observations that biologists have made over the years is that animals tend to find a way to live in places that are in any way habitable, and they tend to adapt to occupy new habitats. Understanding the physiological mechanisms required or used in adaptations to low-oxygen habitats, such as stagnant pools of water, has allowed explanation of some evolutionary changes.

There are several habitat types that undergo hypoxia routinely, and the utilization of the natural resources of those habitats, as well as the effective preservation of the habitats, generally dictates that attention be paid to the effects on the animals. One of the bodies of water that undergoes low-oxygen conditions is the Chesapeake Bay, and the effects on the animals there have been studied.

—*Peter L. deFur*

See also: Circulatory systems of vertebrates; Respiration.

FURTHER READING

Bicudo, J. Eduardo P.W., ed. *The Vertebrate Gas Transport Cascade: Adaptations to Environment and Mode of Life.* Boca Raton, Fla.: CRC Press, 1993. A collection of symposium papers related to gas transport and respiratory mechanisms.

Bryant, Christopher, ed. *Metazoan Life Without Oxygen.* New York: Chapman and Hall, 1991. A collection of essays addressing the question of how ancient animals dealt with low oxygen or interruptions in oxygen supply to vital organs. Traces the evolution of animal species to adapt to the accumulation of oxygen in earth's atmosphere.

Dejours, Pierre. "Mount Everest and Beyond: Breathing Air." In *A Companion to Animal Physiology*, edited by C. Richard Taylor, Kjell Johansen, and Liana Bolis. New York: Cambridge University Press, 1982. This chapter is a scholarly paper that Dejours delivered at a symposium. It is clearly written and has illustrations. Basic principles of respiration are explained; topics such as how animals deal with conditions of low oxygen are discussed. The chapter does not require an extensive science background.

Graham, Jeffrey B. *Air-Breathing Fishes: Evolution, Diversity, and Adaptation.* San Diego, Calif.: Academic Press, 1997. Comprehensive coverage of the puzzling existence of air-breathing fishes. Easy to read for nonspecialists with a background in biology. Numerous tables and illustrations.

Hill, R. W., and G. A. Wyse. *Animal Physiology.* New York: Harper & Row, 1989. The chapter on the physiology of diving in birds and mammals gives an excellent explanation of the problems of limited oxygen supply and metabolic function under such conditions. The chapter discusses responses of two groups of animals that have been well studied. This text also has background material in related topics. The text is understandable, the information is current, and the figures are illustrative.

Hochachka, Peter, and George Somero. *Strategies of Biochemical Adaptation.* Philadelphia: W. B. Saunders, 1973. This book is a monograph that treats the general concepts of biochemical adaptations and gives the logic behind the authors' present explanations. Although the book is more than twenty-five years old, the explanations are first rate; it is clearly written, with sound reasoning. It is not filled with masses of equations, complex graphs, and numerous references, but concentrates on the concepts. The first two chapters are directly related to this topic, and there are other references to low oxygen in the book.

Paganelli, Charles V., and Leon E. Farhi, eds. *Physiological Function in Special Environments.* New York: Springer-Verlag, 1988. This book presents the papers from a symposium on environmental physiology held in Buffalo, New York. The chapters are written by experts in the field for presentation to scientists. The papers are well written and informative, and are suited to the person who is able to master the other entries here and still wishes to pursue the topic. Chapters 1 to 4 on adaptation to altitude, and chapters 12 and 13, on comparative physiology, deal with physiological adaptations related to low-oxygen conditions.

Respiratory system

FIELDS OF STUDY

Anatomy, biochemistry, cell biology, environmental science, histology, immunology, pathology, physiology

SUMMARY

The respiratory system maintains a constant flow of oxygen into the blood while removing carbon dioxide. Other functions of the respiratory system include maintaining blood acid-base balance, reducing body temperature, communicating by means of sounds, and removing inhaled microbes. Animals generally meet their energy needs by oxidation of food, and the respiratory system supplies the oxygen necessary for cell metabolism while removing its waste product, carbon dioxide. Oxygen is available either dissolved in water or as a component of the air, and animals have evolved special organ structures to effectively obtain oxygen from their environment.

PRINCIPAL TERMS

- **alveolus:** the thin-walled, saclike lung structure where gas exchange takes place
- **chemoreceptor:** specialized nervous tissue that senses changes in pH (hydrogen ions) and oxygen
- **countercurrent exchanger:** the process where a medium (air or water) flowing in one direction over a tissue surface encounters blood flowing through the tissue in the opposite direction; this improves the gas diffusion by maintaining a concentration gradient
- **diffusion:** the process by which gas molecules move from a higher to a lower concentration through a medium or across a permeable barrier; the rate at which gases cross a barrier is increased by the surface area, and gas concentration gradient is decreased by the thickness of the barrier; gas solubility determines the amount that crosses the barrier
- **gill:** an evaginated organ structure where the membrane wall turns out and forms an elevated, protruding structure; typically used for water respiration
- **lung:** an invaginated organ structure where the membrane wall turns in and forms a pouch or saclike structure

ORGANS OF GAS EXCHANGE

Single-cell and simple organisms, such as flatworms and protozoa, can obtain sufficient oxygen to meet their energy demands by simple diffusion through their body surface. Some amphibians utilize gas exchange through their skin to supplement their lung respiration, but generally, larger, more complex animals require specialized organ systems with a large surface area for gas exchange and a circulatory system for distribution of oxygen to each cell. The basic mechanism, however, for gas exchange between the environment and the blood and between the blood and cells is by diffusion. The three major types of gas exchange organs are the gill for water respiration, the lung for air and in some special cases water respiration, and the tracheas system of tubules for air respiration in insects.

Gills consist of several gill arches located in the operculum or gill cover on each side of the fish's head. A gill arch contains two rows of gill filaments, and each filament has a row of parallel platelike structures on its surface called lamellae. The lamellae are everted structures that rise up from the filament surface and are only a fraction of a millimeter apart. Water flows between the lamellae, and oxygen diffuses from the water into the lamellar capillary blood. The lamellar blood flows in the opposite direction of the water flow and creates a countercurrent exchanger. The countercurrent maximizes the diffusion of oxygen into the lamellar capillary blood by maintaining a diffusion gradient over its entire length.

Lungs, in contrast to gills, are invaginations, where the surface turns in and forms a hollow or saclike structure. Lungs typically are divided into two functional areas: the conducting zone and the respiratory zone. The conducting zone branches from the trachea to the bronchioles and distributes air to the respiratory zone but is not involved in gas exchange. The respiratory zone comprises the majority of the lung and contains small respiratory bronchioles and ducts that lead to the primary gas exchange area, the alveolus. The alveoli vary from simple saclike structures in a pulmonate land snail to the complex alveolar wall structure of mammals. The alveolar wall is fifty micrometers thick, or about one fiftieth the thickness of a sheet of paper, and is composed of epithelial cells covering the alveolar surface, an interstitial space, and the endothelial cells that make up the capillaries. This thin-wall structure allows for the diffusion of oxygen and carbon dioxide between the air and blood.

The insect tracheas respiratory system is unique because it is both the gas exchange and distribution system. Pairs of openings on the insect's thorax and abdomen called spiracles regulate the movement of air in and out of a tubule system. The spiracles open and close in a pattern that allows unidirectional flow of air through the tubule system. The tubules branch and extend throughout the insect's body and deliver oxygen to the cells independent of the circulatory system.

AIR AND WATER ENVIRONMENTS

Important aspects of the atmosphere for respiration are the barometric pressure and concentration of gases, temperature, and humidity. The atmospheric gases important to animals are oxygen, carbon dioxide, and nitrogen, and the atmosphere is a constant 20.95 percent oxygen, 0.03 percent carbon dioxide, and 79 percent nitrogen (plus other inert gases). The rate of diffusion of oxygen from the inspired air into the circulation depends on the partial pressure of the oxygen. The barometric pressure, however, decreases with increasing altitude, and this decreases the partial pressure of oxygen, which decreases the diffusion of oxygen into the blood. Thus an animal's difficulty in obtaining adequate oxygen at higher altitudes is related to the reduction in atmospheric pressure and not to a change in the percentage of oxygen in the atmosphere.

The temperature and amount of water vapor or humidity in the atmosphere are variable, and during inspiration, the inspired air is warmed to body temperature and saturated with water vapor (100 percent humidity). The heat and moisture come from the airways and can potentially cool and dehydrate an animal. Therefore, a minimal amount of air is inspired to prevent excess heat and water loss. However, heat-stressed animals will use this respiratory heat loss or panting to cool their bodies.

Water poses several challenges for respiration compared to air: a lower oxygen content, slower gas diffusion rate, higher viscosity, and greater weight. The amount of oxygen available in the water is thirty times less than that found in air. Thus, more water has to flow over the gill surface for adequate oxygen delivery. The speed at which oxygen moves through the water is ten thousand times slower than oxygen moving through the air. Thus, the distance between the water and the gill surfaces can only be a fraction of a millimeter apart. In contrast, the lung gas exchange surfaces are a few millimeters apart.

Water's greater viscosity and weight compared to air require more energy to move water over the gill surface. Water-breathing animals compensate for this by having a unidirectional flow through the gill. This avoids water being moved, stopped, and then moved again in the opposite direction, which works well for air, but would be very energy costly for the heavier, more viscous water.

The gill structure depends on water to support and separate the rows of lamellar structures. Thus, when a fish is exposed to air, the gill structure collapses on itself and greatly reduces the surface area available for oxygen diffusion. Thus, the fish will suffocate if not returned to the water.

BREATHING WATER AND BREATHING AIR

Water can be moved through the gill lamellae by either opercular pumping or ram ventilation. Opercular pumping involves the movement of the mouth and opercular covering to create pressure gradients for unidirectional flow of water through the mouth, across the gill surface, and out the opercular covering (unidirectional flow). Ram ventilation takes advantage of the fish's forward speed to flow water through the mouth and gill. Opercular pumping is used from rest to slow swimming speeds, and a fish switches over to ram ventilation when swimming at faster speeds.

For air breathers, inspiration (inflating the lungs) can be accomplished by either positivepressure or negative-pressure breathing. Positivepressure breathing requires air pressure to inflate the lungs, which is similar to inflating a balloon or tire with a compressed air. The pressure is considered positive because it is greater than atmospheric pressure. For example, frogs use positivepressure breathing by closing their mouths and then elevating the floor of the mouth. This compresses and pressurizes the air and forces it into the lung. The elastic lung tissue is stretched like an inflated balloon by the increased volume. The process of the air moving out of the lung is called expiration. When the frog relaxes and opens its mouth the lung elastic recoil forces the air out similar to a balloon deflating.

With negative-pressure breathing, the lung is pulled open by contraction of the diaphragm. The pressure becomes negative (below atmospheric pressure), and air flows into the lung until it equalizes with the atmospheric pressure. If additional inflation is required, such as during exercise, accessory inspiratory muscles lift the ribs to inflate the lungs further. Expiration is accomplished by the relaxation of the inspiratory muscles, and the lung elastic recoil increases airway pressure and air flows out of the lung.

Inspiration is always an active process, whereas expiration results from the passive elastic recoil of the lung tissue. However, active expiration is possible by contracting muscles that pull the ribs down and by using abdominal muscles to push the diaphragm farther into the thoracic (chest) cavity.

SETTING BREATHING RATE

In water-breathing animals, such as fish and lobsters, the level of oxygen sets the ventilation rate (volume of water moved through the gill per minute) such that as oxygen content in the water decreases, the frequency of breathing movements increases. During fast swimming, fish using ram ventilation regulate the mouth opening so that the amount of water flowing over the gills just meets tissue oxygen demand. A wider mouth opening than is necessary increases the fish's frictional drag through the water and thus decreases the energy efficiency. Carbon dioxide is highly soluble in water and easily diffuses from water-breathing animals. Thus, blood carbon dioxide levels in water-breathing animals are very low and not used to regulate respiration rate.

In air-breathing animals, the blood levels of carbon dioxide and oxygen regulate the ventilation rate (air volume moved in and out of the lungs per minute). Carbon dioxide quickly diffuses from the small capillaries in the brain circulation into the fluid surrounding the brain cells (cerebral spinal fluid). Here the carbon dioxide reacts with water and forms carbonic acid. The hydrogen ions released from the carbonic acid stimulate chemoreceptor cells that in turn stimulate the respiratory center in the medulla, located in the brain stem. Higher concentrations of carbon dioxide increase the hydrogen ion concentration and thus increase ventilation rate. Air-breathing animals primarily regulate ventilation rate by carbon dioxide produced from metabolism and not low blood oxygen levels.

However, oxygen can regulate ventilation in animals at high altitudes. Oxygen partial pressure is sensed by chemoreceptors in the aorta and the carotid artery. These peripheral chemoreceptors sense the partial pressure of oxygen in the blood plasma, and as the partial pressure of oxygen in the air decreases, such as with altitude, the partial pressure of oxygen in blood also decreases. This increases ventilation, which then compensates for the lower oxygen partial pressure. In addition to low oxygen partial pressure, the peripheral chemoreceptors are stimulated by blood acidosis. For example, lactic acid released from skeletal muscles during strenuous exercise stimulates the ventilation rate in animals and humans.

—Robert C. Tyler

See also: Anatomy; Bone and cartilage; Brain; Circulatory systems of vertebrates; Digestive tract; Eyes; Noses; Reproductive system of female mammals; Reproductive system of male mammals; Respiration and low oxygen; Skin.

FURTHER READING

Fish, F. E. "Biomechanics and Energetics in Aquatic and Semiaquatic Mammals: Platypus to Whale." *Physiological and Biochemical Zoology* 73, no. 6 (2000): 683-698. An overview of the mechanics of aquatic respiration.

Pough, F., H. Heiser, J. B. Heiser, and W. N. McFarland. *Vertebrate Life.* 5th ed. Upper Saddle River, N.J.: Prentice Hall, 1999. Emphasis on the differences in animals based on the sequences of evolution.

Schmidt-Nielson, Knut. *Animal Physiology: Adaptation and Environment.* 5th ed. New York: Cambridge University Press, 1997. This classic textbook in animal physiology is a standard for many high school and college courses and focuses on physiologic function and adaptation to different environmental conditions.

Weibel, Ewald R. *The Pathway for Oxygen.* Cambridge, Mass.: Harvard University Press, 1984. Ewald Weibel's research presented in this book established many of the concepts and facts currently known about respiratory system structure and function.

Willmer, Pat, Graham Stone, and Ian Johnston. *Environmental Physiology of Animals.* Malden, Mass.: Blackwell Science, 2000. This textbook has simple-to-understand diagrams and explanations of animal organ systems and their adaptation to different environments.

Withers, P. C. *Comparative Animal Physiology.* Fort Worth, Tex.: Saunders, 1992. Gives a broad overview of animal physiology comparing the characteristics of each species.

RESPIRATORY PIGMENTS

The amount of oxygen dissolved into circulatory fluids is insufficient to meet the oxygen requirements for almost all vertebrate animals. Respiratory pigments, colored proteins that contain a metal (usually iron or copper) have evolved to increase the amount of oxygen carried by the blood one hundredfold (20 milliliters of oxygen per 100 milliliters of blood, compared to 0.2 milliliters of oxygen per 100 milliliters of blood). These pigments have the unique property of binding oxygen at the gas exchange surface of the gill or lung, transporting the oxygen in circulation, and then releasing the oxygen to the cells, where it can be utilized for energy metabolism.

The most common respiratory pigment is hemoglobin, which has a positively charged iron atom (divalent) attached to a circular protein ring (porphyrin ring). Oxygen is negatively charged and binds to the positive charge of the iron. This bond is easily broken at the tissue level, where oxygen is released. Copper-containing respiratory pigments are called hemocyanins and are found in mollusks and arthropods. Next to hemoglobin, they are the most widely used respiratory pigment.

Hemoglobin is bright red when it carries oxygen and turns a purpleblue color when it has released some or all of its oxygen molecules. Hemoglobin can be found either circulating free in the plasma, typically in insects called hemolymph, or contained in red bloods cells called erthyrocytes, circulating in the plasma.

AUGUST KROGH

Born: November 15, 1874; Grenå, Jutland, Denmark

Died: September 13, 1949; Copenhagen, Denmark

Fields of study: Anatomy, environmental science, invertebrate biology, physiology, zoology

Contribution: Krogh originally described how animals exchange oxygen and carbon dioxide by diffusion and invented many instruments needed to conduct experiments that led to his conclusions. He received the Nobel Prize in Physiology or Medicine in 1920 for his studies on capillary function and muscle metabolism. In this work, Krogh published the first account of regulation of blood perfusion in muscle and other organs.

In 1897, Schack August Steenberg Krogh began working in the lab of the famous physiology professor, Christian Bohr. Dr. Bohr had studied the solubility of oxygen in different tissues and fluids, as well as the mechanisms of muscle contraction. These experiments greatly influenced Krogh's early studies of gas exchange in snails, frogs, and fishes. In 1899, Krogh published the equivalent of a master's thesis, demonstrating that, in birds, oxygen moved by diffusion through the thin lung membranes into the blood. His dissertation, in 1903, studied gas exchange in the frog and showed that skin respiration remains fairly constant, whereas large variability occurs in lung respiration. Krogh reasoned this was an example of the oxygen secretion hypothesis proposed by Bohr. However, later he would doubt his conclusion and demonstrate that oxygen moves solely by diffusion through tissues.

Krogh participated in an expedition in 1902 to Disko, North Greenland, where he investigated the carbon dioxide and oxygen content in springwater, streams, and the sea. From these studies, Krogh described the important role of the oceans in regulation of atmospheric carbon dioxide. He applied these techniques of measuring dissolved gases in animal physiological studies in 1904.

Krogh won the prestigious Seegen Prize, awarded by the Austrian Academy of Sciences, in 1906 for investigating whether free nitrogen or nitrogenous gases were released as a normal by-product of metabolism. He showed that gaseous nitrogen remained constant by using his unique respiratory gas quantification methods. Krogh determined nitrogen dynamics with gas measurements instead of using the traditional German method of Liebig and Rubner, who measured nitrogen content in ingested food and liquids and excreted nitrogen in feces and urine.

Marie Jörgensen, a medical student and scientist, married August Krogh in 1905. Together they published seven papers on the quantification and diffusion of gases in the blood. This overturned the view held by Dr. Bohr and the scientific establishment that stated that oxygen and carbon dioxide were "secreted" by a glandlike structure in the lung.

In 1908, a special position as Associate Professor in Zoo-Physiology was created for Krogh at the University of Copenhagen, and in 1910 Krogh founded a zoo-physiology (animal) laboratory at the University of Copenhagen. The laboratory was moved and enlarged in 1928 with financing from theRockefeller Foundation. Eight years later, Krogh was promoted to a chair, which he held until his retirement in 1945. The Krogh Institute is still active today.

—*Robert C. Tyler*

RNA/PROTEIN TRANSLATION

FIELDS OF STUDY

Biochemistry; Genetics; Molecular Biology

SUMMARY

The process by which RNA translates protein structures from the genetic code in the DNA molecule is described. The process is controlled by a simple series of three- nucleotide units called codons that determine which amino acids are to be assembled in which order to create every protein and enzyme in the entire biological system.

PRINCIPAL TERMS

- **anticodon:** a sequence of three nucleotide bases in transfer RNA (tRNA) that bonds to a complementary codon in messenger RNA (mRNA).
- **codon:** a sequence of three nucleotide bases that specifies a particular amino acid or control point in the process of protein synthesis.
- **gene expression:** the process by which RNA copies genes, which are specific segments of the DNA molecule, and uses the information to synthesize either proteins or other types of RNA.
- **peptide bond:** a covalent bond that links the carboxyl group of one amino acid to the amine group of another, enabling the formation of proteins and other polypeptides.
- **ribosome complex:** a structure consisting of ribosomal RNA (rRNA) and enzymes that decodes messenger RNA (mRNA) and coordinates the assembly of proteins from amino acids carried by transfer RNA (tRNA).

THE PROTEIN ASSEMBLY LINE

Translation of the primary structure of a protein from the genetic code in the deoxyribonucleic acid (DNA) molecule is a kind of assembly-line manufacturing process that takes place in all biological systems. In essence, ribonucleic acid (RNA) reads the blueprint and assembly instructions from the DNA molecule, delivers the necessary parts to the "assembly line" in the ribosome complex, and assembles the parts into proteins according to the design specified by the DNA molecule.

TRANSCRIPTION AND TRANSLATION

The structure of the DNA molecule consists of a very long backbone of deoxyribose sugar molecules alternating with phosphate groups. To each deoxyribose sugar molecule is attached one of four base molecules called nucleotides: adenine, cytosine, guanine, or thymine (commonly abbreviated A, C, G, and T). A second, equally long strand of DNA has a complementary structure to the first. When the two strands combine in the duplex DNA molecule to form its characteristic double-helix structure, the bases coordinate to each the ribosome complex, and assembles the parts into proteins according to the design specified by the DNA molecule.

The process is, of course, more complicated than that, as it is the result of many different enzyme-mediated chemical reactions that must occur in the proper sequence for each one of the many thousands of different proteins that are synthesized in the routine functioning of a living cell. The DNA molecule itself is composed of very few molecular components, but it is nevertheless subject to modification due to any number of causes. Each modification or perturbation of the normal structure of the DNA molecule has effects that are reflected in the cell's ability to produce the correct proteins for its proper functioning. A mutation of the DNA structure may prevent the production of a necessary protein, or it may cause the production of proteins that are at best unusable and at worst dangerous. Mutations that cause the normal process that halts protein synthesis to be disrupted can result in the formation of cancerous adenine or cytosine in one strand matches to thymine or guanine, respectively, in the other strand. The exact matching of the bases along the entire length of the two strands is required for the formation of the duplex DNA molecule.

To begin the transcription process—the process of copying a segment of DNA onto a segment of RNA— enzymes specific to the process attach to the DNA molecule and temporarily separate the two strands. This exposes the sequence of bases in the nucleotides of the DNA strand and begins the process of gene expression. Other enzymes called RNA polymerases then assemble free nucleotides to match the exposed sequence of DNA nucleotide

RNA/PROTEIN TRANSLATION

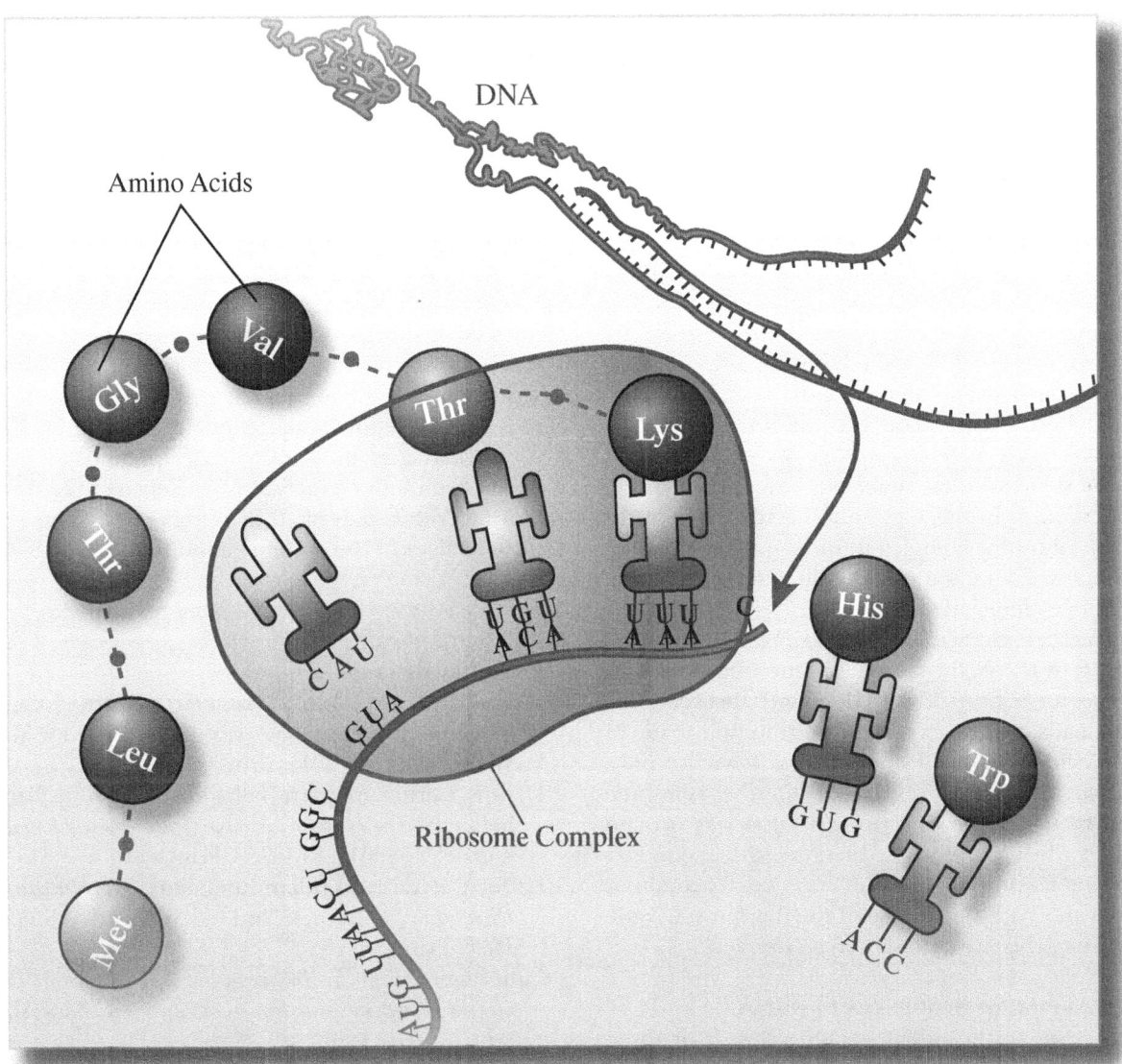

bases. The strand that the RNA attaches to is called the template or non-coding strand, while the other strand, to which the template strand was originally attached, is called the coding strand. Because nucleotides always coordinate to their complementary bases, the sequence of the RNA strand will be identical to that of the DNA coding strand, with one difference: in RNA, the base uracil (U), rather than thymine, is used to complement adenine. Thus, a thymine base in the coding strand will coordinate to an adenine base in the template strand, which will in turn produce a uracil base in the RNA strand.

The formation of this new RNA strand is initiated and terminated by specific sequences of nucleotide bases on the DNA strand. When the "stop" sequence is encountered, the assembly process ends, and another enzyme separates the complementary RNA strand from the DNA strand, releasing it into the cytosol (intracellular fluid) of the cell as a molecule of messenger RNA (mRNA). At this point, transcription is complete and translation begins.

In the cytosol, the molecule of mRNA connects to ribosome complexes, which are themselves composed of another type of RNA called ribosomal RNA (rRNA). The binding of the mRNA to the rRNA occurs by matching nucleotide base sequences in the two strands. This exposes specific three-unit sequences of the mRNA termed codons, which identify the specific amino acid that is to be added onto the protein molecule being assembled at that location.

A third form of RNA, called transfer RNA (tRNA), transports particular amino acids to the site designated by the codon on the mRNA. Each tRNA molecule has a matching anticodon segment in its primary structure that attaches to the codon in the mRNA strand. In this way, amino acids are brought into the ribosome complex in the sequence that had been specified in the nucleotide sequence of the parent DNA molecule.

As each amino acid is brought into the ribosome complex by tRNA, the enzymes in the ribosome catalyze the formation of peptide bonds between the amino acids in the sequence. The resulting string of amino acids is called a polypeptide. After the polypeptide structure is released from the ribosome complex, it undergoes a process known as protein folding, through which it assumes the secondary structural features that characterize it as a protein or an enzyme. With the release of the new protein molecule, the process of translation is complete.

Overlapping of Sequences in mRNA

Each codon of the mRNA strand consists of three nucleotides, and merely shifting over one nucleotide in the sequence creates a new set of three nucleotides and therefore a new codon. For example, the DNA nucleotide series CCTACCTGG codes for the amino acids proline, threonine, and tryptophan. Shifting over two nucleotides creates the sequence TACCTGG, which codes for the amino acids tyrosine and leucine in an entirely different polypeptide chain. Viruses use this codon shifting to generate the various proteins required for their successful infection of a host cell.

The Genetic Code and the Human Genome

The discovery of the molecular structure of DNA in 1943 enabled research that ultimately revealed the genetic code. Through numerous experimental studies that included the use of synthetic nucleotide sequences, researchers were eventually able to identify which groups of nucleotides were codons for which amino acids. Since the number of three-nucleotide combinations that can be formed with four nucleotides (sixty-four) is much greater than the twenty essential amino acids used in protein synthesis, there is significant redundancy in the genetic code, with several different codons specifying the same amino acid. The human genome was finally deciphered at the end of the twentieth century, and in February 2001, the journal *Nature* published its report of the first complete analysis of the human genome.

Richard M. Renneboog, MSc

Further Reading

Berg, Jeremy M., John L. Tymoczko, Gregory J. Gatto, and Lubert Strye. *Biochemistry*. 8th ed. New York: W. H. Freeman, 2015. Print.

Dennis, Carina, Richard Gallagher, and Philip Campbell, eds. *The Human Genome*. Spec. issue of *Nature* 409.6822 (2001): 813–958. Print.

Lafferty, Peter, and Julian Rowe, eds. *The Hutchinson Dictionary of Science*. 2nd ed. Oxford: Helicon, 1998. Print.

Lehninger, Albert L. *Biochemistry: The Molecular Basis of Cell Structure and Function*. 2nd ed. New York: Worth, 1975. Print.

Lodish, Harvey, et al. *Molecular Cell Biology*. 7th ed. New York: Freeman, 2013. Print.

Reece, Jane B., et al. *Campbell Biology*. 10th ed. San Francisco: Cummings, 2013. Print. Watson, James D. *The Double Helix*. New York: Atheneum, 1968. Print.

> **RNA/PROTEIN TRANSLATION SAMPLE PROBLEM**
>
> Given the following sequence of nucleotides in a template strand of DNA, use an RNA codon table to identify the codons and the corresponding amino acids that will be assembled in sequence to form a new protein strand:
>
> TTTCTTACCTTGTTGAGTGATTTT-TACTTCAAACCGCGC
>
> **Answer:**
>
> First, convert the template strand to its complementary mRNA strand by replacing each nucleotide in the sequence with its RNA complement. Remember that U (uracil) is used in RNA instead of T (thymine), while the other nucleotides—C (cytosine), A (adenine), and G (guanine)—remain the same.
>
> template DNA:
> TTTCTTACCTTGTTGAGTGATTTT-TACTTCAAACCG
>
> mRNA:
> AAAGAAUGGAACAACUCACU-AAAAAUGAAGUUUGGC
>
> Divide the mRNA sequence into three-letter groups to isolate the codons:
>
> AAA GAA UGG AAC AAC UCA CUA AAA AUG AAG UUU GGC
>
> Identify the amino acid corresponding to each individual codon, using the standard genetic code chart:
>
> AAA: lysine (Lys)
> GAA: glutamic acid (Glu)
> UGG: tryptophan (Trp)
> AAC: asparagine (Asn)
> UCA: serine (Ser)
> CUA: leucine (Leu)
> AUG: methionine (Met)
> AAG: lysine (Lys)
> UUU: phenylalanine (Phe)
> GGC: glycine (Gly)
>
> Now arrange the amino acids in their proper sequence according to the mRNA codons to determine the primary structure of the resulting polypeptide chain:
>
> Lys-Glu-Trp-Asn-Asn-Ser-Leu-Lys-Met-Lys-Phe-Gly
>
> Note that there are more combinations of the nucleotide bases than there are essential amino acids, and thus the same amino acid can be specified by different codons in the genetic code. In the above problem, both AAA and AAG code for the amino acid lysine.

RNAASE

FIELDS OF STUDY

Cell and Tissue Engineering; Drug Testing; Enzyme Engineering; Metabolic Engineering; Pharmacology; Proteomics; Zymurgy and Zymology

SUMMARY

RNAase is a collective term that refers to a relatively large number of enzymes that act on the many different forms of ribose nucleic acid (RNA). The functions of the RNAase enzymes are very specific. They initiate, control and terminate all stages of the processes of replication and transcription, and are therefore completely self-regulating. One of the RNAase enzymes initiates opening of the DNA molecule from end to end in replication, while another opens the DNA molecule at specific locations only to retrieve the amino acid code for a specific protein. Still other RNAase enzymes control closure of the DNA strands, as well as setting the helical structure of the DNA molecule, and making all of the necessary manipulations of RNA in protein synthesis.

PRINCIPAL TERMS

- **amino acids:** biological molecules that serve as the building blocks of proteins and enzymes. Amino acids are incorporated into proteins by transfer RNA, according to the genetic code contained in DNA. The majority of amino acids have names ending with -ine, and are complex arrangements of atoms of carbon, nitrogen, hydrogen, and oxygen. *See also* enzyme, protein.
- **catalyst:** a substance that makes a chemical reaction between two other substances proceed at a

significantly faster rate, without being consumed in the reaction. (See also enzyme.)

- **enzyme:** a biological catalyst, any protein molecule within a living organism that speeds up biochemical reactions to a rate that will sustain life. The effect may speed up metabolic reactions by a factor of one million, compared with what would occur chemically outside the body. Names and classification of enzymes are regulated by the International Commission on Enzymes. Most enzymes are named by adding -ase to the root of a corresponding substrate, the molecule an enzyme acts upon. Sucrase catalyzes the hydrolysis of sucrose into glucose and fructose. A living cell has a unique set of 3,000 enzymes, each defined by the cell's DNA.

- **protoplasm:** the living substance of a cell, including the content of the cell membrane and the substance within the cell—a transparent gelatinous material composed of inorganic substances (90 percent water with mineral salts and gases such as oxygen and carbon dioxide), and organic substances (proteins, carbohydrates, lipids, nucleic acids, and enzymes). Protoplasm outside the cell nucleus is called cytoplasm.

BASIC PRINCIPLES

No molecular biology or biochemistry textbook will carry a reference to something called 'RNAase'. The term is a 'catch-all' for a number of different enzymes that act on at least as many different types of ribose nucleic acid (RNA). The processes of replication and transcription, for the production of new strands of deoxyribose nucleic acid (DNA) and the many different proteins that are produced and used in the body incorporate several different types of RNA through various stages of those processes. Each different type of RNA requires RNAase enzymes to carry out the specific functions required to assemble the particular type of RNA from component nucleotides, to act as a catalyst to enable the RNA to carry out its specific function, and then to disassemble the RNA molecules into component nucleotides when they are no longer required. The enzymes are assembled from amino acids in the protoplasm of the cell, according to the nucleotide sequence of the parent DNA molecule. Processes involving RNA are accordingly quite complex.

RNA AND PROTEINS

Proteins are produced via the process of transcription according to the blueprint carried in the sequence of nucleotides in the DNA molecule. Transcription is an RNA-mediated process by which the pattern of nucleotides in a segment of DNA specifies the identity and order of the amino acids required for the construction of a specific enzyme or protein. The proteins that are produced include the very RNA-active enzymes that are required for the process. In transcription the DNA molecule is opened at the appropriate location by a specific enzyme. This allows a complementary strand of transfer-RNA, or tRNA, to form complimentary to that segment. Another enzyme brings about the release of the tRNA into the cell protoplasm. There, other enzymes cleave the tRNA into smaller three-unit segments of messenger-RNA, or mRNA. Each unit of mRNA identifies a specific amino acid to be incorporated into the target protein molecule, and the appropriate amino acid becomes attached to its corresponding mRNA codon. This bit of material is carried to the nearest ribosome, where it coordinates to the ribosomal-RNA, or rRNA, for assembly into the protein structure according to the sequence specified by the original sequence of nucleotides obtained from the DNA segment.

RNA AND DNA REPLICATION

The process of replication produces a new DNA molecule. In this process, the DNA molecule is opened up by the corresponding DNA-active enzymes, and duplicated along the entire length of the two complimentary strands by complimentary strands of RNA. In RNA, the deoxyribose sugar backbone of the DNA molecule is replaced by ribose sugar, and uracil instead of thymine is the base compliment to adenosine. The two RNA strands that are produced in this way are subsequently cleaved from their complimentary DNA strand and are replaced nucleotide by nucleotide to form a new complimentary strand of DNA. The end result is that two DNA strands are produced from one.

RNAASES

Throughout the processes of transcription and replication various forms of RNA are required. This also requires that the various types of RNA must be synthesized and manipulated by the various enzymes that collectively may be referred to as RNAase. In addition to tRNA, mRNA and rRNA, forms referred to as gRNA and hnRNA have also been identified, and certainly as molecular biologists and biochemists obtain more detailed information from proteomics regarding the biochemical processes of life other forms of RNA are likely to be identified as integral components of specific processes. The formation of the different known forms of RNA is mediated by three different enzymes termed RNA polymerase I, II and III and by tRNA synthetase. As the names suggest, the polymerases act to build the various RNA strands by polymerization of the appropriate nucleotides. In transcription, the formation of tRNA on the complimentary DNA strand is initiated, controlled and terminated ny tRNA synthetase. Accordingly, it is quite likely that each distinct form of RNA that is required has a corresponding set of RNAase enzymes.

Richard M. Renneboog MSc

FURTHER READING

Stewart, P.R. And Letham, D.S. *The Ribonucleic Acids* 2nd ed., New York, NY: Springer-Verlag, 2012.

Kirby, Lorne T. *DNA Fingerprinting. An Introduction* New York, NY: Oxford University Press, 1993.

Walter, Nils G., Woodson, Sarah A. and Batey, Robert T., eds. *Non-Protein Coding RNAs* New York, NY: Springer-Verlag, 2009.

Dudek, Ronald W. *High-Yield Cell & Molecular Biology* 2nd ed., Baltimore, MD: Lippincott Williams & Wilkins, 2007.

Tropp, Burton E. *Molecular Biology. Genes to Proteins* 3rd ed., Boston, MA: Jones and Bartlett, 2008.

See also: Cell Organelles, Deoxyribose Nucleic Acid.

FASCINATING FACTS ABOUT RNAASE

- The nucleotide sequence of DNA, using A for adenosine, C for cytosine, T for thymine, and G for guanine, is written in the form ATGCTTCAGCG... To write out the complete structure of a strand of human DNA in this form would require about one million closely printed pages. However, the RNA for each amino acid codon can be written with just three letters.

- A different RNAase enzyme is required for each different variation of RNA. This is equivalent to requiring a different hammer and carpenter for each type of nail used in building a house.

- RNAase enzymes control their own formation. They function in and control the processes of replication and transcription. Replication produces exact copies of the DNA molecule that contains the blueprint for all of the enzymes, including the RNAases, and transcription is the process that constructs those enzymes from free amino acids.

- Although similar in function, the RNAase enzymes of one species are not identical to the 'RNAase' enzymes of other species.

SEX DIFFERENCES: EVOLUTIONARY ORIGIN

EVOLUTIONFIELDS OF STUDY

Anatomy, evolutionary science, reproduction science

SUMMARY

In asexual reproduction, genetic material is not exchanged; offspring are genetically identical to the parent. Sexual reproduction involves the exchange of genes. Natural selection favors asexual reproduction in the exploitation of dependable resources, but selection favors sexual reproduction whenever the future is uncertain.

PRINCIPAL TERMS

- **anisogamy:** reproduction using gametes unequal in size or motility asexual reproduction: reproduction in which genes are not exchanged
- **female:** an organism that produces the larger of two different types of gametes
- **gonochorism:** sexual reproduction in which each individual is either male or female, but never both
- **hermaphroditism:** sexual reproduction in which both male and female reproductive organs are present in the same individual, either at the same time or at different times
- **isogamy:** reproduction in which all gametes are equal in size and motility male: an organism that produces the smaller of two different types of gametes
- **parthenogenesis:** asexual reproduction fromunfertilized gametes, producing female offspring only
- **sexual dimorphism:** differences in morphology between males and females
- **sexual reproduction:** reproduction in which genes are exchanged between individuals
- **sexual selection:** selection for reproductive success brought about by the behavioral responses of the opposite sex

BASIC PRINCIPLES

The evolutionary origin of sex differences can be understood only by examining the relative benefits of sexual as compared to asexual reproduction. Those forms of reproduction in which genes are not exchanged are considered asexual. Asexual reproduction may take place from already developed body parts (vegetative reproduction) or fromspecial reproductive tissue. In either case, however, asexual reproduction results in the rapid production of numerous individuals genetically identical to their parents. Because asexual reproduction allows numerous offspring to be produced in a short time, it is favored in situations in which a species can gain an advantage by exploiting an abundant but temporary resource, such as a newly discovered cache of food. There is also a further advantage: The individual that finds a resource that it can effectively exploit, if it can reproduce asexually, is assured that all its offspring will possess the same genotype as itself, and will thus be equally able to exploit the same resource for as long as it lasts. Despite these advantages, asexual reproduction is much less common than sexual reproduction among animals. It is a temporary stage in many species, alternating with sexual reproduction. Asexual reproduction is far more common among microorganisms such as bacteria.

FORMS OF SEXUAL REPRODUCTION

Sexual reproduction may take many forms, but all of them involve the exchange of genes. Some algae and protozoans exchange chromosomes without gametes in a process called conjugation. Most other forms of sexual reproduction use special sex cells called gametes, which exist in different "mating types." Two gametes can combine only if their mating types are different. Some simple organisms, such as the one-celled green alga *Chlamydomonas,* have gametes that are indistinguishable in size or appearance, a condition known as isogamy. Most other organisms have gametes of unequal sizes, a condition called anisogamy. Selection often intensifies the differences between gametes, producing a small, motile sperm and a much larger, immobile egg, laden with stored food (yolk).

Some sexually reproducing organisms have separate sexes, a condition called gonochorism. Individuals producing eggs are called female, while individuals producing sperm are called male. Since sperm are generally small and can be produced in great numbers, males tend to leave more offspring if they reproduce prolifically, indiscriminately, and often. Females, on the other hand, have fewer eggs to offer, and in many species they must also invest nutritional and behavioral energy in the laying of eggs and the care of the resultant offspring. Selection in these species favors females who choose their mates more carefully and take better care of their offspring.

The differing selective forces operating on the two sexes often give rise to sexual dimorphism, or differences in morphology between the sexes. Sexual dimorphism can also be reinforced by competition for reproductive success, a phenomenon first studied by Charles Darwin. Darwin called this type of competition sexual selection. It takes two basic forms—direct competition between members of the same sex, and mate choices made by members of the opposite sex.

Direct male-male competition often takes such spectacular forms as rams or stags fighting in head-to-head combat. Similar fights also occur in many other species, including a variety of turtles, birds, mammals, fishes and invertebrates. Many more species, however, engage in ritual fighting in which gestures and displays substitute for actual combat. Male baboons, for example, threaten each other in a variety of ways, including staring at each other, slapping the ground, jerking the head, or simply walking toward a rival.

Although male-male rivalry has attracted more attention in the past, female-female competition also occurs in many species. Now that more ethologists and sociobiologists are looking for evidence of such direct competition among females, it is being discovered that it is a fairly widespread occurrence which had previously escaped notice only because so few scientists suspected its existence or were interested in looking for it. Female-female competition has been found among langur monkeys, golden lion-marmosets, ichneumon wasps, and several other species.

Sexual Selection

Sexual selection in mating is selection in which reproductive success is determined at least in part by mate choice. No matter what form sexual selection may take, it results in greater reproductive success for those individuals chosen as mates, while those not chosen must try again and again if they are ever to succeed in leaving any offspring at all.

Sexual selection of this kind occurs in nearly all gonochoristic species. In some species, males will attract females by means of a visual display or by various sounds (also called calls or vocalizations). Females in such species will exercise choice by selecting among the available males. For example, male peacocks, lyre-birds, and birds of paradise will court females by showing off their elaborate tail feathers in bright gaudy displays. In other species, the females perform the display and the males do the selecting.

Sometimes, the display will include an object such as a nest constructed by one partner as an attraction to its mate. Bowerbirds, for example, construct elaborate nuptial bowers as a means of attracting their mates. These bowers, which contain a nest in the center, are sometimes adorned with attractive stones, flowers, and other brightly colored objects. In some species of animals, males and females will respond to one another by performing alternating steps; in this manner, each sex selects members of the other.

Many sexually reproducing organisms have both male organs which produce sperm and female organs which produce eggs, a condition known as hermaphroditism. Earthworms and many snails are simultaneous hermaphrodites, meaning that both male and female organs are present at the same time. Hermaphrodites often have their parts so arranged that self-fertilization is difficult or impossible. One system that guarantees cross-fertilization is serial hermaphroditism. In this system, each individual develops the organs of one sex first, then changes into the opposite sex as it matures further.

Some sexually reproducing organisms have become secondarily asexual through a process called parthenogenesis, in which gametes (eggs) develop into new individuals without fertilization. In bees and wasps, males develop parthenogenetically from unfertilized eggs, while females (with twice the chromosome number) develop from fertilized eggs.

The Cost of Sexual Reproduction

Sexually reproducing organisms experience a cost associated with the energy devoted to courtship behavior and to the growing of sexual parts. In addition, the act of courtship usually exposes an individual to a greater risk of predation, and the distractions of mating further increase this risk. In view of these costs, many evolutionists have wondered how sex ever evolved in the first place, or why it is so widespread. Any adaptation so complex and so costly would long ago have disappeared if the organisms possessing it were at a selective disadvantage. The widespread occurrence of sex, and of numerous sexual systems, shows that there must be some advantage to all the various forms of sexual reproduction, and that this advantage is sufficient to overcome the recognized advantages of asexual reproduction in terms of rapid proliferation with relatively low investment of energy.

The answer to this puzzle is based on the fact that asexually produced offspring are all genetically similar to the parent, while sexually produced offspring differ considerably from one another. Organisms exploiting a dependable habitat or food supply often leave more offspring if they produce numerous genetically similar offspring rapidly and asexually. On the other hand, organisms facing uncertain future conditions have a better chance of leaving more offspring if they reproduce sexually and therefore produce a more varied assortment of offspring, at least some of which might have the adaptations needed to survive in the uncertain future. Examination of those species that are capable of reproducing either way confirms this hypothesis: Whenever favorable conditions are likely to persist, they reproduce rapidly and asexually. Faced with conditions of adversity or future uncertainty, however, these same species reproduce sexually. In species that alternate between sexually produced and asexually produced generations, the asexual phases typically occur during the seasons of assured abundance, while the sexual phases are more likely to occur at the onset of harsh or uncertain conditions. Sex, in other words, is a hedge against adversity and against an uncertain future.

Studying Sexual Reproduction

Most biologists who study reproduction are either ecologists, ethologists, or geneticists. Their methods include counting various kinds of offspring and measuring their genetic variability. Reproductive ecologists and ethologists also measure parental investment, or the amount of energy used by individuals of each type (and each sex) in the courting of their mates, in the production of gametes, and in caring for their young. Energy costs of this kind are generally measured by comparing the food consumption of individuals engaged in various types of activity using statistical methods of comparison among large numbers of observations.

The morphology of sex organs in various species is also studied by comparative anatomists and by specialists on particular taxonomic groups such as entomologists (who study insects), helminthologists (who study worms), malacologists (who study snails and other mollusks), and ich- thyologists (who study fishes). In most hermaphroditic species, for example, the organs are so arranged as to make cross-fertilization easier and self-fertilization more difficult.

The above explanation of sexual reproduction as resulting from the greater variability among offspring facing an uncertain future is partially confirmed by studying species that can reproduce either sexually or asexually. Among these species, asexual reproduction is always favored in situations in which an individual discovers a resource (such as a habitat or a food source) too large to exploit by itself. These conditions favor individuals that can reproduce rapidly and asexually produce numerous individuals genetically similar to themselves, who then proceed to exploit the resource. Aphids, for example, produce one or several asexual generations during the spring and early summer, when plant food is abundant. In seasons or situations of great risk or uncertainty, however, the same species often reproduce sexually at somewhat greater energetic cost, leaving a wider variety of offspring but a smaller total number. Under unpredictable conditions (such as those associated with wintering in a cold, temperate climate), the greater energetic costs of reproducing sexually are more than made up by the greater genetic and ecological variability among the offspring. Sexually reproducing individuals leave more offspring (on the average) than asexual individuals under these conditions. Similarly, among hermaphroditic species, cross-fertilization results in more varied offspring than self-fertilization, and is therefore favored under such conditions.

Testing Theories

The several reproductive methods studied by biologists provide a natural laboratory for the testing of several theories. Among these are theories concerned with genetic variability, natural selection, the evolution of sex, and the allocation of re- sources, including the theory of parental investment in the care of their offspring.

In terms of the two most general types of reproductive strategies, those species using a system called the r strategy (reproducing prolifically at small body size) may be either sexual or asexual, or may alternate between these two methods of reproduction. On the other hand, species following the K strategy (reproducing in smaller numbers at larger body size and investing time and energy in parental care) are invariably sexually reproducing and most often gonochoristic as well.

In addition to the theoretical considerations mentioned above, the study of alternative methods of reproduction gives us important insights into the reasons that our species, like other K strategists, is sexually reproducing and gonochoristic. In most species, sexual behavior is largely controlled by instincts, but learned behavior plays a major role among higher primates. Beyond what is necessary in copulation and childbirth, much of sex-specific behavior in humans is culturally defined and may differ from one society to another. This includes the norms of what behavior is appropriate (or inappropriate) for each sex and what personal qualities are considered masculine or feminine. All attempts to redefine sex roles will lead nowhere, unless one is aware of both the biological and the social underpinnings of these roles.

—*Eli C. Minkoff*

See also: Asexual reproduction; Copulation; Gene flow; Genetics; Natural selection; Reproductive systems of female mammals; Reproductive systems of male mammals; Sexual development.

Further Reading

Alcock, John. *Animal Behavior: An Evolutionary Approach.* 7th ed. Sunderland, Mass.: Sinauer Associates, 2001. Perhaps the best overall textbook on the subject of animal behavior, this book takes an evolutionary approach in that it attempts to examine the adaptive reasons behind each behavior pattern. The book is also good in its coverage of a wide variety of organisms, including insects, aquatic invertebrates, fishes, amphibians, reptiles, birds, and mammals. The book has many good black-and-white illustrations and a good, lengthy Bibliography.

Brown, J. L. *The Evolution of Behavior.* New York: W.W. Norton, 1975. Another good review of animal behavior, including an entire chapter on mating systems and sexual selection. A lengthy Bibliography is included.

Campbell, Bernard, ed. *Sexual Selection and the Descent of Man, 1871-1971.* Chicago: Aldine, 1972. A series of eleven articles, each written by a different contributor, outlining the theory of sexual selection as applied to a variety of species.

Campbell, Neil A. *Biology: Concepts and Connections.* 3d ed. Menlo Park, Calif: Benjamin/ Cummings, 2000. An innovative college textbook for students with some biology background.

Clutton-Brock T. H., ed. *Reproductive Success: Studies of Individual Variation in Contrasting Breeding Systems.* Chicago: University of Chicago Press, 1988. A series of twenty-nine individual studies by fifty-three contributors, this book holds a treasure of data on a variety of mating and reproductive systems in both insects and vertebrates (mostly birds and mammals). Most of the articles deal with lifetime measures of reproductive success. The book contains many tables of data, a moderate number of illustrations, and an extensive Bibliography.

Daly, Martin, and Margo Wilson. *Sex, Evolution, and Behavior.* 2d ed. Belmont, Calif.: Wadsworth, 1983. Perhaps the best Summary of the issues related to the advantages and costs of various reproductive strategies, including those related to sex.

McGill, T. E., D. A. Dewsbury, and B. D. Sachs. *Sex and Behavior: Status and Prospectus.* New York: Plenum, 1978. A series of sixteen chapters, each by a different author or authors, on various topics related to sex-related behavior. A large portion of the book is devoted to studies of sex differences in humans. There is a Bibliography at the end of each chapter. There are very few illustrations, mostly in the form of graphs.

Maynard-Smith, John. *Evolution and the Theory of Games.* New York: Cambridge University Press, 1982. This short book presents a more theoretical approach to the problem of evolutionary strategies in general. Sexual reproduction is discussed as a reproductive strategy as is parthenogenesis. There are also discussions of sex ratios and parental care. The book contains a few graphs and many equations. There is a Bibliography, but most of the works listed are articles in technical journals.

Rosenblatt, J. S., and B. R. Komisaruk, eds. *Reproductive Behavior and Evolution.* New York: Plenum, 1977. This book has seven chapters, each by a different author. Included are good chapters on the genetic control of reproductive behavior and reproductive isolation, on mating and child-rearing systems, and on parental care by both mothers and fathers. Most of the discussion centers on mammals. There are black-and-white illustrations of several types, and a Bibliography is included at the end of each chapter.

Sexual development

FIELDS OF STUDY

Anatomy, embryology, genetics, physiology, reproduction science

SUMMARY

The development of an organism into a male or female involves a complex series of interactions, including differential growth, influence of the external or internal environment, and genetic factors.

PRINCIPAL TERMS

- **androgens:** the general term for a variety of male sex hormones, such as testosterone and dihydrotestosterone
- **genital tubercle:** a small swelling or protuberance toward the front of an embryo's
- **genital area:** it is destined to become the penis tip or clitoris
- **genitalia:** the external sex structures
- **gonad:** the structure that produces eggs or sperm cells and sex hormones; the ovary or the testis
- **hermaphrodite:** a single organism that produces both eggs and sperm
- **labial folds:** the paired ridges of tissue on either side of the embryo's genital area, which become penis and scrotum in males and labia in females
- **Müllerian ducts:** the embryonic ducts that will become the female oviducts or Fallopian tubes, uterus, and vagina parthenogenesis: the development of an unfertilized egg
- **urogenital groove:** a slitlike opening behind the genital tubercle that will become enclosed in the penis but remain open in females
- **Wolffian ducts:** an embryonic duct system that becomes the internal accessory male structures that carry the sperm

BASIC PRINCIPLES

While some lower forms of life with no recognizably different sexes exchange genetic material in a form of sexual reproduction, sexual reproduction in most organisms involves individuals with some obviously different physical and behavioral features. Biologically, the real difference between males and females is the type of sex cells they produce—whether large eggs specialized to support embryonic development or tiny sperm specialized for moving to the egg. Eggs and sperm are produced in gonads—the ovaries of females and the testes of males. The gonads of higher animals also produce sex hormones, chemical messengers that affect both embryonic and adult sexual development.

Even these basic sex distinctions are rather flexible in some organisms. Sometimes, sex is determined entirely by the environment. One kind of marine worm becomes a female unless it attaches as a larva to an adult female, whereupon it becomes a male—probably because of hormones secreted by the female. Temperature can control sex development in some animals, such as mosquitoes and amphibians. Sex may also be determined by size. Since it takes more energy to produce eggs than sperm, when food is scarce it may be more adaptive to be male. The European oyster begins adult life as a male, changes to a female as it grows larger, and reverts to being a male after shedding eggs.

In territorial animals, being a large male may be an advantage. A tropical wrasse, or "cleaner fish," travels with a harem of smaller females. If he is removed, the largest female becomes a male within a few days. Many organisms, including earthworms, snails, and some fish, are hermaphrodites— functional males/females that can fertilize themselves or exchange sperm with others. Some insects, worms, crustaceans, goldfish, whip-tailed lizards, and even turkeys lay eggs that can develop without fertilization, a process called parthenogenesis. This strategy is not sound in the long run, since it does not promote genetic diversity. It is an advantage for an organism living under good conditions, however, where an all-female population can exploit the ideal environment most efficiently.

The Origin of Sex Differentiation

Sex differentiation probably originated as differential growth of either the ovary or the testis, mediated in various ways by hormones or other environmental factors. Later in evolution it came under genetic control, which made the process more independent of environment and made possible the development of more complex reproductive structures and behavior.

The genetic sex of an animal is determined by the father at fertilization. In most species, females have two matching X chromosomes, males have an unmatched X and a smaller Y chromosome. If a normal egg with one X chromosome is fertilized by an X-bearing sperm, the XX embryo is genetically female. A Y-bearing sperm will produce a genetic male, XY. In butterflies, fishes, and birds, however, females have XY chromosomes and males have XX. Initially, XX and XY embryos look identical and in a sense are still sexually bipotential. Their gonads are "indifferent," that is, able to form either an ovary or a testis. Each has two sets of undifferentiated sex ducts. One set, the Wolffian ducts, will become the sperm ducts and other male structures. The other set, the Müllerian ducts, form the female oviducts, uterus, and vagina.

Soon, however, genes on the Y chromosome direct the inner part of the indifferent gonad to become a testis, which then produces the male sex hormones (androgens) and Müllerian-inhibiting substance (MIS), which control further events in male development. An androgen called testosterone causes the male duct system to persist and develop, and MIS makes the female duct system degenerate. Testosterone has other developmental effects, as indicated by the fact that in monkeys, male behavior is linked to the length of embryonic exposure to testosterone.

Without the influence of the Y chromosome an XX gonad begins to develop into an ovary. The role of female sex hormones in development is unclear, since in mammals female embryonic development can occur in the absence of female hormones. The mammalian embryo has a tendency to develop in the female direction unless specific influences prevent it. The Wolffian ducts are actually remnants of a drainage system from a temporary embryonic kidney that disappears before birth. Only the presence of male sex hormones will keep these tubes from disintegrating. The Müllerian ducts, on the other hand, tend to persist unless acted upon by the anti-Müllerian substance. In birds, the embryonic ovary is the dominant gonad, and it actively feminizes the reproductive tract. It has been suggested that the early male development in mammals is necessary to allow male differentiation in the female-hormone-rich uterine environment.

Until differentiation begins, both sexes also have the same vaguely female-looking external sex structures or genitalia. In both sexes, a small protuberance called the genital tubercle is found toward the front or belly side of the embryo. Behind the genital tubercle is a slitlike opening, the urogenital groove; it is flanked by two sets of paired folds or swellings, like a river valley paralleled by two sets of ridges on either side. In the female, the genital tubercle will form a small structure called the clitoris. The urogenital groove will remain open, forming a vestibule into which the vagina and the urethra open, which empties the urinary bladder. The folds on either side of the groove will remain relatively unchanged to form labial folds.

In the male, the genital tubercle will become the tip of the penis, and the innermost urogenital folds will fuse together to form the body of the penis; the "scar" of this joining may be seen on the underside of the penis. This fusion closes off the urogenital groove and encloses the male's urethra within the tubelike penis. The outer pair of ridges will fuse to form the scrotum, the sac that encloses the two testes, which descend into the scrotal sac before birth. Another androgen, dihydrotestosterone, may be responsible for the development of these external male structures.

Studying Sexual Development

Many sex-determining mechanisms can be studied by simple modification of the environment. For example, by varying temperature, hormone level, social-group composition, or other environmental factors it is actually possible to reverse the sexes of some invertebrates, fish, and amphibians. Castration experiments are commonly used to study the effects of hormones on the sexual development of birds and mammals. For example, castrated mammals of either sex develop in the female pattern. Since in birds only the left ovary develops, castration of hens may result in the transformation of the right gonad into a testis, with complete functional sex reversal.

Sex development can be studied with naturally occurring hormone imbalances, as in freemartin calves—sterile, masculinized females whose male twin exposed them before birth to male sex hormone. The same masculinizing effect on female fetuses can be achieved by injecting a pregnant mammal with androgens or even by growing an embryonic ovary and testis together in an organ culture outside the body. In each case, the female structures are masculinized by the male hormones.

Sex chromosome mutants in animals as diverse as fruit flies, mice, and birds can be used to study chromosomal influences in sex determination. To use a human example, there are sterile XX men with a tiny piece of the short arm of the Y chromosome attached to one X. There are also XY females who show a deletion of the same short arm of the Y. These observations have led geneticists to think that the testis-determining genes are on the short arm of the Y, since without it an individual—even one with a Y—is female.

The Evolutionary Advantage of Sexual Reproduction

In spite of its great biological costs, sexual reproduction is practiced by almost every kind of living thing. It confers tremendous evolutionary advantage on a species by producing a new individual with the genetic characteristics of two parents but with unique combinations of features that may make the offspring more successful than either parent. The advantage of having separate sexes for reproduction is that it permits the development of extremely specialized reproductive organs for the very different requirements of sperm or egg production, and, when needed, intrauterine support for the embryo. Though some organisms show great sexual flexibility, higher animals and plants have tended toward sexual stability, probably because of the high cost of sex reversal for organisms with highly specialized sexual structures.

The study of sex differentiation helps advance scientific knowledge in many areas. Modes of sex determination often provide clues to evolutionary relationships among groups of organisms. In addition, the development of sex differences makes a good model system for the study of more general questions. For example, one might use the control of sexual size differences to attack the broader question of what makes mice smaller than elephants. The control of sex differentiation by environmental factors, to use another example, might provide geneticists with a way to study how genes are turned on by hormones, temperature, or other external influences.

For embryologists, the stepwise determination and differentiation of the mammalian reproductive system is an excellent general development model; it involves a genetically controlled sequence of events that includes both the preservation of one embryonic structure (the Wolffian duct) and the removal of another (the Müllerian duct). Hormonally controlled events include a wide variety of developmental sequences, from the externally visible large-scale changes involved in the shaping of the external genitalia to the biochemical differentiation that programs the brain hypothalamus for its complex control of the menstrual cycle.

—*Michele Morek*

See also: Embryology; Endocrine systems in vertebrates; Gametogenesis; Reproduction; Reproductive systems of female mammals; Reproductive systems of male mammals.

Further Reading

Carlson, Bruce M. *Patten's Foundations of Embryology.* 6th ed. New York: McGraw-Hill, 1996. Detailed treatment of general anatomy of the reproductive organs and their embryonic development. Though this is an upper-level college embryology textbook and too technical for many general readers, it has unsurpassed illustrations and an extensive Bibliography.

Hickman, Cleveland, et al. *Biology of Animals.* 7th ed. St. Louis: Times Mirror/Mosby, 1998. This freshman college zoology text has a special chapter on reproduction and others on major animal groups. It is beautifully illustrated and has a good glossary, index, and bibliographies.

Hopf, Alice. *Strange Sex Lives in the Animal Kingdom.* New York: McGraw-Hill, 1981. This short book, intended for the general reader, gives a very readable account of the importance of sexual reproduction, unusual types of sex determination, sex switching, hermaphroditism, and parthenogenesis. The Bibliography and index are brief.

Naftolin, Federick, et al. *Science* 211 (March 20, 1981). The entire issue of this magazine is devoted to sexual differences and how they develop. Though difficult for the casual reader, most of the articles were written for a general scientific audience and can be understood by someone with a strong interest in the subject. Each article has a good Bibliography and deals with the genetic or hormonal mechanisms that govern sex organ development.

Rothwell, Norman V. *Understanding Genetics.* 4th ed. New York: Oxford University Press, 1988. This college-level genetics textbook requires little more than a basic familiarity with genetic terminology. Contains a very thorough treatment of sex and inheritance, including information on sex chromosomes and their role in sexual differentiation, and related topics such as sex-linked, sex-influenced, and sex-limited genetic traits in various animals. Good Bibliography and glossary.

Wrangham, Richard W., W. C. McGrew, Frans B. M. DeWaal, and Paul G. Heltne, eds. *Chimpanzee Cultures.* Cambridge, Mass.: Harvard University Press, 1996. Essays by authorities on chimpanzee biology and behavior. Discussions of social relations include sexual development and behavior.

Skin

FIELDS OF STUDY

Anatomy, cell biology, histology, invertebrate biology, physiology

SUMMARY

Skin is the organ that covers the body surface of an animal, and it is composed of cells. The specialized structures and cells associated with the skin are involved in a variety of physiological functions, including protection, communication, regulation of body heat, and respiration.

PRINCIPAL TERMS

- **chromatophores:** pigment-producing cells
- **dermis:** layer beneath the epidermis, primarily connective tissue but also containing nerves and blood vessels
- **epidermis:** surface layer of epithelial cells invertebrate: animal without a backbone
- **mitotic cells:** cells capable of dividing and forming new cells
- **vertebrate:** animal with a backbone made up of individual bones called vertebrae

BASIC PRINCIPLES

Survival in animals requires that the internal body components be separated and protected from the external environment. Most single-cell organisms are separated from the environment only by the plasma membrane (cellular membrane). In multicellular organisms the body surface is covered by a tissue consisting of epithelial cells and connective tissue. The covering is commonly referred to as skin, but the skins of invertebrates and vertebrates have distinct differences. Invertebrates often have a single layer of surface epithelial cells, which is generally referred to as an integument. However some invertebrates, specifically flukes and tapeworms, have a unique, living surface covering called a tegument. In this situation, the epithelial cells have fused and formed a single bag of cellular components called a syncytial epidermis. Thus, the word "skin" is often reserved specifically to describe the surface covering in vertebrates, but the word "integument" is also used.

THE STRUCTURE AND PHYSICAL PROPERTIES OF SKIN

The general structure of the skin is similar in all vertebrates. There are two primary regions, an epidermis and a dermis. The upper region, the epidermis, is made up of multiple layers of epithelial cells. All vertebrate skin has a basal layer (the stratum germinativum) consisting of mitotic cells. These mitotic cells divide to replace the cells closer to the surface as they are worn away, and to heal skin wounds. In most vertebrate species, the outermost layer of cells (the stratum corneum) is dead. The dead cells are filled with a waterproofing protein called keratin, which is produced by keratinocytes, the major type of cells forming the epithelium. An exception can be noted in many species of fish, where the epidermis is composed entirely of living cells and the stratum corneum is absent. The epidermis will differ the most between aquatic (water dwelling) and terrestrial (land dwelling) organisms. When compared to mammals and reptiles, the epidermis of amphibians, birds, and fish is thinner and the stratum corneum may only be one or two cells thick. The epidermis lacks its own blood supply (is avascular). Nourishment reaches the living cells by diffusion from the underlying dermal blood supply. No nerves are present in the epidermis.

The underlying region, the dermis, is primarily composed of connective tissue. Although epidermal and dermal thicknesses vary between groups of animals and thickness may vary along an individual's body surface, the dermal layer is always thicker than the epidermis. The hypodermis, a layer of subcutaneous tissue immediately beneath the dermis, connects the skin with underlying tissues such as muscles and bone. In birds and mammals in particular, this layer often contains a significant amount of fat, which provides insulation and a reserve source of energy.

The protective aspect of skin does not mean that it is entirely impenetrable or that the body is completely isolated from the environment. Materials that are fat-soluble or that disrupt cellular membranes can be absorbed across the skin surface. In some cases, beneficial chemicals cross skin. For example, frogs, which are amphibians, actively take up oxygen and expel carbon dioxide across the skin as well as the surface of the lungs. Amphibian skin is also permeable to water and, in fact, some species absorb amounts comparable to that obtained by drinking in other organisms. In other cases, detrimental materials such as solvents and potential environmental pollutants cross the skin. Acetone present in nail polish remover, methanol which is sometimes used to remove old finish on furniture, and salts of heavy metals such as mercury or lead are some examples.

Skin, like muscle, has the properties of extensibility (stretch) and elasticity (ability to return to the original shape after being stretched). These properties are made possible by the presence of collagen and elastic fibers as major components of the tissue comprising the dermis. When the elastic properties of the skin have been exceeded, white lines known as stretch marks appear.

SPECIALIZED SECRETIONS AND STRUCTURES ASSOCIATED WITH SKIN

Some epithelial cells produce and release protective secretions onto the external surface. Vertebrates and invertebrates both have mucoussecreting cells. On internal surfaces, such as the digestive tract, mucus protects cells from being broken down along with the food. Onthe external surface, mucus may trap bacteria or, as in earthworms, prevent death from desiccation (drying out). Another example of an invertebrate secretion is a covering called a cuticle. In insects this cuticle includes a mixture of proteins that eventually harden and form the exoskeleton.

Vertebrates, including fish, birds, and humans, as well as invertebrates, insects, secrete a group of antimicrobial (bacteria-killing) proteins called defensins (originally called magainins). Species of poison dart frogs have another type of protective secretion which is toxic to potential predators. Some of these secretions are used on poison dart arrows.

Structures with quite varied functions are derived from skin cells, particularly the epithelial cells. The feathers of birds function in flight but they also provide insulation. A bird's beaks and claws provide a method of defense and a way to secure food. Mammalian hair is an epithelial derivative. Body hair provides protection from abrasion and sunlight and has some insulation value. In animals that have them, sweat, oil, and mammary glands are groups of specialized epithelial cells. Reptiles, for example, lack sweat glands. Light organs of deep-water fish are modified epithelial glands. The scales of reptiles, the rattles of snakes, and the claws of turtles are other examples of epithelial derivatives. Geckos are able to walk up walls because they have modified epidermal scales on the tips of digits which serve as suction cups.

Cells of the dermis also are the origin for specialized structures in some organisms. Although there are fewer examples, dermal derivatives include shark teeth, fish scales, and the protective armor plates of an armadillo.

SKIN AND TEMPERATURE REGULATION

The skin is a major organ in controlling body temperature. Mammals and birds are animals that generate internal body heat (warm-blooded or endothermic). Species of reptiles, fish, and amphibians, which are often called cold-blooded (ectothermic), are unable to control their body temperature through internal regulators in the same way that warm-blooded animals can. Both groups of animals depend upon the rich supply of blood vessels in the dermis as one mechanism for maintaining a safe body temperature. When internal body temperatures rise in endotherms, an increase in blood flow carries internal body heat to the surface, where it is lost to the environment. A similar increase in blood flow in ectotherms carries heat from the environment into the body and helps to warm internal organs and tissues. A decrease in blood flow will work in the opposite direction in both groups. In animals with sweat glands, the evaporation of sweat secreted onto the body surface also helps to lower body temperature. One unique feature of birds is a specialized region of skin, the brood patch, located on the ventral (stomach) surface. This area is rich in blood vessels (is highly vascularized) and is used to transmit heat from the female to the eggs or the hatchlings.

Skin Coloration

Vertebrate skin contains pigment-producing cells called chromatophores. Pigment production provides many benefits to animals. Skin pigments help to limit the amount of damaging ultraviolet light or irradiation to which the deoxyribonucleic acid (DNA) in the mitotic cells and the underlying tissues are exposed. Melanocytes located in the epithelium of mammals produce a brown-black pigment called melanin. They are the only pigment- producing cells in most mammals. In addition to epithelial melanocytes, amphibians, fish, reptiles, and birds have other types of pigmentproducing cells which are located in the dermis. Examples are lipophores, which use carotene, a naturally occurring pigment in food, to synthesize yellow, orange, and redpigments, and iridophores, which use molecules called purines to synthesize pigments that are iridescent. The amount of pigment produced, the final location of the pigment in the cells, and the combination of cells producing it result in a range of body and feather coloration. Chromatophores account for the changes in body color that allow chameleons, flounders, and octopuses to easily blend in with different surroundings. Changes in body color also provide a means of communication between individuals.

Skin color, particularly in organisms with a thin, fair surface covering such as humans, is also influenced by the amount of oxygen bound to hemoglobin in the blood. Fully oxygenated hemoglobin is red and it gives a pink coloration to the skin. Hemoglobin that is not fully oxygenated can cause the skin to appear blue or take on a purplish hue, a condition called cyanosis.

—*Robert W. Yost*

See also: Anatomy; Bone and cartilage; Brain; Circulatory systems of vertebrates; Digestive tract; Eyes; Immune system; Noses; Reproductive system of female mammals; Reproductive system of male mammals; Respiratory system.

Further Reading

Hickman, C. P., Jr., L. S. Larsen, and A. Larsen. *Biology of Animals.* 7th ed. Boston: WCB/ McGraw-Hill, 1998. General biology text with sections on invertebrate and vertebrate systems.

Linzey, Donald. *Vertebrate Biology.* Boston: McGraw-Hill, 2001. A good midlevel text outlining structure and function in vertebrates. Excellent use of descriptive features and terminology as part of the text. Good list of related texts in preface.

Miller, S. A., and J. P. Harley. *Zoology.* 4th ed. Boston: McGraw-Hill, 1999. A lower-level text with introductory sections on animal systems.

Solomon, E. P., L. R. Berg, and D.W. Martin. *Biology.* 5th ed. Philadelphia: Saunders College Publishing, 1998. Fairly comprehensive general biology text with sections on invertebrate and vertebrate systems.

Walker, F.W., Jr., and K. F. Liem. "The Integument." In *Functional Anatomy of the Vertebrates: An Evolutionary Perspective.* 4th ed. Philadelphia: Saunders College Publishing, 2001. An upper-level comprehensive text on vertebrate systems. Excellent chapter on comparative anatomy and physiology of vertebrate skin.

> **CHANGING PROTECTIVE LAYERS**
>
> Snakes and other reptiles shed their skin, arthropods shed their exoskeleton, and some worms shed their cuticle in a process called ecdysis. In snakes, the mitotic cells of the basal layer replicate to form a new layer of epidermal cells that will be beneath the cells that are breaking down. There is a layer of cells in a region between the mitotic cells and the outer epidermal layers that is not keratinized and begins to break down. A separation or fission zone occurs at the level where the cells are breaking down. This results in the old epidermal layers breaking away above the newly formed layers. The old skin can now be shed and the new skin will be exposed. The surface of the new skin will be keratinized and scaly just like the old. Most snakes shed their skin in one piece.
>
> Arthropods, which includes the insects, shed their exoskeleton many times during the growth stage. The epidermis detaches from the old exoskeleton and secretes a new epicuticle that will be the new exoskeleton. Molting fluid, which contains enzymes, is secreted into the region between the old and what will be the new exoskeleton. The enzymes degrade the inner region of the old exoskeleton. The old skeleton now splits and the animal emerges with a new exoskeleton. However, this skeleton will remain soft for some time and during that time it is able to stretch to accommodate the larger body size of the arthropod.

> **HAIR-RAISING RESPONSE**
>
> Originating in the superficial dermis and attaching to the hair shaft (long part of the hair) near its base is a small muscle, the arrector pili muscle. This is a smooth muscle, which means that its contraction and relaxation are involuntary responses. The muscle is innervated by a part of the nervous system (autonomic nervous system) that is under subconscious control. One part of the autonomic nervous system regulates specific responses during periods of stress. These responses are generally referred to as "fight-or-flight" responses.
>
> When the muscle attached to a hair is stimulated to contract by the nervous system, the hair will stand up. In many mammals, dogs for example, this phenomenon occurs when a nonfriendly situation is detected. The raised hairs indicate alarm. In other cases, erected body hair serves as a protective method of defense since it will make animals appear larger than normal. In many mammals, elevated body hair traps air and assists in maintaining body warmth when exposed to cold temperatures, or helps in cooling when the core body temperature gets too warm. Hair normally emerges from the body surface at a slight angle. "Goose bumps," noticeable in humans who are frightened or experiencing some type of stress, are the result of a slight elevation of the skin when the hair shaft moves into a vertical position.

SMELL

FIELDS OF STUDY

Anatomy, biochemistry, biophysics, cell biology, developmental biology, genetics, herpetology, histology, human origins, immunology, marine biology, neurobiology, ornithology, pathology, physiology, reproduction science, zoology

SUMMARY

For most animal species, smell is the main sense that is used in locating food and in detecting harmful agents and predators. Smell is used to recognize or exclude members within a social group, such as a pack, herd, or flock, and to find appropriate home sites. In some species, smell is also used to attract and identify mates.

PRINCIPAL TERMS

- **anosmia:** the clinical term for the inability to detect odors
- **chemotaxis:** an oriented response toward or away from chemicals
- **olfaction:** the sense of smell
- **olfactory receptors:** receptor organs which have very high sensitivity and specificity and which are "distance" chemical receptors
- **pheromones:** species-specific compounds (odors) which, acting as chemical stimuli at a distance, have a profound effect on an animal's behavior

BASIC PRINCIPLES

Responses to chemicals are fundamental at all stages in biological organization. Chemotaxis, an oriented response toward or away from chemicals, has been observed in species ranging from single-cell animals such as bacteria and protozoa to very complex multicellular animals including humans. An attraction to a chemical is referred to as positive chemotaxis whereas a rejection or repulsion is called negative chemotaxis. The development of sensitivity (both positive and negative) to particular chemicals is the dominant sense in most animals. In general, receptor organs which have very high sensitivity and specificity, and which are distance chemical receptors, are called olfactory; the receptors of moderate sensitivity, usually found in the mouth, which are associated with feeding and are stimulated by dilute solutions are called taste receptors. Smells can be delivered to the olfactory receptors through air, as is the case with terrestrial animals ranging from insects to humans. On the other hand, smells can be delivered to the olfactory receptors through water, as is the case with aquatic animals such as insects and fish.

SMELL IN INSECTS

In insects, the olfactory (smell) receptors are located on the antennae. Because of the superficial location of their receptors and, more especially, because of their suitability for electrophysiological studies, insects have contributed much basic information about the mechanisms of olfaction. The olfactory receptors of most insects are highly specialized and can detect very trace amounts of compounds that are biologically important to the animal.

Olfaction is an important sensory modality for insects, particularly in mating, egg laying, and food selection. Numerous male insects, such as moths and cockroaches, are attracted by speciesspecific compounds called pheromones that act as chemical stimuli at a distance. Pheromones can be thought of as a language based on the sense of smell. Pheromones are often divided into two categories. Releaser pheromones initiate specific patterns of behavior. For example, they serve as powerful sex attractants, identify territories or trails, signal danger, and bring about swarming or similar types of grouping behavior. Primer pheromones trigger physiological changes in metabolism related to sexual development, growth, or metamorphosis. These changes are usually mediated through the endocrine system.

Male silk moths and gypsy moths may be attracted froma distance of a mile or two by a releaser pheromone from the scent glands of the females. Males will attempt to mate with any object that has touched the female scent gland; however, males deprived of their antennae do not even orient toward the female. Synthetic releaser pheromones are now being used in traps to attract pest insects such as the gypsy moth and the Japanese beetle.

Chemical communication in social insects is used for alarm, attraction, recruitment, and recognition of nest mates and of castes. Ants give off alarm releaser pheromones from mandibular glands and so are able to warn other ants of impending danger. Armyants deposit releaser pheromones on trails to food sources or to nest sites. Primer pheromones-secreted by the queen bee cause the worker bees to cluster and swarm, and they suppress the rearing of other queens in the hive.

Mosquitoes are attracted chemically to warm-blooded animals and are sensitive to several chemicals. Carbon dioxide (the metabolic waste product excreted through the lung) attracts them and they are able to orient themselves and fly to the source of this compound. They also react positively to other mammalian body products. Most common insect repellants work by interfering with the olfactory ability of the mosquito, so that the insect can no longer follow an odor toward its source.

SMELL IN FISH

The olfactory receptors in most fish are located in olfactory sacs in a pit on the head. Chemicals are brought to the receptors while swimming or during respiratory movements. Odors and the olfactory sense play a major role in the life of many species of fish. For example, homing in salmon is controlled mostly by the "smell" of the water in which the fish was born. By following the smell trail composed of the minerals found in the water, salmon are able to return to breed in the same stream in which they were born. Fish can also become rapidly conditioned to odors. For example, once a pike has attacked a school of minnows, the odor of other pike in the water becomes associated with an alarm response in the minnows.

SMELL IN TERRESTRIAL VERTEBRATES

In vertebrates, the olfactory receptor cells are located in the nose along the respiratory airflow path. As a result, when air is brought into the nose either during breathing or sniffing, odorant molecules are delivered to the headspace above the mucus-coated olfactory receptors. The odor molecules then bind to hairlike cilia on the olfactory receptors, producing a signal that is transmitted to the central nervous system. Because they stimulate different receptors, different smells produce different patterns of electrical activity. These odorant-specific patterns are used by the brain in smell identification.

Olfactory receptor cells are primary receptors, with axons running directly to the brain. This makes olfactory receptor cells unique, since most other sensory cells send their signals through processing centers (called synapses) before the message is carried to the brain. In the case of the olfactory receptor cells, all the information recorded by the cell is transmitted to the central nervous system. Once in the brain, the output of the olfactory receptors is sent to the limbic system (a portion of the brain involved with memory), the endocrine system, and throughout the rest of the central nervous system. The connections to the limbic system result in the very strong association that odors have in memory recognition. In humans, smells can often trigger very vivid memories. The rest of the brain also sends messages back to the bulbs, amending the pleasure of a food aroma when the stomach is full. Unlike other neurons, olfactory receptor cells constantly replicate. As a result, after a life span of about thirty days, olfactory receptor cells are replaced.

Odors help bond mothers to their newborn babies. A mother cuddling her infant will invariably brush her nose in the baby's hair to inhale its sweet aroma. She can identify her baby by its smell as much as by its cry. Additionally, one-day-old infants of many species have been shown to be able to recognize the smell of their mothers. A mother rat licks her nipples so that her blind pups can follow the scent of her saliva to the milk. Likewise, a mother kangaroo produces a saliva trail so the newly born and blind babies can follow the trail from the uterus to the mother's pouch. Wash the nipples and eliminate the saliva trail, and the pups are lost.

Female rodents who periodically smell male urine will move more quickly into puberty than females that do not. If a pregnant female mouse smells the urine of a male of another colony, she will immediately terminate her pregnancy. Also, if the olfactory nerves of a newborn rat pup are cut, the rats will never develop sexually.

A diminished sense of smell is termed hyposmia. Hyposmia can occur following a cold or after head trauma, and humans experience some reduction in the sense of smell with age. Also, most conditions that reduce the flow of air through the nose will reduce olfactory acuity. For example, a stuffy nose as a result of an allergy, a cold, or a nasal polyp often creates hyposmia. Anosmia is the complete loss of the ability to detect airborne odorants. Head trauma and severe nasal obstructions can produce anosmia. If the cause of hyposmia or anosmia is related to a blocking in the nasal airflow passageways, then treatment with steroids and/or surgery often can restore the olfactory loss.

Human experience seems to draw a sharp contrast between taste and smell. Taste is the chemical sense related to sampling compounds that come in directed contact with the inside of the mouth whereas smell is the ability of the nose to monitor airborne chemicals, often from distant sources. However, the sensations of taste and smell are not completely independent, since smell can influence taste and vice versa. For example, a lemon smell in the nose can make distilled water appear to "taste" bitter, and a sugar solution in the mouth can affect the perception of a fruit smell such as cherry. Much of what is usually perceived of as being a taste is really a smell. For example, with the nose blocked, it is difficult to tell coffee from bitter water or an onion from a potato. As humans chew, volatile compounds in the food are released into the air in the back of the throat. These compounds then make their way up the back of the nasal cavity, where they stimulate the olfactory receptors, producing a smell sensation that dramatically enriches the perception of the taste. This combination of smell and taste is referred to as flavor. What is often thought of as "taste" is actually a combination of smells and tastes, with additional contributions to the flavor coming from temperature and pain receptors in the nose and mouth.

—*David E. Hornung*

See also: Brain; Communication; Noses; Vision.

FURTHER READING

Association for Chemoreception Sciences. www.achems.org. This Web site contains a description of current work in the field and well as a discussion of smell disorders.

Getchell, T. V., R. L. Doty, L.M. Bartoshuck, and J. B. Snow, eds. *Smell and Taste in Health and Disease*. New York: Raven Press, 1991. Discussion of the clinical aspects of smell problems.

Gibbons, Byron. "The Intimate Sense of Smell." *National Geographic* 170, no. 3 (1986): 321-361. An excellent overview of the anatomy, physiology and psychology of the sense of smell. The making of perfumes, use of dogs for tracking, and the history of smell are all well covered.

Vroon, Piet. *Smell: The Secret Seducer*. Translated by Paul Vincent. New York: Farrar, Straus and Giroux, 1997. A cultural history and compendium of odd facts and a tribute to the sense of smell.

STEM CELL RESEARCH AND TECHNOLOGY

FIELDS OF STUDY

Biotechnology; cell and molecular biology; transplantation; genetic engineering; cell and tissue transplantation; cloning; cell-based therapies; regenerative medicine.

SUMMARY

Stem cell research is the field of science that examines specific cells that have the ability to divide indefinitely in culture and that give rise to specialized cells in order to provide therapy for diseases. There are two main types of stem cells: embryonic and somatic stem cells. Embryonic stem cells are formed in the early stages of embryonic development, and somatic stem cells are adult stem cells found in various tissues in the body. Stem cells have the potential to be used as therapy to replace or repair a person's cells or tissues that are damaged or dysfunctional in the treatment or cure of diseases.

PRINCIPAL TERMS

- **blastocyst:** embryo in a very early stage of development, produced by cell division of a zygote (fertilized egg); consists of about 150 cells in a spherical cell mass of two regions, the inner cell mass and the trophoblast.
- **cloning:** process of producing one or more genetically identical copies of a cell, tissue, or organism, by either natural means, such as cell division (mitosis), or artificial means, such as through an in vitro laboratory setting.
- **differentiation:** process of development in which cells change their complexity to acquire a specialized function.
- **embryonic stem cell:** cell derived from the inner cell mass at the blastocyst stage in the development of an embryo.
- **induced pluripotent stem cell:** cell that is genetically reprogrammed to be induced to express genes and factors to maintain cells in a stem cell line state.
- **multipotent stem cell:** stem cell that has the potential to become many different cell types in an organism's body.
- **pluripotent cell:** stem cell that has the potential to become any type of cell in an organism's body.
- **somatic cell nuclear transfer:** technique that removes the nucleus of a somatic cell, which is then injected or transferred into an egg (that has had its nucleus removed), which will be implanted into the womb of another individual to be born as a clone.
- **somatic stem cell:** stem cell found in the nongermline (egg and sperm) tissues of an individual that remains undifferentiated and can give rise to specialized cell types of the tissue from which it is derived; also known as adult stem cells.

BASIC PRINCIPLES

Stem cells have the basic properties of being undifferentiated cells that can divide indefinitely and have the potential to develop into many different types of cells in a body during early embryogenesis and during growth of an individual. Stem cells are different from other cells in the body in that they can renew themselves through cell division, allowing them to act as a repair mechanism and to replenish cells that are damaged or that die. When each stem

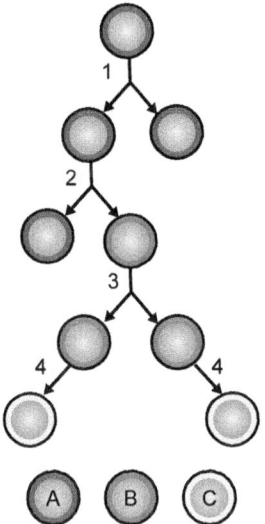

cell divides, it has the potential of either remaining a stem cell or becoming another cell type with a more specialized function.

There are two main types of stem cells: embryonic and somatic (also called adult stem cells). Embryonic stem cells are pluripotent, in that they have the capability of becoming any type of cell in the body. This is because these cells arise from the blastocyst, early in embryogenesis, making up the inner mass of cells. The inner cell mass gives rise to the entire body of an organism, including all the specialized cell types and organs, such as the heart, muscle, brain, skin, and other tissues. Somatic stem cells are considered to be multipotent and are found only in specialized tissues in the body, which are specific populations of cells that are used to generate replacements for cells that are damaged or die through the normal aging process of cells and because of injury or disease.

Stem cell therapy uses stem cells to replace or repair a patient's cells or tissues that are damaged or missing. Stem cell therapy is still experimental, in that it has not yet proven to be effective or safe, but stem cells have the potential to treat many diseases.

Background

Embryonic stem cells were first studied in the mouse in 1981, when scientists discovered ways to derive embryonic stem cells from mouse embryos. This led to the discovery, in 1998, of a method to derive stem cells from human embryos and grow them in the laboratory. However, the use of human embryonic stem cells—taken from embryos that were originally created for reproductive purposes—has been limited because a number of stem cell lines have been allowed to be grown in the laboratory for research purposes. This constraint led to further discoveries of how to derive stem cells from somatic tissues. In 2006, scientists discovered conditions that would allow these specialized tissue stem cells to be reprogrammed genetically to become pluripotent. These stem cells, reprogrammed to express certain genes or maintain these cells in a stem cell-like state, are called induced pluripotent stem cells. In March, 2009, the ban on generating new stem cell lines was lifted by President Barack Obama, making federal funding for embryonic stem cell research available without the previous limits on the stem cell lines generated.

How It Works

To identify stem cells, cells first have to be grown in the laboratory, or cultured. The first step in isolating stem cells is to transfer the inner cell mass of a blastocyst into a cultural medium in a laboratory dish. The culture medium contains nutrients that cells need to grow and divide. Stem cells do not always grow, but when the cells continue to grow and divide, they are then divided into other culture dishes, called subculturing, so that millions of copies of the same stem cell (cloning) can be used for research.

Embryonic Stem Cells. Embryonic stem cells are the easiest of stem cells to divide and reproduce in culture, and they have been shown to live for months without differentiating. When these cells continue in their stem cell state, they are considered pluripotent and have the same genetic makeup as the original stem cells from the inner cell mass. These cells are referred to as an embryonic stem cell line. These cells may be frozen and shipped to other laboratories for further culturing and experimentation.

Somatic Stem Cells. Somatic (adult) stem cells are undifferentiated cells that are found in a tissue or organ that can renew themselves and differentiate to become specialized cells of that tissue or organ. Adult stem cells are used to regenerate or repair the tissue in which they are located. Known somatic stem cells are located in the brain, bone marrow, peripheral blood and blood vessels, muscles, skin, teeth, heart, liver, ovarian epithelium, and testes. To be used as a somatic stem cell, these cells need to demonstrate that they can generate a line of genetically identical cells that can give rise to all the differentiated cell types of that tissue. Once these cells are identified, they can be used to regenerate and repair cells within that tissue. Experiments are ongoing in transdifferentiation, in which certain somatic stem cells are reprogrammed into other cell types or even to become like embryonic stem cells, called induced pluripotent stem cells, with the introduction of embryonic cells.

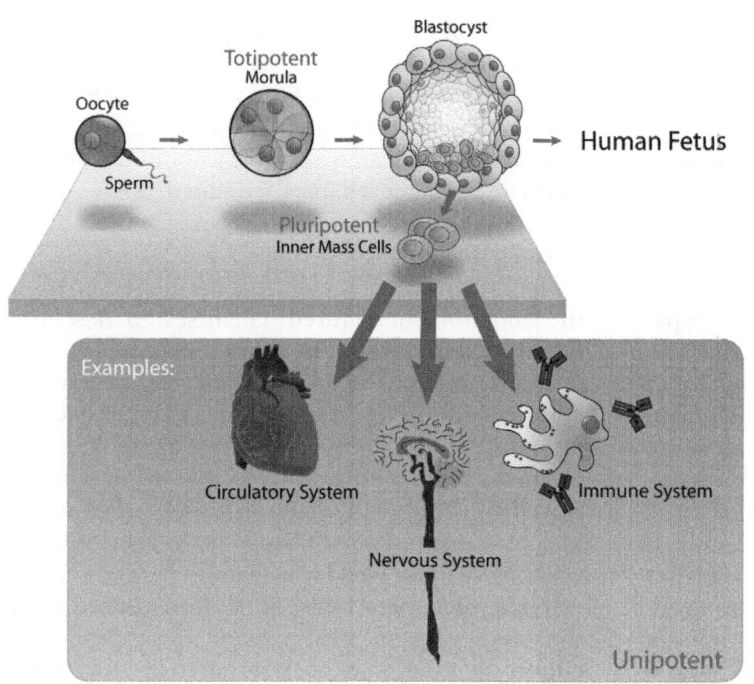

Pluripotent, embryonic stem cells originate as inner cell mass (ICM) cells within a blastocyst. These stem cells can become any tissue in the body, excluding a placenta. Only cells from an earlier stage of the embryo, known as the morula, are totipotent, able to become all tissues in the body and the extraembryonic placenta.

APPLICATIONS

There are several reasons why stem cells are important in science and the advancement of health care.

Cell Specialization and Development. Pluripotent stem cells help scientists understand the complexity of human development and how genes work to make decisions so that cells differentiate to become specialized cells. As development proceeds from an embryo to an individual human, genes turn on and off to give rise to protein expression and cell differentiation. These decision-making genes control the expression of pluripotent stem cells. Scientists know that certain diseases, such as cancer and birth defects, are caused by abnormal cell division and cell specialization. Understanding normal cell development will allow scientists to determine the errors that cause debilitating and often lethal diseases.

Medical Drug Testing. Stem cell research potentially may change the way new medical drugs are developed and tested for safety. These new drugs can be tested on stem cell lines first. Using pluripotent stem lines will expand the cell types that can be tested in the laboratory, before a drug is tested on animals and humans, streamlining the process for drug development.

Cell Therapies. Stem cells have the potential to be used to generate cells and tissues that could replace or regenerate damaged cells and tissues in humans. Such cell therapies could help treat disorders that disrupt cell function or destroy tissues, such as cancer, heart disease, diabetes, spinal cord injury, arthritis, Parkinson's disease, and Alzheimer's diseases. Modern medicine relies on donated organs and tissues to replace destroyed tissue in heart, bone marrow, and kidney transplants. However, the number of people suffering from these disorders far outnumbers the organs and cells available. Stem cells offer a unique opportunity to create a renewable source of replacement cells and tissues to treat these diseases. Another problem in the transplant process is that the recipient's body tends to reject the foreign cells from the donor. With stem cells, research could focus on developing modifications to these cells to minimize tissue incompatibility or to create tissue banks with common tissue type profiles that would be accepted by a large number of individuals.

Somatic Cell Nuclear Transfer. The technique called somatic cell nuclear transfer is still in the research stage, and no human stem cell lines have been created using it. In somatic cell nuclear transfer, the nucleus of virtually any somatic cell is taken from an individual patient and fused with a donor egg cell from which the nucleus has been removed. That cell is then stimulated to develop into a blastocyst and the inner cell mass is taken to create a culture of pluripotent stem cells. These stem cells can be stimulated to develop into specialized cells that are needed to repair damaged tissues or organs. Because the genetic information is taken from the individual patient, these cells theoretically would not be rejected by the patient as they are genetically identical to those of the individual. This type of transplantation would not require immune-suppressing drugs to be successful, and patients would have a far greater chance of survival.

Somatic Stem Cell Therapies. There are disadvantages and advantages to using somatic stem cells for therapies. One disadvantage is that these stem cells are multipotent but not pluripotent, and the types of cells that can be developed are limited. Previously, it was thought that somatic stem cells could develop into only the specialized cells from which they were derived, making it necessary to use only bone marrow stem cells for bone marrow transplantation, liver stem cells for liver diseases, and so on. However, experiments on mice have shown that, for example, when neural stem cells were placed into bone marrow, a variety of blood cell types were produced. So it is possible that even specialized stem cells may be manipulated to be wider reaching in their potential than previously thought. However, the biggest limitation of somatic stem cells to date is that they have not been isolated from all the tissues of the body. So far, it has not been possible to locate adult cardiac stem cells or pancreatic islet stem cells in humans, which would help in heart disease or diabetes, respectively.

Transplantation. One advantage of somatic stem cells is in transplantation. If these cells could be isolated from a patient and directed to divide and specialize in a manner that conveys normal cell function, they could then be transplanted back into the patient without immune rejection. This would also reduce or avoid the need for embryonic stem cells from human embryos or human fetal tissue. However, isolating somatic stem cells and growing them in culture has been difficult. Even if it becomes possible, growing and manipulating them quickly enough to correct a disease state may be impossible. Rigorous research will be required to overcome the obstacles of this type of cell therapy.

FUTURE PROSPECTS

Acceptance of stem cell research has been greatly expanded with publicity regarding the potential benefits of this type of research. Also, there have been numerous scientific publications leading to advances in the field. However, considerable controversy remains regarding the ethical implications of using embryonic stem cells. In the United States, much debate has centered on the use of human embryos and fetal tissue created for reproductive use. Although these embryos are no longer needed, using them for research means they no longer can be used to produce a viable individual. The morals and ethics of this continue to be debated. However, stem cell research is not limited by the availability of embryonic stem cells. Alternative stem cells, such as somatic stem cells and induced stem cells, have been developed, and researchers may be able to use these instead. The competitive nature of scientific endeavors has led to the advancement of all fields of science, and continued work in this field has produced further success in the use and potential of stem cells as a source of eliminating the threat of some of the most deadly human diseases.

Susan M. Zneimer, PhD

FURTHER READING

Fox, Cynthia. *Cell of Cells: The Global Race to Capture and Control the Stem Cell.* New York: W. W. Norton, 2007. Looks at the competition that arises among researchers as they attempt to find applications for stem cell therapy.

Haerens, Margaret, ed. *Embryonic and Adult Stem Cells.* Detroit: Green Haven Press, 2009. Contains a collection of essays arguing the pros and cons of stem cell research.

Humber, James M., and Robert F. Almeder, eds. *Biomedical Ethics Reviews: Stem Cell Research.* Totowa, N.J.: Humana Press, 2004. A collection of objective essays reviewing the principle arguments for and against stem cell research and whether this type of work violates the rights of human embryos.

Panno, Joseph. *Stem Cell Research: Medical Applications and Ethical Controversy.* New York: Facts On File, 2004. Provides information on the technological advances, applications, and issues of stem cell research, including the use of stem cells to repair damaged nerve tissue and the ethical and legal implications of research in this field.

Sell, Stewart, ed. *Stem Cells Handbook.* Totowa, N.J.: Humana Press, 2007. Explains the origins of stem cells and describes how they function and how they can treat illness and disease. Emphasis is placed on the role of stem cells in development, tissue regeneration, repair mechanisms, and carcinogenesis. Also includes technical approaches to obtaining stem cells and manipulating them for therapeutic use.

Wobus, A. M., and K. R. Boheler, eds. *Stem Cells.* New York: Springer, 2006. Presents many novel aspects of stem cell biology, including existing and future applications in research and medicine, particularly uses in drug therapies.

WEB SITES

California Institute for Regenerative Medicine http://cirm.ca.gov

International Society for Stem Cell Research http://www.isscr.org/public

National Institutes of Health Stem Cell Information http://stemcells.nih.gov

See also: Bioengineering; Cloning; Human Genetic Engineering.

FASCINATING FACTS ABOUT STEM CELL RESEARCH AND TECHNOLOGY

- Clinical trials began in 2010 on a stem cell therapy to restore spinal cord function. Oligodendrocyte progenitor cells derived from human stem cells were to be injected directly into the patient's damaged spine.

- In 2010, the biotech company ACT was preparing to enter into clinical trials of retinal cells derived from stem cells in the treatment of Stargardt's macular dystrophy.

- Scientists in 2010 used a combination of three transcription factors to directly reprogram postnatal mouse heart or skin fibroblasts into differentiated cells that shared many features of heart muscle cells, including beating. Scientists hope to use such cells in heart repair.

- Qingdao University began conducting clinical trials in 2010 of umbilical cord mesenchymal stem cells to treat ulcerative colitis.

- Muscle stem cells have proved difficult to grow. However, in 2010, scientists reported that making the culture conditions mimic the physical properties of the muscle improved cell growth.

- University of Rotterdam scientists became the first to film the birth of blood stem cells in 2010. From this knowledge, the Dutch scientists hope to develop a technique to grow blood stem cells in the laboratory.

- Scientists used induced pluripotent stem cells (rather than embryonic stem cells) in 2010 to produce neural stem cells and dopamingeric neurons, which improved the behavior of mice with a condition similar to Parkinson's disease. Neurons from induced pluripotent and embryonic stem cells were similar in their gene expression but not identical.

- In human embryonic stem cells from people with Fragile X Syndrome, the *FMR1* gene is expressed normally until the cells begin to differentiate, when it is silenced. However, in induced pluripotent stem cells from people with Fragile X, the gene is silenced before differentiation, demonstrating that the two types of stem cells can have significant differences.

The Hardy-Weinberg Law of Genetic Equilibrium

FIELDS OF STUDY

Ecology, evolutionary science, genetics

SUMMARY

The Hardy-Weinberg law of genetic equilibrium is one of the foundations of mathematical population genetics. A description of the genetic makeup of a population under ideal conditions, it acts as a benchmark against which the effects of natural selection or other evolutionary forces can be measured.

PRINCIPAL TERMS

- **allele:** one of several alternate forms of a gene; the deoxyribonucleic acid (DNA) of a gene may exist as two or more slightly different sequences, which may result in distinct characteristics allele frequency: the relative abundance of an allele in a population
- **diploid:** having two chromosomes of each type
- **gene:** a section of the DNA of a chromosome, which contains the instructions that control some characteristic of an organism
- **gene pool:** the array of alleles for a gene available in a population; it is usually described in terms of allele or genotype frequencies
- **genotype:** the set of alleles an individual has for a particular gene
- **genotype frequency:** the relative abundance of a genotype in a population
- **haploid:** having one chromosome of each type
- **population:** the individuals of a species that live in one place and are able to interbreed
- **random mating:** the assumption that any two individuals in a population are equally likely to mate, independent of the genotype of either individual

BASIC PRINCIPLES

Genetics began with the study of inheritance in families: Gregor Mendel's laws describe how the alleles of a pair of individuals are distributed among their offspring. Population genetics is the branch of genetics that studies the behavior of genes in populations. The population is the only biological unit that can persist for a span of time greater than the life of an individual, and the population is the only biological unit that can evolve. The two main subfields of population genetics are theoretical (or mathematical) population genetics, which uses formal analysis of the properties of ideal populations, and experimental population genetics, which examines the behavior of real genes in natural or laboratory populations.

Population genetics began as an attempt to extend Mendel's laws of inheritance to populations. In 1908, Godfrey H. Hardy, an English mathematician, and Wilhelm Weinberg, a German physician, each independently derived a description of the behavior of allele and genotype frequencies in an ideal population of sexually reproducing diploid organisms. Their results, now termed the Hardy-Weinberg principle, or Hardy-Weinberg equilibrium, showed that the pattern of allele and genotype frequencies in such a population followed simple rules. They also showed that, in the absence of external pressures for change, the genetic makeup of a population will remain the same, at an equilibrium. Since evolution is change in a population over time, such a population is not evolving. Modern evolutionary theory is an outgrowth of the "New Synthesis" of R. A. Fisher, J. B. S. Haldane, and Sewall Wright, which was done in the 1930's. They examined the significance of various factors that cause evolution by examining the degree to which they cause deviations from the predictions of the Hardy-Weinberg equilibrium.

ASSUMPTIONS AND PREDICTIONS

The predictions of the Hardy-Weinberg equilibrium hold if the following assumptions are true: The population is infinitely large; there is no differential movement of alleles or genotypes into or out of the population; there is no mutation (no new alleles are added to the population); there is random mating (all genotypes have an equal chance of mating with all other genotypes); and all genotypes are equally fit (have an equal chance of surviving to reproduce). Under this very restricted set of assumptions, the following two predictions are true: Allele frequencies will not change from one generation to the next, and genotype frequencies can be determined by a simple equation and will not change from one generation to the next.

The predictions of the Hardy-Weinberg equilibrium represent the working through of a simple set of algebraic equations and can be easily extended to more than two alleles of a gene. In fact, the results were so self-evident to the mathematician Hardy that he at first did not think the work was worth publishing.

If there are two alleles (A, a) for a gene present in the gene pool, let p = the frequency of the A allele and q = the frequency of the a allele. As an example, if $p = 0.4$ (40 percent) and $q = 0.6$ (60 percent), then $p + q = 1$, since the two alleles are the only ones present and the sum of the frequencies (or proportions) of all the alleles in a gene pool must equal 1 (or 100 percent). The Hardy-Weinberg principle states that at equilibrium the frequency of AA individuals will be p_2 (equal to 0.16 in this example), the frequency of Aa individuals will be $2pq$, or 0.48, and the frequency of aa individuals will be q_2, or 0.36.

The basis of this equilibrium is that the individuals of one generation give rise to the next generation. Each diploid individual produces haploid gametes. An individual of genotype AA can make only a single type of gamete, carrying the A allele. Similarly, an individual of genotype aa can make only a gametes. An Aa individual, however, can make two types of gametes, A and a, with equal probability. Each individual makes an equal contribution of gametes, since all individuals are equally fit and there is random mating. Each AA individual will contribute twice as many A gametes as each Aa individual. Thus, to calculate the frequency of A gametes, add twice the number of AA individuals and the number of Aa individuals, then divide by twice the total number of individuals in the population (note that this is the same as the method to calculate allele frequencies). That means that the frequency of A gametes is equal to the frequency of A alleles in the gene pool of the parents.

The next generation is formed by gametes pairing at random (independent of the allele they carry). The likelihood of an egg joining with a sperm is the frequency of one multiplied by the frequency of the other. AA individuals are formed when an A sperm joins an A egg; the likelihood of this occurrence is $p \times p = p_2$ (that is, $0.4 \times 0.4 = 0.16$ in the first example). In the same fashion, the likelihood of forming an aa individual is $q_2 = 0.36$. The likelihood of an A egg joining an a sperm is pq, as is the likelihood of an a egg joining an A sperm; therefore, the total likelihood of forming an Aa individual is $2pq = 0.48$. If one now calculates the allele frequencies (and hence the frequencies of the gamete types) for this generation, they are the same as before: The frequency of the A allele is $p = (2p_2+2pq)/2$ (in the example, $(0.32 + 0.48)/2 = 0.4$), and the frequency of the a allele is $q = (1 - p) = 0.6$. The population remains at equilibrium, and neither allele nor genotype frequencies change from one generation to the next.

IDEAL VERSUS REAL CONDITIONS

The Hardy-Weinberg equilibrium is a mathematical model of the behavior of ideal organisms in an ideal world. The real world, however, does not approximate these conditions very well. It is important to examine each of the five assumptions made in the model to understand their consequences and how closely they approximate the real world.

The first assumption is infinitely large population size, which can never be true in the real world, as all real populations are finite. In a small population, chance effects on mating success over many generations can alter allele frequencies. This effect is called genetic drift. If the number of breeding adults is small enough, some genotypes will not get a chance to mate with one another, even if mate choice does not depend on genotype. As a result, the genotype ratios of the offspring would be different from the parents'. In this case, however, the gene pool of the next generation is determined by those genotypes, and the change in allele frequencies is perpetuated. If it goes on long enough, it is likely that some alleles will be lost from the population, since a rare allele has a greater chance of not being included. Once an allele is lost, it cannot be regained. How long this process takes is a function of population size. In general, the number of generations it would take to lose an allele by drift is about equal to the number of individuals in the population. Many natural populations are quite large (thousands of individuals), so that the effects of drift are not significant. Some populations, however, especially of endangered species, are very small: The total population of California condors is less than twenty-five, all in captivity.

The second assumption is that there is no differential migration, or movement of genotypes into or out of the population. Individuals that leave a population do not contribute to the next generation. If one genotype leaves more frequently than another, the allele frequencies will not equal those of the previous generation. If incoming individuals come from a population with different allele frequencies, they also alter the allele frequencies of the gene pool.

The third assumption concerns mutations. A mutation is a change in the DNA sequence of a gene—that is, the creation of a new allele. This process occurs in all natural populations, but new mutations for a particular gene occur in about one of 10,000 to 100,000 individuals per generation. Therefore, mutations do not, in themselves, play much part in determining allele or genotype frequencies. Yet, mutation is the ultimate source of all alleles and provides the variability on which evolution depends.

The fourth assumption is that there is random mating among all genotypes. This condition may be true for some genes and not for others in the same population. Another common limitation on random mating is inbreeding, the tendency to mate with a relative. Many organisms, especially those with limited ability to move, mate with nearby individuals, which are often relatives. Such individuals tend to share alleles more often than the population at large.

The final assumption is that all genotypes are equally fit. Considerable debate has focused on the question of whether two alleles or genotypes are ever equally fit. Many alleles do confer differences in fitness; it is through these variations in fitness that natural selection operates. Yet, newer techniques of molecular biology have revealed many differences in DNA sequences that appear to have no discernible effects on fitness.

THEORETICAL AND EXPERIMENTAL GENETIC STUDIES

The field of population genetics uses the Hardy-Weinberg equations as a starting place, to investigate the genetic basis of evolutionary change. These studies have taken two major pathways: theoretical studies, using ever more sophisticated mathematical expressions of the behavior of model genes in model populations, and experimental investigations, in which the pattern of allele and genotype frequencies in real or laboratory populations is compared to the predictions of the mathematical models.

Theoretical population genetics studies have systematically explored the significance of each of the assumptions of the Hardy-Weinberg equilibrium. Mathematical models allow one to work out with precision the behavior of a simple, well-characterized system. In this way, it has been possible to estimate the effects of population size or genetic drift, various patterns of migration, differing mutation rates, inbreeding or other patterns of nonrandom mating, and many different patterns of natural selection on allele or genotype frequencies. As the models become more complex, and more closely approximate reality, the mathematics becomes more and more difficult. This field has been greatly influenced by ideas and tools originally devised for the study of theoretical physics, notably statistical mechanics. Some of the most influential workers in this field were trained as mathematicians and view the field as a branch of applied mathematics, rather than biology. As a consequence, many of the results are not easily understood by the average biologist.

Experimental population genetics tests predictions from theory and uses the results to explain patterns observed in nature. The major advances in this field have been determined, in part, by some critical advances in methodology. In order to study the behavior of genes in populations, one must be able to determine the genotype of each individual. The pattern of bands on the giant chromosomes found in the salivary glands of flies such as *Drosophila* form easily observed markers for groups of genes. Since these animals can be easily manipulated in the laboratory, as well as collected in the field, they have been the subjects of much experimental work. Using population cages, one can artificially control the population size, amount of migration, mating system, and even the selection of genotypes, and then observe how the population responds over many generations. More recently, the techniques of allozyme or isozyme electrophoresis and various methods of examining DNA sequences directly have made it possible to determine the genotype of nearly any organism for a wide variety of different genes. Armed with these tools, scientists can address directly many of the predictions from mathematical models. In any study of the genetics of a population, one of the first questions addressed is whether the population is at Hardy-Weinberg equilibrium. The nature and degree of deviation often offer a clue to the evolutionary forces that may be acting on it.

Understanding Genotypes

As the cornerstone of population genetics, the Hardy-Weinberg principle pervades evolutionary thinking. The advent of techniques to examine genetic variation in natural populations has been responsible for a great resurgence of interest in evolutionary questions. One can now test directly many of the central aspects of evolutionary theory. In some cases, notably the discovery of the large amount of genetic variation in most natural populations, evolutionary biologists have been forced to reassess the significance of natural selection compared with other forces for evolutionary change.

In addition to the great theoretical significance of this mathematical model and its extensions, there are several areas in which it has been of practical use. An area in which a knowledge of population genetics is important is agriculture, in which a relatively small number of individuals are used for breeding. In fact, much of the early interest in the study of population genetics came from the need to understand the effects of inbreeding on agricultural organisms. A related example, and one of increasing concern, is the genetic status of endangered species. Such species have small populations and often exhibit a significant loss of the genetic variation that they need to adapt to a changing environment. Efforts to rescue such species, especially by breeding programs in zoos, are often hampered by an incomplete consideration of the population genetics of small populations. A third example of a practical application of population genetics is in the management of natural resources such as fisheries. Decisions about fishing limits depend on a knowledge of the extent of local populations. Patterns of allele frequencies are often the best indicator of population structure. Population genetics, by combining Mendel's laws with the concepts of population biology, gives an appreciation of the various forces that shape the evolution of the earth's inhabitants.

—*Richard Beckwitt*

See also: Evolution: Animal life; Gene flow; Genetics; Natural selection; Reproduction.

Further Reading

Audesirk, Gerald, and Teresa Audesirk. *Biology: Life on Earth*. 5th ed. Upper Saddle River, N.J.: 1999. An introductory college textbook designed for nonscience majors. The chapter on the processes and results of evolution includes a complete explanation of basic population genetics, presented in a nontechnical way. The chapter is well illustrated and includes a glossary and suggestions for Bibliography.

Avers, Charlotte J. *Process and Pattern in Evolution*. New York: Oxford University Press, 1989. A text that introduces modern evolutionary theory to students who already have a background in genetics and organic chemistry. Covers basic population genetics and introduces most of the techniques used in the study of evolution. Includes references to original research, as well as other suggested readings.

Ayala, Francisco J., and John A. Kiger, Jr. *Modern Genetics*. 2d ed. Menlo Park, Calif.: Benjamin/Cummings, 1984. This genetics text assumes an audience that has had college-level biology and some chemistry. It provides a good description of classical as well as molecular genetics. Covers most of the methods and major results of population genetics. Chapters include a Bibliography as well as problem sets, and there is a glossary.

Dobzhansky, Theodosius. *Genetics of the Evolutionary Process*. New York: Columbia University Press, 1970. An older book, this text is an introduction to experimental population genetics, by one of the architects of the field. There are numerous references to the original literature and many examples. The book is suitable for anyone with at least some introduction to biology and provides one of the clearest explanations of the major concepts of modern evolutionary thought.

Futuyma, Douglas J. *Evolutionary Biology*. 3d ed. Sunderland, Mass.: Sinauer Associates, 1998. An advanced text in evolution for students with previous exposure to calculus and a strong biology background, including genetics and various courses in physiology and ecology. The great strength of this book is in the presentation of areas of current research and argument in evolution, rather than a cut-and-dried array of "facts." There are numerous references to original research and a glossary.

Hartl, Daniel L. *A Primer of Population Genetics*. 3d ed. Sunderland, Mass.: Sinauer Associates, 2000. This text is intended for students with a college-level knowledge of biology but does not require prior exposure to genetics, statistics, or higher mathematics. There are examples of the significance of population genetics ideas in many areas of biology and medicine, and each chapter has problem sets with answers. There are numerous references to original research.

Nagylaki, Thomas. *Introduction to Theoretical Population Genetics*. New York: Springer-Verlag, 1992. A college-level textbook, using elementary mathematics to investigate theoretical population genetics. Uses calculus and linear algebra, but does not require a background in genetics.

Starr, Cecie, and Ralph Taggart. *Biology: The Unity and Diversity of Life*. 9th ed. Pacific Grove, Calif.: Brooks/Cole, 2001. A textbook for an introductory college biology course. The chapter on population genetics, natural selection, and speciation covers population genetics and mechanisms of evolution. The book is well provided with examples and many striking photographs.

Svirezhev, Yuri M., and Vladimir P. Passekov. *Fundamentals of Mathematical Evolutionary Genetics*. Translated by Alexey A. Voinov and Dmitrii O. Logofet. Boston: Kluwer Academic Publishers, 1990. Offers a clear exposition of the mathematical material in historical perspective. Part 1 covers deterministic models and part 2 uses stochastic models.

THERMOREGULATION

FIELDS OF STUDY

Biophysics, physiology

SUMMARY

Temperature regulation in animals is a process that may utilize either environmental or physiological sources of heat to maintain conditions conducive to life. By learning about temperature regulation, scientists have gained insights into interactions between animals and their environments or between the functional components of their bodies.

PRINCIPAL TERMS

- **convection:** a transfer of heat from one substance to another with which it is in contact
- **countercurrent mechanism:** a heat exchange system in which heat is passed from fluid moving in
- **ectotherm:** an animal that regulates its body temperature using external (environmental) sources of heat or means of cooling
- **endotherm:** an animal that regulates its body temperature using internal (physiological) sources of heat or means of cooling
- **heliotherm:** an animal that uses heat from the sun to regulate its body temperature homeostasis: the maintenance by an animal of a constant internal environment
- **homeotherm:** an animal that strives to maintain a constant body temperature independent of that of its environment
- **optimum temperature:** the narrow temperature range within which the metabolic activity of an animal is most efficient
- **poikilotherm:** an animal that does not regulate its body temperature, which will be the same as that of its environment
- **thermogenesis:** the generation of heat in endotherms by shivering or increased oxidation of fats

BASIC PRINCIPLES

Body-temperature regulation by animals is essential for life. The maintenance of life relies on the sum of all chemical reactions or metabolic activity in an organism. These reactions are facilitated by catalysts, substances not directly involved in a reaction as either a product or reagent but essential for accelerating the process or allowing the reaction to proceed under conditions compatible with life. For example, a reaction that, in a test tube, might require exceedingly high temperatures will proceed, if catalyzed, at normal body temperatures. Biological catalysts are complex proteins called enzymes. These are fragile molecules and are quite temperature-sensitive.

If exposed to excessively high or low temperatures, they will be denatured and lose their functional properties.

How It Works

Homeostasis is the maintenance of a constant internal environment, one suitable for proper enzymatic activity. Homeostatic mechanisms involve three components: a sensor (or receptor) that reacts to changes in environmental conditions, a coordinator (or integrator) that responds to information from the sensor, and one or more effectors (activated by the coordinator), which elicit appropriate, regulatory responses.

Temperature sensors are scattered throughout the bodies of most animals, but those specifically associated with temperature regulation in vertebrates (animals with backbones) are found in the hypothalamic region of the brain. Coordinators are found within the brain (or its equivalent in simpler animals), again in the hypothalamus of more advanced types. Effectors may be any structure capable of affecting temperature.

Animals generally function at temperatures between 4 and 40 degrees Celsius. Peak metabolic efficiencies, however, exist over a much narrower range, called the optimum temperature. This temperature varies by the animal and its habitat. Optimum temperatures often approach lethal limits, the highest temperature an animal can tolerate. This necessitates precise control of temperature in order to avoid exceeding those limits. Within lower temperature ranges, some animals can alter metabolic requirements in order to adapt to changing temperatures without sacrificing efficiency. This process, which involves complex biochemical and cellular adjustments, is called "temperature compensation." Animals that utilize metabolic mechanisms to maintain constant, relatively high body temperatures are often referred to as being "warm-blooded." Others, whose body temperatures are not regulated or are regulated primarily by behavioral means, are called "cold-blooded." That these terms are imprecise and irrelevant becomes obvious when one considers that the temperature of a desert-dwelling "cold-blooded" lizard or insect may often exceed that of any bird or mammal. On the other hand, the core temperature of some hibernating mammals may be reduced to being anything but "warm."

Most invertebrates (animals lacking backbones) as well as many fishes, amphibians, and some reptiles, do not regulate body temperatures; they are called poikilotherms. They monitor environmental conditions, attempt to seek out areas where temperatures are suitable, and avoid those where they are not. Their temperatures are essentially identical to environmental temperatures. If excessively high temperatures are unavoidable for more than short periods, death may occur. Low temperatures are seldom fatal (unless below freezing) but will result in diminution of metabolic functions, causing the animal to become torpid, or inactive. Since these animals are vulnerable, they will seek shelter, which is why insects, for example, are rarely encountered during colder months.

Ectotherms

Animals that regulate body temperatures fall into two categories. Those that utilize environmental sources of heat are called ectotherms (animals that "heat" their bodies using external sources). Those that utilize physiological temperature control mechanisms are called endotherms (animals that "heat" their bodies using internal sources). Since endotherms (birds and mammals) strive to keep temperatures constant, they may also be called homeotherms (animals that maintain constant temperatures). All regulators must invest considerable energy in the process. To minimize that expenditure, they utilize microhabitats in which regulatory mechanisms are not necessary. Ectotherms use behavior, enhanced by physical or physiological mechanisms, to take advantage of environmental conditions. A principal source of heat for most ectotherms is sunlight; temperature regulators that rely on the sun are called heliotherms (animals that "heat" their bodies using the sun). Lizards from temperate zones (areas with moderate and/or seasonal climates) are the most efficient ectotherms and may serve as models to illustrate the process. Tropical species, which live in constant, warm environments, tend to be poor regulators.

Sunlight and heat may be assimilated directly by basking lizards or indirectly by convection from sun-heated surfaces. Basking occurs when an animal exposes itself to sunlight by seeking unshaded perches. Position and posture are critical. Lizards will orient themselves in order to expose the greatest amount of surface to the sun. This involves a position in which the animal is broadside to the sun. Surface area is further enhanced by flattening the body dorsoventrally (top to bottom). Similarly, animals may absorb heat from the substrate. Lizards flatten themselves against a warm surface to maximize the area through which heat is assimilated. Area is critical in elevating temperatures, either by basking or convection, but does not increase proportionately with volume as animals increase in size. Thus, large ectotherms require disproportionately more energy and time to raise their temperatures than animals with similar proportions but smaller dimensions. This explains why the first animals to emerge in the spring or early morning tend to be small. Also, since dark colors absorb more radiation (heat and light), cold animals will stimulate pigment cells and are invariably much darker than those at optimum temperatures. That these mechanisms work effectively is illustrated by observations of active lizards at near-freezing temperatures at high elevations in the Andes of South America. When these lizards are captured, body temperatures of 31 degrees Celsius are recorded. In another study, lizards active at 4 degrees Celsius have been found to have body temperatures above 10 degrees Celsius. Some investigators have observed lizards, buried in sand during the night, emerging slowly, exposing only their heads. Since many lizards have large blood sinuses in their heads, it has been suggested that they can raise their body temperatures while minimizing exposure to predators. It is unlikely that this is effective, as heat gained would be rapidly lost to the substrate by convection. Only if the ground were warmer than air and only until body temperature reached that of the ground would this mechanism be operative.

In ectotherms, cooling is a much more difficult proposition. Without access to a source of "cold," ectotherms can do little more than minimize heat absorption. Coloration is lighter to increase reflection, orientation is toward the sun, posture involves lateral (side-to-side) compression, and animals will "tiptoe," lifting themselves away from warm substrates. If these are inadequate, animals must seek shelter. Many desert-dwelling lizards exhibit activity cycles that peak twice each day (morning and evening) to avoid cold nights and hot midday periods.

ENDOTHERMS

Endotherms use physiological effectors to raise or lower temperatures. If cold, they will generate heat (thermogenesis) by rapid muscular contractions (shivering) or increased oxidation of fats. Simultaneously, devices minimizing heat loss will be implemented. These include lowered ventilation (breathing) rates; since inhaled air is warmed during passage through the respiratory tract, heat is lost with each expiration. Also, superficial blood vessels narrow (vasoconstriction), reducing flow of warm blood to the skin, from which heat is lost by convection. Attempts to insulate skin are illustrated by "goose bumps." Though ineffective in sparsely haired humans, this reaction to cold is quite effective in mammals with thick body hair or fur. Muscles attached to hair follicles contract and draw hairs into an upright position, and the ends droop, trapping dead air between matted ends and skin. A fine undercoat in many species enhances the process. Dead air is an excellent barrier to heat flow. A similar device affecting feathers exists in birds.

When hot, endotherms keep muscular activity to a minimum, increase ventilation rates (panting), and expand superficial blood vessels (vasodilation). Rates of heat dissipation in some mammals are enhanced by sweating. Sweating and panting rely on evaporative cooling, the same principle involved in using radiators to prevent hot automobile engines from overheating. Endotherms adapted to hot climates produce concentrated urine and dry feces to conserve water, since much is lost in cooling.

Many of these mechanisms are surface-area related. Consequently, endotherms in hot climates, especially large species with relatively poor surface-to-volume ratios, often possess structures, such as elephant's ears, to increase area through which heat may be dissipated. On the other hand, endotherms occupying cold habitats are designed to minimize exposed surfaces. For example, arctic hares have short ears and limbs compared to the otherwise similar jackrabbits of warmer climes. In addition, cold-adapted endotherms may decrease rates of heat loss from poorly insulated appendages by means of countercurrent mechanisms. Heat from blood in arteries flowing into a limb is passed to venous blood returning to the body. This minimizes the amount of heat carried into a limb, whose surface-to-volume ratio is very high. It also functions to warm the returning blood, which prevents cooling of the body core. The appendages themselves are very cold; portions may even be at below-freezing temperatures.

Actual freezing is prevented by special fats in the extremities.

STUDYING THERMOREGULATION

Specific methods vary according to the subject, approach, and discipline in question. Anatomy (study of structure), using both micro- and macroscopic methods, often centers on surface-related phenomena. For example, studies investigating the vascularization (blood supply) of whale flukes, whose physiology is difficult to study, have indicated that these are quite capable of dissipating heat and have led to the knowledge that these animals, even in cold water, because of their large size and poor surface-to-volume ratios, have potential problems with overheating. The role of blubber was reevaluated in this light and is now recognized as being one of fat storage with little to do with insulation. Furthermore, with new technologies in electron microscopy, anatomists have been able to describe, often for the first time, the complex structural components of organs (and even cells) that are active in thermoregulation.

Physiological studies of function are of two major types. One involves measurements of activity under different thermal regimes; for example, patterns of locomotion or digestion (involving specifically neural and muscular or neural, muscular, and glandular entities, respectively) may be observed at different temperatures. Often, these include observations of performance on treadmills or of rates at which food items are processed in controlled laboratory settings. On a different scale, metabolic activity itself might be linked to temperature by measuring rates of oxygen consumption in special metabolic chambers or utilization rates of products necessary for particular chemical reactions. These types of investigations have led to the determination of optimum and lethal temperatures in many species.

A second type of physiological study deals with actual thermoregulation. The ability to monitor body temperatures continuously, even in small animals, by means of radiotelemetry has made possible whole series of experiments in which animals' thermal responses to induced or natural conditions can be evaluated. Investigations of this type have provided insights into, for example, adaptive hypothermia (significantly reduced body temperatures) in small endotherms such as bats and hummingbirds. These species drastically reduce their core temperatures when inactive in order to conserve energy otherwise rapidly lost as heat through their relatively large surface areas.

Since laboratory work often fails to simulate natural conditions adequately, observations of animals in nature have been instituted. These seek to evaluate thermoregulation in the contexts of ethology (the study of behavior) and ecology (the study of organisms' relationships with their environments). These types of studies frequently entail prolonged observations until patterns of behavior or habitat use emerge and can be quantified and evaluated. The use of rapid-reading thermometers or implanted radiothermisters facilitates understanding of the often-subtle modifications in thermoregulatory behavior or microhabitat use characteristic of many animals. Relating recorded temperatures to changes in posture, position, orientation, activity level, and ambient temperatures of substrate and air has, for example, led to an appreciation of how efficiently some ectotherms regulate temperature and the complexity of the mechanisms involved.

APPLICATIONS

Long restricted by concepts of "warm-blooded" versus "cold-blooded" animals, investigators did not begin in-depth explorations of thermoregulation until the twentieth century. Most early efforts grew out of medical studies dealing with dynamics of human temperature regulation, especially in the context of pathological states associated with fever or trauma-induced hypothermia. Monitoring these conditions led to an appreciation of how complex temperature regulation is and how many of the body's systems are involved. These studies, in turn, led to investigations of similar mechanisms in animals. Initially, most dealt with laboratory animals, but pioneering investigations into thermoregulation by animals in natural habitats soon opened whole new vistas. These studies were subsequently extended to "cold-blooded" species, which in turn led to an appreciation of how effective behavioral temperature regulation could be. In the 1970's, suggestions that at least some dinosaurs may have been homeotherms stimulated further interest in this field of study.

Most heat exchange with the environment occurs through skin or respiratory systems; muscular systems generate heat as a by-product of contraction; digestive and urinary systems regulate elimination of wastes, which influences retention or loss of heat-bearing water; cardiovascular systems transport heat; and nervous and endocrine systems regulate the entire complex. In addition, all cells require a proper thermal environment and may affect heat production by altering rates of oxidative metabolism. Therefore, a more complete understanding of thermoregulation has enhanced scientists' awareness of both normal and pathological functions in most body systems. Specific medical applications of these studies include induced hypothermia during surgery-related trauma and treatment of accident-related hypothermia using mechanisms first observed under natural conditions in animals.

Studies of temperature-regulating mechanisms, both behavioral and physiological, have also provided insights into relationships between animals and their environments. Thermoregulatory needs have been used to explain behavioral and ecological phenomena for which causative agents were previously unknown. From a practical perspective, this knowledge is useful in developing management tools to sustain disrupted or endangered ecosystems. Appropriate techniques must be developed with a thorough knowledge of the dynamics in any given system, and this must be based on biological criteria rather than human perceptions. For example, reforested areas have often been managed as crops, with all the attendant problems of monocultures (areas cultivated for plants of only one species). Among these is the lack of biodiversity (variety of life-forms). When efforts began to take into consideration microhabitat requirements, often related to temperature regulation, varieties of plants—many with little or no commercial value in themselves— were planted. This resulted in managed areas becoming capable of supporting many different species.

Finally, a more complete knowledge of structures related to thermoregulation has been applied by paleontologists (scientists who study fossils) to the study of dinosaurs. Long thought to be "sluggish," lizard-like ectotherms, dinosaurs are now thought by many investigators to have been more like mammals and birds in their physiological capabilities. This image is more in tune with their domination of the earth for some hundred million years.

—Robert Powell

See also: Osmoregulation.

FURTHER READING

Avery, Roger A. *Lizards: A Study in Thermoregulation.* Baltimore: University Park Press, 1979. This book effectively summarizes principles of thermoregulation by using lizard models to illustrate adaptations to various habitats. Somewhat technical, but appropriate for advanced high school students. Nicely illustrated; Bibliography.

Bakker, Robert T. *The Dinosaur Heresies: New Theories Unlocking the Mystery of the Dinosaurs and Their Extinction.* New York: Kensington, 1986. A marvelously entertaining book treating dinosaurs as homeotherms. Line drawings shed a whole new light on dinosaurs as active, dynamic animals. The book provides insights into homeothermy and the impact it can have on all aspects of animals' lifestyles. Literature-cited section. Written for a popular audience.

Dukes, H. H. *Dukes' Physiology of Domestic Animals.* 11th ed. Ithaca, N.Y.: Comstock, 1993. A comprehensive textbook for veterinary students, covering all aspects of domestic animal physiology including thermoregulation.

Gans, Carl, and F. Harvey Pough, eds. *Physiology C: Physiological Ecology.* Vol. 12 in *Biology of the Reptilia.* New York: Academic Press, 1982. Technical, but provides very complete coverage of thermoregulation in reptiles, with six articles by different authors covering various aspects of the topic. Extensive literature-cited sections at the end of each article.

Hickman, Cleveland P., Larry S. Roberts, and Frances M. Hickman. *Integrated Principles of Zoology.* 11th ed. Boston: McGraw Hill, 2001. A textbook written with exceptional clarity, one of the best in a field of many good general zoology books. The chapter on homeostasis, osmotic regulation, excretion, and temperature regulation covers general principles of thermoregulation, discusses concepts of ectothermy and endothermy, and describes mechanisms and adaptations in both ectotherms and endotherms. Diagrams illustrate examples. Selected references are given at the end of the chapter. Glossary. College level, but suitable for advanced high school students.

Johnston, Ian A., and Albert F. Bennett, eds. *Animals and Temperature: Phenotypic and Evolutionary Adaptation.* New York: Cambridge University Press, 1996. A comprehensive review of temperature adaptation in animals.

Schmidt-Nielsen, Knut. *Desert Animals: Physiological Problems of Heat and Water.* New York: Dover, 1979. This engaging book provides considerable insights into thermoregulation by animals living in possibly the most inhospitable climate on earth. The use of case studies describing mechanisms by individual species is quite useful. Coverage is fairly technical but suitable for the interested nonscientist. Complete literature-cited section.

———. *How Animals Work.* Cambridge, England: Cambridge University Press, 1972. This well-written text discusses in some detail various mechanisms of temperature regulation in vertebrates. Adequately illustrated. The strength of this work is its integration of temperature regulation into the total context of animal physiology. Complete literature-cited section.

TOOL USE

FIELD OF STUDY

Ethology

SUMMARY

Tools extend an animal's ability to interact with or modify its environment. Most of these interactions involve obtaining food, but animals are known to use tools in many different ways.

PRINCIPAL TERMS

- **echolocation:** the ability of animals to locate objects at a distance by emitting sound waves which bounce off an object and then return to the animal for analysis
- **ectoparasite:** a parasite, such as a tick, that lives on the external surface of the host
- **ethology:** the study of an animal's behavior in its natural habitat
- **insight learning:** using past experiences to adapt and to solve new problems
- **pheromone:** a hormone produced by an animal and then released into the environment
- **predator:** an organism that kills and eats another organism, generally of a different species
- **primates:** a group of mammals including apes, chimpanzees, monkeys, humans, lemurs, and tarsiers

BASIC PRINCIPLES

In general, a tool is considered to be something which is not an integral part of an animal's body but is used by the animal to accomplish a specific task. For example, a lobster may use its claw to crack open shells; however, since the claw is a normal appendage of the lobster, it is not considered to be a tool. When humans use a similar object, a nutcracker, to open shells, the nutcracker serves as a tool. It is difficult to define tools accurately. Examples of tools acceptable under the definition of one scientist may not meet the criteria set down by another investigator. Some scientists expand the definition of tool use to include specialized structures some animals use to extend their capability to locate and capture prey. These capabilities might include echolocation or sonar, electromagnetic fields, and specialized cells used for feeding such as the cnidocytes used by jellyfish. Other scientists consider products produced by an organism to be used to capture food as tools. Under this definition, a spider's web can be considered to be a tool.

Quite often, objects taken directly from the environment, such as stones or sticks, are used as tools without further modification by the animal. Other times, the object may be modified by actions such as stripping the leaves from a stick prior to use. Tools allow the user to complete a task more easily or to accomplish a task that may not have been possible without the advantage provided by the tool. The size, shape, and even texture of tools varies across the animal kingdom. Some animals use trees as tools and others use grains of sand. Some fish use spurts of water as tools. In addition to capturing or obtaining food, tools are also used in grooming, for defense, or even as protection from the elements. Thus, animals that use tools are actively interacting with and even modifying their environment.

Sticks and Stones Used as Tools

Many different species of animals, including insects, fish, birds, mammals, and primates, are known to use tools in some way during their everyday activities. While many different types of tools are used in the animal kingdom, the stick is a common and readily available tool. The use of sticks as tools has been well documented in nonhuman primates, such as chimpanzees, apes, and orangutans. Primates often use insight to solve a problem using tools and the young learn to use tools from either observing or being taught by the adults. A classic example of insight learning leading to multiple tool use in chimpanzees was shown by Wolfgang Köhler, an early twentieth century psychologist. Chimpanzees held in captivity were offered food that had been placed beyond their normal reach. When boxes and sticks were added in the enclosure, the chimpanzees stacked the boxes, climbed them, and then used the sticks to knock down bananas that were hanging overhead. If one stick was not long enough, they would connect them together.

Orangutans and chimpanzees will strip the leaves froma stick and then use it to probe into the nest of insects such as ants or termites. When the stick is removed from the nest, the insects crawling over it can be eaten. Leaves themselves have been used by chimpanzees to gather water for drinking. Birds, too, use sticks to probe for insects and to remove them from crevices in the bark of trees. Some birds, the Galápagos woodpecker finch for example, will use their bill to trim and modify the twig before using it as a probe. Pacific island crows use their beaks to modify sticks as well as leaves before using them as probes. In the absence of sticks or leaves, some animals will use cactus spines as probes. Elephants use trees and sticks in various ways. They will rub against a tree or they may pick up a stick with their trunk to scratch. They have been observed to use tree trunks as levers and to use sticks to remove ectoparasites. When monkeys throw sticks and rocks, they are using these objects as tools for defense.

Stones are another common tool. Sea otters use stones in two different ways. Some otters will carry stones with them when they dive and use the stone as a hammer to free a tightly adhered abalone froma rock. While floating along the surface on their backs, otters use stones to crack open the shells of abalone or of bivalves such as clams, mussels, or oysters, which they also pluck from under the water. Otters may use bottles floating in the water to crack shells. Birds use stones in a similar way. Egyptian vultures pick up stones in their beaks and use them in a pecking fashion, like a hammer, to crack open an ostrich egg. If this method fails, they will fly at the egg while clasping the stone in their talons. Mongooses also use rocks to crack eggs.

Other birds, such as eagles, gulls, and crows, drop shelled animals such as turtles onto the rocks to crack their shells. Vultures are known to drop bones of prey onto rocks to crack them open and expose the marrow. Chimpanzees use stones to crack open nuts, analogous to humans using a hammer and anvil. Even spiders use stones as tools. The trap-door spider, *Stanwellia nebulosa*, uses a stone as a defensive tool. If forced to retreat when being attacked, the spider uses a stone to close off its burrow behind it. In Japan, one species of crow uses a very different tool, a car. It has been reported that these crows use cars as nut crackers by placing the nut on the road and, after a car has run over it, retrieving the nut meat. If the car should miss hitting the shell, the crow may try again.

Tools for Fishing

Humans are not the only species to use tools for fishing. Some green herons are known to drop objects into the water to attract fish looking for food. The herons then consume the curious fish. The archer fish uses jets of water shot from its mouth to knock insects off overhanging branches and into the water. Somescientists do not view this as a tool because the water passes along a specialized region of the mouth. However, it is similar to using a bow and arrow to subdue prey from a distance. Octopuses use water shot from their siphon system as a broom to clean the exoskeletons of eaten invertebrates from its den. An octopus may also use the jet of water to modify the size of the den. Another group of animals that uses a form of liquid tool belongs to the spider family, Scytodidae. These spiders shoot sticky material from modified venom glands to entangle their prey.

Other Tools

Spiders use their webs as tools in various ways. Those species of spiders that construct webs make them with silk produced from modified appendages called spinnerets. Webs are used to ambush animals that happen to enter into them. Some spiders strum their webs and use them as tools for communicating. Others may spin a long single strand of silk that they use as a drag line to find their way back or as a safety line to catch themselves. In some species, young spiders make silk parachutes which trap the air currents and allow them to be dispersed far from the nest. Spiders of the genus *Mastophora* spin a single thread, on the end of which is a sticky globule. By suspending the thread fromone leg, the spider uses the web to "fish" for male moths, which are attracted to the sticky globule containing chemicals similar to the pheromones produced by female moths to lure males for mating.

The jellyfish and the hydra, two members of the phylum Cnidaria, have specialized cells, cnidocytes, concentrated on the surface of their tentacles. Inside these cells is an organelle, the nematocyst, which contains a thread. The nematocyst is stimulated to discharge when prey are near to it. This thread may have a barb on its tip that will penetrate the body surface of the prey, or it may be a lasso that wraps around the prey. The prey is then pulled into the digestive cavity of the cnidarian.

Bats and dolphins are two good examples of animals that use echolocation to locate prey. Since sound waves can travel over great distances, the prey can be well beyond the predator's immediate area. The objects do not need to be large in order to be detected. Bats are able to locate mosquitoes. By analyzing the sound waves returning after bouncing off an object, the bat knows which objects are moving and which are stationary. The moving objects represent potential prey. Some potential prey, moths, have evolved a way to detect that they are being tracked by a bat. Thus, they are able to take evasive action and seek shelter near a stationary object such as a tree, or by landing on the ground. In a similar manner, dolphins use a series of high-frequency clicks to track fish. However, the fish, unlike the moths, are often not aware that they are being followed.

—*Robert W. Yost*

See also: Intelligence.

Further Reading

McFarland, David. *Animal Behavior.* 3d ed. Boston: Longman Science and Technology, 1998. An upper-level textbook on animal behavior. Chapter 27, "Intelligence, Tool Use, and Culture," discusses tool use among animals. This is a good book for readers with some previous background on the subject.

McGrew, W.C. *Chimpanzee Material Culture: Implications for Human Evolution.* New York: Cambridge University Press, 1992. Describes and analyzes the use of tools by chimpanzees in their native habitats through field studies conducted across Africa.

Maier, Richard. *Comparative Animal Behavior: An Evolutionary and Ecological Approach.* Boston: Allyn & Bacon, 1998. An excellent general-audience textbook on various aspects of animal behavior, with sections on tool use in animals.

Sherman, Paul W., and John Alcock, eds. *Exploring Animal Behavior: Readings from "American Scientist."* 2d ed. Sunderland, Mass.: Sinauer Associates, 1998. A good collection of essays on various aspects of animal behavior.

USING TRAPS TO CATCH PREY

Traps are one example of animals using a tool to ambush and capture prey. When a predator is an ambusher, it lies in wait for another animal to happen upon its territory, and then the predator strikes. This technique has been especially perfected by the ant lion. The ant lion is the larval form of hundreds of species of insects in the order Neuroptera, family Myrmeleontidae. Using a series of circular and backward body movements combined with a quick side-to-side motion of the head, the ant lion digs into sandy soil, forming an inverted, cone-shaped impression. When an ant or a small insect crawls along the margin of the cone or happens to fall over the edge, the ant lion vigorously begins to throw grains of sand out of the bottom of the pit. This causes the prey to fall deeper into the pit and into the grasp of the ant lion's two large mandibles. The ant lion thus has used the method of tossing sand grains as a tool to capture food. Natural selection will favor the gene pool of those individuals with the greatest ability to move and throw quantities of sand quickly.

Toxicology

FIELDS OF STUDY

Analytical chemistry; biochemistry; biology; chemistry; clinical chemistry; environmental science; forensics; mathematics; pharmacology; toxicology; veterinary medicine.

SUMMARY

Toxicology involves the study of toxicants, whether biological, chemical, or physical, and how they affect people, animals, and the environment. Toxicologists determine whether these chemicals are actually or potentially harmful by using their knowledge of chemistry and biology and help develop and implement strategies to eliminate, reduce, or control exposure to those harmful substances.

PRINCIPAL TERMS

- **analytical chemistry:** study of the chemical composition of natural and artificial materials.
- **biochemistry:** study of the chemical substances and vital processes occurring in living organisms.
- **chain of custody:** process used to maintain and document the chronological history and person responsible for evidence used in a criminal investigation.
- **environmental science:** science of the interactions between the biological, chemical, and physical components in the environment including the effects of these interactions on all types of organisms.
- **forensics:** use of science, scientific methods, and technology to investigate a crime and establish facts that are admissible in a court of law.
- **pharmacology:** study of drugs and their sources, nature, and properties, and how an organism reacts to them.
- **poison:** toxicant that causes immediate death or illness when experienced in even a small amount.
- **reagent:** substance used in a chemical reaction to detect, measure, and examine a substance or to produce other substances.
- **toxicant:** substance that may produce adverse biological effects of any nature.
- **toxicodynamics:** study of the effects poison has on the body systems and structures.
- **toxicokinetics:** study of how the body processes poisons, including the body systems that are involved.
- **toxin:** specific protein produced by a living organism.
- **xenobiotic:** foreign substance taken into the body.

BASIC PRINCIPLES

Toxicologists study the adverse effects of biological, chemical, or physical agents on living organisms (humans, animals, and plants). Adverse effects can manifest in many forms, ranging from immediate death to subtle changes at a molecular level that do not become known until years later. These effects can also manifest themselves at various levels in the body. For example, some chemical agents affect a certain body organ, others damage a particular type of cell, and even others may interfere with a specific biochemical reaction in the body necessary for life to continue. As medical knowledge has progressed, the understanding of how toxic agents affect the body has changed. A body can be affected on a cellular level by unseen toxins, the damage of which will not be know for many years.

This realization has led to an expansion in the field of toxicology. Toxicologists are now tasked with examining the physical environment to determine whether, how, and at what levels environmental toxins affect humans and other living things. These types of examinations can affect many industries, such as those that emit toxins into the environment and even those that dispose of toxic and hazardous waste and develop agents for biological warfare. Other fields in which toxicology is key is that of animal science (veterinarians who determine treatment for animals who are affected by toxins) and drug development (scientists who determine how certain therapeutic drugs affect the human body and determine safe and effective dosages).

BACKGROUND

Toxicology and the study of poisons has a long and interesting background, possibly beginning with early humans, who recognized poisonous plants and animals and used them in the process of killing, whether for food or in war. Writings as early as 1500 BCE depict substances such as hemlock, opium, and certain metals that were used on arrows to kill animals or humans or even as agents in state execution processes. Stories are told of "poison maidens," beautiful young girls who were fed tiny amounts of poison on a daily basis, causing them to become immune to the effects of the poison, until they became poisonous themselves. They were then sent as gifts to rival kings who died when they touched the poisonous girl.

Poisoning as a method of assassination become more popular in the eighth century, when an Arab chemist discovered how to turn arsenic into an odorless, tasteless, nearly undetectable powder. This substance became an easily available murder weapon, and by the Renaissance period, poison rings, knives, letters, and lipstick were in use for those who wished to do away with a political or amorous rival easily and quickly.

Philippus Aureolus Theophrastus Bombastus von Hohenheim (known more commonly as Paracelsus), a sixteenth-century Swiss physician, was formulating ideas about poisons and toxicology that are still in use. He carefully studied plant and animal poisons and determined that specific chemical compounds, rather than the plant or animal itself, which was immune to the poison it carried, were responsible for toxicity. He documented how the human body responded to those specific chemical compounds and understood that doses of a particular compound could be beneficial or toxic, depending on the amount given (known as the dose-response relationship). A major concept of toxicology, credited to Paracelsus, is that "all substances are poisons; there is none which is not a poison. The right dose differentiates a poison and a remedy." Drug companies continue to use this idea, as many drugs, such as warfarin, were developed from substances that caused immediate death. In the case of warfarin, it began as a type of poison for rats that caused their blood to thin, and they would bleed to death. Therapeutic doses of warfarin help stroke victims (or possible victims) to keep from forming blood clots.

French toxicologist Mathieu Joseph Bonaventure Orfila is referred to as "the father of toxicology" and was the first major proponent of forensic toxicology. In the nineteenth century, he prepared a systematic correlation between chemical and biological properties of poisons. He analyzed autopsy materials to show the effects of poisons on specific organs by showing tissue damage and made chemical analysis a routine part of forensic medicine. Orfila is credited as being one of the first to use a microscope to look for blood and semen stains, and he became an expert witness in the sensational murder trials of his time.

Poisoning is still a relatively major cause of death. In the United States, from 2001 to 2004, there were more than 147,000 deaths related to poison; of these 434 were considered homicides, though more may have been murders as it is sometimes difficult to distinguish a poisoning from a natural death or an accident.

How It Works

Toxicologists work in laboratories, performing tests on substances of different types—often human tissue. They must be familiar with and know how to operate highly sophisticated laboratory equipment and understand the functioning of chemical reagents. They must understand and apply highly sophisticated and exact methodologies to determine reliably the presence or absence of a substance in a sample. Each step of every complicated process must be documented to ensure that procedures have been exactly followed, especially in circumstances involving a chain of custody for criminal cases. Toxicologists must also make informed conclusions about the impact of a certain amount of a specific substance and what effect it would have on a certain individual (based on weight, for example) or what effect a substance would have on a particular environment. These educated opinions are often based on professional, educational, and scientific experience and are sometimes required in court testimony.

Applications

Toxicologists can focus their efforts in a variety of areas. Below are a few of the major areas of specialization.

Forensic Toxicology. These scientists usually work as part of a crime-scene team. They perform tests on bodily fluids and tissues to determine whether any drugs or chemicals in the body may have contributed to a crime, such as alcohol, chemicals, drugs (illegal or prescription), gases, metals, or poisons. Alternatively, a forensic toxicologist may work in drug testing, trying to discover evidence of date-rape or performance-enhancing drugs, or in animal-tissue testing for evidence of wildlife crime or environmental contamination, such as chemical spills.

Environmental Toxicology. These professionals focus on the interaction of chemicals on living systems, including how areas and environments are affected by toxic waste or released industrial chemicals. They may also work with workplace exposure to chemicals and metals and understand principles of toxicodynamics.

Medical Toxicology. This type of toxicologist usually works in a laboratory performing tests on bodily fluid and tissue samples to determine whether there are chemicals present. Though their work is similar to that of a forensic toxicologist and may even involve criminal investigations, this type of toxicologist works more with medical cases, such as chemotherapy adjustments or accidental exposures, rather than criminal cases.

Pharmacological Toxicology. Drug companies use toxicologists to help determine the chemical toxicity of drugs under development. These professionals help determine therapeutic levels for drugs and evaluate whether the proposed drugs build up in tissues or are eliminated from the body to determine maximum safe dosages and durations. They may also help determine under what conditions certain drugs should be avoided by monitoring interactions of drugs with other drugs a patient may be taking or other conditions a patient may have. Their knowledge of toxicokinetics can be helpful in these situations. This knowledge also helps determine age-related effects of certain toxic agents, such as whether a drug affects children and the elderly differently than it affects adults.

FUTURE PROSPECTS

Toxicologists are necessary in many aspects of environmental industry. They are important in many ecological fields, and some professional societies of toxicologists focus exclusively on this area. For example, the Society of Environmental Toxicology and Chemistry concentrates efforts on the study and analysis of environmental problems. It also focuses on environmental education and the management and regulation of natural resources. Its goal is to find solutions to environmental problems that people can live with on a long-term, everyday basis that support sustainable environments and ecosystems. As the field of health care expands, toxicologists have opportunities to become more and more involved. New drugs are constantly being developed, and toxicologists are heavily involved with drug testing, both on animals and on humans. Their knowledge of the human body and how chemicals interact with it is crucial in this field.

Marianne M. Madsen, MS

FURTHER READING

Evans, G. O., ed. *Animal Clinical Chemistry: A Practical Handbook for Toxicologists and Biomedical Researchers.* 2d ed. Boca Raton, Fla.: CRC Press, 2009. Covers pre-analytical and analytical variables along with information on specific-organ toxicity.

Fenton, John Joseph. *Toxicology: A Case-Oriented Approach.* Boca Raton, Fla.: CRC Press, 2002. Includes case studies and information about diagnosis, testing, and treatment.

Hayes, A. Wallace, ed. *Principles and Methods of Toxicology.* 5th ed. Boca Raton, Fla.: CRC Press, 2008. Discusses principles of absorption, distribution, metabolism, and excretion; helps with understanding and using basic experiments in toxicology.

Klaassen, Curtis D. *Casarett & Doull's Toxicology: The Basic Science of Poisons.* 7th ed. New York: McGraw-Hill, 2008. The "gold standard" of toxicology, includes detailed discussions of concepts, principles, and mechanisms of toxicology.

Nelson, Lewis S., et al. *Goldfrank's Toxicologic Emergencies.* 9th ed. New York: McGraw-Hill, 2011. Includes comprehensive references; begins with general principles and moves to detailed discussions of biochemical principles; discusses various exposures— drugs, plants, metals, household products, as well as occupational and environmental.

Osweiler, Gary D., et al., eds. *Blackwell's Five-Minute Veterinary Consult Clinical Companion: Small Animal Toxicology.* Ames, Iowa: Wiley-Blackwell, 2011. Overview of toxicology in veterinary practice; includes color photos and tables in an appendix to help with quick differential diagnoses.

Richards, Ira S. *Principles and Practice of Toxicology in Public Health.* Sudbury, Mass.: Jones and Bartlett, 2008. Introduction to the field of toxicology and its practice in the public-health environment.

Wright, David A., and Pamela Welbourn. *Environmental Toxicology.* Cambridge, England: Cambridge

University Press, 2002. Overview of interaction of chemicals and the environment from molecular to ecosystem levels; includes case studies.

WEB SITES

American College of Medical Toxicology http://www.acmt.net

Society for Environmental Toxicology and Chemistry http://www.setac.org

Society of Toxicology http://www.toxicology.org

United States Department of Health and Human Services National Toxicology Program http://ntp.niehs.nih.gov

See also: Pathology.

FASCINATING FACTS ABOUT TOXICOLOGY

- Toxicology is sometimes called "the science of poisons."
- Most toxicologists, especially forensic toxicologists, work in labs that are part of law-enforcement agencies. Others work with medical examiners to determine cause of death. Private drug-testing facilities or poison-control centers are another source of employment for these scientists.
- Toxicologists work every day with body fluids and tissues. It can be messy, smelly work.
- Toxicologists must be mentally strong. They are often exposed to details of horrific crimes and must make judgments about whether a crime was committed.
- A forensic toxicologist is often called on to testify in court as to the effect a certain amount of a substance would have on a particular person. He or she must explain complicated testing methods in language that a jury can understand.
- Some famous victims of poisoning include Socrates (hemlock) and Cleopatra (snakebite). The Emperor Claudius (Tiberius Claudius Drusus Nero Germanicus) was said to have been poisoned by his wife. Some say she served him poison mushrooms. Another story says that he was suspicious of her and would only eat figs he himself had picked from the tree, so she went into the garden and poisoned figs still on the tree.
- In the fifteenth century, Lucrezia Borgia was one of the Borgia family members famous for poisoning rivals. She was said to have worn a ring that contained poison that she poured into drinks of men and women who were threatening to her family and its status.
- Viktor Yushchenko, a popular Ukrainian politician, was said to have been poisoned by government agents after announcing that he would run for president. After a dinner with Ukrainian officials, his face became pockmarked and disfigured. Toxicologists found that he had more than 1,000 times the normal amount of TCDD dioxin in his body.
- Toxicology is a constantly changing field. Successful toxicologists are constantly learning, keeping pace with new chemicals, methodologies, and technologies. A good toxicology candidate is someone who is fascinated by chemicals and the effect they can have on the human body.

TRIBOLOGY

FIELDS OF STUDY

Applied mathematics; physics; chemistry; material science; mechanical engineering; fluid mechanics; thermodynamics; rheology; polymer chemistry; biomechanics; biophysics

SUMMARY

All physical materials interact where their surfaces interface. The interaction is characterized by friction, abrasion, and the generation of heat. These effects have deleterious effects in every instance. Tribology, the study of those effects, works to eliminate negative effects and to find positive ways to harness them. Tribological effects, primarily friction, play a role in every mechanical aspect of existence. They are of particular significance in the ultrasmall devices of nanotechnology and in the biomechanics of living systems. The development of scanning probe microscopy has made it possible to acquire an understanding of tribology at the atomic scale.

PRINCIPAL TERMS

- **abrasion:** material deformation or removal that results from frictional contact.
- **friction:** the resistance to lateral relative motion of two surfaces in contact with each other.
- **interface:** the surface on which two differentiated materials make contact with each other.
- **lubrication:** the interposition of a third material between two frictional surfaces for the purpose of minimizing the coefficient of friction between them.
- **nanotribology:** the study of friction, abrasion, heat, and lubrication at the scale of the nanometer.
- **oil whirl:** a vibrational instability of the lubricating fluid in high-speed journal bearings.

BASIC PRINCIPLES

The word "tribology" means "study of rubbing." In every practical sense it applies to the study of the interactions of physical matter at an interface (that is, where one surface contacts another). These interactions are characterized by friction, abrasion, and the generation of heat, all of which affect the subsequent behavior of the surfaces and the dimensional characteristics of the material.

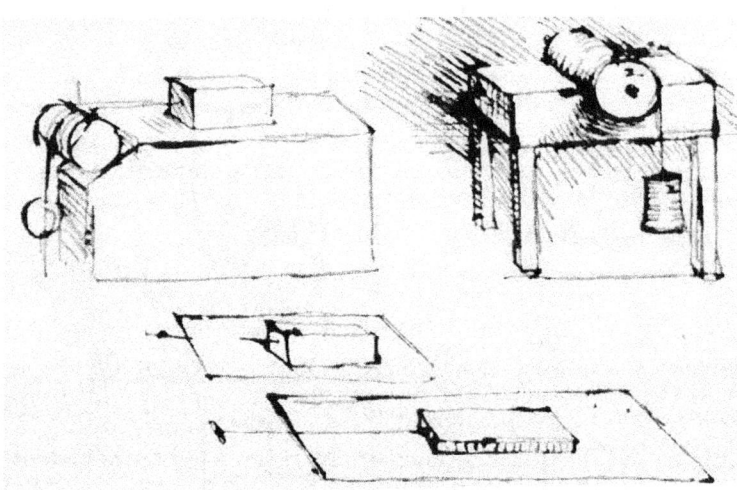

Tribological experiments suggested by Leonardo da Vinci

Friction can be described as the resistance to relative lateral motion between two surfaces, while abrasion describes the deformation and forcible removal of material from one surface by material of another surface. Both of these effects facilitate the release of energy, altering physical surface structure at the atomic and molecular level of the materials. The energy released by the alteration of surface structure by friction and abrasion becomes sensible as heat conducted through the mass of the material.

Tribology examines and quantifies the relationships of friction, abrasion, and heat as they relate to the physical performance of mechanical devices. An especially significant field of research in tribology is the study of the qualities of lubrication, as lubricating materials are used to counteract tribological effects. At the same time, however, the lubricating materials themselves become active contributors to tribological effects, and their study seeks to identify and quantify their corresponding effects.

BACKGROUND

Friction is one of the oldest known technological effects. Archeologists have unearthed implements dating from the Paleolithic Age (early Stone Age) that had been fitted with pieces of antler or bone to act as antifriction bearings. Chariots found in tombs dating from ancient Egypt (from about five thousand years ago) contained the residue of animal fats in the axle-bearing surfaces of their wheels, indicating that the Egyptians of the time understood the value of lubrication. Tomb paintings also indicate that the use of lubrication was an essential component in the movement of large stone blocks used in construction. The physical concept of the coefficient of friction was deduced by Renaissance artist and thinker Leonardo da Vinci in the fifteenth century, but remained generally unknown because his notebooks were not published until some centuries later.

The rules of friction were rediscovered in 1699 by French physicist and inventor Guillaume Amontons and were later verified by French physicist Charles-Augustin de Coulomb. These rules acquired great significance with the mechanization developed during the Industrial Revolution of the eighteenth century. In modern times, the precision with which mechanical devices are built demands that the effects of friction, abrasion, and heat be fully understood, from the atomic scale upward, so that their detrimental effects can be minimized or eliminated.

FRICTION

The classical view of tribology is focused on the study of the causes and effects of friction. The simplest explanation comes from the view that no matter how smooth a surface may appear to be, as the scale of resolution becomes ever smaller, even the smoothest of surfaces becomes more and more irregular.

This process is exemplified by the examination of a billiard ball, a hard and extremely smooth spherical object. If one could expand the scale of the billiard ball to the size of the planet, maintaining the surface irregularities to scale, then the surface of the billiard ball would be covered with bumps and ridges higher than Mount Everest and depressions deeper than the Mariana Trench. In simplest of terms, friction results from the binding of the irregularities of one surface in those of another.

The processes of friction are much more complicated than this simple view, however. Since the development of the current atomic theory and quantum mechanics, it is now known that many other effects play a role in the causes of friction. Researchers in tribological phenomena are only now beginning to acquire an understanding of details of the process at the atomic level, where tribology begins. This new understanding has been made possible by the development of scanning probe microscopy and, particularly, of the atomic force microscope.

Scanning probe microscopes allow examination of surfaces at the atomic scale, with resolutions as fine as 10 picometers (10_{-11} meters). Even the most cursory examination of a surface image from a scanning probe microscope reveals that at the atomic level an assumed perfectly smooth surface consists instead of a series of bumps and depressions reminiscent of what one would observe in a layer of golf balls, marbles, or any other spherical object. Additionally, quantum effects such as van der Waals forces, magnetism, and electronic interactions, and the chemical nature of the material, are important components of friction at the atomic level. The accumulation of effects from the atomic level to the normal size of the object determines how friction is generated between material objects.

The basic principles of friction are deduced from empirical observation. First, the frictional force that resists the sliding of one surface against another is directly proportional to the normal load between them. In other words, the more pressure that is mutually exerted against the two surfaces, the harder it is to slide them across each other. Second, the amount of frictional force does not depend on the size of the area of contact between the two surfaces. This can be examined simply, and cursorily, by sliding an irregularly shaped object across a tabletop, using different surface areas each time. Third, once the sliding motion has begun, the frictional force is independent of the velocity. That is to say, sliding two surfaces against each other at a high velocity requires the same force as it does at a low velocity.

ABRASION

Abrasion is friction to the extreme, resulting in the deformation and displacement of material from one or both interacting surfaces. Abrasion is not the result of matter passing across the surface of other matter. Rather, abrasion is the result of matter physically passing through the same space occupied by other matter. The harder or tougher of the two materials will correspondingly force the other into a new relative position, to the point of separating from the main mass.

It is also possible for material abraded from one surface to transfer to the other surface in a chemical sense and in a physical sense. Research on friction between Teflon and aluminum surfaces, for example, has revealed the formation of a certain amount of aluminum trifluoride on the aluminum surface, a condition made possible only by chemical reaction between the aluminum metal and heat-induced breakdown products of the perfluorinated chemical structure of the Teflon surface.

HEAT

Heat is the third major component of tribological effects, easily examined by rubbing one's hands together briskly, first dry and then wet. The heat produced through friction can be intense, leading to dimensional changes that in turn aggravate both friction and abrasion and perhaps lead to the failure of the mechanism.

Applications

Tribological effects can have both positive and negative effects, both of which are crucial to the functioning of modern machinery. The applied science of tribology is a multi-aspect study that seeks first to identify the genesis of tribological effects. It then seeks to identify the ways in which negative effects can be reduced or eliminated and positive effects used or enhanced.

Friction and wear (abrasion) occur simultaneously in all physical systems. Environmental erosion and skeletal joints obey the same principles of tribology as do steel bearings and internal combustion engines. Examples of positive applications of friction include braking and clutching systems; the drive wheels of trains, cars, and other vehicles; and bolts, nuts, and other devices whose proper function depends on the application of friction. Positive wear or abrasion includes such diverse applications as pencils, pens, and other writing or drawing materials; various machining and polishing techniques; and even a morning shave. Negative friction includes the resulting dimensional changes and physical damage that occur with internal combustion engines, gears, cams, bearings, and seals, and even such minor inconveniences as getting stuck halfway down a playground slide.

Lubrication and Lubricants. The essential principle of lubrication is simply to add a third material to a system to lower the coefficient of friction between them as much as possible. In the worst possible sense, lubrication can have disastrous results, as when water-soaked soil slides under the force of gravity as an avalanche or mudslide. Controlled lubrication, on the other hand, is essential to the long-term functioning of machines and other mobile structures, including biological and biomechanical systems. Lubrication is a surprisingly complex system in its own right, because a lubricating material interacts somewhat differently with each other material in the system. For example, in a system in which an oil is used between steel and aluminum components, the oil molecules will have a different level of adhesion and adsorption to the aluminum surface than to the steel surface, resulting in a dynamic movement of material within the oil that may affect how the system functions over time.

Lubricating materials come in a variety of forms and viscosities, ranging from plain dry air to microgranulated solid particles, such as graphite powder. Typically, the selection of a lubricant depends on the amount of pressure that it must bear in application. Teflon is rather unique in this regard because it is a material that becomes more slippery as the pressure it bears increases; typical lubricants tend to lose their lubricating properties as the pressure placed upon them increases. There are literally as many possible lubricating materials as there are materials and material combinations, presenting an impossible challenge to tribological research. The vast majority of lubricants in common use therefore fall into a few general classes: liquids and semiliquids (such as oils and greases) and solid lubricants (such as graphite). Within these classes there are hundreds of variations. A special class of liquid lubricants are those that function as abrasive carriers as they lubricate. The almost exclusive use of such lubricants is in deep boring operations; for instance, in petroleum and natural gas recovery. The extreme pressures encountered during deep well boring in rock formations demand the use of water as the lubricant, while simultaneously transporting abrasive and abraded material at the drill head to assist the boring process. Various cutting fluids and honing oils used in machine-shop operations for fine grinding and polishing procedures serve a similar function.

Tribological Research and Control Devices. At the lowest end of this technology are grease guns and oil cans for the crude application of lubricants. At the highest end are scanning probe microscopes, enabling researchers to examine the causes and processes of friction at the atomic level. Between the two ends are numerous specially designed devices that test and measure the properties and capabilities of lubricating materials and machine components under operating conditions likely to be encountered in the working environments of those devices. In most cases, such as for a synthetic oil blend in standard roller bearings, this is an almost trivial exercise. In other cases, the working environment is extreme, ranging from the deep ocean floor to deep space, demanding that the materials and designs function flawlessly the first time and for the lifetime of the device.

Future Prospects

In the general scheme of society, tribology plays a large economic role, in both positive and negative ways. Tribological effects are integral parts of the physical world. A world in which friction did not function would be a grand failure at a basic level, given that friction and frictional wear has, for example, enabled humans to walk upright and write meaningful information on materials, whose production, in turn, was possible only because of frictional processes. In modern times, however, devices that function with moving parts call for more ways to defeat the negative effects of friction, abrasion, and heat.

Richard M. Renneboog, MSc

FURTHER READING

Bhushan, Bharat. *Principles and Applications of Tribology.* New York: John Wiley & Sons, 1999. The introduction to this book, also available online, provides an excellent account of the history of tribology.

Bhushan, Bharat, ed. *Measurement Techniques and Nanomechanics.* Nanotribology and Nanomechanics 1. Berlin: Springer, 2011. Provides a concise overview of the conceptual history and principles of tribology.

Donnet, Christophe, and Ali Erdemir. *Tribology of Diamond-like Carbon Films: Fundamentals and Applications.* New York: Springer Science, 2008. This book deals specifically with one specialized material aspect of tribology, demonstrating the breadth and depth of the field.

Gohar, Ramsey, and Homer Rahnejat. *Fundamentals of Tribology.* London: Imperial College Press, 2008. An introductory book for undergraduate engineering specialists that includes heavy mathematical descriptions and relationships for several common situations.

Sinha, Sujeet K., and Brian J. Briscoe. *Polymer Tribology.* London: Imperial College Press, 2009. This book addresses the relatively recent study of the tribology of polymeric materials with their increasing use as dynamic components such as nylon gears and Teflon-coated slides.

See also: Biophysics.

FASCINATING FACTS ABOUT TRIBOLOGY

- Early Stone Age people used boring tools that had been fitted with pieces of bone or antler to act as bearings.
- Chariots found in Egyptian tombs dated to more than five thousand years ago have traces of animal fat, used as a lubricant, on the axle-bearing surfaces of their wheels.
- Leonardo da Vinci deduced the basic physical laws of friction in the late fifteenth century, but these laws remained unknown largely because da Vinci's personal notebooks were not published. The laws were rediscovered by Guillaume Amontons in 1699.
- Atomic force microscopes can examine the causes of friction between single atoms.
- If it were expanded to the size of the planet, the surface of a glass-smooth billiard ball, for example, would have peaks higher than Mount Everest and valleys deeper than the Mariana Trench.
- Images from scanning probe microscopes demonstrate that no surface is perfectly flat, but that all surfaces are made up of bumps and holes, like a layer of golf balls in a box.
- Ignorance of the effects of friction costs about 4 percent of a nation's gross national product, which amounts to about $200 billion per year in the United States.
- About one-third of all energy consumed around the world is used to overcome the force of friction.

U

Urology

FIELDS OF STUDY

Biology; chemistry; anatomy; physiology; biochemistry; endocrinology; embryology; developmental biology; neurology; pharmacology; microbiology; pathology.

SUMMARY

Urology is the study of the anatomy and physiology of the liquid waste removal system in men and women and the associated reproductive system in men. This system includes the kidneys, ureters, urinary bladder, and urethra in both sexes, and the testes, vas deferens, prostate, and penis in men. By understanding the normal structures and functions of the organs involved with the urinary tract, urology can address abnormalities and elective changes such as the correction of birth defects, implantation of a penile prosthesis for erectile dysfunction, sterilization by vasectomy, and sex reassignment surgery (both male-to-female and female-to-male). Urologists typically diagnose and treat more men than women because women generally bring urological concerns to their gynecologists. For men, urologists are comparable specialists for reproductive health.

PRINCIPAL TERMS

- **benign prostatic hyperplasia:** enlargement of the prostate gland.
- **bladder:** muscular organ that distends to hold urine until excretion.
- **calculi:** solid masses of mineral salts; also called kidney stones.
- **catheter:** hollow tube inserted up the urethra into the bladder to drain urine.
- **erectile dysfunction:** inability to achieve or sustain an erection of the penis.
- **kidney:** organ that filters waste products and excess water from the blood.
- **prostate:** organ that secretes an alkaline fluid as a component of semen.
- **ureter:** tube that connects a kidney and the bladder.
- **urethra:** tube through which urine is excreted.
- **vas deferens:** duct that propels sperm from the epididymis to the urethra.

Basic Principles

The urinary tract for the removal of liquid waste from the body consists of several organs. The kidneys, located in the lower back, filter metabolic waste products and excess water from the blood and convert them into urine. Urine passes through tubes called ureters into the bladder, where it is stored until it is excreted through the urethra. In males, the urethra passes through the penis. If metabolic waste products such as urea, uric acid, and creatinine were not removed, they would accumulate in the blood and poison the tissues throughout the body.

The male reproductive system consists of several organs. The testicles, or testes, produce sperm and testosterone as well as other male sex hormones. The sperm mature in the epididymis, which connects the testicle to the vas deferens. The vas deferens propels sperm toward the urethra. Along the way, the sperm mix with alkaline fluid from the prostate and fluid excreted from seminal vesicles. The sperm and semen are expelled through the urethra during ejaculation. Urology studies the structure and function of these related organs. Urologists are physicians who are consulted for the treatment of urinary tract infections, kidney stones, prostate enlargement, and urological cancers; the correction of urogenital birth defects; and the management of stress incontinence (involuntary urination), male infertility, and erectile dysfunction.

Background

Some urological treatments have been in evidence since early times. Catheterization, for example, is one of the first therapeutic interventions. Early civilizations used catheters made of onion stalks, wooden tubes, and metal cylinders to drain painfully distended bladders.

It was not until the twentieth century, however, that urology came into its own as a medical specialization. In 1908, Nobel Prize laureate Paul Ehrlich discovered that salvarsan, an arsenic compound, was an effective treatment for syphilis (a venereal, or sexually transmitted, disease with devastating effects on the urological system) and less toxic than the mercury compounds that had been used previously to treat this disease. This therapy was significant to urologists because they typically treated venereal diseases. In 1910, Hugh Hampton Young developed a novel technique for surgically treating benign prostatic hyperplasia (an enlarged prostate) using a perineal approach. As a result, he became known as the father of American urology. Then, in 1935, Frederick E. B. Foley introduced a rubber balloon catheter that did not require bandages or medical tape to keep it in place.

The overwhelming number of veterans with spinal cord injuries returning home from World War II created a need for advancements in urology and established it as a recognized and respected medical specialty. In 1973, when F. Bradley Scott developed a device to be implanted into a penis and pumped with saline to achieve an erection, urology began to expand into the area of sexual dysfunction as well as the treatment of disease.

In 1986, the U.S. Food and Drug Administration (FDA) approved the first prostate-specific antigen test for prostate cancer screening and monitoring. In 1997, the drug Flomax (tamsulosin) was introduced to improve urination for men with benign prostatic hyperplasia. In 1998, Viagra (sildenafil citrate) was introduced by Pfizer for the treatment of erectile dysfunction. The subsequent popularity of this drug and others designed to treat erectile dysfunction expanded not only the domain of urology but also the validation of disease states as a whole—particularly after major medical insurers began to cover these drugs and thereby to acknowledge the treatment as medically necessary.

How It Works

Urology uses a broad range of technology and processes to diagnose urological conditions. The most important of these are endoscopy and several imaging techniques.

Endoscopy. The development of instruments that could help physicians visualize the urinary tract and thereby enable them to diagnose and treat patients was crucial to the advancement of urology as a medical specialty. In 1805, German physician Philip Bozzini created the Lichtleiter, a tube for viewing the inside of the body, using a wax candle as a light source. This instrument is sometimes referred to as the first laryngoscope.

In 1879, German urologist Maximilian Nitze and Viennese instrument maker Joseph Leiter introduced the cystoscope, a small endoscope for directly viewing urethras and bladders. It consisted of an incandescent platinum wire loop heated by electricity and cooled by ice water for internal illumination and a system of magnifying lenses.

In 1926, in New York City, urologist Maximilian Stern developed the resectoscope, an instrument containing an electrified tungsten wire loop that could be manipulated to clear urinary tract obstructions. The loop was electrified by high-frequency alternating current, and it vaporized cells as it passed through tissue. This design was later modified by American physician Theodore Davis, who used his engineering background to strengthen the parts, enhance the loop, and add insulation. He also introduced a dual-action foot pedal to control the current for either cutting or cauterizing.

The biggest development in this field was the Hopkins rod lens system, introduced in 1959. The inventor was Harold Hopkins, a British professor of applied optics. Hopkins had earlier developed zoom lenses, which can change their degree of magnification without having to be refocused. He used flexible glass rods to transmit light internally from an external source. These rods provided more illumination than a small lightbulb and were even smaller, so that an endoscope using the rod lens system could explore smaller spaces. The resulting endoscopes were maneuverable, sterilizable, and durable, and the light source did not burn the patient.

Imaging Modalities. When direct visualization is not sufficient or possible, urologists rely on other imaging modalities such as ultrasonography, computed tomography (CT), and magnetic resonance imaging (MRI). These imaging modalities use computer reconstruction software programs to generate three-dimensional information from two-dimensional images and thereby function as virtual endoscopies. Imaging provides a noninvasive manner in which to evaluate the urinary tract. It is especially advantageous for watching for the recurrence of bladder cancer as it does not have the risks of repeated cystoscopies: infection, trauma, scarring, and pain. However, imaging lacks the sensitivity to replace cystoscopies altogether.

Although still in trial, wireless capsule endoscopy shows promise for use in urology. This technology consists of a wireless transmitter and a camera contained in a small capsule that the patient swallows. The capsule can be steered by external magnets. Using continuous real-time images, the capsule can be guided to its targeted destination. One potential application of a version of this technology is the delivery of chemotherapy directly to the bladder following bladder cancer surgery.

The application of technologies from seemingly unrelated fields has produced advancements in endoscopy. Miniaturization may lead to microendoscopy, and robotics may make remote control of internal scouts possible. These and further developments in endoscopy technologies are designed to reduce the adverse effects of diagnosis and treatment on patients and improve the course of recovery from urological diseases.

APPLICATIONS

Advances in urology have resulted in a host of technologies and devices to treat urological conditions.

Prosthetic Devices. In urology, a prosthetic device almost always refers to a penile implant, which is a treatment option for men with erectile dysfunction for whom other treatments have failed. About 30 million men in the United States have erectile dysfunction, and about one-third of them do not respond to other treatments. Implantation requires a surgical procedure, which is why other options are tried first. Implants have several advantages. They do not require the doctor visits for prescriptions and monitoring that are necessary with pills, creams, suppositories, and injections, and they are available for immediate use and are aesthetically pleasing because they prevent the penis from contracting when cold. There are two kinds of implants, flexible and inflatable. The less popular choice is a flexible prosthesis, consisting of two firm cylinders made from a silicone elastomer that is able to bend without breaking. These cylinders are surgically implanted within the two erection chambers (corpora cavernosa) of the penis. The implant allows erectile function to return by reducing the cavernosal volume to be engorged, inhibiting venous drainage, and increasing the intrapenile pressure. Although the prosthesis may be bent down for urination and up for sexual intercourse, its constant size and rigidity make it seem less natural.

The inflatable prosthesis consists of two fluid-filled cylinders that are surgically implanted within the two erection chambers of the penis. These cylinders are connected by tubing to a pump that is implanted within the scrotum. To achieve an erection, the pump transfers fluid to the cylinders, increasing the intrapenile pressure. Following ejaculation, the fluid returns to the pump from the cylinders. The adjustable size and rigidity make this prosthesis seem more natural. This type of implant may experience mechanical failure; there is a 10 to 15 percent failure rate in the first five years. If a cylinder leaks, the fluid inside is saline, which is absorbed by the body without harm. However, a failed prosthesis must be replaced or at least removed in a subsequent operation. An implant procedure may be performed on a variety of patients, including young men who are trauma patients, men left impotent after having their prostate removed, and men with age-related erectile dysfunction. One of the biggest considerations is whether the man is otherwise healthy enough for general anaesthesia. Surgeons attempt to minimize the size of the incisions to preserve the sensitivity of the natural tissues. A penile prosthesis is not intended for penile enlargement of an otherwise perfectly functioning organ.

Although rarely used, another urological prosthetic is an artificial testicle, a saline-filled silicone ball that when implanted, mimics the appearance and movement of a natural testicle. It was developed for use in reconstructive surgery on men whose testicle had been removed in the treatment of cancer. It carries a risk of rupture from a sports injury, and one in three cases requires a surgical adjustment within the first year. This prosthetic is not medically necessary and serves only a cosmetic purpose. Perhaps a model will be developed that will provide long-term delivery of testosterone.

Neurourology. Another form of prosthesis under development is an implantable neuroprosthetic device for the treatment of urinary incontinence. An existing treatment for urinary incontinence is electrical stimulation of the nerves and muscles in an attempt to restart them so that they resume their normal function and regulation. However, this stimulation is from an external source and deliberately delivered. As research progresses, scientists hope to create an internal source of electrical stimulation that would be delivered on a regular schedule or in response to a particular signal. Trials have been conducted on subjects, most of whom were not paraplegic but experienced urinary incontinence. Research is also ongoing into rewiring the nerves of patients with spinal cord injuries to help them regain bladder control and eliminate their need for self-catheterization. The application of artificial intelligence, in the form of artificial neural networks, has been successfully used in several fields of medicine and is being applied to urology. Many studies have applied artificial neural networking to the diagnosis and staging of prostate cancer with promising results. It is an analytical way to uncover nonlinear relationships among various clinical parameters, from which to determine optimal therapies and predict outcomes.

Drug Therapies. The pharmaceutical industry has developed numerous prescription drugs for the treatment of urological conditions. Drugs for the treatment of benign prostatic hyperplasia (enlarged prostate) include tamsulosin, Avodart (dutasteride), Jalyn (dutasteride and tamsulosin), and Uroxatral (alfuzosin hydrochloride). Drugs for the treatment of prostate cancer include Jevtana (cabazitaxel), Eulexin (flutamide), Zoladex (goserelin), and Gemzar (gemcitabine hydrochloride). Drugs for the treatment of overactive bladder include Toviaz (fesoterodine), Sanctura (trospium), Vesicare (solifenacin), Detrol LA (tolterodine tartrate), and Ditropan XL (oxybutynin chloride). Drugs for the treatment of cystitis and uncomplicated urinary tract infection include Cipro (ciprofloxacin), Elmiron (pentosan polysulfate), Levaquin (levofloxacin), and Doribax (doripenem). Drugs for the treatment of erectile dysfunction include sildenafil citrate, Cialis (tadalafil), and Levitra (vardenafil).

FUTURE PROSPECTS

Urology will remain a necessary clinical specialty: The development of calculi is a common occurrence in industrialized nations, and prostate disease affects 75 percent of men fifty years of age or older. Neither of these can be prevented with a pharmacological intervention. Because urologists provide both surgical and nonsurgical treatments, they must remain knowledgeable about advances in physiology, pharmacology, and technology.

Urologists are continuing to develop their surgical skills to achieve improvements in appearance, function, and sensation. Internal penile pumps remain a popular choice versus flexible implants, vasodilator pills, or external pumps and creams; more than 250,000 men have already chosen to have such a device implanted. Surgeons will continue to correct birth defects and repair urogenital trauma. Their aim is to restore function or at least preserve it. Scientists are using tissues harvested during surgery for ongoing cancer research, focusing especially on the genes underlying the disease. They are also trying to understand and block the mechanisms of cancer recurrence. Other research groups are studying kidney stones to determine the underlying factors of calculi formation to devise methods of prevention. Diagnostic testing is advancing to facilitate earlier detection of prostate cancer and sexually transmitted diseases. Imaging equipment and technology will continue to develop, offering modalities with stronger clarity and contrast in real time. These will be used for earlier detection of abnormalities as well as closer monitoring of changes and better planning of surgical interventions.

New therapies will continue to be sought. For example, researchers are investigating the use of botulinum toxin (commonly known as Botox) in the treatment of urinary incontinence. New solutions for male infertility and enhancement of sperm conditions to increase the efficiency and effectiveness of fertilization are being studied.

Bethany Thivierge, MPH

FURTHER READING

Chan, Evelyn C. Y., et al. "Informed Consent for Cancer Screening with Prostate-Specific Antigen: How Well Are Men Getting the Message?" *American Journal of Public Health* 93, no. 5 (2003): 779-785. Presents balanced information on why prostate-specific antigen (PSA) screening is controversial and yet beneficial.

Field, Michael J., Carol Pollock, and David Harris. *The Renal System: Systems of the Body Series.* 2d ed. New York: Churchill Livingstone, 2010. Covers the basic anatomy, physiology, and biochemistry of the renal and urogenital system.

Genadry, Rene, and Jacek L. Mostwin. *A Women's Guide to Urinary Incontinence.* Baltimore: The Johns Hopkins University Press, 2007. Provides information about a common yet embarrassing condition and encourages women and their families to seek compassionate treatment.

Gomella, Leonard G. *The Five-Minute Urology Consult.* 2d ed. Philadelphia: Lippincott Williams & Wilkins, 2009. Offers immediate, practical information on a broad range of urological topics.

Tanagho, Emil A., Jack W. McAninch, and Donald R. Smith. *Smith's General Urology.* 17th ed. New York: McGraw-Hill Medical, 2008. This comprehensive textbook contains visual aids, including clinical images.

Walsh, Patrick C., and Janet Farrer Worthington. *The Prostate: A Guide for Men and the Women Who Love Them.* New York: Warner Books, 1997. This comprehensive yet easy-to-read classic explains diagnostic tests and treatments in a reassuring manner suitable for patients and their families.

Web Sites

American Board of Urology http://www.abu.org
American Medical Association http://www.ama-assn.org
American Urological Association http://www.auanet.org
Kidney and Urology Foundation of America http://www.kidneyurology.org
Society for Pediatric Urology http://www.spuonline.org
Urological Sciences Research Foundation http://www.usrf.org

See also: Endocrinology; Reproductive Science and Engineering.

FASCINATING FACTS ABOUT UROLOGY

- An adult's kidneys filter 50 gallons of blood and produce about 0.5 gallon of urine each day.

- Many believe that the children's song "Frère Jacques" was written about Frère Jacques Beaulieu, a seventeenth-century French monk who specialized in surgically removing urinary stones. He performed more than 5,000 lithotomies in thirty years.

- American inventor and statesman Benjamin Franklin designed a silver coil catheter for his brother in 1752.

- In 1831, American physician Philip Syng Physick removed more than 1,000 stones from the bladder of John Marshall, chief justice of the United States Supreme Court. The seventy-six-year-old Marshall recovered and returned to the bench.

- In the nineteenth and early twentieth centuries, men with chronic bladder outlet obstruction commonly self-catheterized with tubes they carried in hatbands, umbrella handles, and hollow walking sticks.

- Railroad magnate James Buchanan "Diamond Jim" Brady donated the funds to found a urological institute at The Johns Hopkins University after urologist Hugh Hampton Young surgically relieved Brady's chronically inflamed prostate on April 12, 1912.

- German aerospace engineer Claude Dornier noticed pitting on the surface of an airplane as it approached the sound barrier; the minute destruction was the result of the shock wave created in front of moisture. This led to the development in 1984 of extracorporeal shock wave lithotripsy, a noninvasive procedure to demolish kidney stones with intense shock waves.

Virology

FIELDS OF STUDY

Microbiology; general biology; cell biology; chemistry; molecular biology; biochemistry; biophysics; microbial genetics; immunology; genetic engineering.

SUMMARY

Virology is a scientific field emphasizing the study of submicroscopic entities known as viruses. Because genetic material within viruses can be easily manipulated and viruses replicate at a relatively rapid rate, research in molecular genetics between the 1940's and 1970's largely consisted of the study of viruses. Scientific application within virology has led to understanding cell processes such as the molecular basis for cancer, as well as developments in the field of biotechnology. Genetic engineering, the process of altering the genetic makeup of organisms using viral vectors, has also provided a means for production of numerous pharmaceuticals.

PRINCIPAL TERMS

- **antibody:** protein produced by lymphocytes following exposure to an antigen or anything perceived by the body as foreign.
- **cell culture:** animal or plant cells grown in the laboratory. used for growth and study of viruses.
- **host range:** animals or specific tissues and cells within the animal that a specific virus is capable of infecting.
- **molecular cloning:** isolation of specific gene or fragment of DNA and its insertion into a viral vector.
- **plaque assay:** method of quantifying viruses in which a viral solution is used to infect a layer of cells growing in a dish. sites of infection by viral particles will appear as "holes" in the layer.
- **restriction enzymes:** enzymes used to cut DNA to allow for insertion of genetic material.
- **transfection:** cellular uptake of "naked" DNA from the environment. Used to introduce DNA vectors without the use of viral capsids, bypassing the problem of host range.
- **virus particle:** genetic material, either DNA or RNA, enclosed within a protein capsid or coat; the capsid may also be surrounded by a membrane or envelope.

BASIC PRINCIPLES

Viruses are intracellular parasites that infect all types of organisms, including bacteria, plants, and animals. Because viruses are commonly associated with human diseases, the public usually views these biological entities only in this context. However, the ability of viruses to infect cells and, in some cases, to integrate within the host genetic material has led to the development of new technologies for insertion or replacement of defective genetic material. Although all cells are capable of being infected by viruses, viruses themselves often exhibit a specific host range. Bacterial viruses infect only specific strains of bacteria, and animal viruses usually are restricted to a specific species or tissues within the species. Viruses contain on their surfaces proteins that determine which cells can be infected. In a manner analogous to a lock and key, viruses attach to specific receptors expressed on the surface of the cell. Viral vaccines induce the body to produce neutralizing proteins, or antibodies, which bind to the surface of the virus and block its adsorption into the cell.

HOW IT WORKS

Viral diseases have probably existed since the early years of human civilization, but viruses were not discovered until the late 1800's. The cause of many human diseases was unknown until the development of the germ theory of disease in the mid-nineteenth century. During the golden age of microbiology, from about 1875 to 1900, scientists were able to determine that microorganisms caused certain diseases. French scientist Louis Pasteur and German physician Robert Koch were prominent among these early microbiologists. The work of Pasteur, Koch, and their associates involved isolating and growing bacteria in the laboratory and demonstrating that these organisms were able to cause specific diseases in animals.

Pasteur, working with rabies in the early 1880's, noted that unlike bacteria, the agent that caused rabies did not grow on culture media and was capable of passing through minute filters. The agent was called a "virus," from the Latin word for poison. During the 1890's and early twentieth century, scientists demonstrated that viruses were capable of replicating, indicating that they were a life-form rather than a poison. A number of human diseases, including smallpox, influenza, and rabies, were shown to be caused by these agents.

The development of the electron microscope during the 1930's allowed viruses to be observed. The first viruses to be extensively studied were bacteriophages, viruses that infect bacteria. Because some bacteriophages kill bacteria, it was at one time thought that they could be used to treat bacterial infections; the plot of Sinclair Lewis's book *Arrowsmith* (1925), in which physician Martin Arrowsmith used "phage" to deal with an outbreak of plague, was predicated on that idea.

Much of the study of viruses has revolved around finding a way to treat viral diseases or prevent them through vaccines. However, as the mechanism by which they infect people became better understood, scientists began to develop ways in which they could use viruses to treat other diseases and conditions.

Viral Vaccines. The first human vaccine against a viral disease was created in the 1790's through a serendipitous discovery by British physician Edward Jenner. He observed that people previously infected with material from lesions on the udders of cattle (cowpox) became immune to smallpox. Although Jenner was unaware of why cowpox provided immunity to smallpox, scientists later determined that the cowpox virus contains proteins that cross-react with those of smallpox and produce immunity in those exposed to it. The process of immunization became known as vaccination, derived from the Latin word *vacca*, which means cow.

Although the smallpox vaccine uses a naturally cross-reactive virus, modern viral vaccines take advantage of the ability of artificially attenuated or killed viruses to provide immunity. Attenuated viruses, such as those used to immunize against rabies, polio (Sabin vaccine), measles, and mumps, are viruses that have had their disease-causing ability reduced through animal or cell culture passage, and they produce immunity to the disease although they are incapable of causing illness. The Salk vaccine against polio uses a virus that has been killed through chemical treatment.

The theory behind any immunization is that components of the vaccine induce the recipient to produce antibodies, which provide protection against any subsequent exposure to that agent. Viruses have their own means to avoid neutralization by antibodies. Influenza virus is prone to two forms of change: Antigenic drift, mutations in the surface hemagglutinin protein that is the target of the antibody response, and antigenic shift, the recombination of the human strain with animal influenza strains to create an entirely new variety of influenza virus. The 2009 swine flu (H1N1 flu) was the result of such a shift. Rhinoviruses, associated with most cases of the common cold, do not mutate significantly. However, more than one hundred strains of the virus exist, and immunity to one does not confer immunity to the other strains.

Genetic Manipulation. Genetic diseases generally have their origin in mutations that affect either the proper expression of specific genes or the function of the gene product. Although in most cases the mutation affects only a single gene, the effect may be significant because many gene products are pleiotropic, producing multiple effects in the organism. An example is the *FBN1* gene, which encodes fibrillin, a protein necessary for proper connective tissue formation. Mutations in the gene result in Marfan syndrome, a weakening in connective tissue throughout the body, which can affect the limbs, aorta, or particular organs.

The principle behind gene therapy is that replacement of the gene containing the mutation with a normal copy may reverse the effects of the genetic error. The idea applies in particular to genetic defects that are associated with a single gene, or monogenic diseases. The challenge has been how to introduce the correct gene into the tissues that are affected in a manner that minimizes side effects yet produces a permanent correction. Viruses are particularly useful in serving as gene vectors because some have the ability to infect multiple cells and multiple types of tissue and, in some cases, integrate into the host genome, becoming a permanent part of the genetic material.

Because viruses by their nature generally alter or kill the cells they infect, they must first be rendered harmless to be used in gene therapy. This is carried out by first deleting the genes necessary for viral replication. For example, to render human adenoviruses harmless, two specific genes, *E1* and *E3*, are deleted to block virus replication. Deletion of an additional gene called *F1*, which encodes the surface fiber that allows the virus to attach, reduces the danger of inflammation following the virus's introduction into a host. Deletion of the fiber also reduces the host range, limiting the variety of cells that can be infected. Cutting of the DNA with the proper restriction enzymes allows the desired therapeutic gene to be inserted at the site once occupied by the *E1* and *E3* genes, creating recombinant DNA. Cell cultures that can provide the necessary functions for duplication of the recombinant DNA are then transfected with the engineered viral DNA.

Anticancer Therapy. The limitation of host range for viral infection has been applied to try to develop anticancer therapies. The theory is that because viruses can infect only certain cells, viruses could be altered to express proteins detrimental to the survival of cancer cells. Gene therapy using viruses has taken a number of approaches. Several clinical trials have involved the introduction into cancer cells of retroviruses carrying genes for cytokines, proteins that stimulate an immune response directed against the neoplasm. A second approach uses viruses that infect only specific types of cells such as nerve cells. For example, viral agents such as the herpesviruses naturally target tissues found in the nervous system. Genetically altered herpesviruses could in theory be used to kill only tumor cells that arise in the brain or spinal cord.

GENE INTRODUCTION AND INTEGRATION

The ability to insert specific genes into viral vectors provides a mechanism for introducing genetic material into individual cells or tissues within organisms. Several viruses, each with its own advantages and disadvantages, have served as such vectors. These viruses have in common the ability to infect a broad range of hosts, allowing for a wide range of applications in the field of genetic engineering. Replacement of defective genes in human cells remains the goal, but only a few therapies have been developed to the point where they could be tested in clinical trials.

The choice of virus is generally based on the size of the genetic material to be introduced and the likelihood of any immune or inflammatory response. For larger segments of DNA, ranging between 7,500 and 35,000 base pairs, enough to encode a protein or proteins with about 2,500 to 10,000 amino acids, the primary choices as vectors are either vaccinia virus or adenovirus. Vaccinia, a large enveloped double-stranded DNA virus, has previously served a role as the vaccine against smallpox. When vaccinia is used as a vector, viral genes necessary for replication are deleted before inserting the desired genetic material, rendering the virus unable to replicate. Unfortunately, vaccinia frequently elicits a significant inflammatory response, limiting its usefulness.

Adenoviruses are relatively large nonenveloped DNA viruses with a protein capsid. Spikes or fibers attached to the capsid determine the host range for infection. There are about fifty serotypes of adenoviruses, some of which are associated with human respiratory and gastrointestinal infections. The most commonly used adenovirus strain for genetic studies is serotype 5. Recombinant adenovirus DNA has been used in limited clinical trials to treat patients with cystic fibrosis and ornithine transcarbamylase (OTC) deficiency. Cystic fibrosis, the most common inherited genetic disease in the West, is associated with a mutation in the gene that encodes a regulator protein necessary for proper transport of ions in and out of cells. OTC deficiency is a metabolic disorder in which urea metabolism is affected. About 25 percent of clinical trials using a viral vector have used adenoviruses. Lentiviruses such as HIV have a vector capacity significantly smaller than that of adenoviruses, but they have the ability to integrate into the host genome, resulting in the recombinant gene becoming part of the host genetic material. In animal trials, lentiviruses have introduced growth factor genes into mouse cells as well as the gene encoding Factor VIII, which is lacking in the most common forms of human hemophilia. Preclinical trials have demonstrated that recombinant HIV can be used to replace defective genes in diseases such as cystic fibrosis and muscular dystrophy. About 25 percent of clinical trials have used retroviruses as the vector, including most of the original clinical trials that used viral vectors.

The ability of lentiviruses to integrate into the host genetic material, one of the desirable features of such viruses, does have its drawbacks. Integration is not random and frequently takes place within the introns (DNA regions in a gene that is not translated into protein) of preexisting genes, some of which are necessary for proper cell function. This can produce severe side effects, such as those that occurred during trials in France that attempted to insert an adenosine deaminase gene to cure a form of severe combined immunodeficiency syndrome (SCID). Although nine patients did show improved immune function, three developed T-cell leukemia, and one person died.

Adeno-associated viruses are naturally defective viruses that require a helper adenovirus to replicate normally. However, they are capable of infecting a range of cells and can integrate at specific sites in human chromatin. Adeno-associated viruses have been used experimentally to kill certain forms of breast, cervical, and prostate cancer cells. As viral vectors, they have been used to introduce a gene for production of insulin and the genes for both Factor VIII and Factor IX, lacking in humans with certain forms of hemophilia (hemophilia A and B, respectively), into mice genes that encode erythropoietin, a glycoprotein that induces the bone marrow to increase red blood cell production. Adeno-associated viruses have been associated with only a small proportion of all clinical trials. Other vectors have used naked plasmid DNA, vaccinia, and other poxviruses, as well as other types of viruses.

Future Prospects

Researchers who attempt to control viral disease must address both the evolving nature of viruses and the new or unusual viral diseases that develop. Many viruses, including influenza and HIV (human immunodeficiency virus), mutate at a rapid rate. Newly emerging viral diseases include Ebola hemorrhagic fever and Lassa fever (both in Africa) and hantavirus (in the United States).

Influenza and HIV provide particular challenges. The potential virulence of influenza was demonstrated during the 1918 worldwide pandemic in which an estimated 50 million people died. Any influenza vaccine that would provide protection against multiple strains would have to address the problem of both genetic drift and genetic shift. The inability to grow the virus in anything but embryonated chicken eggs rather than cell culture also significantly increases the lead time necessary for large-scale vaccine production.

HIV, the etiological agent for AIDS (acquired immunodeficiency syndrome), provides its own challenges. HIV mutates at a high rate immediately following infection of human lymphocytes, producing dozens of varieties in weeks. The high rate of mutation is the primary reason that no AIDS vaccine has thus far proven effective.

The encroachment of human civilization into previously isolated areas has exposed humans to new or unusual forms of viruses. In the 1980's, people in parts of western Africa were exposed for the first time to rodent viruses that produced Lassa fever in people. Luckily, the outbreaks were limited. The movement of people into new areas in both central and western Africa during the 1970's probably brought nonhuman primates infected with HIV in contact with people, and the virus jumped from one species to the other. Although vaccines against newly emerging viruses such as Ebola hemorrhagic fever or Lassa fever could probably be manufactured, the cost of addressing what so far have been localized outbreaks is prohibitive.

Richard Adler, PhD

Further Reading

Allen, Arthur. *Vaccine.* New York: W. W. Norton, 2007. The story of vaccines and the controversies that have long surrounded their use.

Edelstein, Michael, Mohammad Abedi, and Jo Wixon. "Gene Therapy Clinical Trials Worldwide to 2007—An Update." *The Journal of Gene Medicine* 9 (August, 2007): 833-842. Summary of the results of more than 1,300 clinical trials using viral vectors for insertion of replacement genes.

Madigan, Michael, et al. *Brock Biology of Microorganisms.* 12th ed. San Francisco: Pearson Benjamin Cummings, 2009. Several chapters in this textbook of microbiology address the role of viruses in disease as well as in biotechnology.

Strauss, James, and Ellen Strauss. "Virus Vector Systems." In *Viruses and Human Disease.* New York: Elsevier, 2008. Examines the molecular biology of animal viruses and their role in human disease. Addresses the role of viruses as vectors for genetic engineering. Includes numerous illustrations of procedures.

WEB SITES

American Society for Virology http://www.asv.org

Centers for Disease Control and Prevention Vaccines and Immunizations http://www.cdc.gov/vaccines

Pan American Society for Clinical Virology http://www.virology.org/links.html

See also: Immunology and Vaccination; Pathology.

FASCINATING FACTS ABOUT VIROLOGY

- Blossom was the name of the cow from which Edward Jenner isolated cowpox virus, which he used to immunize people against smallpox.
- The association of viruses and cancer was demonstrated by Peyton Rous in 1911, when he used cellfree extracts from a tumor extracted from a Plymouth Rock chicken to induce tumors in healthy chickens. Fifty-five years later, four years before his death at the age of ninety, Rous was awarded a Nobel Prize.
- The Raggedy Ann doll was created in 1915 by a New York illustrator, John Gruelle, for his daughter who became ill following a smallpox vaccination.
- The bacterium *Haemophilus influenzae* does not cause influenza. The species name resulted from its isolation from influenza patients.
- Restriction enzymes, critical to genetic engineering, were discovered during experiments studying the resistance of bacteria to viral infections

---

VISION

FIELDS OF STUDY

Anatomy, biochemistry, cell biology, evolutionary science, neurobiology, physiology

SUMMARY

Vision is the ability of animals to analyze light information in their environment. Given the many unique visual worlds in the animal kingdom, there are many different types of eyes and light gathering mechanisms in them.

PRINCIPAL TERMS

- **accommodation:** changing the shape of the lens in order to keep objects at different distances focused on the retina
- **fovea:** area, often a pit in the retina, of maximal acuity, where each photoreceptor has its own nerve cell, as opposed to many receptors converging on one nerve cell
- **opsin:** a membrane-bound protein or pigment, which absorbs light
- **photon:** a unit used to describe light intensity
- **photoreceptor:** cell containing membranes which house light-sensitive pigments retina: the light-sensitive film at the back of the eye

BASIC PRINCIPLES

Despite the fact that it is often taken for granted, vision is one of the more interesting and complex sense systems. Different animals have, over the course of their evolution, independently fine-tuned their visual systems to adapt to their unique environments and needs. The possession of a good visual system can be the factor dictating a species' survival. Vision is essential for many animal behaviors, such as foraging for food, prey avoidance, and mate choice. Upon considering the many different species in the world, it is evident that there must be many different types of vision; for example, a fish in its unique underwater environment would have a vastly different visual world than an insect in the rainforest.

When animals forage for food, vision is used along with many other senses, especially the senses of smell and hearing, to make a good food choice. Many animals scan a visual field before deciding to forage;

thus, the visual system must be acute, enabling the animal to understand what is in its field of view, often in a very short time. The brain must be able to determine such things as shape, form, and color of objects. Many animals utilize color vision when foraging. Cues as to the suitability of a food choice are often indicated by color, for example, the difference between a poisonous berry and an innocuous one. Eye placement (the exact positioning of eyes on the head) is crucial; appropriate positioning of the eyes from each other, as in humans, makes binocular vision possible. Binocular vision gives the viewer a sense of depth in the field of view, critical for many animals when catching prey. Species with laterally spaced eyes have a smaller amount of binocular vision, and some of their visual field has to be viewed monocularly.

Vision is essential in many animals' communication with each other. Body markings and displays in many species are often used as mating signals. Birds provide a striking example of visual communication in animals, where often specific body markings are used to attract mates. Some birds attract mates by elaborate nest construction; the bird with the most elegant nest attracts a mate and ensures reproduction and the passing of its genes to the next generation.

THE PHYSIOLOGY OF VISION
In simple terms, the visual system takes a signal in the form of light and translates it into a chemical change and later a nervous impulse in the brain; this nervous impulse is what the animal perceives as sight. The two main characteristics of eyes, no matter how complex or simple, are light-sensitive receptors (photoreceptors), and a mechanism to control light. In simple eyes, the light-sensitive receptors make up a layer known as the retina. The nature of a photoreceptor is dictated by opsin, the photosensitive proteinaceous pigment present within the membranes in the photoreceptor. The photoreceptors are variable in size, shape, and content. There are two types in vertebrates: rods sensitive to light and as such makeupthe majority of the retina of nocturnal species, who require as sensitive a system as possible. The photopigment in rods is called rhodopsin. Cones are less sensitive to light, but there are different types which are sensitive to light of different wavelengths (or colors), and can give animals color vision. There are many different types of cones, and hence many types of color vision; possession of these cone types and their specific positioning within the retina is an evolutionary adaptation particular to animals who benefit from color vision.

There are two main types of eye design in the animal kingdom: simple eyes and compound eyes. Simple eyes have a single layer of photoreceptors, which, in the least complicated case, form a cup of photosensitive material. The human eye, with its complex light-focusing apparatus, is still a simple eye. Compound eyes, which are present in most insects, have many separate optical units, called ommatidia. Each ommatidium has a rhabdom, containing a group of up to nine tubular rhabdomeres, with ciliary or microvillar (finger-shaped) photoreceptors. The orientation of groups of photosensitive cilia is structured, often with pairs of rhabdomeres organized at right angles to each other. This is especially key in analysis of polarized light.

Many visual systems have mechanisms to control light. Restriction of the amount of light entering the eye is useful; the opening through which light enters is referred to as an aperture, as in a camera. Many animals have a contractible iris which constricts and dilates to control light entry through the pupil. For example, in dim light conditions the iris can dilate and let in as much light as possible. Some eyes have lenses which enable light to be focused on the retina, allowing for better resolution of objects in the visual field. Many are able to change the shape of the lens in order to bring an image to focus on the retina, a process called accommodation. Other animals use a cornea to bend light onto the retina, although the cornea is rigid and cannot change shape. On the other hand, there are many species with much simpler eyes, which do not possess any kind of light control apparatus.

Upon reception of light by a photoreceptor, a biochemical cascade of events occurs within the photoreceptor itself, which amplifies the original signal received. The result of this cascade is a nervous signal which proceeds through many neural layers to the brain. Throughout most of the retina in the simple eyes of vertebrates, several photoreceptors connect to one neuron (convergence), but there is often an area of the retina where one photoreceptor connects to one neuron. This area is called the fovea and is the part of the retina which has best acuity. The area within the retina which comprises the fovea is variable. Fish possess what is known as a "visual streak" fovea, which gives excellent vision along a horizontal slice of the visual field. This is an ideal adaptation for fish given their particular habitat.

Information is passed through the nervous system in layers of neurons. The retina is, in fact, an extension of the brain, and contains many nerve cells. The exact arrangement and mechanism of action of neural cell types, and the precise pathway to and within the brain, differ greatly from species to species. Phototransduction in vertebrates is different from that of invertebrates, from the arrangement of the retina to the biochemical cascade and the types of neurons involved.

Color Vision

Reflected natural light has a unique property that maybe exploited by the visual system of an animal, namely its wavelength, or color. Color vision can be vital for a species in regard to mate choice and foraging. Photoreceptors can be sensitive to different wavelengths of light, or colors; in vertebrates the color-sensitive receptors are called cones because of their shape. The maximum sensitivity of a photoreceptor is dictated by the nature of the photo-sensitive protein (opsin) within the receptor. Opsins can be classified according to the approximate wavelength of light that stimulates them maximally. Opsins have been studied that are sensitive to light from the ultraviolet region of the visible spectrum all the way to the far red region.

To have the possibility of color vision, an animal must possess at least two photoreceptors with differing sensitivities. The brain must then be able to compare the outputs of both these receptors and discriminate color. Often more than two types of photopigment type are present, as in the fish retina, which results in very complex color vision, including sensitivity to ultraviolet light. Color vision has been shown to exist in many animals within the animal kingdom. Positioning and distribution of the different types of photoreceptor within the retina are also key to the ability to discriminate color. In vertebrates, this aspect of retinal structure is called the cone mosaic, and its nature is often closely related to some behavioral aspect of the animal in question. Color vision also requires a central processing system that can decode the various light signals and turn them into a brain output which is useful to the animal.

—*Lucy A. Newman*

See also: Anatomy; Brain; Eyes; Noses.

Further Reading

Baylor, D. "How Photons Start Vision." *Proceedings of the National Academy of Science* 93 (January, 1996): 560-565. This review explores the biochemistry of phototransduction and is fairly detailed.

Jacobs, G. H. "The Distribution and Nature of Color Vision Among the Mammals." *Biological Reviews* 68 (1993): 413-471. This review covers the topic of color vision and addresses the different types of color vision in many mammalian species.

Lodish, H., A. Berk, S. L. Zipursky, P. Matsudaira, D. Baltimore, and J. Darnell. *Molecular Cell Biology*. 4th ed. New York: W. H. Freeman, 2000. Chapter 21 offers in-depth coverage of phototransduction from a biochemical point of view.

POLARIZED LIGHT VISION

Polarized light plays an important role in the visual world of many animals, including many species of fish and insects. These animals are able to use their visual systems to decode one particular aspect of polarized light, namely the angle of polarization. Light which is polarized has an electric vector which has a specific angle, or orientation. Natural light is unpolarized, but becomes polarized by scattering in air, water, or by reflection off surfaces. Scattering of the ultraviolet (UV) light in sunlight is quite predictable and produces obvious patterns in the sky, invisible to humans without a polarizing filter. These patterns are analyzed by some insects and are used as a kind of map for navigation, for example, to get back to the nesting area after a foraging trip. Similarly, the polarization pattern aids in orientation of the animal. Analysis of the skylight polarization patterns is possible because of the unique positioning of the microvillar photoreceptors within the eye. Pairs of microvillar photoreceptor populations are positioned perpendicular to each other. The light-sensitive molecules within the photoreceptor lie parallel to the axes of the microvilli. Each member of the pair of photoreceptors is maximally sensitive to a different angle of polarized light, and the combined response to a given angle of both members allows the animal to compute the exact angle of polarization. Enhancement of contrast is an advantage of polarized light vision used by fish, although the mechanism of detection of the light is obscure.

ULTRAVIOLET (UV) LIGHT VISION

Sunlight is the natural source of ultraviolet light, whose wavelength runs from less than 280 to 400 nanometers. For the sake of vision, however, one need only be concerned with UV light between 320 and 400 nanometers, since any light below these wavelengths becomes absorbed by air before it reaches earth. Many members of the animal kingdom are able to see in the ultraviolet region of the spectrum, due to their possession of a particular class of opsin molecule which is maximally sensitive to UV light. Humans also possess a UV-sensitive opsin, but are unable to actually see UV because the lens, which is yellow, absorbs UV light. There are certain behavioral advantages for the species who do possess UV vision. For example, fish use UV vision extensively in signaling and communication with members of the same species; many fish and birds have UV markings on their bodies which make it easier for other individuals to see them and which may also be involved in mate choice. There are also very strong UV markings on flower petals, which act as guides for pollinators such as bees.

Marshall, J. "Visual Function: How Spiders Find the Right Rock to Crawl Under." *Current Biology* 9, no. 24 (December, 1999): 918-921. An interesting article on polarization vision in spiders.

Solomon, E., L. Berg, D. Martin, and C. Villee. *Biology*. 3d ed. FortWorth, Tex.: Saunders College Publishing, 1993. Chapter 41 covers image formation and, briefly, phototransduction.

Zoology

FIELD OF STUDY

Zoology

SUMMARY

Zoology is the branch of biology devoted to the study of the animal kingdom. The study of zoology encompasses the analysis of and classification of animals.

PRINCIPAL TERMS

- **comparative anatomy:** the branch of natural science dealing with the structural organization of living things
- **ecology:** the study of the interactions between animals and their environment
- **embryology:** the study of the development of individual animals evolutionary zoology: the study of the mechanisms of evolutionary change and the evolutionary history of animal groups
- **morphology:** the study of structure; includes gross morphology, which examines entire structures or systems, such as muscles or bones; histology, which examines body tissues; and cytology, which focuses on cells and their components
- **phylogenetics:** the study of the developmental history of groups of animals
- **physiology:** the study of the functions, activities, and processes of living organisms
- **systematics:** the delineation and description of animal species and their arrangement into a classification
- **taxonomy:** the classification of organisms in an ordered system that indicates natural relationships
- **zoogeography:** the study of the distribution of animals over the earth

Basic Principles

Attempts at classification are known from documents in the collection of the Greek physician, Hippocrates, as early as 400 BCE. However, the Greek philosopher Aristotle (384-322 BCE) was the first to devise a system of classifying animals that recognized commonalities among diverse organisms. Aristotle arranged groups of animals according to mode of reproduction and habitat. After observing the development of selected animal groups, he noted that general structures appear before specialized ones, and he also distinguished between sexual and asexual reproduction. Aristotle was also interested in form and structure, and concluded that different animals can have similar embryological origins and different structures can have similar functions.

In Roman times, Pliny the Elder (23-79 CE) compiled four volumes on zoology widely read during the Middle Ages. Some scholars have deemed those volumes little more than a collection of folklore, myth, and superstition. One of the more influential figures in the history of physiology, the Greek physician Galen (c. 130-c. 201 CE), dissected farm animals, monkeys, and other mammals and described many features accurately, although scholars have noted that some of these features were then wrongly applied to the human body. His misconceptions, especially with regard to the movement of blood, remained virtually unchanged for hundreds of years. In the seventeenth century, the English physician William Harvey established the true mechanism of blood circulation.

How It Works

Until the Middle Ages, zoology was little more than a collection of folklore and superstition. However, during the twelfth century, zoology began to emerge as a science. The thirteenth century German scholar and naturalist St. Albertus Magnus refuted many of the superstitions associated with biology and reintroduced the work of Aristotle. The anatomical studies of Leonardo da Vinci in the fifteenth century have been noted as being far ahead of their time. His dissections and comparisons of the structure of humans and other animals led him to several important conclusions. For example, Leonardo noted that the arrangement of joints and bones in the leg are similar in both horses and humans, thus embracing the concept of homology, or the similarity of corresponding parts in different kinds of animals, suggesting a common grouping. A Flemish physician of the sixteenth century, Andreas Vesalius, is considered the father of anatomy for establishing the principles of comparative anatomy.

Throughout most of the seventeenth and eighteenth centuries, classification dominated zoology. The Swedish botanist Carolus Linnaeus developed a system of nomenclature still in use today, referred to as the binomial system of genus and species. Linnaeus also established taxonomy as a discipline. His work was built on that of the English naturalist John Ray and relied upon the form of teeth and toes to differentiate mammals and upon beak shape to classify birds. Another leading figure in systematic development of this era was the French biologist Comte Georges-Louis Leclerc de Buffon. The study of comparative anatomy was further developed by men such as Georges Cuvier, who devised a systematic organization of animals based on specimens sent to him from all over the world.

A cell is the smallest structural unit of an organism capable of independent functioning. Although the word "cell" was introduced in the seventeenth century by the English scientist Robert Hooke, it was not until 1839 that two Germans, Matthias Schleiden and Theodor Schwann, proved that the cell is the common structural unit of living things. The concept of the cell provided impetus for progress in embryology and animal physiology, including the concept of homeostasis, referring to the stability of the body's internal environment.

The formation of scientific expeditions in the eighteenth and nineteenth centuries gave scientists the opportunity to study plant and animal life throughout the world. The most famous scientific expedition was the voyage of the HMS *Beagle* in the early 1830's. During this voyage, Charles Darwin observed the plant and animal life of South America and Australia and developed his theory of evolution by natural selection. Although Darwin recognized the importance of heredity in understanding the evolutionary process, he was unaware of the work of a contemporary, the Austrian monk Gregor Mendel, who first formulated the concept of particulate hereditary factors, later called genes. Mendel's work was not widely disseminated until 1900.

Modern Zoology

During the twentieth century, zoology has become more diversified and less confined to such traditional issues as classification and anatomy. Zoology, broadening its span to include such areas of study as genetics, ecology, and biochemistry, has become an interdisciplinary field applying a wide variety of techniques to obtain knowledge about animal kingdom. The current study of zoology has two main focuses, taxonomic groups, and the structures and processes common to these groups. Studies of taxonomy concentrate on the different divisions of animal life. Invertebrate zoology deals with multicellular animals without backbones; its subdivisions include entomology (the study of insects) and malacology (the study of mollusks). Vertebrate zoology, the study of animals with backbones, is divided into ichthyology (the study of fish), herpetology (amphibians and reptiles), ornithology (birds), and mammalogy (mammals).Taxonomic groups also subdivide paleontology, the study of fossils. In each of these fields, researchers investigate the classification, distribution, life cycle, and evolutionary history of the particular animal or group of animals under study. Most zoologists are also specialists in one or more of the related disciplines of morphology, physiology, embryology, and ecology.

Animal behavioral studies have developed along two lines. The first of these, animal psychology, is primarily concerned with physiological psychology and has traditionally concentrated on laboratory techniques such as conditioning. The second, ethology, had its origins in observations of animals under natural conditions, concentrating on courtship, flocking, and other social contacts. One of the important recent developments in the field is the focus on sociobiology, which is concerned with the behavior, ecology, and evolution of social animals such as bees, ants, schooling fish, flocking birds, and humans.

—*Mary E. Carey*

See also: Anatomy; Biology; Demographics; Embryology; Evolution: Animal life; Evolution: Historical perspective; Genetics.

FURTHER READING

Allaby, Michael, ed. *The Concise Oxford Dictionary of Zoology.* 2d ed. New York: Oxford University Press, 1999. This book contains some six thousand entries and covers subjects such as animal behavior, physiology, genetics, cytology, evolution, earth history, and zoogeography, and reflects the current emphasis on ecology in the study of animals. Included are biographical notes on important figures in the history of zoology.

_____, ed. *The Dictionary of Zoology.* 2d ed. New York: Oxford University Press, 1999. A comprehensive review of field of zoology.

Anderson, Donald Thomas. *Invertebrate Zoology.* New York: Oxford University Press, 1999. An up-to-date textbook for undergraduate students studying the biology and evolution of invertebrate animals. Emphasizes function, physiology, and reproductive biology, rather than the more traditional comparative anatomy. Recent advances in the cladistic analysis of invertebrate taxonomy are incorporated into the classifications used in the text.

Griffin, Donald R. *Animal Minds.* Chicago: University of Chicago Press, 1992. The author presents the view that the significance of animal consciousness falls into three categories— philosophical, ethical, and scientific—and discusses examples of each type.

Proctor, Noble S., and Patrick J. Lynch. *Manual of Ornithology: Avian Structure and Function.* New Haven, Conn.: Yale University Press, 1993. A visual guide to the structure and anatomy of birds. The text is informative and written at a level appropriate to undergraduate students and to bird lovers in general.

ELECTRONIC ZOOLOGY RESOURCES

Conservation International: www.conservation.org. This Web site provides information about biodiversity conservation in the world's endangered ecosystems, including a map of global biodiversity hotspots, profiles of hotspots, and many other resources.

International Commission on Zoological Nomenclature (ICZN): www.iczn.org. This Web site describes the official body responsible for providing and regulating the system for ensuring that every animal has a unique and universally accepted scientific name.

International Species Information System (ISIS): www.worldzoo.org. This Web site offers information about this organization that helps zoological institutes manage their living collections by providing software for records keeping and collection management, and then pools this information.

Wildlife Conservation Society: www.wcs.org. This Web site offers information about the organization's conservation activities.

GLOSSARY

Absorption: the movement of nutrients out of the lumen of the gut into the body bile salts: organic compounds derived from cholesterol that are secreted by the liver into the gut lumen and that emulsify fats

Accommodation: changing the shape of the lens in order to keep objects at different distances focused on the retina

Adaptation: the possession by organisms of characteristics that suit them to their environment or their way of life

Adenosine triphosphate (ATP): a molecule produced in the cell that provides energy for cell processes

Aging: A process common to all living organisms, eventually resulting in death or conclusion of the life cycle

Allele: alternative forms of a single gene chromosome: a long strand of DNA with supporting proteins, that contains many genes

Alveoli: the milk-producing areas within the mammary glands

Alveolus: the thin-walled, saclike lung structure where gas exchange takes place

Amino acid: the subunit that makes up larger molecules called proteins

Amplexus: a form of pseudocopulation seen in amphibians, where the male mounts and grasps the female so that their cloacae are aligned, and eggs and sperm are released into the water in close proximity and at the same time

Amygdala: subcortical brain structure related to emotional expression anthropomorphism: attributing human characteristics to animal behavior

Androgens: the general term for a variety of male sex hormones, such as testosterone and dihydrotestosterone

Anisogamy: reproduction using gametes unequal in size or motility asexual

Anosmia: the clinical term for the inability to detect odors

Anterior pituitary gland: the front portion of the pituitary gland, which is attached to the base of the brain; the source of luteinizing hormone (LH) and follicle stimulating hormone (FSH)

Anthropomorphism: attributing human characteristics or states of mind to animals

Antibody: protein produced by lymphocytes, with specificity for a particular antigen

Antigen: chemical that stimulates the immune system to respond in a very specific manner

Aorta: the major arterial trunk, into which the left ventricle of the heart pumps its blood for transport to the body

Apes: large, tailless, semierect anthropoid primates, including chimpanzees, gorillas, gibbons, orangutans, and their direct ancestors—but excluding man and his direct

Aposematic coloration: brightly colored warning coloration that toxic species use to advertise their distastefulness to would-be predators

Archenteron: the primitive gut cavity formed by the invagination of the blastula; the cavity of the gastrula

Arrhenius equation: a mathematical function that relates the rate of a reaction to the energy required to initiate the reaction and the absolute temperature at which it is carried out.

Artery: a blood channel with thick muscular walls which transports blood from the heart to various parts of the body

Articular: pertaining to bone joints bone: the dense, semirigid, calcified connective tissue which is the main component of the skeletons of all adult vertebrates

Atria: the two chambers of the heart, which receive venous blood from the body (via the right atrium) or oxygenated blood from the lungs (left atrium)

Australopithecines: nonhuman hominids, commonly regarded as ancestral to man

Autotomy: the self-induced release of a body part mimicry: a type of defense in which an organism gains protection from predators by looking like a dangerous or distasteful species

Bacteria: single-celled microorganisms that are often the cause of infectious diseases in animals

Binocular vision: the ability to utilize image information from both eyes to form a single image with depth information

Biodiversity: the total of all living organisms in an environment

Blastula: an early stage of an embryo which is shaped like a hollow ball in some animals and a small, flattened disc in others; contains a cavity called the blastocoel

Brainstem: lowest or most posterior portion of the vertebrate brain, including midbrain, pons, and medulla oblongata; controls "housekeeping" functions such as breathing and heartbeat

Calcification: calcium deposition, mostly as calcium carbonate, into the cartilage and other bone-forming tissue, which facilitates its conversion into bone

Capillaries: the very fine vessels in various tissues, which connect arterioles with venules; it is here that the exchange between blood and the extracellular fluid takes place

Carbohydrate: an organic molecule containing only carbon, hydrogen, and oxygen in a 1:2:1 ratio; often defined as a simple sugar or any substance yielding a simple sugar upon hydrolysis

Cardiac output: the amount of blood ejected by the left ventricle into the aorta per minute

Cartilage: elastic, fibrous connective tissue which is the main component of fetal vertebrate skeletons, turns mostly to bone, and remains attached to the articular bone surfaces

Catalyst: a chemical species that initiates or speeds up a chemical reaction but is not itself consumed in the reaction.

Catastrophism: a geological theory explaining the earth's history as resulting from great cataclysms (floods, earthquakes, and the like) on a scale not now observed

Cell-mediated immunity: production of lymphocytes that specifically kill cells with foreign antigens on their surface

Cerebellum: second largest part of the brain, manages fine muscle control and muscle memories

Cerebrum: largest part of most vertebrate brains, with areas that control vocalizations, vision, hearing, smell, and taste, as well as voluntary skeletal muscle movements

Cetaceans: plant-eating marine mammals, such as whales, dolphins, and porpoises marine mammals: part of the class of mammals that adapted to life in the sea

Chemical reaction: a process in which the molecules of two or more chemical species interact with each other in a way that causes the electrons in the bonds between atoms to be rearranged, resulting in changes to the chemical identities of the materials.

Chemoreceptor: specialized nervous tissue that senses changes in pH (hydrogen ions) and oxygen

Chemotaxis: an oriented response toward or away from chemicals

Chorion: the outer cellular layer of the embryo sac of reptiles, birds, and mammals; the term was coined by Aristotle

Chromatophores: pigment-producing cells

Chromophore: the molecule which interacts with opsin; absorption of light changes the interaction and starts the phototransduction cascade

Chromosome: a molecule of deoxyribonucleic acid (DNA) that contains a string of genes, which consist of coded information essential for all cell functions, including the creation of new life

Class: the taxonomic category composed of related genera; closely related classes form a phylum or division

Cleavage: cell division in the early embryo that, unlike division in adults, involves little or no growth between divisions

Cloaca: a common opening for the reproductive, urinary, and digestive systems

Clone: an organism that is genetically identical to the original organism from which it was derived

Cognition: Ability to perceive or understand death: the cessation of all body and brain functions

Cognition: transformation and elaboration of sensory input

Cognitive ethology: scientific study of animal intelligence

Cohort: a group of organisms of the same species, and usually of the same population, that are born at about the same time fecundity: the number of offspring produced by an individual

Collagen: a fibrous protein very plentiful in bone, cartilage, and other connective tissue

Colostrum: the precursor to milk that is formed in the mammary gland during pregnancy and immediately after birth of the young

Comparative anatomy: the branch of natural science dealing with the structural organization of living things

Connective tissue: any fibrous tissue that connects or supports body organs

Controlling site: a sequence of nucleotides generally fifteen to sixty nucleotides long, to which a transcriptional activator or repressor binds

Convection: a transfer of heat from one substance to another with which it is in contact

Corona radiata: the layers of follicle cells that still surround the mammalian egg after ovulation

Cortex: thin layer of gray matter that covers surfaces of the cerebrum and cerebellum

Countercurrent exchanger: the process where a medium (air or water) flowing in one direction over a tissue surface encounters blood flowing through the tissue in the opposite direction; this improves the gas diffusion by maintaining a concentration gradient

cuticle: the outermost layer of a hair, made of scales

Cytoplasm: the living portion of the cell that is contained within the cell membrane deoxyribonucleic acid (DNA): the molecular structure within the chromosomes that carries genetic information

Darwinism: branching evolution brought about by natural selection essentialism (typology): the Platonic-Aristotelian belief that each species is characterized by an unchanging "essence" incapable of evolutionary change

Deme: a local population of closely related living organisms

Deoxyribonucleic acid (DNA): the carrier of all an organism's genetic information

Dermis: layer beneath the epidermis, primarily connective tissue but also containing nerves and blood vessels

Developmental anatomy: the study of the anatomical changes an animal undergoes in the process of growth

Developmental disorders: diseases caused by embryonic or fetal mistakes in normal development

Diastole: relaxation (filling with blood) of the heart chambers

Differentiation: the process during development by which cells obtain their unique structure and function

Diffusion: the process by which gas molecules move from a higher to a lower concentration through a medium or across a permeable barrier; the rate at which gases cross a barrier is increased by the surface area, and gas concentration gradient is decreased by the thickness of the barrier; gas solubility determines the amount that crosses the barrier

Digestion: the process by which larger organic nutrients are broken down to smaller molecules in the lumen of the gut

Diploid: having two chromosomes of each type

Diseases of aging: loss of functions required for health due to age-related degeneration of tissues

Dopamine: neurotransmitter involved in movement and reward systems

Dryopithecines: extinct Miocene-Pliocene apes (sometimes including Proconsul, from Africa) found in Europe and Asia; their evolutionary significance is unclear

Ducts: the tubular structures that carry milk from the alveoli to the outside through the nipple or teat

Duodenum: the first part of the small intestine, where it joins the stomach

Echolocation: the ability of animals to locate objects at a distance by emitting sound waves which bounce off an object and then return to the animal for analysis

Ecology: the study of the interactions between animals and their environment

Ecosystem: a community of organisms in relation to each other and their physical environment

Ectoparasite: a parasite, such as a tick, that lives on the external surface of the host

Ectotherm: an animal that regulates its body temperature using external (environmental) sources of heat or means of cooling

Ediacarian (Ediacaran) fauna: a diverse assemblage of fossils of soft-bodied animals that represents the oldest record of multicellular animal life on the earth eukaryotic cell: a cell that has a nucleus with chromosomes and other complex internal structures; this is the type of cell which makes up all organisms except bacteria

Ejaculation: the process of expelling semen from the male body endocrine glands: glands that produce hormones and secrete them into the blood

Embryo: a young animal that is developing from a fertilized or activated ovum and that is contained within egg membranes or within the maternal body

Embryology: the study of the development of individual animals evolutionary

Endocannibalism: a form of human cannibalism in which members of a related group eat their own dead

Endometrium: an inner, thin layer of cells overlying the muscle layer of the uterus

Endotherm: an animal that regulates its body temperature using internal (physiological) sources of heat or means of cooling

Enterocytes: the cells that line the lumen of the small intestine

Enzyme: a protein that acts as a catalyst under appropriate physiological conditions to break down bonds of a large protein, fat, or carbohydrate

Epidermis: surface layer of epithelial cells invertebrate: animal without a backbone

Erection: the process of enlargement and stiffening of the penis because of increased blood volume within it

Esophagus: the part of the oral cavity (pharynx) that transfers morsels to the stomach; it is usually a long, muscular tube with no digestive function other than transport

Estrus cycle: hormonally controlled changes that make up the female reproductive cycle in most mammals; ovulation occurs during the estrus (heat) period

Ethology: the study of an animal's behavior in its natural habitat

Euryhaline: the ability of an organism to tolerate wide ranges of salinity

Evolution: any cumulative change in the characteristics of organisms or populations over many generations

Exocannibalism: a form of human cannibalism in which unrelated humans are eaten

External genitals: the external reproductive parts of the female

Feedback: in endocrinology, this usually refers to one hormone controlling the secretion of another that stimulates the first, usually in the form of negative feedback, in which the second hormone inhibits the first

Female: an organism that produces the larger of two different types of gametes

Fertilization: the process by which the egg and sperm unite to form the zygote gametes: the haploid cells, ova and spermatozoa, that fuse to form the diploid zygote

Fetus: a mammalian embryo from the stage of its development where its main adult features can be recognized, until birth maternal: referring to the female parent

Field observations: observing behavior in naturalistic settings

Fitness: the relative ability of individuals to pass on genes to subsequent generations

Fossil: a remnant, impression, or trace of an animal or plant of a past geologic age that has been preserved in the earth's crust

Fovea: area, often a pit in the retina, of maximal acuity, where each photoreceptor has its own nerve cell, as opposed to many receptors converging on one nerve cell

Function: Ability, capacity, performance

Gamete: a functional reproductive cell (egg or sperm) produced by the adult male or female

Ganglia: clustered cell bodies of neurons that may form a brain-like center in lower animals

Gaploid: having one chromosome of each type

Gastrula: the stage of development during which the endoderm (gut precursor) and the mesoderm (muscle and connective tissue precursor) are internalized

Gastrulation: the transformation of a blastula into a three-layered embryo, the gastrula; initiated by invagination germ layers: the embryonic layers of cells which develop in the gastrula: ectoderm, mesoderm, and endoderm

Gene flow: the movement of genes from one population to another

Gene pool: the whole body of genes in an interbreeding population that includes each gene at a certain frequency in relation to other genes

Gene: a section of the DNA of a chromosome, which contains the instructions that control some characteristic of an organism

Generalized: not specifically adapted to any given environment; used to describe one group of Neanderthal humans

Genetic diseases: disorders caused by lack of enzymes or structural proteins caused by mutations

Genetic drift: change in gene frequencies in a population owing to chance

Genital area: it is destined to become the penis tip or clitoris

Genital tubercle: a small swelling or protuberance toward the front of an embryo's

Genitalia: the external sex structures

Genome: all of the genetic material of an organism

Genotype frequency: the relative abundance of a genotype in a population

Genotype: the actual genetic makeup of an organism

Genus: the first part of the scientific name of an organism; members of the same genus but different species are closely related, but cannot mate and produce fertile offspring

Geoffroyism: an early theory of evolution in which heritable change was thought to be directly induced by the environment

Gestation: the term of pregnancy hormone: a substance produced by one organ of a multicellular organism and carried to another organ by the blood, which helps the second organ to function

Gill: an evaginated organ structure where the membrane wall turns out and forms an elevated, protruding structure; typically used for water respiration

Gland: a tissue composed of similar cells that produce a hormone

Gonad: the organ responsible for production of gametes—the testis in the male, the ovary in the female

Gonadotropin: a hormone that stimulates the gonads to produce gametes and to secrete other hormones

Gonochorism: sexual reproduction in which each individual is either male or female, but never both

Gopulation: the individuals of a species that live in one place and are able to interbreed

Gracile: slender and light-framed, as opposed to robust

Gray matter: region of the brain or spinal cord that contains cell bodies of neurons, where information processing and storage occur

Growth: the increased body mass of an organism that results primarily from an increase in the number of body cells and secondarily from the increase in the size of individual cells

Haploid: having one of each chromosome; a normal state for animal gametes

Heat: that part of the estral cycle when the female is receptive to male copulatory behavior

Heliotherm: an animal that uses heat from the sun to regulate its body temperature homeostasis: the maintenance by an animal of a constant internal environment

Heritability: the extent to which variation in some trait among individuals in a population is a result of genetic differences

Hermaphrodite: a single organism that produces both eggs and sperm

Hermaphroditism: sexual reproduction in which both male and female reproductive organs are present in the same individual, either at the same time or at different times

Homeotherm: an animal that strives to maintain a constant body temperature independent of that of its environment

Hominid: an anthropoid primate of the family Hominidae, including the genera *Homo* and *Australopithecus*

Hormone: a blood-borne chemical messenger receptor: a protein molecule on or in a cell that responds to the hormone by binding to it and initiating a series of events that compose the response

Humans: hominids of the genus Homo, whether Homo sapiens sapiens (to which all varieties of modern man belong), earlier forms of Homo sapiens, or such presumably related types as Homo erectus and the still earlier (and more problematic) Homo habilis

Humoral immunity: production of antibodies specifically reactive against foreign antigens carried in body fluids (humors)

Hypertonic: describes a solution with a higher osmotic pressure, one containing more osmotically active particles relative to the same volume, than the solution to which it is being compared

Hyperventilation: an increase in the flow of air or water past the site of gas exchange (lung, gill, or skin)

Hypotonic: a solution with a lower osmotic pressure, fewer osmotically active particles relative to the same volume, than the solution to which it is being compared

Hypoxia: from two Latin words, hypo and oxia, meaning "low oxygen"

Immune system: system that produces antibodies and cells that attack foreign substances and pathogens that invade the body

Insight learning: using past experiences to adapt and to solve new problems

Interbreeding: the mating of closely related individuals, which tends to increase the appearance of recessive genes

Intestine: the part of the digestive system involved in completing the process of digestion and absorption of nutrients; usually divided into the small intestine and the large intestine, which opens to the exterior by way of the anus

Invagination: the turning of an external layer into the interior of the same structure; formation of archenteron

Invertebrates: animals lacking a backbone phylogeny: the evolutionary history of a group of species

Isogamy: reproduction in which all gametes are equal in size and motility male: an organism that produces the smaller of two different types of gametes

Isotonic: a solution having the same osmotic pressure, the same number of osmotically active particles relative to the same volume, as the solution to which it is being compared

keratin: a tough fibrous protein, seen in large quantities in epidermal structures such as hair

Labial folds: the paired ridges of tissue on either side of the embryo's genital area, which become penis and scrotum in males and labia in females

Lactation: the process of producing and delivering milk to the young; also, the time period during which milk is produced

Lamarckism: an early evolutionary theory in which voluntary use or disuse of organs was thought to be capable of producing heritable changes scale of being (chain of being): an arrangement of life forms in a single linear sequence from "lower" to "higher"

Larva: a newly hatched form of an organism that looks very different from adults of the species and must undergo metamorphosis to the adult form

Lexigrams: symbols associated with objects or places in keyboard communication experiments with primates

Life cycle: the sequence of development beginning with a certain event in an organism's life (such as the fertilization of a gamete), and ending with the same event in the next generation

Life expectancy: the probable length of life remaining to an organism based upon the average life span of the population to which it belongs

Glossary

Life span: Length of life from birth to death

Life table: a chart that summarizes the survivorship and reproduction of a cohort throughout its life span

Limbic system: brain structures related to the regulation of emotions

Lipid: an organic molecule, such as a fat or oil, composed of carbon, hydrogen, oxygen, and sometimes phosphorus, that is nonpolar and insoluble in water

Longevity: Length of life

Lumen: the central opening through the digestive tract, which is continuous from the mouth to the anus

Lung: an invaginated organ structure where the membrane wall turns in and forms a pouch or saclike structure

Lymphatic vessels: very thin tubes that carry water, proteins, and fats from the gut to the bloodstream

Lymphocyte: white blood cell that produces either cell-mediated or humoral immunity in response to foreign antigens

Macrophage: mature phagocytic cell that works with lymphocytes in destroying foreign antigens

Mammary glands: the milk-producing glands found in all mammals; for example, the cow's udder contains the mammary glands

Mate Competition: competition among members of one sex for mating opportunities with members of the opposite sex

medulla: the innermost layer of a hair

Meiosis: reduction division of the genetic material in the nucleus to the haploid condition; it is the process used by animal cells to form the gametes

Menstrual cycle: a series of regularly occurring changes in the uterine lining of a non-pregnant primate female that prepares the lining for pregnancy

Metabolism: the biochemical action by which energy is stored and used in the body to maintain life

Metabolism: the sum of all of the reactions that take place in an animal allowing it to move, grow, and carry out body functions

Metamorphosis: the form changes in a larva that turn it into the adult form motile: able to move about spontaneously oviparous: born from an externally incubated egg

Migration: the movement of individuals, resulting in gene flow, changing the proportions of genotypes in a population

Migration: the movement of individuals, resulting in gene flow, changing the proportions of genotypes in a population

Milk ejection: also known as milk letdown, this is the reflex response of the mammary gland to suckling of the nipple; the hormone oxytocin mediates this reflex

Mineral: one of the many inorganic elements other than carbon, hydrogen, oxygen, and nitrogen that an organism requires for proper body function

Mitochondria: subcellular structures containing DNA used to estimate the relationships between groups of organisms; the more similar the DNA, the more closely related the groups

Mitosis: the process of cellular division in which the nuclear material, including the genes, is distributed equally to two identical daughter cells

Mitotic cells: cells capable of dividing and forming new cells

Morphology: the scientific study of body shape, form, and composition

Morphology: the study of structure; includes gross morphology, which examines entire structures or systems, such as muscles or bones; histology, which examines body tissues; and cytology, which focuses on cells and their components

Mortality rate: the number of organisms in a population that die during a given time interval

Mortality rate: the percentage of a population dying in a year

Morula: a solid ball or mass of cells resulting from early cleavage divisions of the zygote

Mouth: the anterior part of the digestive system, used for ingesting food; it leads into the oral cavity, which opens into the esophagus

Mucosa: the lining of the inner wall of the gut facing the lumen

Mucus: a secretion of the salivary glands and other parts of the digestive system which lubricates passages

Müllerian ducts: the embryonic ducts that will become the female oviducts or Fallopian tubes, uterus, and vagina parthenogenesis: the development of an unfertilized egg

Multicellular organisms: organisms consisting of more than one cell; there are diverse types of cells, specialized for different functions and generally organized into tissues and organs

Mutation: a change in the nucleotide sequence of a gene or of a controlling site; changes in genes alter the protein, whereas changes in controlling sites determine where and how much of a protein is produced

Myocarditus: inflammation of the heart muscle

Myoepithelial cells: the specialized cells within the mammary gland that surround the alveoli and contract to force milk into the ducts during milk ejection

Natality rate: the number of individuals that are born into a population during a given time interval

National Institutes of Health: the United States' governmental division that monitors and improves public health

Natural selection: any environmental force that promotes reproduction of particular members of the population that carry certain genes at the expense of other members

Neurulation: the process by which the embryo develops a central nervous system notochord: a fibrous rod in an embryo which gives support; a structure that will later be surrounded by vertebrae zygote: the fertilized egg; the first cell of a new organism

Nipple: the raised area on the surface of the skin over the mammary gland that contains the duct openings

Olfaction: the sense of smell

Olfactory receptors: receptor organs which have very high sensitivity and specificity and which are "distance" chemical receptors

Ommatidium: individual unit of the multifaceted compound eye

Oogenesis: gamete formation in the female; it occurs in the female gonads, or ovaries

Opsin: a membrane-bound protein or pigment, which absorbs light

Optic nerve: the main nerve taking information from the eyes to higher processing areas

Optimum Temperature: the narrow temperature range within which the metabolic activity of an animal is most efficient

Organelle: a subcellular structure found within the cytoplasm that has a specialized function

Osmoconformer: an organism whose internal osmotic pressure approximates the osmotic pressure of its environment; such an organism is also referred to as "poikilosmotic"

Osmoregulator: an organism that maintains its internal osmotic pressure despite changes in environmental osmotic pressure; such an organism is also referred to as "euryosmotic"

Osteoblast: a bone cell which makes collagen and causes calcium deposition

Ovary: the female gonad, which produces ova and the hormones estrogen and progesterone

Ovum (pl. ova): the female reproductive cell (gamete); a mature egg cell

Oxytocin: hormone involved with pleasure during bonding

Pacemaker: a specialized group of cardiac muscle cells in the right atrium which initiates the heartbeat; also called the sinoatrial node

Pancreas: an organ derived from the gut that secretes digestive enzymes; it is connected to the gut by a duct through which its secretions enter the gut

Parasites: protozoans, fungi, or animals that survive by obtaining nourishment from a living host, from inside the host or on its surface

Parthenogenesis: a form of asexual reproduction where the young are derived from diploid or triploid eggs produced by the mother without any genetic input from a male

Periosteum: the fibrous membrane which covers all bones except at points of articulation, containing blood vessels and many connections to muscles

Persistent organic pollutants (POPs): chemicals that remain in the environment for a very long time and can be found at long distances from where they are used or released; they are nearly all of human origin

Phenotype: the observable characteristics of an organism (for example, black fur color in a cat)

Pheromone: a hormone produced by an animal and then released into the environment

Pheromones: species-specific compounds (odors) which, acting as chemical stimuli at a distance, have a profound effect on an animal's behavior

Photon: a unit used to describe light intensity

Photoreceptor: cell containing membranes which house light-sensitive pigments

Phylogenetics: the study of the developmental history of groups of animals

Phylum: the taxonomic category of animals and animal-like protists that is contained within a kingdom and consists of related classes

Physiology: the study of the functions, activities, and processes of living organisms

Pinnipeds: flipper-footed marine mammals, such as sea lions, fur seals, true seals, walruses

Pleistocene epoch: the sixth of the geologic epochs of the Cenozoic era; it began about three million years ago and ended about ten thousand years ago

Plexus: a group of nerve cells and their connections to one another

Poikilotherm: an animal that does not regulate its body temperature, which will be the same as that of its environment

Population: a group of individuals of the same species that live in the same location at the same time; a group of individuals that occupy a common area and share a common gene pool

Precambrian eon: the earliest chapter of the earth's history, covering the time interval between the formation of the earth, about 4.6 billion years ago, and the beginning of the Cambrian period, about 570 million years ago

Predation: broadly defined, any interaction in which one organism consumes another living organism, including herbivory (predation on plants), parasitism (predation by small organisms), and familiar predation (where one animal kills and eats another animal)

Predator: an organism that kills and eats another organism, generally of a different species

Primary emotions: emotions related to innate motivations

Primates: a group of mammals including apes, chimpanzees, monkeys, humans, lemurs, and tarsiers

Prions: infectious proteins that cause neurological diseases such as "mad cow disease" viruses: noncellular infectious agents that must enter a host cell to infect it

Prokaryotic cell: a primitive cell that lacks a nucleus, chromosomes, and other well defined internal cellular structures

Protein: a substance made up of amino acids; proteins are the chief building blocks of cellular structures

Protogrammar: word coined to signify the early foundation for grammar development found in primates

Random mating: the assumption that any two individuals in a population are equally likely to mate, independent of the genotype of either individual

Recapitulation: stages of human development reappearing in different animal species

Reproduction: reproduction in which genes are not exchanged

Reproductive Success: the number of offspring produced by one individual relative to other individuals in the same population

Respiratory pigment: a protein that "supercharges" the body fluid (blood) with oxygen; the oxygen can bind to the pigment and then be released

Respiratory surface: the gill, lung, or skin site at which oxygen is taken up from the air or water into the animal, with the release of carbon dioxide at the same time and site

Retina: the light-sensitive film at the back of the eye

Retina: the light-sensitive membrane at the back of the eye

Secondary emotions: emotions with a strong social component

Secondary metabolite: a biochemical that is not involved in basic metabolism, often of unique chemical structure and capable of serving a defensive role for the organism

Semen: fluid produced by the male reproductive system that contains the sperm

Septum: the bony structure that divides the nose into two sections

Sequester: to store a material derived from elsewhere. In defenses, some predators sequester defensive properties from their prey to defend themselves from their own predators

Sexual dimorphism: an observable difference between males and females in morphology, physiology, and behavior

Sexual reproduction: reproduction in which genes are exchanged between individuals

Sexual selection: selection for reproductive success brought about by the behavioral responses of the opposite sex

shaft: the main hair part, made of dead cells arranged in a complex fashion

Sirenians: plant-eating dugongs and manatees

Speciation: the formation of new species as a result of geographic, physiological, anatomical, or behavioral factors

Species: a category of biological classification ranking immediately below the genus or subgenus, comprising related organisms or populations capable of interbreeding

Species: a group of animals capable of interbreeding under normal natural conditions; the smallest major taxonomic category

Spermatogenesis: gamete formation in the male; it occurs in the male gonads, or testes

Spermatogenesis: the structural and functional changes of a spermatid that lead to the formation of a mature sperm cell

Sphincter: a ring of muscle that can close off a portion of the gut

Stenohaline: the inability of an organism to tolerate wide ranges of salinity

Stomach: the part of the digestive system where mechanical breakdown of food is completed and chemical digestion begins

Stratigraphy: in geology, a sequence of sedimentary or volcanic layers, or the study of them—indispensable for dating specimens

Subspecies: the third part of a scientific trinomial, assigned to one of two groups that can mate and produce fertile offspring, but that have some strikingly different characteristics

Survivorship: the pattern of survival exhibited by a cohort throughout its life span

Symbiosis: "living together"; a term that describes the association between two species in which one species typically lives in or on the other species. Parasitism is a common type of symbiosis

Systematics: the delineation and description of animal species and their arrangement into a classification

Systemic: referring to a group of organs that function in a coordinated and controlled manner to accomplish some end, such as respiration

Systole: contraction (emptying of blood) of the heart chambers

Target: cells that contain hormone receptors

Taxon (pl. taxa): group of related organisms at one of several levels such as the family Canidae, the genus *Canis*, or the species *Canis lupus*

Taxonomy: the classification of organisms in an ordered system that indicates natural relationships

Teat: an elongated form of nipple that contains one duct opening

Thermogenesis: the generation of heat in endotherms by shivering or increased oxidation of fats

Triploid: having three of each chromosome; an abnormal state which is unable to produce normal haploid gametes

Turbinates: bony structures that define the internal nasal anatomy

Uniformitarianism: a geological theory explaining the earth's history using processes that can be seen at work today

Urogenital groove: a slitlike opening behind the genital tubercle that will become enclosed in the penis but remain open in females

Uterus: in female mammals, the organ in which the embryo develops

Valves: specialized, thickened groups of muscle cells in the heart chambers, major arterial trunks, arterioles, and veins which prevent backflow of blood

Ventilation: the movement, often by pumping, of air or water to the site of gas exchange; commonly thought of as breathing

Vertebrate: animal with a backbone made up of individual bones called vertebrae

Vitamin: an organic nutrient that an organism requires in very small amounts and which generally functions as a coenzyme

Vitelline envelope: the protective layers that form around the egg while it is still in the ovary

Viviparous: born alive after internal gestation zygote: a fertilized egg

Viviparous: producing young that are active upon birth (often referred to as live birth); the embryo is nurtured within the uterus

White matter: region of neural tissue that contains axons of neurons that carry electrical nerve impulses from one processing center to another

Wolffian ducts: an embryonic duct system that becomes the internal accessory male structures that carry the sperm

Würm glaciation: the fourth and last European glacial period, extending from about seventy-five thousand years ago to twenty-five thousand years ago

Xenotransplantation: the transplantation of organs from one species to another

Zona pellucida: mammalian protective layer analogous to the vitelline envelope

Zoogeography: the study of the distribution of animals over the earth

Zoology: the study of the mechanisms of evolutionary change and the evolutionary history of animal groups

Zygote: the single cell formed when gametes from the parents (ova and sperm) unite, a one-celled embryo

The Last Twenty Years of Nobel Prize Winners in Biological Studies
https://www.nobelprize.org/

Year	Nobel Laureate	Field	Explanation
2016	Yoshinori Ohsumi	Physiology or Medicine	discoveries of mechanisms for autophagy
2015	William C. Campbell Satoshi Ōmura Youyou Tu	Physiology or Medicine	discoveries concerning a novel therapy against infections caused by roundworm parasites & discoveries concerning a novel therapy against Malaria
2014	John O'Keefe May-Britt Moser Edvard I. Moser	Physiology or Medicine	discoveries of cells that constitute a positioning system in the brain
2013	James E. Rothman Randy W. Schekman Thomas C. Südhof	Physiology or Medicine	discoveries of machinery regulating vesicle traffic, a major transport system in our cells
2012	Sir John B. Gurdon Shinya Yamanaka	Physiology or Medicine	discovery that mature cells can be reprogrammed to become pluripotent
2011	Bruce A. Beutler Jules A. Hoffmann Ralph M. Steinman	Physiology or Medicine	discoveries concerning the activation of innate immunity discovery of the dendritic cell and its role in adaptive immunity
2010	Robert G. Edwards	Physiology or Medicine	for the development of in vitro fertilization
2009	Elizabeth H. Blackburn Carol W. Greider Jack W. Szostak	Physiology or Medicine	discovery of how chromosomes are protected by telomeres and the enzyme telomerase
2008	Harald zur Hausen Françoise Barré-Sinoussi Luc Montagnier	Physiology or Medicine	discovery of human papilloma viruses causing cervical cancer discovery of human immunodeficiency virus
2007	Mario R. Capecchi Sir Martin J. Evans Oliver Smithies	Physiology or Medicine	discoveries of principles for introducing specific gene modifications in mice by the use of embryonic stem cells
2006	Andrew Z. Fire Craig C. Mello	Physiology or Medicine	discovery of RNA interference - gene silencing by double-stranded RNA
2005	Barry J. Marshall J. Robin Warren	Physiology or Medicine	discovery of the bacterium Helicobacter pylori and its role in gastritis and peptic ulcer disease
2004	Richard Axel Linda B. Buck	Physiology or Medicine	discoveries of odorant receptors and the organization of the olfactory system
2003	Paul C. Lauterbur Sir Peter Mansfield	Physiology or Medicine	discoveries concerning magnetic resonance imaging

Year	Nobel Laureate	Field	Explanation
2002	Sydney Brenner H. Robert Horvitz John E. Sulston	Physiology or Medicine	discoveries concerning genetic regulation of organ development and programmed cell death
2001	Leland H. Hartwell Tim Hunt Sir Paul M. Nurse	Physiology or Medicine	discoveries of key regulators of the cell cycle
2000	Arvid Carlsson Paul Greengard Eric R. Kandel	Physiology or Medicine	discoveries concerning signal transduction in the nervous system
1999	Günter Blobel	Physiology or Medicine	discovery that proteins have intrinsic signals that govern their transport and localization in the cell
1998	Robert F. Furchgott Louis J. Ignarro Ferid Murad	Physiology or Medicine	discoveries concerning nitric oxide as a signaling molecule in the cardiovascular system
1997	Stanley B. Prusiner	Physiology or Medicine	discovery of Prions - a new biological principle of infection
1996	Peter C. Doherty Rolf M. Zinkernagel	Physiology or Medicine	discoveries concerning the specificity of the cell mediated immune defense

Body Systems

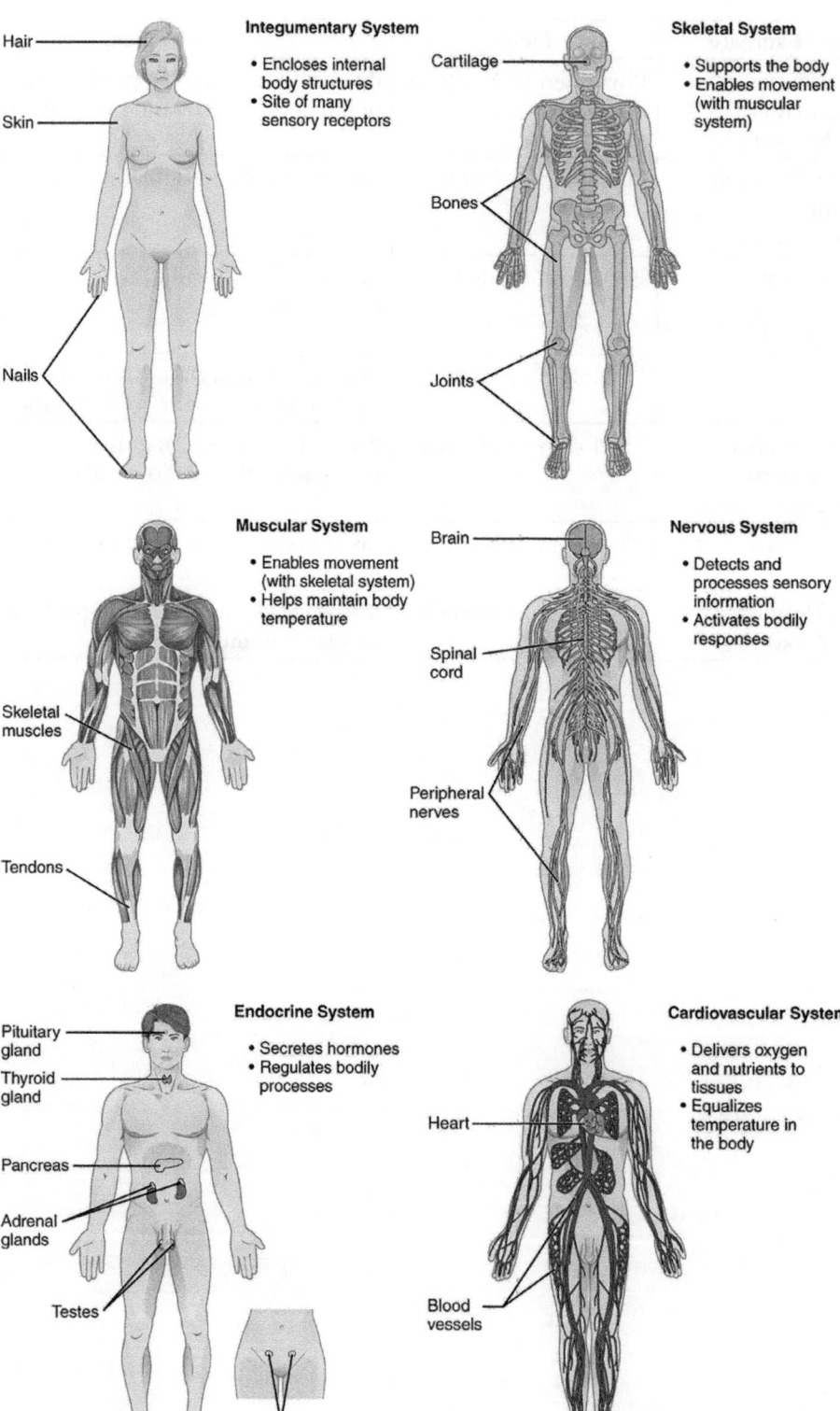

Body Systems

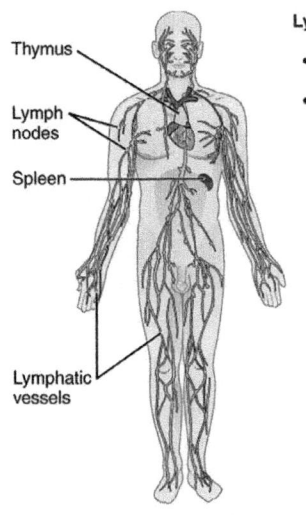

Lymphatic System
- Returns fluid to blood
- Defends against pathogens

Labels: Thymus, Lymph nodes, Spleen, Lymphatic vessels

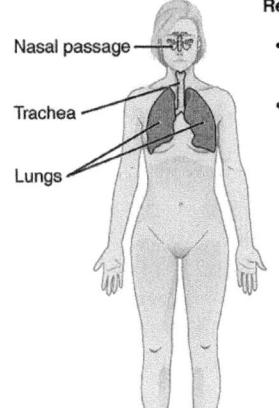

Respiratory System
- Removes carbon dioxide from the body
- Delivers oxygen to blood

Labels: Nasal passage, Trachea, Lungs

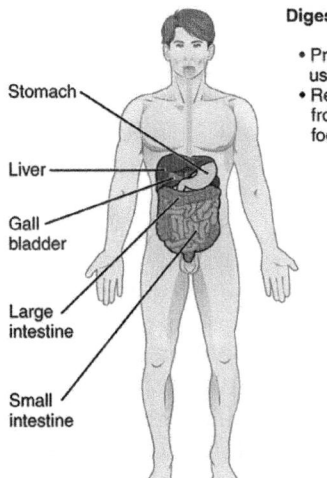

Digestive System
- Processes food for use by the body
- Removes wastes from undigested food

Labels: Stomach, Liver, Gall bladder, Large intestine, Small intestine

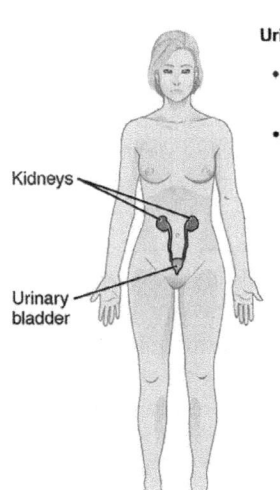

Urinary System
- Controls water balance in the body
- Removes wastes from blood and excretes them

Labels: Kidneys, Urinary bladder

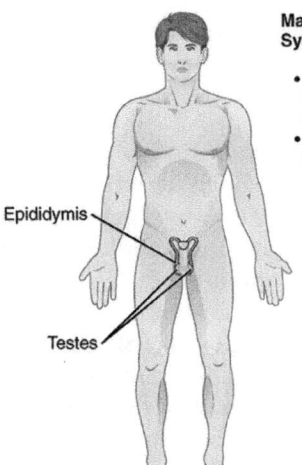

Male Reproductive System
- Produces sex hormones and gametes
- Delivers gametes to female

Labels: Epididymis, Testes

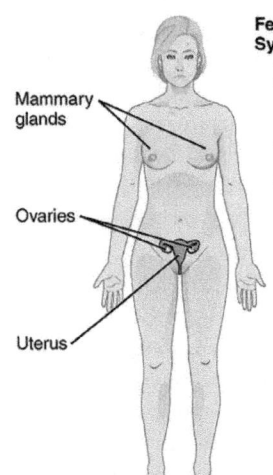

Female Reproductive System
- Produces sex hormones and gametes
- Supports embryo/fetus until birth
- Produces milk for infant

Labels: Mammary glands, Ovaries, Uterus

BIBLIOGRAPHY

Abbott, David. *The Biographical Dictionary of Scientists.* New York: P. Bedrick, 1984. Print.

Acheson, R.M. *An Introduction to the Chemistry of Heterocyclic Compounds.* 3rd ed. New York: Wiley, 2008. Print. Berg,

Adiyodi, K. G., and Rita Adiyodi, eds. *Asexual Propagation and Reproductive Strategies.* Vol. 6, Parts A and B, in *Reproductive Biology of Invertebrates.* New York: John Wiley & Sons, 1993, 1995. Survey of the last one hundred years of research in invertebrate reproduction.

Akazawa, Takeru, Kenichi Aoki, and Ofer Bar-Yosef, eds. *Neandertals and Modern Humans in Western Asia.* New York: Plenum, 1998. A scholarly but understandable group of papers with contributions from many of the major workers in Neanderthal evolution and biology.

Alberts, Bruce, Dennis Bray, Julian Lewis, Martin Raff, Keith Roberts, and James D. Watson. *Molecular Biology of the Cell.* 3d ed. New York: Garland, 1994. This encyclopedic college text is greatly enhanced by its superior diagrams and illustrations. While all aspects of cell biology are covered in great detail, of special interest are the chapters on cellular mechanisms of development and differentiated cells and the maintenance of tissues. Highly recommended for the serious student who wishes to understand developmental processes on a cellular level.

Alberts, Bruce, Dennis Bray, Karen Hopkin, Alexander D. Johnson, Julian Lewis, Martin Raff, Keith Roberts, and Peter Walters. *Essential Cell Biology.* Garland Science, 2013. An easy-to-understand introduction to concepts crucial to the study of cell biology with clear writing and beautiful illustrations.

Alcock, John. *Animal Behavior: An Evolutionary Approach.* 7th ed. Sunderland, Mass.: Sinauer Associates, 2001. An extremely up-to-date textbook, utilizing case studies to illustrate topics. Covers mating and mating behavior in the context of natural selection.

Alexander, Brian. *Rapture: A Raucous Tour of Cloning, Transhumanism, and the New Era of Immortality.* New York: Basic Books, 2004. A reporter examines the fringe groups that support human cloning and genetic enhancement and finds people who want to defeat the effect of entropy and live forever.

Alexander, R. McNeill. *Bones: The Unity of Form and Function.* Reprint. Boulder, Colo.: Westview, 2000. Covers bones, joints, muscle attachment, and related topics in an interesting fashion.

Allaby, Michael, ed. *The Concise Oxford Dictionary of Zoology.* 2d ed. New York: Oxford University Press, 1999. This book contains some six thousand entries and covers subjects such as animal behavior, physiology, genetics, cytology, evolution, earth history, and zoogeography, and reflects the current emphasis on ecology in the study of animals. Included are biographical notes on important figures in the history of zoology.

_____, ed. *The Dictionary of Zoology.* 2d ed. New York: Oxford University Press, 1999. A comprehensive review of field of zoology.

Allen, Arthur. *Vaccine.* New York: W. W. Norton, 2007. Discusses the history of vaccine development, from eighteenth century variolation to modern times, and the controversies that have surrounded vaccination.

Allen, Arthur. *Vaccine.* New York: W. W. Norton, 2007. The story of vaccines and the controversies that have long surrounded their use.

American Diabetes Association. *American Diabetes Association Complete Guide to Diabetes.* 5th ed. Alexandria, Va.: Author, 2011. Provides information to help diabetics manage their disease. Begins with a discussion of the causes and effects of diabetes. Contains a glossary, an appendix on self-monitoring and injection techniques, and a list of resources and organizations.

Ammerman, A. J., and L. L. Cavalli-Sforza. *The Neolithic Transition and the Genetics of Populations in Europe.* Princeton, N.J.: Princeton University Press, 1984. This book is rather technical in its language, but it presents a thoughtful discussion of the influence of migration and gene flow on the Neolithic revolution and the complex workings of gene flow on human evolution.

Anderson, Donald Thomas. *Invertebrate Zoology.* New York: Oxford University Press, 1999. An up-to-date textbook for undergraduate students studying the biology and evolution of invertebrate animals. Emphasizes function, physiology, and reproductive biology, rather than the more traditional comparative anatomy. Recent advances in the cladistic analysis of invertebrate taxonomy are incorporated into the classifications used in the text.

Andrews, Lori B. *The Clone Age: Adventures in the New World of Reproductive Technology.* New York: Henry Holt, 1999. A lawyer specializing in reproductive technology, Andrews examines the legal ramifications of human cloning, from privacy to property rights.

Ankel-Simons, Friderun. *Primate Anatomy.* 2d ed. San Diego, Calif.: Academic Press, 1999. Focuses on all the organ systems of primate species. Many excellent illustrations. Bone, Jesse F. Animal Anatomy and Physiology. 3d ed. Englewood Cliffs, N.J.: Prentice Hall, 1996.Aveterinary textbook that takes a systematic approach to animal anatomy.

Antonucci, Toni, and James Jackson, eds. *Annual Review of Gerontology and Geriatrics: Life-Course Perspectives on Late Life Health Inequalities.* New York: Springer, 2010. Yearly review of the vast field of aging studies. This volume's emphasis is on health disparities and aging.

Arms, Karen, and Pamela S. Camp. *Biology.* 4th ed. FortWorth, Tex.: Saunders College Publishing, 1995.Aclear and well-illustrated general biology text. Presents the functions of the digestive tract in different phyla of animals and specifically of higher animals.

Arnaut, Luís, Sebastião Formosinho, and Hugh Burrows.Chemical Kinetics: From Molecular Structure to Chemical Reactivity. Oxford: Elsevier, 2007. Print.

_____. *Directed Evolution Library Creation: Methods and Protocols.* Totowa, N.J.: Humana Press, 2010. Encyclopedic collection of protocols for generating libraries of randomly mutagenic enzyme genes in bacteria, with tables, graphs, and some figures.

Arnold, Frances H., and George Georgiou, eds. *Directed Enzyme Evolution: Screening and Selection Methods.* Totowa, N.J.: Humana Press, 2010. Laboratory protocol book that describes, in great detail with figures and graphs, some rather ingenious techniques for screening mutant clones of enzyme genes.

Artmann, Gerhard M., and Shu Chien, eds. *Bioengineering in Cell and Tissue Research.* New York: Springer, 2008. Examines bioengineering's role in cell research. Heavily illustrated with diagrams and figures; includes a comprehensive index and references after each section.

Arya, Dev. *Aminoglycoside Antibiotics: From Chemical Biology to Drug Discovery.* New York: Wiley-Interscience, 2007. Describes the design and synthesis of antibiotics and the process of antibiotic resistance.

Askeland, Donald R., Wendelin J. Wright, D. K. Bhattacharya, and Raj P. Chhabra. *The Science and Engineering of Materials.* Boston: Cengage Learning, 2016. Print.

Association for Chemoreception Sciences. http://www.achems.org. ThisWeb site contains a description of current work in the field as well as a discussion of smell disorders. Getchell, T. V., R. L. Doty, L.M. Bartoshuk, and J. B. Snow, eds. *Smell and Taste in Health and Disease.* New York: Raven Press, 1991. Discussion of the clinical aspects of smell problems.

Association for Chemoreception Sciences. www.achems.org. This Web site contains a description of current work in the field and well as a discussion of smell disorders.

Audesirk, Gerald, and Teresa Audesirk. *Biology: Life on Earth.* 5th ed. Upper Saddle River, N.J.: 1999. An introductory college textbook designed for nonscience majors. The chapter on the processes and results of evolution includes a complete explanation of basic population genetics, presented in a nontechnical way. The chapter is well illustrated and includes a glossary and suggestions for Bibliography.

Austad, Steven. *Why We Age.* New York: John Wiley & Sons, 1997.Written by a leading investigator of aging, the book is primarily devoted to human concerns, but there is much discussion of theories of aging, as well as specific biological and environmental influences, pertinent to animals, especially mammals.

Avers, Charlotte J. *Process and Pattern in Evolution.* New York: Oxford University Press, 1989. An intermediate-level college text that lays out the evidence for evolution and the processes that cause it. Chapters 6 and 7 present natural selection and its mechanisms and are a good introduction to the mathematical theories.

Avery, Roger A. *Lizards: A Study in Thermoregulation.* Baltimore: University Park Press, 1979. This book effectively summarizes principles of thermoregulation by using lizard models to illustrate adaptations to various habitats. Somewhat technical, but appropriate for advanced high school students. Nicely illustrated; Bibliography.

Ayala, Francisco J., and James W. Valentine. *Evolving: The Theory and Processes of Organic Evolution*. Menlo Park, Calif.: Benjamin/Cummings, 1979. This book covers the basic principles of evolution and contains a section on the geologic record, which is extremely important in studies of the origin of early life-forms. It is designed for a college-level audience but is clearly written and easy to read.

Ayala, Francisco J., and John A. Kiger, Jr. *Modern Genetics*. 2d ed. Menlo Park, Calif.: Benjamin/Cummings, 1984. This genetics text assumes an audience that has had college-level biology and some chemistry. It provides a good description of classical as well as molecular genetics. Covers most of the methods and major results of population genetics. Chapters include a Bibliography as well as problem sets, and there is a glossary.

Bailey, James E., and David F. Ollis. *Biochemical Engineering Fundamentals*. 2d ed. New York: McGraw-Hill, 2006. Covers all aspects of biochemical engineering in an understandable manner.

Bailey, Jill. *Evolution and Genetics: The Molecules of Inheritance*. New York: Oxford University Press, 1995. An encyclopedia of the current understanding of genetics and evolution.

Bainbridge, William Sims, ed. *Berkshire Encyclopedia of Human-Computer Interaction: When Science Fiction Becomes Fact*. 2 vols. Great Barrington, Mass.: Berkshire, 2004. Contains more than a hundred HCI topics, including gesture recognition, naturallanguage processing, and education. Each article includes Bibliography listings and may contain sidebars, figures, tables, or photographs.

Bakker, Robert T. *The Dinosaur Heresies: New Theories Unlocking the Mystery of the Dinosaurs and Their Extinction*. New York: Kensington, 1986. A marvelously entertaining book treating dinosaurs as homeotherms. Line drawings shed a whole new light on dinosaurs as active, dynamic animals. The book provides insights into homeothermy and the impact it can have on all aspects of animals' lifestyles. Literature-cited section. Written for a popular audience.

Balinsky, B. *An Introduction to Embryology*. 5th ed. Philadelphia: Saunders College Publishing, 1981. A thorough chapter on "Fertilization and the Beginning of Embryogenesis."

Balinsky, B. I. *An Introduction to Embryology*. Philadelphia: Saunders College Publishing, 1981. While somewhat dated, this very well-written text covers the essentials of classical animal embryology while integrating with them the cellular and molecular mechanisms that regulate them. Of specific interest is the chapter on growth, which integrates this topic into the study of developmental biology.

Banks, William J. *Applied Veterinary Histology*. 3d ed. St. Louis: Mosby-Year Book, 1993. A good general reference on animal physiology and organ systems.

Baserga, Renato. *The Biology of Cell Reproduction*. Cambridge, Mass.: Harvard University Press, 1985. This illustrated book can be read by the layperson who is interested in understanding more fully the role of cell division and reproduction in growth. The basics of the cell cycle are covered in detail, as are many of the influences that regulate it. In the latter portion of the book, the author describes the genetic mechanisms involved in growth and development.

Baudrillard, Jean. *The Vital Illusion*. Edited by Julia Witwer. New York: Columbia University Press, 2000. A sociological perspective on what human cloning means to the idea of what it means to be human.

Baylor, D. "How Photons Start Vision." *Proceedings of the National Academy of Science* 93 (January, 1996): 560-565. This review explores the biochemistry of phototransduction, and is fairly detailed.

Beckmann, Charles R. B. *Obstetrics and Gynecology*. 6th ed. Baltimore: Lippincott Williams & Wilkins, 2010. Covers all aspects of the specialty and is endorsed by the American College of Obstetricians and Gynecologists for its compliance with the organization's standards and procedures.

Begon, Michael, Martin Mortimer, and David J. Thompson. *Population Ecology: A Unified Study of Animals and Plants*. 3d ed. Cambridge, Mass.: Blackwell Science, 1996. An up-to-date and readable textbook for those with no background in population ecology. The text is divided into three sections: Single-Species Interactions, Interspecific Interactions, and Synthesis. The first section contains information most relevant to the study of demographics. References.

Bekoff, Marc. *The Smile of a Dolphin*. New York: Discovery Books, 2000. A series of true stories about animal emotions are presented under the headings of love, grief, joy, aggression, anger, and fellow feelings. The research completed by over fifty scientists in the realm of animal emotions is discussed. The book also includes a number of attractive color pictures.

Bell, Donald D., William Daniel Weaver, and Mack O. North. *Commercial Chicken Meat and Egg Production.* 5th ed. Norwell, Mass.: Kluwer Academic, 2002. An essential guide for those interested in the poultry industry. This edition emphasizes managerial aspects.

Bell, Graham. *Selection: The Mechanism of Evolution.* New York: Chapman and Hall, 1997. Clearly explains the processes of natural selection and explores its possible consequences. Offers many examples and gives an extensive review of the literature. Written for nonspecialists with some science background.

Bell, Jerry A.Chemistry: A Project of the American Chemical Society.New York: Freeman, 2005.Print.

Bentley, P. J. *Comparative Vertebrate Endocrinology.* 3d ed. New York: Cambridge University Press, 1998. A very complete treatment of vertebrate hormones from all of the vertebrate classes.

Berg, Jeremy M., John L. Tymoczko, Gregory J. Gatto, and Lubert Strye. *Biochemistry.* 8th ed. New York: W. H. Freeman, 2015. Print.

Berger, Lee. *In the Footsteps of Eve: The Mystery of Human Origins.* Washington, D.C.: Adventure Press/National Geographic Society, 2000. Argues that humans originated in South Africa rather than East Africa, based on the author's own fieldwork with hominid fossils discovered in South African caves.

Berne, Robert M., and Matthew N. Levy. *Principles of Physiology.* 3d ed. St. Louis: C. V. Mosby, 2000. A clearly written textbook at the college level. Emphasis is on cellular mechanisms and regulation. The overview of reproductive function includes an excellent description of interactions between brain, pituitary gland, and gonads. The female system is covered separately. Excellent Summary charts and figures. References provided for each chapter. Index.

Bertino, Anthony J., and Patricia N. Bertino. *Forensic Science: Fundamentals and Investigations.* Mason, Ohio: South-Western Cengage Learning, 2009. Examines the tests and techniques used for the scientific analysis of various evidence types, including hairs and fibers, DNA, handwriting, and soil.

Bhushan, Bharat, ed. *Measurement Techniques and Nanomechanics.* Nanotribology and Nanomechanics 1. Berlin: Springer, 2011. Provides a concise overview of the conceptual history and principles of tribology.

_____. *Principles and Applications of Tribology.* New York: John Wiley & Sons, 1999. The introduction to this book, also available online, provides an excellent account of the history of tribology.

Bickers, D. R., et al. "The Burden of Skin Diseases, 2004: A Joint Project of the American Academy of Dermatology Association and the Society for Investigative Dermatology." *Journal of the American Academy of Dermatology* 55, no. 3 (September, 2006): 490-500. Summary of the well-documented study assessing the prevalence and economic burden of skin diseases and how they effect quality of life.

Bicudo, J. Eduardo P.W., ed. *The Vertebrate Gas Transport Cascade: Adaptations to Environment and Mode of Life.* Boca Raton, Fla.: CRC Press, 1993. A collection of symposium papers related to gas transport and respiratory mechanisms.

Bischof, Marco. "Some Remarks on the History of Biophysics and Its Future." In *Current Development of Biophysics,* edited by Changlin Zhang, Fritz Albert Popp, and Marco Bischof. Hangzhou, China: Hangzhou University Press, 1996. This paper delivered at a 1995 symposium on biophysics in Neuss, Germany, examines how the field of biophysics got its start and predicts future developments.

Bittar, E. Edward, and Neville Bittar, eds. *Evolutionary Biology.* Greenwich, Conn.: JAI Press, 1994. A collection of essays on the structure and dynamics of DNA, RNA, and protein, and more general topics on the origin and cellular basis of life.

Blake, R.D. Informational Biopolymers of Genes and Gene Expression. Properties and Evolution Sausalito, CA: University Science Books, 2005.

Blaustein, A. "Amphibians in a Bad Light." *Natural History* 103 (October, 1994): 32-39. Thorough examination of the role of increased ultraviolet light, which is penetrating a depleted ozone layer, in recent population declines of amphibians.

Blume, Stuart. *The Artificial Ear: Cochlear Implants and the Culture of Deafness.* New Brunswick, N.J.: Rutgers University Press, 2010. Historical study of implant development and implementation.

Boardman, Richard S., Alan H. Cheetham, and Albert J. Rowell, eds. *Fossil Invertebrates.* London: Blackwell Scientific Publications, 1987. A textbook that is designed for use in college courses in invertebrate paleontology. It begins with a number of chapters on basic principles (such as ecology, paleoecology, evolution, classification, and fossil preservation), followed by a run-through of kingdom Protista and twelve invertebrate phyla of kingdom

Animalia that are commonly preserved as fossils. Well illustrated with photographs and line drawings. Will be most useful for those who have had courses in high school biology and historical geology.

Bolognia, Jean, et al., eds. *Dermatology*. 2d ed. 2 vols. St. Louis, Mo.: Mosby Elsevier, 2008. A basic textbook that covers nearly all aspects of dermatology, from cancers to cosmetic procedures.

Borer, Katarina T. *Exercise Endocrinology*. Champaign, Ill.: Human Kinetics, 2003. Looks at the role of hormones in exercise and athletic performance. Topics include regulation of hydration and fuel use during exercise, gender and performance, biological rhythms, and exercise as a stressor.

Bougaze, David, Thomas R. Jewell, and Rodolfo G. Buiser. *Biotechnology. Demystifying the Concepts*. San Francisco: Benjamin/Cummings, 2000. Classical book on biotechnology and bioprocessing.

Bova, Ben. *Immortality*. New York: Avon Books, 1998. Although human life span is his primary topic, Bova discusses animal life spans and the biochemistry of aging in clear, engaging prose for general readers.

Bowden, Douglas M. *Aging in Nonhuman Primates*. New York: Van Nostrand Reinhold, 1979. Discusses aging in monkeys, particularly the effects of disease.

Bowler, Peter J. *Evolution: The History of an Idea*. Rev. ed. Berkeley, Calif.: University of California Press, 1989. A comprehensive history of the evolutionary theory for both specialist and nonspecialist.

_____. *Life's Splendid Drama: Evolutionary Biology and the Reconstruction of Life's Ancestry, 1860-1940*. Chicago: University of Chicago Press, 1996. A history of evolutionary morphology and its relationship with paleontology and biogeography. Covers scientific debates over the emergence of vertebrates, the origins and extinctions of animal species, and the role and influence of Darwin. Biographical appendix, Bibliography.

Bradshaw, Ralph A., and Edward A. Dennis, eds. *Regulation of Organelle and Cell Compartment Signaling: Cell Signaling Collection*. Academic Press, 2011. Contains 55 articles about the function of nuclei and other organelles discussing how they signal each other, respond to stress, and affect cell life and death.

Braga, Newton C. Bionics for the Evil Genius: Twenty-five Build-It-Yourself Projects. New York: McGraw-Hill, 2006. Step-by-step projects that introduce basic concepts in bionics.

Brain, C. K. *The Hunters or the Hunted? An Introduction to African Cave Taphonomy*. Chicago: University of Chicago Press, 1981. In a sophisticated but readable style, Brain reports the evidence about the australopithecines—what they hunted and what hunted them. (Taphonomy is the study of entire bone assemblages.) His study is intended primarily for graduate students and professionals in the field.

Brandon, Robert N. *Concepts and Methods in Evolutionary Biology*. New York: Cambridge University Press, 1996. A collection of essays spanning two decades, addressing problems in the philosophy of biology, particularly the conception of relative adaptedness and the principle of natural selection.

Brennessel, Barbara. "Inborn Error of Metabolism." In *Encyclopedia of Genetics*, edited by Jeffrey A. Knight and Robert McClenaghan. Pasadena, Calif.: Salem Press, 1999. A short history of Sir Archibald Garrod's discoveries and a short discussion of other inborn errors of metabolism in humans.

Brettell, Thomas A., John M. Butler, and José R. Almirall. "Forensic Science." *Analytical Chemistry* 79, no. 12 (2007): 4365-4384. A review of forensic science applications used in common disciplines.

Brewer, Richard. *The Science of Ecology*. 2d ed. Fort Worth, Tex.: Saunders College Publishing, 1994. This clearly written textbook is aimed at upper-level undergraduates and was written by an author who has experience researching both plants and animals. Contains a succinct, well-presented discussion of life tables and survivorship curves. Some of the evolutionary applications of demography are also treated.

Bronson, F. H. *Mammalian Reproductive Biology*. Chicago: University of Chicago Press, 1989. A very complete look at mammalian development. One chapter gives a brief overview of development for each mammalian order.

Brotzman, S. Brent, and Kevin E. Wilk. *Clinical Orthopaedic Rehabilitation*. 2d ed. Philadelphia: Mosby, 2003. Includes chapters on rehabilitation of patients who have had a total knee replacement, lumbar fusion, and knee arthroscopy.

Brown, A. D. *Microbial Water Stress Physiology: Principles and Perspectives*. New York: John Wiley & Sons, 1990. Explores the means by which organisms adapt to survive acute and chronic water stress.

Brown, J. L. *The Evolution of Behavior.* New York: W.W. Norton, 1975. Another good review of animal behavior, including an entire chapter on mating systems and sexual selection. A lengthy Bibliography is included.

Brown, Paula, and Donald Tuzin, eds. *The Ethnography of Cannibalism.* Washington, D.C.: The Society for Psychological Anthropology, 1983. A collection of essays from a 1980 symposium by the Society for Psychological Anthropology.

Brusca, R. C., and G. J. Brusca. *Invertebrates.* Sunderland, Mass.: Sinauer Associates, 1990. A survey of invertebrates, filled with interesting reading and beautiful drawings. Very informative.

Bryant, Christopher, ed. *Metazoan Life Without Oxygen.* New York: Chapman and Hall, 1991. A collection of essays addressing the question of how ancient animals dealt with low oxygen or interruptions in oxygen supply to vital organs. Traces the evolution of animal species to adapt to the accumulation of oxygen in earth's atmosphere.

Budiansky, Stephen. *If a Lion Could Talk.* New York: Free Press, 1998. A good collection of contemporary and historical cases of animal intelligence. The stories cover a wide range of examples seen in various animal species.

Butcher, Graham. *Gastroenterology: An Illustrated Colour Text.* Philadelphia: Elsevier Health Sciences, 2003. Full-color clinical photographs and detailed line drawings illustrate gastroenterological and liver diseases.

Butler, John M. *Fundamentals of Forensic DNA Typing* Burlington, MA: Academic Press, 2010.

Calvin, William H. *The Throwing Madonna: Essays on the Brain.* Updated ed. New York: Bantam Books, 1991. One of the essays describes studies on the brain of *Aplysia*, a sea slug widely used in neurophysiological experimentation on habituation learning.

Campbell, Bernard, ed. *Sexual Selection and the Descent of Man, 1871-1971.* Chicago: Aldine, 1972. A series of eleven articles, each written by a different contributor, outlining the theory of sexual selection as applied to a variety of species.

Campbell, Bernard. *Humankind Emerging.* 8th ed. Boston: Allyn & Bacon, 2000. See especially the chapters "Back beyond the Apes" and "The Behavior of Living Primates." Though a college-level text, this book is useful to everyone because of its comprehensive scope. Older sources should be checked against it.

Campbell, Mary K. and Farrell, Shawn O. *Biochemistry* 8th ed., Stamford, CT: Cengage Learning, 2015.

Cann, R. L., M. Stoneking, and A. C. Wilson. "Mitochondrial DNA and Human Evolution." *Nature* 325 (1987): 31-36. This brief article presents a computer analysis of the mitochondrial DNA studies and makes the conclusion that all living humans descended from one African woman who lived about two hundred thousand years ago.

Capriccioso, Richard P. "Genetic Testing." In *Salem Health: Cancer*, edited by Jeffrey A. Knight. Pasadena, Calif.: Salem Press, 2009. A comprehensive overview of genetic testing covering different types of genetic tests with a review of the science behind the testing. *The Economist.* "Artificial Lifeforms: Genesis Redux." 395, no. 8683 (May 20, 2010): 81-83. Informative article on synthetic biology and the creation of a new form of life in the laboratory.

Carlson, B. *Patten's Foundations of Embryology.* 6th ed. New York: McGraw-Hill, 1996. A comprehensive chapter on fertilization with emphasis on vertebrates.

Carlson, Karen J., Stephanie A. Eisenstat, and Terra Diane Ziporyn. *The New Harvard Guide to Women's Health.* Cambridge, Mass.: Harvard University Press, 2004. Brings together doctors from Harvard Medical School, Massachusetts General Hospital, and Brigham and Women's Hospital to provide complete information on women's health.

Carr, D. E. *The Deadly Feast of Life.* Garden City, N.Y.: Doubleday, 1971. A keen insight on what and how animals eat.

Carradice, Duncan, and Graham J. Lieschke. "Zebrafish in Hematology: Sushi or Science?" *Blood* 111, no. 7 (April, 2008): 3331-3342. Describes the potential of the use of zebrafish for the modeling of hematologic diseases.

Carroll, Sean B., Jennifer K. Grenier, and Scott D. Weatherbee. *From DNA to Diversity: Molecular Genetics and the Evolution of Animal Design.* Malden, Mass.: Blackwell Science, 2001. Covers general principles of the genetic basis of morphological change, including the history of animal evolution, model system developmental genetics, genetic regulatory mechanisms, and case studies of evolutionary change. Color diagrams and images, glossary.

Carter, Carol Sue, I. Izja Lederhendler, and Brian Kirkpatrick, eds. *The Integrative Neurobiology of Affiliation.* Cambridge, Mass.: MIT Press, 1999. A collection of papers from a symposium, with coverage of the neurobiology of affiliation associated with reproductive behaviors.

Cavalli-Sforza, Luigi Luca. *Genes, Peoples, and Languages.* Berkeley: University of California Press, 2000. Although this book focuses on humans and their languages, Cavalli-Sforza's introductory sections on genetics and gene flow are exceptionally clear and well written, accessible to nonspecialists, and applicable to all animal species.

Chan, Evelyn C. Y., et al. "Informed Consent for Cancer Screening with Prostate-Specific Antigen: How Well Are Men Getting the Message?" *American Journal of Public Health* 93, no. 5 (2003): 779-785. Presents balanced information on why prostatespecific antigen (PSA) screening is controversial and yet beneficial.

Chien, Shu, Peter C. Y. Chen, and Y. C. Fung, eds. *An Introductory Text to Bioengineering.* Hackensack, N.J.: World Scientific Publishing, 2008. While definitely written with advanced science students in mind, this text is one of the most basic and yet comprehensive texts available as an introduction to all types of bioengineering.

Chiras, Daniel. *Human Biology: Health, Homeostasis, and the Environment.* Jones and Bartlett, 2002. Discusses basic human biology using the perspective of homeostasis to explain behaviors in response to environment.

Chodzko-Zajko, Wojtek, Arthur F. Kramer, and Leonard W. Poon, eds. *Enhancing Cognitive Functioning and Brain Plasticity.* Champaign, Ill.: Human Kinetics, 2009. A review of recent research supporting the notion that physical activity and brain exercises strengthen the brain. The fairly new concept of neural plasticity, that aging brains can grow and form new neural connections, is discussed extensively.

Ciochon, Russell L., and John G. Fleagle, eds. *The Human Evolution Source Book.* Englewood Cliffs, N.J.: Prentice Hall, 1993. The broad spectrumof human evolution is covered in this book, but a large section titled "The Neanderthal Question and the Emergence of Modern Humans" covers Neanderthals. Several of the most fundamental questions are dealt with in a clear and interesting fashion.

Clancy, Kate. *Greener Eggs and Ham: The Benefits of Pasture-Raised Swine, Poultry, and Egg Production.* Cambridge, Mass.: Union of Concerned Scientists, 2006. The Union of Concerned Scientists looks at egg production, poultry, and pigs and presents an alternative to the intensive production methods in predominant use.

Clark, Jonathan. *Biology: Explaining the Cell: Cell Structure and Organelles, Cell Specialization and Function.* Amazon Digital Services, 2014. Explains the features of the cell including structure and function of organelles and their significance. Generally geared toward students with a question-and-answer format.

Claycomb, James R., and Jonathan Quoc P. Tran. *Introductory Biophysics: Perspectives on the Living State.* Sudbury, Mass.: Jones and Bartlett, 2011. This textbook considers life in relation to the universe. Contains a compact disc that allows computer simulation of biophysical phenomena. Relates biophysics to many other fields and subjects, including fractal geometry, chaos systems, biomagnetism, bioenergetics, and nerve conduction.

Cloud, Preston. *Oasis in Space: Earth History from the Beginning.* New York: W.W. Norton, 1988. This book is part of a series designed to make science more accessible to the general public. It is clearly written, well illustrated, easily understood, and up-to-date. Unfamiliar scientific terms are explained on first mention in the text, and the reader can easily locate definitions by using the index at the back of the book. A two-page time chart indicating major events in geologic history is in the front of the book. Earth history is clearly covered, in addition to many of the basic principles of geology that contribute to the interpretations.

Cloudsley-Thompson, John L. *Tooth and Claw: Defensive Strategies in the Animal World.* London: J. M. Dent&Sons, 1980.Areadable volume that covers defense mechanisms in great detail and with many examples.

Clutton-Brock T. H., ed. *Reproductive Success: Studies of Individual Variation in Contrasting Breeding Systems.* Chicago: University of Chicago Press, 1988.Aseries of twenty-nine individual studies by fifty-three contributors, this book holds a treasure of data on a variety of mating and reproductive systems in both insects and vertebrates (mostly birds and mammals). Most of the articles deal with lifetime measures of reproductive success. The book contains many tables of data, a moderate number of illustrations, and an extensive Bibliography.

Clutton-Brock, Tim H., F. E. Guinness, and S. D. Albon. *Red Deer: Behavior and Ecology of Two Sexes.* Chicago: University of Chicago Press, 1982. Provides an in-depth analysis of a fifteen-year study of the reproductive behavior of male and female red deer. This study has produced some of the very best information on the reproductive biology of a long-lived vertebrate.

Collins, Paul. *Gastroenterology: Crash Course.* Philadelphia: Elsevier Health Sciences, 2008. Offers basic definitions and explanations of all aspects of gastroenterology in an easily understandable manner.

Colomé, Jaime S. "Gene Regulation: Bacteria." In *Encyclopedia of Genetics*, edited by Jeffrey A. Knight and Robert McClenaghan. Pasadena, Calif.: Salem Press, 1999. Regulation of the lactose operon, arabinose operon, tryptophan operon and a flagellin operon is discussed.

Conroy, Glenn C. *Primate Evolution.* New York: W. W. Norton, 1990. A very readable book for students and general readers. Emphasizes the evolution, phylogeny, and classification of hominids, linking the fragmented fossil record with behavior and culture.

Conway-Morris, Simon. *The Crucible of Creation: The Burgess Shale and the Rise of Animals.* Los Angeles: The Getty Center for Education in the Arts, 1999. Afascinating study of one of the richest fossil deposit sites on earth, focusing on what can be learned of evolution from fossil remains.

Costanzo, Linda S. *Physiology: Cases and Problems.* 4th ed. Baltimore: Lippincott, 2012. Print.

Croft, William J. *Under the Microscope: A Brief History of Microscopy.* Hackensack, N.J.: World Scientific, 2006. Traces the microscope from early beginnings to modern instruments, discussing how each works.

Crow, J. F., and Motoo Kimura. *An Introduction to Population Genetics.* New York: Harper & Row, 1970. An excellent starting place for those whose knowledge of migration and gene flow and their influence on evolution is limited and who wish to learn more about the subject. The book is relatively free of the technical jargon that can make some biological texts difficult for the nonbiologist.

Crowley, Leonard V. *An Introduction to Human Disease: Pathology and Pathophysiology Correlations.* 5th ed. Sudbury, Mass.: Jones and Bartlett, 2001. While written for nursing students concerned with the human body, this book discusses many general aspects of diseases and their causes. The latter part of the text covers specific diseases of different organ systems and how they can be treated.

Daly, Martin, and Margo Wilson. *Sex, Evolution, and Behavior.* 2d ed. Belmont, Calif.: Wadsworth, 1983. Perhaps the best Summary of the issues related to the advantages and costs of various reproductive strategies, including those related to sex.

Damjanov, Ivan. *Pathology for the Health Professions.* 4th ed. Maryland Heights, Mo.: Elsevier/Saunders, 2012. A basic overview of pathology with clear pictures and review questions; discusses normal pathology and diseases in various systems.

_____. *Pathology Secrets.* 3d ed. Philadelphia: Mosby/ Elsevier, 2009. A basic review in question-and-answer format, focusing on practical knowledge.

Darwin, Charles R. *On the Origin of Species by Means of Natural Selection: Or, the Preservation of the Favoured Races in the Struggle for Life.* London: John Murray, 1859. This is the original edition, still worth reading. It is better than the more widely reprinted sixth edition, in which Darwin's more forceful statements were toned down as a response to criticism that is no longer greatly valued by biologists. Some knowledge of zoology, geology, and geography would definitely increase any reader's understanding and appreciation of this book. Darwin provided no Bibliography, but some modern editors have supplied one.

Dasheck, William V., and Gurbachan S. Miglani, eds. *Plant Cells and Their Organelles.*Wiley-Blackwell, 2017. Provides a comprehensive overview of plant organelles and their structure and function, describing the differences between these organelles and eukaryotic cell organelles.

Davey, K. G. *Reproduction in the Insects.* San Francisco: W. H. Freeman, 1965. Acompendium of insect reproduction.

Day, Michael H. *Guide to Fossil Man.* 4th ed. Chicago: University of Chicago Press, 1986. Despite some abstruse language (a glossary is included), this authoritative compilation can be of significant help. Organized geographically, it lists and often illustrates all known hominid fossils. Conflicting interpretations are reported but seldom reconciled. This is not, therefore, a book for those seeking the fast answer. Bibliographies accompany each section of the text. For college-level students and professionals, this is a standard work.

De Gray, Aubrey, and Michael Rae. *Ending Aging: The Rejuvenation Breakthroughs That Could Reverse Aging in Our Lifetime.* New York: St. Martin's Griffin, 2008. An investigation of research programs in bioengineering, nutrition, and other fields of medicine that are aimed at prolonging life. Provides interesting coverage of organ transplantation and cellular manipulation.

De Kruif, Paul. *Microbe Hunters.* 1926. Reprint. New York: Harcourt, Brace, 2006. A classic work that introduces some aspects of microbiology and bacteriology and contains some information about parasitology and eukaryotic microbes that cause disease as well as interesting coverage of malaria.

De Luca, Laurival Antonio, Jr., Jose Vanderlei Menani, Alan Kim Johnson, eds. *Neurobiology of Body Fluid Homeostasis: Transduction and Integration.* CRC Press, 2013. A more advanced discussion with examples of homeostasis from the fields of integrative neurobiology and regulatory physiology using animal and human models.

Deems, Eugene F., Jr., and Duane Pursley. *North American Furbearers: A Contemporary Reference.* Baltimore: International Association of Fish and Wildlife Agencies, 1983. Presents a great deal of information on fur and fur-bearing animals.

Dejours, Pierre. "Mount Everest and Beyond: Breathing Air." In *A Companion to Animal Physiology*, edited by C. Richard Taylor, Kjell Johansen, and Liana Bolis. New York: Cambridge University Press, 1982. This chapter is a scholarly paper that Dejours delivered at a symposium. It is clearly written and has illustrations. Basic principles of respiration are explained; topics such as how animals deal with conditions of low oxygen are discussed. The chapter does not require an extensive science background.

Dekkers, Midas. *Birth Day: A Celebration of Baby Animals.* New York: W. H. Freeman, 1995. A book for children, covering conception, pregnancy, and birth in a wide range of animals. Well illustrated.

Dennis, Carina, Richard Gallagher, and Philip Campbell, eds. *The Human Genome.* Spec. issue of *Nature* 409.6822 (2001): 813–958. Print.

DeWaal, Frans B. M. *Good Natured: The Origins of Right and Wrong in Humans and Other Animals.* Cambridge, Mass.: Harvard University Press, 1996. De Waal is a zoologist and ethnologist. His provocative book examines morality in animals.

Dewick, Paul. *Medicinal Natural Products: A Biosynthetic Approach.* New York: John Wiley & Sons, 2009. Comprehensive textbook describing biosynthetic methods and processes, including new techniques in genetic engineering and isolation of genes.

DiLorenzo, Daniel J., and Joseph D. Bronzino, eds. *Neuroengineering.* Boca Raton, Fla.: CRC Press, 2008. Essential review of neuroengineering developments written by leaders in the field.

Dobzhansky, Theodosius. *Genetics of the Evolutionary Process.* New York: Columbia University Press, 1970. An older book, this text is an introduction to experimental population genetics, by one of the architects of the field. There are numerous references to the original literature and many examples. The book is suitable for anyone with at least some introduction to biology and provides one of the clearest explanations of the major concepts of modern evolutionary thought.

Donnet, Christophe, and Ali Erdemir. *Tribology of Diamond-like Carbon Films: Fundamentals and Applications.* New York: Springer Science, 2008. This book deals specifically with one specialized material aspect of tribology, demonstrating the breadth and depth of the field.

Doran, Pauline M. *Bioprocess Engineering Principles.* London: Academic Press, 2009. A solid, basic textbook for students entering the field.

Doudna, Jennifer A. and Samuel H. Sternberg. *A Crack in Creation: Gene Editing and the Unthinkable Power to Control Evolution.* Houghton Mifflin Harcourt, 2017. Written by one of the scientists who discovered this earthshaking technology; focuses on whether or not to actually use this method to change our DNA and the promises and perils of this gene-editing tool.

Dowling, J. E. *The Retina: An Approachable Part of the Brain.* Cambridge, Mass.: Belknap Press of the Harvard University Press, 1987. This work approaches the retina as a part of the brain, and discusses much of the neurobiological research available on this topic.

Dudek, Ronald W. *High-Yield Cell & Molecular Biology* 2nd ed., Baltimore, MD: Lippincott Williams & Wilkins, 2007.

Dukes, H. H. *Dukes' Physiology of Domestic Animals.* 11th ed. Ithaca, N.Y.: Comstock, 1993. A comprehensive textbook for veterinary students, covering all aspects of domestic animal physiology including thermoregulation.

Durand, Dominique M. "What Is Neural Engineering?" *Journal of Neural Engineering* 4, no. 4 (September, 2006). Written by the editor in chief of the journal, who defines NE and its scope.

Eccles, John C. *Evolution of the Brain: Creation of the Self.* London: Routledge, 1989. The chapter on learning and memory uses modern anthropoid apes as a model for prehuman ancestral hominids of modern humans. Comparisons are made of the size of brain areas devoted to learning in apes and humans, as well as other mammals such as monkeys and even rabbits. Emphasis is on the development and characteristics of the neocortex.

Edelstein, Michael, Mohammad Abedi, and Jo Wixon. "Gene Therapy Clinical Trials Worldwide to 2007—An Update." *The Journal of Gene Medicine* 9 (August, 2007): 833-842. Summary of the results of more than 1,300 clinical trials using viral vectors for insertion of replacement genes.

Edmunds, Malcolm. *Defence in Animals.* Burnt Mill, England: Longman, 1974. Technical, comprehensive guide to antipredator defenses, and the evolutionary arms race between predator and prey. Contains photographs, illustrations, and quantitative results from experiments.

Elder, Kay, Doris Baker, and Julie Ribes. *Infections, Infertility, and Assisted Reproduction.* 2004. Reprint. New York: Cambridge University Press, 2010. A detailed, illustrated examination of the microbiology of assisted reproductive technologies. Each chapter includes references, Bibliography suggestions, and frequently appendixes outlining procedures and protocols.

Eldredge, Niles, and Ian Tattersall. *The Myths of Human Evolution.* New York: Columbia University Press, 1982. In fewer than two hundred pages the authors give the nonspecialist a brief, comprehensive look at human evolutionary history, while attempting to show that the once-standard expectations of evolution—slow, steady, gradual progress—are not supported by the evidence. According to the authors, fossil evidence shows human evolution to be the result of long periods of stability interrupted by abrupt change, occurring in smaller populations. The thesis is an important consideration presented in well-written form.

Elgar, Mark A., and Bernard J. Crespi, eds. *Cannibalism: Ecology and Evolution Among Diverse Taxa.* New York: Oxford University Press, 1992. Standard study of the role of cannibalism in animal evolution.

Elit, Laurie, and Jean Chamberlain Froese, eds. *Women's Health in the Majority World: Issues and Initiatives.* New York: Nova Science Publishers, 2006. This first part of the book focuses on health issues that specifically affect women. The second part discusses how agencies such as governments, nongovernmental organizations, and professional societies can work together and improve standards for women.

Elseth, Gerald D., and Kandy D. Baumgardner. *Population Biology.* New York: Van Nostrand, 1981. Intended for graduate students and advanced undergraduates, provides rigorous mathematical treatment of population biology. Discussion of demography includes detailed discussions about age structure, the calculation of population growth rates from demographic data, the evolution of demographic traits, and sex ratios.

Embar-Seddon, Ayn, and Allan D. Pass, eds. *Forensic Science.* 3 vols. Pasadena, Calif.: Salem Press, 2008. Extensive coverage of forensics, including historical events, famous cases, and types of investigations, evidence, and equipment.

Enderle, John D., Susan M. Blanchard, and Joseph D. Bronzino, eds. *Introduction to Biomedical Engineering.* 2d ed. Boston: Elsevier Academic Press, 2005. A broad introductory textbook designed for undergraduates. Each chapter contains an outline, objectives, exercises, and suggested reading.

Endler, John A. *Geographic Variation, Speciation, and Clines.* Princeton, N.J.: Princeton University Press, 1977. Endler's book is valuable primarily because of an excellent chapter on gene flow and its influence on the evolutionary process. Endler sees evolution as a very slow and gradual process in which gene flow and small mutations cause massive change over long periods of time.

_____. *Natural Selection in the Wild.* Princeton, N.J.: Princeton University Press, 1986. This book, although a somewhat more technical presentation, presents the best Summary of the process of natural selection. Chapters 1 and 2 are a nontechnical description of the process and a discussion of the relationship between natural selection and evolution.

Enriquez, Juan and Steve Gullans. *Evolving Ourselves: Redesigning the Future of Humanity—One Gene at a Time.* Current, 2016. Discusses the rapidly changing field of altering human evolution; shows the inner workings of innovative molecular biology and how this will affect who humans become in the future.

Epel, D. "The Program of Fertilization." *Scientific American* 237 (November, 1977): 128-140. A Summary of the events of fertilization and an excellent account of the world of the egg and the sperm. It explains the events that occur after the formation of the gametes. Written for the college-level reader with some background in general biology

Erickson, Milton H. *The Wisdom of Milton H. Erickson.* Compiled by Ronald A. Havens. 1985. Reprint. Williston, Vt.: Crown House, 2003. Erickson's thoughts on hypnosis and psychotherapy.

Esch, Gerald W. *Parasites, People and Places: Essays on Field Parasitology.* New York: Cambridge University Press, 2004. Anecdotal accounts of parasitology illustrate many aspects of the field and cover elements of the history of parasitology as well as medical and evolutionary investigations of parasite behavior.

Evans, David L., and Justin O. Schmidt, eds. *Insect Defenses: Adaptive Mechanisms and Strategies of Prey and Predators.* Albany: State University of New York Press, 1990. An edited volume that examines the many ways that the most successful group of organisms on earth deals with predators.

Evans, G. O., ed. *Animal Clinical Chemistry: A Practical Handbook for Toxicologists and Biomedical Researchers.* 2d ed. Boca Raton, Fla.: CRC Press, 2009. Covers pre-analytical and analytical variables along with information on specific-organ toxicity.

Evans, Randolph W., ed. *Common Neurological Disorders.* Philadelphia: Saunders, 2009. Describes the most common nervous system disorders and their treatment.

Faber, Kurt. *Biotransformations in Organic Chemistry: A Textbook.* 5th ed. New York: Springer-Verlag, 2004. A very clear, useful textbook on the uses of enzymes in chemistry that includes a chapter on engineered enzymes.

Falk, Dean. *Braindance.* New York: Henry Holt, 1992. This study of the evolution of the human brain describes a comparison of the brains and behaviors of humans and other primates, including common chimpanzees, pygmy chimpanzees (bonobos), and monkeys. Areas of particular interest are the visual, motor, and premotor areas of the cerebral cortex, and the concept of brain lateralization in animals as well as humans.

Feher, Gyorgy. Cyclopedia *Anatomicae: More than 1,500 Illustrations of the Human and Animal Figure for the Artist.* New York: Black Dog & Leventhal, 1996. A compendium of anatomical illustrations, focusing on musculoskeletal systems. Focuses on the comparative anatomies of humans, horses, dogs, cats, lions, sheep, cattle, hogs, camels, apes, crocodiles, and seals.

Fenichell, Stephen. *Plastic: The Making of a Synthetic Century.* New York: Harper, 2009. Print.

Fenton, John Joseph. *Toxicology: A Case-Oriented Approach.* Boca Raton, Fla.: CRC Press, 2002. Includes case studies and information about diagnosis, testing, and treatment.

Ferraris, Joan D., and Stephen R. Palumbi, eds. *Molecular Zoology: Advances, Strategies, and Protocols.* New York: Wiley-Liss, 1996. Discusses the uses and methods of molecular biology as related to a wide variety of zoological topics, including reproduction.

Ferri, Fred. *Ferri's Fast Facts in Dermatology: A Practical Guide to Skin Diseases and Disorders.* Philadelphia: Saunders/Elsevier, 2011. A handbook for the diagnosis of dermatological disorders.

Field, Michael J., Carol Pollock, and David Harris. *The Renal System: Systems of the Body Series.* 2d ed. New York: Churchill Livingstone, 2010. Covers the basic anatomy, physiology, and biochemistry of the renal and urogenital system.

Finch, Caleb B., and Rudolph E. Tanzi. "The Genetics of Aging." *Science* 278, no. 5337 (October 17, 1997): 104-109. A concise, well-explained survey of the genetic influences on life span, drawing from studies conducted on animals. Best for readers who understand basic biology and genetics.

Fischman, Josh. "Merging Man and Machine: The Bionic Age." *National Geographic* 217, no. 1 (January, 2010): 34-53. A well-illustrated consideration of the latest advances in bionics, with specific examples of people aided by the most modern prosthetic technologies.

Fish, F. E. "Biomechanics and Energetics in Aquatic and Semiaquatic Mammals: Platypus to Whale." *Physiological and Biochemical Zoology* 73, no. 6 (2000): 683-698. An overview of the mechanics of aquatic respiration.

Fishman, Alfred P., et al. *Fishman's Pulmonary Diseases and Disorders.* 4th ed. New York: McGraw-Hill Professional, 2008. A basic pulmonary textbook and a good source for basic information about procedures for maintaining respiratory health and the history of the science of pulmonary medicine.

Floyd, R. T. *Manual of Structural Kinesiology.* New York: McGraw-Hill, 2009. An introductory text that focuses on movements of the major joints.

Foran, John M. "Thomas Hunt Morgan: 1933." In *The Nobel Prize Winners: Physiology or Medicine,* edited by Frank N. Magill. Pasadena, Calif.: Salem Press, 1991. A history of Morgan's discoveries, in particular his work showing that one of the genes that controls eye color is linked to the X chromosome.

Fornari, Chet S. "Homeotic Genes." In *Encyclopedia of Genetics,* edited by Jeffrey A. Knight and Robert McClenaghan. Pasadena, Calif.: Salem Press, 1999. Discusses the function of homeotic genes.

Fox, Cynthia. *Cell of Cells: The Global Race to Capture and Control the Stem Cell.* New York: W. W. Norton, 2007. Looks at the competition that arises among researchers as they attempt to find applications for stem cell therapy.

Fox, I. S. *Human Physiology.* 6th ed. Boston: WCB/McGraw-Hill, 1999. An excellent treatment of the physiology of nutrition in humans, but applicable to other mammals.

Fox, Michael W. *The Animal Doctor's Answer Book.* New York: Newmarket Press, 1984. This general audience paperback written by a scientific director of the U.S. Humane Society lists questions and answers from his newspaper column. It mainly concerns health and diseases in all sorts of pets and even wild animals, but aspects of animal behavior and careers in animal care are also discussed.

Fox, Renee C., and Judith P. Swazey. *Spare Parts: Organ Replacement in American Society.* New York: Oxford University Press, 1992. Discusses not only the progression of organ transplantation methods but also the emotional significance attached to the human body and its parts as well as the ethical concerns regarding organ replacement.

Fox, Richard G. "Agonistic Science and the Neanderthal Problem." *Current Anthropology,* supp. 39 (June, 1998). All articles in the supplement concern "The Neanderthal Problem and the Evolution of Human Behavior," the supplement's title. The interaction between Neanderthals and modern humans in Europe and in the Middle East and an archaeological consideration of the out-of-Africa model are the subjects of three other articles.

Fukuyama, Francis. *Our Posthuman Future: Consequences of the Biotechnology Revolution.* New York: Picador, 2003. A historian's admonition of the consequences of the biotechnology revolution and its potential to abolish human rights and erode the foundations of liberal democracy.

Furlow, Bryant, and Tara Armijo-Prewitt. "Fly Now, Die Later." *New Scientist* 164, no. 2209 (October 23, 1999): 32-35. The authors describe the age differences between closely related creatures of the same size to support their theory that environmental factors, such as predation, influence the evolution of life spans for species.

Futuyma, Douglas J. *Evolutionary Biology.* 3d ed. Sunderland, Mass.: Sinauer Associates, 1998. A textbook aimed at advanced undergraduates and graduate students, emphasizing the history of life and microevolutionary processes. Comprehensive glossary, index, Bibliography, abundant illustrations.

Ganong, William F. *Review of Medical Physiology.* 19th ed. Stamford, Conn.: Appleton and Lange, 1999. An advanced text, but has well-illustrated and concise explanations of digestive functions and structures. The structures of the digestive tract, their functions, and the control mechanisms are thoroughly covered. The text is very useful because the topics are presented in small sections and subsections. Well indexed for locating information.

Gans, Carl, and F. Harvey Pough, eds. *Physiology C: Physiological Ecology.* Vol. 12 in *Biology of the Reptilia.* New York: Academic Press, 1982. Technical, but provides very complete coverage of thermoregulation in reptiles, with six articles by different authors covering various aspects of the topic. Extensive literature-cited sections at the end of each article.

Gardner, David, and Dolores Shoback. *Greenspan's Basic and Clinical Endocrinology.* 9th ed. New York: McGraw-Hill Medical, 2011. Examines the molecular biology of endocrine glands and discusses metabolic bone disease, pancreatic hormones and diabetes mellitus, hypoglycemia, obesity, geriatric endocrinology, and many other diseases and disorders.

Genadry, Rene, and Jacek L. Mostwin. *A Women's Guide to Urinary Incontinence*. Baltimore: The Johns Hopkins University Press, 2007. Provides information about a common yet embarrassing condition and encourages women and their families to seek compassionate treatment.

Getchell, T. V., R. L. Doty, L.M. Bartoshuck, and J. B. Snow, eds. *Smell and Taste in Health and Disease*. New York: Raven Press, 1991. Discussion of the clinical aspects of smell problems.

Gibbons, Byron. "The Intimate Sense of Smell." *National Geographic* 170, no. 3 (1986): 321-361. An excellent overview of the anatomy, physiology and psychology of the sense of smell. The making of perfumes, use of dogs for tracking, and the history of smell are all well covered.

Gilbert, Scott, F., ed. *A Conceptual History of Modern Embryology*. Baltimore: The Johns Hopkins University Press, 1994. A collection of essays concentrating on historical aspects of research on embryonic induction and the relationship between experimental embryology and genetics.

Gilbert, Scott F. *Developmental Biology*. 11th ed. Sunderland, Mass.: Sinauer Associates, 2016. This is a college-level text written for the upper-level student who has a serious interest in development. The author's approach to development uses a historical experimental analysis of the progress of investigators in the research field. About half of the text deals with specific cell or genetic mechanisms that regulate growth and development. An excellent resource for advanced and accurate information regarding the field of development.

Gilbert, Scott F., and A. M. Raunio. *Embryology: Contructing the Organism*. Sunderland, Mass.: Sinauer Associates, 1997. A college-level textbook on comparative embryology. The first two chapters introduce general terms and concepts, while the following twenty chapters each concentrate on a different animal group.

Gilles, R., E. K. Hoffmann, and L. Bolis, eds. *Volume and Osmolality Control in Animal Cells*. New York: Springer-Verlag, 1991. A comprehensive review and Summary of current literature in the field of osmolality and volume control in animals in terms of both organic and inorganic ions.

Glaser, Roland. *Biophysics*. 5th ed. New York: Springer, 2005. Contains numerous chapters on the molecular structure, kinetics, energetics, and dynamics of biological systems. Also looks at the physical environment, with chapters on the biophysics of hearing and on the biological effects of electromagnetic fields.

Glazer, Alexander N., and Hiroshi Nikaido. *Microbial Biotechnology: Fundamentals of Applied Microbiology*. New York: Cambridge University Press, 2007. Indepth analysis of the application of microorganisms in bioprocessing.

Gliboff, Sander. "Gregor Mendel and Mendelism." In *Encyclopedia of Genetics*, edited by Jeffrey A. Knight and Robert McClenaghan. Pasadena, Calif.: Salem Press, 1999. A well-done history and Summary of Mendel's research with wild-type and mutant pea plants.

Gohar, Ramsey, and Homer Rahnejat. *Fundamentals of Tribology*. London: Imperial College Press, 2008. An introductory book for undergraduate engineering specialists that includes heavy mathematical descriptions and relationships for several common situations.

Goldfarb, Daniel. *Biophysics Demystified*. Maidenhead, England: McGraw-Hill, 2010. Examines anatomical, cellular, and subcellular biophysics as well as tools and techniques used in the field. Designed as a self-teaching tool, this work contains ample examples, illustrations, and quizzes.

Goldman, Laurence R., ed. *The Anthropology of Cannibalism*. Westport, Conn.: Bergin and Garvey, 1999. Provides general textbook-style information

Goldsmith, David, Satish Jayawardene, and Penny Ackland, eds. *ABC of Kidney Disease*. Malden, Mass.: Blackwell Publishing, 2007. This book contains excellent illustrations and clearly written information to provide a greater understanding of renal disease.

Goldsmith, T. "Optimization, Constraint, and History in the Evolution of Eyes." *Quarterly Review of Biology* 65, no. 3 (September, 1990): 281-320. Discusses the evolution and adaptation of eyes and vision.

Gomella, Leonard G. *The Five-Minute Urology Consult*. 2d ed. Philadelphia: Lippincott Williams & Wilkins, 2009. Offers immediate, practical information on a broad range of urological topics.

Gould, James L., and Carol Grant Gould. *Sexual Selection*. 2d ed. New York: Scientific American Library, 1997. Summarizes some of the best research to date on the reproductive behaviors of a variety of animals. It is well written and well illustrated. Provides a small but select Bibliography for more detailed reading.

_____. *The Animal Mind*. New York: Scientific American Library, 1994. A fascinating, well-illustrated inquiry into animal intelligence.

Gould, Stephen J. *Ever Since Darwin*. Reprint. New York: W. W. Norton, 1992.

_____. *The Panda's Thumb*. New York: W.W. Norton, 1980. A compilation of columns from *Natural History* magazine, this book provides a very entertaining and accessible view of evolution and natural selection. This and Gould's other books are the nonscientist's best introduction to the subject.

Graham, Jeffrey B. *Air-Breathing Fishes: Evolution, Diversity, and Adaptation*. San Diego, Calif.: Academic Press, 1997. Comprehensive coverage of the puzzling existence of air-breathing fishes. Easy to read for nonspecialists with a background in biology. Numerous tables and illustrations.

Grant, Verne. *The Evolutionary Process: A Critical Study of Evolutionary Theory*. 2d ed. New York: Columbia University Press, 1991. A comprehesive and critical review of modern evolutionary theory. Focuses on whole organisms and general principles rather than molecular changes and mathematical models.

Green, Ronald Michael. *Babies by Design: The Ethics of Genetic Choice*. 2007. Reprint. New Haven, Conn.: Yale University Press, 2009. A bioethicist tackles moral dilemmas provoked by emerging genetic engineering technologies. Contains a glossary of relevant technical terms.

Greenberg, Arthur, ed. *Primer on Kidney Diseases*. Philadelphia: Saunders Elsevier, 2009. This book is endorsed by the National Kidney Foundation and provides very good background and basics of renal anatomy and physiology. Defines and classifies common kidney disorders and outlines treatment protocols and modalities.

Grendell, James H., Scott L. Friedman, and Kenneth R. McQuaid. *Current Diagnosis and Treatment in Gastroenterology*. 2d ed. New York: Lange Medical Books, 2003. This comprehensive reference discusses all gastroenterological conditions, including hepatic, pancreatic, and biliary conditions.

Griffin, Donald R. *Animal Minds*. Chicago: University of Chicago Press, 1992. This book was written to counteract the behavioristic tradition of John Watson and B. F. Skinner. The book is a key to the understanding of cognitive ethology, and emphasizes the richness of the animal mind.

Griffin, John R., and J. David Grisham. *Binocular Anomalies: Diagnosis and Vision Therapy*. 4th ed. Boston: Butterworth-Heinemann, 2002. Deals with vision therapy as a way to manage some disorders.

Guyton, Arthur C. *Textbook of Medical Physiology*. 10th ed. Philadelphia: W. B. Saunders, 2000. An advanced-level medically oriented physiology text. Strong emphasis on disease mechanisms as well as normal structure and function.

Habif, Thomas P. *Clinical Dermatology*. 5th ed. St. Louis, Mo.: Mosby Elsevier, 2010. Leading manual with excellent photographs, online access, multiple appendixes, and an online differential diagnoses (DDX) mannequin for lesion localization.

Hadley, Mac E., and Jon E. Levine. *Endocrinology*. 6th ed. Upper Saddle River, N.J.: Prentice Hall, 2007. Presents explanations of basic concepts and applications. Focuses on how glands and hormones control physiological processes.

Hadley, Neil F. *Water Relations of Terrestrial Arthropods*. San Diego, Calif.: Academic Press, 1994. Brings together data on the physiology and anatomy of arthropods in the context of their adaptations to terrestrial environments. Written for advanced students and researchers in the field.

Haerens, Margaret, ed. *Embryonic and Adult Stem Cells*. Detroit: Green Haven Press, 2009. Contains a collection of essays arguing the pros and cons of stem cell research.

Hall, Brian J., and John C. Hall. *Sauer's Manual of Skin Diseases*. 10th ed. Philadelphia: Lippincott, Williams & Wilkins, 2010. Accessible textbook includes numerous color photographs, diagnostic algorithms, and a dictionary-index. Has an accompanying Web site.

Halliday, Tim. *Sexual Strategy*. Chicago: University of Chicago Press, 1980. Covers a range of topics related to sexual behavior in a clear, informative manner.

Halter, Jeffrey, et al. *Hazzard's Geriatric Medicine and Gerontology*. 6th ed. New York: McGraw-Hill, 2009. A compendium of evidence-based medicine and clinical applications for treating the aged. Includes 300 illustrations, numerous tables and figures, and additional online resources.

Harmening, Denise M. *Clinical Hematology and Fundamentals of Hemostasis.* 5th ed. Philadelphia: F. A. Davis, 2009. Includes chapters on types of anemia, white blood cell disorders, hemostasis, and laboratory methods.

Hartl, Daniel L. *A Primer of Population Genetics.* 3d ed. Sunderland, Mass.: Sinauer Associates, 2000. This text is intended for students with a college-level knowledge of biology but does not require prior exposure to genetics, statistics, or higher mathematics. There are examples of the significance of population genetics ideas in many areas of biology and medicine, and each chapter has problem sets with answers. There are numerous references to original research.

Hartwell, Leland H., and Ted A. Weinert. "Checkpoints: Controls That Ensure the Order of Cell Cycle Events." *Science* 246 (November 3, 1989): 629-634. An excellent review article describing the biochemical controls of cell division and cell proliferation. Information is a bit technical and is written at the level of the more advanced reader who has some background in biology.

Hartwell, Leland, et al. *Genetics: From Genes to Genomes.* 4th ed. New York: McGraw-Hill, 2011. A comprehensive textbook on genetics, including human genetics discussed in a comparative context.

Hauser, Marc D. *Wild Minds: What Animals Really Think.* New York: Henry Holt, 2000. An exploration of the intellectual and emotional lives of animals and how researchers examine animal skills and cognition.

Hayes, A. Wallace, ed. *Principles and Methods of Toxicology.* 5th ed. Boca Raton, Fla.: CRC Press, 2008. Discusses principles of absorption, distribution, metabolism, and excretion; helps with understanding and using basic experiments in toxicology.

Hayes, Allyson E., ed. *Cryogenics: Theory, Processes and Applications.* Hauppauge, N.Y.: Nova Science Publishers, 2010. Details global research on cryogenics and applications such as genetic engineering and cryopreservation.

Hayes, Karen E. N. *The Complete Book of Foaling: An Illustrated Guide for the Foaling Attendant.* New York: Howell Book House, 1993. A guidebook for horse breeders. Covers the final three weeks of a horse's pregnancy through the first twelve hours of the newborn foal's life, with detailed discussion of the birthing process.

Hayssen, Virginia, Ari van Tienhoven, and Ans van Tienhoven, eds. *Asdell's Patterns of Mammalian Reproduction: A Compendium of Species-Specific Data.* Ithaca, N.Y.: Comstock, 1993. An encyclopedic reference work listing information on the reproductive characteristics of mammalian species, including size and mass of newborns, litter size, age at sexual maturity, length of estrus, gestation, and lactation, and details of frequency and seasons of reproduction.

He, Bin, ed. *Neural Engineering.* New York: Kluwer Academic/Plenum Publishers, 2005. Introductory overview of research in neural engineering.

Heinzle, Elmar, Arno P. Biwer, and Charles L. Cooney. *Development of Sustainable Bioprocesses: Modeling and Assessment.* Hoboken, N.J.: John Wiley & Sons, 2007. Looks at making bioprocesses sustainable by improving them. Includes case studies on citric acid, biopolymers, antibiotics, and biopharmaceuticals.

Hekimi, Siegfried, ed. *The Molecular Genetics of Aging: Results and Problems in Cell Differentiation.* Berlin: Springer, 2000. Examines various genetic aspects of the aging process. Illustrated.

Helander, Martin. *A Guide to Human Factors and Ergonomics.* 2d ed. Boca Raton, Fla.: CRC Press/Taylor & Francis, 2006. Discusses the cognitive and physical aspects of human capabilities that inform product development, and includes an appendix examining the use of checklists to improve safety and effectiveness in human factors design.

Heller, Jacob. *The Vaccine Narrative.* Nashville, Tenn.: Vanderbilt University Press, 2008. Tells how four of the major vaccines of the twentieth century were developed.

Hench, Larry L., and Julian R. Jones, eds. *Biomaterials, Artificial Organs, and Tissue Engineering.* Boca Raton, Fla.: CRC Press, 2005. Provides multiple essays and introductory topics on artificial organs and tissue engineering.

Henschke, Claudia I., Peggy Mcarthy, and Sarah Wernick. *Lung Cancer: Myths, Facts, Choices—and Hope.* New York: W. W. Norton, 2002. This book is a nontechnical examination of the prognosis and treatment of lung cancer for patients and their caregivers.

Herman, Irving P. *Physics of the Human Body.* New York: Springer, 2007. Analyzes how physical concepts apply to human body functions.

Herzenberg, Leonard A., and Leonore A. Herzenberg. "Genetics, FACS, Immunology, and Redox:

A Tale of Two Lives Intertwined." *Annual Review of Immunology* 22 (2004) 1-31. The inventors of flow cytometry describe their lives and work.

Hewitson, Tim D., and Ian A. Darby, eds. *Histology Protocols*. New York: Humana Press, 2010. This laboratory manual looks at tissue preparation and staining, with explanations of complex procedures.

Hickman, Cleveland P., Jr., Larry Roberts, and Frances Hickman. *Integrated Principles of Zoology*. 11th ed. St. Louis: Mosby College Publishing, 2001. A comprehensive textbook, richly illustrated. Chapters on mammals, the reproductive process, and principles of development survey reproductive strategies. A very readable work. A historical perspective and references are included with each chapter. The book has an index, a glossary, and appendices.

Hickman, Cleveland, et al. *Biology of Animals*. 7th ed. St. Louis: Times Mirror/Mosby, 1998. This freshman college zoology text has a special chapter on reproduction and others on major animal groups. It is beautifully illustrated and has a good glossary, index, and bibliographies.

Hildebrand, Milton. *Analysis of Vertebrate Structure*. 4th ed. New York: John Wiley & Sons, 1994. A classic textbook on vertebrate morphology. Uses an organ system approach and relates morphology to evolution. Excellent illustrations, accessible to nonspecialists.

Hill, R. W., and G. A. Wyse. *Animal Physiology*. New York: Harper & Row, 1989. The chapter on the physiology of diving in birds and mammals gives an excellent explanation of the problems of limited oxygen supply and metabolic function under such conditions. The chapter discusses responses of two groups of animals that have been well studied. This text also has background material in related topics. The text is understandable, the information is current, and the figures are illustrative.

Hochachka, Peter, and George Somero. *Strategies of Biochemical Adaptation*. Philadelphia: W. B. Saunders, 1973. This book is a monograph that treats the general concepts of biochemical adaptations and gives the logic behind the authors' present explanations. Although the book is more than twenty-five years old, the explanations are first rate; it is clearly written, with sound reasoning. It is not filled with masses of equations, complex graphs, and numerous references, but concentrates on the concepts. The first two chapters are directly related to this topic, and there are other references to low oxygen in the book.

Hoffman, Shirl J., ed. *Introduction to Kinesiology: Studying Physical Activity*. 3d ed. Champaign, Ill.: Human Kinetics, 2009. Includes an overview of the major areas of kinesiology and a discussion of professions.

Hoffmann, Ary A., and Peter A. Parsons. *Evolutionary Genetics and Environmental Stress*. New York: Oxford University Press, 1991. Another excellent starting place for those whose knowledge of migration and gene flow and their influence on evolution is limited and who wish to learn more about the subject. The book is relatively free of the technical jargon that can make some biological texts difficult for the nonbiologist.

Holler, Teresa. *Cardiology Essentials*. Sudbury, Mass.: Jones and Bartlett Learning, 2007. Presents cardiology with a practical clinical orientation for medical personnel working in a cardiology office.

Holzapfel, William H. and Wood, Brian J.B., eds. *Lactic Acid Bacteria. Biodiversity and Taxonomy* New York, NY: John Wiley & Sons, 2014.

Hopf, Alice. *Strange Sex Lives in the Animal Kingdom*. New York: McGraw-Hill, 1981. This short book, intended for the general reader, gives a very readable account of the importance of sexual reproduction, unusual types of sex determination, sex switching, hermaphroditism, and parthenogenesis. The Bibliography and index are brief.

Horder, T. J., J. A.Witkowski, and C. C.Wylie, eds. *A History of Embryology*. Cambridge, England: Cambridge University Press, 1986. A sourcebook that covers the history of embryology from 1818, when the first human abnormalities were described, until the 1943 production of radioisotopes at Oak Ridge, which are used in marking and tracing development. Describes the contributions of scientists from around the world, with interesting sidelights. Includes an extensive Bibliography on all aspects of embryology.

Houck, Max M., and Jay A. Siegel. *Fundamentals of Forensic Science*. 2d ed. Burlington, Mass.: Academic Press, 2010. An introduction to forensic science and common techniques used for the analysis of physical, biological, and chemical evidence.

Hubel D. H. *Eye, Brain, and Vision*. 2d ed. New York: Scientific American Library, 1995. Approaches vision from a neurobiological standpoint; discusses in detail processing of visual information by the brain.

Huffman, Wallace E., and Robert E. Evenson. *Science for Agriculture: A Long-Term Perspective.* 2d ed. Ames, Iowa: Blackwell, 2006. A history of agricultural engineering research within the United States. Includes a glossary and list of relevant acronyms.

Humber, James M., and Robert F. Almeder, eds. *Biomedical Ethics Reviews: Stem Cell Research.* Totowa, N.J.: Humana Press, 2004. A collection of objective essays reviewing the principle arguments for and against stem cell research and whether this type of work violates the rights of human embryos.

Hung, George K. Biomedical Engineering: Principles of the Bionic Man. Hackensack, N.J.: World Scientific, 2010. Examines scientific bioengineering principles as they apply to humans.

Jacobs, G. H. "The Distribution and Nature of Color Vision Among the Mammals." *Biological Reviews* 68 (1993): 413-471. This review covers the topic of color vision and addresses the different types of color vision in many mammalian species.

James, Stuart H., and Jon J. Nordby, eds. *Forensic Science: An Introduction to the Scientific and Investigative Techniques.* 3d ed. Boca Raton, Fla.: CRC Press, 2009. Discusses mass spectrometry techniques in relation to forensic applications, including forensic toxicology, controlled substance identification, and DNA analysis.

Janig, Wilfrid. *Integrative Action of the Autonomic Nervous System: Neurobiology of Homeostasis.* Cambridge University Press, 2006. Detailed descriptions of how the autonomic nervous system attempts to maintain homeostasis to keep the body running smoothly; covers neurobiological concepts from the cellular and integrative organization aspects.

Jennings, J. B. *Feeding, Digestion, and Assimilation in Animals.* 2d ed. New York: St. Martin's Press, 1972. An excellent comparative approach to nutrition among animals.

Jeremy M., John L. Tymoczko, Gregory J. Gatto, and Lubert Strye. *Biochemistry.* 8th ed. New York: W. H. Freeman, 2015. Print.

Jha, A. R. *Cryogenic Technology and Applications.* Burlington, Mass.: Elsevier, 2006. Deals with most aspects of cryogenics and cryogenic engineering, including historical development and various laws, such as heat transfer, that make cryogenics possible.

Johanson, D., and M. Edey. *Lucy: The Beginnings of Humankind.* New York: Simon & Schuster, 1981. This is the exciting story, told by a world-famous anthropologist, of his discovery of "Lucy," the diminutive three-and-a-half-million-year-old skeleton of one of humankind's australopithecine ancestors. Also gives a readable account of the state of modern research into human evolution, with careful explanations, diagrams of dating methods, and chronologies. Contains many pictures and diagrams, an appendix, and a good Bibliography of books and scientific papers.

Johanson, Donald, and Maitland A. Edey. *Lucy: The Beginnings of Humankind.* Reprint. New York: Simon & Schuster, 1990. Intended for a broad audience, this popular account re-creates the excitement and explains the significance of the finding of an extraordinary specimen, known technically as *Australopithecus afarensis*. It is the oldest hominid whole skeleton extant.

Johanson, Donald, Lenora Johanson, and Blake Edgar. *Ancestors: In Search of Human Origins.* New York: Villard Books, 1994. A companion to the three-part *Nova* television series. Chronological coverage of *Australopithecus*, *Homo habilis*, *Homo erectus*, archaic *Homo sapiens*, the Neanderthals, and *Homo sapiens*. Focuses on the australopithecines and early *Homo*. Lavishly illustrated.

John, David T., et al. *Markell and Voge's Medical Parasitology.* 9th ed. St. Louis, Mo.: Saunders Elsevier, 2006. Covers the basics of medical parasitology from treatments and pharmacology to genomic research. Written for readers with a strong medical or biological background.

Johnson, Leland G., and Rebecca L. Johnson. *Essentials of Biology.* Dubuque, Iowa: Wm. C. Brown, 1986. An introductory-level college text, well illustrated and including outlines, major concepts, key terms, essays, summaries, questions, suggested readings, and a glossary. The chapter on reproduction and development gives a concise Summary of embryonic development in each vertebrate group and has an informative essay on animal cloning.

Johnson, Leonard R., and Thomas Gerwin, eds. *Gastrointestinal Physiology.* 6th ed. St. Louis: C. V. Mosby, 2000. A popular, readable textbook for graduate and medical students. Highlighted key terms, summaries, review questions. Illustrated.

Johnson, Lori. *Cell Function and Specialization (Sci-Hi Life Science).* Raintree, 2009. Discusses how specialized cells function with good photographs.

Johnston, Ian A., and Albert F. Bennett, eds. *Animals and Temperature: Phenotypic and Evolutionary Adaptation.* New York: Cambridge University Press, 1996. A comprehensive review of temperature adaptation in animals.

Jolly, Alison. *Lucy's Legacy: Sex and Intelligence in Human Evolution.* Cambridge, Mass.: Harvard University Press, 1999. Traces four major transitions in human evolution, all based in cooperation, and posits that we are in the process of undergoing a fifth, incorporating species–wide, global communication.

Jones, Richard E., and Kristin H. Lopez. *Human Reproductive Biology.* Rev. ed. London: Academic Press, 2010. A comprehensive introductory textbook, heavily illustrated with diagrams and photographs. Each chapter includes a Summary, Bibliography, and advanced reading list.

Jordan, Paul. *Neanderthal: Neanderthal Man and the Story of Human Origins.* Phoenix Mill, England: Sutton, 1999. A clearly written book for the general reader, highlighting all the ideas involved in the study of human evolution. Describes the discovery of Neanderthal Man, reconstructs the Neanderthal environment and way of life, and traces the emergence of modern humankind.

Jorde, Lynn B., John C. Carey, and Michael J. Bamshad. *Medical Genetics.* 4th ed. Philadelphia: Mosby, 2010. Provides both an introduction to the field of human genetics with chapters on clinical aspects of human genetics, such as gene therapy, genetic screening, and genetic counseling.

Jörres, Achim, Claudio Ronco, and John A. Kellum, eds. *Management of Acute Kidney Problems.* Berlin: Springer-Verlag, 2010. This excellent reference contains information regarding the definition, epidemiology, pathophysiology, and clinical causes of acute kidney failure.

Kalumuck, Karen E. "Drosophila melanogaster." In *Encyclopedia of Genetics*, edited by Jeffrey A. Knight and Robert McClenaghan. Pasadena, Calif.: Salem Press, 1999. A brief review of T. H. Morgan and Sturtevant's work with *Drosophila* followed by a discussion of the use of *Drosophila* to study development.

Kaneko, K. *Life: An Introduction to Complex Systems Biology.* New York: Springer, 2006. Provides an introduction to the field of systems biology, focusing on complex systems.

Kaner, Etta. *Animal Defenses: How Animals Protect Themselves.* Toronto: Kids Can Press, 1999. A colorfully illustrated book aimed at adolescents. It places defenses of animals in context of human behavior.

Kang, Jie *Bioenergetics Primer for Exercise Science* Champagne, IL: Human Kinetics, 2008.

Kang, Manjit S. "One Gene-One Enzyme Hypothesis." In *Encyclopedia of Genetics*, edited by Jeffrey A. Knight and Robert McClenaghan. Pasadena, Calif.: Salem Press, 1999. A short history of the concept that each gene specifies a polypeptide.

Katoh, Shigeo, and Fumitake Yoshida. *Biochemical Engineering: A Textbook for Engineers, Chemists, and Biologists.* Weinheim, Germany: Wiley-VCH Verlag, 2009. A basic, though rather technical, textbook of biochemical engineering by two prominent Japanese biochemical engineers that contains many tables of mathematical symbols and conversions, graphs that illustrate the application of the equations presented, and problems for interested students to solve.

Katz, Bruce F. *Neuroengineering the Future: Virtual Minds and the Creation of Immortality.* Hingham, Mass.: Infinity Science Press, 2008. Fascinating introduction to this field, describing the state of the art and speculating on long-term developments.

Keirl, Andrew William, and Caroline Christie. *Clinical Optics and Refraction: A Guide for Optometrists, Contact Lens Opticians, and Dispensing Opticians.* Oxford, England: Butterworth-Heinemann, 2007. A guide to visual optics, providing straightforward information on clinical optics and refraction.

Kemp, William L., Dennis K. Burns, and Travis G. Brown. *Pathology: The Big Picture.* New York: McGraw-Hill Medical, 2008. Focuses on broad pathology concepts. Contains full-color illustrations, s ummary tables and figures, and questions and answers.

Kendall, Bonnie. *Opportunities in Dental Care Careers.* Rev. ed. New York: McGraw-Hill, 2006. A review of the educational requirements and professional expectations for all specialties of dentistry and dental related careers.

Kinne, Rolf K. H., ed. *Oogenesis, Spermatogenesis, and Reproduction.* New York: Karger, 1991. Covers gametogenesis in fish.

Kirby, Lorne T. *DNA Fingerprinting. An Introduction* New York, NY: Oxford University Press, 1993.

Kirsch, Irving, Steven J. Lynn, and Judith W. Rhue, eds. *The Handbook of Clinical Hypnosis*. 2d ed. Washington, D.C.: American Psychological Association, 2010. Broad introduction to hypnosis covering how it is used clinically to treat conditions and disorders.

Kitchen, Clyde K. *Fact and Fiction of Healthy Vision: Eye Care for Adults and Children*. Westport, Conn.: Praeger, 2007. Starts with anatomy and proceeds to discuss how to maintain healthy eyes. Includes information on many conditions and the types of refractive surgery.

Klaassen, Curtis D. *Casarett & Doull's Toxicology: The Basic Science of Poisons*. 7th ed. New York: McGraw-Hill, 2008. The "gold standard" of toxicology, includes detailed discussions of concepts, principles, and mechanisms of toxicology.

Klavora, P. *Foundations of Kinesiology: Studying Human Movement and Health*. Toronto: Sport Books, 2007. An overview of basic anatomy and physiology, human performance, and the major components of kinesiology.

_____. *Introduction to Kinesiology: A Biophysical Perspective*. Toronto: Sport Books, 2009. A succinct text on basic kinesiology and human movement.

Klitzman, Robert. *The Trembling Mountain: A Personal Account of Kuru, Cannibals, and Mad Cow Disease*. New York: Plenum Trade, 1998. Tells of the author's attempt to show, after months of field study, the relation between endocannibalism and Kuru (a brain disease).

Knobil, Ernst, and Jimmy D. Neill, eds. *The Physiology of Reproduction*. 2 vols. 2d ed. New York: Raven Press, 1994. Covers the entire scope of mammalian reproduction, looking at both human and nonhuman species.

Kobilinsky, Lawrence, Thomas F. Liotti, and Jamel Oeser-Sweat. *DNA: Forensic and Legal Applications*. Hoboken, N.J.: Wiley-Interscience, 2005. Presents an overview of DNA analysis, including the historical perspective, scientific principles, and laboratory procedures.

Kohn, Robert R. *Principles of Mammalian Aging*. Englewood Cliffs, N.J.: Prentice-Hall, 1971. One of the few general surveys of aging in mammalian species.

Koneman, Elmer W., et al. *Color Atlas and Textbook of Diagnostic Microbiology*. 5th ed. Philadelphia: Lippincott, 1997. The first two chapters in this medical text are introductory and cover basic bacteriology and the diagnosis of infectious diseases. Later chapters consider everything about bacteria, viruses, fungi, and parasites that cause diseases in humans and animals. Color plates show what these pathogens look like in culture and in wounds.

Kornberg, Arthur and Baker, Tania A. *DNA Replication* 2nd ed., Sausalito, CA: University Science Books, 2005.

Kornspan, Alan S. *Fundamentals of Sport and Exercise Psychology*. Champaign, Ill.: Human Kinetics, 2009. Discusses basic opportunities and goals in the field.

Kovitz, Kevin L. "Pulmonary and Critical Care: The Unattractive Specialty." *Chest* 127, no. 4 (April, 2005): 1085-1087. A frank discussion of the workforce issues surrounding pulmonary and critical care medicine and an accurate description of the day-to-day activities of a pulmonologist.

Kozubek, James. *Modern Prometheus: Editing the Human Genome with Crispr-Cas9*. Cambridge University Press, 2016. Discusses the potential for gene editing, including ethical and legal implications; tells the story across a 50-year timeline, including stories of the scientists involved in the process.

Krogh, August. *Osmotic Regulation in Aquatic Animals*. Reprint. New York: Dover, 1965. Reprint of the 1939 edition, a classic work in the field. It is clearly written and covers all aspects including techniques used to determine osmotic pressure.

Kumar, Vinay, et al. *Robbins and Cotran Pathologic Basis of Disease*. 8th ed. Philadelphia: Saunders/Elsevier, 2010. Regarded by some as the keystone book and used as the primary text in pathology at many medical schools. Includes illustrations and case studies.

Kumé, Matazo, and Katsuma Dan. *Invertebrate Embryology*. Translated by Jean C. Dan. Belgrade, Yugoslavia: NOLIT Publishing House for the U.S. Department of Health and Human Services, 1968. Extensive compendium of invertebrate development by phylum.

Kurten, Bjorn. *Our Earliest Ancestors*. Translated by Erik J. Friss. New York: Columbia University Press, 1993. A succinct overview of hominid evolution. Focuses on the major changes in anatomy and culture that have marked the development of hominids from *Homo habilis* through *Homo sapiens*, including the emergence of bipedalism, stone toolmaking, and articulate speech. Numerous maps, charts, and drawings.

Lahtinen, Sampo, Ouwehand, Arthur C., Salminen, Seppo and Von Wright, Atte *Lactic Acid Bacteria. Microbiological and Functional Aspects* 4th ed., Boca Raton, FL: CRC Press, 2012.

Lai, Kar Neng, ed. *Practical Manual of Renal Medicine: Nephrology, Dialysis and Transplantation.* Hackensack, N.J.: World Scientific, 2009. This manual provides practical information about dialysis, transplantation, and general nephrology in straightforward language. It describes treatment rationale and kidney disease management.

Lavers, Chris. *Why Elephants Have Big Ears: Understanding Patterns of Life on Earth.* New York: St. Martin's Press, 2001. An encyclopedic exploration of form and function in animal physiology.

Lazo, John, and Peter Wipf. "Combinatorial Chemistry and Contemporary Pharmacology." *The Journal of Pharmacology and Experimental Therapeutics* 293, no. 3 (February, 2000): 705-709. Describes the process of combinatorial chemistry. Includes experimental strategies and flow charts describing the screening of compounds.

Leakey, Maeve, et al. "New Hominin Genus from Eastern Africa Shows Diverse Middle Pliocene Lineages." *Nature* 410 (March 22, 2001): 433-440. A new skeleton, dated to 3.5 million years ago and differing from the roughly contemporary *Autralopithecus afarensis*, throws the lineage of modern man into even further confusion.

Leakey, Mary. *Disclosing the Past.* New York: Doubleday, 1984. This is Mary Leakey's autobiography, but it serves as an unsurpassed history of field research into human origins over the past fifty years, told by an individual who was at the forefront of that story. The book contains photographs, maps, diagrams, and a completely captivating narrative of human evolution. Contains a list of twelve books for Bibliography.

Leakey, Richard E. *The Making of Mankind.* New York: E. P. Dutton, 1981. This book, by the son of Louis and Mary Leakey, is written from the perspective of one who grew up with famous physical anthropologists and scientists who were tracing human origins. The text is complemented with many color photographs. Chapters discussing the development of language and early *Homo sapiens* culture are important. It also distills and synthesizes important ideas of others.

Lebovic, Dan I., John D. Gordon, and Robert N. Taylor. *Reproductive Endocrinology and Infertility: Handbook for Clinicians.* Arlington, Va.: Scrub Hill Press, 2005. A ready reference for endocrinologists treating conditions and disorders related to reproduction. Information from textbooks, articles, and endocrinologists was gathered and analyzed to provide evidence-based approaches and strategies.

Lehninger, Albert L. *Biochemistry: The Molecular Basis of Cell Structure and Function.* 2nd ed. New York: Worth, 1975. Print.

Levine, Herbert M. *Animal Rights.* Austin, Tex.: Steck-Vaughn, 1998. Offers arguments pro and con on many issues concerning animal rights. Mason, Jim, and Peter Singer. *Animal Factories.* New York: Harmony Books, 1990. A stringent and discerning examination of the manufacturing of animals for food and profit.

Lewin, Benjamin. *Genes VIII.* San Francisco: Benjamin Cummings, 2003. In-depth look at genes and molecular biology. Madigan, Michael T., et al. *Brock Biology of Microorganisms.* 12th ed. San Francisco: Benjamin Cummings, 2008. Several chapters of this popular textbook describe microbial metabolism and the application of microorganisms in industry.

Lewin, Roger. *Bones of Contention: Controversies in the Search for Human Origins.* New York: Simon & Schuster, 1987. This lucid but sophisticated treatment reviews scholarly debates regarding the Taung child (*Australopithecus africanus*), *Ramapithecus* (a fossil ape) and "Lucy" (*Australopithecus afarensis*). Too specialized for most high school students, it will appeal primarily to college graduates and professionals. Lewin has also written *Human Evolution: An Illustrated Introduction* (1984) and coauthored three popular books with Richard Leakey: *Origins* (1977), *People of the Lake* (1979), and *The Making of Mankind* (1981).

_____. *The Origin of Modern Humans.* New York: Scientific American Library, 1998. Overview of the key concepts in the origin of modern humans, including the molecular biology and archaeology supporting the out-of-Africa hypothesis versus the multiregional hypothesis of human evolution.

Lewis, Ricki. *Human Genetics: Concepts and Applications.* 9th ed. New York: McGraw-Hill, 2010. This textbook provides a broad overview of human genetics and genomics.

Link, Kurt. *The Vaccine Controversy: The History, Use, and Safety of Vaccinations.* Westport, Conn.: Praeger,

2005. Short synopses of the history behind most major vaccines, from smallpox to acellular vaccines.

Linzey, Donald. *Vertebrate Biology*. Boston: McGraw-Hill, 2001. A good midlevel text outlining structure and function in vertebrates. Excellent use of descriptive features and terminology as part of the text. Good list of related texts in preface.

Lipkin, Steven Monroe and John Luoma. *The Age of Genomes: Tales from the Front Lines of Genetic Medicine*. Beacon Press, 2016. Focuses on the real-life stories of patients who may be helped by this type of gene editing in an easy-to-read and accessible way.

Lodish, H., A. Berk, S. L. Zipursky, P. Matsudaira, D. Baltimore, and J. Darnell. *Molecular Cell Biology*. 4th ed. New York: W. H. Freeman, 2000. Chapter 21 offers in-depth coverage of phototransduction from a biochemical point of view.

Loewy, Ariel G., et al. *Cell Structure and Function: An Integrated Approach*. 3d ed. Philadelphia: Saunders, 1991. A college textbook that covers the major facts and theories of cell biology, as well as genetic manipulation and genetic analysis in their relationship with structural and biochemical approaches to the cell. Suitable for beginning students with a background in elementary organic chemistry.

Long, John A. *The Rise of Fishes: Five Hundred Million Years of Evolution*. Baltimore: The Johns Hopkins University Press, 1996. A detailed history of fish evolution. Color photographs, glossary.

MacKenzie, Dana. "New Clues to Why Size Equals Destiny." *Science* 284, no. 5420 (April 6, 1999): 1607-1608. Discusses theories about the relation of size to life span in a style accessible to general readers.

Macnee, William, and Stephen I. Rennard. *Chronic Obstructive Pulmonary Disease*. London: Health Press, 2009. Spends equal time on the science and clinical treatment of COPD. It is a useful reference to understand the disease and its manifestations at a deeper level.

Madhavan, Guruprasad, Barbara Oakley, and Luis G. Kun, eds. *Career Development in Bioengineering and Biotechnology*. New York: Springer, 2008. An extensive guide to careers in bioengineering, biotechnology, and related fields, written by active practitioners. Covers both traditional and alternative job opportunities.

Madigan, Michael, et al. *Brock Biology of Microorganisms*. 12th ed. San Francisco: Pearson Benjamin Cummings, 2009. Several chapters in this textbook of microbiology address the role of viruses in disease as well as in biotechnology.

Magee, D. J. *Orthopedic Physical Assessment*. 5th ed. St. Louis, Mo.: Saunders 2008. Includes chapters on orthopedic assessments and rehabilitation of the cervical, thoracic, and lumbar spine.

Maier, Richard. *Comparative Animal Behavior: An Evolutionary and Ecological Approach*. Boston: Allyn & Bacon, 1998. An excellent general-audience textbook on various aspects of animal behavior, with sections on tool use in animals.

Malacinski, George M. Essentials of Molecular Biology 4th ed., Sufbury, MA: Jones and Bartlett Education, 2005.

Mann, J., et al. *Natural Products: Their Chemistry and* 2nd ed. London: Chapman & Hall, 1989. Print.

Marguet, Philippe, et al. "Biology by Design: Reduction and Synthesis of Cellular Components and Behavior." *Journal of the Royal Society Interface* 4, no. 15 (2007): 607-623. Review on metabolic engineering and synthetic biology written for the general public.

Margulis, Lynn, and Karlene V. Schwartz. *The Five Kingdoms: An Illustrated Guide to the Phyla of Life on Earth*. 3d ed. New York: W. H. Freeman, 1998. A richly illustrated explanation of taxonomy with detailed information about the biology of the thirty-seven phyla in the kingdom Animalia, including discussion of life cycles and development.

Marieb, Elaine N. *Human Anatomy and Physiology*. 5th ed. San Francisco: Benjamin/ Cummings, 2001. This college text describes the structure and function of neurons in general, mainly dependent on animal studies that can be applied to humans as well. The human brain and its functions are similar in many ways to the general patterns of the mammalian brain and nervous system.

Marshall Graves, J. A., R. M. Hope, and D.W. Cooper, eds. *Mammals From Pouches and Eggs: Genetics, Breeding, and Evolution of Marsupials and Monotremes*. Canberra, Australia: CSIRO, 1990. A collection of papers on the often-overlooked topic of marsupial and monotreme reproduction.

Marshall, Elizabeth L. *The Human Genome Project: Cracking the Code Within Us*. New York: Franklin Watts. 1996. Discusses the goals and structure of the Human Genome Project and its implications. Contains a chapter on the contributions of animal model systems to this project.

Marshall, Elizabeth. *The Hidden Life of Dogs*. Boston: Houghton Mifflin, 1993. As an anthropologist and ethologist, the author provides unique insights into the behavior of dogs. The behavior of a pack of dogs over the course of thirty years is documented.

Martin, Joseph B. "The Integration of Neurology, Psychiatry, and Neuroscience in the Twenty-first Century." *American Journal of Psychiatry* 159 (May, 2002): 695-704. Examines the historical basis for the divergence of neurology and psychiatry and discusses prospects for a potential convergence.

Martin, R. D. *Primate Origins and Evolution: A Phylogenetic Reconstruction*. London: Chapman and Hall, 1990. A wide-ranging book for those interested in mammals and evolutionary biology in general. Covers classification of mammals, the fossil record of mammalian origins, continental drift and evolution, and the problems of phylogenetic reconstruction.

Mason, Jim, and Peter Singer. *Animal Factories*. New York: Harmony Books, 1990. A stringent and discerning examination of the manufacturing of animals for food and profit.

Mason, Robert J., Jay A. Nadel, and John F. Murray. *Murray and Nadel's Textbook of Respiratory Medicine*. 5th ed. Philadelphia: Elsevier Saunders, 2010. The quintessential textbook on pulmonary medicine. Includes chapters on every aspect of pulmonary disease management from basic anatomy of the lungs to lung transplantation.

Masson, Jeffrey M. *When Elephants Weep*. New York: Delacorte Press, 1995. The author is a strong advocate for the recognition of emotions in animals. After discussing the impediments in the scientific community to the serious study of animal emotions, the author presents groupings of emotions expressed by animals. He presents numerous examples to support the wide range of emotions he attributes to animal behavior.

Masterton, William L., Cecile N.Hurley, and Edward Neth.Chemistry: Principles and Reactions. 7th ed. Belmont: Brooks, 2012.Print.

Mataigne, Fen. *Medicine by Design: The Practice and Promise of Biomedical Engineering*. Baltimore: The Johns Hopkins University Press, 2006. An introduction to and investigation of bioengineering and the potential future of the field. Provides discussions of issues such as bioreactors and organ replacements.

Mathews, Willis W. *Atlas of Descriptive Embryology*. 5th ed. Upper Saddle River, N.J.: Prentice Hall, 1998. A paperback manual intended for laboratory work in an intermediate college course. The manual includes a complete series of large photomicrographs of developmental stages.

Maxey, Lisa, and Jim Magnusson, eds. *Rehabilitation for the Postsurgical Orthopedic Patient*. 2d ed. St. Louis, Mo.: Mosby, 2007. Provides detailed descriptions of each orthopedic surgery and addresses patient rehabilitation and how to adapt therapy to geriatric, athletic, and pediatric populations.

Maynard-Smith, John. *Evolution and the Theory of Games*. New York: Cambridge University Press, 1982. This short book presents a more theoretical approach to the problem of evolutionary strategies in general. Sexual reproduction is discussed as a reproductive strategy as is parthenogenesis. There are also discussions of sex ratios and parental care. The book contains a few graphs andmanyequations. There is a Bibliography, but most of the works listed are articles in technical journals.

McClellan, Marilyn. *Organ and Tissue Transplants: Medical Miracles and Challenges*. Berkeley Heights, N.J.: Enslow, 2003. Explores the history of organ transplantation as well as the ensuing medical, ethical, and financial issues.

McClintock, J. Thomas. *Forensic DNA Analysis: A Laboratory Manual*. Boca Raton, Fla.: CRC Press, 2008. Examines the various methods of DNA analysis and DNA fingerprinting.

McClintock, James B., and Bill J. Baker, eds. *Marine Chemical Ecology*. Boca Raton, Fla.: CRC Press, 2001. This technical volume is the most current, comprehensive book on marine chemical ecology. The book provides cellular, physiological, organismal, evolutionary, and applied perspectives creating a high-resolution snapshot of the field at the start of the twenty-first century.

McFarland, David. *Animal Behavior*. 3d ed. Boston: Longman Science and Technology, 1998. An upper-level textbook on animal behavior. Chapter 27, "Intelligence, Tool Use, and Culture," discusses tool use among animals. This is a good book for readers with some previous background on the subject.

McGill, T. E., D. A. Dewsbury, and B. D. Sachs. *Sex and Behavior: Status and Prospectus*. New York: Plenum, 1978. A series of sixteen chapters, each by a different author or authors, on various topics related to sex-related behavior. A large portion of the book is devoted to studies of sex differences in humans. There is a Bibliography at the end of each chapter. There are very few illustrations, mostly in the form of graphs.

McGrew, W.C. *Chimpanzee Material Culture: Implications for Human Evolution.* New York: Cambridge University Press, 1992. Describes and analyzes the use of tools by chimpanzees in their native habitats through field studies conducted across Africa.

McKee, Jeffrey K. *The Riddled Chain: Chance, Coincidence, and Chaos in Human Evolution.* New Brunswick, N.J.: Rutgers University Press, 2000. In contrast to the traditional top-down approach, McKee takes a bottom-up view of human evolution, in which each genetic variation must be tested through successive levels within a species. He argues that the large size of the current human population poises us for a rapid rate of evolution.

McKenzie, Shirlyn B., and J. Lynne Williams. *Clinical Laboratory Hematology.* 2d ed. Upper Saddle River, N. J.: Pearson, 2010. Cover types of anemia, neoplastic hematologic disorders, hemostasis, and hematology procedures. Includes some excellent photomicrographs and illustrations.

McKie, Robin. *Dawn of Man: The Story of Human Evolution.* New York: Dorling Kindersley, 2000. Published in conjunction with the Learning Channel series. Clearly written for a general audience, with many photographs and illustrations.

McMenamin, M. A., and D. L. McMenamin. *The Emergence of Animals: The Cambrian Breakthrough.* New York: Columbia University Press, 1990. An examination of the adaptive radiation that resulted in an explosion of animal forms at the beginning of the Cambrian period.

McNamee, Gregory. *Careers in Renewable Energy: Get a Green Energy Job.* Masonville, Colo.: PixyJack Press, 2008. This highly readable Summary of the renewable- energy job market includes more than just engineering jobs and discusses potential future employment opportunities in alternative energy sources.

Meeks, Robert. *Biology: Science of Life, Cell Theory, Evolution, Genetics, Homeostasis and Energy.* CreateSpace Independent Publishing Platform, 2016. A simple overview of many biological concepts, including homeostasis.

Mellars, Paul, ed. *The Emergence of Modern Humans: An Archaeological Perspective.* Ithaca, N.Y.: Cornell University Press, 1990. A volume of proceedings from a 1987 conference on the origins of modern humans and the adaptations of hominids. Concentrates on the archaeology of hominids in western and central Europe.

Meyr, Ernst. *The Growth of Biological Thought: Diversity, Evolution, and Inheritance.* Reprint. Cambridge, Mass.: Belknap Press of Harvard University Press, 1985. A classic exposition of the history of biology as a field of study and its philosophical evolution.

Miller, S. A., and J. P. Harley. *Zoology.* 4th ed. Boston: McGraw-Hill, 1999. A lower-level text with introductory sections on animal systems.

Millodot, Michel. *Dictionary of Optometry and Visual Science.* Oxford, England: Butterworth-Heinemann, 2008. Provides understandable definitions, tables, and illustrations.

Milo, Ron and Rob Phillips. *Cell Biology by the Numbers.* Garland Science, 2015. Features calculations that investigate key numbers in the study of cell biology such as sizes, concentrations, rates, energies, etc.

Minkoff, Eli C. *Evolutionary Biology.* Reading, Mass.: Addison-Wesley, 1983. A college textbook that deals with evolution in an interdisciplinary way. It is a good reference book that will be most useful to those who have had an introductory course in college-level biology, but it can also be understood by beginners because it clearly explains new terms and has a good index and a glossary.

Mitchell, C. Ben, et al. *Biotechnology and the Human Good.* Washington D.C.: Georgetown University Press, 2007. A distinctly Christian assessment of the application of biotechnology to humans that remains optimistic but cautious and concerned.

Mitchell, Lawrence G., John A. Mutchmor, and Warren D. Dolphin. *Zoology.* Menlo Park, Calif.: Benjamin/Cummings, 1988. This college text covers the nervous system in general, as well as discussing the different individual categories of animals, including their nervous system specializations and behaviors.

Montaigne, Fen. *Medicine By Design: The Practice and Promise of Biomedical Engineering.* Baltimore: The Johns Hopkins University Press, 2006. Bioengineering (including neuroengineering) applications made accessible to the nonspecialist through vignettes and portraits of researchers.

Moore, Janice. *Parasites and the Behavior of Animals.* 2002. Reprint. New York: Oxford University Press, 2005. An interesting look at a variety of parasites and other animals, written for the general reader;

contains information about medical, agricultural, and ecological parasitology.

Morell, V. "Life on a Grain of Sand." *Discover* 16 (April, 1995): 78-86. A close look at the sand beneath shallow waters, home to incredibly diverse microscopic creatures.

Morgan, T. H. "Sex Limited Inheritance in *Drosophila*." *Science* 32 (1910): 120-122. Morgan demonstrates how a spontaneous mutation (sport) affecting eye color in *Drosophila* is passed to succeeding generations. He showed that a mutated form of the gene is recessive to the normal form of the gene and that the mutated and normal gene are linked to the X chromosome. Genes that are linked to the X chromosome are now known as sex-linked genes.

Morrison, Robert T, Robert N. Boyd, and Saibal K. Bhattacharjee. *Organic Chemistry*. Noida: Dorling Kindersley, 2013. Print.

Mosier, Nathan S., and Michael R. Ladisch. *Modern Biotechnology: Connecting Innovations in Microbiology and Biochemistry to Engineering Fundamentals*. Hoboken, N.J.: Wiley-AIChE, 2009. A very practical and richly illustrated and referenced guide to the advances in molecular biology for aspiring biochemical engineers.

Moss, Cynthia. *Elephant Memories*. New York: William Morrow, 1988. An interesting account of thirteen years of field observations concerning the behavior of elephants in the Amboseli National Park in Kenya.

Moussaieff Masson, Jeffrey, and Susan McCarthy. *When Elephants Weep: The Emotional Lives of Animals*. New York: Delacorte Press, 1995. Full of engaging anecdotes, this scholarly, insightful book is a delight to read.

Mousseau, T. A., B. Sinervo, J. A. Endler, eds. *Adaptive Genetic Variation in the Wild*. New York: Oxford University Press, 1999. Provides excellent background and current information for readers interested in population genetics and evolution in natural systems.

Mueller, Richard L., and Timothy A. Sanborn. "The History of Interventional Cardiology: Cardiac Catheterization, Angioplasty, and Related Interventions." *American Heart Journal* 129, no. 1 (January, 1995): 146-172. Contains plenty of names, dates, and details of interest in cardiology history.

Mumenthaler, Marco, and Heinrich Mattle. *Fundamentals of Neurology: An Illustrated Guide*. New York: Thieme, 2006. A well-written, logically ordered textbook for the novice in neuroscience and neurology.

Murphy, Joseph G. *Mayo Clinic Cardiology: Concise Textbook*. 3d ed. London: Informa Healthcare Communications, 2006. This easy-to-read textbook features information on all aspects of cardiology.

Murphy, Kenneth M., Paul Travers, and Mark Walport. *Janeway's Immunobiology*. 7th ed. Oxford, England: Taylor & Francis, 2007. A standard immunology textbook that has an excellent section on the therapeutic use of antibodies.

Murray, Patrick D. F. *Bones: A Study of the Development and Structure of the Vertebrate Skeleton*. Cambridge, England: Cambridge University Press, 1936. Reprint. New York: Cambridge University Press, 1985. Covers many aspects of the development and anatomy of the skeletons of the vertebrates.

Myers, Richard. *The Basics of Chemistry*. Westport: Greenwood, 2003. Print.

Naftolin, Federick, et al. *Science* 211 (March 20, 1981). The entire issue of this magazine is devoted to sexual differences and how they develop. Though difficult for the casual reader, most of the articles were written for a general scientific audience and can be understood by someone with a strong interest in the subject. Each article has a good Bibliography and deals with the genetic or hormonal mechanisms that govern sex organ development.

Nagylaki, Thomas. *Introduction to Theoretical Population Genetics*. New York: Springer-Verlag, 1992. A college-level textbook, using elementary mathematics to investigate theoretical population genetics. Uses calculus and linear algebra, but does not require a background in genetics.

Nakamura, Yusuke. "DNA Variations in Human and Medical Genetics: Twenty-five Years of My Experience." *Journal of Human Genetics* 54 (2009): 1-8. A historic perspective on the progression of DNA analysis techniques with a particular emphasis on human disease characterization.

Nalbandov, A. V. *Reproductive Physiology of Mammals and Birds: The Comparative Physiology of Domestic and Laboratory Animals and Man*. San Francisco: W. H. Freeman, 1976. This book is written for beginning students of reproductive physiology, and it focuses on the essentials. As the title implies, the author takes a comparative perspective, surveying reproduction in different species.

Nash, Michael R., and Amanda J. Barnier, eds. *The Oxford Handbook of Hypnosis: Theory, Research and Practice*. New York: Oxford University Press, 2008. Presents both academic theory and practical approaches. Contains name and subject indexes.

National Agricultural Statistics Service. *U.S. Broiler and Egg Production Cycles*. Washington, D.C.: USDA National Agricultural Statistics Service, 2005. A governmental document providing information on egg production cycles and chickens for those in the poultry industry.

Nebel, Bernard J., and Richard T. Wright. *Environmental Science: Towards a Sustainable Future*. 10th ed. Englewood Cliffs: Prentice Hall, 2008. Describes several bioprocesses used in waste treatment and pollution control.

Nelson, Lewis S., et al. *Goldfrank's Toxicologic Emergencies*. 9th ed. New York: McGraw-Hill, 2011. Includes comprehensive references; begins with general principles and moves to detailed discussions of biochemical principles; discusses various exposures— drugs, plants, metals, household products, as well as occupational and environmental.

Nemerow, Nelson Leonard, et al., eds. *Environmental Engineering*. 3 vols. 6th ed. Hoboken, N.J.: John Wiley & Sons, 2009. Discusses topics such as food protection, soil management, waste management, water supply, and disease control. Each section includes references and a Bibliography.

Norman, A. W., and G. Litwack. *Hormones*. 2d ed. San Diego, Calif.: Academic Press, 1997. An excellent source of information on the biochemistry and molecular biology of the endocrine system.

Norris, June, ed. *Diseases*. 2d ed. Springhouse, Pa.: Springhouse, 1997. Written for nursing and allied health professions students, this text examines infections, trauma, neoplasms, and other disorders of every body system. Many tables cover causes, assessment, diagnostic tests, and treatments for these disorders.

Norwitz, Errol R., et al., eds. *Oxford American Handbook of Obstetrics and Gynecology*. New York: Oxford University Press, 2007. Covers all aspects of obstetrics and gynecology as well as diagnosis and treatment options.

Oatis, Carol A. *Kinesiology: The Mechanics of Human Movement*. Baltimore: Lippincott Williams & Wilkins, 2009. Covers biomechanics, movement at the major joints, and posture.

Oppenheimer, Steven B,, and Edward J. Carroll. *Introduction to Embryonic Development*. Upper Saddle River, N.J.: Pearson Education, 2004. Print. An intermediate-level college text which gives extensive coverage to the embryological stages in primitive chordate and vertebrate classes. Molecular and cellular aspects of development are emphasized, and the discussion of molecular genetics is informative. Extensive coverage of the topics of cleavage, gastrulation, and neurulation. Illustrated, glossary, references.

Ostergaard, Simon, Lisbeth Olsson, and Jens Nielsen. "Metabolic Engineering of *Saccharomyces cerevisiae*." *Microbiology and Molecular Biology Reviews* 64, no. 1 (2000): 34-50. Describes metabolic engineering techniques using *S. cerevisiae* as an example.

Osweiler, Gary D., et al., eds. *Blackwell's Five-Minute Veterinary Consult Clinical Companion: Small Animal Toxicology*. Ames, Iowa: Wiley-Blackwell, 2011. Overview of toxicology in veterinary practice; includes color photos and tables in an appendix to help with quick differential diagnoses.

Ovalle, William K., and Patrick C. Nahirney. *Netter's Essential Histology*. Philadelphia: Saunders/Elsevier, 2008. This atlas covers cells and tissues and the major bodily systems. Features a great collection of images by Frank H. Netter.

Owen, Denis. *Survival in the Wild: Camouflage and Mimicry*. Chicago: University of Chicago Press, 1980. A look at animals that appear to be something other than what they are. Some try to look like their background to avoid detection, others try to appear like a dangerous animal, while others are brightly colored to advertise nastiness. Easy reading with numerous illustrations and photographs.

Paganelli, Charles V., and Leon E. Farhi, eds. *Physiological Function in Special Environments*. New York: Springer-Verlag, 1988. This book presents the papers from a symposium on environmental physiology held in Buffalo, New York. The chapters are written by experts in the field for presentation to scientists. The papers are well written and informative, and are suited to the person who is able to master the other entries here and still wishes to pursue the topic. Chapters 1 to 4 on adaptation to altitude, and chapters 12 and 13, on comparative physiology, deal with physiological adaptations related to low-oxygen conditions.

Page, George. *Inside the Animal Mind.* New York: Doubleday, 1999. The author begins with a historical account of the popular and scientific views about animal intelligence. He provides good details about the various attempts to communicate with primates by teaching them sign language or through the use of keyboards.

Pagliarulo, Michael A., ed. *Introduction to Physical Therapy.* 3d ed. St. Louis, Mo.: Mosby, 2007. Examines the history of physical therapy, discusses financial and legal aspects, and contains a chapter on physical therapy for musculoskeletal conditions.

Pahl, Greg. *Biodiesel: Growing a New Energy Economy.* 2d ed. White River Junction, Vt.: Chelsea Green, 2008. A popular guide to the advances in biodiesel technology and biofuel industries that also examines the food-for-fuel controversy and the issues surrounding genetically modified crops.

Palmore, Erdman B. *The Facts on Aging Quiz: A Handbook of Uses and Results.* New York: Springer, 1988. Contains several quizzes that test common misconceptions of the aging process.

Panksepp, Jaak. *Affective Neuroscience: The Foundation of Human and Animal Emotions.* New York: Oxford University Press, 1998. This book presents a very thorough and scientific exploration of the neurochemicals associated with emotions. The author shows how the basic emotions and motivational processes are controlled by brain chemistry.

Panno, Joseph. *Stem Cell Research: Medical Applications and Ethical Controversy.* New York: Facts On File, 2004. Provides information on the technological advances, applications, and issues of stem cell research, including the use of stem cells to repair damaged nerve tissue and the ethical and legal implications of research in this field.

Park, Sheldon J., and Jennifer R. Cochran, eds. *Protein Engineering and Design.* Boca Raton, Fla.: CRC Press, 2010. Covers the broader field of protein engineering—methods of developing altered proteins for novel applications—in two sections: one on experimental protein engineering and the other on computational design. Includes discussion of enzyme engineering using both rational and combinatorial approaches.

Pasternak, Jack J. *An Introduction to Human Molecular Genetics: Mechanisms of Inherited Diseases.* 2d ed. Hoboken, N.J.: Wiley-Liss, 2005. Discusses treatment advances, fundamental molecular mechanisms that govern human inherited diseases, the interactions of genes and their products, and the consequences of these mechanisms on disease states in major organ systems such as muscles, the nervous system, and the eyes. Also addresses cancer and mitochondrial disorders.

Patlak, Margie. "Targeting Leukemia: From Bench to Bedside." *The Federation of American Societies for Experimental Biology* 16, no. 3 (March, 2002): 273. An interesting review of leukemia diagnosis and treatment.

Pattie, Frank A. *Mesmer and Animal Magnetism: A Chapter in the History of Medicine.* Hamilton, N.Y.: Edmonston, 1994. Mesmer specialist Pattie provides a thoughtful biography of Mesmer that includes the probable source of his thought.

Patton, Kevin T. and Thibodeau, Gary A. *Anatomy & Physiology* 9th ed., St. Louis, MO: Elsevier, 2016.

Paul, William E. *Fundamental Immunology.* 4th ed. Philadelphia: Lippincott-Raven, 1999. Despite the name, this medical text covers all aspects of human immune function in over 1,500 pages. Regulation of the immune system and its functions in fighting disease are discussed extensively.

Pelczar, Michael J., Jr., E. C. S. Chan, and Noel R. Krieg. *Microbiology: Concepts and Applications.* New York: McGraw, 1993. Print.

Pereira, Filipe, Joao Carneiro, and Antonio Amorim. "Identification of Species with DNA-Based Technology: Current Progress and Challenges." *Recent Patents on DNA and Gene Sequence* 2 (2008): 187-200. Contains an excellent table comparing methods of DNA analysis, along with a helpful flowchart. Also contains clear descriptions of each method, including diagrams.

Pettit, George. *Biosynthetic Products for Cancer Chemotherapy.* Vol. 5 London: Elsevier Science, 1985. A discussion of the fundamental processes involved with screening for antitumor agents.

Phenice, Terrell W. *Hominid Fossils.* Dubuque, Iowa: Wm.C. Brown, 1973. One of the best introductory illustrated keys of hominid fossil remains of the Pleistocene epoch. The drawings are all oriented the same way, and each fossil illustration is listed with its measurements and with information concerning tools found at the respective sites. This book (and similar types of atlases) is very helpful to the nonspecialist.

Phillips, Anthony. *The Optometrist's Practitioner-Patient Manuel.* Oxford, England: Butterworth-Heinemann, 2008. Information and illustrations of procedures and conditions pertaining to eye health.

Picard, Alyssa. *Making the American Mouth: Dentists and Public Health in the Twentieth Century.* New Brunswick, N.J.: Rutgers University Press, 2009. Presents a history of dentistry as well as essays on issues such as dental hygiene, dental economics, and the American diet.

Pilla, Louis. "Cosmetic Versus Medical Dermatology: A Widening Gap?" *Skin and Aging* 11, no. 6 (June, 2003). Analysis of the interplay between medical and cosmetic dermatology in modern practices.

Pine, Stanley H. *Organic Chemistry.* 5th ed. New York: McGraw, 1987. Print.

Pinney, Chris C. *Veterinary Guide for Dogs, Cats, Birds, and Exotic Pets.* Blue Ridge Summit, Pa.: Tab Books, 1992. Holds much useful data on pet-keeping, breeding, and parturition

Pintar, Judith, and Steven Jay Lynn. *Hypnosis: A Brief History.* Malden, Mass.: Wiley-Blackwell, 2008. Offers a compact but critical overview of the origins and the history of hypnosis, including accounts on debates within psychology, references and index.

Plotkin, Stanley A., Walter A. Orenstein, and Paul A. Offit, eds. *Vaccines.* 5th ed. Philadelphia: Saunders/ Elsevier, 2008. An excellent source for understanding the history of vaccine development against most major agents.

Pollard, Thomas, D. MD, William C. Earnshaw, Jennifer Lippincott-Schwartz PhD, and Graham Johnson MA PhD, CMI. *Cell Biology.* Elsevier, 2016. Cover key principles of cellular function with clear, concise text and visually interesting illustrations, diagrams, and charts.

Potter, Daniel A., and Jennifer S. Hanin. *What to Do When You Can't Get Pregnant: The Complete Guide to All the Technologies for Couples Facing Fertility Problems.* New York: Marlowe, 2005. A thorough guide for couples with fertility problems.

Pough, F.,H. Heiser, J. B. Heiser, andW. N. McFarland. *Vertebrate Life.* 5th ed. Upper Saddle River, N.J.: Prentice Hall, 1999. Emphasis on the differences in animals based on the sequences of evolution.

Prescott, David M. *Cells: Principles of Molecular Structure and Function.* Boston: Jones and Bartlett, 1988. This college-level textbook on classic cell biology is one of the best written on the topic. Cytology is covered well, as is other molecular biology information. Chapters dealing with cellular evolution and specialized cells are especially well done and will be of interest to the reader who wants further information.

Princeton Review. *High School Biology Unlocked: Your Key to Understanding and Mastering Complex Biology Concepts.* Princeton Review, 2016. Straightforward and easy-to-understand lessons on biology concepts; includes practice questions with complete explanation of answers to enhance understanding of concepts such as homeostasis.

Proctor, Noble S., and Patrick J. Lynch. *Manual of Ornithology: Avian Structure and Function.* New Haven, Conn.: Yale University Press, 1993. Avisual guide to the structure and anatomy of birds. The text is informative and written at a level appropriate to undergraduate students and to bird lovers in general.

Prosser, C. Ladd. *Adaptational Biology: Molecules to Organisms.* New York: JohnWiley & Sons, 1986. This text, by the leader in the field of adaptational physiology, is one of the most comprehensive in the field.

Provine, William B. *Sewall Wright and Evolutionary Biology.* Chicago: University of Chicago Press, 1986. Sewall Wright was one of the most important figures in the development of evolutionary biology in the twentieth century. Besides describing his life, this book places his work within the context of the development of evolutionary theory. Chapters 7, 8, and 9 present an excellent overview of theories of evolution and natural selection and are understandable by a general audience.

Pyle, Marsha, et al. "The Case for Change in Dental Education." *Journal of Dental Education* 70, no. 9 (September, 2006): 921-924. The American Dental Education Association's Commission on Change and Innovation in Dental Education examines the need for change in dental education. It takes into account the financial expense of a dental education and the professional responsibilities of meeting all individual and public health needs.

Raup, D. M., and D. Jablonski, eds. *Patterns and Processes in the History of Life.* New York: Springer-Verlag, 1986. Raup and Jablonski's book is a compilation of articles from the Dahlem workshop, "Patterns and Processes in the History of Life," held in Berlin in 1985. The articles in the book discuss the evidence concerning the role of migration and gene flow in the evolutionary process. Thought-provoking.

Raven, P. H., and G. B. Johnson. *Biology*. 4th ed. Boston: McGraw-Hill, 1996. An introductory biology text for college students. The chapter on circulation provides a good description of circulatory systems in animals, with helpful diagrams and illustrations. Good for the beginner.

Raymond, Kenneth W. *General, Organic, and Biological Chemistry: An Integrated Approach*. 4th ed. Hoboken: Wiley, 2014. Print.

Reader, John. *Missing Links: The Hunt for Earliest Man*. Boston: Little, Brown, 1981. Organized chronologically, chapters cover the discoveries of the major hominids. Outstanding original photographs by the author are featured. Though other treatments have been more detailed (and more technical), Reader's agreeable combination of history and science makes his book irresistible. For readers at all levels.

Reece, Jane B, and Neil A. Campbell. *Biology: Concepts & Connections*. Boston: Pearson, 2015. Print.

Regan, Tom. *The Case for Animal Rights*. Berkeley: University of California Press, 1983. Regan's book describes his philosophy, as an animal rights leader, that nonhuman animals have moral rights and that recognition of these rights requires changes in how we treat them.

Rennie, J. "Living Together." *Scientific American* 266 (January, 1992): 122. The fascinating interactions between parasites and their hosts provide insights into life on earth.

Restak, Richard M. *The Modular Brain*. New York: Charles Scribner's Sons, 1994. Discussion of the workings of the human brain is accompanied by descriptions of animal brain and behavior experimentation as well as observations on humans. Brain activities in cats, dogs, rats, and monkeys are examined and described.

Richards-Kortum, Rebecca. *Biomedical Engineering for Global Health*. New York: Cambridge University Press, 2010. Examines the potential of biomedical engineering to treat diseases and conditions throughout the world. Examines health care systems and social issues.

Richards, Ira S. *Principles and Practice of Toxicology in Public Health*. Sudbury, Mass.: Jones & Bartlett, 2008. An overview of toxicology in public health practice, written in an easy-to-read style. Includes glossary and index.

Richardson, S. "The Benefits of Virgin Birth." *Discover* 17 (March, 1996): 33. A brief summary of the advantages of asexuality in Pacific geckos.

Rijnberk, A., and H.W. De Vries. *Medical History and Physical Examination in Companion Animals*. Translated by B. E. Belshaw. Boston: Kluwer, 1995. A handbook for veterinarians, includes coverage of the reproductive systems of animals that are commonly found as pets.

Robbins, Clarence R. *Chemical and Physical Behavior of Human Hair*. 3d ed. New York: Springer-Verlag, 1994. The first chapter of this text covers the morphological and macromolecular structure of hair.

Roberts, Michael, Michael Jonathan Reiss, and Grace Monger. *Advanced Biology*. Cheltenham: Nelson,

Robertson, James, ed. *Forensic Examination of Hair*. London: Taylor & Francis, 2000. This book on the forensics of human hair has two solid, basic chapters on hair physiology and growth and on its microscopic examination.

Robinson, T. F., et al. "The Heart as a Suction Pump." *Scientific American* 254 (June, 1986): 84-91. An excellent introduction to the heart and its functioning, for high school students, college freshmen, and general readers. Provides basic information at a nontechnical level.

Roitt, Ivan, Jonathan Brostoff, and David Male. *Immunology*. 5th ed. Philadelphia: C. V. Mosby, 1998. Both vertebrate and invertebrate immune functions are discussed in Chapter 15, "Evolution of Immunity." Good color photos and diagrams help explain cellular and molecular aspects of immunology.

Romano, Amy. *Cell Specialization and Reproduction: Understanding How Cells Divide and Differentiate (Library of Cells)*. Rosen Publishing Group, 2005. Gives both a high-level overview and a deep look into how specialized cells, tissues, and organs develop; covers both mitosis and meiosis in depth.

Romundstad, Liv Bente, et al. "Effects of Technology or Maternal Factors on Perinatal Outcome After Assisted Fertilisation: A Population-based Cohort Study." *The Lancet* 372, no. 9640 (August, 2008): 737-743. Finds that adverse outcomes associated with births following assisted reproduction are not caused by technological factors. Includes several tables.

Roper, Stephan M., and Owatha L. Tatum. "Forensic Aspects of DNA-Based Human Identity Testing."

Journal of Forensic Nursing 4 (2008): 150-156. A straightforward description of all pertinent methods and applications of DNA analysis, including simple diagrams as well as a glossary of terms.

Rose, Michael A. *Darwin's Spectre: Evolutionary Biology in the Modern World.* Princeton, N.J.: Princeton University Press, 1998. Written for the general reader. Outlines the fundamental ideas of evolutionary biology and its influence in other fields such as agriculture, medicine, and eugenics.

Rose, Nickolas. *The Politics of Life Itself: Biomedicine, Power, and Subjectivity in the Twenty-first Century.* Princeton, N.J.: Princeton University Press, 2006. An introduction to the moral, ethical, and political issues that surround medical engineering, genetic manipulation, and bioengineering. Addresses several prominent fields in cell and tissue engineering.

Rosen, Vicki, and R. Scott Theis. *The Cellular and Molecular Basis of Bone Formation and Repair.* Austin, Tex.: R. G. Landes, 1995. This very nice book covers many topics related to bone growth, bone regeneration, other remodeling, and growth factors. Included are a great many bibliographical references and useful illustrations.

Rosenfeld, Israel, Ziff, Edward and Van Loon, Borin *DNA. A Graphic Guide to the Molecule That Shook the World* New York, NY: Columbia University Press, 2011.

Ross, Michael H., and Pawlina Wojciech. *Histology: A Text and Atlas—With Correlated Cell and Molecular Biology.* 6th ed. Philadelphia: Wolters Kluwer/Lippincott Williams & Wilkins Health, 2011. Contains great illustrations, easy to follow diagrams. Recommended for medical, health professions, and undergraduate biology students

Rossomando, Edward F., and Mathew Moura. "The Role of Science and Technology in Shaping the Dental Curriculum." *Journal of Dental Education* 72, no. 1 (January, 2008): 19-25. Offers a history of the changing dental school curricula in the United States and offers perspectives for the future of dentistry education.

Rothwell, Norman V. *Understanding Genetics.* 4th ed. New York: Oxford University Press, 1988. This college-level genetics textbook requires little more than a basic familiarity with genetic terminology. Contains a very thorough treatment of sex and inheritance, including information on sex chromosomes and their role in sexual differentiation, and related topics such as sex-linked, sex-influenced, and sex-limited genetic traits in various animals. Good Bibliography and glossary.

Rowland, Lewis P., Thomas A. Pedley, and H. Houston Merritt, eds. *Merritt's Neurology.* 12th ed. Philadelphia: Lippincott Williams & Wilkins, 2010. Provides the essential facts about common and rare diseases. Includes chapters on neurological symptoms, diagnostic tests, and rehabilitation.

Rubin, Raphael, and David S. Strayer, eds. *Rubin's Pathology: Clinicopathologic Foundations of Medicine.* 6th ed. Philadelphia: Lippincott Williams & Wilkins, 2011. Award-winning book taking a clinical approach to pathology. Includes full-color, enhanced illustrations.

Rudin, Norah, and Keith Inman. *An Introduction to Forensic DNA Analysis.* 2d ed. Boca Raton, Fla: CRC Press, 2002. Discusses forensic DNA analysis from both the medical and legal standpoints. Examines the advantages and limitations of the various techniques.

Sacks, Oliver. *Awakenings.* 1973. Reprint. New York: Vintage Books, 1999. Exceptional case history and literary work, describing the outcome of treating encephalitis lethargica patients institutionalized since World War I.

Sadava, David E. *Cell Biology: Organelle Structure and Function.* Boston: Jones and Bartlett, 1993. A textbook on the biochemistry, molecular biology, and structure of eukaryotic cells and organelles. For students with college-level introductory biology and chemistry. Illustrated.

Sadler, R.M. *The Reproduction of Vertebrates.* New York: Academic Press, 1973. A survey of reproductive patterns and mechanisms of reproductive control in the major vertebrate groups. Useful to the student who wants a comparative presentation of all aspects of reproductive events.

Saferstein, Richard. *Criminalistics: An Introduction to Forensic Science.* 10th ed. Upper Saddle River, N.J.: Prentice Hall, 2011. Provides an introduction to forensic science, detailing the techniques to analyze physical, biological, and chemical evidence.

San Krieg. *Microbiology: Concepts and Applications.* New Francisco: Cummings, 2013. Print. York: McGraw, 1993. Print.

Savage-Rumbaugh, Sue, Stuart G. Shanker, and Talbot J. Taylor. *Apes, Language, and the Human Mind.* New York: Oxford University Press, 1998. A book written from an academic and scientific perspective about the ability of chimpanzees to communicate by means of symbols and a keyboard display.

Savage-Rumbaugh, Sue. *Kanzi: The Ape at the Brink of the Human Mind.* New York: John Wiley&Sons, 1994. This book presents an apparent breakthrough in the communication with chimpanzees using symbols and a keyboard. It includes a number of incidences of the spontaneous construction of sentences by the primates.

Savageau, Michael. *Biochemical Systems Analysis: A Study of Function and Design in Molecular Biology.* New York: CreateSpace, 2010. Detailed textbook describing the immune system and gene regulation.

Sayre, Anne. *Rosalind Franklin and DNA.* New York: W.W. Norton, 1975. Details the invaluable contribution of Dr. Rosalind Franklin to the discovery of the structure of the DNA molecule.

Schaie, K. Warner, and Laura L. Carstensen, eds. *Social Structures, Aging, and Self-Regulation in the Elderly.* New York: Springer, 2006. Examines the evolution of personal and social roles in aging with particular emphasis on familial changes, immigration, and increased life span.

Scheindlin, Stanley. "Clinical Enzymology: Enzymes As Medicine." *Molecular Interventions* 7, no. 1 (February, 2007): 4-8. An absorbing and readable Summary of the use of engineered enzymes in clinical diagnoses and treatments.

Schmidt-Nielsen, Knut. *Animal Physiology: Adaptation and Environment.* New York: Cambridge University Press, 1997. A well-regarded college-level textbook on animal physiology, which covers aging in the context of the whole life of the animal.

_____. *Desert Animals: Physiological Problems of Heat and Water.* New York: Dover, 1979. This engaging book provides considerable insights into thermoregulation by animals living in possibly the most inhospitable climate on earth. The use of case studies describing mechanisms by individual species is quite useful. Coverage is fairly technical but suitable for the interested nonscientist. Complete literature-cited section.

_____. *How Animals Work.* Cambridge, England: Cambridge University Press, 1972. This well-written text discusses in some detail various mechanisms of temperature regulation in vertebrates. Adequately illustrated. The strength of this work is its integration of temperature regulation into the total context of animal physiology. Complete literature-cited section.

Schrier, Robert W. *Manual of Nephrology.* 7th ed. Philadelphia: Lippincott, Williams & Wilkins, 2009. An excellent clinical reference guide for the advanced student.

Schwadron, Terry. "Hot Sounds From a Cold Trumpet? Cryogenic Theory Falls Flat." *New York Times*, November 18, 2003. Explains how two Tufts University researchers studied cryogenic freezing of trumpets and determined the cold did not improve the sound.

Scuderi, Giles R., and Peter D. McCann, eds. *Sports Medicine: A Comprehensive Approach.* 2d ed. Philadelphia: Mosby-Elsevier, 2005. Provides informative chapters on each body part with a focus on sports medicine injuries and rehabilitation.

Sears, Andrew, and Julie A. Jacko, eds. *The Human-Computer Interaction Handbook: Fundamentals, Evolving Technologies, and Emerging Applications.* 2d ed. New York: Lawrence Erlbaum Associates, 2008. Prominent researchers in the field address issues in design, development, testing, and evaluation. Contains hundreds of explanatory figures and tables.

Selander, Robert K., Andrew G. Clark, and Thomas S. Whittam, eds. *Evolution at the Molecular Level.* Sunderland, Mass.: Sinauer Associates, 1991. A collection of essays focused on four themes: bacteria and viruses, organelles, "selfish" genes, and nuclear multigene families.

Sell, Stewart, ed. *Stem Cells Handbook.* Totowa, N.J.: Humana Press, 2007. Explains the origins of stem cells and describes how they function and how they can treat illness and disease. Emphasis is placed on the role of stem cells in development, tissue regeneration, repair mechanisms, and carcinogenesis. Also includes technical approaches to obtaining stem cells and manipulating them for therapeutic use.

Setchell, B. P. *The Mammalian Testis.* Ithaca, N.Y.: Cornell University Press, 1978. Although written for a scientific audience, the work is noteworthy because of its completeness. Contains detailed information about all aspects of the testis, scrotum, spermatogenesis, and hormonal control, including historical perspectives of the research in male reproduction. The book is also useful because it covers all mammals, not only humans, so the reader is able to appreciate similarities and differences among the mammals.

Shanks, Pete. *Human Genetic Engineering: A Guide for Activists, Skeptics, and the Very Perplexed.* New York: Nation Books, 2005. A helpful explication of the science behind cloning, coupled with stern warnings against it, by a noted social activist.

Sharp, Heken, Yvonne Rogers, and Jenny Preece. *Interaction Design: Beyond Human-Computer Interaction.* 2d ed. Hoboken, N.J.: John Wiley & Sons, 2007. Each chapter on interaction design contains an outline, Summary, subsections, and text boxes highlighting case studies and other points of focus. Includes a comprehensive Bibliography.

Sharp, Lesley A. *Bodies, Commodities, and Biotechnologies: Death, Mourning, and Scientific Desire in the Realm of Human Organ Transfer.* New York: Columbia University Press, 2008. Explores how organ transplantation and artificial organs have changed cultural attitudes toward the body.

Sheldrake, Rupert. *Dogs That Know When Their Owners Are Coming Home.* New York: Crown, 1999. The author is a scientist and philosopher who presents numerous incidents of unusual abilities found in dogs. The information on animal empathy is particularly informative.

Sherman, Paul W., and John Alcock, eds. *Exploring Animal Behavior: Readings from "American Scientist."* 2d ed. Sunderland, Mass.: Sinauer Associates, 1998. A good collection of essays on various aspects of animal behavior.

Short, R. V., and E. Balaban, eds. *The Differences Between the Sexes.* New York: Cambridge University Press, 1994. A collection of symposium papers covering a wide variety of topics related to sex differences in animals.

Shostak, Stanley. *Becoming Immortal: Combining Cloning and Stem-Cell Therapy.* Albany: State University of New York Press, 2002. Examines the question of whether human beings are equipped for potential immortality.

Shreeve, J. *The Neanderthal Enigma.* New York: William Morrow, 1995. The author wrote this book after spending many months interviewing leading anthropologists. The book presents the controversies over Neanderthal interpretation in an exciting and readable fashion. One may disagree with the author's hypothesis, but he tells a truly fascinating story. Contains an extensive Bibliography of both research articles and books.

Siebel, Markus J., Simon P. Robins, and John P. Bilezikian, eds. *Dynamics of Bone and Cartilage Metabolism.* San Diego, Calif.: Academic Press, 1999. Covers many aspects of bone and cartilage metabolism and diseases. References and illustrations.

Sieglaff, D. "Most Spectacular Mating." In *University of Florida Book of Insect Records*, edited by T. J.Walker. Gainesville: University of Florida Department of Entomolgy and Nematology, 1999. Also at theWeb site http://gnv.ifas.ufl.edu/~tjw/recbk.htm. An electronically published paper on bee mating.

Silbey, Robert J., Robert A. Alberty, and Moungi G. Bawendi. *Physical Chemistry.* 5th ed. Hoboken: Wiley, 2012. Print.

Silver, Lee. *Challenging Nature: The Clash Between Biotechnology and Spirituality.* New York: Harper Perennial, 2006. A Princeton stem cell scientist explains the science behind biotechnology and stem cells. He offers some rather harsh critiques of more conservative thinkers who do not agree with his optimistic views of genetic enhancement and embryonic stem cells.

Singer, Peter. *Writings on an Ethical Life.* New York: HarperCollins, 2000. Includes extracts from this Australian philosopher's scholarly writings on animal liberation.

Sinha, Sujeet K., and Brian J. Briscoe. *Polymer Tribology.* London: Imperial College Press, 2009. This book addresses the relatively recent study of the tribology of polymeric materials with their increasing use as dynamic components such as nylon gears and Teflon-coated slides.

Skinner, Harry B., ed. *Current Diagnosis and Treatment in Orthopedics.* 4th ed. New York: Lange Medical Books/McGraw-Hill Medical Publishing Division, 2006. An accessible resource for diagnosis of orthopedic conditions, covering trauma, sports medicine, oncology, surgery, amputations, and rehabilitation

Skugor, Mario, and Jesse Bryant Wilder. *The Cleveland Clinic Guide to Thyroid Disorders.* New York: Kaplan, 2009. Skugor, an endocrinologist, teamed with writer Wilder to present detailed information on thyroid diseases and treatment options.

Slater, P. J. B. *Essentials of Animal Behavior.* New York: Cambridge University Press, 1999. A basic introduction to ethology, which considers the effects of aging on behavior.

Slobodkin, Lawrence B. *Growth and Regulation of Animal Populations.* New York: Holt, Rinehart and Winston, 1961. A classic text on population analysis; covers the effects of aging on population demographics.

Smith, C. G. *Ancestral Voices: Language and the Evolution of Human Consciousness.* Englewood Cliffs, N.J.: Prentice-Hall, 1985. This book is written for the general reader and presents a fusion of anthropology, linguistics, and neuroscience to describe the final emergence of the human species from its animal background. It emphasizes the importance of cultural evolution in shaping modern humans. Contains many illustrations, a list of references, and a list of suggested readings.

Smith, Homer W. "The Kidney." *Scientific American* 188 (January, 1973): 40-48. A classic exploration of the structure and function of the kidney and its role in osmoregulation.

Smith, Marquard, and Joanne Morra, eds. *The Prosthetic Impulse: From a Posthuman Present to a Biocultural Future.* Cambridge, Mass.: MIT Press, 2007. Examines the developments in prosthetic devices and addresses the social aspects, including what it means to be human.

Smith, R. L. ed. *Sperm Competition and the Evolution of Animal Mating Systems.* Orlando, Fla.: Academic Press, 1984. A thorough look at mating patterns in animals.

Smith, Robert Leo. *Elements of Ecology.* 4th ed. San Francisco, Calif.: Benjamin/Cummings, 2000. This textbook does an excellent job of covering the breadth of topics found in modern ecology. The treatment of the topics is unusually complete and provides a close linkage between theoretical principles and numerous examples from specific ecological situations. Covers population age structure, as well as natality, mortality, and survivorship.

Solecki, R. S. *Shanidar, the First Flower People.* New York: Alfred A. Knopf, 1971. Solecki was the anthropologist who excavated a Neanderthal skeleton that pollen analysis revealed to have been laid on a bed of flowers. This is a poetic book, with interesting speculation about the "humanness" of the Neanderthals, and a very good story about the process of excavation. Solecki has been an authority in the field of late human evolution and gives a good account of the state of knowledge in the 1960's. Bibliography, pictures, and diagrams.

Solomon, E. P., L. R. Berg, and D.W. Martin. *Biology.* 5th ed. Philadelphia: Saunders College Publishing, 1998. Fairly comprehensive general biology text with sections on invertebrate and vertebrate systems.

Spar, Debora. *The Baby Business: How Money, Science and Politics Drive the Commerce of Conception.* Boston: Harvard Business School Press, 2006. A critical overview of the fertility industry and how reproductive technologies are used in the marketplace. Includes numerous tables and extensive end notes listing sources.

Spaulding, C. E., and Jackie Clay. *Veterinary Guide for Animal Owners: Sheep, Poultry, Rabbits, Dogs, Cats.* Emmaus Pa.: Rodale Press, 1998. Provides much interesting information on keeping animals, including gestation, parturition and related problems

Spearman, R. I. C., and P. A. Riley, eds. *The Skin of Vertebrates.* New York: Academic Press, 1980. The proceedings of an international congress, including sound coverage of skin and hair.

Spentzos, Dimitri. "Gene Expression Signature with Independent Prognostic Significance in Epithelial Ovarian Cancer." *Journal of Clinical Oncology* 22, no. 23 (December, 2004): 4648-4658. The research article describes the diagnosis of ovarian cancer and the use of biomarkers for detection.

Sponenberg, D. P. *Equine Color Genetics.* Ames: Iowa State University Press, 1996. A fascinating, in-depth look at the genetic mechanisms that cover coat color in horses and donkeys.

Staines, Norman, Jonathan Brostoff, and Keith James. *Introducing Immunology.* 2d ed. St. Louis: C. V. Mosby, 1993. For readers with no previous knowledge of immunology, this primer covers the complex language without using jargon and explains the information clearly.

Stanford Environmental Law Society. *The Endangered Species Act.* Stanford, Calif.: Stanford University Press, 2001. Describes endangered species law and legislation, with a solid Bibliography.

Stanforth, Stephen. *Natural Product Chemistry at a Glance.* New York: Wiley-Blackwell, 2006. An introductory textbook that describes much of the organic chemistry involved in biosynthesis.

Stanley, Steven M. *Earth and Life Through Time.* 2d ed. New York: W. H. Freeman, 1989. A textbook for introductory college historical geology classes that can be readily understood by those without previous training in geology. It clearly covers the origin of life and rise of multicellular organisms during the Precambrian. It also presents a clear picture of the interrelationship between the physical and biological history of the earth. It is amply illustrated and has a glossary, comprehensive index, and Summary of the classification of major fossil groups in the back.

Starr, Cecie, Ralph Taggart, Christine A, Evers, and Lisa Starr. *Biology : The Unity and Diversity of Life.* 14th ed. N.p.: n.p., 2016. Print. An introductory-level college text that uses the principles of evolution and energy flow as a conceptual framework for each chapter. Clear writing style and color illustrations on every page make this an attractive and informative text. Gives a concise overview of the early embryological stages and describes experiments that have led to the understanding of mechanisms of development.

Stedelman, William, and Owen Cotterill. *Egg Science and Technology.* 4th ed. New York: Haworth Press, 1995. Long recognized as the most comprehensive handbook on the egg-processing industry.

Stephanopoulos, Gregory N., Aristos A. Aristidou, and Jens Nielsen. *Metabolic Engineering: Principles and Methodologies.* San Diego: Academic Press, 1998. Classic text on metabolic engineering.

Stewart, P.R. And Letham, D.S. *The Ribonucleic Acids* 2nd ed., New York, NY: Springer-Verlag, 2012.

Strange, Kevin, ed. *Cellular and Molecular Physiology of Cell Volume Regulation.* Boca Raton, Fla.: CRC Press, 1994. A textbook on cellular osmoregulation and membrane transport physiology.

Strauss, James, and Ellen Strauss. "Virus Vector Systems." In *Viruses and Human Disease.* New York: Elsevier, 2008. Examines the molecular biology of animal viruses and their role in human disease. Addresses the role of viruses as vectors for genetic engineering. Includes numerous illustrations of procedures.

Sturtevant, A. H. "The Linear Arrangement of Six Sex-Linked Factors in *Drosophila*, as Shown by Their Mode of Association." *Journal of Experimental Zoology* 14 (1913): 43- 59. Sturtevant provides evidence that genes are arranged in a linear sequence along the X chromosome. He was able to map the genes relative to each other and determine a distance between them.

Suhowatsky, Gary. "The Role of Trapping in Wildlife Disease." http://articles.animal concerns.org/arvoices/archive/trapping_disease.html. Testimony delivered before the New York State Assembly Subcommittee on Wildlife in March, 1977.

Suzuki, David T., Antony Griffiths, Jeffrey Miller, and Richard Lewontin. *An Introduction to Genetic Analysis.* 4th ed. New York: W. H. Freeman, 1989. An intermediate-level college text that presents the experimental basis of the understanding of genetic processes. Chapter 23 focuses on natural selection and evolution and is useful for gaining a more mathematical understanding of the phenomenon.

Svirezhev, Yuri M., and Vladimir P. Passekov. *Fundamentals of Mathematical Evolutionary Genetics.* Translated by Alexey A.Voinov and Dmitrii O. Logofet. Boston: Kluwer Academic Publishers, 1990. Offers a clear exposition of the mathematical material in historical perspective. Part 1 covers deterministic models and part 2 uses stochastic models.

Szaly, Frederick S., and Eric Delson. *Evolutionary History of the Primates.* New York: Academic Press, 1979. Though not intended for beginners and somewhat dated in places, this remains a standard reference.

Tanagho, Emil A., Jack W. McAninch, and Donald R. Smith. *Smith's General Urology.* 17th ed. New York: McGraw-Hill Medical, 2008. This comprehensive textbook contains visual aids, including clinical images.

Tanford, Charles and Reynolds, Jacqueline Nature's Robots: A History of Proteins New York, NY: Oxford University Press, 2003.

Tattersall, Ian, and Jeffrey H. Schwartz. "Hominids and Hybrids: The Place of Neanderthals in Human Evolution." *Proceedings of the National Academy of Science* 96 (June, 1999): 7. This commentary is a reply to the paper by Duarte, Trinkaus, and their colleagues (pages 7604-7609 of this same issue) in which they report the description of a 24,500-year-old fossil with characteristics of both Neanderthals and early modern humans. The two positions are clearly laid out in the article and the commentary.

_____. *Extinct Humans.* Boulder, Colo.: Westview Press, 2000. Tattersall excels at explaining complex paleoanthropological topics for the general reader. This book presents the idea of a "bushy" human evolutionary history, with many branches, as opposed to the now-obsolete notion that humankind evolved in a single linear path. Discusses the similarities and differences among *Australopithecus, Paranthropus,* and *Homo,* and the importance of tools in hominid mental evolution.

_____. *The Fossil Trail: How We Know What We Think We Know About Human Evolution.* New York: Oxford University Press, 1995. The reconstruction of human evolution on the basis of interpreting fossil remains, including the Laetoli footprints, Lucy, Turkana Boy, and others. Good coverage of the development of dating techniques such as fluorine analysis and mitochondrial DNA.

_____. *The Last Neanderthal.* New York: Macmillan, 1995. An extensively illustrated account of what is known about Neanderthals and how it was learned. The prologue outlines two interesting, if imaginative, characterizations of the "last Neanderthal."

Tauber, Alfred. "Metchnikoff and the Phagocytosis Theory." *Molecular Cell Biology* 4 (November, 2003): 897-901. Discussion of Metchnikoff's discovery of phagocytosis. Includes original illustrations.

Telford, Ira R., and Charles F. Bridgman. *Introduction to Functional Histology.* 2d ed. Grand Rapids, Mich.: HarperCollins, 1995. This introductory histology text with excellent diagrams and photographs shows the relationships between individual animal cells and the tissues they form. Beginning with basic animal cell biology, the text moves to cover the major organ systems. The frequent use of line drawings to complement electron microscope photographs makes their interpretation much more apparent.

Temes, Roberta. *The Complete Idiot's Guide to Hypnosis.* 2d ed. Indianapolis, Ind.: Alpha Books, 2004. Provides the basics of hypnosis. Contains a glossary, an index, and suggestions for Bibliography.

Thacker, Holly. *The Cleveland Clinic Guide to Menopause.* New York: Kaplan, 2009. Thacker, a physician at the Center for Specialized Women's Health at the Cleveland Clinic, offers safe treatments for the menopause and explains myths and facts regarding hormonal replacement therapy.

Thatcher, Jim, et al. *Web Accessibility: Web Standards and Regulatory Compliance.* New York: Springer, 2006. Presents an overview of laws, policies, and technical standards for creating accessible Web Sites.

Theunissen, Berg. *Eugène Dubois and the Ape-Man from Java: The History of the First "Missing Link" and Its Discoverer.* Boston: Kluwer Academic, 1989. Java man, discovered by Dubois in 1891 (with one important discovery the previous year), was the first example of *Homo erectus* to be recovered. Theunissen's account of this historic addition to our knowledge of early man is detailed but nontechnical, suitable for readers at the high school level and beyond.

Thompson, James N., and R. C. Woodruff. "Mutation and Mutagenesis." In *Encyclopedia of Genetics,* edited by Jeffrey A. Knight and Robert McClenaghan. Pasadena, Calif.: Salem Press, 1999. Discusses the processes of mutation and mutagenesis.

Thro, Ellen. *Genetic Engineering: Shaping the Material of Life.* New York: Facts On File, 1995. Print.

Topol, Eric J., ed. *Textbook of Cardiovascular Medicine.* 3d ed. Philadelphia: Lippincott Williams & Wilkins, 2006. A complete, well-organized, user-friendly reference book, complete with audio and visual aids.

Tortora, Gerard J., and Bryan Derrickson. *Principles of Anatomy and Physiology.* 12th ed. Hoboken, N.J.: John Wiley & Sons, 2009. Excellent anatomy and physiology textbook with great illustrations and clinical references.

Travis, Simon P. L., et al. *Gastroenterology.* 3d ed. Malden, Mass.: Blackwell, 2005. A concise, informative manual on gastroenterology with a global perspective.

Trinkaus, Erik, and Pat Shipman. *The Neanderthals: Changing the Image of Mankind.* New York: Alfred A. Knopf, 1992. An interesting and well-written account of the Neanderthals and the scientists who study them.

Tropp, Burton E. *Molecular Biology. Genes to Proteins* 3rd ed., Boston, MA: Jones and Bartlett, 2008.

Tufte, Edward R. *The Visual Display of Quantitative Information.* 2d ed. Cheshire, Conn.: Graphics Press, 2007. A seminal work on ways in which complex information can be clearly and simply displayed in visual forms such as graphics, maps, charts, and tables.

Valentinuzzi, Max. *Understanding the Human Machine: A Primer for Bioengineering.* Hackensack, N.J.: World Scientific Publishing, 2004. An accessible reference designed to give students much of the biological knowledge needed to pursue studies in bioengineering. Also provides useful information about the nature, goals, and development of the bioengineering field.

Van Blerkom, Jonathan, and Pietro M. Motta, eds. *Ultrastructure of Reproduction: Gametogenesis, Fertilization, and Embryogenesis.* Boston: Kluwer, 1984. A collection of essays focusing on the use of electron microscopy to understand reproductive processes such as gametogenesis.

Van Tienhoven, Ari. *Reproductive Physiology of Vertebrates.* 2d ed. Ithaca, N.Y.: Cornell University Press, 1983. This is a good reference for anyone interested in reproduction among nonmammalian vertebrates. The text covers many subjects not often found in one volume, including effects of the environment, immunological aspects of reproduction, puberty, and animals with intersex characteristics.

Vasic-Racki, Durda. "History of Biotransformations: Dreams and Realities." *Industrial Biotransformations*, edited by Andreas Liese, Karsten Seelbach, and Christian Wandrey. Weinheim, Germany: Wiley-VCH Verlag, 2000. This essay chronicles the history of using microorganisms and enzymes to synthesize commercially valuable products and the rise of biochemical engineering as an inevitable consequence of these developments.

Vaughan, Janet Maria. *The Physiology of Bone.* 3d ed. Oxford, England: Clarendon Press, 1981. Covers the basis of the physiology of bone.

Ventura, Gugliemo, and Lara Risegari. *The Art of Cryogenics: Low- Temperature Experimental Techniques.* Burlington, Mass.: Elsevier, 2008. Comprehensive discussion of various aspects of cryogenics from heat transfer and thermal isolation to cryoliquids and instrumentation for cryogenics, such as the use of magnets.

Vieira, Ernest R. *Elementary Food Science.* 4th ed. Gaithersburg: Aspen, 1999. Print.

Vroon, Piet. *Smell: The Secret Seducer.* Translated by Paul Vincent. New York: Farrar, Straus and Giroux, 1997. A cultural history and compendium of odd facts and a tribute to the sense of smell.

Waechter, John. *Man Before History.* Oxford, England: Elsevier-Phaidon, 1976. This is a concise, readily accessible volume on the evolutionary history of humans. It contains some of the finest visual and graphic representations (photographs and illustrations) on the subject found in any single work on the evolutionary journey of the human species. It also contains an extensive glossary of terms with numerous illustrations. Particularly strong in its discussion and representation of late prehistoric culture (especially art), it is an excellent reference work geared toward the nonspecialist.

Walford, Roy L. *Maximum Life Span.* New York: W.W. Norton, 1983. Delightfully written and lucid, this book, written by a pioneering researcher in aging, is somewhat out of date and focuses primarily upon human aging, but it still supplies solid information about the biological and ecological limits on animal life spans.

Walker, F.W., Jr., and K. F. Liem. "The Integument." In *Functional Anatomy of the Vertebrates: An Evolutionary Perspective.* 4th ed. Philadelphia: Saunders College Publishing, 2001. An upper-level comprehensive text on vertebrate systems. Excellent chapter on comparative anatomy and physiology of vertebrate skin.

Walker, Richard. *Animal Anatomy on File Collection.* New York: Facts on File, 1990. A guide to the external and internal structure of animals, grouped thematically into eight sections (introduction, lower groups, annelids and mollusks, arthropods and echinoderms, fish, amphibians and reptiles, birds, and mammals).

Walker, Sharon. *Biotechnology Demystified.* New York: McGraw-Hill Professional, 2006. This is an introduction to the basics of and latest advances in molecular biology and the latest applications of these concepts to a range of discoveries, including new drugs and gene therapies.

Walsh, Patrick C., and Janet Farrer Worthington. *The Prostate: A Guide for Men and the Women Who Love Them.* New York: Warner Books, 1997. This comprehensive yet easy-to-read classic explains diagnostic tests and treatments in a reassuring manner suitable for patients and their families.

Walter, Nils G., Woodson, Sarah A. and Batey, Robert T., eds. *Non-Protein Coding RNAs* New York, NY: Springer-Verlag, 2009.

Wang, Zhaocai, Jian Tan, Dongmei Huang, Yingchao "A Biological Algorithm to Computation." *Applied Mathematics and Computation.* Volume 244 (2014), pp. 183-190.

Warsh, Cheryl Lynn Krasnick. *Prescribed Norms: Women and Health in Canada and the United States Since 1800.* Toronto: University of Toronto Press, 2010. Examines the history of medical health treatment of women in the United States and Canada, examining menstruation, the menopause, childbirth, and medical education, including nursing.

Watson, James D., and Andrew Berry. *DNA: The Secret of Life.* New York: Alfred A. Knopf, 2006. This comprehensive introduction to DNA has the famous biologist Watson as one of its authors.

Weibel, Ewald R. *The Pathway for Oxygen*. Cambridge, Mass.: Harvard University Press, 1984. Ewald Weibel's research presented in this book established many of the concepts and facts currently known about respiratory system structure and function.

Weindrach, R. "Caloric Restriction and Aging." *Scientific American* 274 (1996): 46-52. This article gives a very good discussion of how organisms from protists to mammals live longer on well-balanced but low-calorie diets.

Wessels, Norman K., and Janet L. Hopson. *Biology.* New York:RandomHouse, 1988. The chapter on animal development gives an overview of classic embryology, including the role of growth. The chapter on developmental mechanisms and differentiation takes the reader through the major cellular and genetic regulatory mechanisms involved in development. This extremely well written introductory college text is easily read; it is well illustrated with color photographs and self-explanatory diagrams.

Whitford, David Proteins: Structure and Function New York, NY: John Wiley & Sons, 2013.

Wiley, R. Haven. "Lek Mating System of the Sage Grouse." *Scientific American* 238 (May, 1978): 114-125. Gives additional information on the reproductive behavior of this fascinating species and on the general phenomenon of lekking behavior.

Wilkins, Robert L., and James R. Dexter. *Respiratory Disease: A Case Study Approach to Patient Care*. Philadelphia: F. A. Davis, 2006. Describes different pulmonary conditions by examining specific case studies of different pulmonary patients.

Williams, Tony. *The Pox and the Covenant*. Naperville, Ill.: Sourcebooks, 2010. Story of Cotton Mather and variolation in Boston during the 1720's.

Willmer, Pat, Graham Stone, and Ian Johnston. *Environmental Physiology of Animals*. Malden, Mass.: Blackwell Science, 2000. This textbook has simple-to-understand diagrams and explanations of animal organ systems and their adaptation to different environments.

Wills, Christopher, and Jeffrey Bada. *The Spark of Life: Darwin and the Primeval Soup*. Cambridge, Mass.: Perseus, 2000. Describes theories of the origins of terrestrial life and its evolution. Written for the general reader.

Wilmut, Ian, Keith Campbell, and Colin Trudge. *The Second Creation: Dolly and the Age of Biological Control*. New York: Farrar, Straus and Giroux, 2000. The two researchers who made Dolly team up with a noted British science writer to give a personal but rigorous explanation and thoughtful examination of cloning. Contains a helpful glossary of terms.

Wilson, Edward O. *On Human Nature*. 1978. Cambridge, Mass.: Harvard University Press, 2004. A look at the significance of biology and genetics for the way people understand human behaviors, including aggression, sex, and altruism, and the institution of religion.

Wilson, Edward O., and William Bossert. *A Primer of Population Biology*. Sunderland, Mass.: Sinauer Associates, 1977. This classic handbook has been used extensively by students wishing to master the basics of population biology. Demography is discussed in a way that emphasizes the estimation of population growth rates from survivorship and reproduction data, as well as the determination of age distributions.

Withers, P. C. *Comparative Animal Physiology*. Fort Worth, Tex.: Saunders, 1992. Gives a broad overview of animal physiology comparing the characteristics of each species.

Wobus, A. M., and K. R. Boheler, eds. *Stem Cells*. New York: Springer, 2006. Presents many novel aspects of stem cell biology, including existing and future applications in research and medicine, particularly uses in drug therapies.

Wrangham, Richard W., W. C. McGrew, Frans B. M. DeWaal, and Paul G. Heltne, eds. *Chimpanzee Cultures*. Cambridge, Mass.: Harvard University Press, 1996. Essays by authorities on chimpanzee biology and behavior. Discussions of social relations include sexual development and behavior.

Wright, David A., and Pamela Welbourn. *Environmental Toxicology*. Cambridge, England: Cambridge University Press, 2002. Overview of interaction of chemicals and the environment from molecular to ecosystem levels; includes case studies.

Wuethrich, B. "The Asexual Life: Why Sex? Putting Theory to the Test." *Science* 281, no. 25 (September, 1998): 1980-1982. A comparison of the advantages of a sexual versus an asexual lifestyle.

Wynbrandt, James. *The Excruciating History of Dentistry: Toothsome Tales and Oral Oddities from Babylon to Braces*. New York: St. Martin's Press, 1998. An entertaining history of the development of the dental profession, offering humorous anecdotes and macabre tales of the profession.

Yang, Shang-Tian. *Bioprocessing for Value-Added Products from Renewable Resources: New Technologies and Applications.* Amsterdam: Elsevier, 2007. Reviews the techniques for producing products through bioprocesses and lists suitable organisms, including bacteria and algae, and describes their characteristics.

Yen, S. C., R. B. Jaffe, and R. L. Barbieri. *Endocrinology: Physiology, Pathophysiology, and Clinical Management.* 4th ed. Philadelphia: W. B. Saunders, 1999. Integrates normal endocrinology, the diseases of the hormones, and their treatments.

Zaher, Aiman. *Pathology Made Ridiculously Simple.* Miami: MedMaster, 2007. A book of mnemonics and cartoons to aid in memory and retention.

Zenios, Stefanos, Josh Makower, and Paul Yock, eds. *Biodesign: The Process of Innovating Medical Technologies.* New York: Cambridge University Press, 2010. Covers the biomedical industry, with a particular focus on the process of creating and marketing medical technology. Also provides information about the future of biotechnology and bioengineered products.

Zimmer, Carl. *Parasite Rex: Inside the Bizarre World of Nature's Most Dangerous Creatures.* New York: Simon and Schuster, 2001. An account of interesting parasites, their ecology, evolution, and effect on culture and society; covers medical issues, parasite ecology, and other facets of the field.

Zuk, Marlene. *Riddled with Life: Friendly Worms, Ladybug Sex, and the Parasites That Make Us Who We Are.* Fort Washington, Pa.: Harvest Books, 2008. This introduction to parasites that affect humans in both negative and negligible ways focuses on the evolutionary aspect of parasitism and also contains information on medical and agricultural parasitology.

SUBJECT INDEX

A

abrasion 511–512
absolute temperature 2
absolute zero 129
absorption 166
  of nutrients 17
abzymes 216
acacia-ant mutualism 142
accessibility 308
accessory organs 455
accommodation 525
acellular vaccine 324
acetylcholine 99
Acheulian culture 297
acquired immunodeficiency syndrome (AIDS) 323, 327
action 333
activation energy
  in biological systems 3
  in chemical reactions 1
  defined 1
  diagram 2
  of reaction 3
  and reaction rates 1–3
active immunization 325
active site, enzymes 213
active transport 4–5
  ATP and 7
  cell walls and lipid bilayers 5–6
  functions 6
  ions transfer 6–7
  mechanics of 5
activities of daily living 263
adaptation 224
  and evolution 365, 368
adenosine monophosphate (AMP) 99
adenosine triphosphate (ATP) 4, 7, 60, 70, 101, 104, 108, 184, 333, 335, 392
adensosine diphosphate (ADP) 7
adrenal glands 206
adult stem cells 490
Advanced Research Projects Agency (ARPA) 380
Aegyptopithecus 23
aerobic exercise 333, 336
aerobic metabolism 465
aerobic respiration 108
age regression and progression 316
age-related macular degeneration (AMD) 381
age-specific approach 145–146
age structures and sex ratios 145
aging
  death, cause of 9–10
  definition 8
  effects of 9–10
  theories 263–264
agricultural and veterinary parasitology 420
air and water environments, respiratory system 467
alcoholic beverages 47
alkaloids 351
alkaptonuria 363
allele 255, 258, 360, 364, 495–496
allozyme 497
altered state of consciousness (ASC) 316
alveoli 333, 339, 444
alveolus 466
Alzheimer's disease 49, 263
American Academy of Physical Education 334
American Association of Opticians 402
American Optometric Association 405
amino acids 104, 152–153, 213, 430–435, 473–474
  formation of proteins and enzymes 12–13
  genetic code into proteins, translating 13
  mRNA Codon 13–14
  nature of 10–12
  and proteins 12–13
α-amino acids 12
amino group 10
amphibians (amphibia) 21
Amphipithecus 23
amplexus 125
amygdala 198
amylase 439
amyotrophic lateral sclerosis (ALS) 388
anabolic pathway determination 363
anabolic processes 6, 45
anaerobic respiration 108
analytical chemistry 507
anatomic pathology 424–425
anatomy
  circulatory system 16
  defined 15
  digestive system 17
  endocrine system 17
  immune system 16

integumentary system 17
musculoskeletal system 15–16
nervous system 16
reproductive system 17
respiratory system 16
androgens 481–482
anemia 273
  treatment 377
anemones 19
anestrus 455
angiogram 87
angiography 89
angioplasty 87, 89
animal
  fighting, animal farming, and sport hunting 136
  kingdom 18–22
  and plant cell cultures 67
Animal Minds (1992) 329
Animal Thinking (1984) 329
anisogamy 477
anosmia 487, 489
antagonist 333
anterior pituitary gland 454
anthropomorphism 328
antibiotics 67
  resistance 175
antibodies 214, 321, 324, 521
  classes 323
  production 71
antibody-directed enzyme prodrug therapy (ADEPT) 217
anticancer therapy 523
anticholinergic agent 384, 387
anticodon 12, 432, 434, 470
antidiuretic hormone (ADH) 203, 412
antigen 324
  preparation 71
  presentation and receptors 322–323
anti-inflammatory agent 67
aorta 87, 112
apes 22, 284
apes to hominids 22–23
  ancestors 23–24
  ancient fossils and modern primates 25–26
  fossil apes 25–26
  higher primate fossil record 23
  theorizing from fossils 24–25
aphids (phylum Arthropoda) 32–33
aposematic coloration 138

appliance 148
applied kinesiology 333
archenteron 115
arithmetic skills in primates 331
Arrhenius equation 1–2, 4
arrhythmia 87
arterioles 16, 113
artery 112
arthritis 406, 410
arthropods 20, 80
arthroscope 406
arthroscopy 406
artificial hearts 28, 90
artificial insemination 448–450, 452
artificial organs 27, 31
  applications 29–30
  miniaturization 30
  working of 28–29
artificial retinas (AR) 381
asexual reproduction 32
  parthenogenesis 32–33
Aspergillus terreus 351
assisted reproductive technology (ART) 207, 449
associational defenses 140
Asthma and Allergy Foundation of America 444
ataxia 384–385, 388
atherosclerosis 346
atomic arrangements 61
atresia 88
atria 112
atrioventricular valve 114
atrium 87
atrophy 423
auscultation 439
  and percussion 441–442
australopithecines 22, 284
Australopithecus afarensis 223
autocrine signaling 96
autoimmune disease 326
automated cell counter 273
automated cell-counting instruments 275
automated perimetry 403
autonomic nervous system 16
autopsy 277
autotomy 138, 141

B

Bacillus thuringiensis 69
bacteria 171

bacterial diseases 172–173
bacterial infections 175
bacterium 419
balloon angioplasty 88
basal ganglia 384–385
basal metabolic rate (BMR) 393
base analogues 361
base pair 176
benign prostatic hyperplasia 515
bevacizumab (Avastin) 39
bindins 234
binocular vision 230, 401–402, 526
bioartificial device 91
bioartificial liver (BAL) 95
biochemical engineering 35–37
  work 37–40
biochemical macromolecules 436
biochemical processes of metabolism 6
biochemicals
  carbohydrates 42–44
  lipids 44–45
  nucleic acids 44
  proteins 44
biochemistry 436, 507
  basics 42
  biochemicals
    carbohydrates 42–44
    lipids 44–45
    nucleic acids 44
    proteins 44
  metabolic processes 45
biocompatibility 27
biocompatible material 46
biodiesel production 352
biodiversity 53
bioengineering 46–48, 91
  applications
    agriculture 50
    biomedical applications 49
    bionanotechnology 49–50
    environmental and ecological 50–51
    geoengineering 51
    military 50
  biochemical engineering 48
  electrical engineering 48–49
  materials science 48
biofilters 67
biofuels 35, 67
  production 40

biohybrid 27
bioinformatics 46, 300, 304
bioinstrumentation 49
biological catalysts 499–500
biologics 55, 300
biology 53–54
biomass 35, 65
  production 67
biomaterials 27, 29, 55, 58, 91
biomechanics 46
biomedical devices 57
biomicroscopy 403
biomolecular structures 62
biomolecule 42
bionanotechnology 55
bionics 55–57
  biomaterials 58
  biomedical devices 57
  cloning 57
  cloning and stem cells 58
  equipment and machinery 57–58
  imaging systems 58
  medical devices 57
  medications 58
  recombinant DNA 57
  restorative bionics 57
biophysics 60–61
  biomolecular structures 62
  CD spectroscopy 62
  drug discovery 63
  electron microscopy 61–62
  membrane structure and transport 62–63
  nanobiology 63
  NMR spectroscopy 62
  synthetic biology 63
  X-ray crystallography 62
biopolymers 431
  formation 432
bioprocess engineering 65–66
  applications
    animal and plant cell cultures 67
    antibiotics 67
    biofuels 67
    biomass production 67
    chemicals 68
    environmental applications 67
    enzymes 67
    food 66–67
biopsy 277, 400, 423

bioreactors 35, 37, 46, 65, 95, 349–351
  landfills 40
bioremediation 46, 65
biosensors 71
biosynthesis 69–70
  antibody production 71
  antigen preparation 71
  applications
    biosensors 71
    biosynthetic temporary skin substitute 72
    disposable micropumps 71
    gene expression databases 72
    high-throughput screening 71
    nanoparticles 72
    needle-free drug delivery systems 72
    protein biomarker assays 71
    therapeutic proteins 71
  combinatorial chemistry 71
  general process 70–71
biosynthetic temporary skin substitute 72
birds (Aves), diversity of 21–22
birth 74
  in cattle 76
  control pills 399, 450
  oviparous 74–75
  ovoviviparous 75
  shark 76
  viviparous 74
bite 148
black cherts 357
bladder 515
blastocyst 490
blastula 115
blockage of coronary vessels 88
blood-brain barrier 384, 389
blood flow and oxygen delivery 462–463
blood salt concentrations 412
blood sugar levels 283
B lymphocytes 324–325
body-temperature regulation 499
Boltzmann fraction 2
bombardier beetles 139
bone and cartilage 76–77
  composition, development, and remodeling 78
  physical characteristics 77–78
bone fractures 406
bone marrow
  smear 424
  transplant 276

bonobo chimpanzee, sign language 330–331
Botox 518
botulinum toxin 155, 518
bovine serum albumin (BSA) 71
bovine spongiform encephalopathy (BSE) 136
bradycardia 87
brain 79
  evolutionary development of vertebrate 81–82
  invertebrates with and without 79–81
  primate 82–83
  and reproduction 457–458
brain-controlled interfaces (BCI) 48, 381–382
brain-machine interface (BMI) systems 382
brainstem 79, 83, 384–385
breast examination 397, 399
breathing
  rate, respiratory system 468
  water and breathing air 467–468
brewing 65
bronchodilators 442
Brownian motion 160
buccal cavity 168

C

calcification 76
calculi 515
calyx 374–375
cancerous cysts 400
candling 191
cannibalism 85–86
  human 86
capillaries 16, 112
capillary beds 114
carbohydrates 42–44, 392, 437–438
  -carbon 12
carbon dioxide, osmoregulation 411–412
carboxyl group 5–6, 10
carboxylic acid group 10
cardiac catheterization 87
cardiac output 112
cardiology 87, 91
  angiography 89
  angioplasty 89
  artificial hearts 90
  blockage of coronary vessels 88
  catheters 89
  defibrillators 89
  electrical irregularities 88
  electrocardiograms 89

faulty heart valves 88
heart valves 89–90
pacemakers 89
stents 89
stethoscopes 89
cardiovascular rehabilitation 336
caries 148
carotenoids 351
cartilage 76
bone 77
cartilaginous fishes (Chondrichthyes) 21
Cas-9 127
catabolic pathway determination 363
catabolic processes 6, 98
catabolic reaction 163
catalyst 1–2, 10, 214, 473–474
catalytic antibodies 215
catalyzed activation energy 3
catarrhini 23
catastrophism 224
catheterization 515–516
catheters 87, 89, 515
CD spectroscopy 62
cell and tissue engineering 91–94
bioartificial organs 94–95
cell matrices 94
cell body 79
cell communication 96–98
basics 98
functions and methods 98–99
receptors in 99
types of signaling 99
cell culture 521
cell division (mitosis) 345
cell-mediated immunity 321
cell membrane 4, 211
cell metabolism and meiosis 234–235
cell organelle 100–101
eukaryotic organelles 101
endoplasmic reticulum 101
golgi apparatus 101
mitochondria 101
nucleus 101
cell specialization 102–103
cell-surface glycoproteins 322
cell therapy 91
cell-to-cell connections 106
cell types 104
nucleus and its contents 105

shapes and types 106–107
size and function 105–106
structure 104–105
studying cells 107
cellular biology 53
cellular defenses 321
cellular immunity 324
cellularization 357
cellular metabolism 392
nutrient requirements 392
cellular respiration 108–109
anaerobic respiration in action 110–111
electron transport chain 110–111
glycolysis 110
Krebs cycle 110
cellulose 431
cellulosic biomass 350
cell walls and lipid bilayers 5–6
centenarian 263
Centers for Disease Control and Prevention 327
central nervous system 384–385, 387
cephalopod, *Nautilus pompilius* 230
cercopithecoidea (Old World monkeys) 23
cerebellum 79
cerebrospinal fluid (CSF) 82, 384, 387–388
cerebrum 79
cesarean sections 397, 399
cetaceans 135
chain of custody 507
chalaza 191
changing protective layers 487
chemical communication 488
chemical defenses 139–140
chemical pathology 425
chemical plaques 345
chemical reaction 1
chemical reactor 36
chemicals 68
chemoreceptor 466
chemotaxis 487–488
chemotherapy 276
Chicago-born Belding Scribner 376
chickens 191
chimeric antibodies 39
Chlamydomonas 477
cholecystokininpancreozymin (CCKPZ) 203
chordae tendinae 114
chordata 18
chorion 427

chorionic gonadotropin 429
choroid plexuses 82
chromatophores 484, 486
chromophore 230–231
chromosome 127, 153, 457
chronic conditions 263
chronic kidney disease detection 376
chronic obstructive pulmonary disease (COPD) 440, 442
chyme 17
cinematography 334
circular dichroism (CD) 60
circulatory systems of vertebrates 112–113
  of birds and mammals 113–114
  blood volume and blood vessels 114
  of fish, amphibians, and reptiles 113
  vertebrate circulation 114–115
citric acid 69
class 18
  evidence 236
cleaning, oral 148
cleavage 115–116, 195
clinical genetics 305–306
clitoris 482
cloaca 125
cloned genes 121
clones 32, 55, 119, 300
cloning 57, 119–121, 449, 490
  molecular 121
  reproductive 121–122
  and stem cells 58
  therapeutic 122–123
Clustered Regularly Interspaced Short Palindromic Repeats (CRISPR) 127
cluster of differentiation (CD) proteins 327
cnidarians 19
coaches, kinesiology 336
cochlea 379–381
cochlear implants 48, 381
coding strand 471
codon 432, 470
cofactors 214
cognition 8
cognitive ethology 328–329
cohesion 312
cohort 143
coitus. *see* copulation
cold-blooded 500
collagen 76

fibers 77
colligative properties, osmoregulation 411
colon 251
colonoscopy 251
colorectal cancer screening 253
color vision 527
colostrum 339
combinatorial chemistry 71
Combined DNA Index System (CODIS) 176
comparative anatomy 15, 529
competition and reproductive success 447
complement 324
complementary strand 182, 186
compound eyes 231
compression of morbidity 263
computational fluid dynamics (CFD) 28
computed tomography (CT) 58, 386
concentration gradient 4, 160, 414
conjugated monoclonal antibody 327
connective tissue 76
contact lenses 404
continuous culture systems 37
contraception and sterilization 452
controlling reproduction 460
controlling site 360, 362
convection 499–501
conveyance cells 107
copulation 124–125
  copulatory behaviors 126
  copulatory organs 125–126
copulatory organs 125–126
cornea 402, 404–405
corona radiata 233
cortex 79, 242–243, 374–375
cortical 385–386, 388
Corynebacterium glutamicum 69
countercurrent exchanger 466
countercurrent mechanism 499, 501
crackles 439
cranial electrotherapy stimulation 382–383
cranial nerve palsy 403
Creutzfeld-Jacob disease 173
criminalistics 236
CRISPR-Cas9 127
  applications 127–128
critical care medicine 440
Cro-Magnon man 292
cross-fertilization 478
cross-linking during polymer 432

cryocooler 129
cryogenic processing 129
cryogenics 128–130
  applications 131
    aerospace industry 132
    automotive 132
    food and beverage 131–132
    health care 131
    superconductivity 131
    tools, equipment, and instrument 132
  creating low temperatures 130–131
  cryogenic refrigeration 131
  processing and tempering 131
cryogenic tempering 129
cryopreservation 129, 449
Cushing's syndrome (overactive adrenal glands) 207
cuticle 242
cyclic adenosine monophosphate (cAMP) 202
cylindrical errors 403
cystic fibrosis 451
cytogenic analysis 275–276
cytology 278
cytoplasm 104–105
cytosol (intracellular fluid) 12
cytotoxic T cells (T) 322

D
daily food pyramid 392
damage sensors 346
Darwin, Charles 365–369
  evolution, animal life 221
  revolution 225–226
Darwinism 224
death and dying 135
  organized animal fighting, animal farming, and sport hunting 136
  pollutants 135–136
  scientific experiments 136–137
  viral and bacterial disease 136
decomposition reaction 163
deep brain stimulation (DBS) 380, 382–383
defense mechanisms 138
  associational defenses 140
  autotomy 141
  chemical defenses 139–140
  defensive behaviors 140–141
  nutritional defenses 141
  structural defenses 139
defensins 485

defensive behaviors 140–141
defibrillators 87, 89
degeneration 423
deme 255
dementias 387–388
demographic parameters 143–144
demography 143
  age-specific approach 145–146
  age structures and sex ratios 145
  demographic parameters 143–144
  patterns of reproduction 144–145
  patterns of survival 144
  time-specific approach 146
  uses of 146–147
demyelinating diseases 388
De Novo design 215
dental public health 151
dental therapy and devices 149–150
dental tools 149
dentistry 148–149
  dental public health 151
  dental therapy and devices 149–150
  dental tools 149
  endodontics 151
  general dentistry 150
  oral and maxillofacial
    pathology 151
    radiology 151
    surgery 150
  orthodontics 150
  pedodontics 150
  periodontics 150
  prosthodontics 150–151
deoxygenated blood 113
deoxyribonucleic acid (DNA) 42, 53, 101–102, 127, 152–153, 176, 182, 367, 370, 470
  forensic science 236
  genetics 258
  human evolution analysis 296
  human genetic engineering 300
  and proteins 153–154
  replication and transcription 154
  unique character 154
dermatology and dermatopathology 155–156
  applications 157
    infections 157
    inflammatory diseases 157
    melanocytes 157–158
    papulosquamous diseases 157

Subject Index

  surgical and cosmetic procedures 158
  treatment modalities 158
  tumors 157
 diagnosis 156
 management 156–157
 treatment 156
dermis 17, 484–485
designer baby concept 453
developmental anatomy 15
developmental biology 53
developmental disorders 171, 408
diabetes 208, 377
diabetes mellitus 17
dialysis 374, 377
diamond 432
diastole 112
diencephalon 83
differentiation 91, 104, 268, 490
diffusion 5, 415, 466
 applications 162
 atoms and molecules 160–162
 coefficient 162
 defined 160
 of nuclear isotopes 162
 and osmotic pressure 416
 visualizing 160
digestion 163–164
 enzyme activity 165
 forms of 164–165
digestive hormones 203–204
digestive tract 166–167
 complex animals 167–168
 mucous layers 168–169
 nerve and muscle layers 169
 simple animals 167
 small intestine 169–170
 studying 170
digital fundus photography 404
dilatation and curettage (D&C) 400
diploid 32, 247, 495–496
direct manipulation 308
diseases 171–172
 of aging 171
 bacterial diseases 172–173
 genetic and congenital diseases 174
 inflammation and immunity 172
 metabolic, neoplastic, and degenerative diseases 174
 prion and fungal diseases 173–174
 viral diseases 173

disposable micropumps 71
disruptive selection 366
disseminated intravascular coagulation (DIC) 275
DNA
 analysis 176–178
  crime scenes 179–180
  DNA polymorphism 178–179
  medical applications 180
  PCR, limitations 180
  probes and primers 178
 fingerprinting 185
 polymorphism 178–179
 and proteins 431–432
 recombination 326
 sequencing 181, 349–350
  and genetic engineering 349–350
 strands 360
DNA/RNA
 synthesis 182–184
  amplifying DNA for Analysis 185
  discovery and analysis 184
  formation of DNA and RNA 184–185
 transcription 186
  genetic code 188
  process 187–188
  structure of DNA Versus RNA 186–187
Dolly (sheep) 303
dopamine 198
dopaminergic 385, 387
Down syndrome 451
dried egg products 194
Drosophila 361–362, 364, 497
drug
 discovery 63
 and gene delivery 92
 therapies, urology 518
dryopithecines 22, 284
Dryopithecus nyanzae 23
Dual energy X-ray absortiometry (DEXA) 208
ducts 339–340
Dunaliella tertiolecta 40
duodenum 166
dysmorphology 300

E

earthworms *(Lumbricus terrestris)* 125
echinoderms 20
echocardiogram 87
echolocation 504, 506

erectile dysfunction 515
erection 457
ergonomic, kinesiology 336
Escherichia coli 173, 350
essentialism (typology) 224
estrogen 455–456
estrus cycle 454
ethnography 308
ethology 504
eukaryotes and prokaryotes 218–219
eukaryotic cell 100, 218–219
eukaryotic fossil cells 354–355
eukaryotic organelles 101
  endoplasmic reticulum 101
  golgi apparatus 101
  mitochondria 101
  nucleus 101
eukaryotic organisms 354–355
Eunice viridis 233
euryhaline 410
euryosmotic 411
evaporative cooling 129
evoked potentials (EPs) test 386–387
evolution 53
  animal life 220–221
    adaptation to environment 222
    australopithecines 223
    beginnings 221
    Charles Darwin's explanation 221
    genes and evolution 221–222
    time frame 222–223
  historical perspective 224–225
    Darwin's revolution 225–226
    struggle to conceptualize 225
evolutionary theories 19
executive functions 263
exercise
  physiology 335
  specialists 336
exocannibalism 85–86
exocytosis 212
expiration 468
external genitals 454–455
extracellular matrix 92
extracorporeal membrane oxygenation (ECMO) 29
extracorporeal systems 27
eyeglasses 404
eyes 230
  compound eyes 231
  optimization 232
  photoreceptors 231
  simple eyes 230

F

fatty acids 5
faulty heart valves 88
feedback, endocrinology 201
female 477
fermentation 47, 65, 218, 342–343
  by microorganisms 39
fermenters 37, 47, 65
fertilization 74, 124, 195, 233, 268, 457
  cell metabolism and meiosis 234–235
  and development 196–198
  penetration 233–234
  proximity, eggs and sperm 233
  in vitro 235
fetal monitoring 399
fetus 427–429
fibroid 400
field observations 198
fighting for mates 446–447
"fight or flight" responses 487
fitness 365–367
flagellate algae 356
flatworms 19, 167
flavonoids 352
flow cytometer 273
flow cytometry 155, 275
fluid volume 283
fluorescence-activated cell sorter (FACS) 274
fluorescence in situ hybridization (FISH) 424
follicle-stimulating hormone (FSH) 249, 449–450, 454, 458
food
  allergies 254
  bioprocess 66–67
  engineering 39
forensic 507
  application 153
  genetics 300, 306
  pathology 425
  toxicology 508–509
forensic science 236–238
  electrophoresis 240–241
  gas chromatography-mass spectrometry 239–240
  infrared spectroscopy 238
  ultraviolet/visible microspectrophotometry 238–239

ecological parasitology 421
ecology 529
ecosystem 85
ectoparasite 419, 504
ectotherms 499–501
Ediacarian (Ediacaran) fauna 353
egg production 191–193
  breeding stock 193
  dried egg products 194
  ducks for 193
  from egg to layer 192
  laying stock 193
  liquid egg products 194
  shell eggs 193–194
ejaculation 457
electrical irregularities 88
electrical stimulators 382
electrocardiogram (EKG) 87, 89
electrode 380–383
electrodesiccation and curettage 158
electroencephalogram (EEG) 49, 386
electromyograms (EMs) 387
electron microscopy 61–62
electron transport chain 108
electrophoresis 240–241
embryo 427
embryology 118, 195–196, 278, 529
  fertilization and development 196–198
  gametogensis 196
embryonic stem cells 119, 490–492
ementia 390
emergency medicine 408
emotions 198
  biology of 200
  defining and communicating 198–199
  primary and secondary 199–200
enamel 148
endocannibalism 85–86
endocrine control systems 202
endocrine signaling 96
endocrine systems of vertebrates 201–202
  digestive hormones 203–204
  endocrine control systems 202
  hormones controlling growth 202–203
  reproductive hormones 204
  water and salt balance control 203
endocrinology 205–208
  diabetes 208
  infertility 209

  medical laboratory 208
  menopause 209
  osteoporosis 208–209
  performance-enhancing hormones 209
  thyroid disease 208
endocytosis 212
  and exocytosis 211–213
endodontics 151
endometrial cancer 400
endometrium 427, 455–456
endoparasite 419
endoplasmic reticulum (ER) 101, 105
endoscopy 251–253, 516
endosomes 211–212
endotherms 499, 501–502
energy
  balance, nutrient requirements 393
  and metabolism 343
enterocytes 166
entomology 420
enucleation 119
environmental applications 67
Environmental Protection Agency (EPA) 135
environmental science 507
environmental toxicology 509
enzyme engineering 213–214
  applications
    abzymes 216
    enzyme immobilization 216
    as medicines 216
    organic solvents, making enzymes soluble in 216
    pharmaceutical production 215–216
  catalytic antibodies 215
  De Novo design 215
  directed evolution 215
  rational design 215
  semirational design 215
  semisynthetic enzymes 214–215
Enzyme-linked immunosorbent assay (ELISA) 71
enzymes 3, 35, 44, 65, 67, 127, 153, 163, 186, 214, 342–343, 349–350, 430–432, 435, 474
  activity in digestion 165
  immobilization 216
  proteins 154
epidermis 155, 242, 484–485
epididymus and vas deferens 459
epithelio-chorial placenta 428–429
equilibrium 160, 415
equipment and machinery 57–58

fossil 255, 353
  apes 24
    importance of 25–26
fovea 525
fracture 406
  care 407
Fragile X Syndrome 494
free-energy change 6–7
freezing point, osmoregulation 413
freshwater organisms, osmoregulation 411
freshwater *versus* saltwater environments 413–414
friction 511–512
fructose 42
function 8
functional electrical stimulation (FES) 380, 382
functional group 436
fundus examination 404
fungus *Penicillium* 65
fur and hair 242
  growth and replacement 243
  origin and function 243
  protecting endangered fur-bearers 243–244
  structure 243

G

gamete 104, 247, 268, 445, 457
gamete intrafallopian transfer (GIFT) 451
gametogenesis 196, 247
  eggs 248–249
    and RNA 249
  sperm 247–248
  studying 249–250
gamma-amino butyric acid (GABA) 99
gamma rays 361
ganglia 79
gas chromatography-mass spectrometry 239–240
gastric and peptic ulcers 253
gastroenterology 250–252
  patient care 252
    colorectal cancer screening 253
    conditions treated 252
    endoscopy 252–253
    food allergies 254
    gastric and peptic ulcers 253
    gastroesophageal reflux disease 253
    imaging techniques 252
    long-term and end-of-life care 254
    management duties 252
    obesity 254

    parenteral nutrition 253–254
    pediatric gastroenterology 253
    research 254
gastroesophageal reflux disease (GERD) 252–253
gastrointestinal tract 251
gastrula 195–196
gastrulation 115–117
gene 301, 360, 434, 495
  expression 182, 186, 470
    databases 72
  flow 220, 255–256
    and gene exchange 256
    hybridization 256–257
  introduction and integration 523–524
  pool 289, 495
  testing 304–305
  therapy 301, 305
  transfer 301
generalized humans 290
genetically engineered organisms 35
genetic and congenital diseases 174
genetic code 435
genetic diseases 171
genetic drift 220
genetic engineering 261–262, 349–350
genetic manipulation 350, 522–523
  of metabolic pathways 350–351
genetic pathology 425
genetics 258–259, 360, 450
  genes and chromosomes 260
  genetic engineering 261–262
  Mendelian genetics 259–260
  molecular genetics 260–261
genetic screening 301
genetic testing 301, 423, 426
genetic tools 350
genital area 481
genitalia 481
genital tubercle 481
genome 104, 119, 153, 220, 258
  and chromosomes 260
  pool 255
genomic and proteomic research 377
genomics 301, 377
genotype 224, 258, 495
  frequency 495
genus 370–371
geoengineering 46
Geoffroyism 224

geriatrics and gerontology 262–263
  aging, theories 263–264
  app 264
  applications
    assistive technology 266
    education 267
    medicine 265–266
    pharmacology 266
  in field 264–265
gestation 74, 197, 427
gestational diabetes 398
gestation periods 429–430
gigantism 210
Gigantopithecus ("giant ape") 24
gill 466–467
gingiva 148
gland 201
glomerular filtration rate 376
gluconeogenesis 202
glucose 44
glycerol 42, 352
glyceryl palmitate 439
glycolysis 343
glycosaminoglycan (GAG) 39
glycosylation 346
GnRH. *See* gonadotropin-releasing hormone (GnRH)
Goldmann field examinations 403
golgi apparatus 101
Golgi complex 212
gonad 454, 457, 481
gonadotropin 457
gonadotropin-releasing hormone (GnRH) 249, 458
gonadotropins 204
gonads 482
gonochorism 477–478
goose bumps 487
G- protein-coupled receptors (GPCRs) 99
gracile 296
graft rejection 322
graphical user interface (GUI) 308
gray matter, brain 79
green algae 351
group selection 366–367
growth 268–269
  animal growth and development 269
  developmental biology 271
  factor 92
  postnatal development 270
  reproduction 269–270
  studying 270–271
gynecology 398, 401
  applications 399–400
  breast examination 399
  future prospects 400
  infertility 397–398
  obstetrician 398
  precancerous conditions 399

H

hair-raising response 487
haploid 32, 196, 247, 495
Hayflick finite doubling potential phenomenon 345
heart
  artificial 28
  human 88
  valves 89–90
heat 125, 512–513
  conduction 129
heliotherm 499
helminth 419
helminthology 420
hematology 273–274, 278, 425
  applications
    automated cell-counting instruments 275
    bone marrow transplant 276
    chemotherapy 276
    cytogenic analysis 275–276
    flow cytometry 275
    molecular-targeted therapy 276
    stem cell transplant 276
  disorders
    of platelets and hemostasis 275
    of red blood cells 274–275
    of white blood cells 275
hematopoiesis 273
hematopoietic stem cell 273
hemocyanin 463
hemodialysis 27, 377
hemodynamics 28
hemoglobin 273, 333, 486
hemoglobinopathy 273
hemoperfusion 27
hemophilia 217
hemostasis 273
hepatitis 251
heritabilities 366
hermaphrodite 481
hermaphroditism 477

herpes simplex virus 1 (HSV1) 157
heterologous graft 92
heuristics 308
high-energy electromagnetic radiation and particles 361
high-throughput screening 71
histochemistry 278
histology 277–279
  applications
    clinical and educational applications 281
    diagnosis of diseases 281
    fluorescence microscopy 280
    sample preparation 280–281
  fluorescent staining 280
  microscopy 279
  tissue culture 279
  tissue samples 279
histopathology 278
hollow-fiber membrane bioreactor (HFMB) 35
homeostasis 282–283, 500
  blood sugar levels 283
  fluid volume 283
  stomach pH 283
homeotherm 499
hominids 22, 284–285, 290, 296
  ancestors and origins 287–288
  australopithecines 285–286
  evolution into homo 286
  fossils 286–287
  humans and primates 284–285
Hominoidea 23
Homo habilis 223, 287
Homo sapiens 26, 289
Homo sapiens and human diversification 289–291
  Cro-Magnon man 292
  emergence of modern humans 292–293
  human evolution in context of animal evolution 294–295
  Neanderthal man 291
  paleoanthroplogy 293
Homo sapiens neanderthalensis 297
Homo sapiens 290
hormonal regulation of osmotic pressure 412
hormonal therapy 399
hormones 201
  control 206
  controlling growth 202–203
host 419
  range 521

human 22, 284
  brain 385
  cannibalism 86
  embryonic stem cells 491
  evolution analysis 296
    evolutionary controversies 297–298
    fossils and paleoanthropology 298–299
    human origins 299
    tool use and culture 296–297
  evolution in context of animal evolution 294–295
  genetic engineering 55, 300–304
    bioinformatics 304
    clinical genetics 305–306
    early medical applications 304
    forensic genetics 306
    gene testing 304–305
    gene therapy 305
    pharmacogenomics 306
    potential for human cloning 306
    proteomics 304
    toxicogenomics 304
  monoclonal antibodies 39
human chorionic gonadotropin (hCG) 449–451
human-computer interaction 307–309
  applications
    computers 310
    consumer appliances 310–311
    education and training 311
    workplace information systems 311
  input and output devices 309
  perceptual-motor interaction 309–310
human immunodeficiency virus (HIV) 179, 323, 327, 421
human-insulin-producing genes 55
human menopausal hormone (HMG) 209
human papillomaviruses (HPVs) 157
humoral immunity 321–322, 324
hyaline cartilage 77
hybridoma 35
hydra 19
hydrocarbon
  chains 6
  molecules 431
hydrochloric acid 164
hydrogen bond 312
hydrophilic and hydrophobic materials 311–313
  biological systems 314–315
  nature 313–314
  solvent partitioning and phase-transfer catalysis 314

hymen 455
hyperosmotic 410, 413
hyperplasia 423
hyperthyroidism 206
hypertonic 415
hypertrophy 423
hyperventilation 461–462
hypnosis 316–318
   altered state and dissociation theories 318
   hypnotherapy, psychotherapy, and counseling 318
   hypnotic analgesia 318
   nonclinical applications 318
   social role and response expectancy theories 318
hypnotherapy 316
hypodermis 485
hypoosmotic 410
hypotension 377
hypothalamic control over reproduction 458
hypothalamus 206, 282
hypotonic 415
   solutions 411
hypoxia 461–462
hysterectomy 397, 400

I

idiopathic thrombocytopenic purpura (ITP) 275
illicit drug 236
imaging systems 58
imaging techniques 252
immune system 171–172, 321
   antibodies, classes 323
   antigen presentation and receptors 322–323
   immunity 321–322
   nonspecific defenses 321
   primary and secondary immune responses 323
immunity 321–322
immunization 327
immunohistochemistry 155, 278
immunology 425
   and vaccination 324–325
     active immunization 325
     autoimmune disease 326
     cellular basis 325
     cellular mechanism 325
     future prospects 327
     monoclonal antibodies (mAB) 327
     vaccine production 325–327
     variolation 325
     viral vaccines 326

immunomodulation 27
impression evidence 236
individualizing evidence 236
induced pluripotent stem cells 490–492
induction 316
infections 157
infertility 209
inflammation 423
   and immunity 172
inflammatory diseases 157
Influenza viruses 327
information architecture 308
infraorder Catarrhini 23
infrared spectroscopy 238
in-line production 191
insertion 333
insight learning 504
inspiration 467–468
insulin production 65
integral proteins 104
intelligence 329
   animals who might think 329–330
   cognitive ethology, theories 329
   primates and sign language 330–331
interbreeding 220
interface 511
internal naris 390
interstitial lung disease 440
interventional neurology 389
intracellular communication 98
intracellular fluid 12
intracellular processes 6
intracytoplasmic sperm injection (ICSI) 449–450
intrafallopian transfer 451–452
intrauterine devices (IUDs) 399
invagination 115
invertebrates 18
in vitro engineering 92, 94
in vitro fertilization 397–398, 400, 451
in vivo engineering 92, 94
involuntary movements, neurology 385
ions transfer 6–7
isogamy 477
isokinetic machine 333
isomerization 164
isoprenoids 351
isosmotic 410
isotonic 415
   solution 411

J

Jarvik-7 total artificial heart 91
jellyfish 19
joint pain and dysfunction 407
joint prosthesis 406
Joule-Thomson effect 129
juxtacrine signaling 96

K

kehole limpet hemocyanin (KLH) 71
Kelvin temperature scale (K) 129
keratin 106, 242, 485
keratinocyte 155
kidney 374–375, 515
   disease, diagnosing 376
   exchange programs 378
   stone detection 376–377
   structure 375
kinesiology 337
   behavioral control 335
   coaching 336
   component 333
   contributors to 334
   ergonomics 336
   field of 334
   future prospects 337
   health promotion and injury prevention 336
   movement analysis 334–335
   physiological function 335
   rehabilitation 335–336
Krebs (citric acid) cycle 108

L

labial folds 481
lactation 339–340
   definition 339
   mammary glands 339–340
   milk production 340
   suckling 340–341
lactic acid fermentation 39, 344
   in absence of oxygen 343
   energy and metabolism 343
   vs. respiration 343
lactogenesis 340
Lamarckism 224
lancelets 21
laparoscopy 397, 400
larva 74
latent print 236

leg prostheses 410
leukemia 273
leukocytes 273
levodopa (L-dopa) 385, 387
lexigrams 329
Leydig cells 458
life cycle 344
life expectancy 344
life form date of emergence 223
life span 8, 344
life spans
   animal kingdom 344
   limiters and extenders 345–346
   meanings 344
   size and 346
   theories of 346–347
life-support systems 49
life table 143
Likert scale 308
limbic system 198
limiters and extenders 345–346
Limnopithecus 23
line 191
Linnaeus, Carl 21
lipids 42, 44–45, 392–393, 437–439
   membrane layer 72
liquid egg products 194
Locked-in syndrome 389
longevity 8
long-term and end-of-life care 254
loop electrosurgical excision procedure (LEEP) 399
low-density lipoprotein (LDL) 209
low-oxygen conditions 461–465
Lowry assay 63
low vision 402, 404–405
lubrication 511
lubrication and lubricants 513
lumbar puncture and cerebrospinal fluid studies 387
lumen 166
lungs 440, 466
   cancer 444
      therapy 443
   transplant 443
lung volume reduction surgery (LVRS) 443
luteinizing hormone (LH) 454, 458
Lyme disease 419
lymphatic vessels 166
lymphocytes 321–323
lymphoma 273

M

macromolecule 436
macrophage 321, 325
macule 155
mad cow disease. *see* bovine spongiform encephalopathy (BSE)
magainins 485
magnetic resonance imaging (MRI) 49, 58, 386, 406
major histocompatibility complex (MHC) 322–323
malaria 419, 421
male infertility 450–451
mammary glands 339–340
marsupials 427
mass spectrometry 424
mass transfer efficiency 29
mate competition 445
  and mate choice 446
maternal-fetal medicine 397
mating types 477
maxillofacial pathology 151
maxillofacial radiology 151
maxillofacial surgery 150
mechanical ventilation technology 440
medical devices 57
medical drug testing 492
medical laboratory 208
medical parasitology 420
medical toxicology 509
medications 58
medulla 242
meiosis 247
melanocytes 103, 155, 157–158, 486
membrane
  bone 77
  and cartilage bone 79
  structure and transport 62–63
memory-storage cards 266
Mendelian genetics 259–260
meninges 385
menopause 209
menstrual cycle 454, 456
menstrual flow 399
messenger RNA (mRNA) 12, 472
metabolic, neoplastic, and degenerative diseases 174
metabolic adaptation 463
metabolic alteration, respiration and low oxygen 463
metabolic engineering 353
  altering nutrient uptake 351
  alternatives 349
  applications
    chemicals 352
    fuels 352
    pharmaceuticals 351–352
  DNA sequencing and genetic engineering 349–350
  future prospects 352
  genetic tools 350
  host and host design 350
  metabolic pathways 349
    genetic manipulation of 350–351
  microbial metabolism 350
  phases 349
  production systems 351
  substrates 350
  tasks of 349
metabolic engineers 350
metabolism 6, 45, 333, 342–343, 349–351, 461
metabolomics 69
metamorphosis 74
metazoan origin, theories 356–357
Methylophilus methylotrophus 217
MHC. *See* major histocompatibility complex (MHC)
Microalgae 40
microbe 218, 342–343
microbial metabolism 350
microbial production 350
microbiology 53, 425
micromanipulation 449–450
microphage 325
microscopy 278
microtome 278
migraines 388
migration 220
milk
  ejection 339–340
  production 340
Miltefosine 421
mimicry 140
mineral 392–393
minute-volume 114
misalignment 403
  of teeth 149
mitochondria 101, 218–219, 296, 371
mitosis 268
mitotic cells 484
molecular cloning 121, 521
molecular genetics 60, 260–261
molecular hydrogen (H_2) production 351
molecular-targeted therapy 276

molluscs 20
monoclonal antibodies (mAB) 35, 38–39, 324, 327
monocytes 324–325
monomer 436
monotremes 427
morphine 351
morphology 290, 529
mortality rate 143, 344
morula 115
motile cells 74
motility 251
motor 333
   cortex 379–380, 382
movement disorders 387
MRI. *See* magnetic resonance imaging (MRI)
mRNA 434, 439
mucosa 166
mucus 390–391
Müllerian ducts 481–482
Müllerian-inhibiting substance (MIS) 482
multicellular animals 355–356
   life 353
multicellularity
   advantages and origins of 354
   eukaryotic organisms 354–355
   evidence from rocks 357–358
   in evolutionary process 358–359
   metazoan origin, theories 356–357
   multicellular animals 355–356
   multicellular kingdoms 354
   multicellular life, theories of the origin of 356
   multicellular organisms 353–354
multilayer insulation (MLI) 130
multinucleate sphere unknown (MSX) 414
multiple sclerosis 388
multipotent stem cell 490
municipal solid waste (MSW) 40
muscle balance 336
muscle cells 103, 107
muscle stem cells 494
muscular contractions 16
musculoskeletal system 406
mutants 360
mutations 220, 255, 258
   anabolic pathway determination 363
   catabolic pathway determination 363
   causes
      chemical 360–361
      heat 360
      high-energy electromagnetic radiation and particles 361
   characterization 360
   in development and evolution 361–362
   DNA strands 360
   genetics 360
   mutants 360
   nucleotides 360
   usefulness 361
myelin 385, 387–388
myocarditus 135
myoelectric arm 410
myoepithelial cells 339–340

N
nanobiology 63
nanoparticles 72
nanotribology 511
nasal swell spaces 390
natality rate 143
National Institutes of Health 53
natural products, biosynthesis 351
natural selection 220, 290, 445–446
   adaptation and evolution 368
   group selection 366–367
   importance 365
   measurements 367–368
   process of 365–366
   slow and holistic process 366
Neanderthal man 291
Neanderthals
   late twentieth century 372–373
   significance 373
   structure and behavior 371
   taxonomic relationship to modern humans 371–372
needle-free drug delivery systems 72
negative chemotaxis 488
negative-pressure breathing 467–468
nematodes or roundworms 80
neocortex 82
nephrolithiasis 375–376
nephrologist 375
nephrology 378. *See also* kidney
   applications
      anemia treatment 377
         dialysis 377
         genomic and proteomic research 377
   chronic kidney disease detection 376

future prospects 377–378
kidney 375–378
nephrologist 375
patient pathology 375–376
nephron 375
neural engineering 380, 384
  applications
    brain-controlled interfaces 382
    cochlear implants 381
    deep brain stimulation (DBS) 382–383
    functional electrical stimulation (FES) 382
    retinal bioengineering 381–382
  future prospects 383
  microsystems 381
  neural prostheses (NP) 381
  neuromodulation and neuroaugmentation 380
  neuromuscular stimulation 380–381
  neuron types 380
  neuroscience and neurophysiology 379
neural microsystem 381
neural prostheses (NP) 381
neuroaugmentation 380
neuroendovascular therapy 389
neurological disorders 386
neurological examination 386
neurology 385–386, 389
  applications
    dementias 387–388
    demyelinating diseases 388
    movement disorders 387
    paroxysmal disorders 388
    spinal cord diseases 388
    vascular diseases 387
  clinical history 386
  diagnostic tools 387
  electromyograms (EMs) 387
  electrophysiological studies 386–387
  future prospects 388–389
  human brain 385
  imaging studies 386
  involuntary movements 385
  lumbar puncture and cerebrospinal fluid studies 387
  neurological disorders 386
  neurological examination 386
  neurons 386
neuromodulation 380
neuromuscular stimulation 380–381
neurons 386
neuroscience and neurophysiology 379

neurourology 517–518
neurulation 116–118
neutrophiles 324–325
nicotinamide adenine dinucleotide phosphate (NADPH) 60
nipple 339–340
nitrogenous wastes, osmoregulation 411
nitrous acid 360–361
NMR spectroscopy 62
nocebo effect 316
noncarbohydrate molecules 202
non-Hodgkin's lymphoma 38
nonspecific defenses 321
nonspontaneous reaction 1
nonsteroidal anti-inflammatory drugs (NSAIDs) 407
normal delivery 397–398
noses
  ementia 390
  heat, humidity and purification 390–391
  in mammals 390
nucleic acid amplification testing (NAAT) 179
nucleic acids 42, 44, 437–438
nucleotides 176, 182, 360, 470
  sequences 42
nucleus (cell) 101, 219
nutrient molecules, nutrient requirements 393–394
nutrient requirements
  activities and 392–393
  animal 392
  basal metabolic rate (BMR) 393
  cellular metabolism 392
  daily food pyramid 392
  energy balance 393
  essential minerals and vitamins, functions 394
  feeding mechanisms 394
  nutrient molecules 393–394
nutritional defenses 141

O

obesity 254
obligate aerobes 343
obstetrics 398, 401
  applications 399–400
  breast examination 399
  future prospects 400
  infertility 397–398
  obstetrician 398
  precancerous conditions 399
occupational therapist 406

ocular motility examination 403
off-line production 191
oil whirl 511
olfaction 390, 487–488
olfactory receptors 487
  cells 489
ommatidium 230
oncology 278
"one gene-one enzyme" hypothesis 363
oogenesis 247, 249
opercular pumping 467
ophthalmologist 402, 405
ophthalmoscopy 404
opsin 230, 525
optician 402, 405
optic nerve 230
optimum temperature 499
optometrists 402
optometry 402, 405
  applications 404–405
  definition 402
  fundus examination 404
  future prospects 405
  ocular motility examination 403
  optometrists 402
  refraction 403
  slit lamp examination 403–404
  tonometry 404
  visual acuity examination 403
  visual field examination 403
oral contraceptives 399
orbital floor fracture 403
organelle 104
organ failure 27
organic chemistry
  and biochemistry 45
organic compound 436
organic eggs 193
organic solvents, making enzymes soluble in 216
organs of gas exchange 466–467
organ-specific diseases include type 1 diabetes 326
origin 333
orthodontics 150
orthopedic prosthesis 408
orthopedics 92
  assessment 407
  bone fractures 406
  developmental disorders 408

  emergency medicine 408
  fracture care 407
  future prospects 409
  joint pain and dysfunction 407
  orthopedic prosthesis 408
  pediatric orthopedic problems 406–407
  pharmacotherapy and medical devices 409
  spinal stabilization 408–409
  sports medicine 408
  surgeons 406
  X-ray 407
orthopedic surgeon 406
orthoptist 402
osmoconformer 410–411
osmoregulation
  carbon dioxide 411–412
  challenges 411
  colligative properties 411
  definition 411
  freshwater organisms 411
  freshwater *versus* saltwater environments 413–414
  hormonal regulation of osmotic pressure 412
  hypotonic solutions 411
  isotonic solution 411
  measuring 413
  nitrogenous wastes 411
  osmosis 411
  osmoticity 411
  salt loss 412
  solutions characteristics 411
osmoregulator 411
osmosis 160, 411, 417
  in biological systems 417
  concept 415–416
  demonstration 416–417
  diffusion and osmotic pressure 416
osmotic concentrations 413
osmoticity 411
osmotic pressure 415
osteoblast 76
osteoporosis 208–209
ovaries 206, 454–455
ovary 454
oviparous birth 74–75
ovoviviparous birth 75
ovulation 17, 456
ovum 427–428, 454
oxytocin 17, 198

P

Pacemakers 49, 57, 87, 89, 112
paleoanthroplogy 293
pancreas 166, 206
Pap smear 397, 399, 424–425
papule 155
papulosquamous diseases 157
paracrine signaling 96
parasites 171
  prevention 421
parasitism 419
parasitology 419
  agricultural and veterinary 420
  applications
    malaria 421
    parasite prevention 421
  ecological 421
  future prospects 421–422
  impact on other medical fields 421
  medical 420
  parasitism 416
parathyroid glands 206
parathyroid hormone (PTH) 202
parenteral nutrition 253–254
paroxysmal disorders 388
parthenogenesis 32–33, 74, 120, 477, 482
passive transport 5
pathology 427
  applications
    anatomic 424–425
    chemical 425
    forensic 425
    genetic 425
    hematology 425
    immunology 425
    microbiology 425
    in research and development 425
  bone marrow smear 424
  future prospects 425–426
  genetic testing 423, 426
  microscopic technology 424
  molecular level technology 424
  tissues or body fluids donation 423
patterns of reproduction 144–145
patterns of survival 144
Pearson, Karl 61
pediatric gastroenterology 253
pedodontics 150
penetration, fertilization 233–234

penis 459
peptide bond 10, 44, 470
  formation 13
percussion 440–442
performance-enhancing hormones 209
periodontics 150
periosteum 76–77
peripheral nervous system 385
peritoneal dialysis 377
permanent teeth 148
persistent organic pollutants (POPs) 135–136
petroleum-based transportation fuels 349
phagocytosis 211, 324–325
pharmaceuticals 38
pharmacogenomics 301, 306
pharmacological toxicology 509
pharmacology 507
pharmacotherapy and medical devices, orthopedics 409
pharming 120
phenotype 258
pheromones 98, 487–488, 504
phoropter 403
phospholipid 44
photobioreactor 35
photochromic lenses 404
photodiode 379
photon 525
photoreceptors 230–231, 525–527
photosynthesis 42, 354, 358, 430–431
phylogenetics 529
phylum 18
  annelida 20
  chordata 21
  mollusca 20
physical rehabilitation 406
physical therapy 406
physiological digestion 165
physiology 251, 529
physiotherapy 387
pinnipeds 135
pinocytosis 211
pituitary gland 206
placebo effect 316
placental mammals
  gestation 427
  gestation periods 429–430
  marsupials 427
  monotremes 427

placenta
  structure 428
  types 428–429
planuloids 357
plaque 148, 155
  assay 521
Platyrrhini (flat-nosed New World monkeys) 23
Pleistocene epoch 290
plexus 166
Pliopithecus 23
pluripotent 120
  cell 490
pneumococcus bacterium 184
poikilosmotic organisms 412
poikilotherms 499–500
poison 507
poisoning 508
polarity 313
polarized light 60
  vision 527
pollutants 135–136
polychaete annelid 233
polycystic ovary syndrome (PCOS) 450
polymerase chain reaction (PCR) 121, 176, 178, 182, 424
  limitations 180
polymeric and non-polymeric biomolecules 438–439
polymerization reaction 431
polymers and monomers
  biopolymer formation 432
  biopolymers 431
  cellulose 431
  cross-linking during polymer 432
  diamond 432
  DNA and proteins 431–432
  hydrocarbon molecules 431
  polymerization reaction 431
polypeptide 472
Pondaungia 23
population 143, 365–367, 495
positive chemotaxis 488
positive predictive value 423
positron emission tomography (PET) 58
potential for human cloning 306
precambrian eon 353
predation 138
predator-prey relationships 10
predators 139, 504
pre-exponential factor 2
preimplantation genetic diagnosis (PGD) 449

preventable diseases 263
primary and secondary immune responses 323
primary emotions 198–200
primary producers 354
primary teeth 148
primates 22, 284, 504
primates and sign language 330–331
prion and fungal diseases 173–174
prions 171, 389
Proconsul 23
progesterone 455–456
programmed death theories 346–347
prokaryotic cell 100–101, 218–219, 353
prolactin 17, 340
Propionibacterium acnes 157
Propliopithecus 23
prostate 515
prosthesis 55
prosthetic devices 46, 517
prosthetic leg 47
prosthodontics 150–151
protease 214
protein 10, 42, 44, 102, 104, 392–393, 431–432
  biomarker assays 71
  and enzymes 436–437
  like hormones 202
  synthesis 45, 432
    applications 435
    mRNA, tRNA, and rRNA 434
    protein structure and function 434–435
    structure 435
    visualizing 433–434
proteinogenic amino acids 12
proteomics 301, 304
protogrammar 329
protoplasm 153, 474
protozoan 419
protozoology 420
proximity, eggs and sperm 233
pullet 191
pulmonary 440
pulmonary function tests 442
pulmonary medicine 444
  applications
    bronchodilators 442
    lung cancer therapy 443
    lung transplant 443
    lung volume reduction surgery (LVRS) 443
    ventilators 442–443

auscultation and percussion 441–442
definition 440
future prospects 443
imaging studies 442
mechanical ventilation technology 440
practice 440
pulmonary examination 440–441
pulmonary function tests 442
pulmonology 440
pulp 148
pump 5
pupil 402, 404
pustule 155
pyramid, kidney 375

Q
quantum mechanics 60

R
radioactive carbon-14 298
radioimmunoassay (RIA) techniques 456
radiotelemetry 502
radiotherapy 207
rales 439
Ramapithecus 24, 296
random mating 495
rapport 316
reaction rate 1
reagent 423, 507
recapitulation 329
receptor 96
recombinant DNA 55, 57, 301
recombinant human erythropoietin (rHuEpo) 377
red blood cell 103
refraction 402–403
refractive errors 403
regenerative medicine 46, 92
rejection (immune response) 92
renal artery 375
renal vein 375
reproduction
  competition and reproductive success 447
  field research and laboratory studies 447–448
  fighting for mates 446–447
  mate competition and mate choice 446
  one-celled organisms 445
  reproductive success 445–446
reproductive cloning 121–122
reproductive endocrinology 204, 397

reproductive genetics 449
reproductive hormones 204
reproductive pharmacology 450
reproductive science and engineering 453
  applications
    artificial insemination 452
    contraception and sterilization 452
    intrafallopian transfer 451–452
    surrogacy 452
    in vitro fertilization 451
  artificial insemination 448, 450
  future prospects 452–453
  genetics 450
  micromanipulation 449–450
  reproductive pharmacology 450
  screening tests 451
reproductive success 445–446
reproductive system
  of female mammals
    accessory organs 455
    examination 456
    female reproductive system 456
    menstrual cycle 456
    ovaries 454–455
    sexual maturity 455
  of male mammals
    brain and reproduction 457–458
    controlling reproduction 460
    epididymus and vas deferens 459
    penis 459
    studying 459–460
    testes 458–459
repulsion 313
respiration 16
respiration and low oxygen
  blood flow and oxygen delivery 462–463
  evolution, metabolism and ecology 464–465
  hypoxia 461–462
  metabolic alteration 463
  respiratory and circulatory systems 463–464
respiratory and circulatory systems 463–464
respiratory pigments 461, 469
respiratory surface 461
respiratory system 440
  air and water environments 467
  breathing rate 468
  breathing water and breathing air 467–468
  organs of gas exchange 466–467
  respiratory pigments 469

restorative bionics 57
restriction endonucleases (REs) 120–121
restriction enzymes 521
restriction fragment length polymorphism (RFLP) 176, 178
resulting 1
retina 230, 379, 402–404
retinal bioengineering 381–382
retinitis pigmentosa (RP) 381
retinoids 155
retinoscope 403
reverse osmosis 415–416
Rh disease 398–399
ribonucleic acid (RNA) 42, 101–102, 105, 182, 470
ribosomal RNA (rRNA) 12, 472
ribosome 12, 249
  complex 432, 470
RNAase 473, 474–475, 475
  replication and transcription 474
  RNA and DNA replication 474
  RNA and proteins 474
RNA polymerases 186–187, 470–471
RNA/protein translation
  genetic code and human genome 472
  sequences in mRNA, overlapping of 472
  transcription and translation 470–472
Robert Hooke's innovative Micrographia 54
root 148
Rosembaum chart 403
Rotifers (phylum Rotifera) 32
rough endoplasmic reticulum (rER) 101, 105
Roundworm *(Caenorhabditis elegans)* 262
roundworms (phylum Nematoda) 20
rRNA 434
ruminants 17

S
Sabin poliovirus strains 326
Saccharomyces cerevisiae 39, 350–351
salmonella 191
salpingo-oophorectomy 400
salt loss 412
sarcoplasmic reticulum 7
scaffolds 94
scale of being (chain of being) 224–225
scanning electron microscope (SEM) 62
scorpions 20
screening tests, reproductive science and engineering 451

scrotum 458
secondary emotions 198–200
secondary metabolite 138
self-fertilization 478
self-neutralization 12
semen 125, 457
semipermeable membrane 160, 415
semisynthetic enzymes 214–215
sensitivity 423
separation, bioreactors 37
septum 390
sequences in mRNA, overlapping of 472
sequester 138
sequestration 142
serostim 38
sex chromosomemutants 483
sex differences, evolutionary origin
  asexual reproduction 477
  sexual reproduction
    cost of 478–479
    forms of 477–478
    studying 479
  sexual selection 478
  testing theories 479–480
sex differentiation, origin of 482
sexual development
  evolutionary advantage 483
  hormone imbalances 483
  parthenogenesis 482
  sex chromosomemutants 483
  sex differentiation, origin of 482
sexual dimorphism 445, 477
sexual intercourse. *see* copulation
sexual maturity 455
sexual reproduction 477
sexual selection 445, 477
shaft 242
shikonin 67
short tandem repeat (STR) 176
short-term hypoxia 464
sigmoidoscopy 251
sign language 330–331
simple eyes 230
single-celled organisms 213
single nucleotide polymorphism (SNP) 176, 178
sirenians 135
Sivapithecus 24
skeletonized faunas 355–356
skeletons 16

skillful man 223
skin
  changing protective layers 487
  coloration 486
  hair-raising response 487
  single-cell organisms 484
  specialized secretions and structures associated with 485
  structure and physical properties of 484–485
  syncytial epidermis 484
  and temperature regulation 485–486
slit lamp examination 403–404
slow and holistic process, natural selection 366
small intestine 169–170
smell
  in fish 488
  in insects 488
  positive chemotaxis 488
  responses to chemicals 487–488
  taste receptors 488
  in terrestrial vertebrates 488–489
smooth endoplasmic reticulum (sER) 101, 105
sodium dodecyl sulfate polyacrylamide gel electrophoresis (SDS-PAGE) 63
solubility 313
solute 415
solutions characteristics, osmoregulation 411
solvent 415
somatic cell nuclear transfer (SCNT) 120–121, 490, 492–493
somatic stem cells 490–491
  therapies 493
somatotropin (STH) 202
speciation 255
species 18, 220, 255, 365–367, 371–373
specificity 423
sperm 74
  bank 449
  cells 103
spermatogenesis 247
spermatogonia 458
spermiogenesis 247
spherical errors 403
sphincter 166
spiders 20
spinal cord diseases 388
spinal stabilization 408–409
spiracles 467
sponges (phylum porifera) 19, 79

spontaneous mutations 360
sports medicine 408
sport sociology 335
Staphylococcus aureus 157
stem cell 92, 301
  research and technology 490, 494
    cell specialization and development 492
    cell therapies 492
    embryonic stem cells 491–492
    future prospects 493
    human embryonic stem cells 491
    induced pluripotent stem cells 491
    medical drug testing 492
    somatic cell nuclear transfer 492–493
    somatic stem cells 491
    somatic stem cell therapies 493
    therapy 491
    transplantation 493
    types 491
  transplant 276
stenohaline 411
stenosis 88
stents 87, 89
sterilization 38
steroids 67
  hormones 202
sterols 352
stethoscopes 87, 89
stink bugs 139
stomach pH 283
stratigraphy 22
stretching program 336
stromatolites 354
structural defenses 139
Suborder Prosimii (lower primates) 22–23
subspecies 371–372
substrates 214, 349–352
suckling 340–341
superconducting device 129
superconducting magnet 129
superconductivity 129
superfluidity 129
surgeons, orthopedics 406
surgical and cosmetic procedures 158
surgical pathology 424–425
surrogacy 452
survivorship 143
susceptibility 316
symbiosis 138

synapse 385
syncytial epidermis 484
synesthesia 389
synthetic biology 63
systematics 18, 529
systemic, respiration and low oxygen 461
systems biology 46
systole 112, 114

T
tachycardia 87
target, endocrinology 201
tartar 148
tautomeric shifts 360
Taxol 351
taxonomic relationship to modern humans 371–372
taxonomy 18, 371–372, 529
taxon (pl. taxa) 220
teat 339–341
Teflon 376
telomeres 345–346
temperature
  compensation 500
  sensors 500
testes 206, 458–459
testosterone 458, 482
test-tube baby 450
thalamus 379
The Descent of Man (1871) 329
The Hardy-Weinberg law of genetic equilibrium
  assumptions and predictions 495–496
  genotypes 498
  ideal *vs.* real conditions 496–497
  population genetics 495
  theoretical and experimental genetic studies 497–498
therapeutic cloning 122–123
therapeutic proteins 71
thermite reaction 3
thermodynamics 60
thermogenesis 499
thermoregulation
  applications 502–503
  biological catalysts 499–500
  body-temperature regulation 499
  cold-blooded 500
  ectotherms 500–501
  endotherms 501–502
  function 502
  homeostasis 500
  poikilotherms 500
  radiotelemetry 502
  temperature compensation 500
  temperature sensors 500
  warm-blooded 500
thyroid disease 208
thyroid gland 206
thyroid-stimulating hormone (TSH) 202, 206
thyrotropin 202
thyrotropin-releasing hormone (TRH) 206
time-specific approach 146
tissue engineering 39
T lymphocyte 324–325
tonometer 402, 404
tonometry 404
tool use 504, 506
  for fishing 505
  sticks and stones used as 505
  traps 506
toothbrushes 149
tositumomab 38
toxicant 507
toxicodynamics 507
toxicogenomics 301, 304
toxicokinetics 507
toxicology 236, 510
  applications
    environmental toxicology 509
    forensic toxicology 508–509
    medical toxicology 509
    pharmacological toxicology 509
  concept 508
  dose-response relationship 508
  future prospects 509
  poisoning 508
  toxicologists 507–509
toxin 507
trace evidence 236
trance 316
transcranial magnetic stimulation 383
transcription
  factor 186
  process 470
  and translation 470–472
transcutaneous electrical nerve stimulation (TENS) 382
transesterification 352
transfection 521

transfer RNA (tRNA) 12, 303, 472
transgenic foods 51
transgenic organism 46, 120
transition state 1, 214
translation 432–433
transmembrane proteins 105
transmission electron microscope (TEM) 62, 107
transmitter 96
transplant 92
transplantation, stem cell research and technology 493
traps 506
treatment modalities 158
tribology 510, 514
  abrasion 512
  applications
    lubrication and lubricants 513
    tribological research and control devices 513
  friction 511–512
  future prospects 513–514
  heat 512–513
trimethylaminoxide (TMAO) 412
trinucleotide repeat expansion disorders 180
triploid 32
tRNA 434
tubal ligation 452
tumors 157
tunicates 21
turbinates 390
Tyrannosaurus rex 85

U

ubiquitous computing 308
ultrasound 399–400
ultraviolet (UV) light 361, 421
  vision 528
ultraviolet/visible microspectrophotometry 238–239
unconscious 316
uniformitarianism 224
ureter 375–376, 515
urethra 515
urine 375
urogenital groove 481
urology 515–516, 519
  applications
    drug therapies 518
    neurourology 517–518
    prosthetic devices 517
  endoscopy 516
  future prospects 518

  imaging modalities 516–517
usability, HCI 308
user-friendly microorganism hosts 350
uterine mouth 74
uterus 427, 454

V

vaccine production 325–327
vaginal delivery 397
vagus nerve 379
valves 112
vapor pressure determinations 413
variable number tandem repeats (VNTRs) 176, 178–179
variolation 325
vascular diseases 387
vas deferens 515
vasectomy 452
vector 60, 419
vena cava 87
ventilation 461
ventilators 440, 442–443
ventricle 87
ventricular assist device (VAD) 28
  supplement 29
venules 113
vertebrate 18, 21, 484
  circulation 114–115
vertical integration 191
vesicle 211
Viagra 516
viral and bacterial disease 136
viral diseases 173
viral vaccines 325–326, 522
virology 525
  anticancer therapy 523
  future prospects 524
  gene introduction and integration 523–524
  genetic manipulation 522–523
  viral vaccines 522
virulent strains 367
virus particle 521
viscosity 467
vision 525
  color vision 527
  physiology of 52–527
  polarized light vision 527
  ultraviolet (UV) light vision 528
vision therapy 402, 404
visual acuity 402–403

visual cortex 379
visual field
  defects 403
  examination 403
vitamin 392–394
vitelline envelope 233–234
viviparous 427
viviparous birth 74
vomeronasal organ 98

W
warm-blooded 500
waste management 40
water and salt balance control 203
wave bioreactor 35
wayfinding 308
white matter, brain 79
widgct 308
wire frame 308
Wolffian ducts 481–482

worms, diverse forms 19–20
Würm glaciation 290

X
xenobiotic 507
xenotransplantation 53
X-ray 361, 406–407
  crystallography 62

Y
yogurt 39

Z
zona pellucida 233
zoogeography 529
zoology 529–530
  electronic zoology resources 531
  modern 530
zygote 74, 196, 247, 268
zygote intrafallopian transfer (ZIFT) 451–452